T0320919

ANALYTIC ELEMENT METHOD

Analytic Element Method

Complex Interactions of Boundaries and Interfaces

David R. Steward

Professor of Civil and Environmental Engineering
North Dakota State University

OXFORD
UNIVERSITY PRESS

Great Clarendon Street, Oxford, OX2 6DP,
United Kingdom

Oxford University Press is a department of the University of Oxford.
It furthers the University's objective of excellence in research, scholarship,
and education by publishing worldwide. Oxford is a registered trade mark of
Oxford University Press in the UK and in certain other countries

First Edition published in 2020

Impression: 2

Published in the United States of America by Oxford University Press
198 Madison Avenue, New York, NY 10016, United States of America

British Library Cataloguing in Publication Data
Data available

Library of Congress Control Number: 2020943752

ISBN 978–0–19–885678–8

Printed and bound by
CPI Group (UK) Ltd, Croydon, CR0 4YY

To my wife Jill and our children Jacob and Kaitlin

Inspiration and faith reveal
the pathway.
Steadfastness to purpose propels
the process.
Companionship with love sustains
the journey.

Preface

This text on the *Analytic Element Method* (AEM) assembles a broad range of mathematical and computational approaches for solving important problems in engineering and science. As the subtitle *Complex Interactions of Boundaries and Interfaces* suggests, problems are partitioned into sets of elements and methods are formulated to solve conditions along their boundaries and interfaces. Presentation will place an element within its landscape, formulate its interactions with other elements using linear series of influence functions, and then solve for its coefficients to match its boundary and interface conditions. Computational methods enable boundary and interface conditions of closely interacting elements to be matched with nearly exact precision, commonly to within 8–12 significant digits. Comprehensive solutions provide elements that collectively interact and shape the environment within which they exist.

Aims: The ***purpose*** of this book is to illustrate how the methods of the AEM may be employed to solve important problems. Readers will become immersed within a ***paradigm*** viewing problems as collections of elements with mathematical properties and processes that reflect their physical environment. The ***goals*** of this book are:

- to introduce readers to the basic principles of the AEM,
- to provide a template for those interested in pursuing these methods, and
- to empower readers to extend the AEM paradigm to an even broader range of problems.

It does not attempt to solve all known problems, but instead focuses on the most important methods and their application to broadly relevant problems.

Motivation: This work is grounded in a wide range of ***foundational studies***, which are presented comprehensively within their fields of study in Chapter 1. Their authors developed methods used to analyze ***exact solutions*** to important boundary value problems. However, the computational capacity of their times limited solutions to idealized problems, commonly involving a single isolated element within a uniform regional background. With the advent of modern computers, such mathematically based methods were passed over by many, in the pursuit of discretized domain solutions using finite element and finite difference methods. Yet, the elegance of the mathematical foundational studies remains, and the ***rationale*** for the AEM was inspired by the realization that ***computational advances could also lead to advances in the mathematical methods*** that were unforeseeable in the past.

Seminal Literature: This book articulates the ***Analytic Element Method paradigm*** in Chapter 2, which comprehensively employs a background of existing methodology using complex functions, separation of variables, and singular integral equations. The AEM was originally developed for problems in groundwater mechanics, with early applications including the Netherlands National Groundwater Model (de Lange, 2006) as well as capture zone analysis for municipal wells (Haitjema, 1995). The pioneering developments by Strack (1989) supported these endeavors using conformal mappings and singular integral equations. I became involved in their extension to curvilinear elements (Steward et al., 2008) and three-dimensional flows beginning with Steward (1998, 1999). Such analytic solutions form a basis for the methods of ***complex functions*** (Chapter 3) and ***singular integral equations*** (Chapter 5). Analytic elements may also be formulated using ***separation of variables*** (Chapter 4), expanding upon early studies of two- and three-dimensional vector fields in groundwater flow conducted by Barnes, and Janković (1999) and Janković and Barnes (1999*b*). For readers, convenience, an overview at the beginning of each chapter summarizes its content and emphasizes important concepts.

General Approach: While the AEM was first developed to study groundwater flow, it provides a ***unifying framework*** that may be employed to solve transdisciplinary problems. The first chapter illustrates the types of solutions that are available across wide-ranging fields of study. The second chapter gathers the processes and properties of disciplinary perspective within a comprehensive foundation to solve these problems using the AEM. And the remaining chapters develop a broad range of solution methods and applications. A philosophy of ***just in time mathematics*** is adopted throughout, which introduces the methods of the AEM as needed to solve a particular problem.

Scope: This book is organized in ***sections with an appropriate content*** to be covered within a ***typical class period***. Each section contains ***examples*** with figures that illustrate applications of the mathematical methods. ***Problems*** require students to solve an application similar to these examples, and enough problem sets are provided to avoid repeating answers over several offerings of a course.

Potential Readership: This text puts forth new methods for solving important problems across engineering and science, and has a tremendous potential to broaden perspective and change the way problems are formulated. Methods utilize basic mathematical procedures that ***students who have completed a sequence in calculus*** should be prepared to master. Presentation begins with basic principles and moves through to their implementation in computational methods. Important ***examples have been implemented*** in Scilab (http://www.scilab.org) and will be made available through a companion website with the publisher. Students with computational acumen should be able to easily adapt examples to solve the problem sets. ***Solutions to alternate problem sets are provided*** in Appendix B, and a complete set is available to assist with grading.

Chapter 1 and Section 2.1 will be made publicly available. This freely introduces use of the AEM to a broad range of engineers and scientists. It also demonstrates the ease by which important problems may be formulated within the AEM paradigm. This is expected to ***motivate a broad range of engineers, scientists, and students to become introduced to the concepts in the remaining chapters***, where they can learn and master the AEM methodology.

The AEM worldview for solving problems has the potential to become widely accepted, adapted, and utilized.

Acknowledgements: Most importantly, I dedicate this book to my wife, Jill, and our children, Jacob and Kaitlin. Thank you for your constant encouragement and support. Special thanks to Professor Otto D. L. Strack at the University of Minnesota and the wide range of AEM colleagues whom I have had the privilege of knowing and collaborating with over the past decades. Thanks to North Dakota State University and Kansas State University, the friends and collaborators with whom I have learned interdisciplinary engineering and science, and all the students whom I have learned to teach and hopefully inspire.

Contents

Analytic Element Method across Fields of Study

1

- This chapter introduces the philosophical perspective for solving problems with the Analytic Element Method, organized within three common types of problems: gradient driven flow and conduction, waves, and deformation by forces.
- These problems are illustrated by classic, well-known solutions to problems with a single isolated element, along with their extension to complicated interactions occurring amongst collections of elements.
- Analytic elements are presented within fields of study to demonstrate their capacity to represent important processes and properties across a broad range of applications, and to provide a template for transcending solutions across the wide range of conditions occurring along boundaries and interfaces.
- While the mathematical and computational developments necessary to solve each problem are developed in later chapters, each figure documents where its solutions are presented.

1.1 Philosophical Perspective

The Analytic Element Method (AEM) conceptualization is quite intuitive. It is founded in the mathematical representations employed across fields of study in science and engineering, which describe a disciplinary perspective of important processes and properties. For many problems, the physical environment may be represented as a collection of elements with distinct properties. Each element interacts with its adjacent environment, and methods are formulated along an element to solve its conditions with nearly exact precision. Collectively, sets of elements interact to transform their environment, and these synergistic interactions are expanded upon for three common types of problems.

Problem 1. Flow and Conduction

The first type of problem studies a vector field that is directed from high to low values of a function as illustrated in Fig. 1.1. The function is visualized using a gray color pallet that shades the domain from lighter tones where the value of the function is larger to darker tones where it is smaller, and solid black lines lie along ***equipotential*** contours that have constant function value. The vector field is computed at discrete locations and arrows are drawn from these points in the direction of the vector with lengths scaled by the vector's magnitude. Many solutions may also be represented using ***streamlines***, which are everywhere tangent to the vector field and drawn as white lines.

Computer code is available at the publisher's companion website to illustrate implementations of important examples.

Analytic Element Method: Complex Interactions of Boundaries and Interfaces. David R. Steward, Oxford University Press (2020).

DOI: 10.1093/oso/9780198856788.001.0001

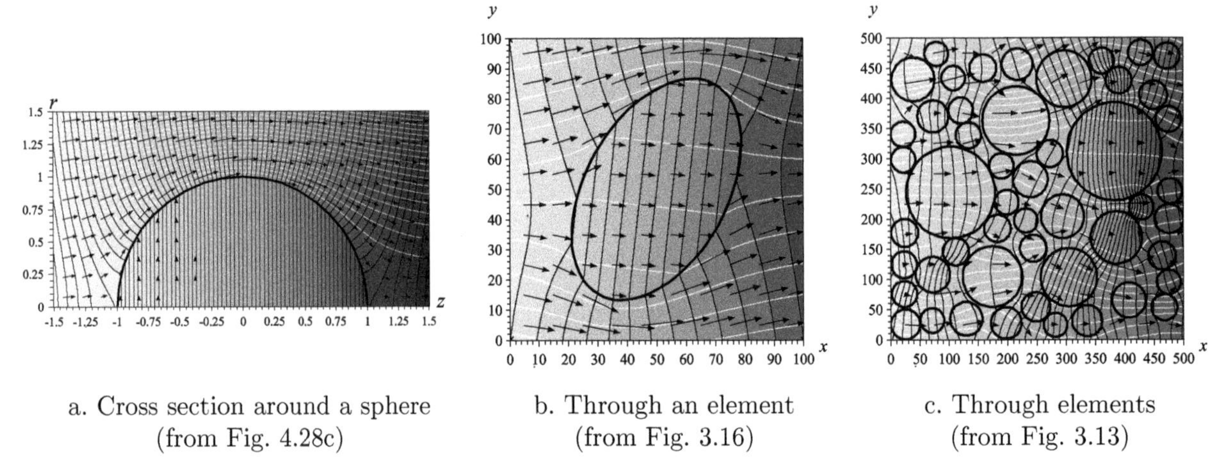

Figure 1.1 *Example problems with flow and conduction in a background uniform vector field, with figure numbers indicating where solution methods are presented.*

The use of analytic elements to representation solutions is presented across broad fields of study. ***Groundwater flow*** is formulated in Section 1.2.1 with equipotential surfaces of uniform groundwater elevation (head), the vector field points in the direction of flow, and streamlines represent the trajectories of water particles. A similar representation is used to study flow through a ***vadose zone*** in Section 1.2.2 using equipotentials with uniform pressure, and vector fields are developed for ***incompressible fluid flow*** in Section 1.2.3. Analytic elements are formulated for ***thermal conduction*** in Section 1.2.4 using equipotentials with uniform temperature and the vector field represents conduction of heat. ***Electrostatics*** in Section 1.2.5 represent solutions for electric current using electric potentials related to voltage.

A ***common representation of solutions*** are formulated using relevant properties and processes from these disciplinary perspectives, but with ***cross-disciplinary interpretations of results*** (Bulatewicz et al., 2010, 2013). For example, the solution in Fig. 1.1a was used to study fluid flowing about an impermeable sphere by Lamb (1879, sec.102), and also represents the electric field near an object with dielectric constant equal to zero (Sommerfeld, 1952, p.62). The solution in Fig. 1.1b illustrates electric current through a dielectric inhomogeneity with the geometry of an ellipse or a prolate spheroid (Moon and Spencer, 1961*a*, sec.6), and has related applications in heat conduction (Carslaw and Jaeger, 1959, sec.16.4). The example in Fig. 1.1c illustrates a vector field for flow and conduction with many interacting elements that collectively shape the solution. A ***general framework*** is presented in Section 1.2 for such problems with flow and conduction to achieve a ***consistent representation with broadly applicable solutions***.

Problem 2. Periodic Waves

A second type of problem studies the ***interactions of analytic elements with waves***. This is illustrated in Fig. 1.2 for a background field of plane waves traveling from left to right that encounter and interact with elements. Wave fields are visualized using two plots for each solution, where the upper panels illustrate the ***amplitude*** of waves, and the lower panels illustrate the ***phase*** of a wave field. The amplitude shown in Fig. 1.2 is scaled by the

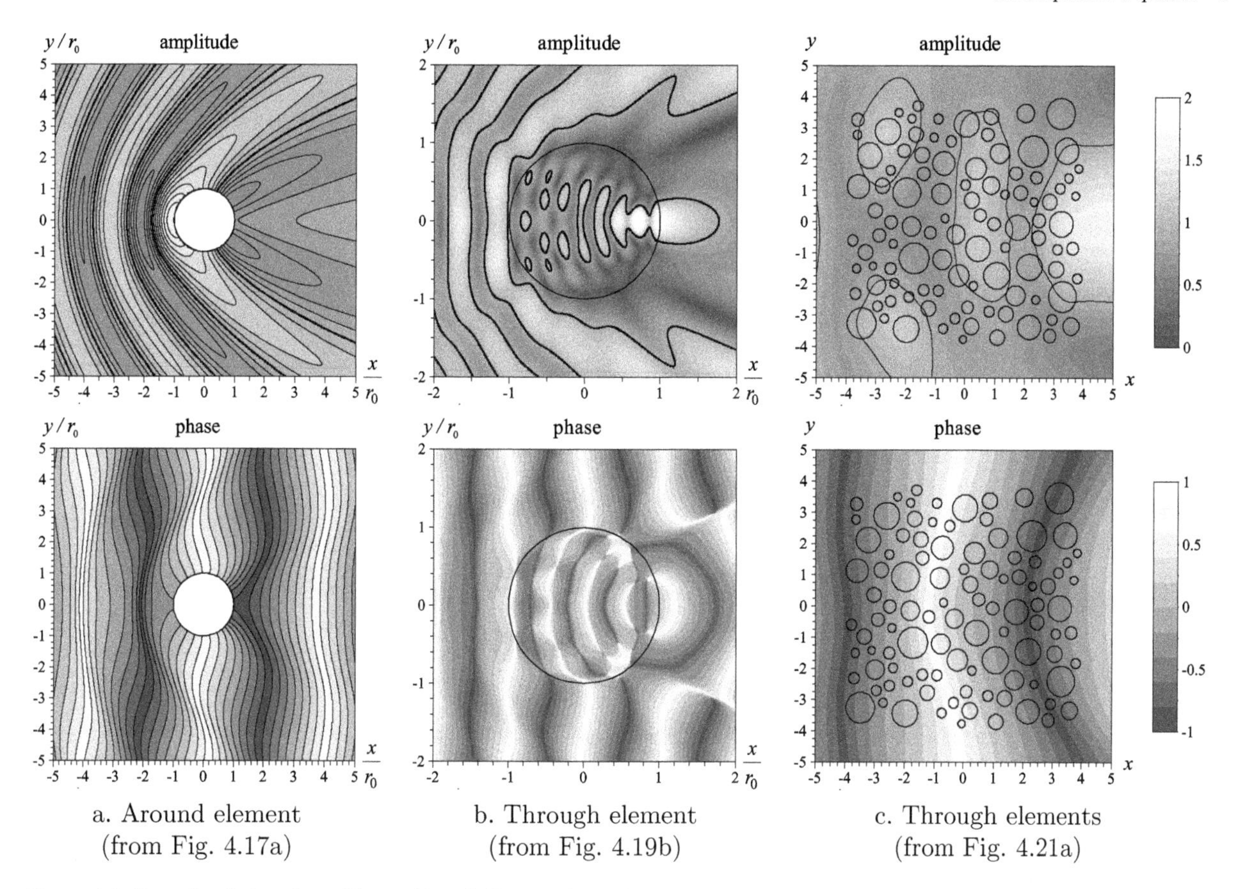

Figure 1.2 *Example solutions for problems with periodic waves in a field of plane waves. (Reprinted from Steward, 2018, Wave resonance and dissipation in collections of partially reflecting vertical cylinders,* Journal of Waterways, Ports, Coastal, and Ocean Engineering, *Vol. 144(4), Fig. 1, with permission from ASCE.) (Reprinted from Steward, 2020, Waves in collections of circular shoals and bathymetric depressions,* Journal of Waterways, Ports, Coastal, and Ocean Engineering, *Vol. 146, Figs. 2, 4, preproduction version DOI 10.1061/(ASCE)WW.1943-5460.0000570, with permission from ASCE.)*

amplitude of the background wave field and thicker lines are drawn where the amplitude would be equal to the background if the elements did not exist. The phase is contoured at equal position on a wave; for example, if the darker tones represent the highest elevation at the crest of a wave field at a specific time, then the lighter tones represent the lowest elevation at the trough of the wave field at the same time.

Analytic elements are developed for ***water waves*** in Section 1.3.1 to illustrate the amplitude and phase of the surface of a sea. The analytic elements for ***acoustics*** in Section 1.3.2 illustrate vibrations traveling through solids, liquids, or gases. Boundary conditions are developed for wave ***reflection*** where waves are scattered, for example, by the fully reflective element in Fig. 1.2a. This solution was developed by MacCamy and Fuchs (1954) to study the formation of partially standing waves on the windward side of an island, and wave ***diffraction*** occurs on the leeward side with calmer water. Analytic elements may also be formulated to study wave ***refraction*** as illustrated in Fig. 1.2b to represent solutions when water waves travel across a shoal with shallower water, or when acoustic waves travel through an inhomogeneity with a wavelength shorter than that of the surrounding media. Wave refraction also occurs in Fig. 1.2c, where many elements interact to collectively shape the wave field. Analytic elements are ***formulated for waves*** in Section 1.3 for problems with ***reflection, diffraction, and refraction***.

Problem 3. Deformation by Forces

A third type of problem studies the ***interactions of analytic elements with stresses and displacements***. The examples in Fig. 1.3 illustrate stress fields occurring as analytic elements interact with a background field generated by body forces. These solutions are visualized using two panels for each solution: the top panels show ***principle stresses*** and mean stress, and the lower panels show ***displacement*** and maximum shear. The principle stresses are computed at equally spaced positions and lines are drawn from these locations to indicate direction and magnitude, and ***isostress*** contours with uniform mean stress have darker tones, indicating a larger compression. Displacements are shown by vectors scaled

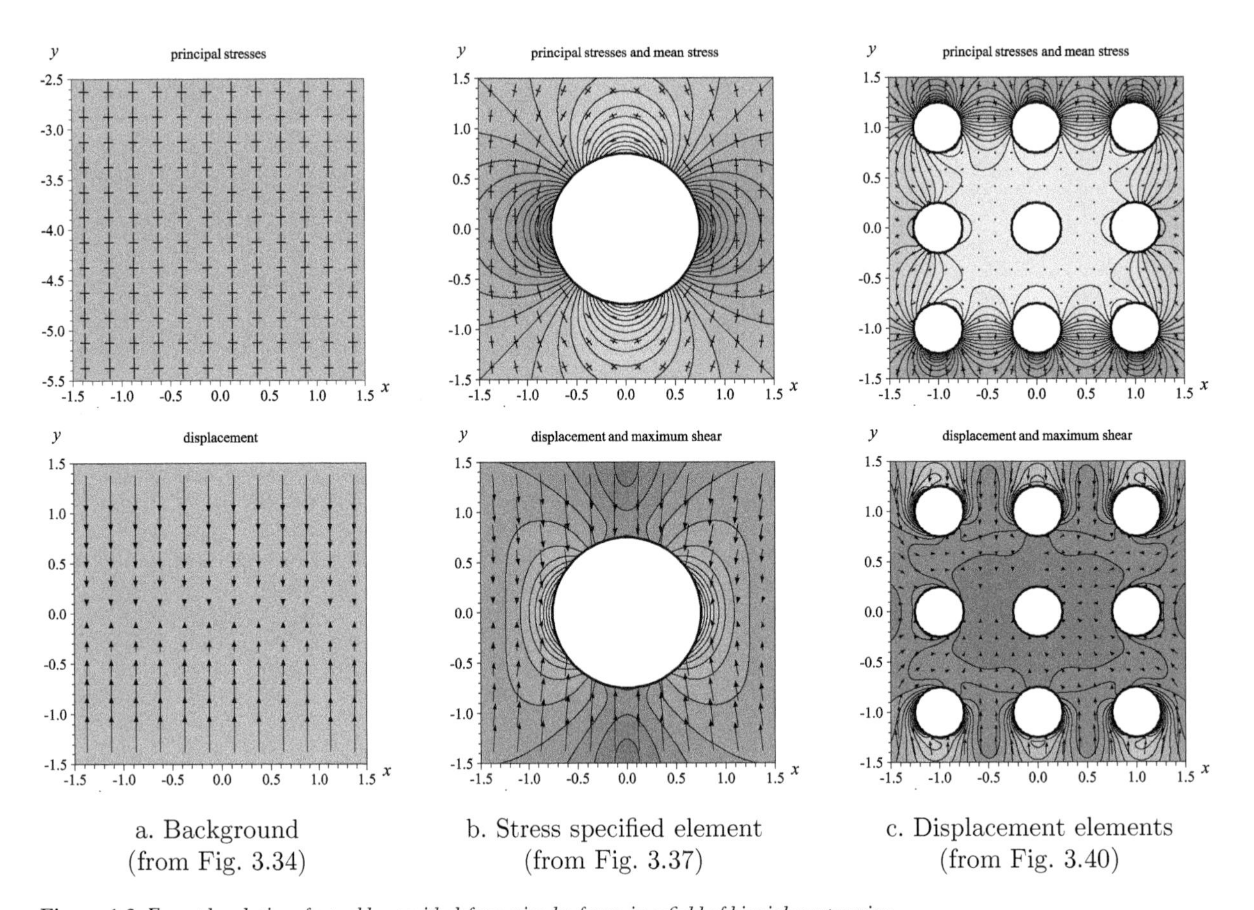

Figure 1.3 *Example solutions for problems with deformation by forces in a field of biaxial compression.*

by its magnitude, and ***isoshear*** contours of uniform maximum shear show lighter tones where it has a larger value.

The use of ***analytic elements to solve problems with deformation by forces*** is formulated in Section 1.4. The example in Fig. 1.3a illustrates the background stress field in geomechanics generated at a depth within a soil column by the ***body forces*** of the overlying material. The solution in Fig. 1.3b from Jaeger (1969) illustrates the stress field and displacement formed around a tunnel placed in this background field, where the boundary is ***traction free with no normal or shear stresses***. The example in Fig. 1.3c shows a set of elements in a plate that satisfy a boundary condition of ***zero displacement*** where their fixed locations cause a distribution of the stress field. Solutions are developed in Section 1.4.1 for analytic elements with boundary conditions of specified stress or specified displacement. While the examples are shown for stress–strain studies of ***elasticity in structures and geomechanics***, the methods are extensible to the studies of thermal stress and inviscid fluid flow.

These three types of problems each compute the value of a function, and a vector or stress field across a region of interest. This chapter formulates such boundary value problems across a broad range of fields of study in engineering and science. For each type of problems, the mathematical tools necessary to address problems are first presented to establish a consistent notation and common equations. Each field of study introduces the properties and processes of its perspectives and presents examples of analytic elements used to solve its important boundary and interface conditions. A comprehensive foundation assembles these methods in Chapter 2, the symbols and variables are collected in Appendix A and the Index, and references are gathered at the end of each chapter.

1.2 Studies of Flow and Conduction

A common problem in engineering and science is to compute the value of a function and an associated vector field across a region of interest, where the vector field is directed from higher to lower values of the function (as in Fig. 1.1). For example, the function might represent groundwater elevation or temperature, with associated vector fields of groundwater flow or thermal conduction. Such problems are formulated in terms of a ***potential function*** Φ and a ***vector field*** $\mathbf{v}$, and vector calculus provides the basic operations to manipulate these functions.

The ***gradient***, ∇, operates on the scalar function and produces a vector with components equal to the partial derivative of the function. The gradient of a three-dimensional $\Phi(x, y, z)$ has components

$$\boxed{\nabla\Phi = \frac{\partial\Phi}{\partial x}\hat{\mathbf{x}} + \frac{\partial\Phi}{\partial y}\hat{\mathbf{y}} + \frac{\partial\Phi}{\partial z}\hat{\mathbf{z}}} \tag{1.1}$$

where $\hat{\mathbf{x}}$, $\hat{\mathbf{y}}$, and $\hat{\mathbf{z}}$ are unit vectors pointing in the x, y, and z coordinate directions. The gradient is directed normal to surfaces of constant Φ and directed towards increasing Φ, as illustrated in Fig. 1.4a. A second vector operator is the ***divergence***, $\nabla\cdot$, which is used to quantify the net flux of a vector field through a control volume as illustrated in Fig. 1.4b. The divergence of $\mathbf{v}$ with Cartesian components (v_x, v_y, v_z) is obtained by multiplying the normal components of the vector field directed out minus those directed into the control volume times the surface area of each face of the parallelepiped with sides of length Δx, Δy, and Δz, and dividing by the volume:

$$\begin{aligned}\nabla \cdot \mathbf{v} = &\lim_{\Delta x \to 0} \frac{v_x(x+\frac{\Delta x}{2}, y, z)\Delta y \Delta z - v_x(x-\frac{\Delta x}{2}, y, z)\Delta y \Delta z}{\Delta x \Delta y \Delta z} \\ &+ \lim_{\Delta y \to 0} \frac{v_y(x, y+\frac{\Delta y}{2}, z)\Delta x \Delta z - v_y(x, y-\frac{\Delta y}{2}, z)\Delta x \Delta z}{\Delta x \Delta y \Delta z} \\ &+ \lim_{\Delta z \to 0} \frac{v_z(x, y, z+\frac{\Delta z}{2})\Delta x \Delta y - v_z(x, y, z-\frac{\Delta z}{2})\Delta x \Delta y}{\Delta x \Delta y \Delta z}\end{aligned} \tag{1.2}$$

Evaluating the limit of this central difference equation gives

$$\boxed{\nabla \cdot \mathbf{v} = \frac{\partial v_x}{\partial x} + \frac{\partial v_y}{\partial y} + \frac{\partial v_z}{\partial z}} \tag{1.3}$$

A third vector operator is the ***curl***, $\nabla\times$, which may be written in Cartesian coordinates for the vector $\mathbf{v}$:

$$\boxed{\nabla \times \mathbf{v} = \left(\frac{\partial v_z}{\partial y} - \frac{\partial v_y}{\partial z}\right)\hat{\mathbf{x}} + \left(\frac{\partial v_x}{\partial z} - \frac{\partial v_z}{\partial x}\right)\hat{\mathbf{y}} + \left(\frac{\partial v_y}{\partial x} - \frac{\partial v_x}{\partial y}\right)\hat{\mathbf{z}}} \tag{1.4}$$

The curl is used to quantify infinitesimal rotation as illustrated in Fig. 1.4c, where the component of the curl in the z-direction, $\frac{\partial v_y}{\partial x} - \frac{\partial v_x}{\partial y}$, is zero when the angles $\vartheta_1 = \vartheta_2$ in this figure are equal in the limit as Δx and $\Delta y \to 0$, and the vector field is irrotational.

There are two functions commonly utilized to solve problems with flow and conduction, a ***scalar potential function***, Φ, and a ***vector potential function***, $\boldsymbol{\Psi}$. The ***Helmholtz theorem*** states that any vector field, $\mathbf{v}$, may be decomposed into a linear combination of the gradient of this scalar potential and the curl of this vector potential:

$$\boxed{\mathbf{v} = -\nabla\Phi - \nabla \times \boldsymbol{\Psi}} \tag{1.5}$$

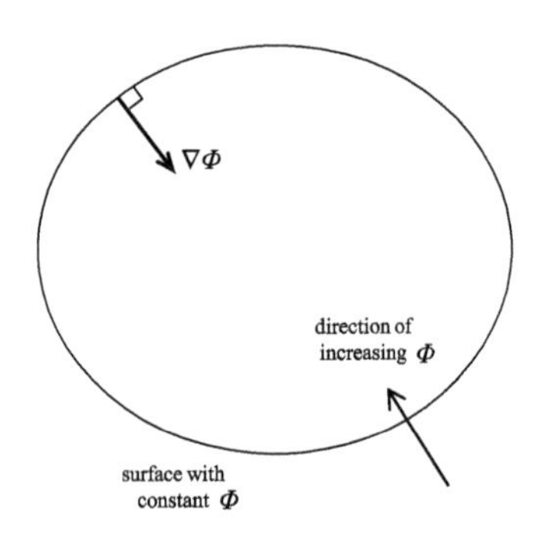

a. Gradient of Φ is normal to surfaces of constant Φ

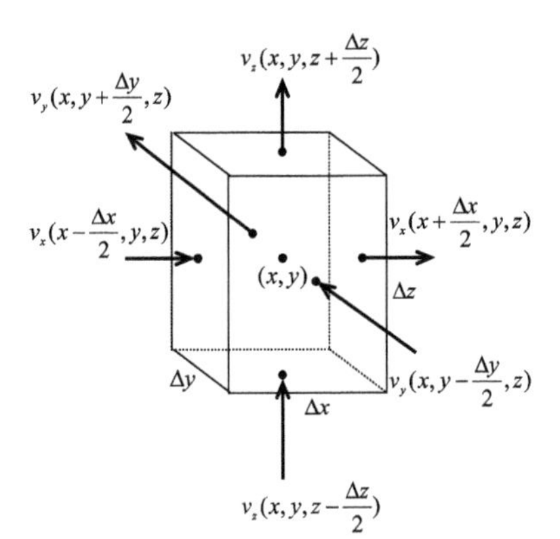

b. Divergence of $\mathbf{v}$ for flux through a control block

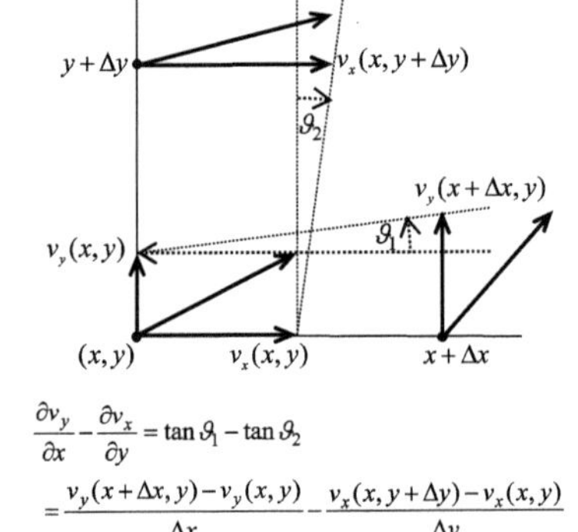

c. The component of curl in the z-direction

Figure 1.4 *Vector operators: gradient, divergence, and curl.*

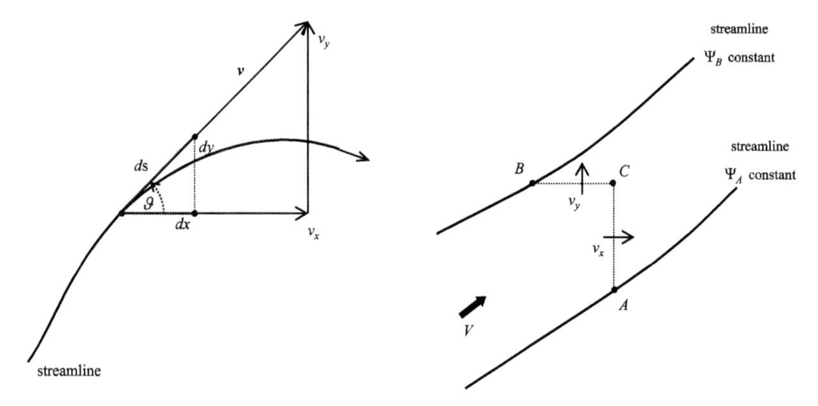

Figure 1.5 *Two-dimensional vector fields and the stream function.*

The ***first corollary*** to the Helmholtz theorem states that a vector field may be represented using the gradient of the scalar potential if and only if the field is irrotational:

$$\boxed{\mathbf{v} = -\nabla\Phi \iff \nabla\times\mathbf{v} = 0} \tag{1.6}$$

and most problems will be formulated in terms of Φ. This sign convention aligns a vector field in the direction of decreasing potential. The ***second corollary*** to the Helmholtz theorem states that a vector field may be represented using the curl of the vector potential function if and only if the field is divergence-free:

$$\boxed{\mathbf{v} = -\nabla\times\boldsymbol{\Psi} \iff \nabla\cdot\mathbf{v} = 0} \tag{1.7}$$

and use of vector potentials for three-dimensional problems is presented in Section 5.7.2. Vector fields that have both divergence and curl require use of both the scalar and vector potential, while those that are divergence-free and irrotational may be formulated using either function.

A simpler representation exists for divergence-free problems with a two-dimensional vector field that lies in a plane (e.g., $v_z = 0$) and is invariant in the direction normal to this plane (i.e., $\frac{\partial v_x}{\partial z} = \frac{\partial v_y}{\partial z} = 0$). A ***streamline*** in such a flow is defined to be everywhere tangent to the vector field, and this property is satisfied in Fig. 1.5a when

$$\tan\vartheta = \frac{v_y}{v_x} = \frac{dy}{dx} = \frac{\frac{dy}{ds}}{\frac{dx}{ds}} \quad\rightarrow\quad v_y\frac{dx}{ds} - v_x\frac{dy}{ds} = 0 \tag{1.8}$$

where the incremental step ds along the streamline has components dx and dy in the coordinate directions. The ***Lagrange (1781) stream function***, Ψ, is defined to be constant along a streamline in the s-direction, and the chain rule for differentiation gives

$$\frac{d\Psi}{ds} = \frac{\partial\Psi}{\partial x}\frac{dx}{ds} + \frac{\partial\Psi}{\partial y}\frac{dy}{ds} = 0 \tag{1.9}$$

Together, the last two equations give a relation between the partial derivatives of Ψ and the components of the velocity vector:

$$\boxed{v_x = -\frac{\partial \Psi}{\partial y}, \quad v_y = +\frac{\partial \Psi}{\partial x}} \tag{1.10}$$

A vector potential for a two-dimensional vector field is given by multiplying the Lagrange stream function times $\hat{\mathbf{z}}$, a unit vector pointing in the positive z-direction (Steward, 2002),

$$\boldsymbol{\Psi} = \Psi\hat{\mathbf{z}} = (0, 0, \Psi) \tag{1.11}$$

and the curl of this vector in (1.7) reproduces the two-dimensional relationship (1.10). Consequently, any two-dimensional divergence-free vector field may be studied using a Lagrange stream function Ψ.

The stream function is important because contours of constant Ψ may be visualized to illustrate streamlines. The stream function also provides the flux occurring between two points, as illustrated in Fig. 1.5b for the two streamlines that pass through the points A and B. This flux is obtained from the integral of the normal component of the vector field along the vertical and horizontal lines connecting A to B,

$$V = \int_A^C v_x \mathrm{d}y + \int_B^C v_y \mathrm{d}x \tag{1.12}$$

Substituting the relationship between the vector field and stream function (1.10), integrating, and evaluating the stream function at the limits of integration gives

$$\begin{aligned} V &= \int_A^C -\frac{\partial \Psi}{\partial y}\mathrm{d}y + \int_B^C \frac{\partial \Psi}{\partial x}\mathrm{d}x = -\int_A^C \mathrm{d}\Psi + \int_B^C \mathrm{d}\Psi \qquad \rightarrow \qquad \boxed{V = \Psi_A - \Psi_B} \\ &= -(\Psi_C - \Psi_A) + (\Psi_C - \Psi_B) \end{aligned} \tag{1.13}$$

where Ψ_A and Ψ_B are the value of the stream function at points A and B. Thus, ***the flux between two streamlines is equal to their difference in stream function***.

The potential and stream functions both exist for two-dimensional vector fields that are irrotational and divergent-free. For this case, the relations between the vector field and the potential function (1.6) and the stream function (1.10) together give the ***Cauchy–Riemann equations***

$$\boxed{\begin{aligned} v_x &= -\frac{\partial \Phi}{\partial x} = -\frac{\partial \Psi}{\partial y} \\ v_y &= -\frac{\partial \Phi}{\partial y} = +\frac{\partial \Psi}{\partial x} \end{aligned}} \quad \rightarrow \quad \begin{aligned} \frac{\partial^2 \Phi}{\partial x^2} + \frac{\partial^2 \Phi}{\partial y^2} &= 0 \\ \frac{\partial^2 \Psi}{\partial x^2} + \frac{\partial^2 \Psi}{\partial y^2} &= 0 \end{aligned} \tag{1.14}$$

Furthermore, such functions Φ and Ψ each satisfy the Laplace equation, which will be discussed further at (2.2a). A ***flow net*** exists for irrotational and divergent-free vector fields, where ***families of equipotentials*** with constant Φ and ***families of streamlines*** with constant Ψ are ***mutually orthogonal***.

These scalar and vector potentials and the stream function will be formulated next across a broad range of fields of study. Each field begins with the unique variables of interest to its perspective, as well as the properties of the problem domain. While the examples illustrate applications of the Analytic Element Method for specific problems, many of the examples have applications within closely aligned problems in other areas. Later chapters then develop methods for solving these problems.

Problem 1.1 Develop expressions for the vector field (components v_x and v_y) associated with the following potential and stream functions, where $r = \sqrt{x^2 + y^2}$, and $\theta = \arctan(y/x)$. The equation numbers are listed for where these functions occur later.

A. $\Phi = \frac{Q}{2\pi} \ln r$, (3.1)

B. $\Phi = -\frac{\Gamma}{2\pi}\theta$, (3.3)

C. $\Psi = \frac{\Gamma}{2\pi} \ln r$, (3.3)

D. $\Psi = \frac{Q}{2\pi}\theta$, (3.1)

E. $\Phi = -\frac{r^2}{4}$, (3.114a)

F. $\Psi = \frac{r^2}{4}$, (3.114b)

G. $\Phi = \frac{\sinh 2\pi n \frac{x_{max}-x}{y_{max}-y_{min}}}{\sinh 2\pi n \frac{x_{max}-x_{min}}{y_{max}-y_{min}}} \cos 2\pi n \frac{y-y_{min}}{y_{max}-y_{min}}$, (4.26)

H. $\Phi = \frac{\sinh 2\pi n \frac{x_{max}-x}{y_{max}-y_{min}}}{\sinh 2\pi n \frac{x_{max}-x_{min}}{y_{max}-y_{min}}} \sin 2\pi n \frac{y-y_{min}}{y_{max}-y_{min}}$, (4.26)

I. $\Phi = \cos 2\pi n \frac{x-x_{min}}{x_{max}-x_{min}} \frac{\sinh 2\pi n \frac{y_{max}-y}{x_{max}-x_{min}}}{\sinh 2\pi n \frac{y_{max}-y_{min}}{x_{max}-x_{min}}}$, (4.26)

J. $\Phi = \sin 2\pi n \frac{x-x_{min}}{x_{max}-x_{min}} \frac{\sinh 2\pi n \frac{y_{max}-y}{x_{max}-x_{min}}}{\sinh 2\pi n \frac{y_{max}-y_{min}}{x_{max}-x_{min}}}$, (4.26)

K. $\Phi = \frac{\sinh 2\pi n \frac{x-x_{min}}{y_{max}-y_{min}}}{\sinh 2\pi n \frac{x_{max}-x_{min}}{y_{max}-y_{min}}} \cos 2\pi n \frac{y-y_{min}}{y_{max}-y_{min}}$, (4.26)

L. $\Phi = \frac{\sinh 2\pi n \frac{x-x_{min}}{y_{max}-y_{min}}}{\sinh 2\pi n \frac{x_{max}-x_{min}}{y_{max}-y_{min}}} \sin 2\pi n \frac{y-y_{min}}{y_{max}-y_{min}}$, (4.26)

M. $\Phi = \cos 2\pi n \frac{x-x_{min}}{x_{max}-x_{min}} \frac{\sinh 2\pi n \frac{y-y_{min}}{x_{max}-x_{min}}}{\sinh 2\pi n \frac{y_{max}-y_{min}}{x_{max}-x_{min}}}$, (4.26)

N. $\Phi = \sin 2\pi n \frac{x-x_{min}}{x_{max}-x_{min}} \frac{\sinh 2\pi n \frac{y-y_{min}}{x_{max}-x_{min}}}{\sinh 2\pi n \frac{y_{max}-y_{min}}{x_{max}-x_{min}}}$, (4.26)

1.2.1 Groundwater Flow: Head and Discharge

Studies of groundwater flow examine the movement of water through geological media, and the important properties and processes are illustrated in Fig. 1.6. Groundwater enters the flow domain as the portion of precipitation that seeps downward through the vadose zone, where pores between soil particles are partially filled with water and partially with air, to the groundwater zone, where porous and fractured media become fully filled with water. Destinations for groundwater include water pumped by wells, baseflow from groundwater to surface water, and phreatophyte root uptake by plants that can directly tap groundwater. The geologic media are separated into layers through which water moves readily called aquifers (e.g., coarse sands and gravels), and aquifer layers may be separated by aquitards with low flow rates (e.g., clays and shales), as illustrated by the confining layer in Fig. 1.6. It is important to know whether an unsaturated zone with unconfined conditions exists at its top or whether an aquifer layer is completely saturated with groundwater and is confined, and these conditions are quantified by the

$$\text{saturated depth} = \begin{cases} h - B & \text{unconfined } (0 < h - B \leq H) \\ H & \text{confined } (H < h - B), \end{cases} \tag{1.15}$$

where B is the base elevation and H is the vertical thickness of the aquifer layer. Different approaches are employed to formulate the equations for saturated groundwater flow in

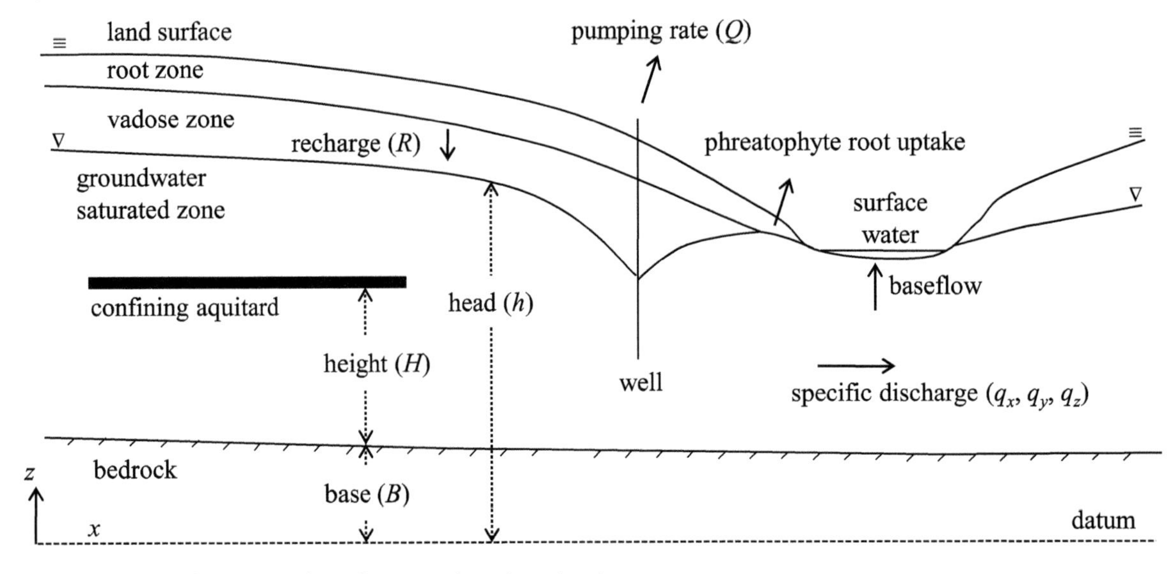

Figure 1.6 *Groundwater properties and processes shown in section view.*

this section (where horizontal flow predominates) and the unsaturated vadose zone flow in Section 1.2.2 (where vertical flow predominates).

Solutions to groundwater flow problems are presented in terms of head, h, which is the level to which water will rise in a pipe and is a measure of potential energy. Groundwater moves from regions with high head to low head, which is represented in ***Darcy's law*** (Darcy, 1856) as a constitutive relation between the specific discharge, (q_x, q_y, q_z), and changes in head:

$$q_x = -K\frac{\partial h}{\partial x}, \; q_y = -K\frac{\partial h}{\partial y}, \; q_z = -K\frac{\partial h}{\partial z} \qquad K = \frac{\kappa \rho g}{\mu} \tag{1.16}$$

This equation uses the hydraulic conductivity K, which is a property of the fluid (density, ρ, and absolute viscosity, μ) and the porous media (intrinsic permeability, κ), and g is the acceleration of gravity. The example solutions in Fig. 1.7 illustrate the distribution of groundwater level (head) and flow rates. The ***Dupuit assumption*** (Dupuit, 1863) approximates head and the horizontal components of flow as uniform in the vertical direction within an aquifer later, which enables the specific discharge to be vertically integrated across the saturated depth to give the two-dimensional discharge per unit width with components Q_x and Q_y:

$$\begin{aligned} Q_x &= \begin{cases} (h-B)q_x \\ Hq_x \end{cases}, \quad & q_x &= \begin{cases} \frac{Q_x}{(h-B)} & (0 < h - B \leq H) \\ \frac{Q_x}{H} & (H < h - B) \end{cases} \\ Q_y &= \begin{cases} (h-B)q_y \\ Hq_y \end{cases}, \quad & q_y &= \begin{cases} \frac{Q_y}{(h-B)} & (0 < h - B \leq H) \\ \frac{Q_y}{H} & (H < h - B) \end{cases} \end{aligned} \tag{1.17}$$

It is convenient to adopt a representation where equations take on a common form for both unconfined and confined flow. This is accomplished by substituting Darcy's law, (1.16), into

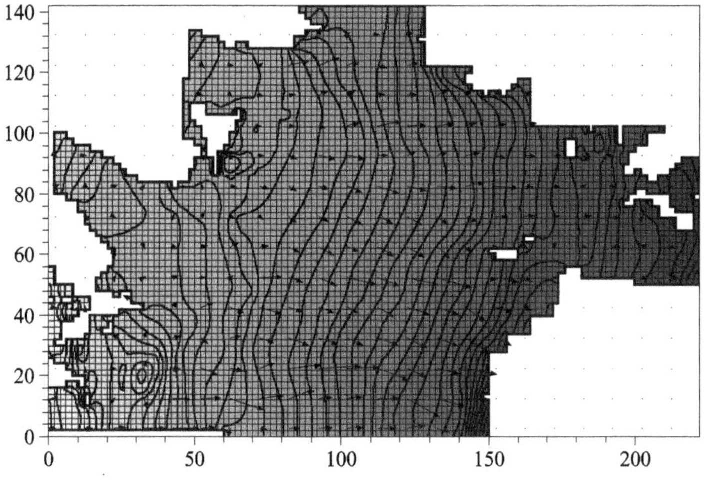

a. Regional groundwater flow (Fig. 4.11b)

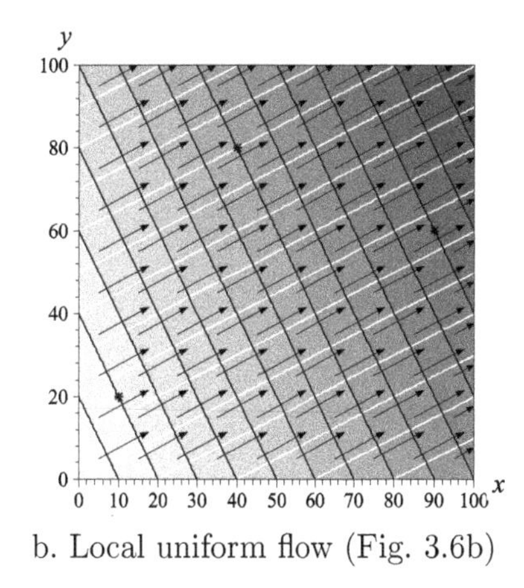

b. Local uniform flow (Fig. 3.6b)

Figure 1.7 *A regional model of groundwater, and approximating the regional flow for another problem as uniform in a local model. (Reprinted from Steward and Allen, 2013, The Analytic Element Method for rectangular gridded domains, benchmark comparisons and application to the High Plains Aquifer,* Advances in Water Resources, *Vol. 60, Fig. 5, with permission from Elsevier.)*

the Dupuit form of discharge per width, (1.17), ***and formulating the problem using the discharge potential, Φ, from Strack (1989)***:

$$\boxed{\mathbf{Q} = -\nabla\Phi}\ , \quad \boxed{\Phi = \begin{cases} \frac{K(h-B)^2}{2} & (0 < h - B \leq H) \\ KH(h-B) - \frac{KH^2}{2} & (H < h - B) \end{cases}}$$
$$\leftrightarrow \quad \boxed{h = B + \begin{cases} \sqrt{\frac{2\Phi}{K}} & \left(0 < \Phi \leq \frac{KH^2}{2}\right) \\ \frac{\Phi}{KH} + \frac{H}{2} & \left(\frac{KH^2}{2} < \Phi\right) \end{cases}} \tag{1.18}$$

when K, B, and H are piecewise constant in an aquifer layer.

Solutions to groundwater problems must also satisfy ***conservation of mass***, which, together with (1.18), leads to the ***Laplace equation***, (2.2a), for divergence-free flows:

$$\nabla \cdot \mathbf{Q} = \frac{\partial Q_x}{\partial x} + \frac{\partial Q_y}{\partial y} = 0 \quad \rightarrow \quad \boxed{\nabla^2\Phi = \frac{\partial^2\Phi}{\partial x^2} + \frac{\partial^2\Phi}{\partial y^2} = 0} \tag{1.19}$$

The examples in Fig. 1.8 illustrate the use of discharge potentials that satisfy this equation to reproduce the ***drawdown associated with wells***, and the interactions between groundwater and ***surface water bodies*** such as lakes. Boundary conditions are utilized in their construction by specifying the pumping rate of the wells, and by specifying the head h^* at locations (x_m, y_m) along the boundary of a lake that is directly connected to the groundwater:

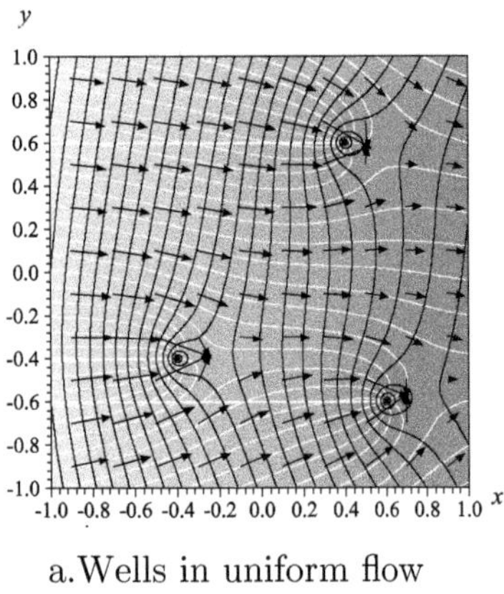

a. Wells in uniform flow (Fig. 3.1c)

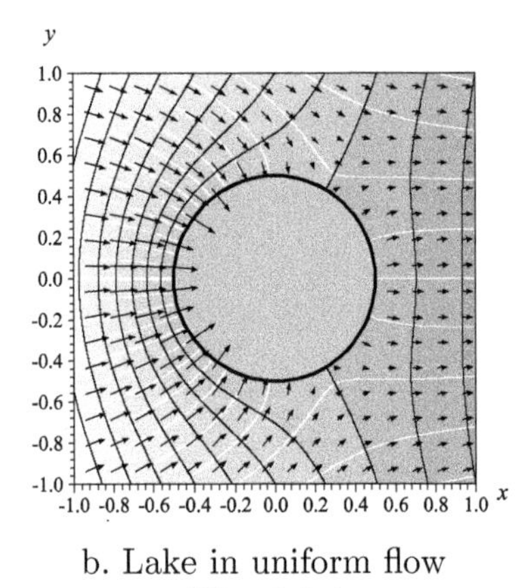

b. Lake in uniform flow (Fig. 3.9a)

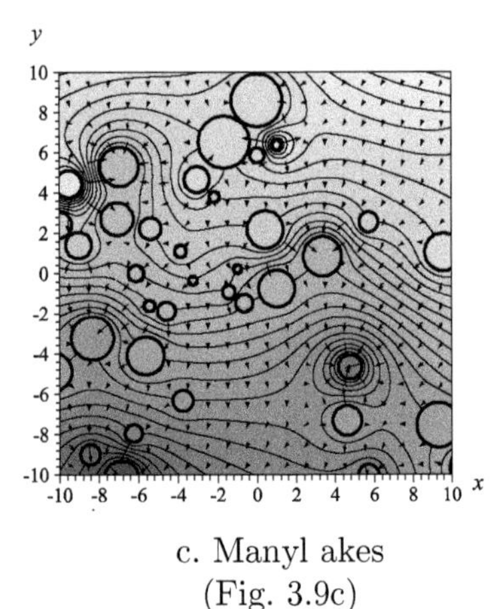

c. Manyl akes (Fig. 3.9c)

Figurc 1.8 *Groundwater interactions with wells and lakes.*

$$h(x_m, y_m) = h^* \quad \rightarrow \quad \Phi(x_m, y_m) = \Phi^* \tag{1.20}$$

Solutions to the Laplace equation with a discharge vector that is both irrotational and divergence-free may be studied using a stream function, (1.14), that is constant along streamlines. These streamlines illustrate the zones where water is captured by a well, as well as those portions of the boundary of a lake fed by groundwater and those where surface water moves from a lake to groundwater. The straight white line to the left of each element is formed by branch cuts, which are discussed later in Section 3.1.

A primary source of groundwater is ***recharge*** from the terrestrial system. Conservation of mass requires the divergence of the discharge per width to equal the recharge rate R, which leads to the ***Poisson equation***, (2.2b):

$$\nabla \cdot \mathbf{Q} = \frac{\partial Q_x}{\partial x} + \frac{\partial Q_y}{\partial y} = R \quad \rightarrow \quad \boxed{\nabla^2 \Phi = \frac{\partial^2 \Phi}{\partial x^2} + \frac{\partial^2 \Phi}{\partial y^2} = -R} \tag{1.21}$$

The recharge illustrated in Fig. 1.9 is modeled using analytic elements that satisfy conservation of energy where the head (and the discharge potential) has the same values outside and inside the recharge area, and conservation of mass where the normal derivative in the n-direction is also continuous at locations (x_m, y_m) on the polygon's boundary is

$$\Phi^+(x_m, y_m) = \Phi^-(x_m, y_m), \quad Q_n^+(x_m, y_m) = Q_n^-(x_m, y_m) \tag{1.22}$$

The notation used to designate sides of interfaces outside an element (the + side) and inside an element (the − side) is further elaborated for the Riemann–Hilbert problem in Section 2.3.1. The uniform recharge within a polygon in Fig. 1.9a is extended to model the ***extraction of groundwater by a plant's, roots*** over a circular area in Fig. 1.9b, and by many plants in Fig. 1.9c. This formulation is also extensible to studying the interactions occurring between aquifer layers interconnected by leaky aquitards (Strack, 1989, sec. 14).

The groundwater and surface water exchanges occurring through ***rivers*** are often approximated using line elements. A river may be ***directly connected to groundwater***, with head h_m specified at a set of M points located at (x_m, y_m) along the river,

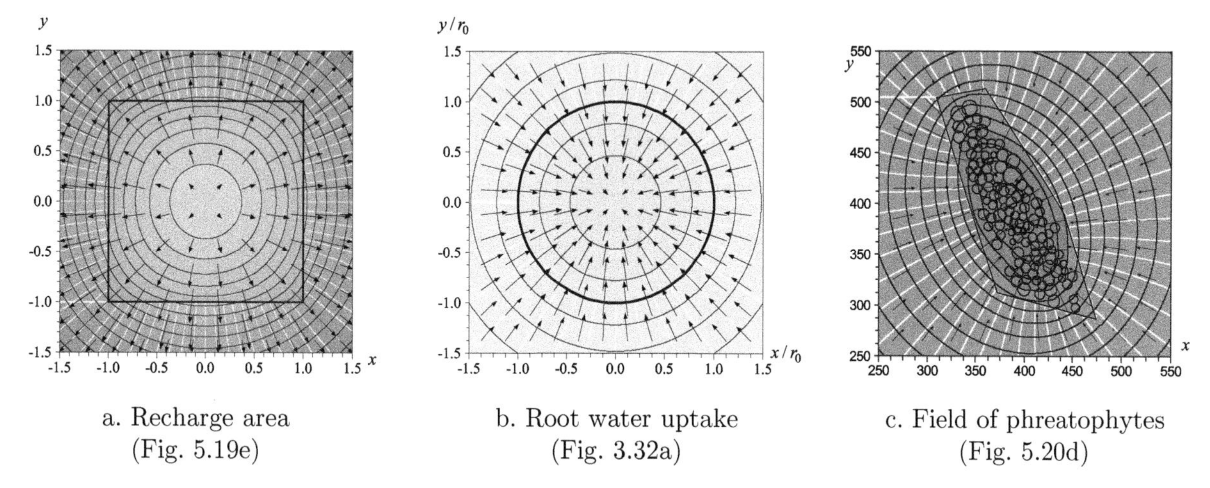

a. Recharge area (Fig. 5.19e)

b. Root water uptake (Fig. 3.32a)

c. Field of phreatophytes (Fig. 5.20d)

Figure 1.9 *Recharge to groundwater and root water uptake by plants (phreatophytes). (Reprinted from Steward and Ahring, 2009, An analytic solution for groundwater uptake by phreatophytes spanning spatial scales from plant to field to regional,* Journal of Engineering Mathematics, *Vol. 64(2), Figs. 2,5, used with permission from Springer Nature; distributed under Creative Commons license (Attribution-Noncommercial), https://creativecommons.org.)*

$$h(x_m, y_m) = h_m \quad \rightarrow \quad \Phi(x_m, y_m) = \Phi_m \qquad (m = 1, \cdots, M) \tag{1.23}$$

where the potential Φ_m are obtained from h_m using (1.18). This boundary condition is illustrated in Figs. 1.10a and 1.10b, where head varies linearly along each river segment. A river may also be separated from groundwater by riverbed sediments of hydraulic conductivity K^* and vertical thickness H^* that cause a difference in head between the surface water h^* and the groundwater h. In this case, the discharge per length of the river flowing into the aquifer, ΔQ_n, is equal to minus the vertical component of flow through the riverbed times the width of the river w^*,

$$Q_n^+ - Q_n^- = \Delta Q_n = -w^* q_z = -w^* K^* \frac{(h - h^*)}{H^*} \tag{1.24}$$

using Darcy's law (1.16). The riverbed in Fig. 1.10c has sediments that satisfy boundary conditions at points m expressed in terms of the ***resistance of the river bed sediments***, c^*,

$$(\Delta Q_n)_m = -w^* c^* (h_m - h^*), \qquad c^* = \frac{K^*}{H^*} \tag{1.25}$$

where this head may be expressed in terms of the discharge potential, (1.18).

Groundwater flow is influenced by the geological medium through which water flows. Porous media may be grouped into ***heterogeneity zones*** where jumps in an aquifer layer may occur in the base elevation (B^+ to B^-), thickness (H^+ to H^-), and/or hydraulic conductivity (K^+ to K^-), across the boundary of a zone. Heterogeneities may either be isolated (like the polygon elements in Figs. 1.11a and 1.11b) or form adjacent elements that tile a domain (like the rectangle elements in Fig. 1.11c). Each interface satisfies ***conservation of energy*** and ***conservation of mass***, like the recharge zone in (1.22):

$$h^+(x_m, y_m) = h^-(x_m, y_m), \quad Q_n^+(x_m, y_m) = Q_n^-(x_m, y_m) \tag{1.26a}$$

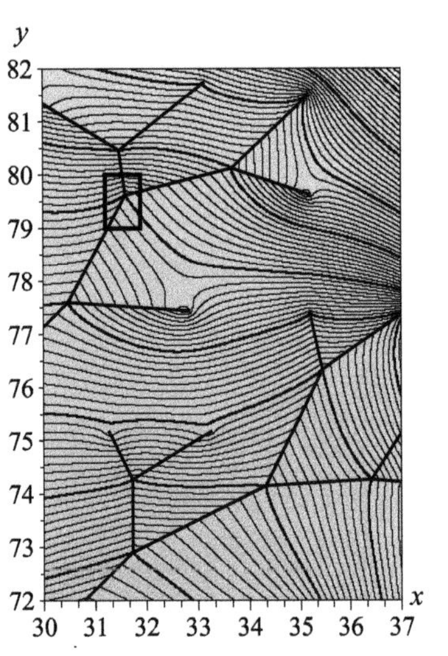

a. Interconnected river
(Fig. 3.26b)

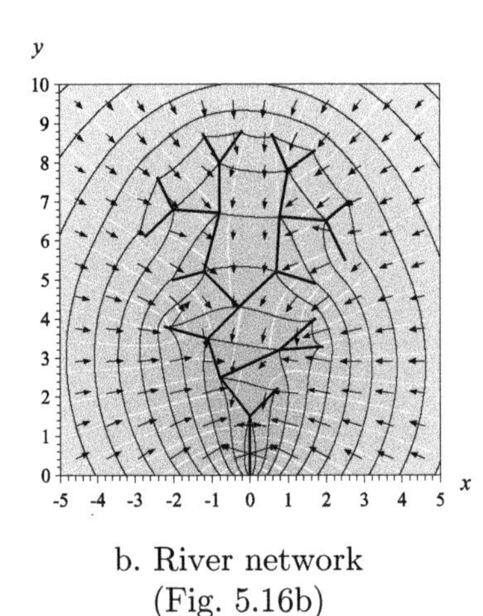

b. River network
(Fig. 5.16b)

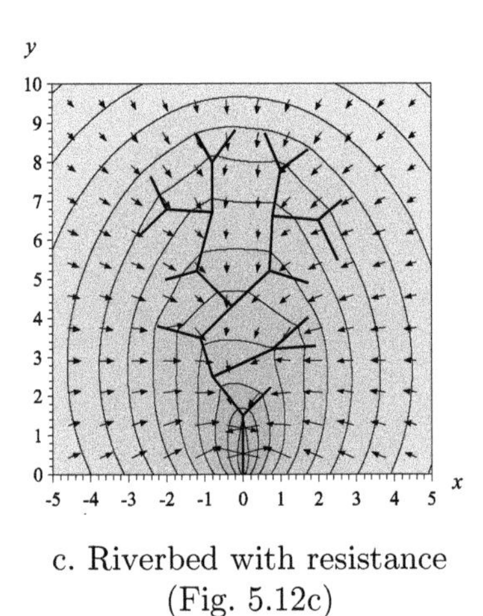

c. Riverbed with resistance
(Fig. 5.12c)

Figure 1.10 *River networks with specified head or a riverbed with resistance. (Reprinted from Steward, 2015, Analysis of discontinuities across thin inhomogeneities, groundwater/surface water interactions in river networks, and circulation about slender bodies using slit elements in the Analytic Element Method,* Water Resources Research, *Vol. 51(11), Fig. 6, with permission from John Wiley and Sons. Copyright 2015 by the American Geophysical Union.)*

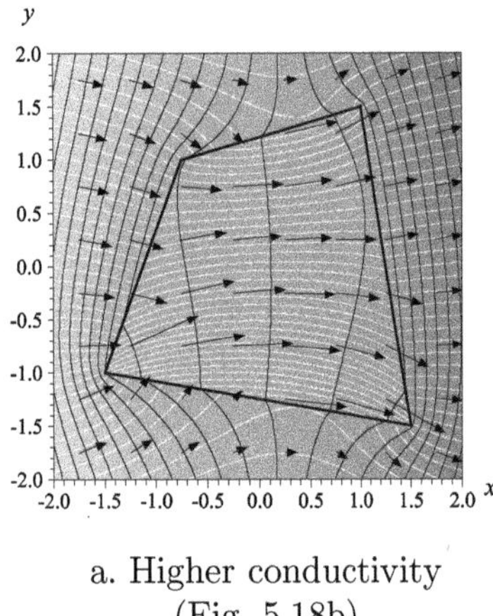

a. Higher conductivity
(Fig. 5.18b)

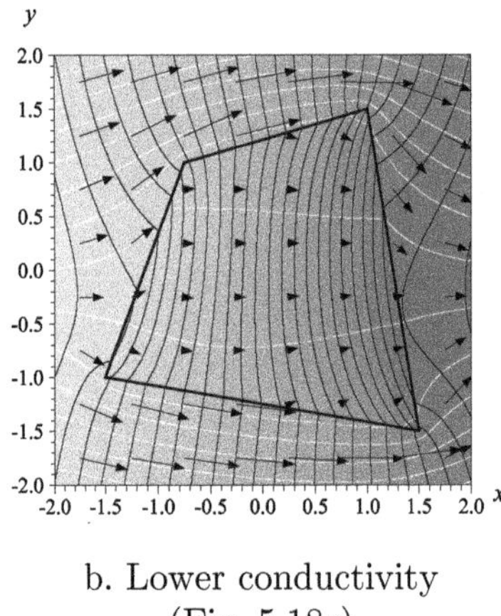

b. Lower conductivity
(Fig. 5.18c)

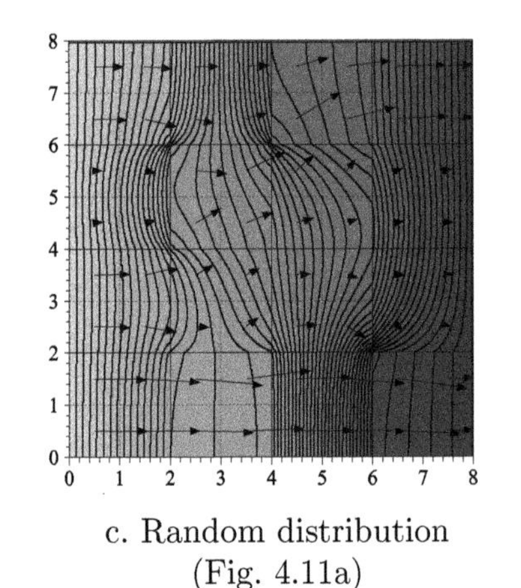

c. Random distribution
(Fig. 4.11a)

Figure 1.11 *Heterogeneity with the geometry of a polygon or interconnected rectangles. (Reprinted from Steward and Allen, 2013, The Analytic Element Method for rectangular gridded domains, benchmark comparisons and application to the High Plains Aquifer,* Advances in Water Resources, *Vol. 60, Fig. 3, with permission from Elsevier.)*

Heterogeneities with a jump only in hydraulic conductivity lead to the following condition for the potential,

$$\frac{\Phi^+(x_m, y_m)}{K^+} = \frac{\Phi^-(x_m, y_m)}{K^-} = 0 \qquad B^+ = B^-,\ H^+ = H^- \tag{1.26b}$$

and conditions for discontinuous base and thickness follow from Steward (2007). The contours of head and the vector field in Fig. 1.11 illustrate the continuous head and normal component of flow across the interfaces of heterogeneities.

Thin geologic features with a small cross-sectional width w^* and properties different than those in the surrounding media form heterogeneities that may be approximated using line elements. A ***thin heterogeneity with higher conductivity*** K^* than K in the surrounding media provides a conduit through which groundwater travels more readily than in the surrounding media, with examples in Fig. 1.12. Such an element has a continuous potential across the element, and the discharge per width in the $\mathfrak{s}$-direction tangential to the heterogeneity may be expressed in terms of the stream function or the partial derivative of the potential by

$$\Phi^+ = \Phi^-, \quad Q_{\mathfrak{s}} = \frac{K}{K^* w^*}\left(\Psi^- - \Psi^+\right) = -w^* \frac{K^*}{K}\frac{\partial \Phi}{\partial \mathfrak{s}} \tag{1.27a}$$

In the limiting case of an infinitely conductive heterogeneity, the potential is uniform along the element:

$$K^* \to \infty \quad \to \quad \frac{\partial \Phi}{\partial \mathfrak{s}} = 0 \tag{1.27b}$$

Analytic elements may also be developed to satisfy the boundary conditions of ***thin heterogeneities with lower conductivity*** than the surrounding media. The flow through

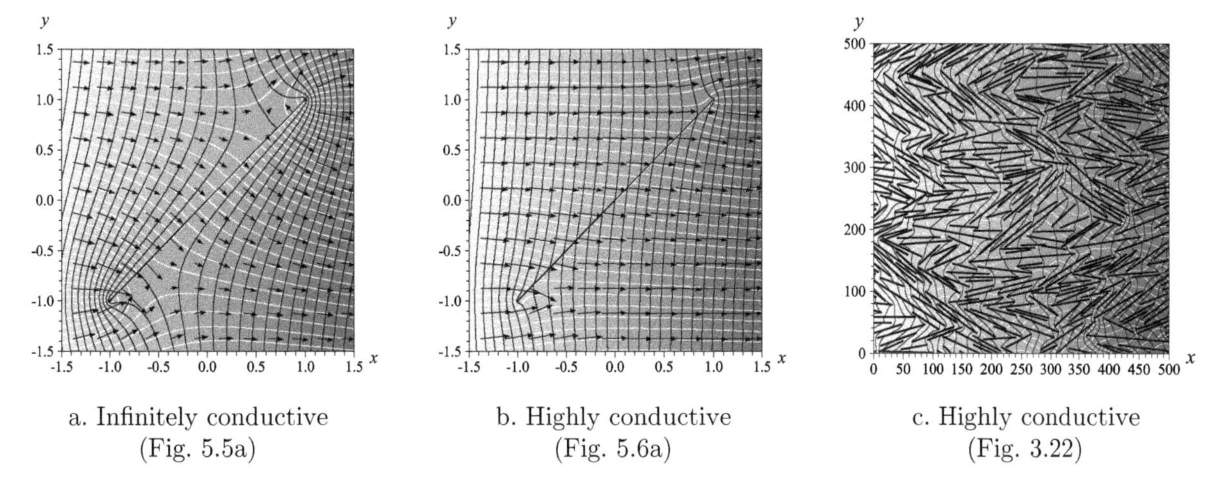

Figure 1.12 *Thin groundwater features with high conductivity. (Reprinted from Steward, 2015, Analysis of discontinuities across thin inhomogeneities, groundwater/surface water interactions in river networks, and circulation about slender bodies using slit elements in the Analytic Element Method,* Water Resources Research, *Vol. 51(11), Fig. 4, with permission from John Wiley and Sons. Copyright 2015 by the American Geophysical Union.)*

such a leaky element is related to the jump in discharge potential across the element, and the stream function is continuous across the element:

$$Q_n = -\frac{K^*}{Kw^*}\left(\Phi^+ - \Phi^-\right), \quad \Psi^+ = \Psi^- \tag{1.28a}$$

In the limiting case of zero conductivity, there is no normal component of flow, and the stream function is uniform along the element

$$K^* = 0 \quad \rightarrow \quad Q_n = \frac{\partial \Psi}{\partial n} = 0 \tag{1.28b}$$

Such thin geologic features provide a barrier that impedes groundwater flow, as illustrated in Fig. 1.13.

Analytic elements may also be developed for aquifer features that generate ***three-dimensional flow***. The specific discharge vector for such problems satisfy Darcy's law, (1.16), which is represented here in terms of a three-dimensional specific discharge potential $\Phi(x, y, z)$, and this together with conservation of mass gives the Laplace equation:

$$\begin{aligned} \mathbf{q} &= -\nabla\Phi, \quad \boxed{\Phi = Kh} \\ \nabla \cdot \mathbf{q} &= \frac{\partial q_x}{\partial x} + \frac{\partial q_y}{\partial y} + \frac{\partial q_z}{\partial z} = 0 \end{aligned} \quad \rightarrow \quad \nabla^2\Phi = \boxed{\frac{\partial^2 \Phi}{\partial x^2} + \frac{\partial^2 \Phi}{\partial y^2} + \frac{\partial^2 \Phi}{\partial z^2} = 0} \tag{1.29}$$

The average groundwater velocity $\mathbf{v}$ is related to the specific discharge by

$$\boxed{v_x = \frac{q_x}{n}, \quad v_y = \frac{q_y}{n}, \quad v_z = \frac{q_z}{n}} \tag{1.30}$$

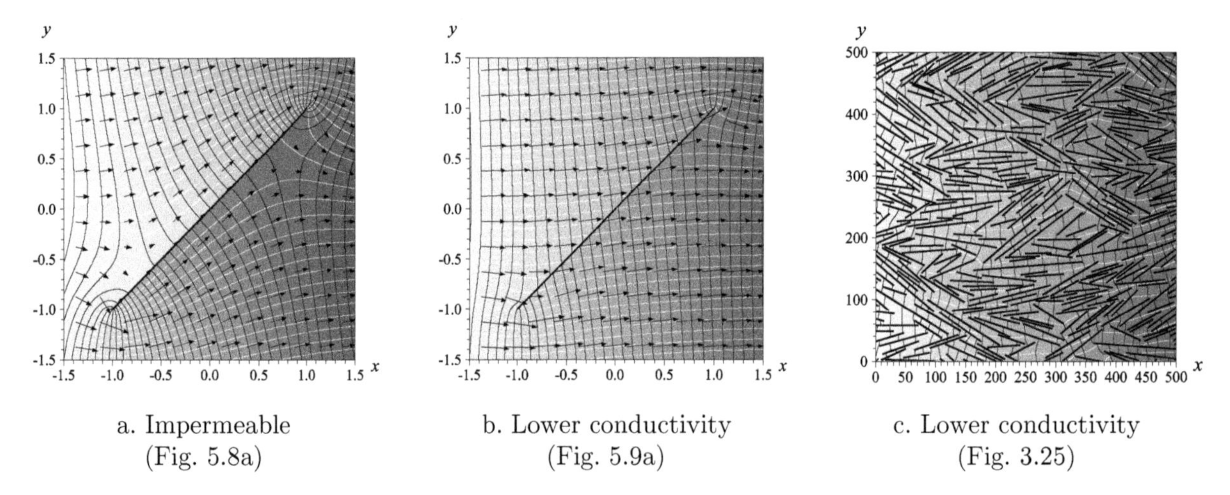

Figure 1.13 *Thin groundwater features with low conductivity. (Reprinted from Steward, 2015, Analysis of discontinuities across thin inhomogeneities, groundwater/surface water interactions in river networks, and circulation about slender bodies using slit elements in the Analytic Element Method,* Water Resources Research, *Vol. 51(11), Fig. 5, with permission from John Wiley and Sons. Copyright 2015 by the American Geophysical Union.)*

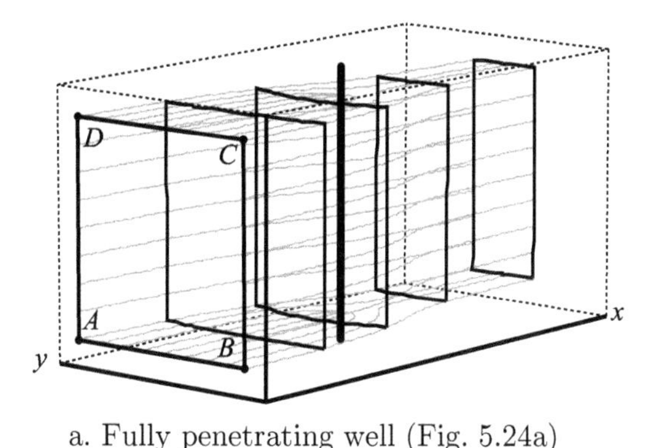

a. Fully penetrating well (Fig. 5.24a)

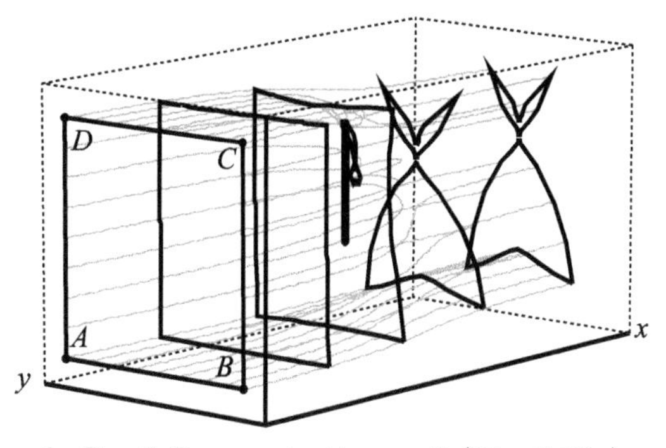

b. Partially penetrating well (Fig. 5.25a)

Figure 1.14 *Three-dimensional streamlines along stream surfaces connecting points A, B, C, and D near fully penetrating and partially penetrating wells. (Reprinted from Steward, 1998, Stream surfaces in two-dimensional and three-dimensional divergence-free flows,* Water Resources Research, *Vol. 34(5), Figs. 3,5, with permission from John Wiley and Sons. Copyright 1998 by the American Geophysical Union.)*

where the porosity, n, is a ratio of the volume of voids between soil particles divided by the total volume. The examples in Fig. 1.14 illustrate three-dimensional ***pathlines*** existing around wells (Steward and Jin, 2001), which are obtained by tracking particles that originate along the straight lines connecting points A, B, C, and D and travel with the groundwater velocity $\mathbf{v}$.

Three-dimensional pathlines occur as groundwater travels ***through and around heterogeneities in geologic media***. This is illustrated in Fig. 1.15 for inclusions, where the conductivity inside the three-dimensional element is higher than that of the surrounding soil. Groundwater becomes transported faster in such heterogeneity, resulting in a lateral shift in the relative position of water particles as observed by deformations to the geometry of upgradient and downgradient stream surfaces. This phenomena is called ***advective mixing*** and contributes towards transverse dispersion in contaminant transport (Janković et al., 2009).

A ***vertical component of flow for quasi three-dimensional flow***, q_z, is also generated for the two-dimensional formulation of the discharge potential, (1.18), due to recharge R occurring into the top of an unconfined aquifer and due to the gradient of head (Anderson, 2005). This component is obtained by vertically integrating the three-dimensional conservation of mass equation (1.29) from the base of an aquifer layer to the point z where q_z is to be evaluated (Strack, 1984)

$$q_z(x, y, z) = -\int_B^z \left(\frac{\partial q_x}{\partial x} + \frac{\partial q_y}{\partial y} \right) d\tilde{z} = -(z - B) \left(\frac{\partial q_x}{\partial x} + \frac{\partial q_y}{\partial y} \right) \tag{1.31a}$$

where the Dupuit assumption (1.17) requires q_x and q_y to remain constant in the z-direction. These partial derivatives are expressed in terms of the discharge per width using (1.17)

$$\frac{\partial q_x}{\partial x} + \frac{\partial q_y}{\partial y} = \frac{1}{h - B} \left(\frac{\partial Q_x}{\partial x} + \frac{\partial Q_y}{\partial y} \right) - \frac{1}{(h - B)^2} \left[Q_x \frac{\partial (h - B)}{\partial x} - Q_y \frac{\partial (h - B)}{\partial y} \right] \tag{1.31b}$$

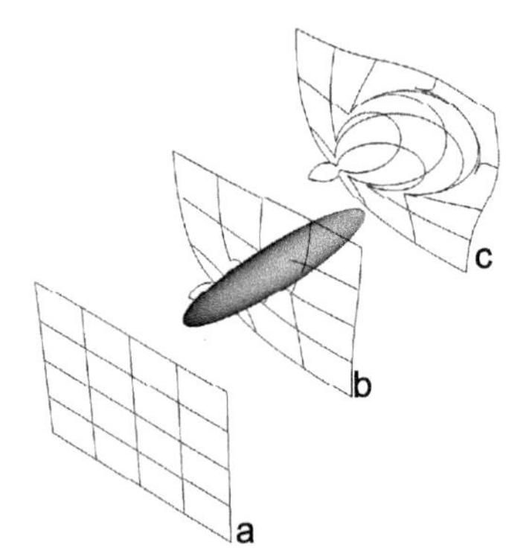

a. Prolate spheroids (Fig. 4.32)

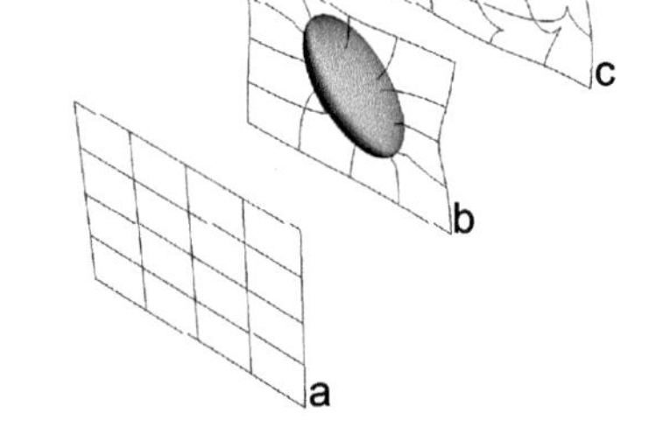

b. Oblate spheroids (Fig. 4.35)

Figure 1.15 *Three-dimensional stream surfaces near a highly conductive heterogeneity. (Reprinted from Janković, Steward, Barnes, and Dagan, 2009, Is transverse macrodispersivity in three-dimensional transport through isotropic heterogeneous formation equal to zero? A counterexample,* Water Resources Research, *Vol. 45(8), Fig. 2, with permission from John Wiley and Sons. Copyright 2009 by the American Geophysical Union.)*

These terms may be simplified using conservation of mass, (1.19), and rearranging the relations between discharge potential, head and discharge per width (1.18):

$$\frac{\partial Q_x}{\partial x} + \frac{\partial Q_y}{\partial y} = R, \qquad \begin{aligned} Q_x \frac{\partial (h-B)}{\partial x} &= \frac{Q_x}{K(h-B)} \frac{\partial \Phi}{\partial x} = -\frac{{Q_x}^2}{K(h-B)} \\ Q_y \frac{\partial (h-B)}{\partial y} &= \frac{Q_y}{K(h-B)} \frac{\partial \Phi}{\partial y} = -\frac{{Q_y}^2}{K(h-B)} \end{aligned} \tag{1.31c}$$

Together, the equations in (1.31) give the vertical component of flow:

$$q_z = -\frac{z-B}{h-B}\left(R + \frac{{Q_x}^2 + {Q_y}^2}{K(h-B)^2}\right) \tag{1.32}$$

Groundwater computations:

- head, h, is obtained from Φ by (1.18)
- discharge per width, $\mathbf{Q}$ with components (Q_x, Q_y), is obtained for two-dimensional flow from minus the gradient of Φ in (1.18);
- horizontal components of the specific discharge, (q_x, q_y), are obtained by dividing $\mathbf{Q}$ by the depth of the saturated zone in (1.17), and the vertical component of the specific discharge, q_z, is obtained from (1.32); or $\mathbf{q}$ is obtained for three-dimensional flow from minus the gradient of Φ in (1.32);
- velocity, $\mathbf{v}$, with components (v_x, v_y, v_z), is obtained by dividing $\mathbf{q}$ by the porosity n in (1.30)

The three-dimensional streamlines associated with groundwater uptake by phreatophytes are illustrated in Fig. 1.16.

Example 1.1 This example illustrates computation of head and flow in unconfined and confined aquifer layers for the flow in Fig. 1.17. with a thin aquitard separating zones 2 and 3. The discharge potential at the boundaries of the four zones in Fig. 1.17 may be obtained from the specified heads h_A and h_D and the computed heads:

$$h_B = \sqrt{\frac{R}{K} L_1(L_2 + L_4) + \frac{L_2 + L_4}{L_1 + L_2 + L_4} {h_A}^2 + \frac{L_1}{L_1 + L_2 + L_4} {h_D}^2}$$

$$h_C = \sqrt{\frac{R}{K} L_4(L_1 + L_2) + \frac{L_4}{L_1 + L_2 + L_4} {h_A}^2 + \frac{L_1 + L_2}{L_1 + L_2 + L_4} {h_D}^2}$$

Within each zone, the general solution to the Poisson equation, (1.21), is

$$\frac{d^2\Phi}{dx^2} = -R \quad \rightarrow \quad \begin{aligned} \Phi &= -\frac{R}{2}x^2 + c_1 x + c_0 \\ Q_x &= Rx - c_1 \end{aligned}$$

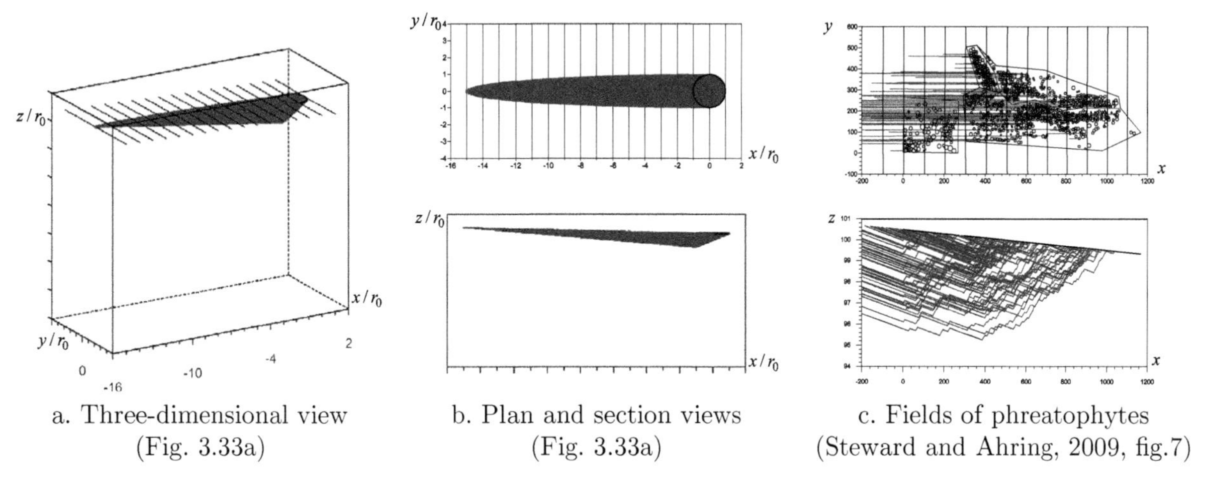

a. Three-dimensional view (Fig. 3.33a)

b. Plan and section views (Fig. 3.33a)

c. Fields of phreatophytes (Steward and Ahring, 2009, fig.7)

Figure 1.16 *Quasi three-dimensional capture zones for root water uptake from groundwater by a field of trees, tracing pathlines backwards from tree locations. (Reprinted from Steward and Ahring, 2009, An analytic solution for groundwater uptake by phreatophytes spanning spatial scales from plant to field to regional,* Journal of Engineering Mathematics, *Vol. 64(2), Figs. 6,7, used with permission from Springer Nature; distributed under Creative Commons license (Attribution-Noncommercial), https://creativecommons.org.)*

where the coefficients c_0 and c_1 are adjusted to give the correct value of discharge potential and head at the edges of each zone. This gives the following for zone 1 ($0 \leq x < L_1$ and $B_1 \leq z \leq h$)

$$\begin{aligned} \Phi &= -\frac{R}{2}x(x - L_1) - \frac{\Phi_{A1} - \Phi_{B1}}{L_1}x + \Phi_{A1} \\ Q_x &= R\left(x - \frac{L_1}{2}\right) + \frac{\Phi_{A1} - \Phi_{B1}}{L_1} \end{aligned}, \qquad \begin{aligned} \Phi_{A1} &= \frac{K}{2}{h_A}^2 \\ \Phi_{B1} &= \frac{K}{2}{h_B}^2 \end{aligned}$$

for zone 2 ($L_1 \leq x < L_1 + L_2$ and $B_2 \leq z \leq h$)

$$\begin{aligned} \Phi &= -\frac{R}{2}(x - L_1)(x - L_1 - L_2) - \frac{\Phi_{B2} - \Phi_{C2}}{L_2}(x - L_1) + \Phi_{B2} \\ Q_x &= R\left(x - L_1 - \frac{L_2}{2}\right) + \frac{\Phi_{B2} - \Phi_{C2}}{L_2} \end{aligned}, \qquad \begin{aligned} \Phi_{B2} &= \frac{K}{2}(h_B - B_2)^2 \\ \Phi_{C2} &= \frac{K}{2}(h_C - B_2)^2 \end{aligned}$$

for zone 3 ($L_1 \leq x < L_1 + L_2$ and $B_3 \leq z \leq H_3$)

$$\begin{aligned} \Phi &= -\frac{\Phi_{B3} - \Phi_{C3}}{L_2}(x - L_1) + \Phi_{B3} \\ Q_x &= \frac{\Phi_{B3} - \Phi_{C3}}{L_2} \end{aligned}, \qquad \begin{aligned} \Phi_{B3} &= KH_3h_B - \frac{K{H_3}^2}{2} \\ \Phi_{C3} &= KH_3h_C - \frac{K{H_3}^2}{2} \end{aligned}$$

and for zone 4 ($L_1 + L_2 \leq x \leq L_1 + L_2 + L_4$ and $B_4 \leq z \leq h$)

$$\begin{aligned} \Phi &= -\frac{R}{2}(x - L_1 - L_2)(x - L_1 - L_2 - L_4) - \frac{\Phi_{C4} - \Phi_{D4}}{L_4}(x - L_1 - L_2) + \Phi_{C4} \\ Q_x &= R\left(x - L_1 - L_2 - \frac{L_4}{2}\right) + \frac{\Phi_{C4} - \Phi_{D4}}{L_4} \end{aligned}, \qquad \begin{aligned} \Phi_{C4} &= \frac{K}{2}{h_C}^2 \\ \Phi_{D4} &= \frac{K}{2}{h_D}^2 \end{aligned}$$

similar to Strack (1981).

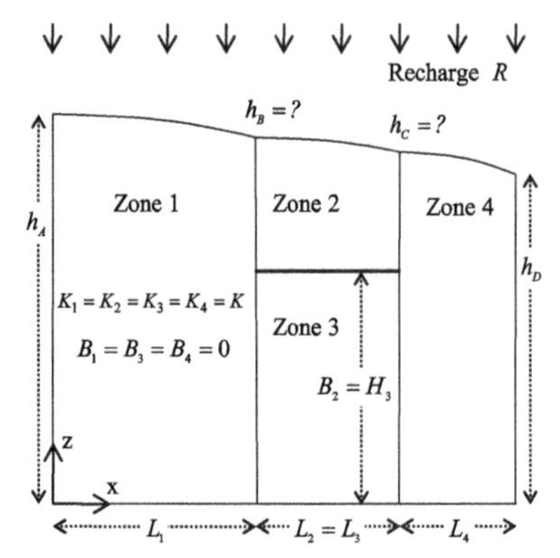

a. Ground water properties

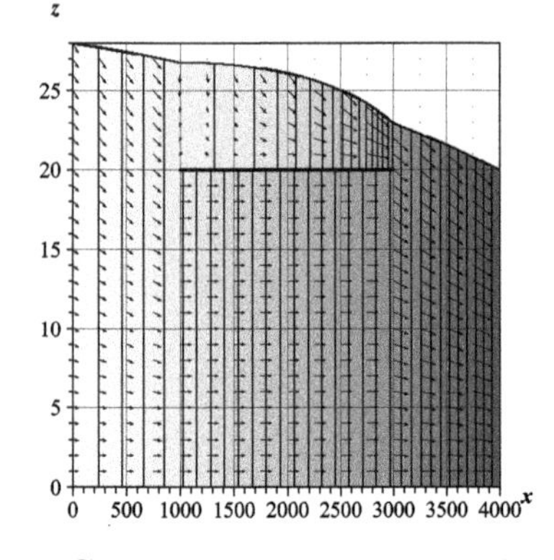

b. Groundwater elevation and velocit

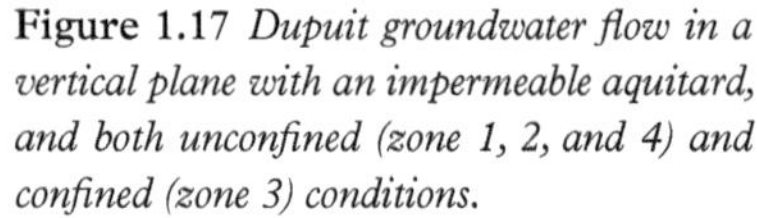

Figure 1.17 *Dupuit groundwater flow in a vertical plane with an impermeable aquitard, and both unconfined (zone 1, 2, and 4) and confined (zone 3) conditions.*

This groundwater head and velocity is illustrated in Fig. 1.17b for the variables $L_1 = 1000$ m, $L_2 = 2000$ m, $L_4 = 1000$ m, $B_2 = H_3 = 20$ m, $K = 30$ m/day, $R = 0.0003$ m/day, $n = 0.25$, $h_A = 28$ m, and $h_D = 20$ m. Note that recharge R enters the top of zones 1, 2, and 4, and no recharge seeps through the confining layer into zone 3. Conservation of energy is explicitly satisfied by requiring that the head be equal to h_B and h_C across respective zones. Conservation of mass may be verified at $x = L_1$ by computing the discharge per width on the left side in zone 1 and on the right side in zones 2 and 3:

$$\begin{aligned} Q_x^- &= Q_x|_{\text{Zone 1}} = 1.14 \text{ m}^2/\text{day} \\ Q_x^+ &= Q_x|_{\text{Zone 2}} + Q_x|_{\text{Zone 3}} = 1.14 \text{ m}^2/\text{day} \end{aligned} \qquad (x = L_1)$$

Likewise, continuity of flow may be verified at $x = L_1 + L_2$ on the left and right sides of the interface

$$\begin{aligned} Q_x^- &= Q_x|_{\text{Zone 2}} + Q_x|_{\text{Zone 3}} = 1.74 \text{ m}^2/\text{day} \\ Q_x^+ &= Q_x|_{\text{Zone 4}} = 1.74 \text{ m}^2/\text{day} \end{aligned} \qquad (x = L_1 + L_2)$$

The values of Φ and Q_x may be computed at any point to provide solutions for head (1.18), specific discharge (1.17) and (1.32), and velocity (1.30).

Problem 1.2 Compute the head (h in m) and the horizontal and vertical components of the velocity (v_x, v_z in m/day) for the aquifer in example 1.1 at the following locations:

A. $x = 0$ m and $z = h$ (zone 1)
B. $x = 0$ m and $z = h/2$ (zone 1)
C. $x = 500$ m and $z = h$ (zone 1)
D. $x = 1000^-$ m and $z = h$ (zone 1)
E. $x = 1000^+$ m and $z = h$ (zone 2)
F. $x = 2000$ m and $z = h$ (zone 2)
G. $x = 3000^-$ m and $z = h$ (zone 2)
H. $x = 1000^+$ m and $z = H_3$ (zone 3)
I. $x = 2000$ m and $z = H_3$ (zone 3)
J. $x = 3000^-$ m and $z = H_3$ (zone 3)
K. $x = 3000^+$ m and $z = h$ (zone 4)
L. $x = 3500$ m and $z = h$ (zone 4)
M. $x = 4000$ m and $z = h$ (zone 4)
N. $x = 4000$ m and $z = h/2$ (zone 4)

1.2.2 Vadose Zone Flow: Pressure and Seepage Velocity

Studies of vadose zone flow examine the movement of water from the land surface to groundwater as depicted earlier in Fig. 1.6. In the vadose zone, water seeps downward through pores between soil particles that are partially filled with water and partially filled with air. Solutions are presented in terms of the ***pressure head*** p from ***Bernoulli's equation*** (Bernoulli, 1738)

$$h = p + z\,, \qquad p = \frac{P}{\rho g} \tag{1.33}$$

where head is partitioned into pressure, P, and elevation terms with the z-axis directed upward against gravity, and the kinetic velocity head $|\mathbf{v}|^2/(2g)$ is ignored since it is small. For example, the pressure head distribution is shown in Fig. 1.18 for three different soils experiencing the same rate of recharge above a saturated groundwater table. While $p = 0$ at $z = 0$, it decreases asymptotically at higher elevations towards a background suction pressure head (Raats and Gardner, 1974).

The governing equations for vadose zone processes contain non-linear relations since water moves more readily through moist soil (higher p) than dry soil (lower p). This is observed in ***Darcy's law***, (1.16),

$$q_x = -K(p)\frac{\partial p}{\partial x}, \quad q_y = -K(p)\frac{\partial p}{\partial y}, \quad q_z = -K(p)\frac{\partial p}{\partial z} - K(p) \tag{1.34}$$

with hydraulic conductivity expressed as a function of pressure head. The ***Richards equation*** (Richards, 1931) is obtained by substituting this form of Darcy's law into conservation of mass

$$-\nabla \cdot \mathbf{q} = \frac{\partial}{\partial x}\left[K(p)\frac{\partial p}{\partial x}\right] + \frac{\partial}{\partial y}\left[K(p)\frac{\partial p}{\partial y}\right] + \frac{\partial}{\partial z}\left[K(p)\frac{\partial p}{\partial z}\right] + \frac{\partial K(p)}{\partial z} = 0 \tag{1.35}$$

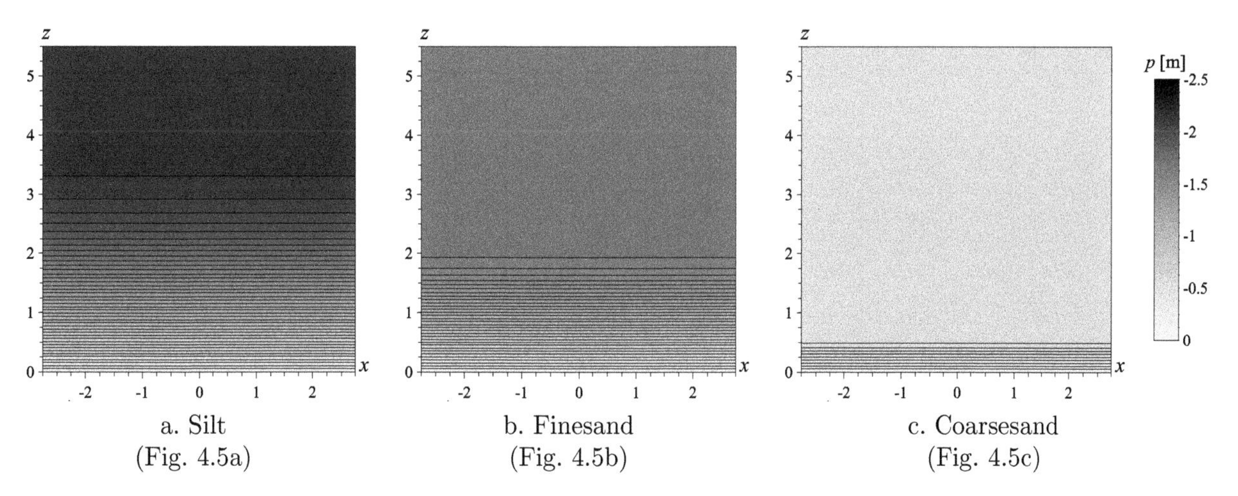

Figure 1.18 *Pressure head distribution above saturated groundwater.*

Table 1.1 *Representative saturated hydraulic conductivity K_s and sorptive number a for the Gardner equation, (1.38), matric flux potential at saturation F_s, (1.40), and modified Helmholtz coefficient k, (1.39) from Steward (2016, table 1).*

Soil type	K_s [m/day]	a [1/m]	F_s [m^2/day]	k [1/m]
Silt	0.02	2	0.01	1
Fine sand	1	5	0.2	2.5
Coarse sand	35	25	1.4	12.5

This equation may be rearranged using the matric flux potential, denoted F, defined by the ***Kirchhoff transformation*** (Kirchhoff, 1894) as

$$F = \int_{-\infty}^{p} K(\tilde{p})\,\mathrm{d}\tilde{p} \quad \rightarrow \quad \frac{\partial F}{\partial x} = K\frac{\partial p}{\partial x}, \quad \frac{\partial F}{\partial y} = K\frac{\partial p}{\partial y}, \quad \frac{\partial F}{\partial z} = K\frac{\partial p}{\partial z} \tag{1.36}$$

to provide a linear form of the Richards equation

$$\nabla^2 F = \frac{\partial^2 F}{\partial x^2} + \frac{\partial^2 F}{\partial y^2} + \frac{\partial^2 F}{\partial z^2} = -\frac{\mathrm{d}K}{\mathrm{d}p}\frac{1}{K}\frac{\partial F}{\partial z} \tag{1.37}$$

where the last term was rearranged using the chain rule for differentiation $\frac{\partial K}{\partial z} = \frac{\mathrm{d}K}{\mathrm{d}p}\frac{\partial p}{\partial z}$, with $\frac{\partial p}{\partial z}$ from (1.36).

A solution to flow in the vadose zone requires specification of the soil properties relating the hydraulic conductivity $K(p)$ to the pressure head. While many functional forms exist, it is convenient to use the ***Gardner equation*** (Gardner, 1958), which provides ***the "quasi-linear" formulation of the Richards equation*** (Philip, 1968; Pullan, 1990),

$$\boxed{K(p) = K_s \mathrm{e}^{ap}} \quad \rightarrow \quad \frac{\mathrm{d}K}{\mathrm{d}p}\frac{1}{K} = a \quad \rightarrow \quad \nabla^2 F = -a\frac{\partial F}{\partial z} \tag{1.38}$$

where a is the sorptive number, and the saturated conductivity K_s occurs at $p = 0$. The first derivative on the right-hand side of the last equation may be rearranged by changing variables to a potential function Φ and constant k (Wooding, 1968; Knight et al., 1989), giving the ***modified Helmholtz equation***, (2.2d):

$$\begin{matrix} F = \Phi \mathrm{e}^{-kz} \\ \boxed{a = 2k} \end{matrix} \quad \rightarrow \quad \boxed{\nabla^2 \Phi = \frac{\partial^2 \Phi}{\partial x^2} + \frac{\partial^2 \Phi}{\partial y^2} + \frac{\partial^2 \Phi}{\partial z^2} = k^2 \Phi} \tag{1.39}$$

The Gardner equation (1.38) also provides a relation between the matric flux potential and pressure head by integrating the Kirchhoff transformation (1.36):

$$F = \int_{-\infty}^{p} K_s \mathrm{e}^{a\tilde{p}}\,\mathrm{d}\tilde{p} = F_s \mathrm{e}^{ap}, \quad \boxed{F_s = \frac{K_s}{a}} \tag{1.40}$$

where F_s is the matric flux potential at $p = 0$. The representative values for soil properties in Table 1.1 were used to construct Fig. 1.18.

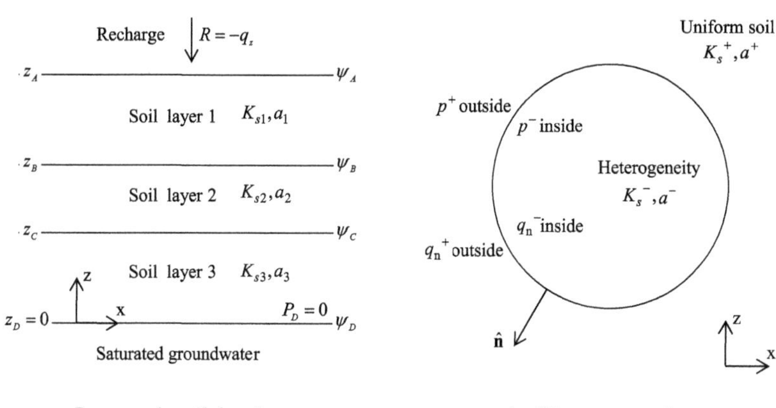

Figure 1.19 *Typical vadose zone problems of seepage through horizontal soil layers and through embedded heterogeneities.*

While solutions to vadose zone problems are formulated in terms of the potential Φ, solutions are represented in terms of pressure head h and flow. Expressions relating pressure head and potential are obtained by equating the expressions for F in the last two equations, and rearranging terms as,

$$\boxed{\Phi = F_s e^{k(z+2p)}} \quad \leftrightarrow \quad \boxed{p = \frac{1}{2k}\ln\frac{\Phi e^{-kz}}{F_s} = -\frac{z}{2} + \frac{1}{2k}\ln\frac{\Phi}{F_s}} \tag{1.41}$$

Likewise, expressions that provide the specific discharge from Φ are obtained from Darcy's law (1.34) with the Kirchhoff transformation (1.36),

$$\begin{aligned} q_x &= -\frac{\partial F}{\partial x} && = -e^{-kz}\frac{\partial \Phi}{\partial x} \\ q_y &= -\frac{\partial F}{\partial y} && = -e^{-kz}\frac{\partial \Phi}{\partial y} \\ q_z &= -\frac{\partial F}{\partial z} - aF && = -e^{-kz}\left(\frac{\partial \Phi}{\partial z} + k\Phi\right) \end{aligned} \tag{1.42}$$

using derivative of the matric flux, (1.39).

Soils may be placed in horizons with layered soils of different types as depicted in Fig. 1.19a. Such problems require the pressure head to be continuous across the interface of adjacent soil layers, and solutions are shown in Fig. 1.20 for horizontal layers of alternating soil types with the same recharge rate as the single soil in Fig. 1.18. The continuous pressure head and vertical component of flow between soil layers is explored shortly in Example 1.2.

Soils may also contain heterogeneities with one soil type with properties K_s^- and a^- embedded within a uniform soil with properties K_s^+, a^+ as illustrated in Fig. 1.19b. Such interfaces satisfy conservation of energy, with the same value of pressure head outside and inside the heterogeneity, which leads to a non-linear boundary condition

$$p^+ = p^- \quad \rightarrow \quad \left(\frac{e^{-k^+z}}{F_s^+}\Phi^+\right)^{\frac{1}{2k^+}} = \left(\frac{e^{-k^-z}}{F_s^-}\Phi^-\right)^{\frac{1}{2k^-}} \tag{1.43}$$

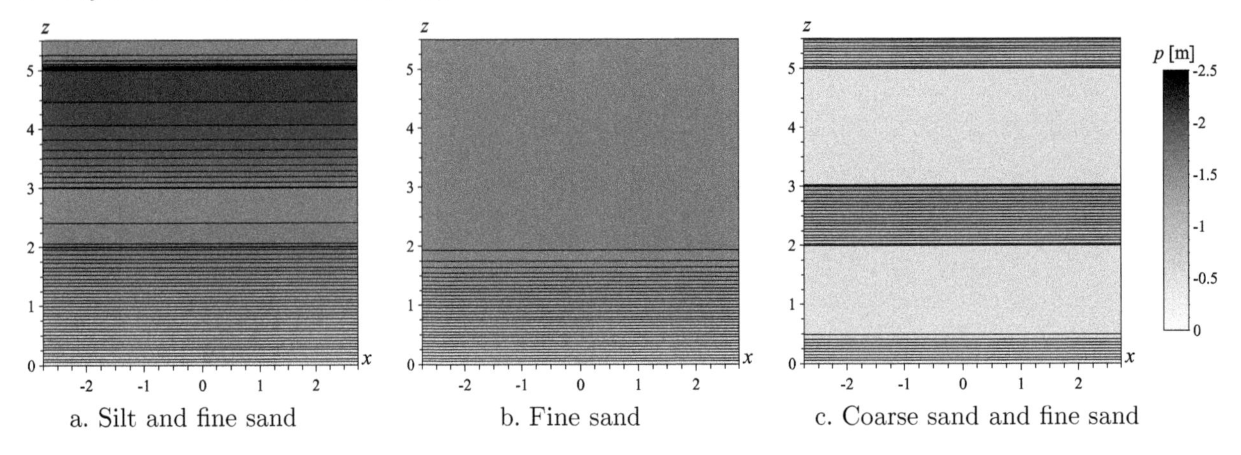

Figure 1.20 *Pressure head distribution in soil layers with interface conditions.*

in terms of the potential function outside and inside the heterogeneity, Φ^+ and Φ^-, using (1.41). Conservation of mass is satisfied if the normal component of the specific discharge is continuous across the boundary

$$q_n{}^+ = q_n{}^- \quad \rightarrow \quad e^{-k^+z}\left(\frac{\partial \Phi^+}{\partial n} + n_z k^+ \Phi^+\right) = e^{-k^-z}\left(\frac{\partial \Phi^-}{\partial n} + n_z k^- \Phi^-\right) \tag{1.44}$$

which is expressed in terms of the potential using (1.42). Solutions to these two interface conditions are illustrated for heterogeneities formed by silt embedded within a fine sand in Fig. 1.21, and for a coarse sand embedded within a fine sand in Fig. 1.22. Note that

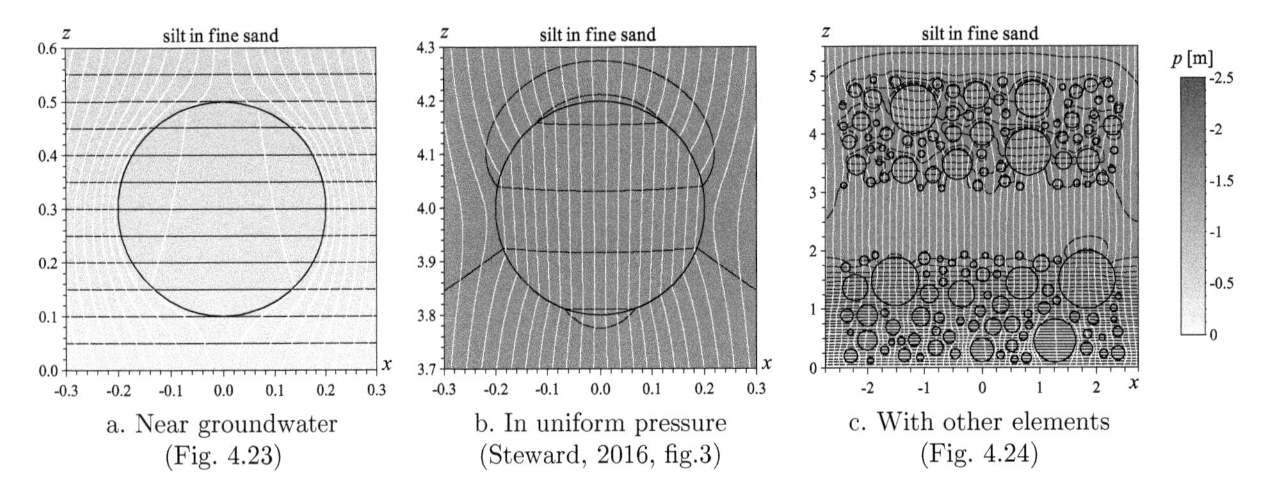

Figure 1.21 *Vadose zone seepage through silt heterogeneities embedded in a fine sand. (Reprinted from Steward, 2016, Analysis of vadose zone inhomogeneity toward distinguishing recharge rates: Solving the nonlinear interface problem with Newton method,* Water Resources Research, *Vol. 52(11), Figs. 3,4,5, with permission from John Wiley and Sons. Copyright 2016 by the American Geophysical Union.)*

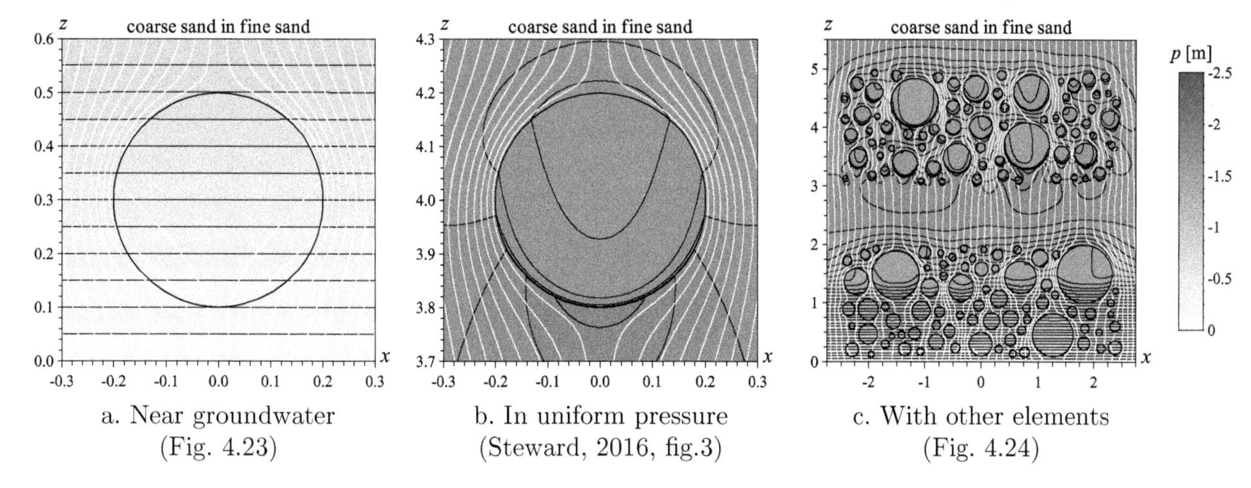

Figure 1.22 *Vadose zone seepage through coarse sand heterogeneities embedded in a fine sand. (Reprinted from Steward, 2016, Analysis of vadose zone inhomogeneity toward distinguishing recharge rates: Solving the nonlinear interface problem with Newton method,* Water Resources Research, *Vol. 52(11), Figs. 3,4,5, with permission from John Wiley and Sons. Copyright 2016 by the American Geophysical Union.)*

the recharge flowing through these layers of embedded heterogeneities is the same as that flowing through the continuous horizontal layers in Fig. 1.20.

An example is presented next to illustrate how the equations for the vadose zone are used to compute pressure and velocity, as well as application of the conditions of conservation of energy and mass across the interfaces between soils with different properties. The velocity is computed from the specific discharge in the vadose zone by

$$v_x = \frac{q_x}{\theta}, \qquad v_y = \frac{q_y}{\theta}, \qquad v_z = \frac{q_z}{\theta} \tag{1.45}$$

where the volumetric moisture content θ is a ratio of the volume of pores filled with water to the total volume. When the soil moisture content $\theta(p)$ is approximated using the same form as $K(p)$ in the Gardner equation, this leads to an expression for the specific moisture capacity

$$\theta = \theta_r + (\theta_s - \theta_r)e^{ap} \tag{1.46}$$

where θ_r is the residual (irreducible) volumetric water content and θ_s is the value at saturation (Warrick, 1974; Basha, 1999).

Vadose zone computations:

- matric flux potential, $F(\Phi)$, is given by (1.39).
- pressure head, $p(\Phi)$, is given by (1.41).
- specific discharge, (q_x, q_y, q_z), is obtained from Φ and its gradient in (1.42).
- velocity, (v_x, v_y, v_z), is obtained by dividing the specific discharge by the moisture content using (1.45) and (1.46).

Example 1.2 Vertical seepage of water through a vadose zone with horizontal soil layers is shown in Fig. 1.20. This solution is obtained from separation of variables, (4.19),

$$p = \frac{1}{2k} \ln \frac{c_1 e^{-2kz} + c_2}{F_s}, \qquad q_z = -2kc_2$$

with coefficients, (4.20), obtained by matching specified values of pressure head and specific discharge at elevation z_0:

$$\begin{matrix} p(z_0) = p_0 \\ q_z(z_0) = -R \end{matrix} \quad \rightarrow \quad c_1 = \left(F_s e^{2kp_0} - \frac{R}{2k}\right) e^{2kz_0}, \quad c_2 = \frac{R}{2k}$$

In the limit as z increases in each layer the pressure approaches

$$p \rightarrow \frac{1}{2k} \ln \frac{R}{2kF_s} = \frac{1}{a} \ln \frac{R}{K_s} \tag{1.47}$$

The pressure head distribution in a stacked set of layers may be computed successively following Raats and Gardner (1974), beginning by computing c_1 and c_2 at the the bottom of the lowest soil layer, where $p = 0$ and $z = 0$. The pressure head from this layer is computed at the interface with the next higher layer, which gives the coefficients c_1 and c_2 in that layer, and so on. The solution in Fig. 1.20a was constructed for recharge $R = 0.1$m/year. The pressure head at the interfaces are:

$$\begin{matrix} z_A = 0 \text{ m} \\ z_B = 2 \text{ m} \\ z_C = 3 \text{ m} \\ z_D = 5 \text{ m} \end{matrix}, \quad \begin{matrix} K_{s1} = 0.02 \text{ m/day} \\ K_{s2} = 1 \text{ m/day} \\ K_{s3} = 0.02 \text{ m/day} \end{matrix}, \quad \begin{matrix} a_1 = 2/\text{m} \\ a_2 = 5/\text{m}, \\ a_3 = 2/\text{m} \end{matrix} \quad \begin{matrix} p_A = 0.0 \text{ m} \\ p_B = -1.72 \text{ m} \\ p_C = -1.64 \text{ m} \\ p_D = -2.13 \text{ m} \end{matrix}$$

Problem 1.3 Compute the pressure head at elevations z_A, z_B, z_C, and z_D for Example 1.2, with the following recharge rates:

A. $R = 0.01$ m/yr
B. $R = 0.02$ m/yr
C. $R = 0.03$ m/yr
D. $R = 0.04$ m/yr
E. $R = 0.05$ m/yr
F. $R = 0.06$ m/yr
G. $R = 0.07$ m/yr
H. $R = 0.08$ m/yr
I. $R = 0.09$ m/yr
J. $R = 0.011$ m/yr
K. $R = 0.012$ m/yr
L. $R = 0.013$ m/yr

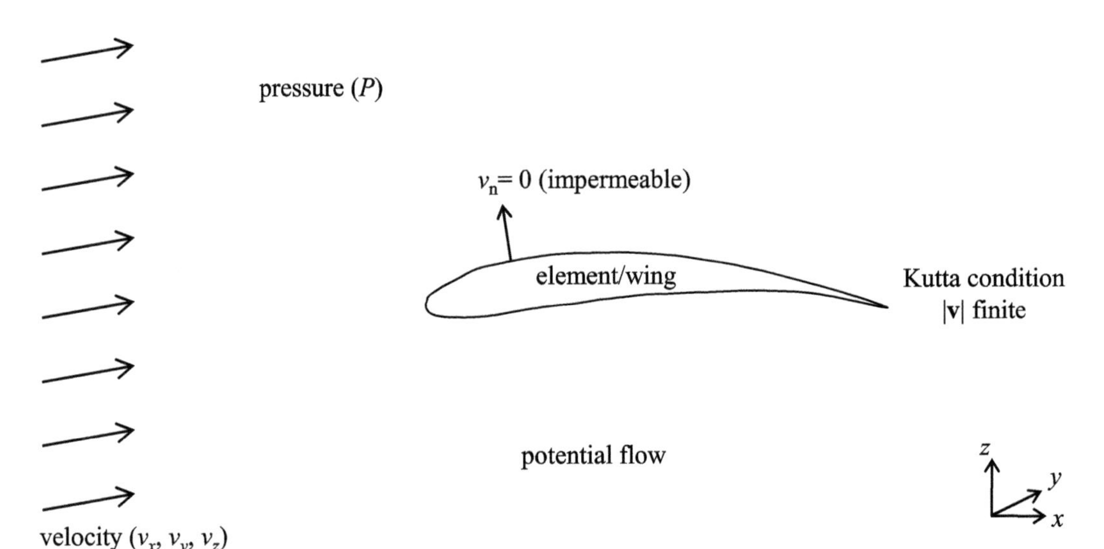

Figure 1.23 *Incompressible fluid flow properties and processes: Wing shown in section view.*

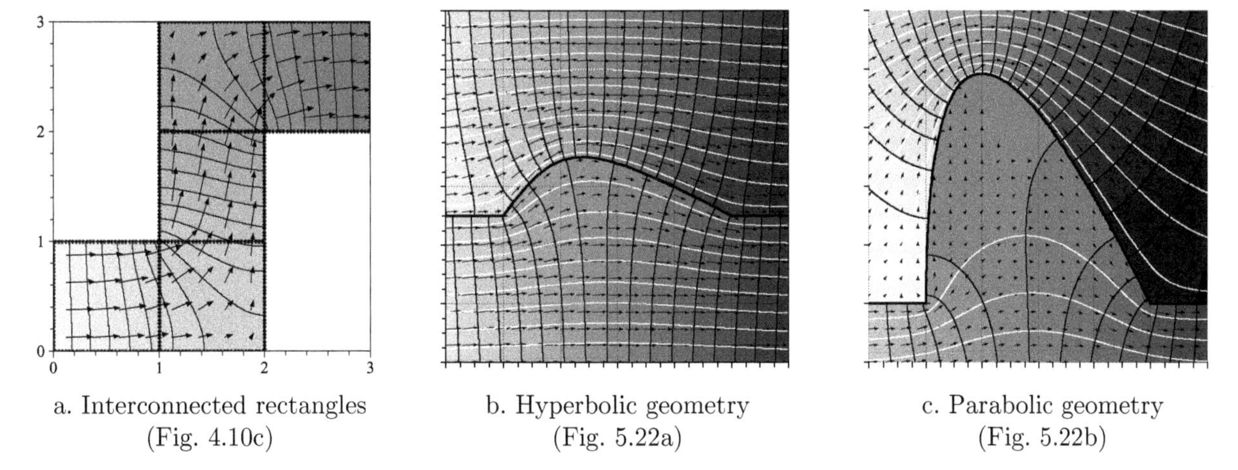

a. Interconnected rectangles (Fig. 4.10c) b. Hyperbolic geometry (Fig. 5.22a) c. Parabolic geometry (Fig. 5.22b)

Figure 1.24 *Fluid flow through a duct comprised of interconnected rectangles and around curvilinear elements. (Reprinted from Steward, Le Grand, Janković, and Strack, 2008, Analytic formulation of Cauchy integrals for boundaries with curvilinear geometry,* Proceedings of The Royal Society of London, Series A, *Vol. 464, Fig. 3, used with permission of The Royal Society; permission conveyed through Copyright Clearance Center, Inc.)*

1.2.3 Incompressible Fluid Flow: Pressure and Velocity

The study of fluid flowing around an element with velocity $\mathbf{v}$ is illustrated in Fig. 1.23. A wing imposes boundary conditions along the element, where the the normal component v_n is zero and fluid flows parallel to its surface. Wings are designed so the fluid moves at with a higher magnitude of velocity $|\mathbf{v}_1|$ at point 1 along the top surface than $|\mathbf{v}_2|^2$ at point 2 along the bottom. This creates a lower pressure P along the top that may be quantified using ***the Bernoulli (1738) equation***

$$\frac{P_1}{\rho g} + \frac{|\mathbf{v}_1|^2}{2g} + z_1 = \frac{P_2}{\rho g} + \frac{|\mathbf{v}_2|^2}{2g} + z_2 \tag{1.48}$$

where z is the elevation and the density ρ is uniform for incompressible fluids. The difference in pressure generates a lift force that is easily observed by placing your hand in a wind and varying this force by cupping ones hand and changing the attack angle relative to the background air velocity. Lift forces exist in slender bodies such as airfoils, hydrofoils, and helicopters; and are used in the motion of birds and fish.

Incompressible fluid flow with constant density may be studied using a velocity potential Φ (Lamb, 1879, sec.23)

$$\mathbf{v} = -\nabla\Phi \tag{1.49}$$

For example, fluid flowing through an impermeable duct is shown in Fig. 1.24a where the fluid particles travel from higher to lower values of this velocity potential. Such flows satisfy conservation of mass and may be formulated using the ***Laplace equation***, (2.2a):

$$\nabla\cdot\mathbf{v} = 0 \quad \rightarrow \quad \boxed{\nabla^2\Phi = \frac{\partial^2\Phi}{\partial x^2} + \frac{\partial^2\Phi}{\partial y^2} + \frac{\partial^2\Phi}{\partial z^2} = 0} \tag{1.50}$$

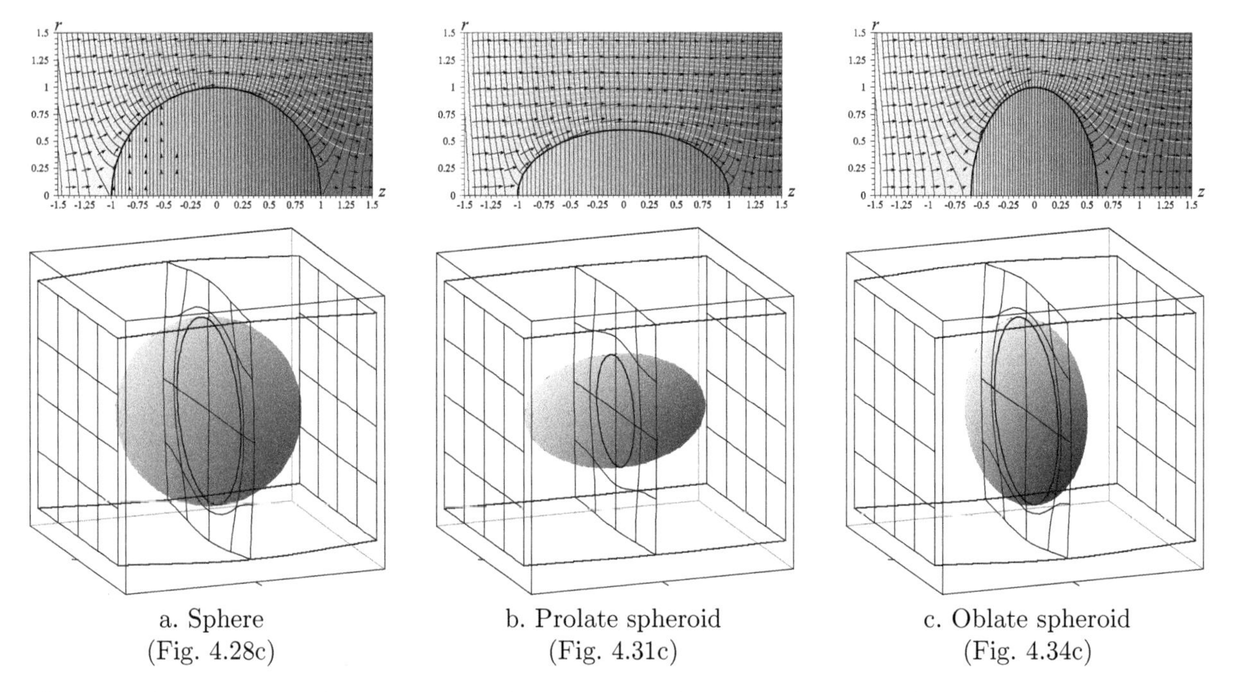

Figure 1.25 *Streamlines in half-plane and three-dimensional breakthrough planes near an impermeable element.*

Two-dimensional problems with incompressible fluid flow may also be formulated using the Lagrange stream function (Lamb, 1879, sec.69), presented earlier in (1.10)

$$v_x = -\frac{\partial \Psi}{\partial y}, \quad v_y = \frac{\partial \Psi}{\partial x}, \quad v_z = 0 \tag{1.51}$$

Such solutions are illustrated by streamlines with constant Ψ in Figs. 1.24b and 1.24c for fluid flow around obstructions on a flat surface and potential flow in cavities.

Analytic elements may be formulated to study the three-dimensional flow of incompressible fluid around impermeable elements as illustrated in Fig. 1.25. The upper panels show the velocity potential Φ and vector field around a sphere, a prolate spheroid, and an oblate spheroid; where each three-dimensional element is formed by rotation of a circle or ellipse about the z-axis. These ***axisymmetric flows may be represented by streamlines with uniform Stokes (1842) stream function***, obtained for components of the vector field in cylindrical (r, θ, z) coordinates using

$$v_r = -\frac{1}{r}\frac{\partial \Psi}{\partial z}, \quad v_\theta = 0, \quad v_z = \frac{1}{r}\frac{\partial \Psi}{\partial r} \tag{1.52}$$

The three-dimensional visualization in the lower panels of this figure illustrate stream surfaces formed by tracing the trajectories of fluid particles released along the lines at the left-most edge. These pathlines are traced to the right-most edge and ***stream surfaces are illustrated by connecting the locations where neighboring streamlines intersect breakthrough planes*** at the middle and edges of the displayed region.

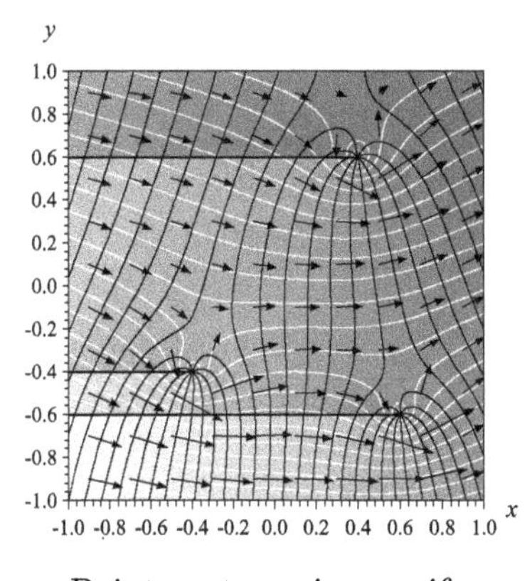

a. Point vortexes in a uniform vector field (Fig. 3.2c)

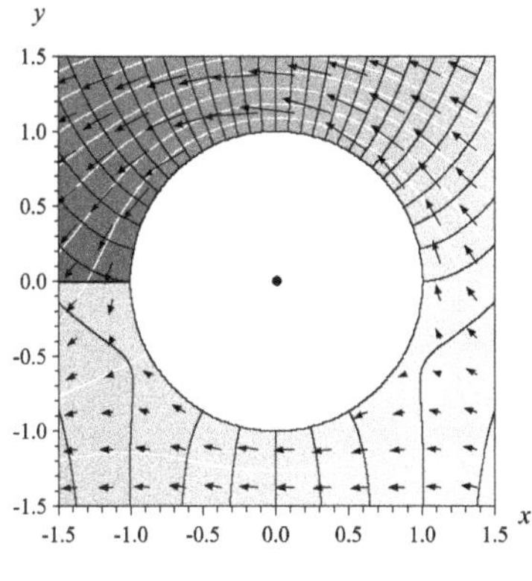

b. Rotating cylinder (Fig. 3.5c)

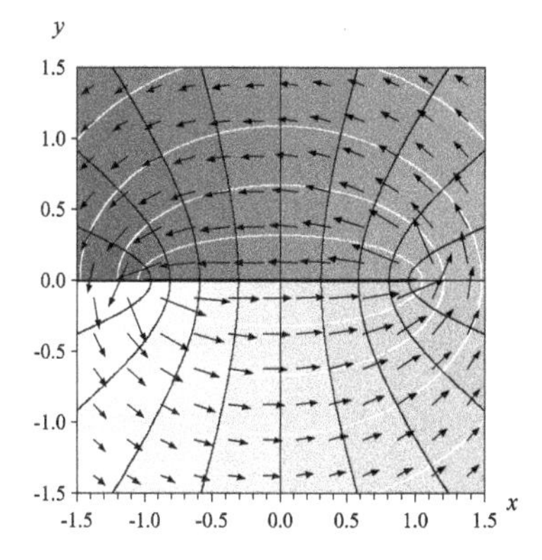

c. Circulation around a line (Fig. 3.19b)

Figure 1.26 *Point vortexes and circle and line elements with circulations.*

The velocity potential and stream function may be used to study fluid flow with circulation as shown in Fig. 1.26. These examples illustrate circulation about isolated points, about a infinitely long cylinder perpendicular to this plane, and about a two-dimensional line element. For elements that generate circulation, a discontinuity in the velocity potential occurs along a branch cut, which has been aligned to occur on the left side of these elements. A ***branch cut is related to the net circulation*** Γ ***generated by an element*** and will be analyzed in detail beginning in Section 3.1. The use of circulation to study flight is described next.

The geometry of a wing that is commonly studied was developed by Joukowsky (1912), and is illustrated in Fig. 1.27. The vector field in Fig. 1.27a satisfies the far-field condition of uniform vector field approaching the wing, and the impermeable condition that the vector field must be aligned tangent to the wing. Another vector field is illustrated in Fig. 1.27b where flow circulates about an impermeable wing, but with no flow at large distances from the element. However, both solutions contain a singularity at the sharp trailing edge where the vector field becomes infinite. The ***Kutta condition requires flow to remain finite at the trailing edge***, and this condition is satisfied when the wing generates exactly the correct circulation so that the two singularities cancel. This gives the fluid flow in Fig. 1.27c where the far-field velocity passes around a wing that satisfies this Kutta condition.

Fluid flow computations:

- velocity, $\mathbf{v}$ with components (v_x, v_y, v_z), is obtained from the potential, (1.49), or stream function, (1.51).
- pressure, P, is obtained from the Bernoulli equation, (1.48)

Examples of flow about wings in Fig. 1.28 illustrate the capacity to study fluid flow around flat airfoils, curved wings, and the Joukowsky airfoil. While some flows may be described mathematically using a single analytic element, more complicated geometries required the wing to be discretized into a set of adjacent elements. Development of the AEM for solving these problems is founded on the panel method for flow about airplane wings developed by Hess and Smith (1967) and Hess (1990).

Example 1.3 The vector field for flow about the rotating cylinder in Fig. 1.26b is given by

$$v_x = v_{x0} - v_{x0}\frac{x^2 - y^2}{(x^2+y^2)^2} - \frac{\Gamma}{2\pi}\frac{y}{x^2+y^2}, \quad v_y = -v_{x0}\frac{2xy}{(x^2+y^2)^2} + \frac{\Gamma}{2\pi}\frac{x}{x^2+y^2} \tag{1.53}$$

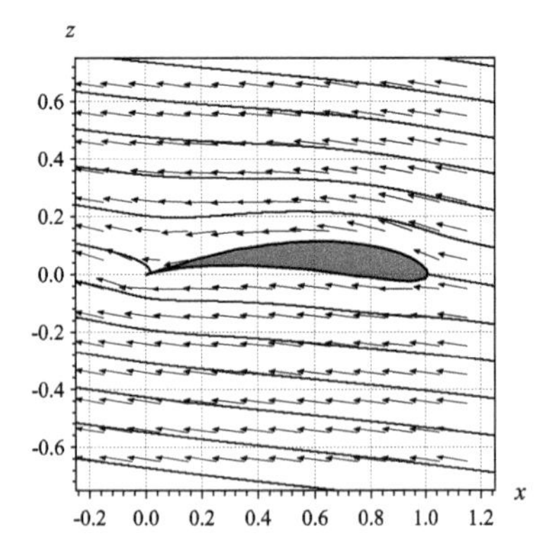

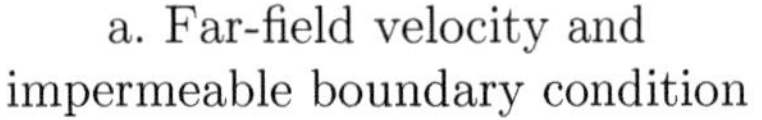
a. Far-field velocity and impermeable boundary condition

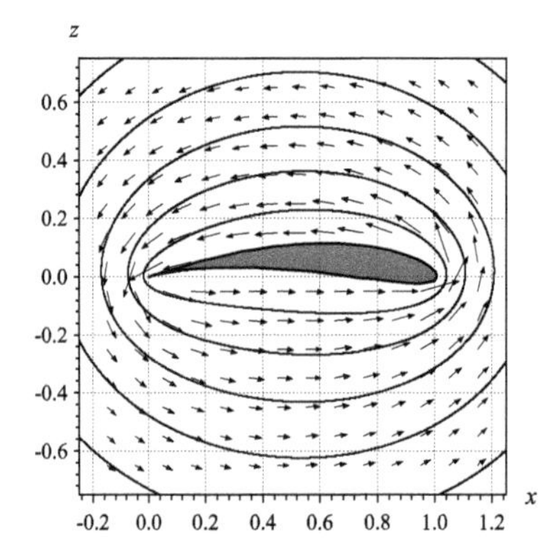

b. Circulation to satisfy the Kutta condition at the trailing edge

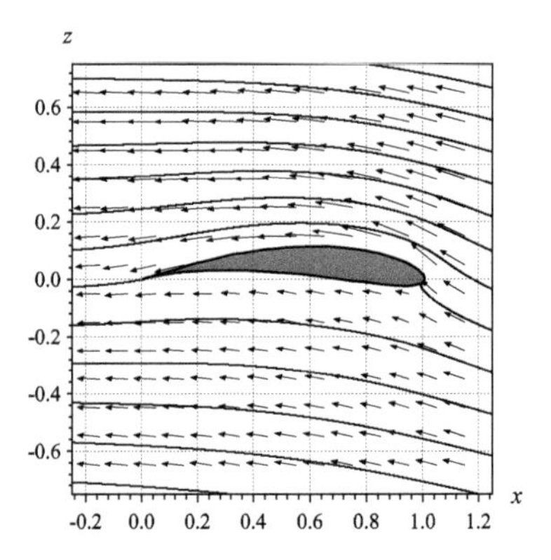

c. Vector field about a wing (7.5° attack angle)

Figure 1.27 *Flight boundary conditions:* $v_n = 0$ *and Kutta condition, from Section 3.5.*

where v_{x0} is the background vector field in the x-direction and Γ is the circulation, adapted from (3.3) and (3.7). Note that the vertical z-axis is uniform for this problem and pressure is related to the fluid velocity by Bernoulli's equation (1.48).

Problem 1.4 Compute the difference in pressure in units of N/m^2 between points at the top (x=0, y=1 m) and bottom (x=0, y=-1 m) of a rotating cylinder in water ρ=1 g/cm^3 for the specified background velocity v_{x0} and circulation Γ:

A. $v_{x0} = -1$ m/s, $\Gamma = 0.25\pi$/s
B. $v_{x0} = -2$ m/s, $\Gamma = 0.25\pi$/s
C. $v_{x0} = -3$ m/s, $\Gamma = 0.25\pi$/s
D. $v_{x0} = -1$ m/s, $\Gamma = 0.5\pi$/s
E. $v_{x0} = -2$ m/s, $\Gamma = 0.5\pi$/s
F. $v_{x0} = -3$ m/s, $\Gamma = 0.5\pi$/s
G. $v_{x0} = -1$ m/s, $\Gamma = \pi$/s
H. $v_{x0} = -2$ m/s, $\Gamma = \pi$/s
I. $v_{x0} = -3$ m/s, $\Gamma = \pi$/s

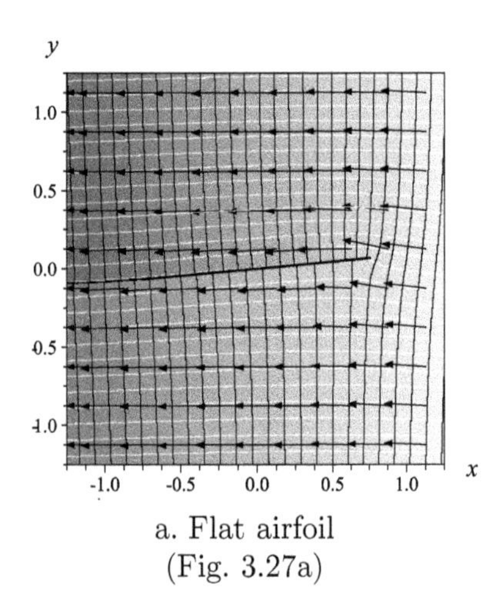

a. Flat airfoil (Fig. 3.27a)

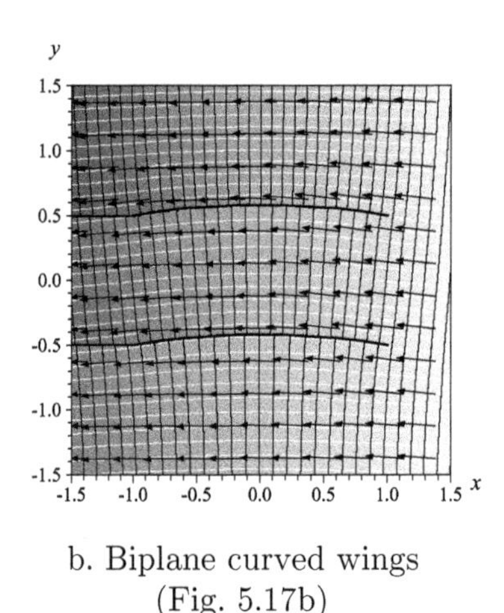

b. Biplane curved wings (Fig. 5.17b)

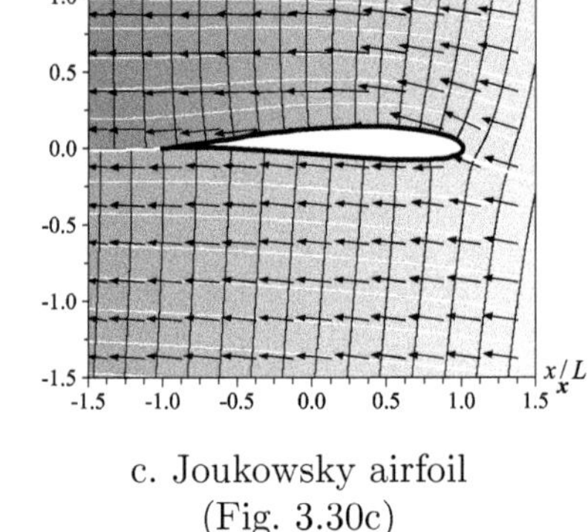

c. Joukowsky airfoil (Fig. 3.30c)

Figure 1.28 *Wings with Kutta condition at left ends (shown in x-y plane).*

1.2.4 Thermal Conduction: Temperature and Heat Flux

Studies of thermal conduction examine the propagation of heat through media as illustrated in Fig. 1.29. Temperature gradients are generated as heat propagates from sources to sinks, and these processes satisfy Fourier's law (Fourier, 1878):

$$\mathbf{q} = -K\nabla T \quad \rightarrow \quad \boxed{\mathbf{q} = -\nabla\Phi, \quad \Phi = KT} \tag{1.54}$$

where the thermal conduction $\mathbf{q}$, which is rate of heat flow per unit time per unit area, varies linearly with the gradient of temperature T, and K is the thermal conductivity. Problems with piecewise uniform thermal conductivity may be formulated in terms of the ***thermal potential*** Φ. Such problems are governed by the ***heat equation***, which results from the conservation law for heat conduction together with Fourier's law (Carslaw and Jaeger, 1959):

$$\nabla\cdot\mathbf{q} = -\rho c\frac{\partial T}{\partial t} \quad \rightarrow \quad \boxed{\nabla^2\Phi = \frac{1}{\mathscr{D}}\frac{\partial\Phi}{\partial t}, \quad \mathscr{D} = \frac{K}{\rho c}} \tag{1.55}$$

where ρ is the density, c is the specific heat, and $\mathscr{D}$ is the thermal diffusivity. For ***steady heat conduction***, the transient term on the right-hand side of the heat equation is zero, leading to the ***Laplace equation***, (2.2a):

$$\nabla^2\Phi = \boxed{\frac{\partial^2\Phi}{\partial x^2} + \frac{\partial^2\Phi}{\partial y^2} + \frac{\partial^2\Phi}{\partial z^2} = 0} \tag{1.56}$$

Use of the AEM to formulate solutions for thermal conduction builds upon foundational studies by Carslaw and Jaeger (1959), Moon and Spencer (1961*a*), and MacRobert (1967).

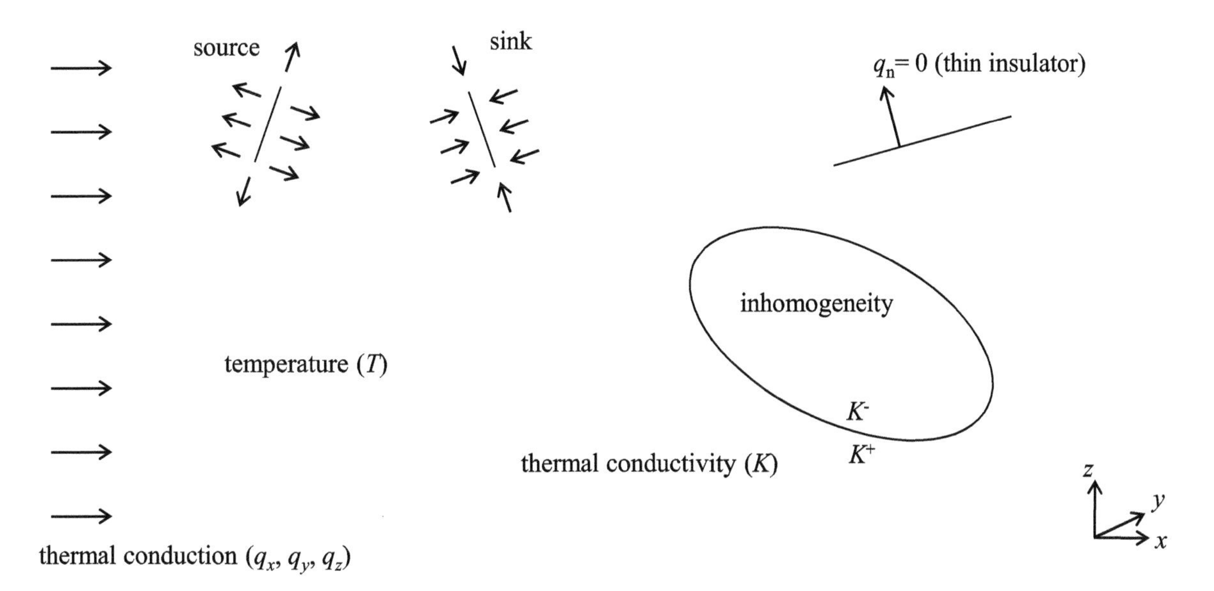

Figure 1.29 *Thermal conduction properties and processes shown in section view.*

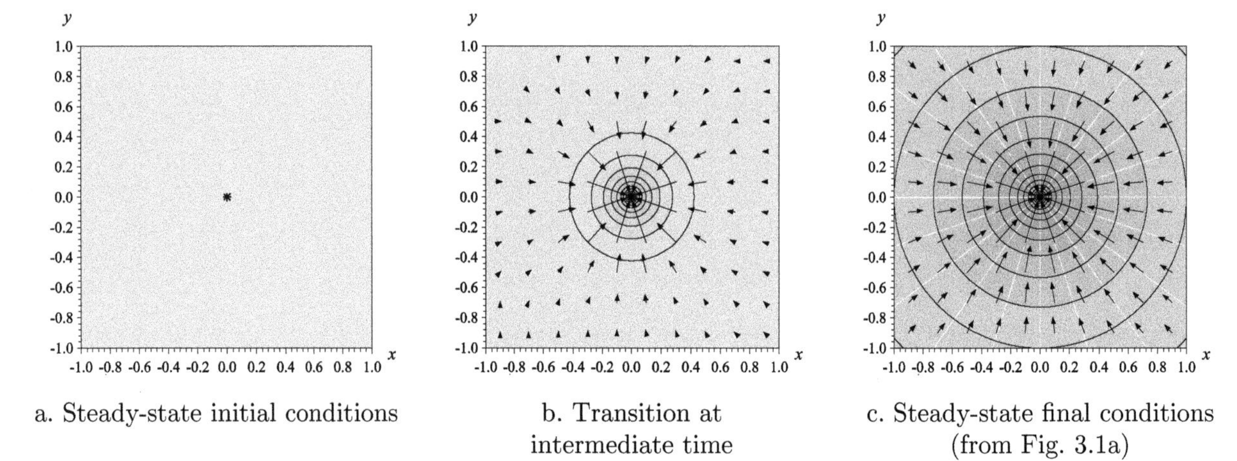

Figure 1.30 *Thermal conduction: temperature and flux near a cooling rod.*

The temporal process of heat condition for a heat sink is illustrated in Fig. 1.30. For this problem, an infinitely long cooling rod oriented normal to this picture begins to withdraw heat from the domain from initial conditions of uniform temperature as shown in Fig. 1.30a. The temperature becomes lower in the vicinity of the sink in Fig. 1.30b, and at larger times the thermal condition $\mathbf{q}$ approaches the final steady-state conditions in Fig. 1.30c. This is the same problem as groundwater flow toward a pumping well studied by Theis (1935), which approaches the final steady-state conditions from Thiem (1906). The ***transition from initial to final steady states*** is explored later in Example 1.4 and Problem 1.5.

Extensions to problems for ***heat generation along elements with more complex geometries*** are illustrated next. For example, heat is conducted from four rods in a cylinder to its boundary of uniform temperature in Fig. 1.31a. Such solutions are extensible to problems where ***heat is generated across the domain*** using functions that satisfy the Poisson equation (Moon and Spencer, 1961*a*, p.427):

$$\nabla^2\Phi = \frac{\partial^2\Phi}{\partial x^2} + \frac{\partial^2\Phi}{\partial y^2} + \frac{\partial^2\Phi}{\partial z^2} = -R \tag{1.57}$$

where R is the rate of heat production. Such solutions may be used to represent, for example, propagation of the heat of hydration within a concrete cylinder inserted into a water bath. Heat production is also illustrated for elements with the geometry of a Bezier curve in Fig. 1.31b or a B-spline in Fig. 1.31c.

Analytic elements may be used to study ***three-dimensional problems with heat production***, as illustrated in Fig. 1.32, where each plate is bounded on the top and bottom surfaces by insulators that hold heat within the domain. A heat source is embedded within the domain in Fig. 1.32a, and the three-dimensional trajectories of thermal conduction illustrate pathways as heat moves away from the source to further regions within the plate. ***Trajectories of thermal conduction*** are also illustrated for elements that apply heat to the boundary of the plate along a line in Fig. 1.32b, and across an area on the top boundary in Fig. 1.32c.

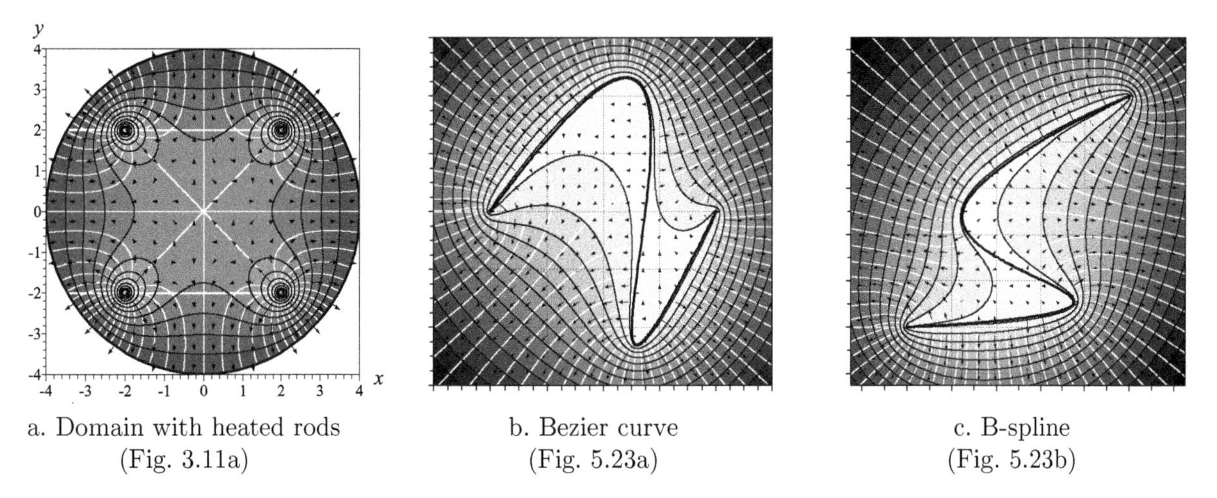

Figure 1.31 *Thermal conductors with uniform temperature. (Reprinted from Steward, Le Grand, Janković, and Strack, 2008, Analytic formulation of Cauchy integrals for boundaries with curvilinear geometry,* Proceedings of the Royal Society of London, Series A, *Vol. 464, Fig. 4, used with permission of The Royal Society; permission conveyed through Copyright Clearance Center, Inc.)*

Thermal conduction around thin insulators is illustrated in Fig. 1.33 for both ***isolated elements and strings of connected elements***. A ***perfect insulator through which heat cannot penetrate*** satisfies a boundary condition where the normal component of conduction is zero:

$$\boxed{q_{\rm n} = 0, \quad \Psi \text{ uniform}} \tag{1.58}$$

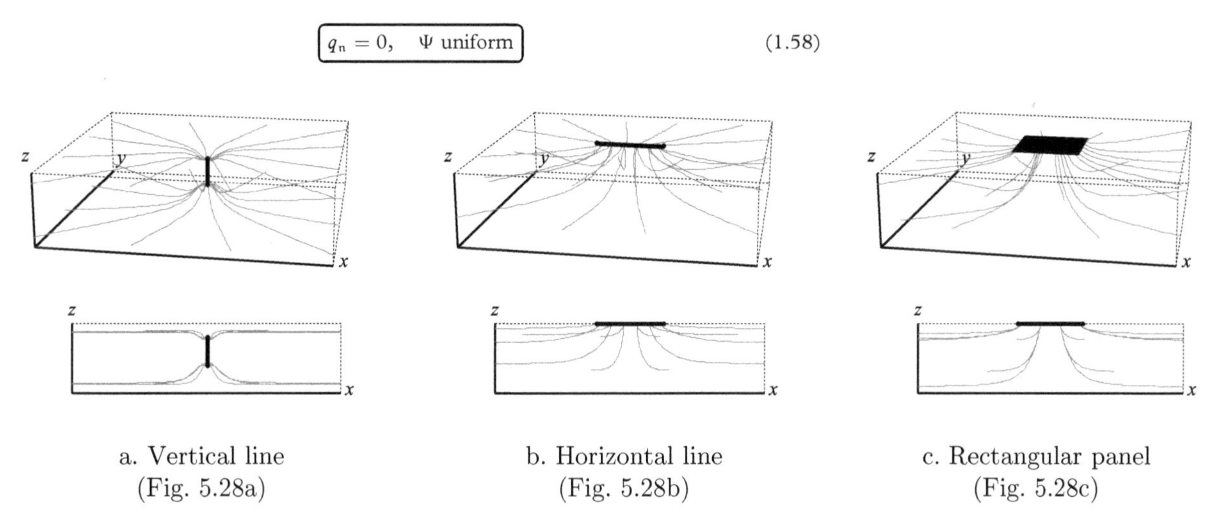

Figure 1.32 *Three-dimensional thermal conduction near a source or sink in a plate. (Reprinted from Steward, 1999, Three-dimensional analysis of the capture of contaminated leachate by fully penetrating, partially penetrating, and horizontal wells,* Water Resources Research, *Vol. 35(2), Fig. 2, with permission from John Wiley and Sons. Copyright 1999 by the American Geophysical Union.)*

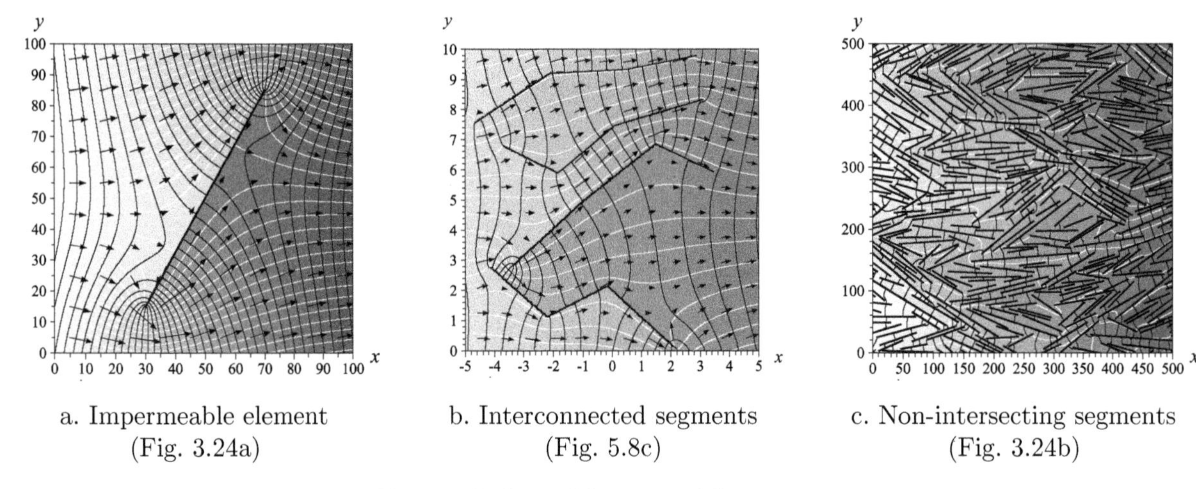

Figure 1.33 *Thermal conduction around impermeable lines and interconnected line segments. (Reprinted from Steward, 2015, Analysis of discontinuities across thin inhomogeneities, groundwater/surface water interactions in river networks, and circulation about slender bodies using slit elements in the Analytic Element Method,* Water Resources Research, *Vol. 51(11), Fig. 5, with permission from John Wiley and Sons. Copyright 2015 by the American Geophysical Union.)*

Such a boundary also has a continuous stream function along the element, which is evident in the Fig. 1.33 solutions. Analytic elements will also be formulated for ***thin features with lower thermal conductivity***, K^*, than tthe surrounding media, K. Such elements satisfy a boundary condition where the normal component of conduction is related to the partial derivative of the thermal potential,

$$q_n = -K^* \frac{\partial T}{\partial n} = -\frac{K^*}{K}\frac{\partial \Phi}{\partial n} = -\frac{K^*}{K}\left(\frac{\Phi^+ - \Phi^-}{w^*}\right) \tag{1.59}$$

which is approximated as the jump in thermal potential across the thin feature of width w^*. This gives a Robin boundary condition,

$$\boxed{q_n = -\frac{1}{\delta}\left(\Phi^+ - \Phi^-\right), \quad \delta = \frac{Kw^*}{K^*}} \tag{1.60}$$

and in the limit as $\delta \to \infty$ the normal component become zero, (1.58).

Thermal conduction is influenced by changes in the media through which heat propagates. For example, the circular element in Fig. 1.34a is ***impermeable to heat***, and the object satisfies boundary conditions of a perfect insulator, (1.58). Interface conditions may be formulated for ***inhomogeneities with a different thermal conductivity,*** K^-, ***than the background,*** K^+ into which it is inserted. Such elements satisfy continuity of the normal component of the temperature gradient at each point m on the interface

$$\boxed{\begin{aligned} q_n^+(x_m, y_m, z_m) &= q_n^-(x_m, y_m, z_m), \\ \Psi^+(x_m, y_m, z_m) &= \Psi^-(x_m, y_m, z_m) \end{aligned}} \tag{1.61a}$$

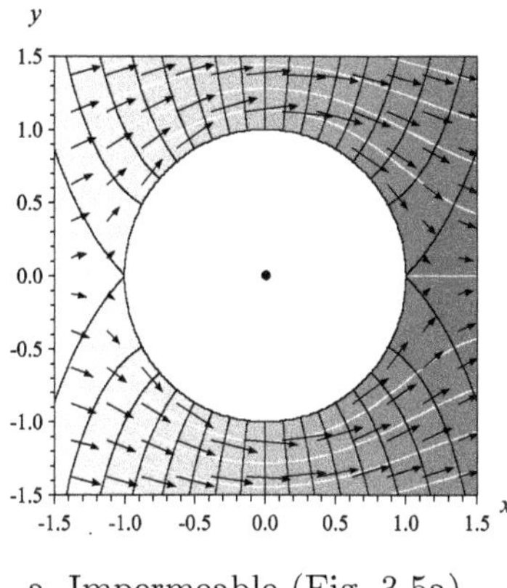

a. Impermeable (Fig. 3.5a)

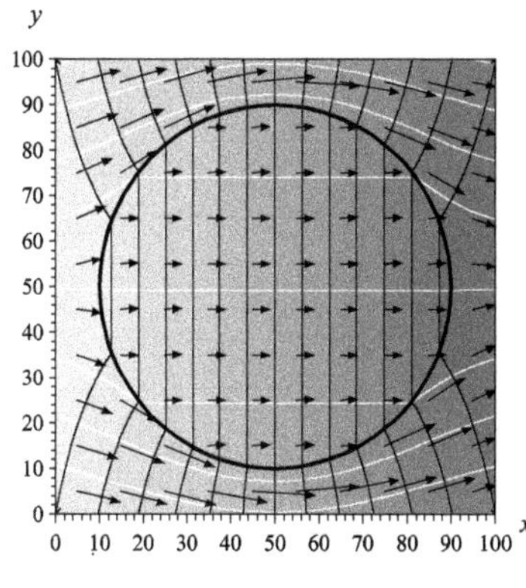

b. Low conductivity (Fig. 3.13a)

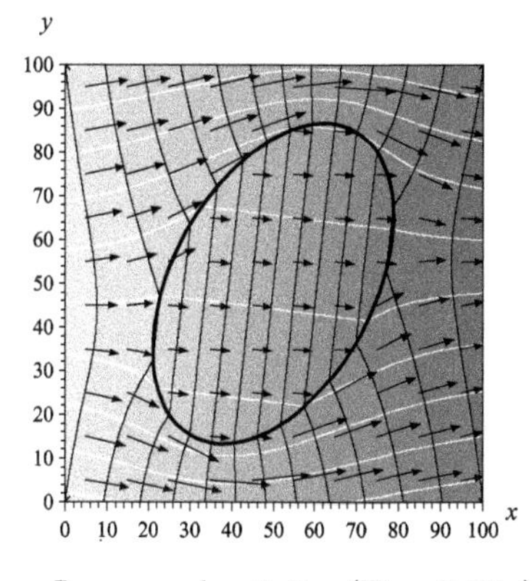

c. Low conductivity (Fig. 3.16a)

Figure 1.34 *Thermal conduction around elements with low thermal conduction in a uniform vector field (showing contour intervals with uniform temperature).*

a condition that is also satisfied if the stream function is continuous at adjacent points across the interface. Additionally, the temperature must be continuous, which provides conditions for the potential function

$$T^+(x_m, y_m, z_m) = T^-(x_m, y_m, z_m)$$
$$\rightarrow \boxed{\frac{1}{K^+}\Phi^+(x_m, y_m, z_m) = \frac{1}{K^-}\Phi^-(x_m, y_m, z_m)} \tag{1.61b}$$

Thermal conduction through and around a circular or elliptical element with lower thermal conductivity than the background is illustrated in Fig. 1.34.

These methods for solving problems with an inhomogeneity embedded into a uniform background may be extended to problems involving many elements. For example, the thermal conduction near circular elements is shown in Fig. 1.35a, and elliptical elements in Fig. 1.35b. These examples illustrate inhomogeneities with randomly distributed thermal conductivity, as evident by thermal trajectories diverging around elements with lower conductivity and converging within elements with higher conductivity. The interface conditions for inhomogeneities, (1.61), may also be applied to interconnected regions with different thermal conductivity as illustrated in Fig. 1.35c. These formulations facilitate studies of heterogeneous media ***with two different conceptualizations: embedded inhomogeneities within a uniform background, or a domain filled with interconnected regions each with unique thermal properties***.

Analytic elements may also be formulated to study ***three-dimensional conduction near inhomogeneities***. Such solutions are depicted in Fig. 1.36 by the trajectories of thermal conduction near spheres. Each element satisfies the interface conditions for an inhomogeneity, (1.61), and results are shown as the thermal conductivity varies from conditions of impermeable to infinitely conductive.

Thermal conduction computations:

- temperature, T, is obtained from Φ by (1.54)
- heat flow, $\mathbf{q}$, is obtained from minus the gradient of Φ, (1.54)

Example 1.4 The thermal potential and conduction for a point-sink in Fig. 1.30 that removes heat flux Q may be computed at distance r from the sink at time t using

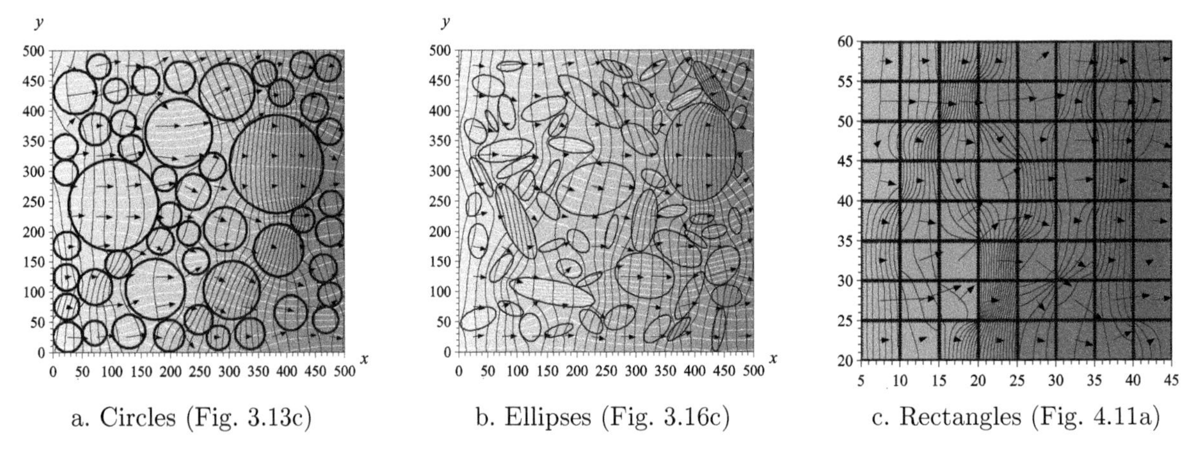

Figure 1.35 *Thermal conduction through heterogeneity with elements of randomly distributed conductivity embedded within a uniform background or with interconnected edges.*

$$\begin{array}{l} \Phi = -\dfrac{Q}{4\pi}E_1(u) + \Phi_0 \\ q_r = -\dfrac{Q}{2\pi}\dfrac{e^{-u}}{r} \end{array}, \quad u = \frac{r^2}{4\mathscr{D}t}, \quad E_1(u) = \int_u^\infty \frac{e^{-\tilde{u}}}{\tilde{u}}\mathrm{d}\tilde{u} \tag{1.62}$$

where $\Phi_0 = KT_0$ represents the thermal potential at the initial condition, and the exponential integral, $E_1(u)$, and may be evaluated using Abramowitz and Stegun (1972, eqns 5.1.53, 5.1.56). Final steady-state conditions are given by

$$\Phi = \frac{Q}{2\pi}\ln r + \text{constant}, \quad q_r = -\frac{\partial \Phi}{\partial r} = -\frac{Q}{2\pi}\frac{1}{r} \tag{1.63}$$

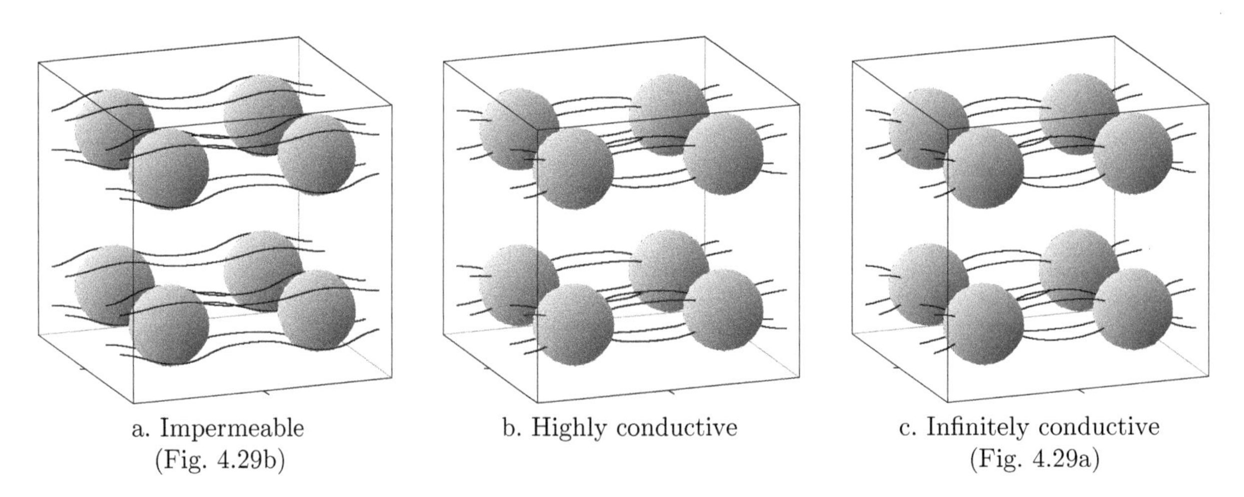

Figure 1.36 *Thermal conduction near impermeable, highly conductive, or infinitely conductive spheres.*

and extensions to transient problems with more complicated geometries are developed by Steward (1991), Zaadnoordijk and Strack (1993), Furman and Neuman (2003), Kuhlman and Neuman (2009), and Bakker (2013).

Problem 1.5 Compute thermal conduction, q_r, toward a heat sink with unit flux $Q = 1$ watt/m and thermal diffusivity at distance r from a point-sink in the specified material at initial conditions $t = 0$, at $t = 10^{-6}$sec, at $t = 1$ sec, and at the final steady-state conditions. Thermal diffusivity values from (Moon and Spencer, 1961*a*, p.213).

A. copper, $\mathscr{D} = 113$ m^2/sec, $r = 1$ mm

B. copper, $\mathscr{D} = 113$ m^2/sec, $r = 1$ cm

C. cast iron, $\mathscr{D} = 13.3$ m^2/sec, $r = 1$ mm

D. cast iron, $\mathscr{D} = 13.3$ m^2/sec, $r = 1$ cm

E. concrete, $\mathscr{D} = 0.58$ m^2/sec, $r = 1$ mm

F. concrete, $\mathscr{D} = 0.58$ m^2/sec, $r = 1$ cm

G. dry soil, $\mathscr{D} = 0.31$ m^2/sec, $r = 1$ mm

H. dry soil, $\mathscr{D} = 0.31$ m^2/sec, $r = 1$ cm

1.2.5 Electrostatics: Voltage and Electric Fields

Studies of electric fields examine the propagation of electric currents through a conducting medium as illustrated in Fig. 1.37. Electrostatics may be formulated in terms of the ***electric field strength*** **E**, which is obtained from minus the gradient of the ***electric potential*** Φ. This, together with conservation of charge, provides solutions in terms of the Poisson equation (Sommerfeld, 1952, p.38):

$$\boxed{\mathbf{E} = -\nabla\Phi} \quad \nabla\cdot\mathbf{E} = \frac{\rho}{\varepsilon} \quad \rightarrow \quad \boxed{\nabla^2\Phi = -\frac{\rho}{\varepsilon}} \tag{1.64}$$

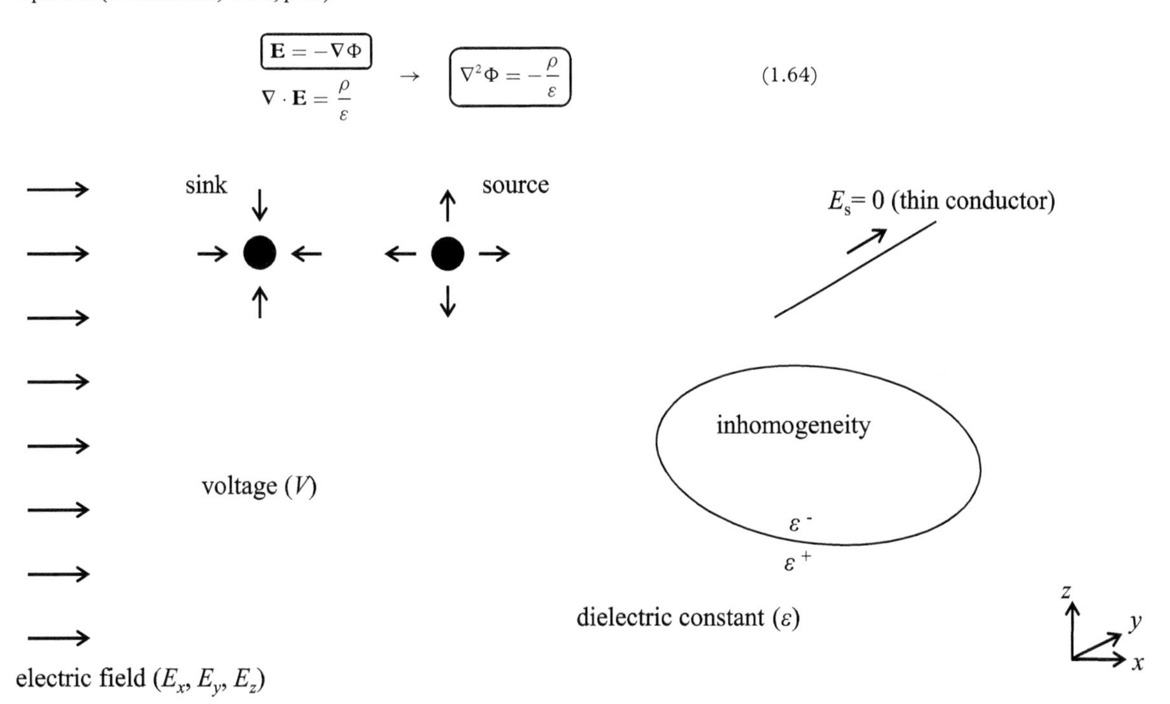

Figure 1.37 *Electric field properties and processes.*

where ρ is charge density, and ε is the dielectric constant. Problems where steady charges propagate through the domain from sinks and sources along boundaries reduce to solutions of the Laplace equation, (2.2a): (Moon and Spencer, 1960)

$$\nabla^2\Phi = \frac{\partial^2\Phi}{\partial x^2} + \frac{\partial^2\Phi}{\partial y^2} + \frac{\partial^2\Phi}{\partial z^2} = 0 \tag{1.65}$$

This formulation is grounded in ***Ohm's law*** (Ohm, 1827), where the current I is equal to voltage V divided by resistance R:

$$I = \frac{V}{R}, \quad I = \int_A \kappa\mathbf{E}\cdot dA, \quad V_{12} = \Phi_1 - \Phi_2 \tag{1.66}$$

and these quantities are related to $\mathbf{E}$ and Φ as follows. The current through a surface A may be obtained from the surface integral of electric current density $\mathbf{J} = \kappa\mathbf{E}$, where κ is the electric conductivity (Moon and Spencer, 1960, p.87). The voltage between points 1 and 2 is equal to the differences in electric potential at these points (Sommerfeld, 1952, p.39). The following solutions for problems in electrostatics build upon examples presented in the foundational works of Sommerfeld (1952), Moon and Spencer (1960, 1961*a*,*b*), and MacRobert (1967).

Electric currents generated by charged elements are illustrated in Fig. 1.38. The solution in Fig. 1.38a illustrates an electric field $\mathbf{E}$ directed towards a long charged wire, where the electric potential Φ decreases in the direction of this sink. This solution may be extended to the problem of two charged rods with opposite charge in Fig. 1.38b, and the field is oriented from the positively to negatively charged rods. Each rod satisfies a boundary condition along the interface between the rod and the problem domain where the ***potential is uniform*** along its boundary

$$\Phi \text{ uniform along boundary} \tag{1.67}$$

In the limit as two charged rods with opposite strength become very thin and approach one another, the field takes the form of a ***point-dipole*** in Fig. 1.38c.

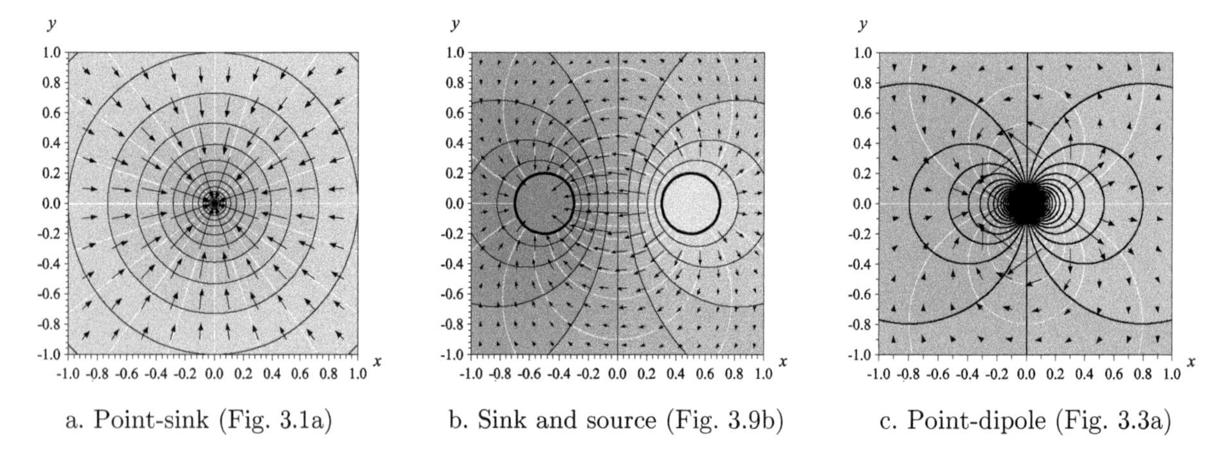

Figure 1.38 *Sources and sinks of electric current, and a dipole.*

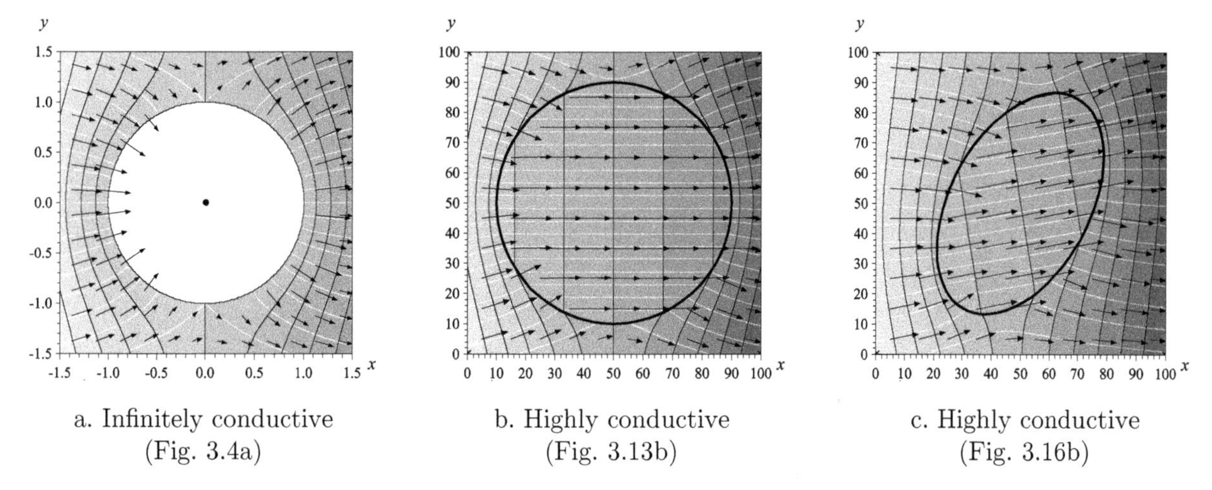

Figure 1.39 *Electric conduction through infinitely conductive and highly conductive circle and ellipse.*

The ***point-dipole is used in developing solutions for electric fields near inhomogeneities***, where elements with a dielectric constant ε^- are inserted into a uniform background with a different dielectric constant ε^+. For example, the solution in Fig. 1.39a illustrates conduction through an infinitely conductive element in the limit as $\varepsilon^- \to \infty$. Solutions are also shown for ***highly conductive elements*** for a circle in Fig. 1.39b (Sommerfeld, 1952, p.62) and for an ellipse in Fig. 1.39c (Moon and Spencer, 1961*a*, p.259). These elements have a larger dielectric constant inside the element than outside, $\varepsilon^- > \varepsilon^+$, but not large enough to approach the limiting value in Fig. 1.39a. These inhomogeneities satisfy conditions where ***voltage is continuous across its interface***, so the electric potential is also continuous at point m on the interface:

$$\Phi^+(x_m, y_m, z_m) = \Phi^-(x_m, y_m, z_m) \tag{1.68}$$

They also satisfy a condition where the ***normal component of the displacement, D, is continuous*** across the interface:

$$\mathbf{D} = \varepsilon\mathbf{E}, \quad \begin{matrix} D_n^+ = \varepsilon^+ E_n^+ \\ D_n^+ = \varepsilon^+ E_n^+ \end{matrix} \quad \to \quad \varepsilon^+ \frac{\partial \Phi^+}{\partial n} = \varepsilon^- \frac{\partial \Phi^-}{\partial n} \tag{1.69}$$

This, together with the relation between the electric field and the gradient of the potential, (1.64), requires the normal component of the gradient of Φ to jump across the interface (Sommerfeld, 1952, p.60). Note that the solutions in Fig. 1.39 visualize surfaces of constant Φ and the displacement vector.

The electric fields near ***thin conducting elements*** with the geometry of a two-dimensional line are illustrated in Fig. 1.40. Each element satisfies a boundary condition of uniform potential, (1.68), similar to the infinitely conductive circle in Fig. 1.39a. This condition also requires that the electric field be oriented normal to each element and its tangential component is zero along the boundary

$$E_s(x_m, y_m, z_m) = 0 \tag{1.70}$$

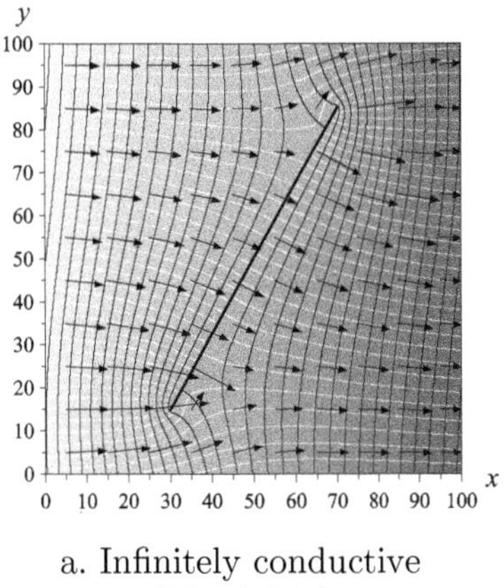

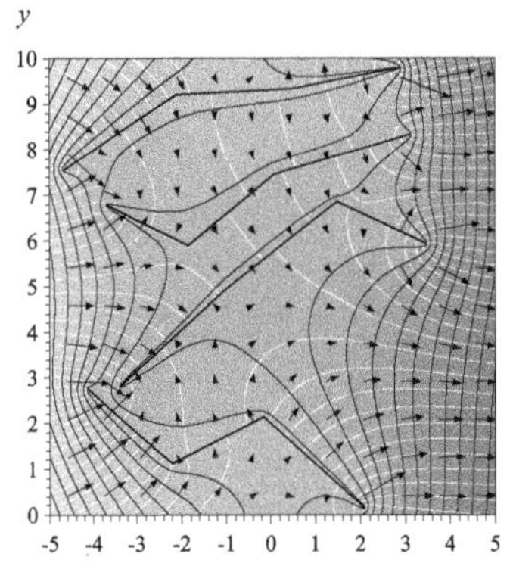

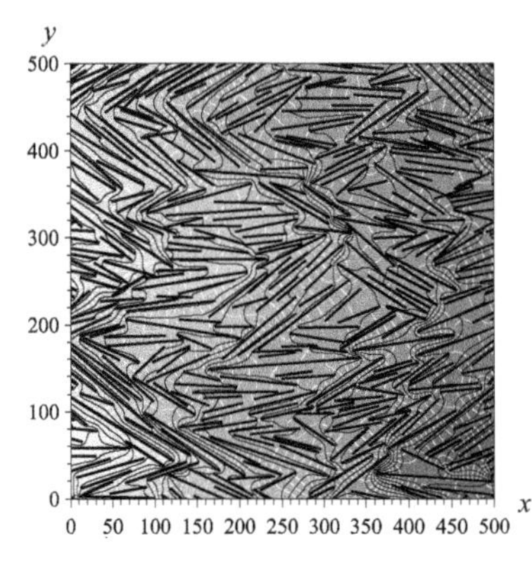

Figure 1.40 *Electric conduction through thin conductors and interconnected line segments. (Reprinted from Steward, 2015, Analysis of discontinuities across thin inhomogeneities, groundwater/surface water interactions in river networks, and circulation about slender bodies using slit elements in the Analytic Element Method,* Water Resources Research, *Vol. 51(11), Fig. 4, with permission from John Wiley and Sons. Copyright 2015 by the American Geophysical Union.)*

Electrostatic computations:

- electric field strength, **E**, is obtained from minus the gradient of the electric potential Φ, (1.64)
- displacement **D** is obtained from the electric field strength, **E**, and dielectric constant, ε, using (1.69)

Examples are shown for an isolated element, for connected elements that ***transmit current along a string of elements***, and for a group of non-connecting elements that collectively shape the electric field. It is evident that the potential is uniform along the thin conductors in each solution, and the vector field is directed normal to their boundaries.

The electrostatic formulation for studying steady electric fields and potential is extended to ***three-dimensional problems*** in Fig. 1.41. These examples illustrate inhomogeneities where the dielectric constant is larger inside the element than in the surrounding media, $\varepsilon^- > \varepsilon^+$, similar to the two-dimensional solutions in Fig. 1.39. The electric field near the sphere in Fig. 1.41a follows Moon and Spencer (1961*a*, pp.224–8), and solutions illustrate the pathways for current in half-planes. The ***intersection of these pathways with breakthrough planes*** is also shown for three planes, one at the left side of the review area, one at the right, and one midway between these. The electric field near a prolate spheroid in Fig. 1.41b follows (Moon and Spencer, 1961*a*, p.244–59), and has been used to study conduction through a grounding rod. The electric field is also illustrated in Fig. 1.41c for an object with the geometry of an oblate spheroid.

Example 1.5 The electric potential for the solution in Fig. 1.39a may be obtained from (3.6)

$$\Phi = -E_{x0}x + \frac{E_{x0}{r_0}^2 x}{x^2 + y^2} \tag{1.71}$$

where E_{x0} is the background electric field in the x-direction, and the element is of radius $r_0 = 1$ m and centered at the origin.

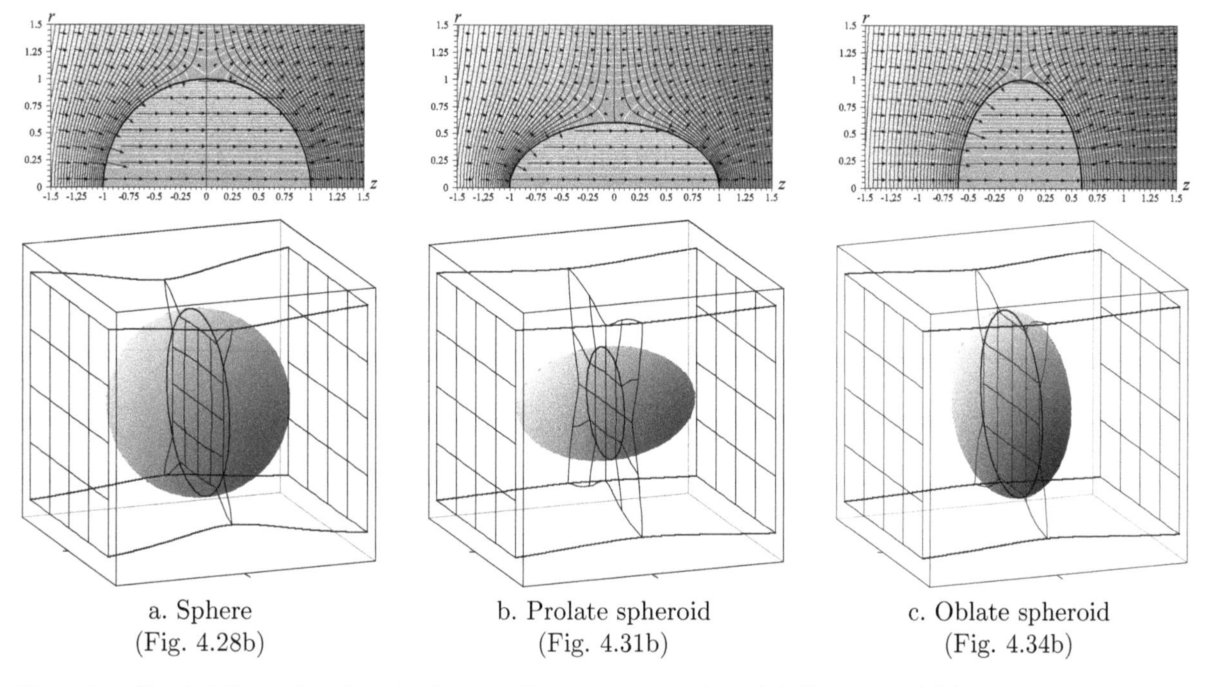

Figure 1.41 *Electric fields near three-dimensional elements illustrating current pathways in half-planes and their intersections with breakthrough planes.*

Problem 1.6 Compute the electric field **E** for the field in Example 1.5 at the specified locations for unit $E_{x0} = 1$ with $r_0 = 1$ cm.

A. $(x, y) = (-4, -1)$cm, $(-4, 0)$cm, and $(-4, 1)$cm

B. $(x, y) = (-3, -1)$cm, $(-3, 0)$cm, and $(-3, 1)$cm

C. $(x, y) = (-2, -1)$cm, $(-2, 0)$cm, and $(-2, 1)$cm

D. $(x, y) = (-1, -1)$cm, $(-1, 0)$cm, and $(-1, 1)$cm

E. $(x, y) = (1, -1)$cm, $(1, 0)$cm, and $(1, 1)$cm

F. $(x, y) = (2, -1)$cm, $(2, 0)$cm, and $(2, 1)$cm

G. $(x, y) = (3, -1)$cm, $(3, 0)$cm, and $(3, 1)$cm

H. $(x, y) = (4, -1)$cm, $(4, 0)$cm, and $(4, 1)$cm

1.3 Studies of Periodic Waves

The methods used to formulate solutions for wave fields and their interactions with analytic elements are presented next. The propagation of waves may be studied using the wave equation, (2.2g):

$$\frac{\partial^2 \Phi}{\partial x^2} + \frac{\partial^2 \Phi}{\partial y^2} + \frac{\partial^2 \Phi}{\partial z^2} = \frac{1}{c^2} \frac{\partial^2 \Phi}{\partial t^2} \tag{1.72}$$

where the constant c is related to the wave velocity (Airy, 1845; Lamb, 1879). The separation of variables subdivides the spatial and temporal variables in Φ for periodic waves into functions that repeat with ***angular frequency*** ω and a complex function φ that satisfies the ***Helmholtz equation***, (2.2c):

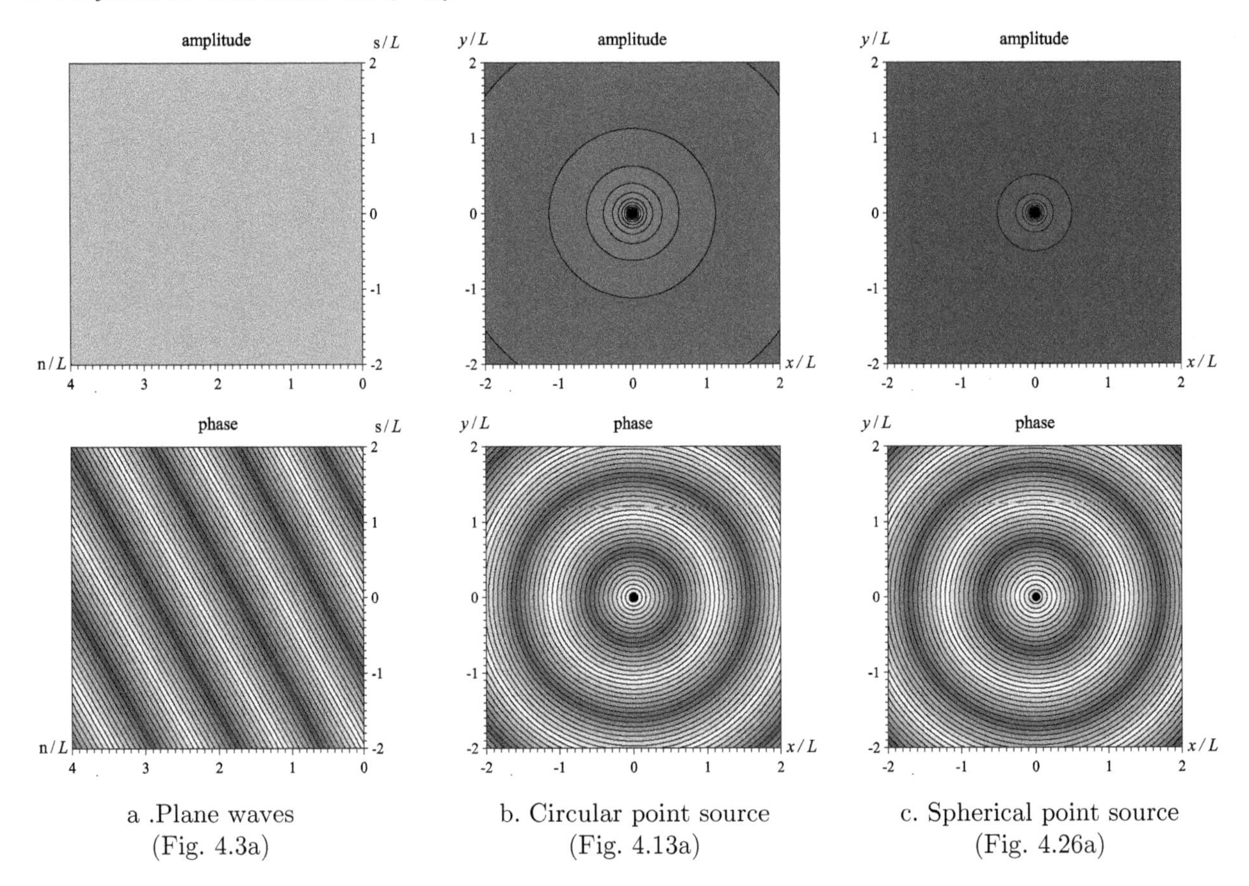

Figure 1.42 *Background wave fields of plane waves and waves emanating from a point.*

$$\boxed{\frac{\partial^2 \varphi}{\partial x^2} + \frac{\partial^2 \varphi}{\partial y^2} + \frac{\partial^2 \varphi}{\partial z^2} = -k^2 \varphi}, \quad \Phi(x, y, z, t) = \Re\left[\varphi(x, y, z)e^{-\mathrm{i}\omega t}\right] \tag{1.73}$$

where k is the ***wave number*** (Courant and Hilbert, 1962). This formulation utilizes complex variables with the imaginary number $\mathrm{i} = \sqrt{-1}$, and an introduction to complex functions with real and imaginary parts, $\varphi = \Re(\varphi) + \mathrm{i}\Im(\varphi)$, is provided in Section 2.3.2. This formulation of the Helmholtz equation is used to study periodic waves in acoustics (Morse and Ingard, 1968, p.728), electromagnetism (Sommerfeld, 1972, p.159), optics (Goodman, 1996, p.55), and surface water (Lamb, 1879, p.195).

The use of the complex function φ for studying periodic wave fieldsis illustrated in Fig. 1.42 for the following examples:

$$\begin{aligned} \varphi &= \varphi_0 e^{\mathrm{i}k(x\cos\theta_0 + y\sin\theta_0)} && \text{plane waves from (4.9)} \\ \varphi &= \frac{S}{2\pi} H_0^{(1)}(kr) && \text{cylindrical point source from (4.42)} \\ \varphi &= \frac{S}{4\pi} h_0^{(1)}(kr) && \text{spherical point source from (4.105)} \end{aligned} \tag{1.74}$$

and the exponential and Bessel functions are introduced later with the specified equations. These solutions are illustrated using lines of constant amplitude of the wave field, $|\varphi|$, and lines of constant cosine of the phase, $\cos[\arg(\varphi)]$. The plane waves in Fig. 1.42a have uniform amplitude $|\varphi_0|$, and travel in the θ_0=30° direction. The waves originating from points in Figs. 1.42b and 1.42c have decreasing amplitude as they travel away from these sources. The ***wavelength***, L, is related to the wave number and the ***wave period***, T, is related to angular frequency by

$$\begin{array}{ll} L = \dfrac{2\pi}{k}, & k = \dfrac{2\pi}{L} \\ T = \dfrac{2\pi}{\omega}, & \omega = \dfrac{2\pi}{T} \end{array} \quad \to \quad C = \frac{L}{T} = \frac{\omega}{k} \tag{1.75}$$

which provides an expression for the celerity, C, or wave velocity. The wavelength is evident in Fig. 1.42 by identifying locations with the same value of the cosine of the phase on neighboring waves.

Waves are transformed by their interactions at the boundaries and interfaces of elements. Wave ***reflection*** results as incoming waves approach a boundary in Fig. 1.43a and part of the energy is reflected back into the domain at a reduced amplitude factored by the reflection coefficient R. This leads to a ***Robin boundary condition***

$$\boxed{\frac{\partial \varphi}{\partial \mathrm{n}} + \alpha \varphi = 0} \tag{1.76}$$

where the coefficient α is developed later as a function of the wave number k and the reflection coefficient R. Wave fields also also transformed by ***diffraction*** as waves bend around elements in Fig. 1.43b This requires the scattered waves to satisfy a radiation condition at large distances from the circular boundary condition (Sommerfeld, 1949, Section 28)

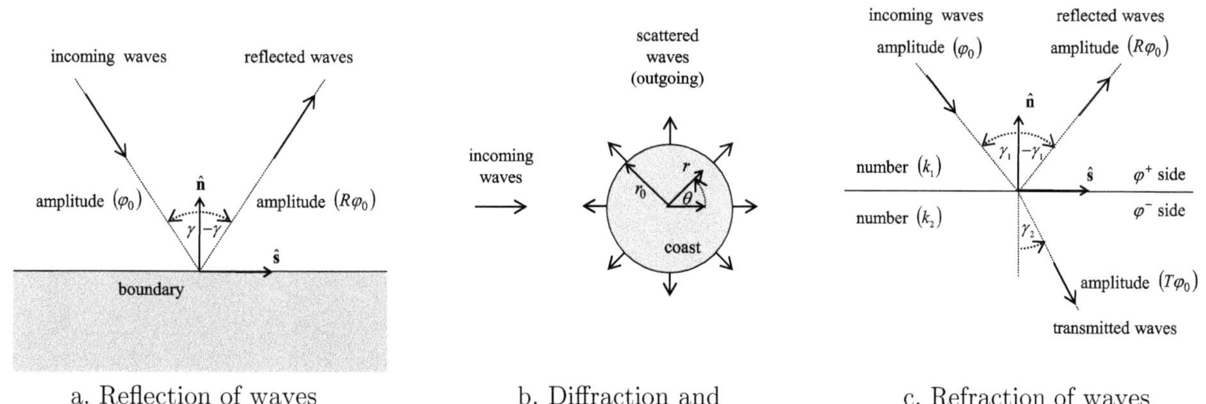

Figure 1.43 *Wave interactions at boundaries and interfaces.*

$$\boxed{\lim_{r\to\infty} \sqrt{r}\left(\frac{\partial \varphi}{\partial r} - ik\varphi\right) = 0} \tag{1.77}$$

so the scattered waves are outgoing and move away from the element. The wave field in Fig. 1.44 illustrates the reflected waves in front of the circle element and the calmer zone behind it. As the refection coefficient decreases, the wave ***absorption*** increases since less energy is reflected from the boundary.

A third mechanism for transformation of waves is ***refraction***, where waves travel between regions with different wave numbers. This is illustrated in Fig. 1.43c for incoming waves that move across an interface normal to the $\mathfrak{n}$-direction. Two conditions must be satisfied at this interface: continuity of the complex wave function, and a condition relating the normal derivatives across the interface:

$$\boxed{\varphi^+ = \varphi^-}, \quad \boxed{\beta^+ \frac{\partial \varphi^+}{\partial \mathfrak{n}} = \beta^- \frac{\partial \varphi^-}{\partial \mathfrak{n}}} \tag{1.78}$$

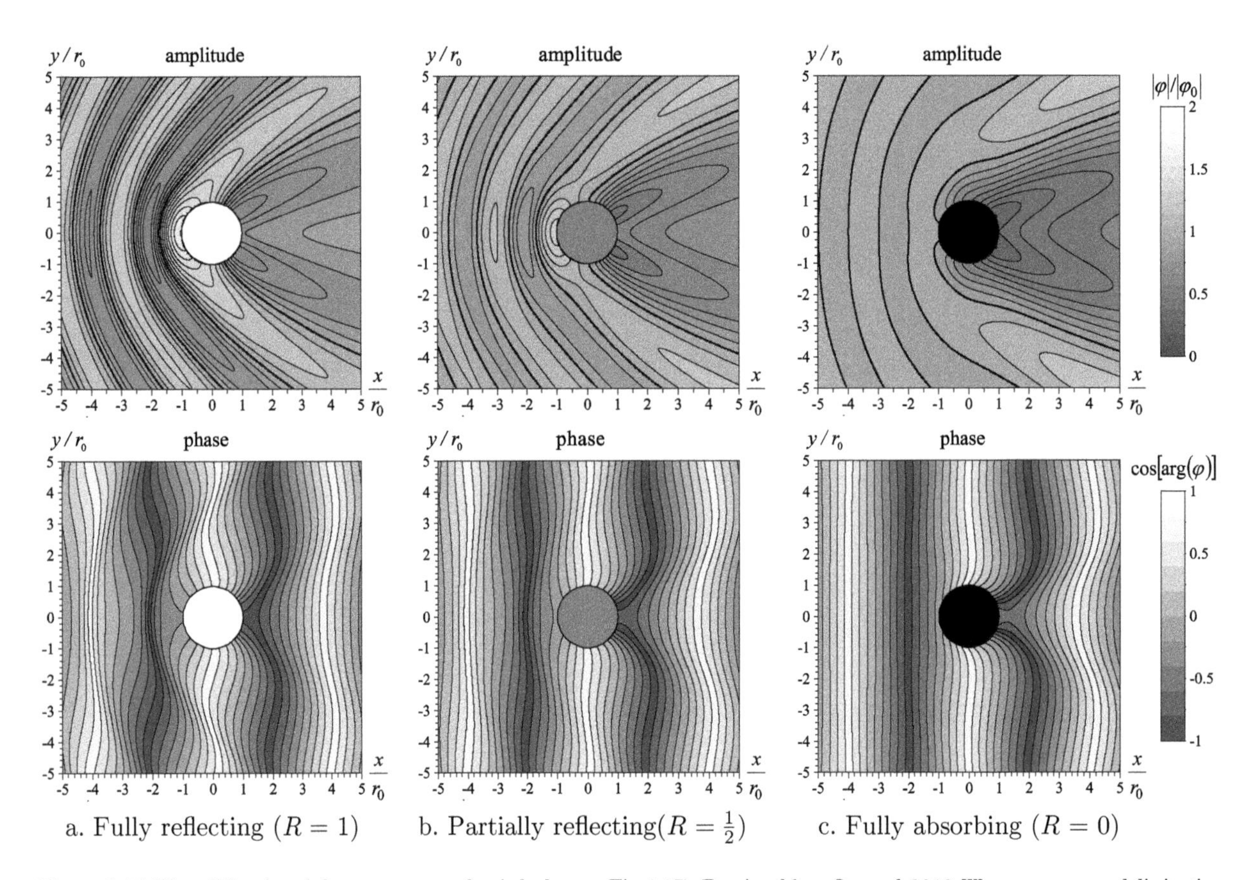

Figure 1.44 *Wave diffraction of plane waves around a circle element (Fig. 4.17). (Reprinted from Steward, 2018, Wave resonance and dissipation in collections of partially reflecting vertical cylinders,* Journal of Waterways, Ports, Coastal, and Ocean Engineering, *Vol. 144(4), Fig. 1, with permission from ASCE.)*

The coefficients for this interface are developed later by satisfying Snell's law

$$k_1 \sin \gamma_1 = k_2 \sin \gamma_2 \tag{1.79}$$

relating the change in wave direction to the wave number on each side of the interface.

Problem 1.7 Compute the variable φ, its magnitude, $|\varphi|$, and the cosine of its phase, $\cos[\arg(\varphi)]$ for the plane waves in Fig. 1.42a and Eq. (1.74) at the specified locations: $(x_1, y_1) = (0, 0)$m, $(x_2, y_2) = (10, 10)$m, and $(x_3, y_3) = (20, 20)$m. This wave field is specified for $\varphi_0 = 10$ m^2/s, $k = 0.25$/m and the following direction θ_0:

A. $\theta_0 = 0°$
B. $\theta_0 = 10°$
C. $\theta_0 = 20°$
D. $\theta_0 = 30°$
E. $\theta_0 = 40°$
F. $\theta_0 = 50°$
G. $\theta_0 = 60°$
H. $\theta_0 = 70°$
I. $\theta_0 = 80°$
J. $\theta_0 = 90°$

1.3.1 Water Waves: Amplitude and Phase

Waves are generated by winds blowing over an open sea that cause water particles in deep water to move in circular paths with the size of the orbits decreasing with depth, as illustrated in Fig. 1.45. These motions generate wave trains moving with a common period T in the same direction, and wave envelopes may travel hundreds to thousands of kilometers from their source. Waves are impacted by bathymetry as they move into coastal regions with shallower depth h, and the motion of water particles takes on the form of elliptical orbits. The wavelength L decreases and the amplitude A increases near the coastal boundary until the maximum slope of the water surface reaches $\frac{1}{7}$, at which point surface tension can no longer hold the wave together and it breaks and dissipates energy (Bascom, 1980).

The circular orbits of water particles are illustrated in Fig. 1.46 using plots of the water surface profile and water particle velocity shown at specific time intervals. Solutions are formulated by separating the spatial coordinates of the scalar potential in (1.73) into a real function Z of the z-coordinate and a complex function φ of x and y:

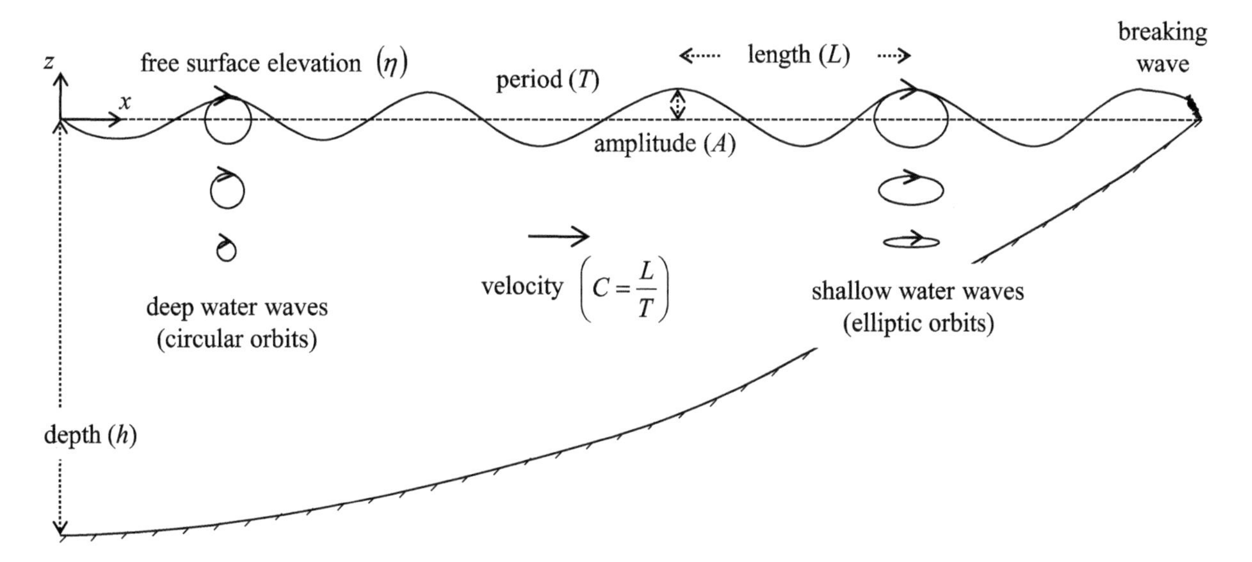

Figure 1.45 *Water waves properties and processes: The periodic motion of water particles generates a wave train.*

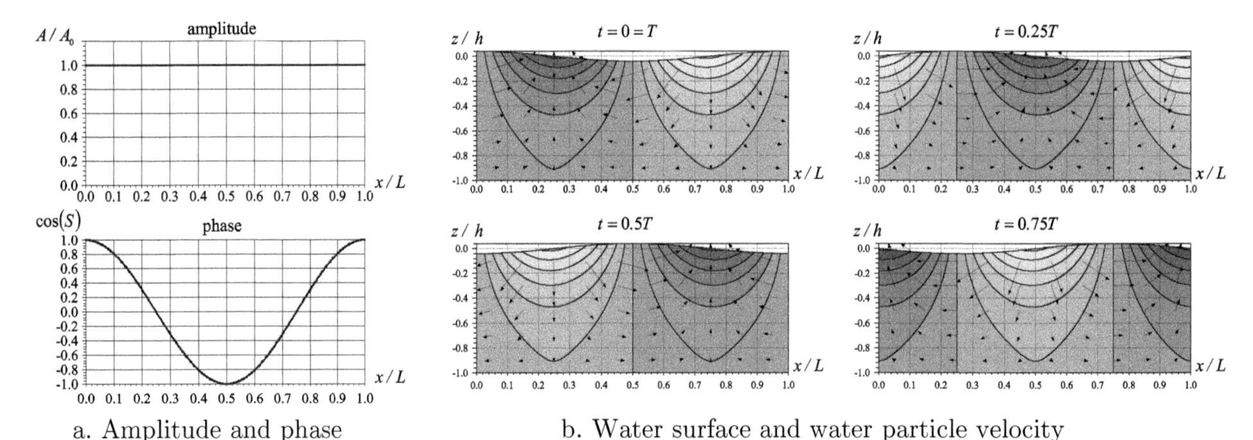

Figure 1.46 *Monochromatic waves in water of constant depth.*

$$\mathbf{v} = -\nabla\Phi, \quad \Phi(x,y,z,t) = Z(z)\Re\left[\varphi(x,y)e^{-i\omega t}\right] \\ = Z(z)\left[\Re(\varphi)\cos\omega t + \Im(\varphi)\sin\omega t\right] \tag{1.80}$$

and $\mathbf{v}$ is the water particle velocity. This gives solutions to the Helmholtz equation for φ, and an ordinary differential equation of Z

$$\boxed{\frac{\partial^2\varphi}{\partial x^2} + \frac{\partial^2\varphi}{\partial y^2} = -k^2\varphi}, \\ \frac{d^2 Z}{dz^2} = k^2 Z \quad \rightarrow \quad Z = \frac{\cosh k(z+h)}{\cosh kh} \tag{1.81}$$

where the solution for Z is given for waves of small amplitude in water of uniform depth (Dean and Dalrymple, 1984). Water waves satisfy a dispersion relation from Airy (1845, p. 290)

$$\boxed{\omega^2 = kg\tanh kh} \tag{1.82}$$

which fixes the wave number k for a specific angular frequency and water depth. This gives expressions for the wave velocity (1.75), and the group velocity C_g, which is velocity at which energy gets propagated by waves (Dean and Dalrymple, 1984, p. 98)

$$C = \frac{\omega}{k} = \sqrt{\frac{g}{k}\tanh kh}, \quad C_g = \frac{\partial\omega}{\partial k} = \frac{C}{2}\left(1 + \frac{2kh}{\sinh 2kh}\right) \tag{1.83}$$

The hyperbolic functions in these relations approach asymptotic limits for shallow water (where $\tanh kh \rightarrow kh$ and $\sinh 2kh \rightarrow 2kh$) and deep water (where $\tanh kh \rightarrow 1$ and $\sinh 2kh \rightarrow e^{2kh}/2$) that are summarized in Table 1.2.

Waves are transformed by ***reflection of wave energy by the coast back to the sea***. The incoming plane waves in Fig. 1.43a are directed towards a coastline at angle γ, and it is natural to express their potential, (1.74), in terms of the tangential and normal coordinates

Table 1.2 *Water wave properties for water depth that is shallow, intermediate, or deep*

Variable	Shallow	Intermediate	Deep
Range of wavelength	$20h < L$	$2h < L < 20h$	$L < 2h$
Range of wave number	$kh < \pi/10$	$\pi/10 < kh < \pi$	$\pi < kh$
Dispersion relation, (1.82)	$\omega^2 = k^2 gh$	$\omega^2 = kg \tanh kh$	$\omega^2 = kg$
Wavelength, (1.75)	$L = \frac{2\pi\sqrt{gh}}{\omega}$	$L = \frac{2\pi}{k}$	$L = \frac{2\pi g}{\omega^2}$
Wave velocity, (1.83)	$C = \sqrt{gh}$	$C = \sqrt{\frac{g}{k} \tanh kh}$	$C = \sqrt{\frac{g}{k}}$
Group velocity, (1.83)	$C_g = C$	$C_g = \frac{C}{2}\left(1 + \frac{2kh}{\sinh 2kh}\right)$	$C_g = \frac{C}{2}$

$\mathfrak{s}$-$\mathfrak{n}$ along the coast with incoming wave amplitude A_0 proportional to φ_0. These ***incoming waves are superimposed with reflected waves*** directed at angle $-\gamma$ away from the coast with amplitude $A_0 R$,

$$\varphi = \varphi_0 \left[\mathrm{e}^{\mathrm{i}k(\mathfrak{s} \sin\gamma - \mathfrak{n} \cos\gamma)} + R\mathrm{e}^{\mathrm{i}k(\mathfrak{s} \sin\gamma + \mathfrak{n} \cos\gamma + \beta)} \right], \quad \varphi_0 = \frac{gA_0}{\omega} \tag{1.84}$$

where the reflection coefficient R takes on values from $R = 0$ for a coast that fully absorbs wave energy to $R = 1$ for full reflection of wave energy, and β is a phase shift between incoming and reflected waves at the boundary $\mathfrak{n} = 0$ (Berkhoff, 1976). A coastal boundary condition may be developed by evaluating the partial derivative with respect to the normal direction on the boundary

$$\begin{aligned} \frac{\partial\varphi}{\partial\mathfrak{n}} &= \mathrm{i}k\cos\gamma\,\varphi_0 \left[-\mathrm{e}^{\mathrm{i}k(\mathfrak{s} \sin\gamma - \mathfrak{n} \cos\gamma)} + R\mathrm{e}^{\mathrm{i}k(\mathfrak{s} \sin\gamma + \mathfrak{n} \cos\gamma + \beta)} \right] \\ &= -\mathrm{i}k\cos\gamma \frac{1 - R\mathrm{e}^{\mathrm{i}k\beta}}{1 + R\mathrm{e}^{\mathrm{i}k\beta}} \varphi \qquad (\mathfrak{n} = 0) \end{aligned} \tag{1.85}$$

It is common to approximate the ***coastal boundary condition*** as

$$\boxed{\frac{\partial\varphi}{\partial\mathfrak{n}} + \alpha\varphi = 0, \quad \alpha = \mathrm{i}k\frac{1 - R}{1 + R} \qquad (\mathfrak{n} = 0)} \tag{1.86}$$

with incoming waves normal to the coast and no phase shift (Steward and Panchang, 2000). Wave reflection is illustrated for a fully absorbing coast with $\gamma = 30°$ in Fig. 1.47a, for partial wave reflection in Fig. 1.47b, and for a fully reflective coast in Fig. 1.47c. It is observed that standing waves form with increased coastal reflection and generate a sea with amplitude varying between 0 and twice that of the incoming waves, and phase shifts occur with full reflection across locations where $|\varphi| = 0$.

Diffraction of water waves occur when the coastal boundary condition (1.86) is satisfied along elements, and the scattered waves travel away from the element, (1.77). For example, the envelope of incoming water waves that travel around a circular island in Fig. 1.44a is extended to multiple islands in Fig. 1.48. These elements interact to form a complicated sea with some regions experiencing wave amplification and others with reduction in amplitude (Steward, 2018). The boundary condition (1.86) may be separated into its amplitude and phase parts (Berkhoff, 1976):

$$\frac{\partial|\varphi|}{\partial\mathfrak{n}} = 0, \qquad \frac{\partial \arg\varphi}{\partial\mathfrak{n}} = -k\frac{1 - R}{1 + R} \tag{1.87}$$

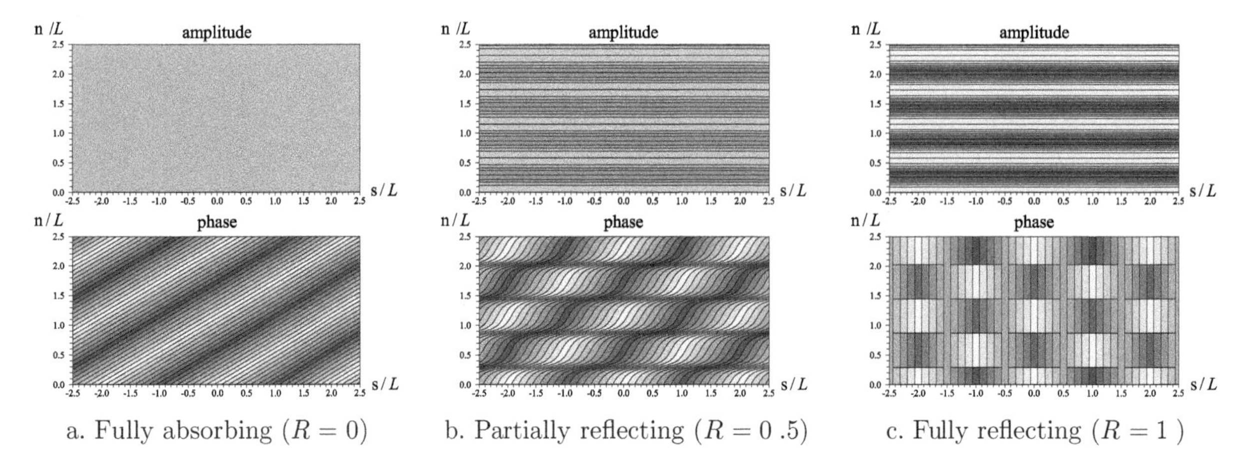

Figure 1.47 *Water wave reflection with incoming waves at an oblique angle* $\gamma = 30°$ *(Fig. 4.3).*

and the close-up figures illustrate that lines of constant amplitude and phase correctly intersect the elements orthogonally along their fully reflected boundaries.

Wave refraction occurs as waves propagate across changes in depth as illustrated in Fig. 1.43c for a region of uniform depth h_1 on the + side of an interface to a depth of h_2 on the − side. ***Continuity of the free surface elevation*** requires that the potential φ takes on the same values across this interface,

$$\boxed{\varphi^+ = \varphi^-} \tag{1.88}$$

A second condition is developed from the component of the vector for water particle movement (1.80) normal to the interface with (1.81)

$$v_n = -\frac{\partial \Phi}{\partial n} = \frac{\cosh k(z+h)}{\cosh kh}\frac{\partial}{\partial n}\Re\left(\varphi e^{-i\omega t}\right) \tag{1.89}$$

which is vertically integrating over the depth of the water column to give

$$V_n = \int_{-h}^{0} v_n \, dz = \int_{-h}^{0} \frac{\cosh k(z+h)}{\cosh kh} dz \frac{\partial}{\partial n}\Re\left(\varphi e^{-i\omega t}\right) = \frac{\sinh kh}{k\cosh kh}\frac{\partial}{\partial n}\Re\left(\varphi e^{-i\omega t}\right) \tag{1.90}$$

Conservation of mass is satisfied when

$$V_n{}^+ = V_n{}^- \rightarrow \boxed{\beta^+\frac{\partial \varphi^+}{\partial n} = \beta^-\frac{\partial \varphi^-}{\partial n}} \tag{1.91a}$$

with coefficients

$$\begin{aligned} \beta^+ &= \frac{\sinh k_1 h_1}{k_1 \cosh k_1 h_1} \\ \beta^- &= \frac{\sinh k_2 h_2}{k_2 \cosh k_2 h_2} \end{aligned}, \quad \begin{aligned} \beta^+ &= h_1 \quad \left(k_1 h_1 < \frac{\pi}{10}\right) \\ \beta^- &= h_2 \quad \left(k_2 h_2 < \frac{\pi}{10}\right) \end{aligned} \tag{1.91b}$$

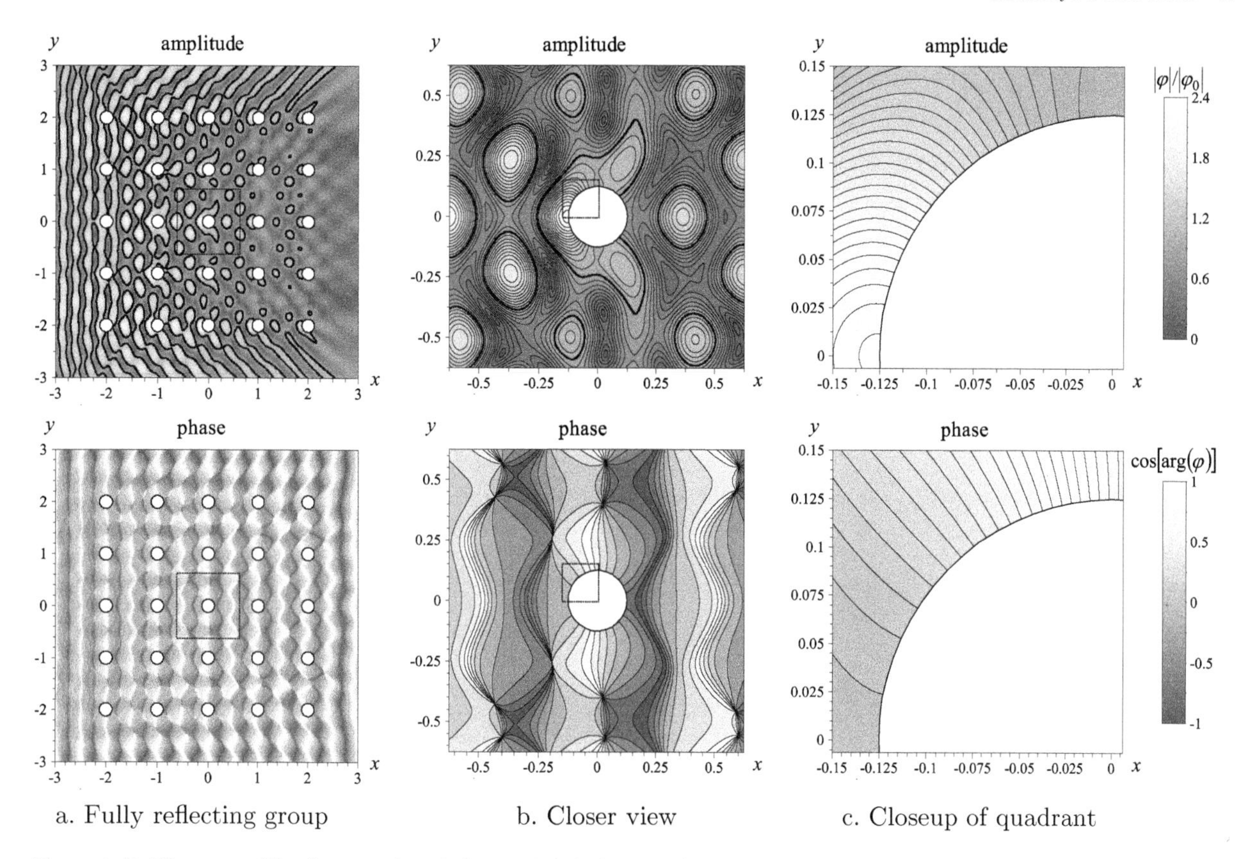

Figure 1.48 *Water wave diffraction around regularly spaced circle elements with fully reflective boundaries (Fig. 4.18). (Reprinted from Steward, 2018, Wave resonance and dissipation in collections of partially reflecting vertical cylinders,* Journal of Waterways, Ports, Coastal, and Ocean Engineering, *Vol. 144(4), Figs. 5,10, with permission from ASCE.)*

that take on simpler forms for shallow water waves where $\sinh kh \to kh$ and $\cosh kh \to 1$. Examples of wave refraction are illustrated in Fig. 1.49 with the same incident wave angle $\gamma_1 = 30°$ as shown for reflection in Fig. 1.47.

Water waves computations:

- frequency ω is related to period T, (1.75).
- wave number k is obtained from ω and water depth h in (1.82).
- free surface amplitude, $A = |\varphi|\omega/g$, is obtained from the magnitude of the wave function φ, (1.84).
- phase diagrams illustrate the cosine of the phase, $S = \arg(\varphi)$
- velocity (v_x, v_y, v_z) of water particles is obtained from minus the gradient of Φ, (1.80).

Example 1.6 Surface water waves with period $T = 4$ sec and amplitude A_0=0.4 m in water of depth $h = 10$ m are shown in Fig. 1.46. The frequency (1.75) is $\omega = 1.5708$/sec, and the wave number k may be determined iteratively by applying Newton's method, (2.56) to the dispersion relation (1.82),

$$k|_{l+1} = k|_l + \frac{\omega^2 - k|_l g \tanh k|_l h}{\frac{k|_l gh}{\cosh^2 k|_l h} + g \tanh k|_l h} \quad \rightarrow \quad \begin{aligned} k|_0 &= \frac{\omega^2}{g} = \frac{(1.5708/\text{sec})^2}{9.80665\text{m}/\text{sec}^2} = 0.2516/\text{m} \\ k|_1 &= 0.2547/\text{m} \\ k|_2 &= 0.2547/\text{m} \end{aligned} \tag{1.92}$$

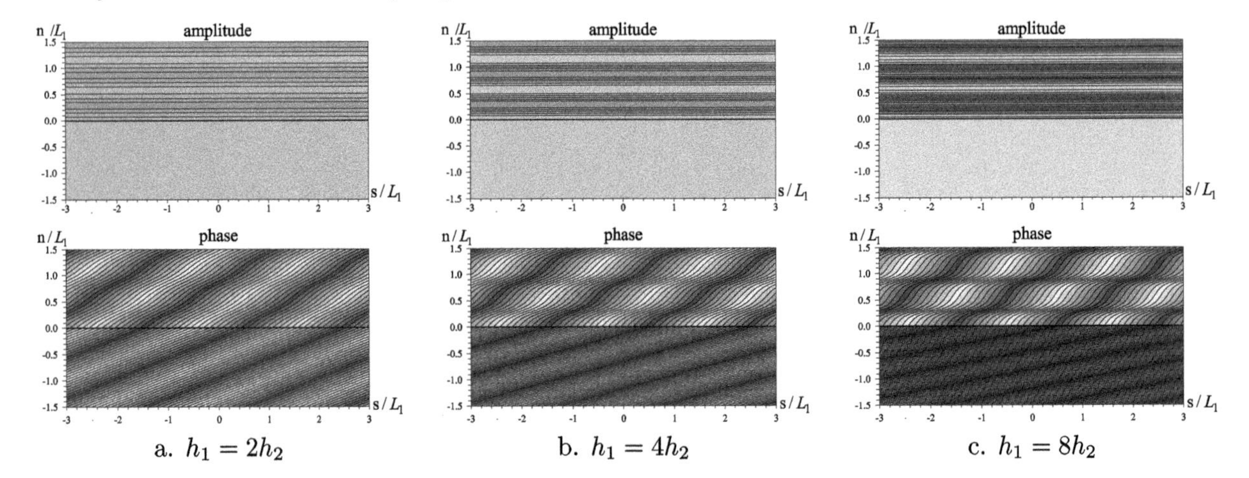

Figure 1.49 *Water wave refraction by a change in depth for shallow water waves (Fig. 4.4).*

where $k|_l$ is the current estimate for k and $k|_{l+1}$ is the next estimate for k, and $k|_0$ is chosen to be the wave number for deep water given in Table 1.2. The wavelength (1.75) is $L = 24.67$ m, and the wave velocity and group velocity (1.83) are $C = 6.167$ m/sec and $C_g = 3.276$ m/sec. The potential and velocity of water particles are

$$\begin{aligned} \Phi &= -\frac{\cosh k(z+h)}{\cosh kh}\frac{g}{\omega}A_0 \sin(kx-\omega t) \\ v_x &= -\frac{\partial \Phi}{\partial x} = \frac{\cosh k(z+h)}{\cosh kh}\frac{gk}{\omega}A_0 \cos(kx-\omega t) \\ v_z &= -\frac{\partial \Phi}{\partial z} = \frac{\sinh k(z+h)}{\cosh kh}\frac{gk}{\omega}A_0 \sin(kx-\omega t) \end{aligned} \tag{1.93}$$

giving maximum velocities at the water surface $\max v_x = 0.6361$ m/sec and $\max v_z = 0.6283$ m/sec.

Problem 1.8 Compute the frequency (ω in 1/sec), wave number (k in 1/m), wavelength (L in m), phase velocity (C in m/sec), group velocity (C_g in m/sec), and maximum horizontal and vertical components of the velocity of water particles ($\max v_x$ and $\max v_z$ in m/sec) for water waves with the following period, amplitude, and water depth:

A. $T = 3$ sec, $A_0 = 0.04$ m, and $h = 0.1$ m
B. $T = 3$ sec, $A_0 = 0.04$ m, and $h = 1$ m
C. $T = 3$ sec, $A_0 = 0.04$ m, and $h = 10$ m
D. $T = 5$ sec, $A_0 = 0.1$ m, and $h = 0.5$ m
E. $T = 5$ sec, $A_0 = 0.1$ m, and $h = 5$ m
F. $T = 5$ sec, $A_0 = 0.1$ m, and $h = 50$ m
G. $T = 10$ sec, $A_0 = 0.5$ m, and $h = 2$ m
H. $T = 10$ sec, $A_0 = 0.5$ m, and $h = 20$ m
I. $T = 10$ sec, $A_0 = 0.5$ m, and $h = 200$ m
J. $T = 20$ sec, $A_0 = 1$ m, and $h = 5$ m
K. $T = 20$ sec, $A_0 = 1$ m, and $h = 50$ m
L. $T = 20$ sec, $A_0 = 1$ m, and $h = 500$ m

1.3.2 Acoustics: Intensity and Vibration

The field of acoustics studies the mechanical propagation of waves through gases, liquids, and solids. Acoustics may be formulated using the wave equation for pressure waves (Airy, 1871) and separation of variables leads to solutions in terms of the Helmholtz equation (Moon and Spencer, 1961*a*, Chap.16)

$$\nabla^2\varphi = \frac{\partial^2\varphi}{\partial x^2} + \frac{\partial^2\varphi}{\partial y^2} + \frac{\partial^2\varphi}{\partial z^2} = -k^2\varphi \tag{1.94}$$

Waves may originate at a point-source as shown in Fig. 1.50, or they may be generated a large distance from the study region and be treated as plane waves. The reflection, diffraction, and refraction processes formulated for water waves are extended in this presentation of acoustics to illustrate solutions to wave propagation through and around inhomogeneities with different wave properties than those of the surrounding media.

The reflection of acoustic waves by a fully reflective boundary at $x = 0$ is illustrated in Fig. 1.51. This extends the solution for plane waves and point-sources in free space from Fig. 1.42 and Eq. (1.74) using the following expressions for the complex potential

$$\begin{aligned}
\varphi &= \varphi_0 \mathrm{e}^{\mathrm{i}k(x\cos\theta_0 + y\sin\theta_0)} + \varphi_0 \mathrm{e}^{\mathrm{i}k(-x\cos\theta_0 + y\sin\theta_0)}, \quad \text{plane waves from (4.11)} \\
\varphi &= \frac{S}{2\pi} H_0^{(1)}\left(k\sqrt{(x-d)^2+y^2}\right) \\
&\quad + \frac{S}{2\pi} H_0^{(1)}\left(k\sqrt{(x+d)^2+y^2}\right), \quad \text{cylindrical point source from (4.43)} \\
\varphi &= \frac{S}{2\pi} h_0^{(1)}\left(k\sqrt{(x-d)^2+y^2}\right) \\
&\quad + \frac{S}{2\pi} h_0^{(1)}\left(k\sqrt{(x+d)^2+y^2}\right), \quad \text{spherical point source from (4.106)}
\end{aligned} \tag{1.95}$$

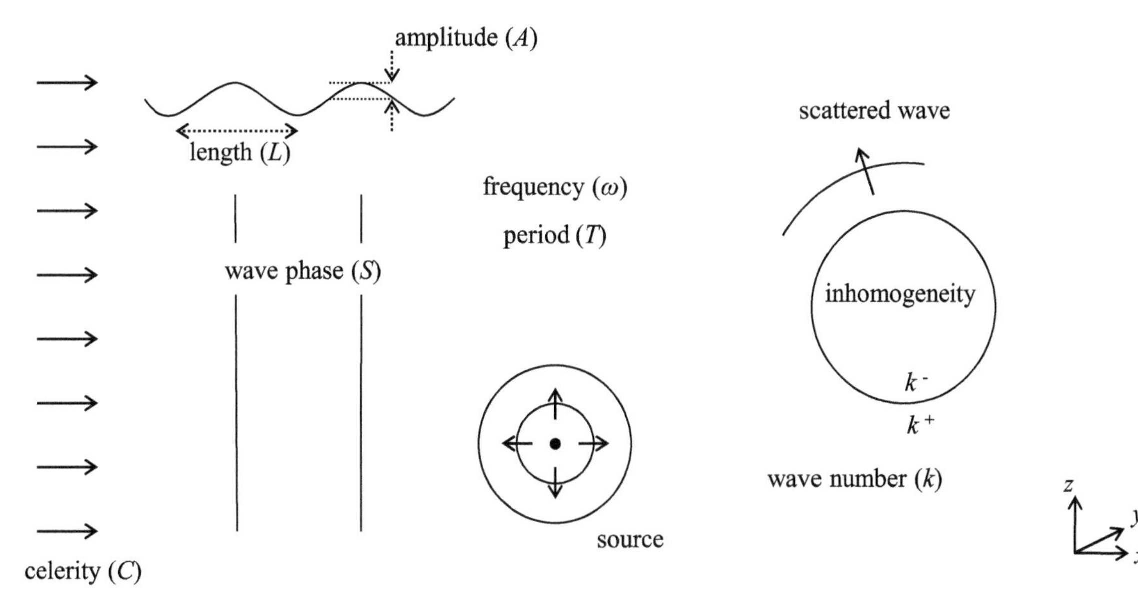

Figure 1.50 *Acoustics properties and processes: waves translated by the periodic motion of particles.*

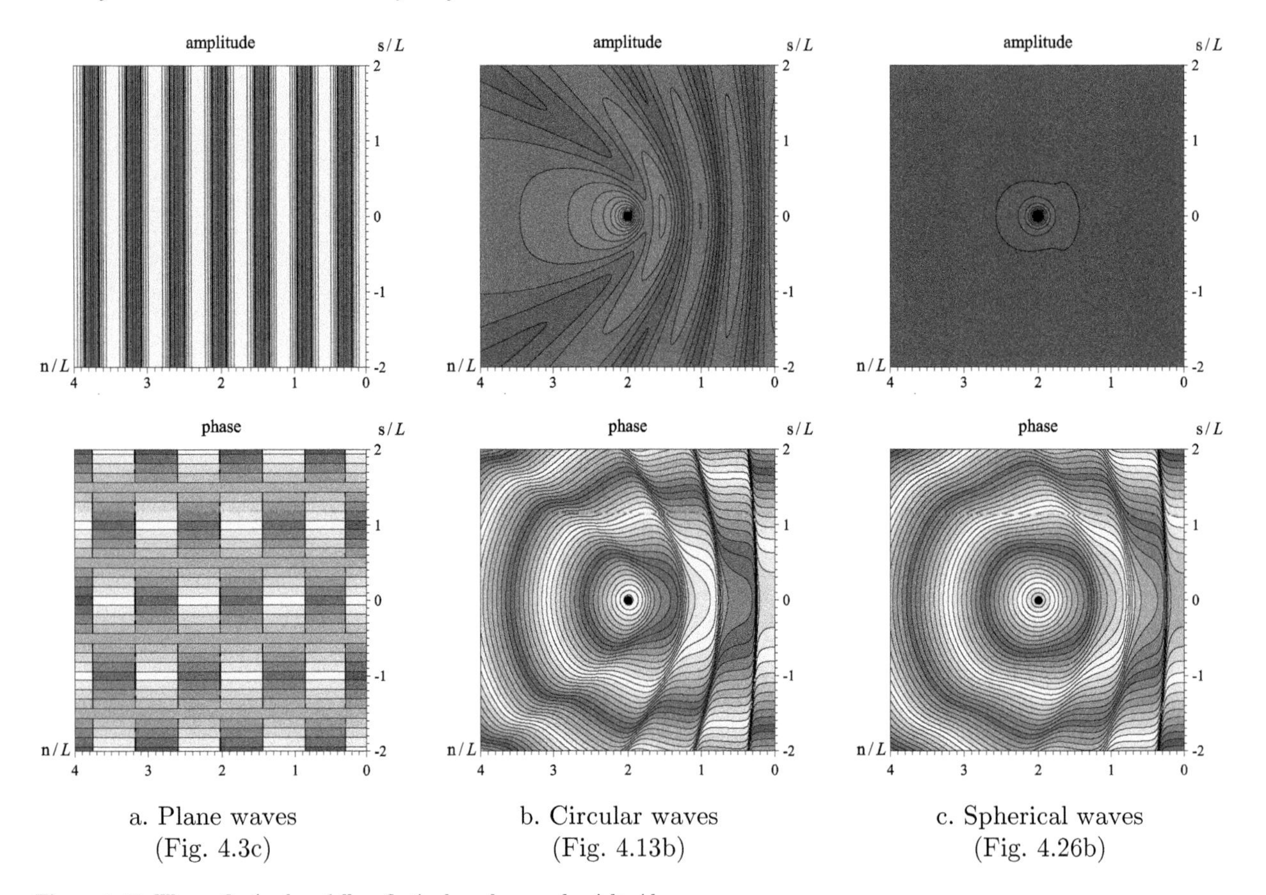

Figure 1.51 *Wave reflection by a fully reflective boundary on the right side.*

which are developed later. Each solution satisfies a ***boundary condition for fully reflective surfaces*** with $R = 1$ in

$$\frac{\partial \varphi}{\partial \mathrm{n}} + \alpha \varphi = 0, \quad \alpha = ik\frac{1-R}{1+R} \quad (\mathrm{n} = 0) \tag{1.96}$$

where the normal component of the derivative is zero at the boundary where $x = 0$ in this figure.

The diffraction produced by collection of partially reflecting elements is illustrated in Fig. 1.52. Each element reproduces the partially reflective boundary condition (1.96) with the specified reflection coefficient R, and the scattered waves generated by each element travels in the outward direction, (1.77). The wavelength for an acoustic field may be computed from (1.75)

$$L = \frac{2\pi C}{\omega} \tag{1.97}$$

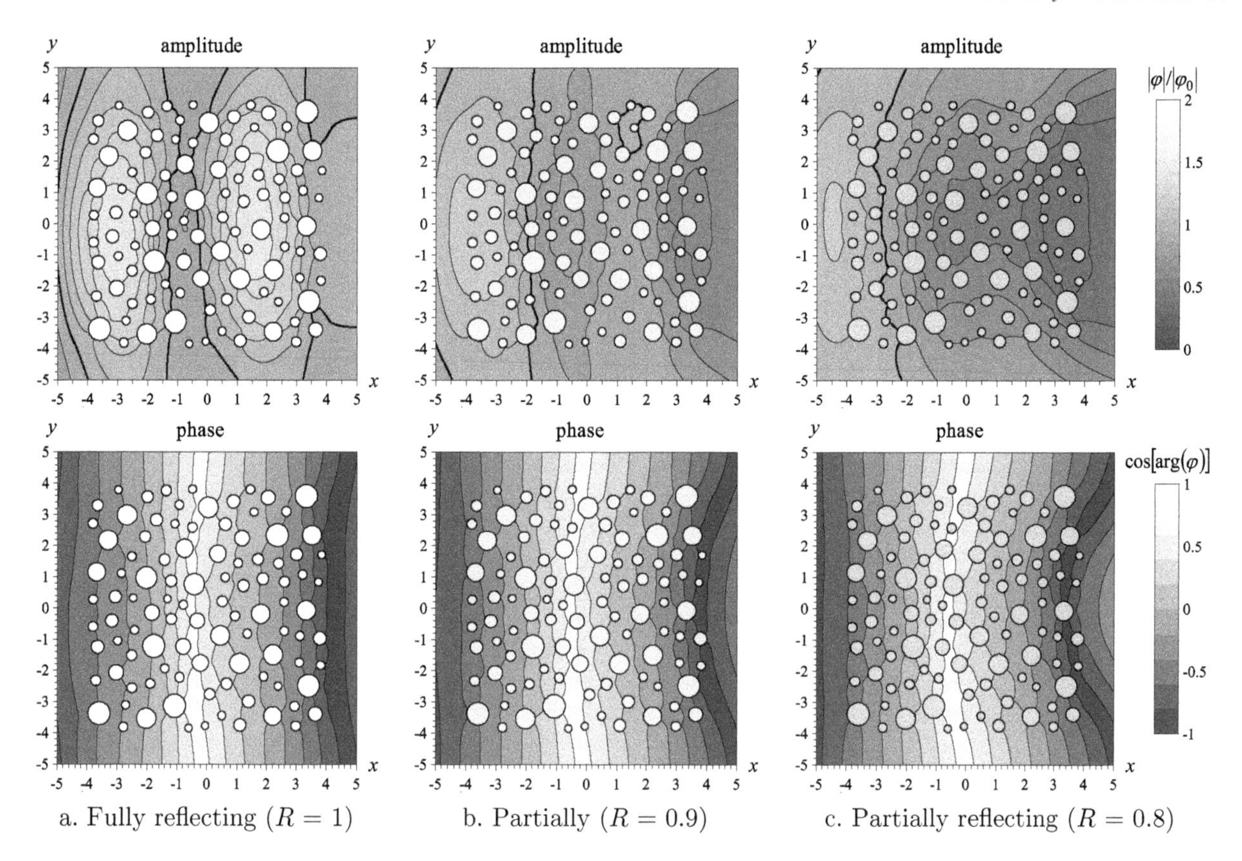

Figure 1.52 *Randomly distributed circle elements, from Fig. 4.18. (Reprinted from Steward, 2018, Wave resonance and dissipation in collections of partially reflecting vertical cylinders,* Journal of Waterways, Ports, Coastal, and Ocean Engineering, *Vol. 144(4), Fig. 9, with permission from ASCE.)*

for waves with specified frequency, ω, where the speed of wave propagation, C, is specified for the media through which waves propagate (Morse, 1936; Morse and Ingard, 1968). These results illustrate the resonance occurring within a collection of elements, where regions exist with amplitudes higher than the incoming waves, and dissipation occurs in other areas with lower amplitude. They also show the impact of reflection coefficient, where resonance is larger for more reflective objects.

The transmission of waves through elements is illustrated in Figs. 1.53 and 1.54. These elements require that waves have the same amplitude and phase across their interface

$$\boxed{\varphi^+ = \varphi^-}, \qquad \boxed{\beta^+ \frac{\partial \varphi^+}{\partial n} = \beta^- \frac{\partial \varphi^-}{\partial n}} \tag{1.98}$$

similar to (1.88) for water wave transmission across changes in water depth. The additional constraints on the normal component of the partial derivative across an interface is similar to (1.91). The coefficients β^+ and β^- may be chosen to satisfy continuity of the gradient

Acoustics computations:

- wave intensity is obtained from the magnitude of the wave function φ.
- phase diagrams illustrate the cosine of the phase, $S = \arg(\varphi)$.
- wavelength L and wave number k are related by (1.75).

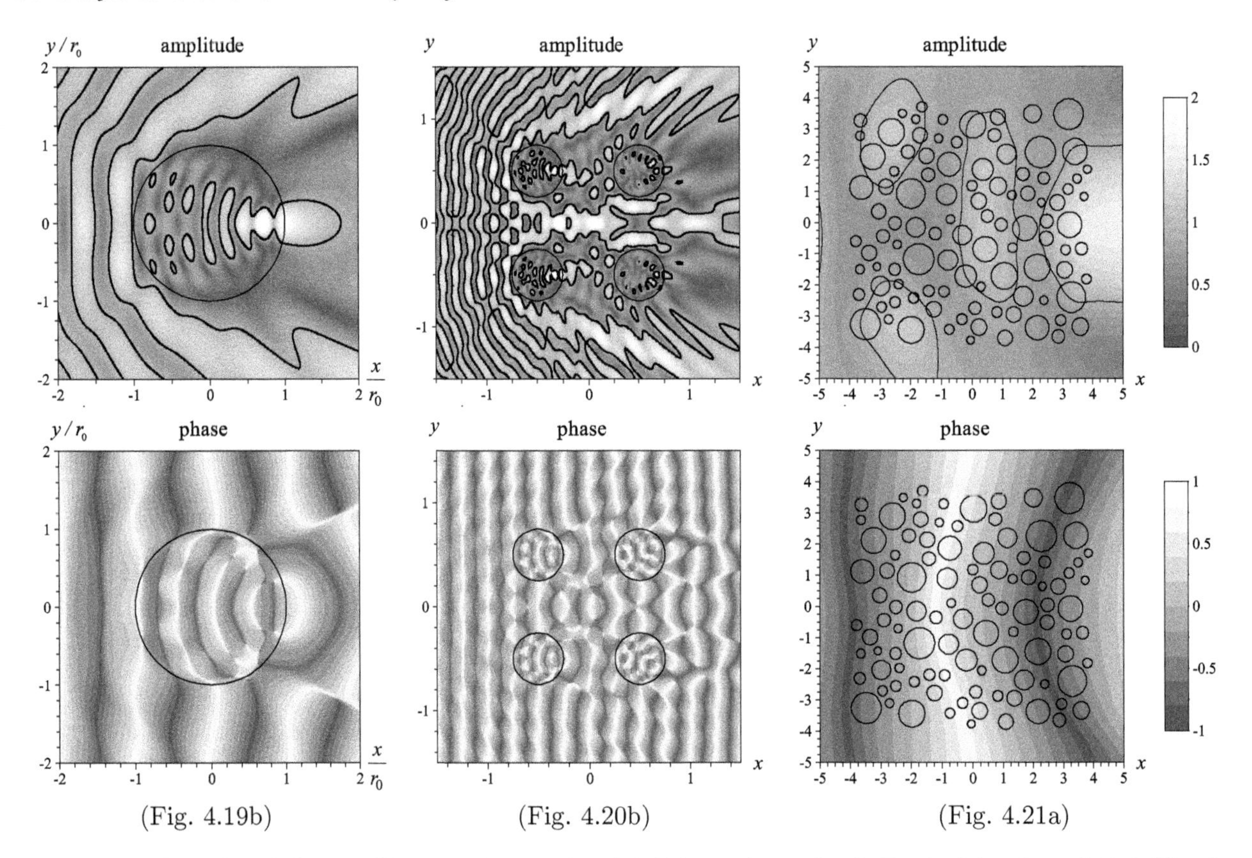

Figure 1.53 *Waves through elements with wavelengths approximately half the exterior wavelength. (Reprinted from Steward, 2020, Waves in collections of circular shoals and bathymetric depressions,* Journal of Waterways, Ports, Coastal, and Ocean Engineering, *Vol. 146, Figs. 2,3,4, preproduction version DOI 10.1061/(ASCE)WW.1943-5460.0000570, with permission from ASCE.)*

of the potential in the wave equation (1.72) following Moon and Spencer (1961*a*). The transmission of waves through inhomogeneities is illustrated in Fig. 1.53 for elements with a shorter wavelength than that on the outside, and for larger wavelengths inside the element in Fig. 1.54. Individually, the interactions of an element with a field of plane waves generate locations within and outside the inhomogeneity with higher and lower amplification of the waves. Collectively, groups of inhomogeneities have the capacity to create zones of resonance and dissipation of wave energy, which are influenced by the acoustic properties of the medium.

Problem 1.9 Compute the intensity, $|\varphi|$, for the reflected plane waves of Fig. 1.51a and Eq. (1.95) at the specified locations: $(x_1, y_1) = (0, 0)$ m, $(x_2, y_2) = (-25, 0)$ m, and $(x_3, y_3) = (-50, 0)$ m. These waves are specified for unit φ_0=1 with the following wavelength and orientation.

A. $L = 20$ m, $\theta_0 = 0°$

B. $L = 20$ m, $\theta_0 = 30°$

C. $L = 20$ m, $\theta_0 = 60°$

D. $L = 20$ m, $\theta_0 = 90°$

E. $L = 70$ m, $\theta_0 = 0°$

F. $L = 70$ m, $\theta_0 = 30°$

G. $L = 70$ m, $\theta_0 = 60°$

H. $L = 70$ m, $\theta_0 = 90°$

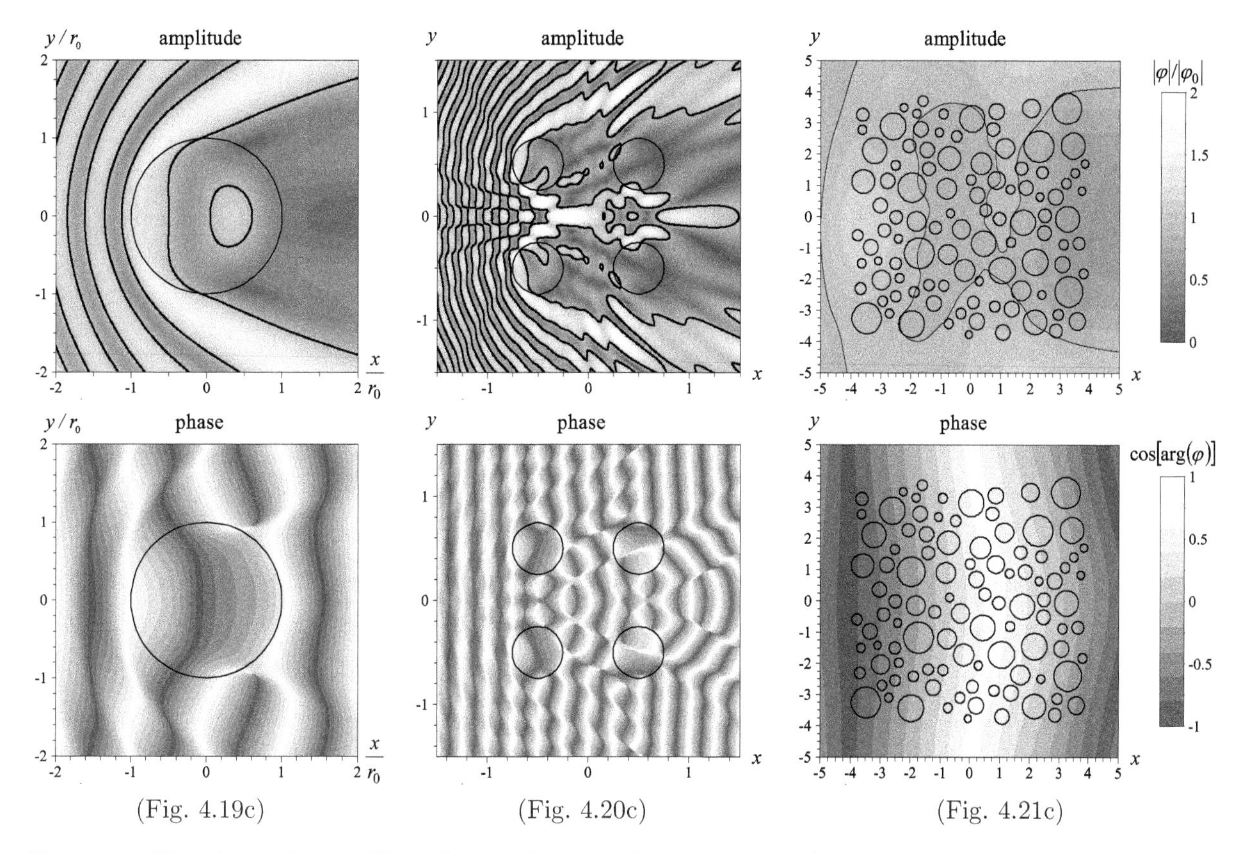

Figure 1.54 *Waves through elements with wavelengths twice the exterior wavelength. (Reprinted from Steward, 2020, Waves in collections of circular shoals and bathymetric depressions,* Journal of Waterways, Ports, Coastal, and Ocean Engineering, *Vol. 146, Figs. 2,3,4, preproduction version DOI 10.1061/(ASCE)WW.1943-5460.0000570, with permission from ASCE.)*

1.4 Studies of Deformation by Forces

The deformation of material by forces is formulated in terms of the stress (force per area) and displacement, as illustrated by the three-dimensional control block in Fig. 1.55. Surface forces at the boundary with stress components σ_x, σ_y, and σ_z act normal to planes of constant x, y, and z, and a sign convention is adopted where stress is positive for tension and negative for compression. Shear stresses act tangentially to these planes (e.g., τ_{xy} is the y component of shear stress acting on the x-face). Body forces with components F_x, F_y, and F_z in the directions of the coordinate axes may also act internally to deform the material. The displacements u_x, u_y, and u_z generate strain (displacement per length) with normal strain ε_x, ε_y, and ε_z, and shear strain with components that act in the same direction as shear stress (e.g., γ_{xy} is the shear strain generated by shear stress τ_{xy}).

A set of equations that enables solutions are given by ***Newton's first law***, where the sum of forces in each of the coordinate directions is zero, together with stress–strain relations for elastic deformations from a generalized ***Hooke's law*** (Hooke, 1678),

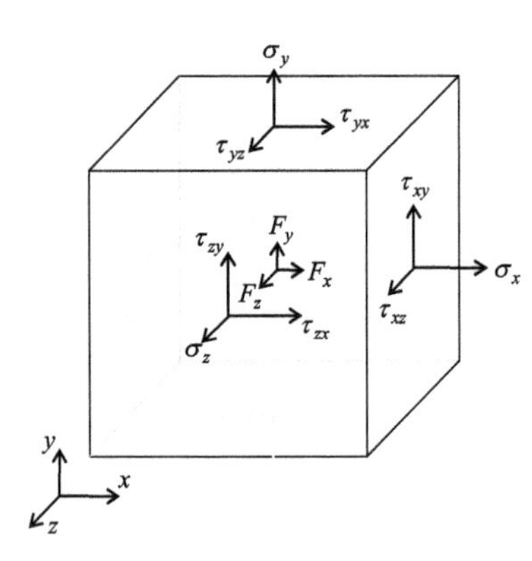

a. Normal and shear stress, and body force

b. Normal and shear strain, and displacement

Figure 1.55 *Stresses generated by surface and body forces result in displacement and strain.*

$$\begin{aligned}
&\frac{\partial \sigma_x}{\partial x}+\frac{\partial \tau_{xy}}{\partial y}+\frac{\partial \tau_{xz}}{\partial z}+F_x=0\,, \quad \varepsilon_x=\frac{1}{E}\left[\sigma_x-\nu\left(\sigma_y+\sigma_z\right)\right], \quad \gamma_{xy}=\frac{\tau_{xy}}{\mu}\\
&\frac{\partial \sigma_y}{\partial y}+\frac{\partial \tau_{xy}}{\partial x}+\frac{\partial \tau_{yz}}{\partial z}+F_y=0\,, \quad \varepsilon_y=\frac{1}{E}\left[\sigma_y-\nu\left(\sigma_z+\sigma_x\right)\right], \quad \gamma_{yz}=\frac{\tau_{yz}}{\mu}\\
&\frac{\partial \sigma_z}{\partial z}+\frac{\partial \tau_{xz}}{\partial x}+\frac{\partial \tau_{yz}}{\partial y}+F_z=0\,, \quad \varepsilon_z=\frac{1}{E}\left[\sigma_z-\nu\left(\sigma_x+\sigma_y\right)\right], \quad \gamma_{xz}=\frac{\tau_{xz}}{\mu}
\end{aligned} \tag{1.99a}$$

where E is the modulus of elasticity, ν is the Poisson ratio, and

$$\mu=\frac{E}{2\left(1+\nu\right)} \tag{1.99b}$$

is the modulus of rigidity (or shear modulus). An additional set of moment equilibrium conditions about the center of the control block in Fig. 1.55 requires shear stress and strain components in coordinate planes to be equal:

$$\begin{aligned}
\tau_{xy}&=\tau_{yx} & & \gamma_{xy}=\gamma_{yz}\\
\tau_{yz}&=\tau_{zy} & \Rightarrow \quad & \gamma_{yz}=\gamma_{zy}\\
\tau_{zx}&=\tau_{xz} & & \gamma_{zx}=\gamma_{xz}
\end{aligned} \tag{1.100}$$

The remaining components of strain are related to the displacement by

$$\begin{aligned}
\varepsilon_x&=\frac{\partial u_x}{\partial x}\,, & \gamma_{xy}&=\frac{\partial u_x}{\partial y}+\frac{\partial u_y}{\partial x}\\
\varepsilon_y&=\frac{\partial u_y}{\partial y}\,, & \gamma_{yz}&=\frac{\partial u_y}{\partial z}+\frac{\partial u_z}{\partial y}\\
\varepsilon_z&=\frac{\partial u_z}{\partial z}\,, & \gamma_{xz}&=\frac{\partial u_z}{\partial x}+\frac{\partial u_x}{\partial z}
\end{aligned} \tag{1.101}$$

Together, Eqs (1.99) and (1.101) provide ***15 conditions*** that must be satisfied along with appropriate boundary conditions prescribed in terms of stress and/or displacement to solve for the ***6 stress, 6 strain, and 3 displacement components***.

Two-dimensional problems in the $x - y$ plane may be studied using two different assumptions that reduce the equilibrium conditions (1.99) to

$$\begin{aligned} \frac{\partial \sigma_x}{\partial x} + \frac{\partial \tau_{xy}}{\partial y} + F_x = 0 \\ \frac{\partial \sigma_y}{\partial y} + \frac{\partial \tau_{xy}}{\partial x} + F_y = 0 \end{aligned} \tag{1.102}$$

One assumption is that of ***plane strain***, where displacements occur in a plane and there is no strain in the z-direction. This leads to the following stress–strain relations in Hooke's law (1.99) that, together with the compatibility conditions (1.101), directly relate displacements to stress:

$$\begin{aligned} &\varepsilon_z = 0 \rightarrow \sigma_z = \nu\left(\sigma_x + \sigma_y\right) \\ &\gamma_{yz} = 0 \rightarrow \gamma_{yz} = 0 \\ &\gamma_{xz} = 0 \rightarrow \gamma_{xz} = 0 \end{aligned} \quad \Rightarrow \quad \begin{aligned} \frac{\partial u_x}{\partial x} &= \varepsilon_x = \frac{1}{2\mu}\left[\sigma_x - \nu\left(\sigma_x + \sigma_y\right)\right] \\ \frac{\partial u_y}{\partial y} &= \varepsilon_y = \frac{1}{2\mu}\left[\sigma_y - \nu\left(\sigma_x + \sigma_y\right)\right] \\ \frac{\partial u_x}{\partial y} + \frac{\partial u_y}{\partial x} &= \gamma_{xy} = \frac{1}{\mu}\tau_{xy} \end{aligned} \tag{1.103a}$$

Another assumption is that of ***plane stress***, where the stresses act in a plane and there is no stress in the z-direction,

$$\begin{aligned} &\sigma_z = 0 \rightarrow \varepsilon_z = -\frac{\nu\left(\sigma_x + \sigma_y\right)}{2\mu(1+\nu)} \\ &\tau_{yz} = 0 \rightarrow \gamma_{yz} = 0 \\ &\tau_{xz} = 0 \rightarrow \gamma_{xz} = 0 \end{aligned} \quad \Rightarrow \quad \begin{aligned} \frac{\partial u_x}{\partial x} &= \varepsilon_x = \frac{1}{2\mu}\left[\sigma_x - \frac{\nu}{1+\nu}\left(\sigma_x + \sigma_y\right)\right] \\ \frac{\partial u_y}{\partial y} &= \varepsilon_y = \frac{1}{2\mu}\left[\sigma_y - \frac{\nu}{1+\nu}\left(\sigma_x + \sigma_y\right)\right] \\ \frac{\partial u_x}{\partial y} + \frac{\partial u_y}{\partial x} &= \gamma_{xy} = \frac{1}{\mu}\tau_{xy} \end{aligned} \tag{1.103b}$$

Together, Eq. (1.102) and either form of (1.103a) or (1.103b) provide ***five equations*** to solve for the ***five unknown*** components of stress ($\sigma_x, \sigma_y, \tau_{xy}$), and displacement ($u_x, u_y$).

Example 1.7 Body forces are important for many problems, and their solutions provide a background into which other solutions may be superimposed to study stress and displacement around elements. For example, gravity force directed in the negative y-direction in a semi-infinite dry soil below y_s with bulk density ρ_s (Jaeger, 1969, p.120) is illustrated in Fig. 1.56 and contains the following body force, stress, and displacements

$$\begin{aligned} &F_x = F_z = 0 \\ &F_y = -\rho_s g \end{aligned} \quad \begin{aligned} &\rightarrow \begin{cases} \sigma_x &= \sigma_z = \frac{\nu}{1-\nu}\rho_s g(y - y_s) \\ \sigma_y &= \rho_s g(y - y_s) \\ \tau_{xy} &= \tau_{yz} = \tau_{xz} = 0 \end{cases} \qquad (y \le y_s) \\ &\rightarrow \begin{cases} u_x &= u_z = 0 \\ u_y &= \frac{1}{2\mu}\frac{1-2\nu}{1-\nu}\rho_s g\frac{(y-y_s)^2}{2} \end{cases} \end{aligned} \tag{1.104}$$

where these stress components satisfy the equilibrium equations for two-dimensional plane strain in (1.102) and the relations between displacement and stress in (1.103a) are satisfied.

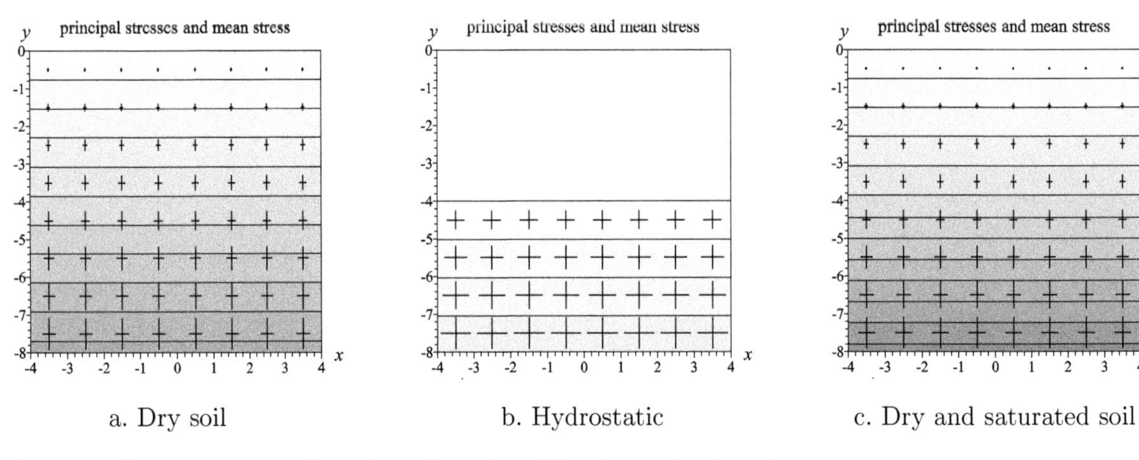

Figure 1.56 *Body force due to gravity for dry and wet soil conditions, showing the principal stress components.*

This solution may be adapted to elasticity problems with hydrostatic pressure using $\nu = \frac{1}{2}$ (Jaeger and Cook, 1976, p.112), where the density of water ρ_w occurs below y_w, giving

$$\begin{aligned} F_x = F_z = 0 \\ F_y = -\rho_w g \end{aligned} \quad \rightarrow \quad \begin{cases} \sigma_x &= \sigma_y = \sigma_z = \rho_w g(y - y_w) \\ \tau_{xy} &= \tau_{yz} = \tau_{xz} = 0 \end{cases} \quad (y \le y_w) \qquad (1.105)$$
$$\rightarrow \quad u_x = u_y = u_z = 0$$

Individually, these solutions for body force may be applied to the elasticity problems in saturated soils where either dry soil lies above a saturated zone in $y_{\text{sat}} < y \le y_s$ or water lies above the saturated zone in $y_{\text{sat}} < y \le y_w$. This gives a plane strain solution in this saturated domain where water fills the voids between soil particles with porosity n beneath y_{sat}, with Poisson ratio of ν_s for the body forces due to the soil and $\nu_w = \frac{1}{2}$ for the hydrostatic forces, giving

$$\begin{aligned} F_x &= F_z = 0 \\ F_y &= -(\rho_s + \mathsf{n}\rho_w)\, g \end{aligned} \qquad (y \le y_{\text{sat}})$$
$$\rightarrow \begin{cases} \sigma_x &= \sigma_z = \left[\frac{\nu_s(\rho_s - \rho_w + \mathsf{n}\rho_w)}{1-\nu_s} + \rho_w\right] g(y - y_w) + \sigma_{x\text{sat}} \\ \sigma_y &= (\rho_s + \mathsf{n}\rho_w)\, g\,(y - y_0) + \sigma_{y\text{sat}} \\ \tau_{xy} &= u_x = u_z = 0 \end{cases} \qquad (1.106)$$
$$\rightarrow \quad u_y = \frac{1}{2\mu}\frac{1 - 2\nu_s}{1 - \nu_s}(\rho_s - \rho_w + \mathsf{n}\rho_w)\, g\frac{(y - y_s)^2}{2} + u_{y\text{sat}}$$

where $\sigma_{x\text{sat}}$, $\sigma_{y\text{sat}}$, and $u_{y\text{sat}}$ are the values occurring at $y = y_{\text{sat}}$ due to overlying dry soil or water.

Problem 1.10 Compute the principal stresses and maximum shear stress at the specified locations for a soil below $y_s = 0$ with porosity $\mathsf{n} = 0.30$, density $2650(1\text{-}\mathsf{n})\text{kg/m}^3$, Poisson ration $\nu = 0.3$, coefficient of elasticity $E = 80 \times 10^6 \text{kg/(ms}^2)$ and water below $y_w = y_0 = -4m$ with density 1000kg/m^3.

A. $y = -1$ m
B. $y = -2$ m
C. $y = -3$ m
D. $y = -4$ m
E. $y = -5$ m
F. $y = -6$ m
G. $y = -7$ m
H. $y = -8$ m
I. $y = -9$ m
J. $y = -10$ m

Two-dimensional problems may be further simplified to a single partial differential equation using the ***Airy (1863) stress function***, $\mathcal{F}$, defined as

$$\boxed{\sigma_x = \frac{\partial^2 \mathcal{F}}{\partial y^2}, \quad \sigma_y = \frac{\partial^2 \mathcal{F}}{\partial x^2}, \quad \tau_{xy} = -\frac{\partial^2 \mathcal{F}}{\partial x \partial y}} \tag{1.107}$$

Displacement is related to the Airy stress function by substituting these terms into (1.103a) for plane strain,

$$\begin{aligned} \frac{\partial u_x}{\partial x} &= \frac{1}{2\mu}\left[\frac{\partial^2 \mathcal{F}}{\partial y^2} - \nu \nabla^2 \mathcal{F}\right] \\ \frac{\partial u_y}{\partial y} &= \frac{1}{2\mu}\left[\frac{\partial^2 \mathcal{F}}{\partial x^2} - \nu \nabla^2 \mathcal{F}\right] \end{aligned}, \qquad \frac{\partial u_x}{\partial y} + \frac{\partial u_y}{\partial x} = -\frac{1}{\mu}\frac{\partial^2 \mathcal{F}}{\partial x \partial y} \tag{1.108a}$$

or into (1.103b) for plane stress,

$$\begin{aligned} \frac{\partial u_x}{\partial x} &= \frac{1}{2\mu}\left(\frac{\partial^2 \mathcal{F}}{\partial y^2} - \frac{\nu}{1+\nu} \nabla^2 \mathcal{F}\right) \\ \frac{\partial u_y}{\partial y} &= \frac{1}{2\mu}\left(\frac{\partial^2 \mathcal{F}}{\partial x^2} - \frac{\nu}{1+\nu} \nabla^2 \mathcal{F}\right) \end{aligned}, \qquad \frac{\partial u_x}{\partial y} + \frac{\partial u_y}{\partial x} = -\frac{1}{\mu}\frac{\partial^2 \mathcal{F}}{\partial x \partial y} \tag{1.108b}$$

Integration of $\mathcal{F}$ in these equations provides the displacement. A governing equation for $\mathcal{F}$ is obtained by substituting either of these sets of displacement expressions into the the equilibrium conditions (1.102),

$$\frac{\partial^2}{\partial y^2}\left(\frac{\partial u_x}{\partial x}\right) + \frac{\partial^2}{\partial x^2}\left(\frac{\partial u_y}{\partial y}\right) = \frac{\partial^2}{\partial x \partial y}\left(\frac{\partial u_x}{\partial y} + \frac{\partial u_y}{\partial x}\right) \tag{1.109}$$

to give the ***biharmonic equation***, (2.2e):

$$\boxed{\nabla^4 \mathcal{F} = \frac{\partial^4 \mathcal{F}}{\partial x^4} + 2\frac{\partial^4 \mathcal{F}}{\partial x^2 \partial y^2} + \frac{\partial^4 \mathcal{F}}{\partial y^4} = 0} \tag{1.110}$$

This formulation of the Airy stress function is used in the study of elasticity (Muskhelishvili, 1953*b*), thermal stress (Westergaard, 1952), and viscous fluid flow (Jaeger, 1969).

The biharmonic equation may be formulated using complex functions following Muskhelishvili (1953*b*). Mathematics are expressed in terms of the complex variable z and its conjugate $\overline{z}$, and readers unfamiliar with complex functions are referred to Section 2.3.2, and specifically Section 3.6 where complex functions of both z and $\overline{z}$ are presented. The biharmonic equation is expressed in a complex form using (3.111),

$$\nabla^4 \mathcal{F} = 16\frac{\partial^4 \mathcal{F}}{\partial z^2 \partial \overline{z}^2} = 0 \quad \rightarrow \quad \mathcal{F} = \frac{\overline{z}\phi + z\overline{\phi} + \chi + \overline{\chi}}{2} = \Re\left(\overline{z}\phi + \chi\right) \tag{1.111}$$

where this arrangement of complex functions $\phi(z)$ and $\chi(z)$ with their complex conjugates $\overline{\phi}$ and $\overline{\chi}$ provides a real function $\mathcal{F}$ and satisfies the biharmonic equation. The ***Kolosov–Muskhelishvili formulas*** are obtained through rearrangement of these functions with $\psi = \chi'$ to provide stress and displacement related to these functions by

$$\boxed{\begin{aligned} \frac{\sigma_x + \sigma_y}{2} &= 2\Re\phi' \\ \frac{\sigma_y - \sigma_x}{2} + \mathrm{i}\tau_{xy} &= \overline{z}\phi'' + \psi' \\ u_x + \mathrm{i}u_y &= \frac{1}{2\mu}\left(\kappa\phi - z\overline{\phi'} - \overline{\psi}\right) \end{aligned}} \quad \Rightarrow \quad \begin{aligned} \sigma_x &= \Re\left(2\phi' - \overline{z}\phi'' - \psi'\right) \\ \sigma_y &= \Re\left(2\phi' + \overline{z}\phi'' + \psi'\right) \\ \tau_{xy} &= \Im\left(\overline{z}\phi'' + \psi'\right) \end{aligned} \tag{1.112}$$

The constant κ is related to the Poisson ratio as follows (Muskhelishvili, 1953*b*, p.112):

$$\kappa = \begin{cases} 3 - 4\nu & \text{plane strain} \\ \frac{3-\nu}{1+\nu} & \text{plane stress} \end{cases} \tag{1.113}$$

Note that derivation of the Kolosov–Muskhelishvili formulas follows from insertion of the complex form of the Airy stress function (1.111) into the relations (1.107) and (1.108) and rearranging terms, and a detailed formulation is presented by Muskhelishvili (1953*b*). ***Solutions for stress and displacement are developed either in terms of $\mathcal{F}$, or of ϕ and ψ.***

Visualization of these solutions requires the use of eigenvalue σ and eigenvector $\mathbf{v}$ for the the stress tensor, $\mathbf{A}$, which satisfies the fundamental equation

$$\mathbf{A}\mathbf{v} = \sigma\mathbf{v}, \quad \mathbf{A} = \begin{bmatrix} \sigma_x & \tau_{xy} \\ \tau_{xy} & \sigma_y \end{bmatrix} \quad \rightarrow \quad (\mathbf{A} - \sigma\mathbf{I})\mathbf{v} = 0, \quad \mathbf{I} = \begin{bmatrix} 1 & 0 \\ 0 & 1 \end{bmatrix} \tag{1.114}$$

where $\mathbf{I}$ is the identity matrix, defined to have ones on the diagonal and zeros elsewhere. Thus, matrix multiplication of $\mathbf{A}$ times the eigenvector $\mathbf{v}$ produces the same result as scaling $\mathbf{v}$ by the factor σ. The determinant of $\mathbf{A} - \sigma\mathbf{I}$ must be zero since the columns in this matrix are linearly dependent, and this gives the characteristic equation

$$\det(\mathbf{A} - \sigma\mathbf{I}) = \begin{vmatrix} \sigma_x - \sigma & \tau_{xy} \\ \tau_{xy} & \sigma_y - \sigma \end{vmatrix} = (\sigma_x - \sigma)\left(\sigma_y - \sigma\right) - \tau_{xy}^2 = 0 = 0 \tag{1.115}$$

where the determinant for the two-dimensional stress matrix is given by Cramer's rule. The solution to this second-order polynomial gives the two principal stresses

$$\begin{aligned} \sigma_1 &= \frac{\sigma_x + \sigma_y}{2} + \sqrt{\left(\frac{\sigma_x - \sigma_y}{2}\right)^2 + \tau_{xy}^2} \\ \sigma_2 &= \frac{\sigma_x + \sigma_y}{2} - \sqrt{\left(\frac{\sigma_x - \sigma_y}{2}\right)^2 + \tau_{xy}^2} \end{aligned} \tag{1.116}$$

The principal axes along which the principal stresses operate lie in the direction of the eigenvectors of the stress matrix. These eigenvalues may be substituted into (1.114) to give a set of equations for the two eigenvectors, $\mathbf{v}_1$ and $\mathbf{v}_2$:

$$\begin{aligned} (\mathbf{A} - \sigma_1\mathbf{I})\,\mathbf{v}_1 &= \begin{bmatrix} \sigma_x - \sigma_1 & \tau_{xy} \\ \tau_{xy} & \sigma_y - \sigma_1 \end{bmatrix}\mathbf{v}_1 = 0 \\ (\mathbf{A} - \sigma_2\mathbf{I})\,\mathbf{v}_2 &= \begin{bmatrix} \sigma_x - \sigma_2 & \tau_{xy} \\ \tau_{xy} & \sigma_y - \sigma_2 \end{bmatrix}\mathbf{v}_2 = 0 \end{aligned} \tag{1.117}$$

Each set of equations is linearly dependent, and thus provide the directions of the eigenvectors but not their length. It is common to scale eigenvectors to have length of 1, and equations for the unit eigenvectors are given by

$$\begin{aligned}\hat{\mathbf{v}}_1 &= \pm\begin{bmatrix}\frac{-\tau_{xy}}{\sqrt{\tau_{xy}{}^2+(\sigma_x-\sigma_1)^2}}\\ \frac{\sigma_x-\sigma_1}{\sqrt{\tau_{xy}{}^2+(\sigma_x-\sigma_1)^2}}\end{bmatrix} = \pm\begin{bmatrix}\frac{\sigma_y-\sigma_1}{\sqrt{\tau_{xy}{}^2+(\sigma_y-\sigma_1)^2}}\\ \frac{-\tau_{xy}}{\sqrt{\tau_{xy}{}^2+(\sigma_y-\sigma_1)^2}}\end{bmatrix}\\ \hat{\mathbf{v}}_2 &= \pm\begin{bmatrix}\frac{-\tau_{xy}}{\sqrt{\tau_{xy}{}^2+(\sigma_x-\sigma_2)^2}}\\ \frac{\sigma_x-\sigma_2}{\sqrt{\tau_{xy}{}^2+(\sigma_x-\sigma_2)^2}}\end{bmatrix} = \pm\begin{bmatrix}\frac{\sigma_y-\sigma_2}{\sqrt{\tau_{xy}{}^2+(\sigma_y-\sigma_2)^2}}\\ \frac{-\tau_{xy}}{\sqrt{\tau_{xy}{}^2+(\sigma_y-\sigma_2)^2}}\end{bmatrix}\end{aligned} \tag{1.118}$$

Example 1.8 Determine eigenvalues and eigenvectors for the two-dimensional matrix with coefficients

$$\mathbf{A} = \begin{bmatrix}13 & 3\\ 3 & 5\end{bmatrix}$$

Eigenvalues are given by (1.116)

$$\sigma_1 = \frac{13+5}{2} + \sqrt{\left(\frac{13-5}{2}\right)^2 + 3\times 3} = 9+5 = 14$$

$$\sigma_2 = \frac{13+5}{2} - \sqrt{\left(\frac{13-5}{2}\right)^2 + 3\times 3} = 9-5 = 4$$

Unit eigenvectors are given by (1.118)

$$\hat{\mathbf{v}}_1 = \pm\begin{bmatrix}\frac{-3}{\sqrt{3^2+(13-14)^2}}\\ \frac{13-14}{\sqrt{3^2+(13-14)^2}}\end{bmatrix} = \pm\begin{bmatrix}\frac{5-14}{\sqrt{3^2+(5-14)^2}}\\ \frac{-3}{\sqrt{3^2+(5-14)^2}}\end{bmatrix} = \pm\begin{bmatrix}\frac{3}{\sqrt{10}}\\ \frac{1}{\sqrt{10}}\end{bmatrix}$$

and

$$\hat{\mathbf{v}}_2 = \pm\begin{bmatrix}\frac{-3}{\sqrt{3^2+(13-4)^2}}\\ \frac{13-4}{\sqrt{3^2+(13-4)^2}}\end{bmatrix} = \pm\begin{bmatrix}\frac{5-4}{\sqrt{3^2+(5-4)^2}}\\ \frac{-3}{\sqrt{3^2+(5-4)^2}}\end{bmatrix} = \pm\begin{bmatrix}\frac{-1}{\sqrt{10}}\\ \frac{3}{\sqrt{10}}\end{bmatrix}$$

Problem 1.11 Compute the eigenvalues and unit eigenvectors for the following matrices.

A. $\begin{bmatrix}1 & 2\\ 2 & 3\end{bmatrix}$ C. $\begin{bmatrix}3 & 4\\ 4 & 5\end{bmatrix}$ E. $\begin{bmatrix}5 & 6\\ 6 & 7\end{bmatrix}$ G. $\begin{bmatrix}7 & 8\\ 8 & 9\end{bmatrix}$ I. $\begin{bmatrix}9 & 10\\ 19 & 11\end{bmatrix}$

B. $\begin{bmatrix}2 & 3\\ 3 & 4\end{bmatrix}$ D. $\begin{bmatrix}4 & 5\\ 5 & 6\end{bmatrix}$ F. $\begin{bmatrix}6 & 7\\ 7 & 8\end{bmatrix}$ H. $\begin{bmatrix}8 & 9\\ 9 & 10\end{bmatrix}$ J. $\begin{bmatrix}10 & 11\\ 11 & 12\end{bmatrix}$

1.4.1 Elasticity: Stress and Displacement

Solutions to boundary value problems in elasticity may be represented using the principal stresses and the maximum shear stress. The ***principal stresses*** follow directly from the eigenvectors of the Cauchy stress tensor, in (1.116):

$$\sigma_1 = \frac{\sigma_x + \sigma_y}{2} + \sqrt{\left(\frac{\sigma_x - \sigma_y}{2}\right)^2 + {\tau_{xy}}^2}$$
$$\sigma_2 = \frac{\sigma_x + \sigma_y}{2} - \sqrt{\left(\frac{\sigma_x - \sigma_y}{2}\right)^2 + {\tau_{xy}}^2} \tag{1.119}$$

These are illustrated near a point where a force is applied in Figs. 1.57a and 1.57b, where regions in tension are indicated by light eigenvectors while compression uses dark eigenvectors. This figure also illustrates a set of ***isostress*** contour lines with uniform mean stress, $(\sigma_1 + \sigma_2)/2$. The ***maximum shear stress*** is given by

$$\tau_{\max} = \frac{\sigma_1 - \sigma_2}{2} = \sqrt{\left(\frac{\sigma_x - \sigma_y}{2}\right)^2 + {\tau_{xy}}^2} \tag{1.120}$$

This is illustrated using ***isoshear*** contour lines with uniform maximum shear stress $\tau_{\max}$, which are shown along with the displacement vector with components u_x and u_y.

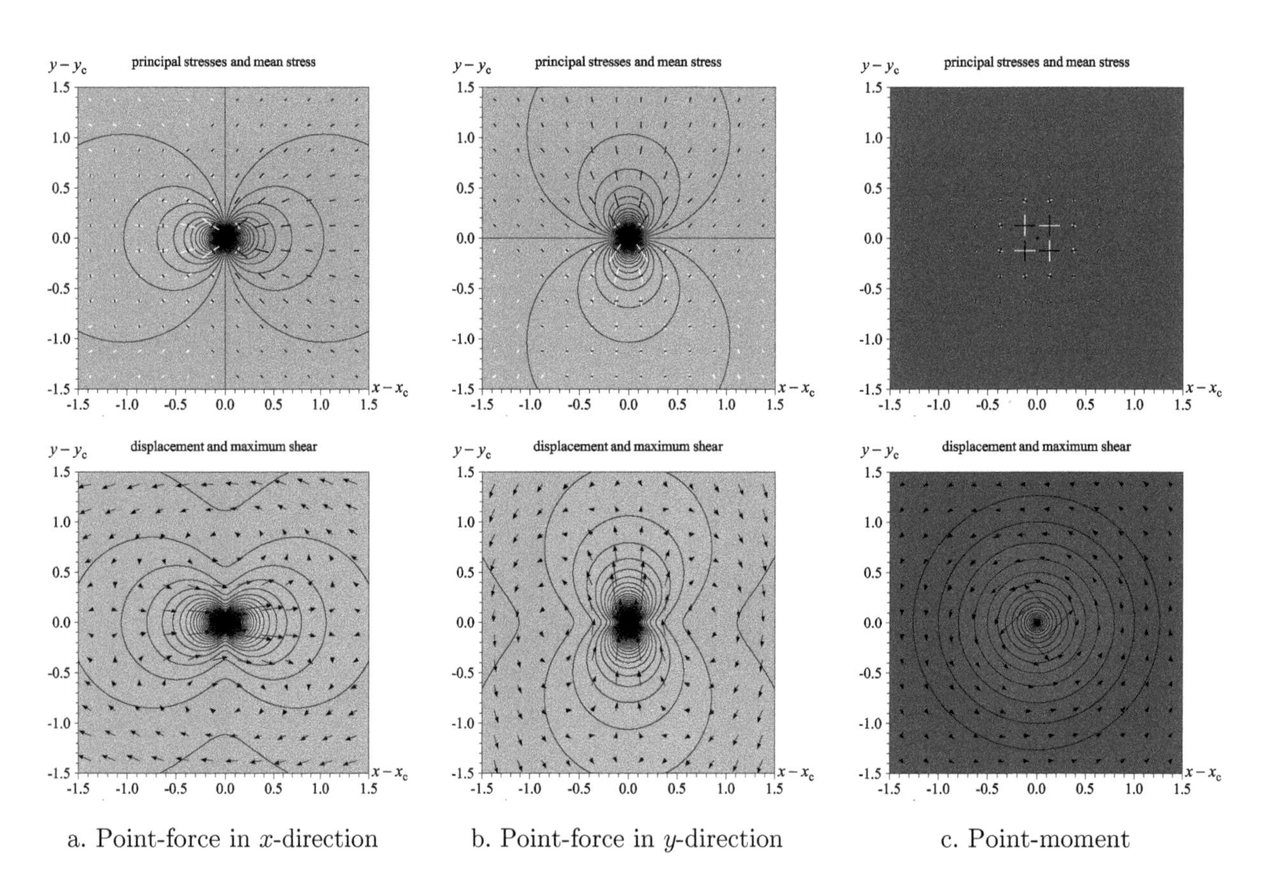

Figure 1.57 *Stress and displacement near a point force and a point moment (Fig. 3.35).*

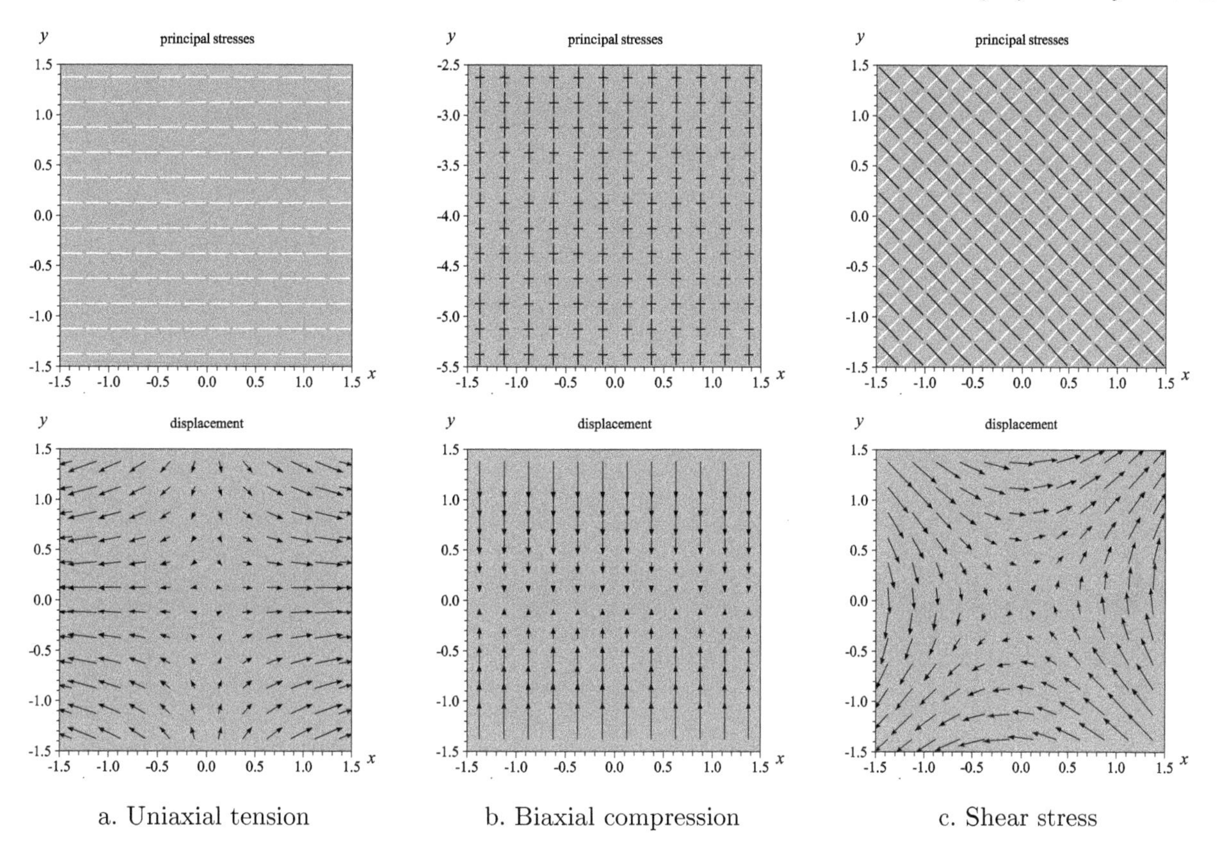

Figure 1.58 *Background field of uniform stress (Fig. 3.34).*

A background stress field into which analytic elements are placed is illustrated in Fig. 1.58 for uniaxial tension, biaxial compression, and shear stress. Each solution is developed later at the locations indicated in the figure using the complex Kolosov functions (1.112), with the visualized stress and displacement components obtained directly from ϕ and ψ using (1.112). These background solutions may be superimposed with the Kolosov formulas for an analytic element to study the influence of an element in the distribution of stress and displacement; for example, the circular element in Fig. 1.60 is inserted into the settings shown in Fig. 1.58. Each element will satisfy a boundary condition of specified stress, (1.121a), where the normal and shear stress components are both zero on the boundary and the only non-zero principle stress component will act in the tangent direction.

Solutions to elastic deformation require specification of boundary conditions, as illustrated by the cantilever beam in Fig. 1.59 (with stress and displacement computed later in Example 1.9). A boundary condition may ***specify the normal and shear stress*** components

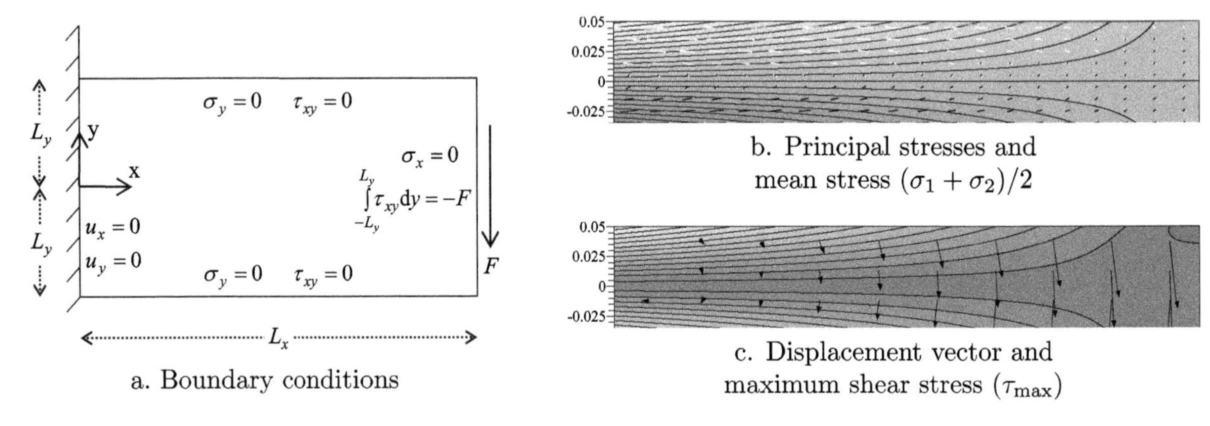

Figure 1.59 *Elasticity: Cantilever beam with a force applied to the free end.*

$$\text{shear specified} \rightarrow \begin{array}{c} \sigma_{\mathfrak{n}}(\mathfrak{s}) = \sigma_0 \\ \tau_{\mathfrak{n}\mathfrak{s}}(\mathfrak{s}) = \tau_0 \end{array} \quad \mathfrak{s} \in \partial D \tag{1.121a}$$

It may ***specify displacement***:

$$\text{displacement specified} \rightarrow \begin{array}{c} u_x(\mathfrak{s}) = u_{x0} \\ u_y(\mathfrak{s}) = u_{y0} \end{array} \quad \mathfrak{s} \in \partial D \tag{1.121b}$$

where $\mathfrak{s}$ and $\mathfrak{n}$ are the tangential and normal directions to the boundary. Or, it may ***specify elastic support*** where a linear relation exists between the the normal and shear stresses and the displacement in these directions

$$\text{elastic support specified} \rightarrow \begin{array}{c} \sigma_{\mathfrak{n}}(\mathfrak{s}) = ku_{\mathfrak{n}0} \\ \tau_{\mathfrak{n}\mathfrak{s}}(\mathfrak{s}) = ku_{\mathfrak{s}0} \end{array} \quad \mathfrak{s} \in \partial D \tag{1.121c}$$

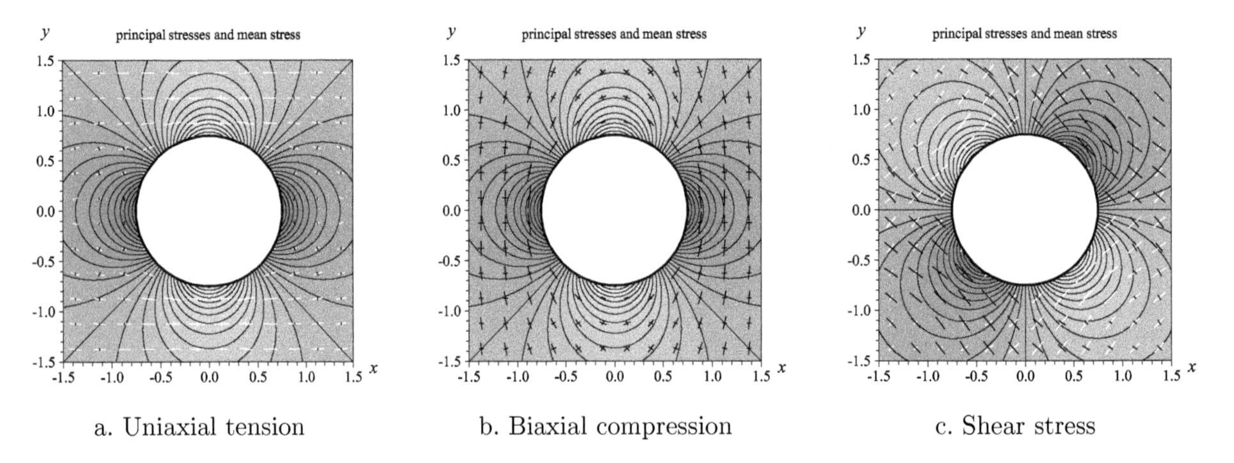

Figure 1.60 *Circle with zero traction in uniform stress (Fig. 3.37).*

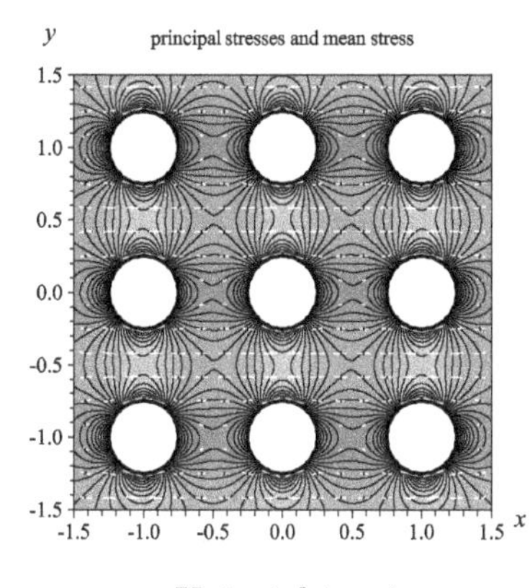

a. Uniaxial tension

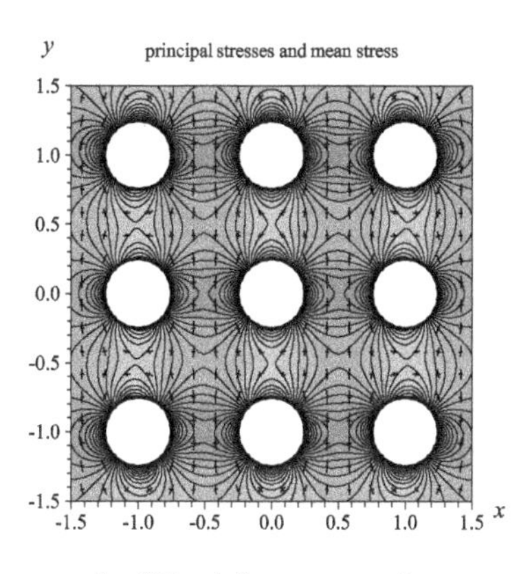

b. Biaxial compression

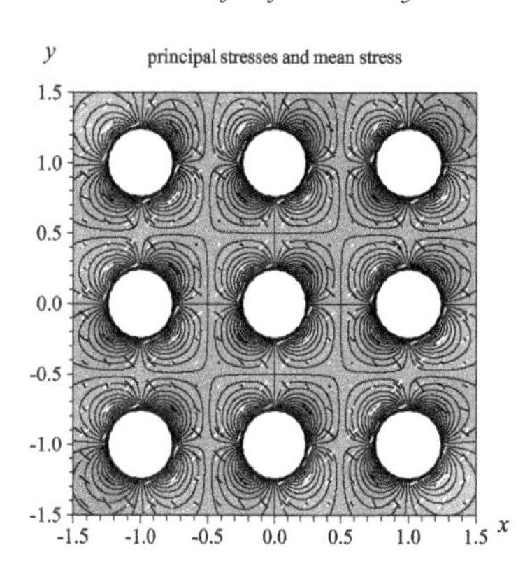

c. Shear stress

Figure 1.61 *Circles with zero traction in uniform stress (Fig. 3.38).*

and k is the stiffness. For example, the left end of the beam in Fig. 1.59 has specified displacement and stresses are specified along the other boundaries.

Boundary conditions along circle elements are formulated by adapting the Kolosov–Muskhelishvili functions (1.112) to solve problems with boundary conditions specified along boundaries at an angle θ. This is accomplished using conservation of momentum in the rotated two-dimensional coordinates which gives the ***Mohr's representation of stress*** in the r and θ directions: (Muskhelishvili, 1953*b*, Eq. 8.7)

$$\begin{aligned} \sigma_r &= \frac{\sigma_x + \sigma_y}{2} + \frac{\sigma_x - \sigma_y}{2} \cos 2\theta + \tau_{xy} \sin 2\theta \\ \sigma_\theta &= \frac{\sigma_x + \sigma_y}{2} - \frac{\sigma_x - \sigma_y}{2} \cos 2\theta - \tau_{xy} \sin 2\theta \\ \tau_{r\theta} &= - \frac{\sigma_x - \sigma_y}{2} \sin 2\theta + \tau_{xy} \cos 2\theta \end{aligned} \tag{1.122}$$

The components of the displacement vector in the rotated coordinates are related to those in the x–y plane by

$$u_r + \mathrm{i}u_\theta = \left(u_x + \mathrm{i}u_y\right) \mathrm{e}^{-\mathrm{i}\theta} \tag{1.123}$$

A form of the Kolosov–Muskhelishvili formulas in rotated coordinates is given by substituting (1.112) into these equations to give (Jaeger, 1969, p.197) (Muskhelishvili, 1953*b*, p.138)

$$\boxed{\begin{aligned} \frac{\sigma_r + \sigma_\theta}{2} &= 2\Re\phi' \\ \frac{\sigma_\theta - \sigma_r}{2} + \mathrm{i}\tau_{r\theta} &= \mathrm{e}^{2\mathrm{i}\theta} \left(\overline{z}\phi'' + \psi'\right) \\ u_r + \mathrm{i}u_\theta &= \frac{\mathrm{e}^{-\mathrm{i}\theta}}{2\mu} \left(\kappa\phi - z\overline{\phi'} - \overline{\psi}\right) \end{aligned}} \tag{1.124}$$

Subtracting the first two equations gives an expression

$$\sigma_r - \mathrm{i}\tau_{r\theta} = 2\Re\phi' - \mathrm{e}^{2\mathrm{i}\theta}\left(\overline{z}\phi'' + \psi'\right) \tag{1.125}$$

that enables the solution to boundary value problems with specified radial and tangential components along a circular boundary.

Boundary conditions for analytic elements may be specified in terms of the value of the complex functions ϕ and ψ along the boundary. A boundary may ***specify the normal and shear stress*** components σ_0 and τ_0 using (1.124)

$$2\Re\phi'(\mathfrak{s}) - \left[\overline{z}\phi''(\mathfrak{s}) + \psi'(\mathfrak{s})\right]\mathrm{e}^{2\mathrm{i}\theta(\mathfrak{s})} = \sigma_0 - \mathrm{i}\tau_0 \tag{1.126}$$

where θ is oriented so the radial and tangential components are aligned in the normal and tangential directions. The boundary may ***specify displacement*** with components u_{x0} and u_{y0} by evaluating the functions in (1.112) at locations along the boundary:

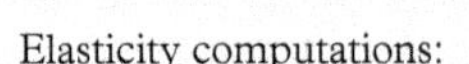

Elasticity computations:

- normal and shear stresses (σ_x, σ_y and τ_{xy}) from the Airy stress function $\mathcal{F}$, (1.107), or from ϕ and ψ in the Kolosov formula, (1.112).
- principal stresses (σ_1 and σ_2) from (1.119), and principal directions ($\hat{\mathbf{v}}_1$ and $\hat{\mathbf{v}}_2$) from (1.118).
- maximum shear stress τ_{max} using (1.120).
- displacement (u_x and u_y) by integrating partial derivatives of the Airy stress function using (1.108) or from the Kolosov formula, (1.112).

$$\frac{1}{2\mu}\left[\kappa\phi(\mathfrak{s}) - z\overline{\phi'(\mathfrak{s})} - \overline{\psi(\mathfrak{s})}\right] = u_{x0} + \mathrm{i}u_{y0} \tag{1.127}$$

This solution is shown in Fig. 1.62 for one circle, and in Fig. 1.63 for many circles. A Robin boundary condition may ***specify elastic support*** as a linear relation between the the normal and shear stresses and the displacement in these directions

$$\text{elastic support specified} \rightarrow \quad \begin{matrix} \sigma_{\mathrm{n}}(\mathfrak{s}) = ku_{\mathrm{n}0} \\ \tau_{\mathrm{n}\mathfrak{s}}(\mathfrak{s}) = ku_{\mathfrak{s}0} \end{matrix} \qquad \mathfrak{s} \in \partial D \tag{1.128}$$

and k is the stiffness. The analytic elements in this section illustrate the application of these boundary conditions, and their functions ϕ and ψ are formulated later in Section 3.7.

Example 1.9 The Airy stress function is applied to the plane stress problem of a cantilever beam in Fig. 1.59 with a fixed left side and a net force per unit width F applied on the right.

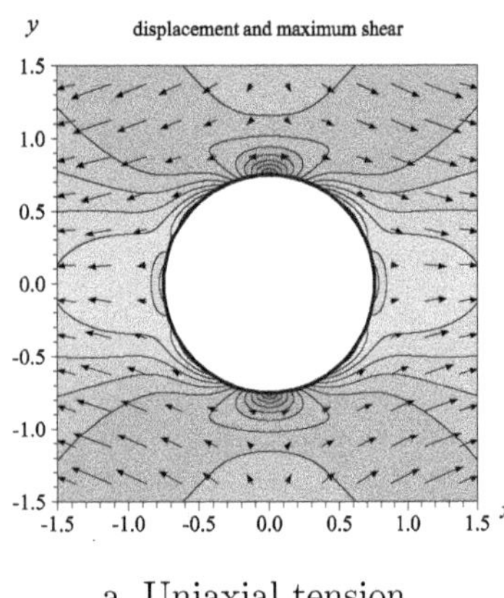

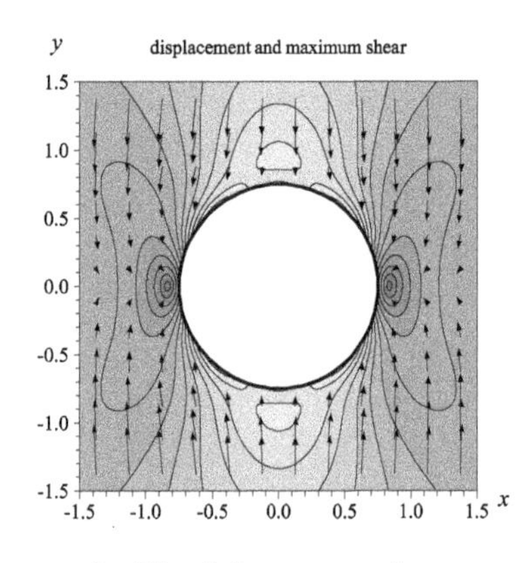

b. Biaxial compression

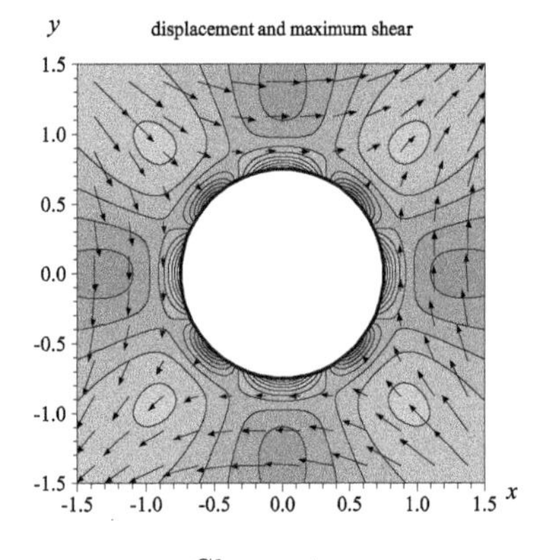

c. Shear stress

Figure 1.62 *Circle with zero displacement in uniform stress (Fig. 3.39).*

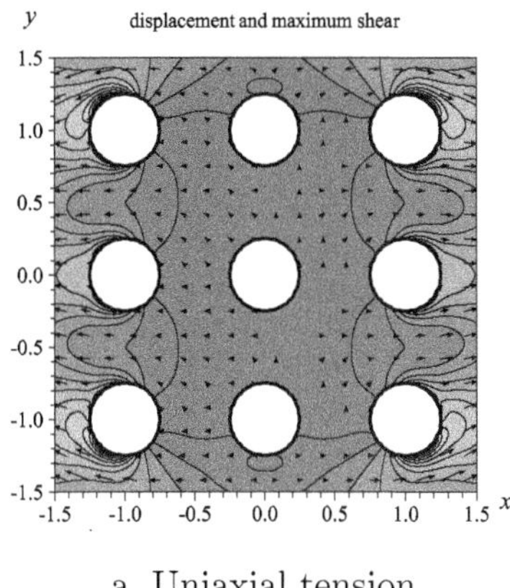

a. Uniaxial tension

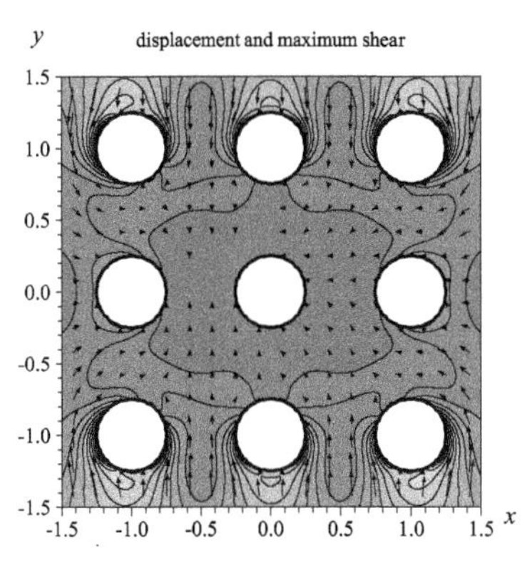

b. Biaxial compression

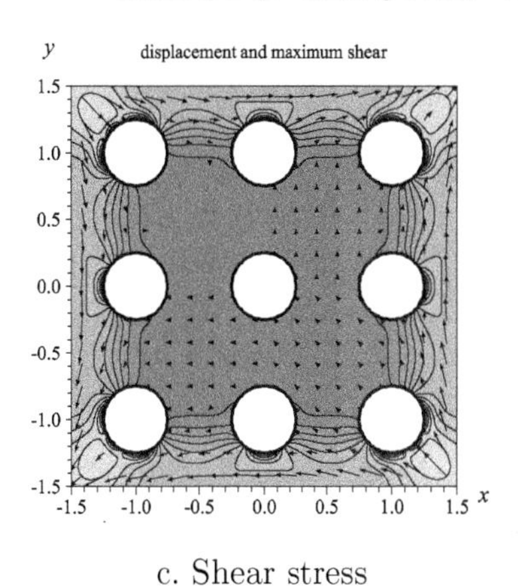

c. Shear stress

Figure 1.63 *Circles with zero displacement in uniform stress (Fig. 3.40).*

A solution from separation of variables (Chapter 4) follows with stress components obtained by evaluating the derivatives in (1.107)

$$\mathcal{F} = -\frac{F}{2I}(x - L_x)\left(\frac{y^3}{3} - L_y{}^2 y\right) \quad \Rightarrow \quad \begin{aligned} \sigma_x &= \frac{\partial^2 \mathcal{F}}{\partial y^2} = -\frac{F}{I}(x - L_x)y \\ \sigma_y &= \frac{\partial^2 \mathcal{F}}{\partial x^2} = 0 \\ \tau_{xy} &= -\frac{\partial^2 \mathcal{F}}{\partial x \partial y} = \frac{F}{2I}\left(y^2 - L_y{}^2\right) \end{aligned}$$

with $I = 2/3L_y{}^3$ being the y moment of inertia per width in the z-direction. This satisfies the traction boundary condition along the top and bottom, where the normal σ_y and shear τ_{xy} components are zero at $y = \pm L_y$, and along the right-hand side, where the normal stress is $\sigma_x = 0$ and the integral of shear stress is equal to the prescribed force: $\int_{-L_y}^{L_y} \tau_{xy} dy = -F$. The displacement is obtained by substituting these partial derivatives of $\mathcal{F}$ into (1.108) and integrating

$$\begin{aligned} \frac{\partial u_x}{\partial x} &= -\frac{F}{EI}(x - L_x)\,y \\ \frac{\partial u_y}{\partial y} &= \frac{F}{EI}\nu\,(x - L_x)\,y \end{aligned} \quad \rightarrow \quad \begin{aligned} u_x &= -\frac{F}{EI}\left(\frac{x^2}{2} - L_x x\right)y + f_1(y) \\ u_y &= \frac{F}{EI}\nu\,(x - L_x)\frac{y^2}{2} + f_2(x) \end{aligned}$$

where use has been made of $2\mu(1 + \nu) = E$ from (1.99). The functions f_1 and f_2 may be obtained fro this problem by substituting u_x, u_y, and $\mathcal{F}$ into the third equation in (1.108)

$$\frac{\partial u_x}{\partial y} + \frac{\partial u_y}{\partial x} = -\frac{1}{\mu}\frac{\partial^2 \mathcal{F}}{\partial x \partial y} \quad \rightarrow \quad -\frac{F}{EI}\left(\frac{x^2}{2} - L_x x\right) + \frac{\partial f_1}{\partial y} + \frac{F}{EI}\nu\frac{y^2}{2} + \frac{\partial f_2}{\partial x} = \frac{F}{EI}(1 + \nu)\left(y^2 - L_y{}^2\right)$$

then separating terms contain x and y, and integrating the separated terms

$$\frac{\partial f_1}{\partial y} - \frac{F}{EI}(2+\nu)\frac{y^2}{2} = a_1 \qquad\qquad f_1 = \frac{F}{EI}(2+\nu)\frac{y^3}{6} + a_1 y + b_1$$

$$\frac{\partial f_2}{\partial x} - \frac{F}{EI}\left(\frac{x^2}{2} - L_x x\right) = a_2 \quad \rightarrow \quad f_2 = \frac{F}{EI}\left(\frac{x^3}{6} - L_x\frac{x^2}{2}\right) + a_2 x + b_2$$

$$a_1 + a_2 = -\frac{F}{EI}(1+\nu)L_y^2$$

The coefficients in f_1 and f_2 require three additional conditions that may be chosen by approximating the fixed conditions at $x = y = 0$ with displacement $u_x = u_y = 0$ (so $b_1 = b_2 = 0$). The additional constant may be chosen as $a_2 = 0$ so $\partial u_y/\partial x = 0$ and there is no rotation of the body about the origin. Substituting these coefficients and f_1 and f_2 into u_x and u_y gives the displacement vector

$$u_x = -\frac{F}{EI}\left(\frac{x^2}{2} - L_x x\right)y + \frac{F}{EI}(2+\nu)\frac{y^3}{6} - \frac{F}{EI}(1+\nu)L_y^2 y$$

$$u_y = \frac{F}{EI}\nu\,(x - L_x)\frac{y^2}{2} + \frac{F}{EI}\left(\frac{x^3}{6} - L_x\frac{x^2}{2}\right)$$

The solution in Fig. 1.59 is illustrated for a lifting hook with dimensions $L_x = 0.5$ m and $Ly = 0.05$ m, coefficients for steel $E = 200 \times 10^9 \text{N/m}^2$ and $\nu = 0.3$, and force per width in the direction normal to the plane $F = 1 \times 10^6 \text{N/m}$. The top half of the bar is in tension, the bottom half is in compression, and the maximum shear stress occurs along the left side.

Problem 1.12 For the problem in Example 1.9 and Fig. 1.59, compute the stresses σ_x, σ_y, τ_{xy}, σ_1, σ_2, and $\tau_{\max}$ in units of N/m^2, and the displacement u_x and u_y in units of m at the following points:

A. $x = 0, y = -L_y$
B. $x = 0, y = -L_y/2$
C. $x = 0, y = 0$
D. $x = 0, y = L_y/2$
E. $x = 0, y = L_y$
F. $x = L_x/2, y = -L_y$
G. $x = L_x/2, y = -L_y/2$
H. $x = L_x/2, y = 0$
I. $x = L_x/2, y = L_y/2$
J. $x = L_x/2, y = L_y$
K. $x = L_x, y = -L_y$
L. $x = L_x, y = -L_y/2$
M. $x = L_x, y = 0$
N. $x = L_x, y = L_y/2$
O. $x = L_x, y = L_y$

Further Reading

Section 1.2.1, Groundwater Flow

- Anderson (2005)
- Aravin and Numerov (1965)
- Bear (1972)
- Bulatewicz, Yang, Peterson, Staggenborg, Welch, and Steward (2010)
- Bulatewicz, Allen, Peterson, Staggenborg, Welch, and Steward (2013)
- Dagan (1981)
- Darcy (1856)
- Dupuit (1863)
- Freeze and Cherry (1979)
- Haitjema (1995)
- Janković, Steward, Barnes, and Dagan (2009)
- Muskat (1937)

- Polubarinova-Kochina (1962)
- Steward (1998)
- Steward and Jin (2001)
- Steward (2002)
- Steward (2007)
- Steward and Ahring (2009)
- Steward and Allen (2013)
- Steward (2015)
- Strack (1981)
- Strack (1984)
- Strack (1989)
- Theis (1935)
- Thiem (1906)
- Verruijt (1982)

Section 1.2.2, Vadose Zone Flow

- Bakker and Nieber (2004*a*)
- Basha (1999)
- Bernoulli (1738)
- Gardner (1958)
- Kirchhoff (1894)
- Knight, Philip, and Waechter (1989)
- Mualem (1976)
- Philip (1968)
- Pullan (1990)
- Raats (1970)
- Raats (1971)
- Raats and Gardner (1974)
- Richards (1931)
- Steward (2016)
- van Genuchten (1980)
- Warrick (1974)
- Wooding (1968)

Section 1.2.3, Incompressible Fluid Flow

- Abbott and von Doenhoff (1949)
- Anderson, Jr. (1978)
- Batchelor (1967)
- Bernoulli (1738)
- Chou and Pagano (1967)
- Corke (2003)
- Hess and Smith (1967)
- Hess (1990)
- Jaeger (1969)
- Joukowsky (1912)
- Kraus (1978)
- Lagrange (1781)
- Lamb (1879)
- Lamb (1916)
- Maxwell (1870)
- Selvadurai (2000)
- Steward (2002)
- Steward, Le Grand, Janković, and Strack (2008)
- Stokes (1842)
- Stokes (1880)
- von Mises (1945)

Section 1.2.4, Thermal Conduction

- Abramowitz and Stegun (1972)
- Bakker (2013)
- Carslaw and Jaeger (1959)
- Davies (1978)
- Fourier (1878)
- Furman and Neuman (2003)
- Goodier and Hodge, Jr. (1958)
- Kuhlman and Neuman (2009)
- MacRobert (1967)
- Maxwell (1904*a*)
- Moon and Spencer (1961*a*)
- Muskhelishvili (1953*b*)
- Planck (1903)
- Steward (1991)
- Steward (1999)
- Steward, Le Grand, Janković, and Strack (2008)
- Steward (2015)
- Theis (1935)
- Thiem (1906)
- Westergaard (1952)

- Zaadnoordijk (1998)

Section 1.2.5, Electrostatics

- Choudhury (1989)
- Grant and Phillips (1990)
- Hammond (1986)
- MacRobert (1967)
- Maxwell (1890*a*)
- Maxwell (1890*b*)
- Maxwell (1904*b*)
- Muskat (1932)
- Moon and Spencer (1960)
- Moon and Spencer (1961*a*)
- Moon and Spencer (1961*b*)
- Ohm (1827)
- Slater and Frank (1947)
- Sommerfeld (1952)
- Steward (2015)
- Thomson (1904)
- Wait (1970)

Section 1.3.1, Water Waves

- Airy (1845)
- Bascom (1980)
- Berkhoff (1972)
- Berkhoff (1976)
- Bernoulli (1738)
- Boussinesq (1871)
- Dean and Dalrymple (1984)
- Ehrenmark (1998)
- Homma (1950)
- Jonsson, Skovgaard, and Brink-Kjaer (1976)
- Lamb (1879)
- Liu, Lin and Shankar (2004)
- Liu and Li (2007)
- Longuet-Higgins (1967)
- MacCamy and Fuchs (1954)
- Mei (1989)
- Sarpkaya and Isaacson (1981)
- Smith and Sprinks (1975)
- Sommerfeld (1949)
- Steward and Panchang (2000)
- Steward (2018)
- Steward (2020)
- Stokes (1847)
- Xu, Panchang, and Demirbilek (1996)

Section 1.3.2, Acoustics

- Airy (1871)
- Brekhovskikh and Godin (1990)
- Byerly (1893)
- Courant and Hilbert (1962)
- Goodman (1996)
- Kuttruff (2007)
- Lamb (1879)
- Morse (1936)
- Morse and Ingard (1968)
- Moon and Spencer (1961*a*)
- Morse and Ingard (1968)
- Sommerfeld (1949)
- Sommerfeld (1972)
- Stephens and Bate (1966)
- Stewart and Lindsay (1930)
- Steward (2018)
- Steward (2020)

Section 1.4.1, Elasticity

- Airy (1863)
- Boresi (1965)
- Goodier and Hodge, Jr. (1958)
- Goursat (1898)
- Green and Zerna (1968)
- Hooke (1678)
- Jaeger (1969)
- Jaeger and Cook (1976)
- Love (1927)
- Mogilevskaya and Crouch (2001)
- Muskhelishvili (1953*b*)
- Sanford (2003)
- Selvadurai (2000)
- Westergaard (1952)

Foundation of the Analytic Element Method

2

- The Analytic Element Method provides a foundation for solving boundary value problems commonly encountered in engineering and science, where problems are structured around elements to organize mathematical functions and methods.
- While this text mostly adheres to a "*just in time mathematics*" philosophy, whereby mathematical approaches are introduced when they are first needed, a comprehensive paradigm is presented in Section 2.1 as four steps necessary to achieve solutions.
- Likewise, Section 2.2 develops general solution methods, and Section 2.3 presents a consistent notation and concise representation to organize analytic elements across the broad range of disciplinary perspectives introduced in Chapter 1.

2.1 The Analytic Element Method Paradigm

The Analytic Element Method (AEM) organizes problems within a common sequence of steps (Steward and Allen, 2013):

1. The problem domain is partitioned into analytic elements with prescribed geometry.
2. Influence functions are developed for each element as a linear superposition of coefficients times mathematical expressions that satisfy governing equations.
3. Boundary conditions are prescribed and solved by adjusting these coefficients.
4. A comprehensive solution is collectively provided by the influence functions for all elements.

Each step in this algorithm is further developed to provide the theoretical framework for solving problems with the AEM.

2.1.1 Principle 1: Partitioning a Problem Domain into Elements

Boundary value problems are specified over the ***domain*** of a problem, which is the region over which the problem is to be solved. For example, Fig. 2.1 illustrates domains where the problem is formulated outside a circle in Fig. 2.1a, inside an rectangle in Fig. 2.1b, or along a line in Fig. 2.1c. The ***boundary*** of the domain D is designated ∂D, and two vectors are

Analytic Element Method: Complex Interactions of Boundaries and Interfaces. David R. Steward, Oxford University Press (2020).
DOI: 10.1093/oso/9780198856788.001.0001

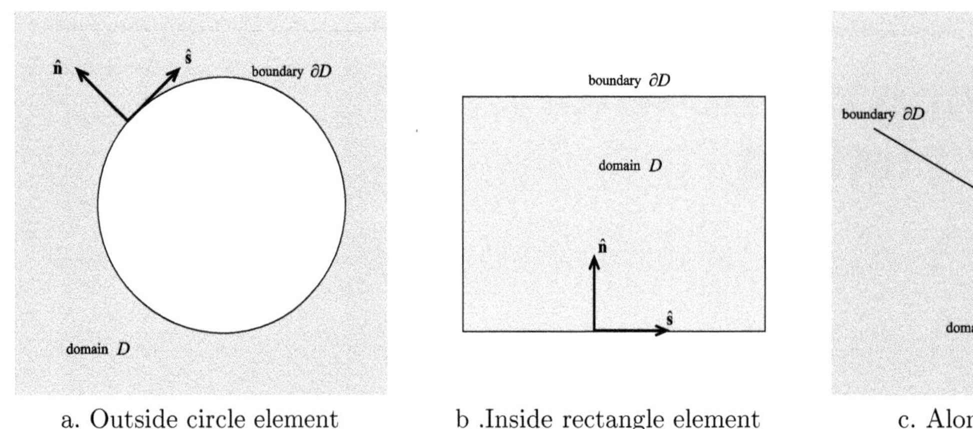

Figure 2.1 *Specifying the boundary and domain of an analytic element.*

specified along the boundary. A ***unit tangential vector*** $\hat{s}$ is tangent to this boundary, and a ***unit normal vector*** $\hat{n}$ is normal to this boundary and chosen to point into the domain D. The ^ (hat) symbol on $\hat{s}$ and $\hat{n}$ denotes that a vector is of unit length. Note that this two-dimensional representation is extended to three-dimensional problems in Chapter 4 and Section 5.7.

Solution to boundary value problems requires mathematical expressions that describes how important functions vary across the domain. They may be represented using the ***Laplacian*** operator

$$\nabla^2 f = \frac{\partial^2 f}{\partial x^2} + \frac{\partial^2 f}{\partial y^2} + \frac{\partial^2 f}{\partial z^2} \tag{2.1}$$

which is written here for the function f in terms of the x, y, and z axes in Cartesian coordinates. Formulation of conservation laws and constitutive equations across fields of study in Chapter 1 provides a common set of partial differential equations for representing important properties and processes:

- ***Laplace equation***:

$$\nabla^2 f = 0 \tag{2.2a}$$

- ***Poisson equation***:

$$\nabla^2 f = -d(x, y, z) \tag{2.2b}$$

where d is a given function, which is commonly equal to the divergence, (1.3), of a vector field.

- ***Helmholtz equation***:

$$\nabla^2 f = -k^2 f \tag{2.2c}$$

where k is the wave number.

- ***Modified Helmholtz equation***:

$$\nabla^2 f = k^2 f \tag{2.2d}$$

where $1/k$ is the leakage factor for groundwater and $2k$ is the sorptive number for the vadose zone.

- ***Biharmonic equation***:

$$\nabla^4 f = \nabla^2 \nabla^2 f = \frac{\partial^4 f}{\partial x^4} + 2\frac{\partial^4 f}{\partial x^2 \partial y^2} + \frac{\partial^4 f}{\partial y^4} = 0 \tag{2.2e}$$

- ***Heat equation***:

$$\nabla^2 f = \frac{1}{\mathscr{D}}\frac{\partial f}{\partial t} \tag{2.2f}$$

where $\mathscr{D}$ is the diffusivity and t is time.

- ***Wave equation***:

$$\nabla^2 f = \frac{1}{c^2}\frac{\partial^2 f}{\partial t^2} \tag{2.2g}$$

where c is the wave velocity (celerity).

This common nomenclature organizes mathematical and numerical techniques for solving the same partial differential equation across a variety of applications.

2.1.2 Principle 2: Linear Superposition of Influence Functions

Analytic elements are represented using mathematical functions formulated as a set of ***influence functions***, $\overset{\text{inf}}{f}_n$ that each individually satisfy the governing equation. Solutions to boundary value problems are developed through ***linear superposition*** of these influence functions:

$$f(x,y) = \sum_{n=0}^{N} c_n \overset{\text{inf}}{f}_n(x,y) \tag{2.3}$$

where the $N+1$ coefficients c_n are adjusted to achieve a solution. For example, the influence functions in Fig. 2.2 were used to obtain solutions to the Laplace equation in Fig. 1.1, to the Helmholtz equation in Fig. 1.2, and to the biharmonic equations in Fig. 1.3. ***The significant mathematical challenge of developing tractable forms for influence functions is a primary focus of Chapters 3, 4, and 5.***

Note that the mathematical form of (2.3) provides continuously varying functions that may be evaluated at any location from the summation of influence functions. Solutions may be computed at points $\mathfrak{s}$ along the boundary of an element, $\overset{\text{inf}}{f}_n(\mathfrak{s})$, and these expansions commonly take on the forms of either

- Power series: $f(\mathfrak{s}) = \sum_{n=0}^{N} c_n \mathfrak{s}^n$, (2.15), (2.25), and (2.36), or
- Fourier series: $f(\mathfrak{s}) = c_0 + \sum_{n=1}^{N} \overset{\cos}{c}_n \cos\frac{2\pi n \mathfrak{s}}{T} + \overset{\sin}{c}_n \sin\frac{2\pi n \mathfrak{s}}{T}$. (2.40)

These linear representations relating the ***coefficients*** c_n with the function f will be studied further in Section 2.2.

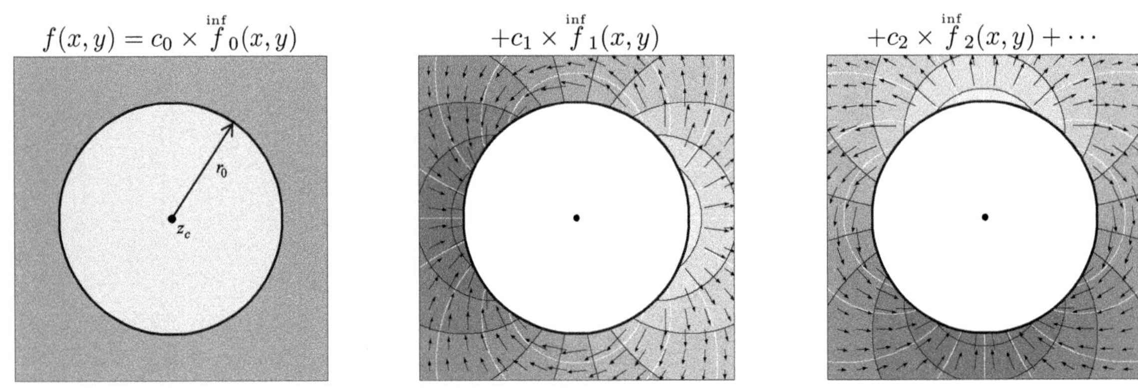

a. Influence functions for the Laplace equation used in Fig. 1.1 (from Figs. 3.8, 3.12)

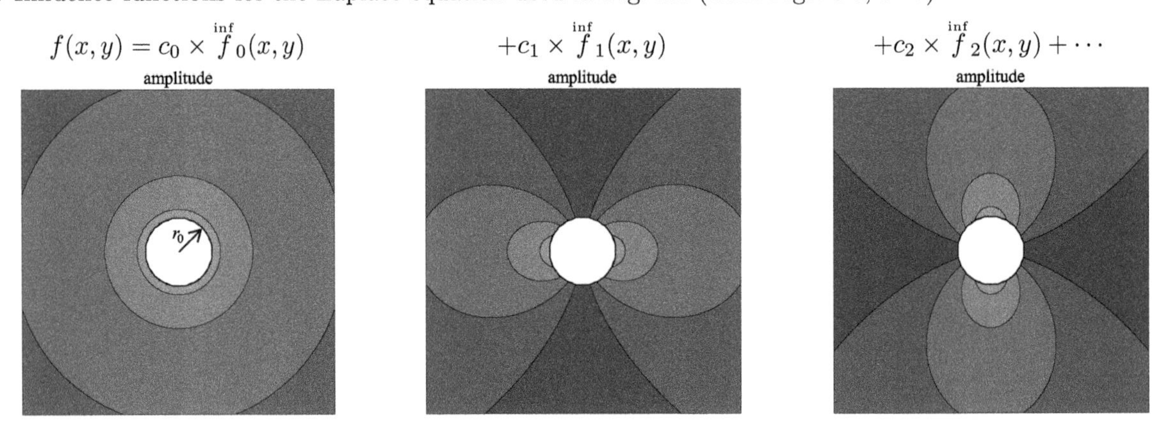

b. Influence functions for the Helmholtz equation in Fig. 1.2 amplitude (from Fig. 4.16)

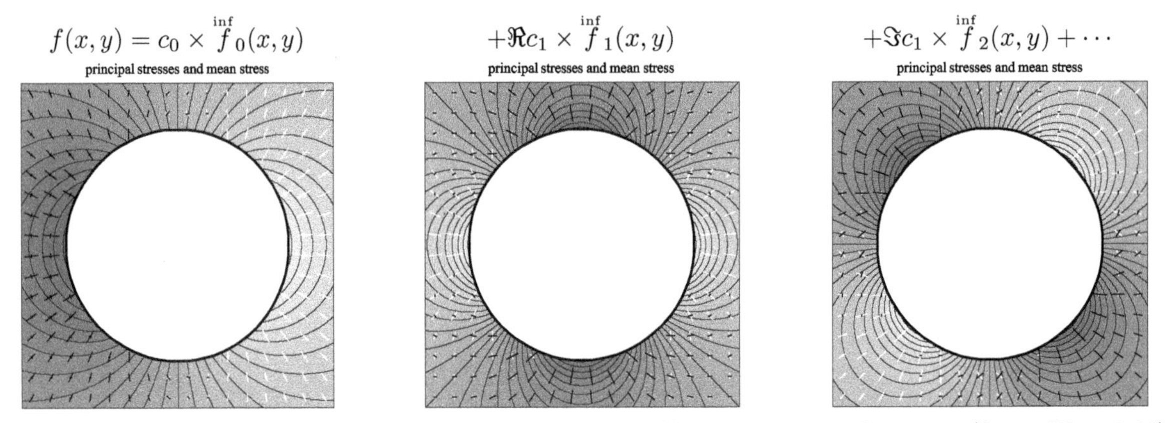

c. Influence functions for the biharmonic equation in Fig. 1.3 principal stress (from Fig. 3.36)

Figure 2.2 *Influence functions for sample analytic elements.*

2.1.3 Principle 3: Conditions at Control Points

Boundary value problems require specification of conditions at a set of control points along an analytic element that must be solved to achieve a solution. For example, the locations of sets of control point locations are identified in Fig. 2.3 along the boundaries of circle, ellipse, and line elements. The index m is used to denote the mth condition f_m at the location $\mathfrak{s}_m$, where m takes on values of 1 for the first data point to M for the last point. These values are tabulated as

index m	location $\mathfrak{s}_m$	boundary value f_m
1	$\mathfrak{s}_1$	f_1
2	$\mathfrak{s}_2$	f_2
⋮	⋮	⋮
M	$\mathfrak{s}_M$	f_M

While this set of data is expressed in terms of f, this function is used to represent the broad range of variables used to delineate boundary conditions occurring across fields of study, as developed later in Section 2.3.

A system of equations may be developed to organize the linear relations between the boundary values and the influence functions. Evaluating the function f, (2.3), at the location $\mathfrak{s}_m$ of each control point gives the expression

$$f(\mathfrak{s}_m) = \sum_{n=0}^{N} c_n \overset{\text{inf}}{f}_n(\mathfrak{s}_m) = f_m \qquad (m = 1, \cdots, M) \tag{2.4}$$

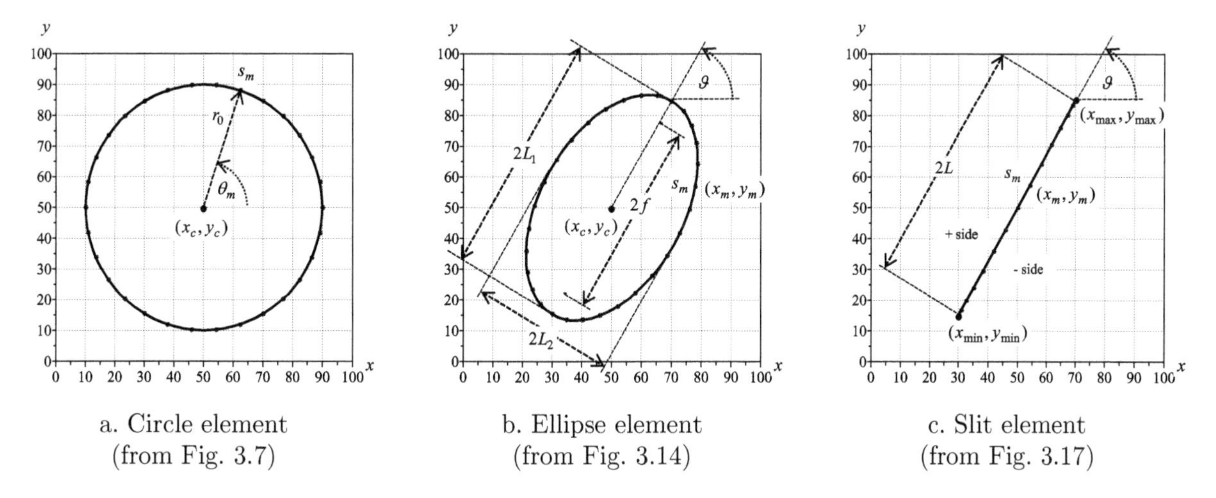

Figure 2.3 *Control point m on the boundary of an analytic element. (Reprinted from Steward, 2015, Analysis of discontinuities across thin inhomogeneities, groundwater/surface water interactions in river networks, and circulation about slender bodies using slit elements in the Analytic Element Method,* Water Resources Research, *Vol. 51(11), Fig. 2, with permission from John Wiley and Sons. Copyright 2015 by the American Geophysical Union.)*

This provides a system of M equations with $N + 1$ unknowns that may be represented in the matrix form $\mathbf{Ac} = \mathbf{b}$ with

$$\mathbf{A} = \begin{bmatrix} \overset{\text{inf}}{f}_0(\mathfrak{s}_1) & \overset{\text{inf}}{f}_1(\mathfrak{s}_1) & \cdots & \overset{\text{inf}}{f}_N(\mathfrak{s}_1) \\ \overset{\text{inf}}{f}_0(\mathfrak{s}_2) & \overset{\text{inf}}{f}_1(\mathfrak{s}_2) & \cdots & \overset{\text{inf}}{f}_N(\mathfrak{s}_2) \\ \cdots & \cdots & \cdots & \cdots \\ \overset{\text{inf}}{f}_0(\mathfrak{s}_M) & \overset{\text{inf}}{f}_1(\mathfrak{s}_M) & \cdots & \overset{\text{inf}}{f}_N(\mathfrak{s}_M) \end{bmatrix}, \quad \mathbf{c} = \begin{bmatrix} c_0 \\ c_1 \\ \vdots \\ c_N \end{bmatrix}, \quad \mathbf{b} = \begin{bmatrix} f_1 \\ f_2 \\ \vdots \\ f_M \end{bmatrix} \tag{2.5}$$

Methods for solving such systems of equations to determine the coefficients c_n that satisfy the prescribed conditions are presented in Section 2.2.

2.1.4 Principle 4: Collective Solutions of Interacting Elements

Problems with analytic elements are formulated by embedding collections of elements into a common setting. For example, the setting introduced in Fig. 1.1 is a regional gradient from left to right (from Fig. 3.6), in Fig. 1.2 it is monochromatic plane waves traveling from left to right (from Fig. 4.17), and in Fig. 1.3 it is biaxial compression (from Fig. 3.34). The functions used to compute important parameters throughout the problem domain are obtained by linear superposition of the background setting into which elements are placed with influence functions, (2.3), for all elements

$$f(x,y) = \overset{\text{set}}{f}(x,y) + \sum_{i=1}^{I} \sum_{n=0}^{N} \underset{i}{c_n} \overset{\text{inf}}{\underset{i}{f}}_n(x,y) \tag{2.6}$$

This uses notation with underscript i to designate that these coefficients and influence functions are associated with element i, and the function is summed across the I elements in the problem domain.

It is convenient to separate the set of coefficients and functions associated with element i from all others, and this is achieved using the ***additional function*** (Steward, 2015):

$$\begin{gathered} f(x,y) = \left[\sum_{n=0}^{N} \underset{i}{c_n} \overset{\text{inf}}{\underset{i}{f}}_n(x,y) \right] + \overset{\text{add}}{\underset{i}{f}}(x,y) \\ \boxed{\overset{\text{add}}{\underset{i}{f}}(x,y) = \overset{\text{set}}{f}(x,y) + \sum_{j \neq i}^{I} \sum_{n=0}^{N} \underset{j}{c_n} \overset{\text{inf}}{\underset{j}{f}}_n(x,y)} \end{gathered} \tag{2.7}$$

where the additional function contains the contributions from the setting and all other elements. The equations for element i may be evaluated at each control point, similar to (2.4),

$$f(\underset{i}{\mathfrak{s}_m}) = \left[\sum_{n=0}^{N} \underset{i}{c_n} \overset{\text{inf}}{\underset{i}{f}}_n(\underset{i}{\mathfrak{s}_m}) \right] + \overset{\text{add}}{\underset{i}{f}}(\underset{i}{\mathfrak{s}_m}) = \underset{i}{f_m} \qquad (m = 1, \cdots, M) \tag{2.8}$$

where $\underset{i}{\mathfrak{s}_m}$ is the location of control point m for element i, and $\underset{i}{f_m}$ is the boundary condition at this point. This provides a system of M equations with $N + 1$ for element i gathered in the matrices $\mathbf{Ac} = \mathbf{b}$ with

$$\mathbf{A} = \begin{bmatrix} \overset{\text{inf}}{\underset{i}{f}}_0(\underset{i}{s}_1) & \overset{\text{inf}}{\underset{i}{f}}_1(\underset{i}{s}_1) & \cdots & \overset{\text{inf}}{\underset{i}{f}}_N(\underset{i}{s}_1) \\ \overset{\text{inf}}{\underset{i}{f}}_0(\underset{i}{s}_2) & \overset{\text{inf}}{\underset{i}{f}}_1(\underset{i}{s}_2) & \cdots & \overset{\text{inf}}{\underset{i}{f}}_N(\underset{i}{s}_2) \\ \cdots & \cdots & \cdots & \cdots \\ \overset{\text{inf}}{\underset{i}{f}}_0(\underset{i}{s}_M) & \overset{\text{inf}}{\underset{i}{f}}_1(\underset{i}{s}_M) & \cdots & \overset{\text{inf}}{\underset{i}{f}}_N(\underset{i}{s}_M) \end{bmatrix}, \quad \mathbf{c} = \begin{bmatrix} \underset{i}{c}_0 \\ \underset{i}{c}_1 \\ \vdots \\ \underset{i}{c}_N \end{bmatrix}, \quad \mathbf{b} = \begin{bmatrix} \underset{i}{f}_1 - \overset{\text{add}}{\underset{i}{f}}(\underset{i}{s}_1) \\ \underset{i}{f}_2 - \overset{\text{add}}{\underset{i}{f}}(\underset{i}{s}_2) \\ \vdots \\ \underset{i}{f}_M - \overset{\text{add}}{\underset{i}{f}}(\underset{i}{s}_M) \end{bmatrix} \tag{2.9}$$

This formulation separates the influence functions and unknown coefficients for element i from all other functions. Yet, the elements interact, and so the solution of each element impacts the solutions of the others.

Collective solutions for all analytic elements in a problem are determined through an ***iterative solution method*** that fixes the coefficients for all other elements in the additional function, and then solve the system of equations for the element i. Gauss–Seidel iteration is performed across elements by sequencing through the $i = 1, \cdots, I$ elements and updating coefficients for each element as the solve algorithm progresses through the iterate (Janković and Barnes, 1999*a*). Once the coefficients for all I elements have been solved, the process is repeated. This continues until only small variations occur in the values of f across all M control points for all I elements between successive iterates (examples in the book stop when f has at least 10 significant digits at every control point for every element).

The computations associated with the additional functions may be organized as matrix multiplication and addition. This is accomplished by organizing terms in the additional function for element i

$$\overset{\text{add}}{\underset{i}{f}}(\underset{i}{s}_m) = \overset{\text{set}}{\underset{i}{f}}(\underset{i}{s}_m) + \sum_{j \neq i}^{I} \sum_{n=0}^{N} \underset{j}{c}_n \overset{\text{inf}}{\underset{j}{f}}_n(\underset{i}{s}_m) \tag{2.10}$$

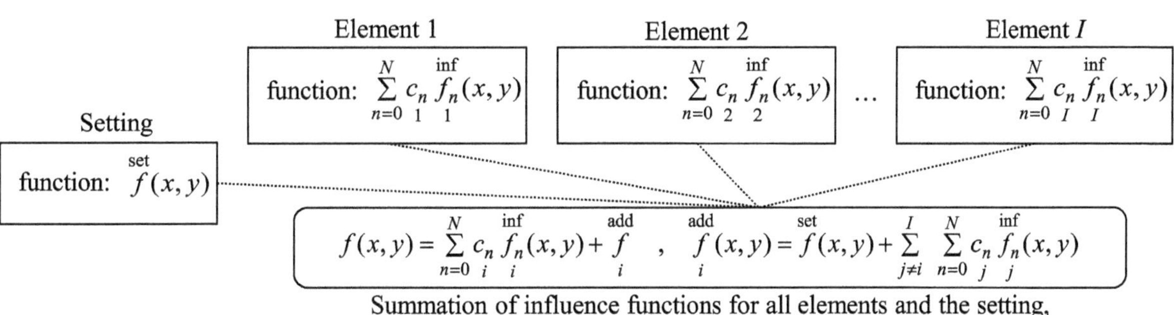

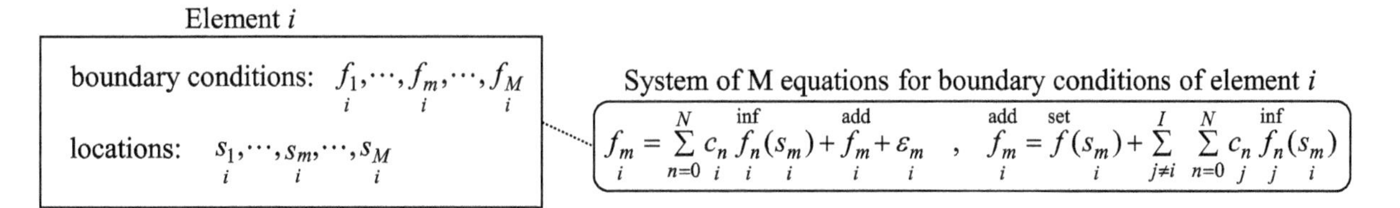

Figure 2.4 *Partitioning functions, boundary conditions and control point locations per element.*

within the matrices

$$\underset{i}{\overset{\text{add}}{\mathbf{f}}} = \begin{bmatrix} \underset{i}{\overset{\text{add}}{f}}(\underset{i}{\mathfrak{s}_1}) \\ \underset{i}{\overset{\text{add}}{f}}(\underset{i}{\mathfrak{s}_2}) \\ \vdots \\ \underset{i}{\overset{\text{add}}{f}}(\underset{i}{\mathfrak{s}_M}) \end{bmatrix}, \quad \underset{i}{\overset{\text{set}}{\mathbf{f}}} = \begin{bmatrix} \overset{\text{set}}{f}(\underset{i}{\mathfrak{s}_1}) \\ \overset{\text{set}}{f}(\underset{i}{\mathfrak{s}_2}) \\ \vdots \\ \overset{\text{set}}{f}(\underset{i}{\mathfrak{s}_M}) \end{bmatrix}, \quad \underset{ij}{\overset{\text{inf}}{\mathbf{f}}} = \begin{bmatrix} \underset{j}{\overset{\text{inf}}{f}}{}_0(\underset{i}{\mathfrak{s}_1}) & \underset{j}{\overset{\text{inf}}{f}}{}_1(\underset{i}{\mathfrak{s}_1}) & \cdots & \underset{j}{\overset{\text{inf}}{f}}{}_N(\underset{i}{\mathfrak{s}_1}) \\ \underset{j}{\overset{\text{inf}}{f}}{}_0(\underset{i}{\mathfrak{s}_2}) & \underset{j}{\overset{\text{inf}}{f}}{}_1(\underset{i}{\mathfrak{s}_2}) & \cdots & \underset{j}{\overset{\text{inf}}{f}}{}_N(\underset{i}{\mathfrak{s}_2}) \\ \cdots & \cdots & \cdots & \cdots \\ \underset{j}{\overset{\text{inf}}{f}}{}_0(\underset{i}{\mathfrak{s}_M}) & \underset{j}{\overset{\text{inf}}{f}}{}_1(\underset{i}{\mathfrak{s}_M}) & \cdots & \underset{j}{\overset{\text{inf}}{f}}{}_N(\underset{i}{\mathfrak{s}_M}) \end{bmatrix}, \quad \underset{j}{\mathbf{c}} = \begin{bmatrix} \underset{j}{c_0} \\ \underset{j}{c_1} \\ \vdots \\ \underset{j}{c_N} \end{bmatrix} \tag{2.11}$$

This organization provides a direct relation between the coefficients of each analytic element and their influence on the boundary condition of the other elements:

$$\underset{i}{\overset{\text{add}}{\mathbf{f}}} = \underset{i}{\overset{\text{set}}{\mathbf{f}}} + \sum_{j \neq i}^{I} \underset{ij}{\overset{\text{inf}}{\mathbf{f}}} \underset{j}{\mathbf{c}} \tag{2.12}$$

This is advantageous, since the matrices $\underset{ij}{\overset{\text{inf}}{\mathbf{f}}}$ may be evaluated once for a set of analytic elements, and changes in f at element i due to adjusting coefficients in element j are then computed with matrix multiplication rather than evaluating functions.

A ***notation convention*** is adopted throughout this text to drop the underscript for element i, but with the understanding that computations are performed on an element-by-element basis.

$$\sum_{n=1}^{N} c_n \overset{\text{inf}}{f}{}_n(\mathfrak{s}_m) = f_m - \overset{\text{add}}{f}{}_m \tag{2.13}$$

and $\overset{\text{add}}{f}{}_m$ is the additional function for control point m containing the contributions from the setting and all other elements. Thus, the matrices in the system of Eqs (2.9) are expressed as

$$\mathbf{A} = \begin{bmatrix} \overset{\text{inf}}{f}{}_0(\mathfrak{s}_1) & \overset{\text{inf}}{f}{}_1(\mathfrak{s}_1) & \cdots & \overset{\text{inf}}{f}{}_N(\mathfrak{s}_1) \\ \overset{\text{inf}}{f}{}_0(\mathfrak{s}_2) & \overset{\text{inf}}{f}{}_1(\mathfrak{s}_2) & \cdots & \overset{\text{inf}}{f}{}_N(\mathfrak{s}_2) \\ \cdots & \cdots & \cdots & \cdots \\ \overset{\text{inf}}{f}{}_0(\mathfrak{s}_M) & \overset{\text{inf}}{f}{}_1(\mathfrak{s}_M) & \cdots & \overset{\text{inf}}{f}{}_N(\mathfrak{s}_M) \end{bmatrix}, \quad \mathbf{c} = \begin{bmatrix} c_0 \\ c_1 \\ \vdots \\ c_N \end{bmatrix}, \quad \mathbf{b} = \begin{bmatrix} f_1 \\ f_2 \\ \vdots \\ f_M \end{bmatrix} - \begin{bmatrix} \overset{\text{add}}{f}{}_1 \\ \overset{\text{add}}{f}{}_2 \\ \vdots \\ \overset{\text{add}}{f}{}_M \end{bmatrix} \tag{2.14}$$

The next section shows how to solve for variation in f found along the boundary of analytic elements.

2.2 Solving Systems of Equations to Match Boundary Conditions

Please visit the Oxford University Press companion website for instructional examples of computer code that implement equations for many of the figures in this book.

Analytic elements contain mathematical functions that were just formulated as a linear superposition of influence functions multiplied by coefficients. This section develops methods for adjusting these coefficients in the function $f(\mathfrak{s})$ by matching boundary conditions

f_m at locations $\mathfrak{s}_m$ along the boundary. Solutions are obtained for functions that will commonly appear later, including local power series in Section 2.2.1, and Fourier series in Section 2.2.2. For some problems, boundary conditions will be specified as functions with a non-linear relation between coefficients, and methods for solving such problems are developed in Section 2.2.3.

2.2.1 Linear Regression: Minimizing a Least Squares Objective Function

A common problem from linear regression is to determine the coefficients c_0 and c_1 in a ***linear approximation*** given by the function

$$\boxed{f(\mathfrak{s}) = c_0 + c_1 \mathfrak{s}} \tag{2.15}$$

to match a set of specified conditions. For example, a set of $M = 11$ conditions are plotted in Fig. 2.5a, where f_m is specified at the location $\mathfrak{s}_m$, along with the linear approximation with intercept c_0 and slope c_1. Clearly, the linear function is not capable of exactly passing through all of the specified values; instead, there is a residual error ε_m at the mth point, which is the difference between the approximate function evaluated at $\mathfrak{s}_m$ and the measurement:

$$f(\mathfrak{s}_m) = f_m + \varepsilon_m \qquad (m = 1, M) \tag{2.16}$$

The linear function (2.15) and the last equation give a system of M equations with the two unknowns c_0 and c_1:

$$\begin{aligned} c_0 + c_1 \mathfrak{s}_1 &= f_1 + \varepsilon_1 \\ c_0 + c_1 \mathfrak{s}_2 &= f_2 + \varepsilon_2 \\ \cdots\cdots&\cdots\cdots \\ c_0 + c_1 \mathfrak{s}_M &= f_M + \varepsilon_M \end{aligned} \tag{2.17}$$

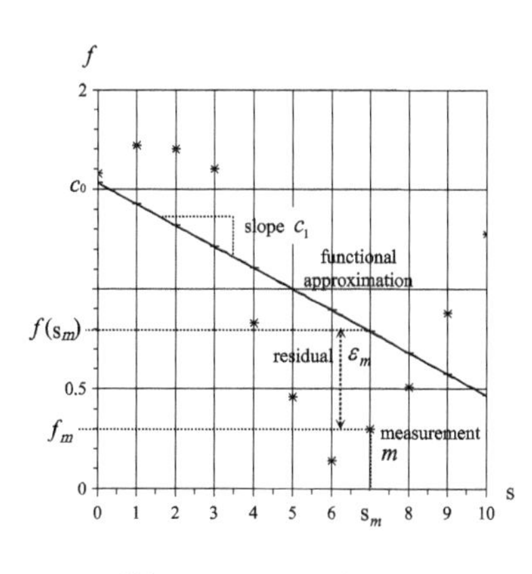

a. Linear approximation, $f(\mathfrak{s}) = c_0 + c_1 \mathfrak{s}$

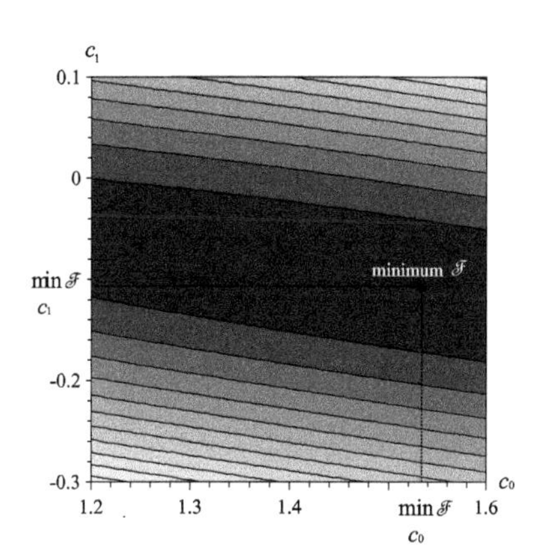

b. Objective function $\mathscr{F}$ for linear approximation, and c_0 and c_1 at its minimum

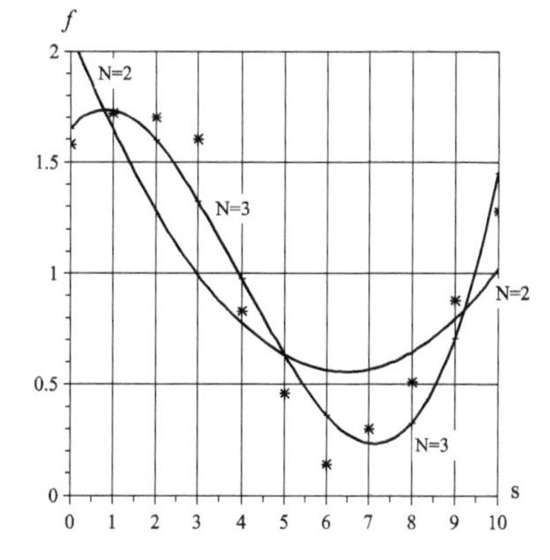

c. Higher-order approximation, $f(\mathfrak{s}) = \sum_{n=0}^{N} c_n \mathfrak{s}^n$

Figure 2.5 *Matching boundary conditions using a power series.*

A good estimate of the function f is obtained by adjusting the coefficients c_0 and c_1 to minimize the residuals.

A computationally attractive method for minimizing residual errors is obtained using an objective function $\mathscr{F}$ equal to the sum of the square of the residual errors divided by the number of measurements:

$$\boxed{\mathscr{F} = \frac{1}{M}\sum_{m=1}^{M}(\varepsilon_m)^2 = \frac{1}{M}\sum_{m=1}^{M}(c_0 + c_1\mathfrak{s}_m - f_m)^2} \tag{2.18}$$

This objective function is plotted for a range of values of c_0 and c_1 in Fig. 2.5b, and the value of the coefficients where $\mathscr{F}$ has its minimum for this data is identified. This minimum occurs at the location where the partial derivatives with respect to c_0 and c_1 are both equal to zero, which gives two equations:

$$\begin{aligned}\frac{\partial\mathscr{F}}{\partial c_0} &= \frac{2}{M}\sum_{m=1}^{M}(c_0 + c_1\mathfrak{s}_m - f_m) = 0\\ \frac{\partial\mathscr{F}}{\partial c_1} &= \frac{2}{M}\sum_{m=1}^{M}\mathfrak{s}_m(c_0 + c_1\mathfrak{s}_m - f_m) = 0\end{aligned} \tag{2.19}$$

These equations may be rearranged to obtain the normal equations:

$$\begin{aligned}Mc_0 + \left(\sum_{m=1}^{M}\mathfrak{s}_m\right) c_1 &= \sum_{m=1}^{M} f_m\\ \left(\sum_{m=1}^{M}\mathfrak{s}_m\right)c_0 + \left[\sum_{m=1}^{M}(\mathfrak{s}_m)^2\right] c_1 &= \sum_{m=1}^{M}\mathfrak{s}_m f_m\end{aligned} \tag{2.20}$$

The solution to this linear system of two equations provides the coefficients

$$c_0 = \frac{\sum_{m=1}^{M}(\mathfrak{s}_m)^2\sum_{m=1}^{M}f_m - \sum_{m=1}^{M}\mathfrak{s}_m\sum_{m=1}^{M}\mathfrak{s}_m f_m}{M\sum_{m=1}^{M}(\mathfrak{s}_m)^2 - \left(\sum_{m=1}^{M}\mathfrak{s}_m\right)^2}, \quad c_1 = \frac{M\sum_{m=1}^{M}\mathfrak{s}_m f_m - \sum_{m=1}^{M}\mathfrak{s}_m\sum_{m=1}^{M}f_m}{M\sum_{m=1}^{M}(\mathfrak{s}_m)^2 - \left(\sum_{m=1}^{M}\mathfrak{s}_m\right)^2} \tag{2.21}$$

and this linear approximation was used in Fig. 2.5a.

This system of equations may also be obtained by formulating the system of M equations in (2.17) in matrix form as

$$\mathbf{Ac} = \mathbf{b}, \qquad \mathbf{A} = \begin{bmatrix}1 & \mathfrak{s}_1\\ 1 & \mathfrak{s}_2\\ \cdots & \cdots\\ 1 & \mathfrak{s}_M\end{bmatrix}, \quad \mathbf{c} = \begin{bmatrix}c_0\\ c_1\end{bmatrix}, \quad \mathbf{b} = \begin{bmatrix}f_1\\ f_2\\ \vdots\\ f_M\end{bmatrix} \tag{2.22}$$

The least squares solution to this system of equations may be expressed as follows,

$$\boxed{\mathbf{A}^{\mathrm{T}}\mathbf{A}\mathbf{c} = \mathbf{A}^{\mathrm{T}}\mathbf{b}} \quad \rightarrow \quad \begin{bmatrix} 1 & 1 & \cdots & 1 \\ \mathfrak{s}_1 & \mathfrak{s}_2 & \cdots & \mathfrak{s}_M \end{bmatrix} \begin{bmatrix} 1 & \mathfrak{s}_1 \\ 1 & \mathfrak{s}_2 \\ \cdots & \cdots \\ 1 & \mathfrak{s}_M \end{bmatrix} \begin{bmatrix} c_0 \\ c_1 \end{bmatrix} = \begin{bmatrix} 1 & 1 & \cdots & 1 \\ \mathfrak{s}_1 & \mathfrak{s}_2 & \cdots & \mathfrak{s}_M \end{bmatrix} \begin{bmatrix} f_1 \\ f_2 \\ \vdots \\ f_M \end{bmatrix} \qquad (2.23)$$

where $\mathbf{A}^{\mathrm{T}}$ is the transpose of $\mathbf{A}$, obtained by interchanging rows and columns. Matrix multiplication gives

$$\begin{bmatrix} M & \sum_{m=1}^{M} \mathfrak{s}_m \\ \sum_{m=1}^{M} \mathfrak{s}_m & \sum_{m=1}^{M} (\mathfrak{s}_m)^2 \end{bmatrix} \begin{bmatrix} c_0 \\ c_1 \end{bmatrix} = \begin{bmatrix} \sum_{m=1}^{M} f_m \\ \sum_{m=1}^{M} \mathfrak{s}_m f_m \end{bmatrix} \qquad (2.24)$$

which is the same system of equations as (2.20).

It is desirable to develop robust functional approximations that approximate conditions more accurately than the linear approximation in Fig. 2.5a. A logical improvement is to extend the linear function (2.15) to a functional approximation with the form of a ***higher-order power series*** of order N. This may be written as

$$\boxed{f(\mathfrak{s}) = \sum_{n=0}^{N} c_n \mathfrak{s}^n = c_0 + c_1 \mathfrak{s} + c_2 \mathfrak{s}^2 + \cdots + c_N \mathfrak{s}^N} \qquad (2.25)$$

Substituting the power series approximation (2.25) into the relation between the specified conditions and residuals (2.16) gives a set of M equations for the $N + 1$ unknowns

$$\begin{aligned} c_0 + c_1 \mathfrak{s}_1 + c_2 \mathfrak{s}_1^2 + \cdots + c_N \mathfrak{s}_1^N &= f_1 + \varepsilon_1 \\ c_0 + c_1 \mathfrak{s}_2 + c_2 \mathfrak{s}_2^2 + \cdots + c_N \mathfrak{s}_2^N &= f_2 + \varepsilon_2 \\ \cdots\cdots\cdots\cdots\cdots\cdots\cdots\cdots & \\ c_0 + c_1 \mathfrak{s}_M + c_2 \mathfrak{s}_M^2 + \cdots + c_N \mathfrak{s}_M^N &= f_M + \varepsilon_M \end{aligned} \qquad (2.26)$$

A solution that minimizes the residual errors may be obtained using the objective function

$$\mathscr{F} = \frac{1}{M} \sum_{m=1}^{M} (\varepsilon_m)^2 = \frac{1}{M} \sum_{m=1}^{M} \left(\sum_{n=0}^{N} c_n \mathfrak{s}^n - f_m \right)^2 \qquad (2.27)$$

for an ***overspecified*** set of equations where the number of measurements M is greater than or equal to the number of unknown coefficients $N + 1$. This system of equations relating the unknown coefficients to the known conditions may be expressed in matrix form as

$$\mathbf{A}\mathbf{c} = \mathbf{b}\,, \quad \mathbf{A} = \begin{bmatrix} 1 & \mathfrak{s}_1 & \mathfrak{s}_1^2 & \cdots & \mathfrak{s}_1^N \\ 1 & \mathfrak{s}_2 & \mathfrak{s}_2^2 & \cdots & \mathfrak{s}_2^N \\ \cdots & \cdots & \cdots & \cdots & \cdots \\ 1 & \mathfrak{s}_M & \mathfrak{s}_M^2 & \cdots & \mathfrak{s}_M^N \end{bmatrix}, \quad \mathbf{c} = \begin{bmatrix} c_0 \\ c_1 \\ c_2 \\ \vdots \\ c_N \end{bmatrix}, \quad \mathbf{b} = \begin{bmatrix} f_1 \\ f_2 \\ \vdots \\ f_M \end{bmatrix} \qquad (2.28)$$

The least squares solution, $\mathbf{A}^{\mathrm{T}}\mathbf{Ac} = \mathbf{A}^{\mathrm{T}}\mathbf{b}$, which minimizes the objective function, (2.27), is given by

$$\begin{bmatrix} M & \sum_{m=1}^{M} \mathfrak{s}_m & \cdots & \sum_{m=1}^{M} (\mathfrak{s}_m)^N \\ \sum_{m=1}^{M} \mathfrak{s}_m & \sum_{m=1}^{M} (\mathfrak{s}_m)^2 & \cdots & \sum_{m=1}^{M} (\mathfrak{s}_m)^{N+1} \\ \cdots & \cdots & \cdots & \cdots \\ \sum_{m=1}^{M} (\mathfrak{s}_m)^N & \sum_{m=1}^{M} (\mathfrak{s}_m)^{N+1} & \cdots & \sum_{m=1}^{M} (\mathfrak{s}_m)^{2N} \end{bmatrix} \begin{bmatrix} c_0 \\ c_1 \\ \vdots \\ c_N \end{bmatrix} = \begin{bmatrix} \sum_{m=1}^{M} f_m \\ \sum_{m=1}^{M} \mathfrak{s}_m f_m \\ \vdots \\ \sum_{m=1}^{M} (\mathfrak{s}_m)^N f_m \end{bmatrix} \tag{2.29}$$

This linear system of equations may be solved using readily available computational methods, and provide the functions for $N > 1$ in Fig. 2.5c.

Example 2.1 The values of the variable and data in Fig. 2.5 follow, where data is sorted by control points with adjacent locations $\mathfrak{s}_m$, and the variables S_m and θ_m will be used later.

m	$\mathfrak{s}_m$	f_m	S_m	θ_m
1	0	1.58	−1.0	0.000
2	1	1.72	−0.8	0.571
3	2	1.70	−0.6	1.142
4	3	1.60	−0.4	1.713
5	4	0.83	−0.2	2.285
6	5	0.46	0.0	2.856
7	6	0.14	0.2	3.427
8	7	0.30	0.4	3.998
9	8	0.51	0.6	4.569
10	9	0.88	0.8	5.141
11	10	1.28	1.0	5.712

The linear approximation in Fig. 2.5a has coefficients and objective function

$$\begin{bmatrix} 11 & 55 \\ 55 & 385 \end{bmatrix} \begin{bmatrix} c_0 \\ c_1 \end{bmatrix} = \begin{bmatrix} 11.00 \\ 43.28 \end{bmatrix} \quad \rightarrow \quad \begin{bmatrix} c_0 \\ c_1 \end{bmatrix} = \begin{bmatrix} 1.5327 \\ -0.1065 \end{bmatrix}$$
$$\mathscr{F} = 0.2126$$

The coefficients and objective function for $N = 2$ are

$$\begin{bmatrix} 11 & 55 & 385 \\ 55 & 385 & 3025 \\ 385 & 3025 & 25333 \end{bmatrix} \begin{bmatrix} c_0 \\ c_1 \\ c_2 \end{bmatrix} = \begin{bmatrix} 11.00 \\ 43.28 \\ 299.36 \end{bmatrix} \quad \rightarrow \quad \begin{bmatrix} c_0 \\ c_1 \\ c_2 \end{bmatrix} = \begin{bmatrix} 2.0845 \\ -0.4744 \\ 0.0368 \end{bmatrix}$$
$$\mathscr{F} = 0.1071$$

and the solution for $N = 3$ is

$$\begin{bmatrix} 11 & 55 & 385 & 3025 \\ 55 & 385 & 3025 & 25333 \\ 385 & 3025 & 25333 & 220825 \\ 3025 & 25333 & 220825 & 1978405 \end{bmatrix} \begin{bmatrix} c_0 \\ c_1 \\ c_2 \\ c_3 \end{bmatrix} = \begin{bmatrix} 11.00 \\ 43.28 \\ 299.36 \\ 2484.92 \end{bmatrix} \quad \rightarrow \quad \begin{bmatrix} c_0 \\ c_1 \\ c_2 \\ c_3 \end{bmatrix} = \begin{bmatrix} 1.6524 \\ 0.2121 \\ -0.1432 \\ 0.0120 \end{bmatrix}, \quad \mathscr{F} = 0.0262$$

The objective function becomes smaller as N increases; however, the matrix coefficients also become larger.

Problem 2.1 Compute the coefficients c_0 and c_1 for a linear approximation, and the objective function $\mathscr{F}$ that minimizes the sum of the square of the errors for the following data. (For Problem 2.1A, use the data corresponding to the column with top row containing A; for Problem 2.1B, use the data corresponding to the column with top row containing B; etc.)

		A.	B.	C.	D.	E.	F.	G.	H.	I.	J.
m	s_m	f_m	f_m	f_m	f_m	f_m	f_m	f_m	f_m	f_m	f_m
1	0	1.57	1.64	1.60	1.82	1.76	1.57	1.63	1.59	1.80	1.62
2	1	1.91	1.89	1.88	1.80	1.91	1.90	1.76	1.86	1.81	1.73
3	2	1.93	1.83	1.71	1.71	1.74	1.72	1.70	1.70	1.89	1.93
4	3	1.44	1.51	1.44	1.57	1.63	1.46	1.38	1.52	1.52	1.38
5	4	1.00	0.83	0.83	1.00	0.87	0.83	0.83	0.82	0.95	0.82
6	5	0.36	0.56	0.32	0.34	0.39	0.54	0.38	0.40	0.34	0.34
7	6	0.29	0.12	0.26	0.13	0.18	0.10	0.20	0.32	0.33	0.34
8	7	0.28	0.24	0.18	0.19	0.36	0.29	0.18	0.18	0.30	0.19
9	8	0.47	0.61	0.53	0.61	0.64	0.64	0.70	0.49	0.51	0.48
10	9	1.01	0.90	1.01	0.88	0.96	0.96	0.88	0.97	1.11	0.98
11	10	1.28	1.39	1.32	1.45	1.28	1.29	1.50	1.36	1.45	1.37

Problem 2.2 Compute the coefficients c_0, c_1, c_2, and c_3 for $N = 3$, and the objective function $\mathscr{F}$ that minimizes the sum of the square of the errors using the data set specified in Problem 2.1.

Lagrange Multipliers to Match Exact Constraints

For some problems, it may be desirable to be able to satisfy some of the conditions exactly. This may be accomplished by first subdividing the system of equations in two groups of matrices: one where M conditions in $\mathbf{A}$ and $\mathbf{b}$ are to be approximately satisfied, and the other where $\widetilde{M}$ conditions in $\widetilde{\mathbf{A}}$ and $\widetilde{\mathbf{b}}$ are to be satisfied exactly:

$$\begin{aligned} \mathbf{A}\mathbf{c} = \mathbf{b}\,, \quad & \sum_{n=1}^{N} a_{mn} c_n + \varepsilon_m = b_m \quad (m = 1, M) \\ \widetilde{\mathbf{A}}\mathbf{c} = \widetilde{\mathbf{b}}\,, \quad & \sum_{n=1}^{N} \widetilde{a}_{mn} c_n = \widetilde{b}_m \quad (m = M+1, M+\widetilde{M}) \end{aligned} \tag{2.30}$$

An objective function that satisfies the overspecified conditions in the least squares sense and the exact conditions is given by

$$\mathscr{F} = \frac{1}{M} \sum_{m=1}^{M} \left(\sum_{n=1}^{N} a_{mn} c_n - b_m \right)^2 + \frac{2}{M} \sum_{m=M+1}^{M+\widetilde{M}} \lambda_{m-M} \left(\sum_{n=1}^{N} \widetilde{a}_{mn} c_n - \widetilde{b}_m \right) \tag{2.31}$$

where λ_m are Lagrange multipliers. The objective function is minimal where the partial derivatives with respect to the unknown coefficients c_n and the Lagrange multipliers λ_m are zero. The derivatives with respect to c_n give the N conditions

$$\frac{\partial \mathscr{F}}{\partial c_1} = \frac{2}{M}\sum_{m=1}^{M} a_{m1}\left(\sum_{n=1}^{N} a_{mn}c_n - b_m\right) + \frac{2}{M}\sum_{m=M+1}^{M+\widetilde{M}} \lambda_{m-M}\widetilde{a}_{m1} = 0$$
$$\cdots \tag{2.32}$$
$$\frac{\partial \mathscr{F}}{\partial c_N} = \frac{2}{M}\sum_{m=1}^{M} a_{mN}\left(\sum_{n=1}^{N} a_{mn}c_n - b_m\right) + \frac{2}{M}\sum_{m=M+1}^{M+\widetilde{M}} \lambda_{m-M}\widetilde{a}_{mN} = 0$$

and the derivatives with respect to λ_m give the $\widetilde{M}$ exact conditions in (2.30),

$$\frac{\partial \mathscr{F}}{\partial \lambda_1} = \frac{2}{M}\left[\sum_{n=1}^{N} \widetilde{a}_{(M+1)n}c_n - \widetilde{b}_{M+1}\right] = 0$$
$$\cdots \tag{2.33}$$
$$\frac{\partial \mathscr{F}}{\partial \lambda_{\widetilde{M}}} = \frac{2}{M}\left[\sum_{n=1}^{N} \widetilde{a}_{(M+\widetilde{M})n}c_n - \widetilde{b}_{M+\widetilde{M}}\right] = 0$$

This system of $N + \widetilde{M}$ equations with $N + \widetilde{M}$ unknowns may be written in symbolic notation as

$$\begin{bmatrix} \mathbf{A}^{\mathrm{T}}\mathbf{A} & \widetilde{\mathbf{A}}^{\mathrm{T}} \\ \widetilde{\mathbf{A}} & \mathbf{0} \end{bmatrix} \begin{bmatrix} \mathbf{c} \\ \boldsymbol{\lambda} \end{bmatrix} = \begin{bmatrix} \mathbf{A}^{\mathrm{T}}\mathbf{b} \\ \widetilde{\mathbf{b}} \end{bmatrix} \tag{2.34}$$

where $\boldsymbol{\lambda}$ is a vector containing the Lagrange multipliers, and $\mathbf{0}$ is a matrix with $\widetilde{M}$ rows and $\widetilde{M}$ columns containing zeros.

Problem 2.3 For the data from Problem 2.1, an additional constraint is placed on the quadratic function that it must exactly pass through the point $f(\mathfrak{s} = 0) = 0$. Formulate a system of equations with three unknown coefficients c_0, c_1, and the Lagrange multiplier for this problem. Solve the system of equations to determine the unknowns. Also compute the value of the approximation f for each measurement, and the objective function $\mathscr{F}$.

Local Power Series

Limitations on the functional approximations that can be solved via least squares are imposed by the capacity of computers to evaluate mathematical expressions. In particular, the least squares solution for power series contains terms of order $\sum \mathfrak{s}_m{}^{2N}$ in the least squares matrix $\mathbf{A}^{\mathrm{T}}\mathbf{A}$, which become quite large for many problems, such as example 2.1. This problem may be resolved by scaling the variable $\mathfrak{s}$ to a local coordinate system where the variable S varies between -1 and $+1$ over all data values. This is accomplished using (Steward et al., 2008)

$$S = \frac{\mathfrak{s} - \mathfrak{s}_c}{L}, \quad \mathfrak{s}_c = \frac{\mathfrak{s}_{\max} + \mathfrak{s}_{\min}}{2}, \quad L = \frac{\mathfrak{s}_{\max} - \mathfrak{s}_{\min}}{2} \tag{2.35}$$

where $\mathfrak{s}_{\min}$ and $\mathfrak{s}_{\max}$ are the minimum and maximum values of $\mathfrak{s}$ across data values, $\mathfrak{s}_c$ is the center, and L is 1/2 the data interval. A ***local power series*** relating the function f and the local variable S is given by

$$f(S) = \sum_{n=0}^{N} c_n S^n \tag{2.36}$$

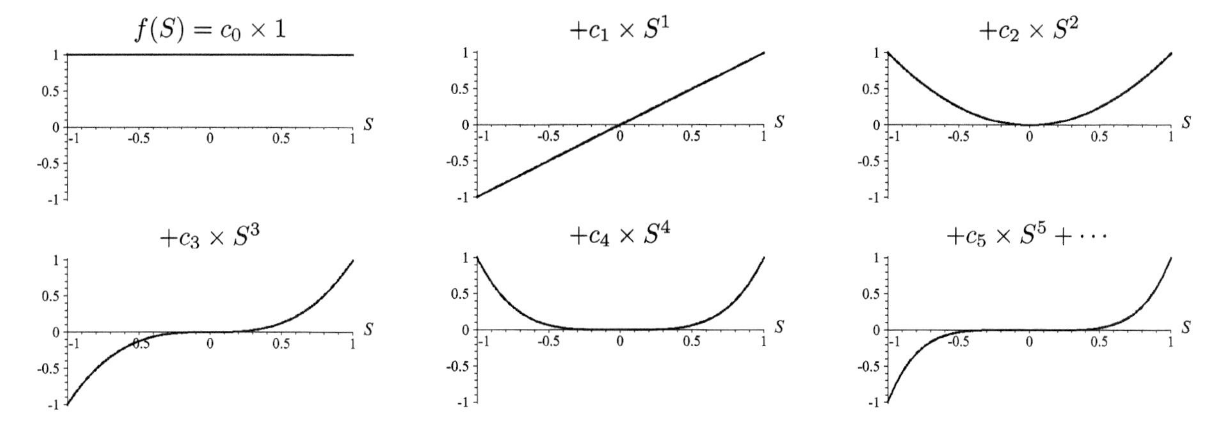

Figure 2.6 *Functional approximations composed of coefficients times powers of a local variable, from (2.36).*

and is illustrated by summation of the series in Fig. 2.6. This represents a linear combination of coefficients times local power series that each varies between ± 1 over the interval.

The solution that minimizes the residual error between the conditions f_m and the function evaluated at $S_m = S(\mathfrak{s}_m)$ is given by

$$\sum_{n=0}^{N} S_m{}^n c_n = f_m + \varepsilon_m \qquad (m = 1, M) \tag{2.37}$$

The solution that minimizes the error in the least squares sense is obtained using the matrices

$$\mathbf{A} = \begin{bmatrix} 1 & S_1 & S_1{}^2 & \ldots & S_1{}^N \\ 1 & S_2 & S_2{}^2 & \ldots & S_2{}^N \\ \cdots & \cdots & \cdots & \cdots & \cdots \\ 1 & S_M & S_M{}^2 & \ldots & S_M{}^N \end{bmatrix}, \quad \mathbf{c} = \begin{bmatrix} c_0 \\ c_1 \\ c_2 \\ \vdots \\ c_N \end{bmatrix}, \quad \mathbf{b} = \begin{bmatrix} f_1 \\ f_2 \\ \vdots \\ f_M \end{bmatrix} \tag{2.38}$$

Each column in the matrix $\mathbf{A}$ in (2.38) contains the value of S^n evaluated at a set of M points along the nth curve in Fig. 2.6. Since these all vary between ± 1, the matrix $\mathbf{A}^{\mathrm{T}}\mathbf{A}$ is comprised of terms all with absolute value less than or equal to 1 irregardless of the order N. The least squares solution for the coefficients c_n is given by

$$\mathbf{A}^{\mathrm{T}}\mathbf{A}\mathbf{c} = \mathbf{A}^{\mathrm{T}}\mathbf{b} \tag{2.39}$$

The approximate function is obtained by finding values of the coefficients c_n by which each function S^n is multiplied, and the function f is approximated by summing the scaled function $c_n S^n$ over all n. The next example illustrates the solutions in Fig. 2.7 for M=11 conditions as the number of coefficients $N + 1$ increases.

As the number of coefficients approaches the number of conditions, $N + 1 \to M$, the objective function decreases. However, the solution may exhibit spurious variations between control points, where f takes on values well outside the limits of the conditions f_m. The

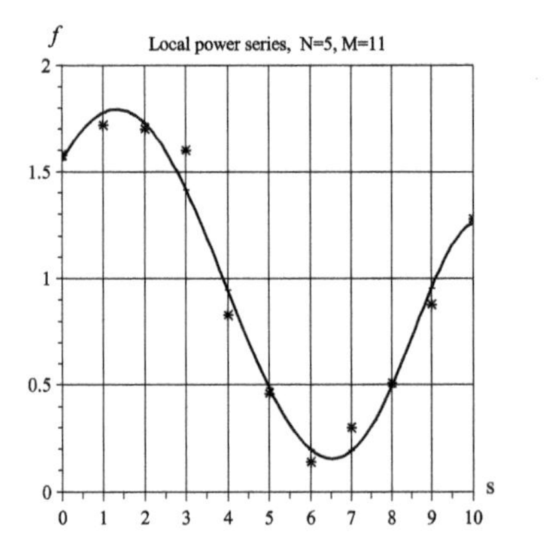

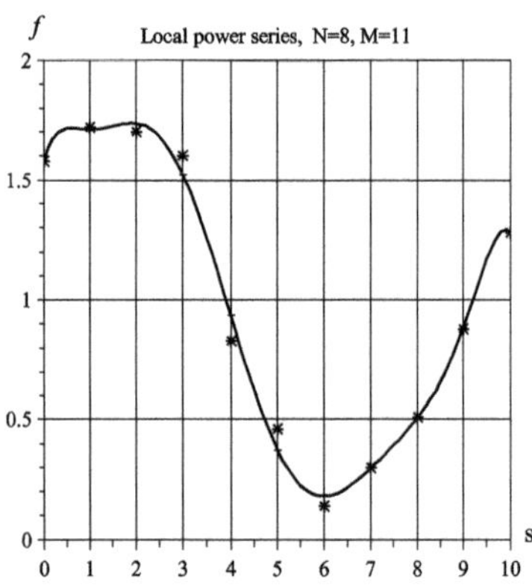

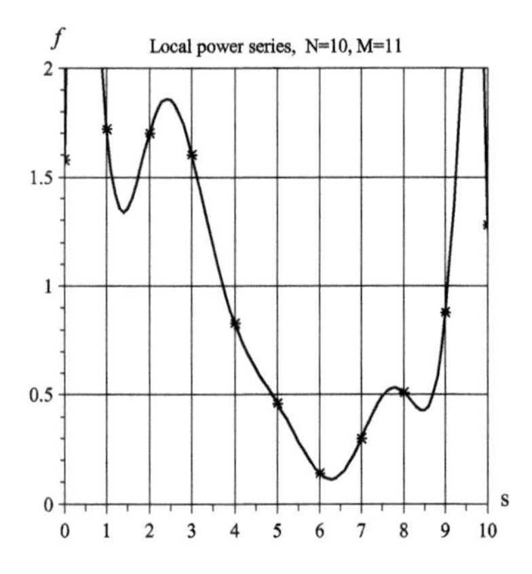

Figure 2.7 *Matching boundary conditions using a local power series.*

overspecification principle of Janković and Barnes (1999*a*) will be adopted throughout this text, whereby the number of conditions M will be approximately 1.5 times the number of unknown conditions. This provides a function with continuous variation that passes through the specified conditions.

Example 2.2 The solution for the data in example 2.1 is shown in Fig. 2.7 for $N = 5$. The least squares solution for these coefficients and the objective function are given by

$$\begin{bmatrix} 11 & 0 & 4.4 & 0 & 3.1328 & 0 \\ 0 & 4.4 & 0 & 3.1328 & 0 & 2.6259 \\ 4.4 & 0 & 3.1328 & 0 & 2.6259 & 0 \\ 0 & 3.1328 & 0 & 2.6259 & 0 & 2.3704 \\ 3.1328 & 0 & 2.6259 & 0 & 2.3704 & 0 \\ 0 & 2.6259 & 0 & 2.3704 & 0 & 2.2270 \end{bmatrix} \begin{bmatrix} c_0 \\ c_1 \\ c_2 \\ c_3 \\ c_4 \\ c_5 \end{bmatrix} = \begin{bmatrix} 11 \\ -2.344 \\ 5.6624 \\ -1.0758 \\ 4.2616 \\ -0.6813 \end{bmatrix} \rightarrow \begin{bmatrix} c_0 \\ c_1 \\ c_2 \\ c_3 \\ c_4 \\ c_5 \end{bmatrix} = \begin{bmatrix} 0.4861 \\ -2.0026 \\ 2.1871 \\ 3.1684 \\ -1.2675 \\ -1.3171 \end{bmatrix}$$

$$\mathscr{F} = 0.0068$$

Figure 2.7 also shows the solution for $N = 8$ with $\mathscr{F} = 0.0026$, and for $N = 10$ with $\mathscr{F} = 0.0000$.

Problem 2.4 Compute the coefficients c_n for $n = 0 \cdots N$ for a local power series with $N = 5$, and the objective function $\mathscr{F}$ that minimizes the sum of the square of the errors for the specified data set in Problem 2.1. Also, plot the data and the approximate function.

2.2.2 Fourier Series: Orthogonal Periodic Solutions

Many influence functions for analytic elements may be approximated along a boundary as a function that repeats over a period 2π using a ***Fourier series***:

$$f(\theta) = c_0 + \sum_{n=1}^{N} \overset{\cos}{c}_n \cos n\theta + \overset{\sin}{c}_n \sin n\theta \tag{2.40}$$

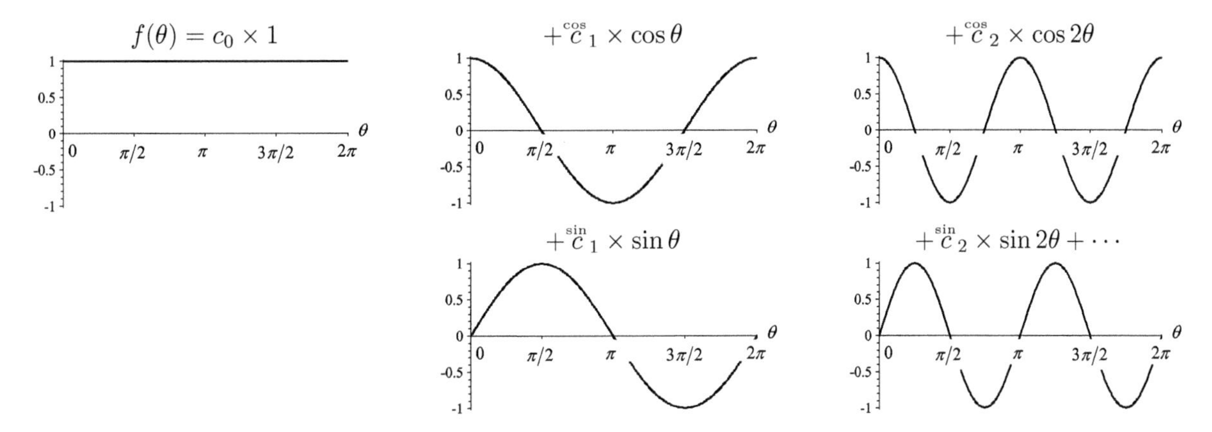

Figure 2.8 *Functional approximations composed of coefficients times Fourier cosine and sine functions, from (2.40).*

where $\overset{\cos}{c}_n$ and $\overset{\sin}{c}_n$ are coefficients to be determined by matching conditions. For example, the location $\mathfrak{s}$ in Example 2.1 may be scaled to occur within the interval $\theta \in [0, 2\pi)$ using

$$\theta = 2\pi \frac{\mathfrak{s} - \mathfrak{s}_1}{M \Delta \mathfrak{s}} \tag{2.41}$$

for M equally spaced intervals $\Delta\mathfrak{s}$. The linear superposition of these repeating functions to approximate functions along a boundary is illustrated in Fig. 2.8.

The coefficients may be determined in a least square sense by minimizing the objective function

$$\mathscr{F} = \frac{1}{M} \sum_{m=1}^{M} \left[c_0 + \sum_{n=1}^{N} \left(\overset{\cos}{c}_n \cos n\theta_m + \overset{\sin}{c}_n \sin n\theta_m \right) - f_m \right]^2 \tag{2.42}$$

where the M conditions f_m occur at the specified locations $\mathfrak{s}_m$, and the Fourier series is truncated at $2N + 1 < M$ terms. This system of equations may be arranged using the matrices (Steward, 2015)

$$\mathbf{A} = \begin{bmatrix} 1 & \cos\theta_1 & \sin\theta_1 & \cdots & \cos N\theta_1 & \sin N\theta_1 \\ 1 & \cos\theta_2 & \sin\theta_2 & \cdots & \cos N\theta_2 & \sin N\theta_2 \\ \cdots & \cdots & \cdots & \cdots & \cdots & \cdots \\ 1 & \cos\theta_M & \sin\theta_M & \cdots & \cos N\theta_M & \sin N\theta_M \end{bmatrix}, \quad \mathbf{c} = \begin{bmatrix} c_0 \\ \overset{\cos}{c}_1 \\ \overset{\sin}{c}_1 \\ \vdots \\ \overset{\cos}{c}_N \\ \overset{\sin}{c}_N \end{bmatrix}, \quad \mathbf{b} = \begin{bmatrix} f_1 \\ f_2 \\ \vdots \\ f_M \end{bmatrix} \tag{2.43}$$

and the least squares solution is given by

$$\mathbf{A}^{\mathrm{T}} \mathbf{A} \mathbf{c} = \mathbf{A}^{\mathrm{T}} \mathbf{b} \tag{2.44}$$

This solution takes on a simple form for series where f_m is specified at constant intervals with

$$\theta_m = 2\pi \frac{m-1}{M} \tag{2.45}$$

In this case, the orthogonality of the Fourier series results in a diagonal matrix $\mathbf{D}$

$$\mathbf{D} = \mathbf{A}^{\mathrm{T}}\mathbf{A} = \begin{bmatrix} M & 0 & 0 & \cdots & 0 \\ 0 & \frac{M}{2} & 0 & \cdots & 0 \\ 0 & 0 & \frac{M}{2} & \cdots & 0 \\ \cdots & \cdots & \cdots & \cdots & \cdots \\ 0 & 0 & 0 & \cdots & \frac{M}{2} \end{bmatrix} \quad \rightarrow \quad \mathbf{D}\mathbf{c} = \mathbf{A}^{\mathrm{T}}\mathbf{b} \tag{2.46}$$

with terms equal to M for the first diagonal entry and $M/2$ for the others. This gives a solution where the unknown coefficients may be computed directly in summation form as

$$c_0 = \frac{1}{M}\sum_{m=1}^{M} f_m\,, \quad \overset{\cos}{c}_n = \frac{2}{M}\sum_{m=1}^{M} f_m \cos n\theta_m\,, \quad \overset{\sin}{c}_n = \frac{2}{M}\sum_{m-1}^{M} f_m \sin n\theta_m \tag{2.47}$$

Example 2.3 The solution for the data in Example 2.1 is shown for Fourier series in Fig. 2.9. The coefficients for the least squares solutions and the objective function for $N = 1$, 3, and 5 are

$$\begin{bmatrix} c_0 \\ \overset{\cos}{c}_1 \\ \overset{\sin}{c}_1 \end{bmatrix} = \begin{bmatrix} \frac{1}{11} & 0 & 0 \\ 0 & \frac{2}{11} & 0 \\ 0 & 0 & \frac{2}{11} \end{bmatrix} \begin{bmatrix} 11 \\ 3.5595 \\ 2.5534 \end{bmatrix} = \begin{bmatrix} 1.0000 \\ 0.6472 \\ 0.4642 \end{bmatrix}, \quad \mathscr{F} = 0.0090$$

$$\begin{matrix} N = 3\,, & c_0 = 1.0 \\ \mathscr{F} = 0.0028 & \end{matrix}\,, \quad \begin{bmatrix} \overset{\cos}{c}_1 \\ \overset{\cos}{c}_2 \\ \overset{\cos}{c}_3 \end{bmatrix} = \begin{bmatrix} 0.6472 \\ -0.0989 \\ 0.0203 \end{bmatrix}, \quad \begin{bmatrix} \overset{\sin}{c}_1 \\ \overset{\sin}{c}_2 \\ \overset{\sin}{c}_3 \end{bmatrix} = \begin{bmatrix} 0.4642 \\ 0.0028 \\ -0.0470 \end{bmatrix}$$

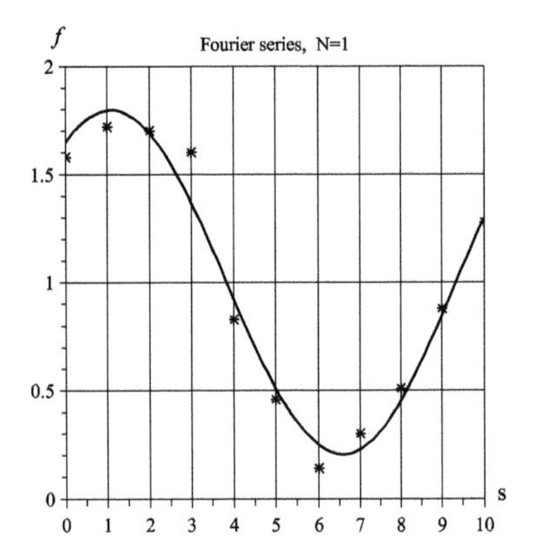

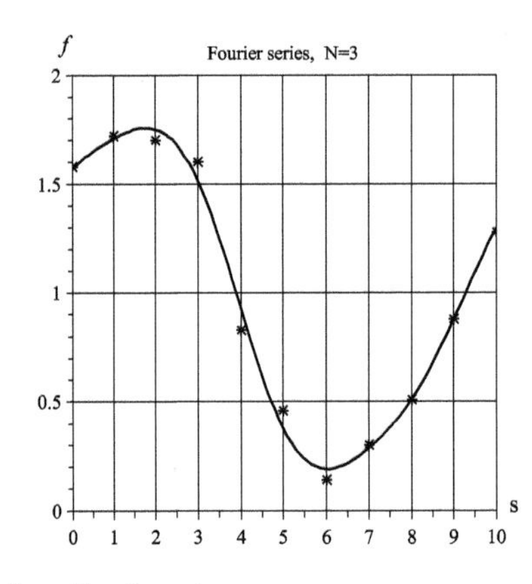

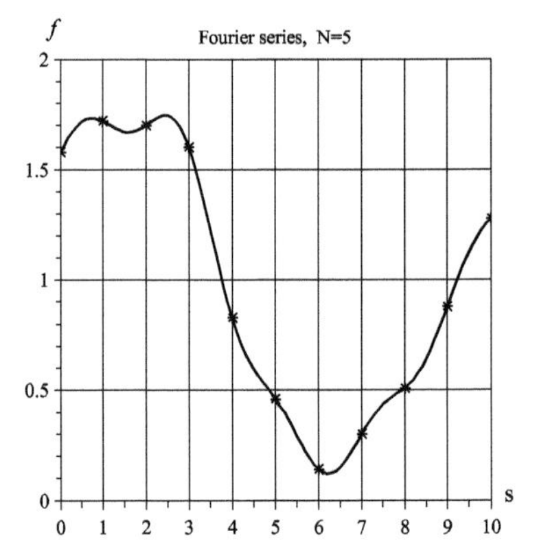

Figure 2.9 *Matching boundary conditions using a Fourier series.*

$$N = 5, \quad c_0 = 1.0 \qquad \mathscr{F} = 0.0000, \quad \begin{bmatrix} \overset{\cos}{c}_1 \\ \overset{\cos}{c}_2 \\ \overset{\cos}{c}_3 \\ \overset{\cos}{c}_4 \\ \overset{\cos}{c}_5 \end{bmatrix} = \begin{bmatrix} 0.6472 \\ -0.0989 \\ 0.0203 \\ 0.0342 \\ -0.0229 \end{bmatrix}, \quad \begin{bmatrix} \overset{\sin}{c}_1 \\ \overset{\sin}{c}_2 \\ \overset{\sin}{c}_3 \\ \overset{\sin}{c}_4 \\ \overset{\sin}{c}_5 \end{bmatrix} = \begin{bmatrix} 0.4642 \\ 0.0028 \\ -0.0470 \\ -0.0057 \\ 0.0616 \end{bmatrix}$$

Problem 2.5 Compute the Fourier coefficients c_0, $\overset{\cos}{c}_n$, and $\overset{\sin}{c}_n$ with $N = 3$ and the objective function $\mathscr{F}$ that minimizes the sum of the square of the errors for the specified data set in Problem 2.1. Also, plot the data and the approximate function.

2.2.3 Non-linear Boundary Conditions

Boundary conditions exist in the AEM where the function $f(\mathfrak{s})$ cannot be written as a linear combination of coefficients times influence functions. An objective function may still be written as the sum of squares of differences between the function evaluated at the M control points and the boundary conditions:

$$\mathscr{F} = \frac{1}{M} \sum_{m=1}^{M} [f(\mathfrak{s}_m) - f_m]^2 \tag{2.48}$$

The solution for the coefficients $\mathbf{c}$ that minimizes this non-linear objective function may be determined using iterative techniques.

The method of ***steepest descent*** identifies the minimum of an objective function by successively traveling a short distance in the direction of minus the gradient of this function. The gradient vector of the objective function $\mathscr{F}$ with respect to the unknown coefficients has components

$$\mathbf{g} = \begin{bmatrix} \frac{\partial \mathscr{F}}{\partial c_0} \\ \frac{\partial \mathscr{F}}{\partial c_1} \\ \vdots \\ \frac{\partial \mathscr{F}}{\partial c_N} \end{bmatrix}, \quad \frac{\partial \mathscr{F}}{\partial c_n} = \frac{2}{M} \sum_{m=1}^{M} \frac{\partial f(\mathfrak{s}_m)}{\partial c_n} [f(\mathfrak{s}_m) - f_m] \tag{2.49}$$

Iteration begins by choosing an initial estimate of the coefficients $\mathbf{c}|_0$, and evaluating the gradient vector for these values. A next estimate of the coefficients $\mathbf{c}|_1$ is found by stepping a small distance in the direction of decreasing $\mathscr{F}$:

$$\mathbf{c}|_1 = \mathbf{c}|_0 - \Delta c\, \mathbf{g}|_{\mathbf{c}|_0} \quad \rightarrow \quad \mathbf{c}|_2 = \mathbf{c}|_1 - \Delta c\, \mathbf{g}|_{\mathbf{c}|_1} \quad \rightarrow \quad \cdots \tag{2.50}$$

This estimate of the coefficients, $\mathbf{c}|_1$ is used to evaluate the gradient vector, and obtain the next iterate, and iteration occurs until only small differences in the objective function $\mathscr{F}$ occur between successive iterates. The derivatives of $\mathscr{F}$ in the gradient vector, (2.49), for iterate l:

$$\mathbf{c}|_{l+1} - \mathbf{c}|_l = -\Delta c\, \mathbf{J}^{\mathrm{T}} \mathbf{f}\Big|_{\mathbf{c}|_l} \tag{2.51}$$

may be organized using the ***Jacobian*** matrix $\mathbf{J}$ and the vector $\mathbf{f}$:

$$\mathbf{J} = \begin{bmatrix} \frac{\partial f(\mathfrak{s}_1)}{\partial c_0} & \frac{\partial f(\mathfrak{s}_1)}{\partial c_1} & \cdots & \frac{\partial f(\mathfrak{s}_1)}{\partial c_N} \\ \frac{\partial f(\mathfrak{s}_2)}{\partial c_0} & \frac{\partial f(\mathfrak{s}_2)}{\partial c_1} & \cdots & \frac{\partial f(\mathfrak{s}_2)}{\partial c_N} \\ \cdots & \cdots & \cdots & \cdots \\ \frac{\partial f(\mathfrak{s}_M)}{\partial c_0} & \frac{\partial f(\mathfrak{s}_M)}{\partial c_1} & \cdots & \frac{\partial f(\mathfrak{s}_M)}{\partial c_N} \end{bmatrix}, \quad \mathbf{f} = \begin{bmatrix} f(\mathfrak{s}_1) - f_1 \\ f(\mathfrak{s}_2) - f_2 \\ \vdots \\ f(\mathfrak{s}_M) - f_M \end{bmatrix} \tag{2.52}$$

Newton's method uses second derivatives of the objective function to iteratively find the coefficients that minimize the objective function. This may be written in a form similar to steepest descent for the lth iterate, (2.51), as

$$\mathbf{H}|_{\mathbf{c}|_l} \left(\mathbf{c}|_{l+1} - \mathbf{c}|_l\right) = -\left.\mathbf{J}^{\mathrm{T}}\mathbf{f}\right|_{\mathbf{c}|_l} \tag{2.53}$$

where the ***Hessian matrix***, $\mathbf{H}$,

$$\mathbf{H} = \begin{bmatrix} \frac{\partial^2 \mathscr{F}}{\partial c_0 \partial c_0} & \frac{\partial^2 \mathscr{F}}{\partial c_0 \partial c_1} & \cdots & \frac{\partial^2 \mathscr{F}}{\partial c_0 \partial c_N} \\ \frac{\partial^2 \mathscr{F}}{\partial c_1 \partial c_0} & \frac{\partial^2 \mathscr{F}}{\partial c_1 \partial c_1} & \cdots & \frac{\partial^2 \mathscr{F}}{\partial c_1 \partial c_N} \\ \cdots & \cdots & \cdots & \cdots \\ \frac{\partial^2 \mathscr{F}}{\partial c_N \partial c_0} & \frac{\partial^2 \mathscr{F}}{\partial c_N \partial c_1} & \cdots & \frac{\partial^2 \mathscr{F}}{\partial c_N \partial c_N} \end{bmatrix} \tag{2.54}$$

containing terms

$$\frac{\partial^2 \mathscr{F}}{\partial c_i \partial c_j} = \frac{2}{M} \sum_{m=1}^{M} \frac{\partial f(\mathfrak{s}_m)}{\partial c_i} \frac{\partial f(\mathfrak{s}_m)}{\partial c_j} + \frac{\partial^2 f(\mathfrak{s}_m)}{\partial c_i \partial c_j} \left[f(\mathfrak{s}_m) - f_m\right] \tag{2.55}$$

is evaluated using the estimates of coefficients for iterate l. It is common to drop the second derivative terms in the Hessian, giving a least squares form

$$\left.\mathbf{J}^{\mathrm{T}}\mathbf{J}\right|_l \left(\mathbf{c}|_{l+1} - \mathbf{c}|_l\right) = -\mathbf{J}^{\mathrm{T}}\mathbf{f}|_l \tag{2.56}$$

The ***Levenberg–Marquardt method*** (Levenberg, 1944; Marquardt, 1963) combines these two methods to use the steepest descent method at large distances from a solution and Newton's method locally. This is accomplished using

$$\left(\mathbf{J}^{\mathrm{T}}\mathbf{J}|_l + \lambda \mathbf{I}\right) \left(\mathbf{c}|_{l+1} - \mathbf{c}|_l\right) = -\mathbf{J}^{\mathrm{T}}\mathbf{f}|_l \tag{2.57}$$

where the identity matrix, $\mathbf{I}$, has values of 1 along the diagonal and 0 elsewhere:

$$\mathbf{I} = \begin{bmatrix} 1 & 0 & \cdots & 0 \\ 0 & 1 & \cdots & 0 \\ \cdots & \cdots & \cdots & \cdots \\ 0 & 0 & \cdots & 1 \end{bmatrix} \tag{2.58}$$

The coefficient λ is adjusted for each iterate following Marquardt (1963):

- pick $\lambda = 0.01$,
- compute the objective function $\mathscr{F}$,
- solve for the next iterate and recompute $\mathscr{F}$,

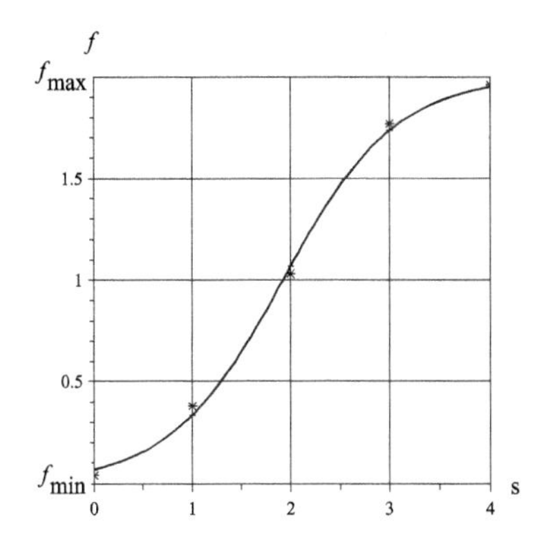

a. Nonlinear approximation

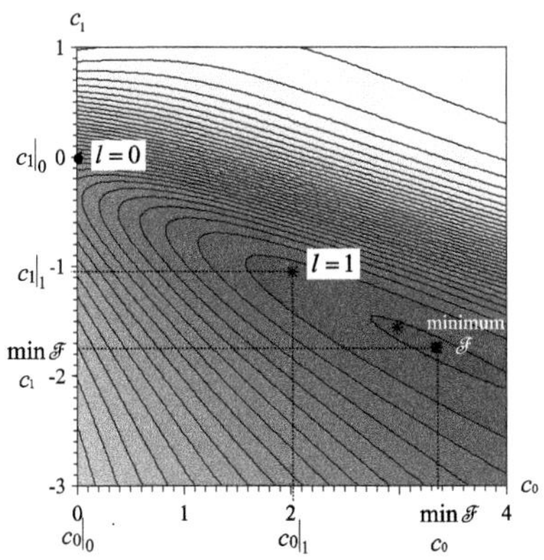

b. Objective function $\mathscr{F}$ and iterates

Figure 2.10 *Matching boundary conditions using Newton's method for nonlinear functions.*

- if the objective function increases, set $\lambda = \lambda\nu$; otherwise, set $\lambda = \lambda/\nu$ where $\nu > 1$, and
- continue solving and adjusting λ for each iterate until the difference between $\mathbf{c}|_l$ and $\mathbf{c}|_{l+1}$ is small.

Example 2.4 Non-linear optimization is illustrated by estimating coefficients in the logistic equation that best fit a set of conditions using Newton's method. The logistic equation forms an S-shaped curve that was developed by Verhulst (1838) to smoothly transition between a minimum f_{min} and maximum f_{max} over changes in $\mathfrak{s}$, and the function and its derivatives with respect to the coefficients are

$$f = f_{min} + \frac{f_{max} - f_{min}}{1 + e^{(c_0 + c_1 \mathfrak{s})}} \,, \quad \frac{\partial f}{\partial c_0} = -\frac{(f_{max} - f_{min})\, e^{(c_0 + c_1 \mathfrak{s})}}{\left[1 + e^{(c_0 + c_1 \mathfrak{s})}\right]^2} \,, \quad \frac{\partial f}{\partial c_1} = -\frac{(f_{max} - f_{min})\, \mathfrak{s} e^{(c_0 + c_1 \mathfrak{s})}}{\left[1 + e^{(c_0 + c_1 \mathfrak{s})}\right]^2}$$

The following conditions and functional approximation are illustrated in Fig. 2.10, along with the objective function, and the iterative progression of coefficients towards the minimum.

m	$\mathfrak{s}_m$	f_m
1	0	0.04
2	1	0.38
3	2	1.03
4	3	1.77
5	4	1.96

The first iterate $l = 0$ in Fig. 2.10 with $f_{min} = 0$ and $f_{max} = 2$ gives

$$\begin{aligned} c_0|_0 &= 0 \\ c_1|_0 &= 0 \\ \mathscr{F} &= 0.5643 \end{aligned} \,, \quad \mathbf{J} = \begin{bmatrix} -0.5 & 0.0 \\ -0.5 & -0.5 \\ -0.5 & -1.0 \\ -0.5 & -1.5 \\ -0.5 & -2.0 \end{bmatrix} \,, \quad \mathbf{f} = \begin{bmatrix} 0.96 \\ 0.62 \\ -0.03 \\ -0.77 \\ -0.96 \end{bmatrix}$$

This gives an estimate for iterate $l = 1$, and progressive iteration leads to the final solution

$$\rightarrow \begin{bmatrix} 1.25 & 2.5 \\ 2.5 & 7.5 \end{bmatrix} \left(\mathbf{c}|_1 - \begin{bmatrix} 0 \\ 0 \end{bmatrix} \right) = - \begin{bmatrix} 0.09 \\ 2.795 \end{bmatrix} \quad \rightarrow \quad \begin{matrix} c_0|_1 = 2.02 \\ c_1|_1 = -1.046 \\ \mathscr{F} = 0.0325 \end{matrix} \quad \rightarrow \quad \cdots \quad \rightarrow \quad \begin{matrix} c_0 = 3.3498 \\ c_1 = -1.7439 \\ \mathscr{F} = 0.0011 \end{matrix}$$

Problem 2.6 Compute the coefficients c_0 and c_1 for the logistic curve with $f_{\min} = 0$ and $f_{\max} = 2$ using estimates for the first iterate $c_0|_0 = 0$ and $c_1|_0 = 0$. Also compute the objective function at the minimum, and plot the function and data.

		A.	B.	C.	D.	E.	F.	G.	H.	I.	J.
m	$\mathfrak{s}_m$	f_m	f_m	f_m	f_m	f_m	f_m	f_m	f_m	f_m	f_m
1	0	0.04	0.13	0.04	0.24	0.12	0.36	0.32	0.13	0.04	0.15
2	1	0.52	0.26	0.23	0.55	0.44	0.56	0.68	0.24	0.54	0.34
3	2	1.03	1.45	1.08	1.00	1.01	1.23	1.11	1.07	1.00	1.09
4	3	1.81	1.94	1.76	1.82	1.77	1.76	1.76	1.78	1.77	1.87
5	4	1.96	1.96	1.96	1.96	1.96	1.96	1.96	1.96	1.96	1.96

2.3 Consistent Notation for Boundary Value Problems

This section develops a common representation for problems and their solutions, which provides a framework for organizing the fields of study in Chapter 1, as well as the mathematical development of analytic elements using complex variables (Chapter 3), separation of variables (Chapter 4), and singular integral equations (Chapter 5).

2.3.1 Boundary and Interface Conditions

A unique solution to a boundary value problem may be obtained by specifying the value of a function and/or its partial derivative at control points along analytic elements. These boundary and interface conditions are presented comprehensively here in terms of the general function f, and examples of their application follow later. The following ***boundary conditions*** are common:

- ***Dirichlet condition*** specifies the value of a function along the boundary

$$f \text{ specified on } \partial D \tag{2.59a}$$

For some problems, a boundary condition exists where the function is know to be ***uniform*** along a boundary but its value is not known until after the solution is obtained:

$$f \text{ uniform on } \partial D \tag{2.59b}$$

This boundary condition is the same as specifying the change of f in the $\mathfrak{s}$-direction tangent to the boundary

$$\frac{\partial f}{\partial \mathfrak{s}} \text{ specified on } \partial D \tag{2.59c}$$

where this derivative is zero if f is uniform along the boundary.

- ***Neumann condition*** contains a ***specified partial derivative of f in the normal $\mathfrak{n}$-direction*** along the boundary

$$\frac{\partial f}{\partial \mathfrak{n}} \text{ specified on } \partial D \tag{2.60}$$

- ***Robin condition*** specifies a linear combination of f and its normal derivative along the boundary:

$$\alpha f + \beta \frac{\partial f}{\partial \mathfrak{n}} \text{ specified on } \partial D \tag{2.61}$$

- ***Cauchy condition*** specifies both a function f and its normal derivative:

$$f \text{ specified and } \frac{\partial f}{\partial \mathfrak{n}} \text{ specified} \tag{2.62}$$

Boundary value problems may contain boundaries where different conditions exist along different parts of the boundary. A ***mixed boundary value problem*** occurs when the boundary is partitioned into elements and one of the above boundary conditions is specified along each element. Time-dependent boundary value problems may specify ***initial conditions*** at locations within the domain:

$$f(\mathbf{x}, t_0) \text{ specified} \qquad \forall \mathbf{x} \in D \tag{2.63}$$

for the initial time t_0,

Problem 2.7 Are the following boundary conditions Dirichlet, Neumann, Robin, and/or Cauchy? Briefly justify your answer.

Groundwater flow (Section 1.2.1)

A. River with specified head

B. Discharge specified into a well

C. Highly conductive slit

Vadose zone flow (Section 1.2.2)

D. Edge with given pressure

E. Edge with given inflow

Incompressible fluid flow (Section 1.2.3)

F. Given pressure

G. Impermeable wall

Thermal conduction (Section 1.2.4)

H. Given temperature

I. Perfect insulator

Electric conduction (Section 1.2.5)

J. Voltage specified

K. Current into domain specified

Water waves (Section 1.3.1)

L. No reflection from coast

M. Free surface boundary

N. Given wave amplitude

Acoustics (Section 1.3.2)

O. Intensity specified

P. Impedance boundary

Elasticity (Section 1.4.1)

Q. Displacement specified

R. Given displacement and derivative of displacement in the normal direction

S. Given force/stress

T. Elastic support

Boundaries also exist at the ***interface of two-sided elements*** as shown in Fig. 2.11 for a line element and an ellipse element. These elements contain two sides, a + side and − side, that each lie within the domain. A sign convention is adopted for line elements, where the + side lies at points immediately above the line in $\mathfrak{s}$-$\mathfrak{n}$ coordinates and the − side is below

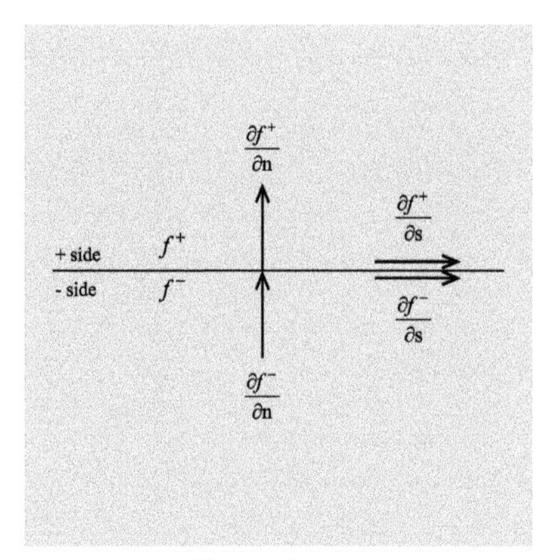

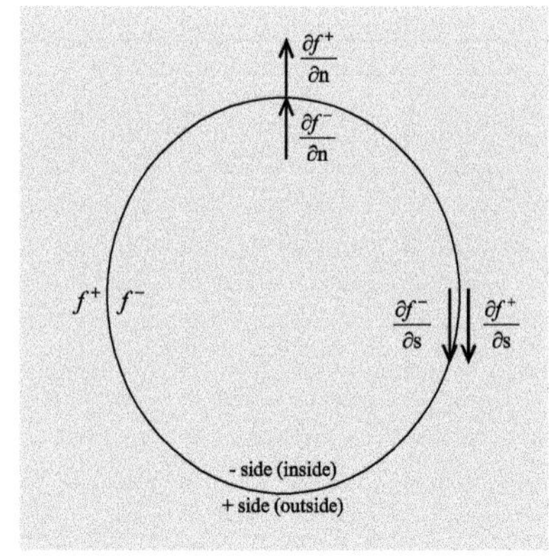

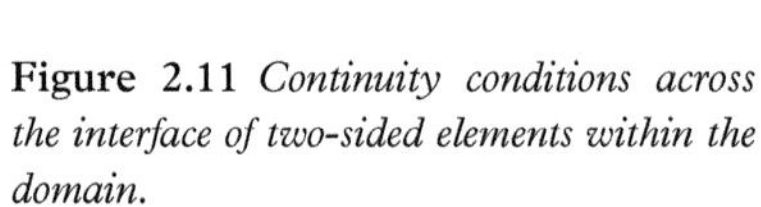

Figure 2.11 *Continuity conditions across the interface of two-sided elements within the domain.*

the line. For elements with outside and inside regions, the $+$ side is outside and the $-$ side is inside the element. Interface conditions are formulated as a ***Riemann–Hilbert problem*** that specifies a relation between the value of a function and its partial derivative across the two sides of an interface. ***Continuity conditions*** for analytic elements may be specified in terms of parameters α, β, and γ:

- ***Continuity of a function***:

$$\alpha^+ f^+ - \alpha^- f^- = \gamma \tag{2.64a}$$

- ***Continuity of the normal derivative of a function***:

$$\beta^+ \frac{\partial f^+}{\partial n} - \beta^- \frac{\partial f^-}{\partial n} = \gamma \tag{2.64b}$$

- ***Continuity of the tangential derivative of a function***:

$$\beta^+ \frac{\partial f^+}{\partial s} - \beta^- \frac{\partial f^-}{\partial s} = \gamma \tag{2.64c}$$

- ***Continuity of a Robin condition***:

$$\alpha^+ f^+ + \beta^+ \frac{\partial f^+}{\partial n} - \left(\alpha^- f^- + \beta^- \frac{\partial f^-}{\partial n} \right) = \gamma \tag{2.64d}$$

Note that the values of the parameters $\alpha^+, \alpha^-, \beta^+, \beta^-$, and γ are developed from the specific properties occurring within each field of study. For some problems, the continuity conditions contain parameters were $\alpha^+ = \alpha^-$ and $\gamma = 0$, and these conditions reduce to ***continuous conditions*** across an interface:

- ***Continuous function***:

$$f^+ = f^- \tag{2.65a}$$

- Continuous normal derivative of the potential or ***continuous normal component of the vector field***:

$$\frac{\partial f^+}{\partial \mathfrak{n}} = \frac{\partial f^-}{\partial \mathfrak{n}}, \quad v_\mathfrak{n}{}^+ = v_\mathfrak{n}{}^- \tag{2.65b}$$

- Continuous tangential derivative of the potential or ***continuous tangential component of the vector field***:

$$\frac{\partial f^+}{\partial \mathfrak{s}} = \frac{\partial f^-}{\partial \mathfrak{s}}, \quad v_\mathfrak{s}{}^+ = v_\mathfrak{s}{}^- \tag{2.65c}$$

In some cases, alternate forms may be used to specify related continuity conditions. For example, the continuous function in (2.65a) would be satisfied for an element with continuous function at some point along the element with a continuous derivative in the tangential direction in (2.65c).

Two-sided interfaces may also specify the normal or tangential component of the vector field in terms of the discontinuity in potential or stream function across the interface.

- The ***Robin interface condition for the normal component of the vector field*** specifies the vector field in the direction normal to the line and a linear relation between the jump in a function across the interface:

$$v_\mathfrak{n} = -\frac{f^+ - f^-}{\delta} \tag{2.66a}$$

where the parameter δ is developed in the fields of study.

- The ***Robin interface condition for the tangential component of the vector field*** specifies a linear relation between the component of the vector field in the direction tangential to the line and the jump in a function across the interface:

$$v_\mathfrak{s} = -\frac{f^+ - f^-}{\delta} \tag{2.66b}$$

Problem 2.8 Do the following interface conditions specify a jump in the function, the normal derivative, and/or tangential derivatives across the interface? Briefly justify your answer.

Groundwater flow (Section 1.2.1)

A. Line element with low conductivity

B. Line element with high conductivity

C. Area of recharge

Vadose zone flow (Section 1.2.2)

D. Jump in conductivity

E. Jump in sorptive number

Incompressible fluid flow (Section 1.2.3)

F. Jump in viscosity

G. Jump in fluid depth

Thermal conduction (Section 1.2.4)

H. Line element with low conductivity

I. Line element with high conductivity

J. Area of heat production

Electric conduction (Section 1.2.5)

K. Line with low resistance

L. Line with high resistance

Water waves (Section 1.3.1)

M. Jump in water depth

N. Harbor entrance loss

Acoustics (Section 1.3.2)

O. Transmission into a softer medium

P. Transmission into a harder medium

Elasticity (Section 1.4.1)

Q. Jump in coefficient of elasticity

R. Jump in Poisson ratio

2.3.2 Mathematical Representation using Complex Variables

It is convenient to represent functions of two variables x and y in terms of a ***complex variable*** z. This variable lies in the plane shown in Fig. 2.12a with a ***real axis*** corresponding to the variable x and an ***imaginary axis*** corresponding to y. A complex variable

$$z = x + \mathrm{i}y \tag{2.67}$$

is obtained by adding the component along the x-axis to the component along the y-axis multiplied by the ***imaginary number***

$$\boxed{\mathrm{i} = \sqrt{-1}} \tag{2.68}$$

A notation is adopted where $\Re z$ represents the ***real part*** of z and $\Im z$ represents the ***imaginary part***:

$$z = \Re z + \mathrm{i}\Im z \tag{2.69}$$

A complex number may also be expressed in terms of polar (r, θ) coordinates where $x = r\cos\theta$ and $y = r\sin\theta$ as shown in Fig. 2.12b:

$$z = r\cos\theta + \mathrm{i}r\sin\theta \tag{2.70}$$

The notation is adopted where the ***modulus***, $|z|$, is the distance from the origin of the complex plane, and the ***argument***, $\arg(z)$, is the angle measured in the θ-direction. This gives

$$z = |z|\mathrm{e}^{\mathrm{i}\arg(z)} \tag{2.71}$$

where use has been made of the ***Euler formula***:

$$\boxed{\mathrm{e}^{\mathrm{i}\theta} = \cos\theta + \mathrm{i}\sin\theta} \tag{2.72}$$

and e is the Euler number corresponding to $\ln(\mathrm{e}) = 1$.

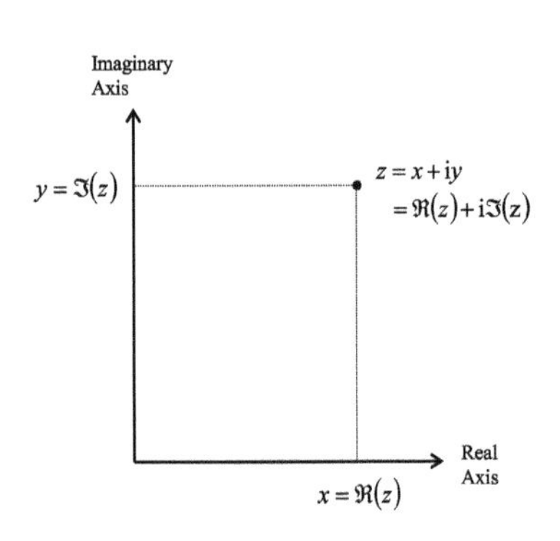

a. Cartesian x-y coordinates

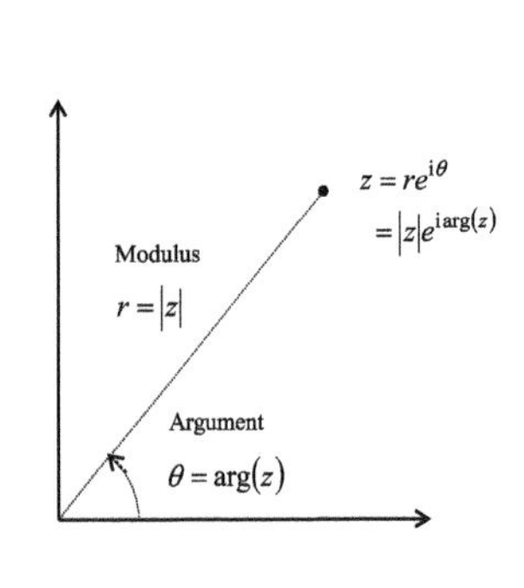

b. Polar r-θ coordinates

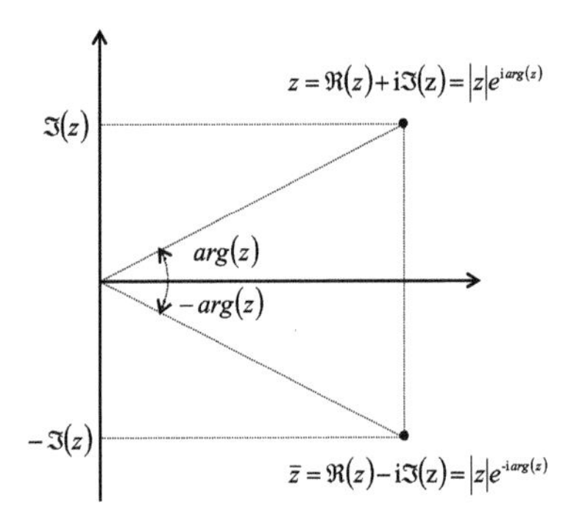

c. Complex conjugate

Figure 2.12 *Complex numbers in Cartesian and polar coordinates.*

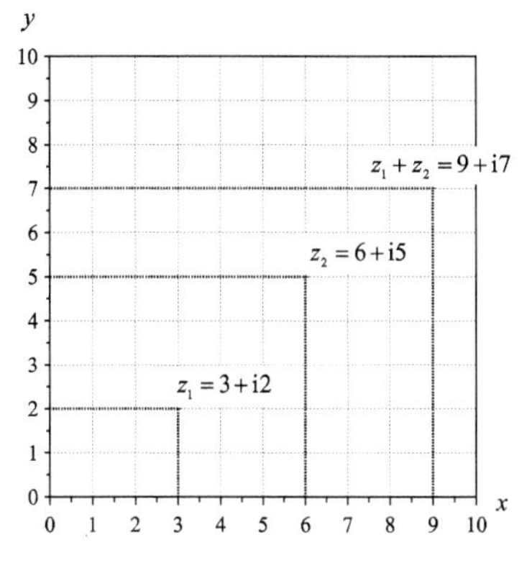

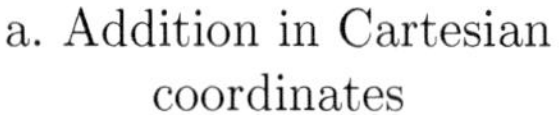

a. Addition in Cartesian coordinates

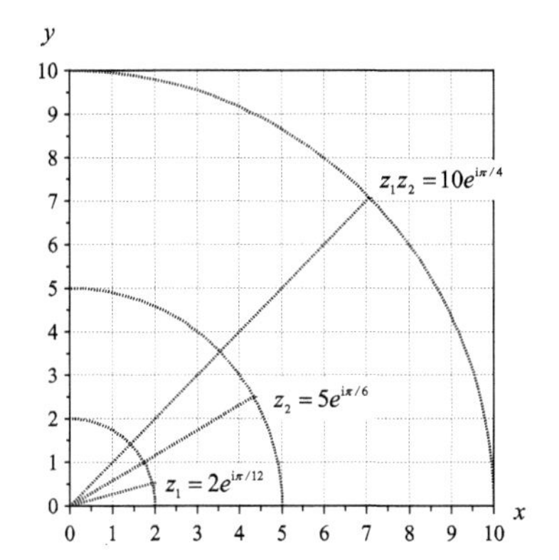

b. Multiplication in polar coordinates

Figure 2.13 *Algebra with complex variables.*

Arithmetic operations using complex variables can be performed using either real and imaginary components or modulus and argument components. Addition of complex variables is easiest to visualize when variables are expressed in their real and imaginary components as shown in Fig. 2.13a (i.e., $z_1 = x_1 + iy_1$ and $z_2 = x_2 + iy_2$). Grouping terms into real and imaginary components gives

$$\Re(z_1 + z_2) = \Re(z_1) + \Re(z_2), \quad \Im(z_1 + z_2) = \Im(z_1) + \Im(z_2) \tag{2.73}$$

Multiplication and division of complex variables is easiest to visualize when variables are expressed using their modulus and argument as shown in Fig. 2.13b (i.e., $z_1 = r_1 e^{i\theta_1}$ and $z_2 = r_2 e^{i\theta_2}$). Multiplication gives $z_1 z_2 = r_1 r_2 e^{i(\theta_1+\theta_2)}$. Thus,

$$|z_1 z_2| = |z_1||z_2|, \quad \arg(z_1 + z_2) = \arg(z_1) + \arg(z_2) \tag{2.74}$$

Division gives $z_1/z_2 = (r_1/r_2)e^{i(\theta_1-\theta_2)}$. Thus,

$$\left|\frac{z_1}{z_2}\right| = \frac{|z_1|}{|z_2|}, \quad \arg(\frac{z_1}{z_2}) = \arg(z_1) - \arg(z_2) \tag{2.75}$$

The conjugate of a complex function is formed by reflection across the real axis as shown in Fig. 2.12c. Thus, the ***complex conjugate*** of z (denoted $\overline{z}$) is

$$\boxed{\overline{z} = x - iy = re^{-i\theta}} \tag{2.76a}$$

The complex conjugate is useful for computing the modulus of a complex variable, since

$$|z| = (z\overline{z})^{1/2} \tag{2.76b}$$

Example 2.5 Addition using complex variables is illustrated in Fig. 2.13a, where

$$\begin{aligned} z_1 &= 3 + \mathrm{i}2 \\ z_2 &= 6 + \mathrm{i}5 \end{aligned} \quad \rightarrow \quad \begin{cases} \Re z_1 = 3 \,, \; \Im z_1 = 2 \\ \Re z_2 = 6 \,, \; \Im z_2 = 5 \end{cases}$$

$$\rightarrow \quad \begin{cases} \Re(z_1 + z_2) = 3 + 6 = 9 \\ \Im(z_1 + z_2) = 2 + 5 = 7 \end{cases}$$

$$\rightarrow \quad z_1 + z_2 = 9 + \mathrm{i}7$$

Multiplication is illustrated in Fig. 2.13b, where

$$\begin{aligned} z_1 &= 2\mathrm{e}^{\mathrm{i}\pi/12} \\ z_2 &= 5\mathrm{e}^{\mathrm{i}\pi/6} \end{aligned} \quad \rightarrow \quad \begin{cases} |z_1| = 2 \,, \; \arg z_1 = \frac{\pi}{12} \\ |z_2| = 5 \,, \; \arg z_2 = \frac{\pi}{6} \end{cases}$$

$$\rightarrow \quad \begin{cases} |z_1 z_2| = 2 \times 5 = 10 \\ \arg(z_1 z_2) = \frac{\pi}{12} + \frac{\pi}{6} = \frac{\pi}{4} \end{cases}$$

$$\rightarrow \quad z_1 z_2 = 10\mathrm{e}^{\mathrm{i}\pi/4}$$

A ***complex function*** Ω may be expressed as a function of z in much the same way as a real function may be expressed as a function of x and y. Just as the complex variable $z = x + \mathrm{i}y$ may be decomposed into its real (x) and imaginary (y) parts, the complex function

$$\boxed{\Omega(z) = \Phi(z) + \mathrm{i}\Psi(z)} \tag{2.77}$$

may be decomposed into a ***real part*** Φ and an ***imaginary part*** Ψ.

It may be shown that Φ is the ***scalar potential*** and Ψ is the ***stream function*** for a two-dimensional vector field as follows. The total derivative of Ω in (2.77) with respect to z may be evaluated in the x-direction (where $z = x$):

$$\frac{\mathrm{d}\Omega}{\mathrm{d}z} = \frac{\mathrm{d}\Omega}{\mathrm{d}x} = \frac{\partial\Omega}{\partial x}\frac{\mathrm{d}x}{\mathrm{d}x} + \frac{\partial\Omega}{\partial y}\frac{\mathrm{d}y}{\mathrm{d}x} = \frac{\partial\Omega}{\partial x} = \frac{\partial\Phi}{\partial x} + \mathrm{i}\frac{\partial\Psi}{\partial x} \tag{2.78a}$$

or in the y-direction (where $z = \mathrm{i}y$):

$$\frac{\mathrm{d}\Omega}{\mathrm{d}z} = \frac{1}{\mathrm{i}}\frac{\mathrm{d}\Omega}{\mathrm{d}y} = \frac{1}{\mathrm{i}}\left(\frac{\partial\Omega}{\partial x}\frac{\mathrm{d}x}{\mathrm{d}y} + \frac{\partial\Omega}{\partial y}\frac{\mathrm{d}y}{\mathrm{d}y}\right) = -\mathrm{i}\frac{\partial\Omega}{\partial y} = -\mathrm{i}\frac{\partial\Phi}{\partial y} + \frac{\partial\Psi}{\partial y} \tag{2.78b}$$

where use has been made of the chain rule for partial derivatives. The condition that $\mathrm{d}\Omega/\mathrm{d}z$ is continuous means that the last two equations are equal, and the real and imaginary terms in these equations give the ***Cauchy–Riemann equations*** (1.14):

$$\begin{aligned} v_x &= -\frac{\partial\Phi}{\partial x} = -\frac{\partial\Psi}{\partial y} \\ v_y &= -\frac{\partial\Phi}{\partial y} = +\frac{\partial\Psi}{\partial x} \end{aligned} \tag{2.79}$$

A ***complex vector field*** v with real part v_x and imaginary part v_y is given by

$$\boxed{v = v_x + \mathrm{i}v_y = -\overline{\frac{\mathrm{d}\Omega}{\mathrm{d}z}}} \tag{2.80}$$

which is minus the complex conjugate of the derivatives in (2.78).

Complex functions $\Omega(z)$, which satisfy the Cauchy–Riemann equations, are called ***analytic*** or ***holomorphic*** functions. The real part $\Phi = \Re\Omega$ is a scalar ***potential function,*** and, by the first corollary to the Helmholtz theorem, (1.6), the vector field with components v_x and v_y is irrotational. The imaginary part $\Psi = \Im\Omega$ is a ***stream function,*** (1.10), which exists for vector fields that are divergence-free. Thus, vector fields generated by complex functions using $v = -\overline{\mathrm{d}\Omega/\mathrm{d}z}$ are both ***divergence-free and irrotational.*** Furthermore, the equipotential lines along which the value of Φ is constant are perpendicular to this vector field and streamlines along which the value of Ψ is constant are tangent to it. Thus, ***equipotential lines and streamlines obtained from the real and imaginary parts of $\Omega(z)$ form two mutually orthogonal families of curves, with one perpendicular to and the other tangent to the vector field.***

Example 2.6 Two exact solutions for boundary value problems using complex variables are shown in Fig. 2.14. The complex function $\Omega = -z^{1/2}$ in Fig. 2.14a satisfies a Dirichlet condition along the negative x-axis, where the potential function is constant when $\theta = \pm\pi$:

$$\Omega = -z^{1/2} = -r^{1/2}\mathrm{e}^{\mathrm{i}\theta/2} = -\sqrt{r}\left(\cos\frac{\theta}{2} + \mathrm{i}\sin\frac{\theta}{2}\right) \quad \rightarrow \quad \Phi = -\sqrt{r}\cos\frac{\theta}{2} = 0$$

and a singularity exists where the vector field becomes infinite at the endpoint of this semi-infinite line. The complex function $\Omega = -z^2$ in Fig. 2.14b satisfies a Neumann condition along the x- and y-axes where the normal component of the vector field is zero:

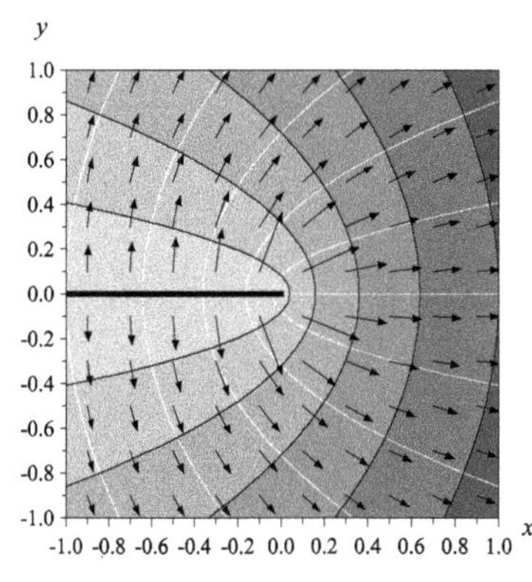

a. $\Omega = -z^{1/2}$ with Dirichlet condition

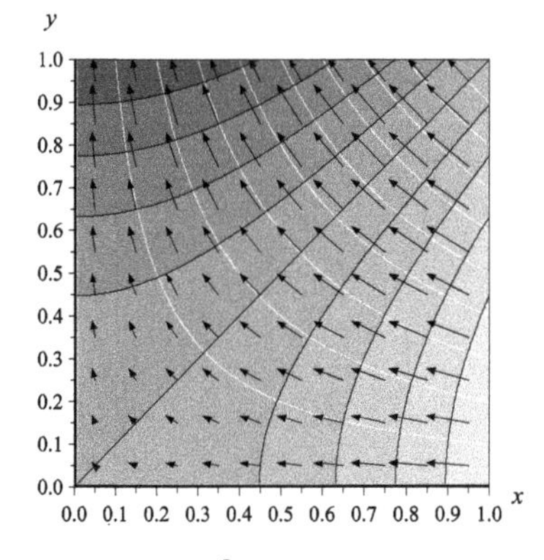

b. $\Omega = z^2$ with Neumann condition

Figure 2.14 *The equipotential lines, streamlines, and vector field for complex functions.*

$$v = -\overline{\frac{\mathrm{d}(z^2)}{\mathrm{d}z}} = -2\overline{z} = -2x + \mathrm{i}2y \quad \rightarrow \quad \begin{cases} v_x = -2x = 0 & (x = 0) \\ v_y = 2y = 0 & (y = 0) \end{cases}$$

and a stagnation point occurs where the vector field becomes zero inside the 90° corner.

Problem 2.9 Compute the real and imaginary parts and the modulus and argument (in units of radians) of z_1, z_2, $z_1 + z_2$, $z_1 - z_2$, $z_1 \times z_2$, and z_1/z_2 for the following complex numbers:

A. $z_1 = 1 + \mathrm{i}1$, $z_2 = 2 + \mathrm{i}2$

B. $z_1 = 3 + \mathrm{i}3$, $z_2 = -4 + \mathrm{i}4$

C. $z_1 = 5 + \mathrm{i}5$, $z_2 = 6 - \mathrm{i}6$

D. $z_1 = 7 + \mathrm{i}7$, $z_2 = -8 - \mathrm{i}8$

E. $z_1 = 1\mathrm{e}^{\mathrm{i}\pi/2}$, $z_2 = 2\mathrm{e}^{\mathrm{i}2\pi/4}$

F. $z_1 = 3\mathrm{e}^{\mathrm{i}3\pi/2}$, $z_2 = 4\mathrm{e}^{\mathrm{i}4\pi/4}$

G. $z_1 = 5\mathrm{e}^{\mathrm{i}5\pi/2}$, $z_2 = 6\mathrm{e}^{\mathrm{i}6\pi/4}$

H. $z_1 = 7\mathrm{e}^{\mathrm{i}7\pi/2}$, $z_2 = 8\mathrm{e}^{\mathrm{i}8\pi/4}$

Problem 2.10 Develop expressions for Φ, Ψ, v_x, and v_y for the following complex functions. Evaluate each of these functions at $z = 1 + \mathrm{i}$.

A. $\Omega = z$

B. $\Omega = z^2$

C. $\Omega = z^3$

D. $\Omega = z^4$

E. $\Omega = z^{-1}$

F. $\Omega = z^{-2}$

G. $\Omega = z^{-3}$

H. $\Omega = z^{=4}$

I. $\Omega = z^{1/2}$

J. $\Omega = z^{1/3}$

K. $\Omega = z^{1/4}$

L. $\Omega = z^{1/5}$

M. $\Omega = z^{-1/2}$

N. $\Omega = z^{-1/3}$

O. $\Omega = z^{-1/4}$

P. $\Omega = z^{-1/5}$

Further Reading

Section 2.1, The Analytical Element Method Paradigm

- de Lange (2006)
- Haitjema (1995)
- Janković and Barnes (1999*a*)
- Kraemer (2007)
- Steward and Bernard (2006)
- Steward and Allen (2013)
- Steward (2015)
- Steward (2016)
- Steward (2018)
- Strack (1989)
- Strack (1999)
- Strack (2003)

Section 2.2, Solving Systems of Equations

- Janković and Barnes (1999*a*)
- Carslaw (1921)
- Levenberg (1944)
- Marquardt (1963)
- Press, Teukolsky, Vetterling, and Flannery (1992)
- Steward, Le Grand, Janković, and Strack (2008)
- Steward, Bruss, Yang, Staggenborg, Welch, and Apley (2013)
- Steward and Allen (2013)
- Steward (2015)
- Steward and Allen (2016)
- Strang (1980)
- Strang (1986)
- Verhulst (1838)

Section 2.3, Consistent Notation

- Abramowitz and Stegun (1972)
- Clebsch (1857)
- Duschek and Hochrainer (1950)

- Einstein (1995)
- Feynman, Leighton, and Sands (1965)
- Gauss (1809)
- Goursat (1904)
- Gradshteyn and Ryzhik (1980)
- Goursat (1904)
- Helmholtz (1881)
- Jeffreys and Jeffreys (1956)
- Lagrange (1768)
- Lagrange (1781)
- Moon and Spencer (1961*b*)
- Morse and Feshbach (1953)
- Sampson (1891)
- Selby (1975)
- Sokolnikoff and Sokolnikoff (1941)
- Sommerfeld (1949)
- Stehfest (1970)
- Stevenson (1954)
- Steward (2002)
- Stokes (1842)
- Strang (2007)
- Thomson (1904)
- Watson (1914)

Section 2.3.2, Complex Variables

- Churchill and Brown (1984)
- Hille (1982)
- Kellogg (1929)
- Muskhelishvili (1953*a*)
- Remmert (1991)
- Watson (1914)
- Weatherburn (1960)

Analytic Elements from Complex Functions

3

- The mathematical functions associated with analytic elements may be formulated using a complex function $\Omega(z)$ of a complex variable $z = x + iy$. This function was introduced in Section 2.3.2 as a solution of the Laplace equation, and it was shown to provide a concise mathematical foundation since the potential function, Φ, stream function, Ψ, and complex vector field $v = v_x + iv_y$ may be obtained directly from Ω.
- Complex formulation of analytic elements is introduced in Section 3.1 for exact solutions obtained by embedding point elements that generate divergence, circulation, or velocity within a uniform vector field.
- Influence functions for analytic elements with circular geometry are obtained using Taylor and Laurent series expansions in Section 3.2, and conformal mapping extends this formulation to analytic elements with the geometry of ellipses (Section 3.3).
- The Courant's Sewing Theorem is employed in Section 3.4 to develop solutions for interface conditions across straight line segments, and the Joukowsky transformation extends methods to circular arcs and wings (Section 3.5), which satisfy a Kutta condition of non-singular vector fields at their trailing edges.
- Vector fields with spatially distributed divergence and curl are formulated using the complex variable z with its complex conjugate $\overline{z}$ in Section 3.6, and the complex conjugate is further employed in the Kolosov formulas (Section 3.7) to solve force deformation problems for analytic elements with traction or displacement specified boundary conditions.

3.1 Point Elements in a Uniform Vector Field

Complex functions with singularities at points form a basis for the development of analytic elements. A ***point-sink*** represents a singularity that withdraws a flux Q at the point $z_p = x_p + iy_p$ and is illustrated in Fig. 3.1a. The complex potential is visualized using potential and stream functions:

$$\Omega = \frac{Q}{2\pi}\ln(z - z_p) \quad \rightarrow \quad \begin{cases} \Phi = \frac{Q}{2\pi}\ln|z - z_p| = \frac{Q}{2\pi}\ln\sqrt{(x - x_p)^2 + (y - y_p)^2} \\ \Psi = \frac{Q}{2\pi}\arg(z - z_p) = \frac{Q}{2\pi}\arctan\frac{y-y_p}{x-x_p} \end{cases} \tag{3.1a}$$

where the complex logarithm has been separated into real and imaginary parts, (2.77), using $\ln z = \ln|z|e^{i\arg(z)} = \ln|z| + i\arg(z)$. A point-sink satisfies a condition of uniform potential

Analytic Element Method: Complex Interactions of Boundaries and Interfaces. David R. Steward, Oxford University Press (2020). © David R. Steward.
DOI: 10.1093/oso/9780198856788.001.0001

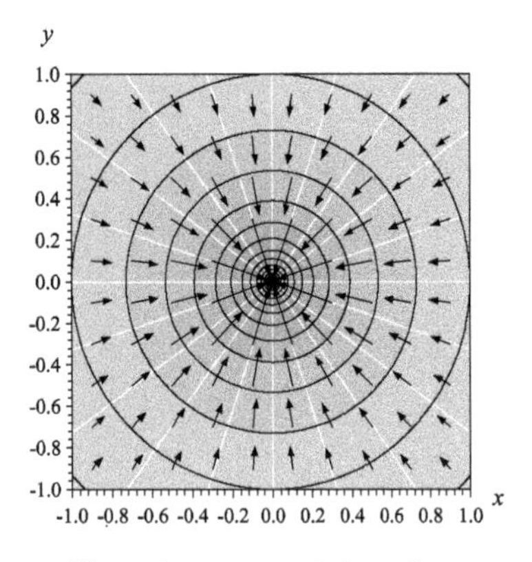

a. Complex potential and vector field

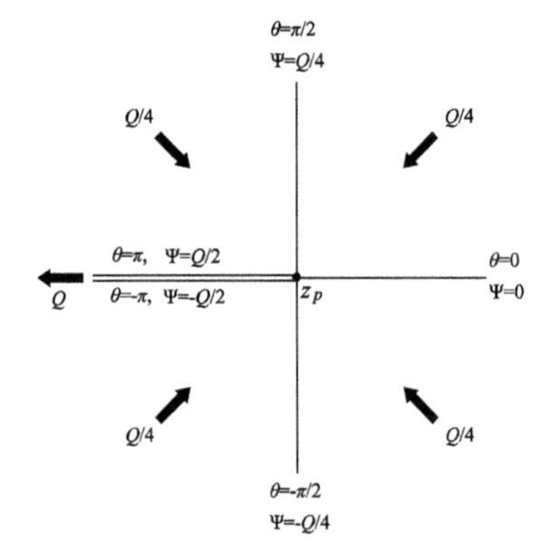

b. Branch cut with jump in stream function

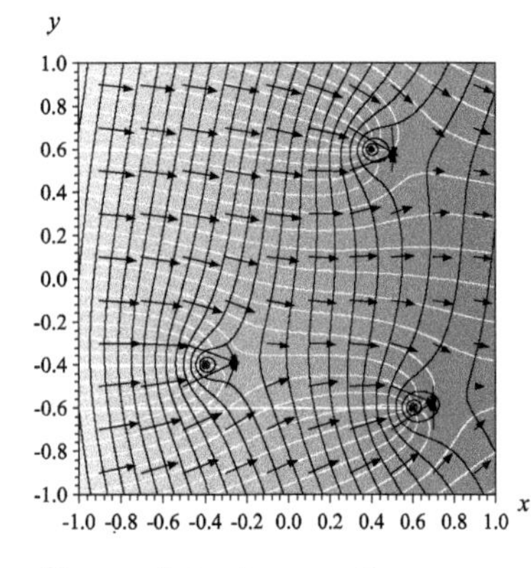

c. Point-sinks in a uniform vector field

Figure 3.1 *Point-sink: removing flux at a point.*

on a circle while withdrawing a flux Q through this boundary, and streamlines form rays about the point-sink. Its vector field, (2.80), is given by

$$v = -\overline{\frac{d\Omega}{dz}} = -\frac{Q}{2\pi}\frac{1}{\overline{z - z_p}} \quad \rightarrow \quad \begin{cases} v_x = -\frac{Q}{2\pi}\frac{\Re(z-z_p)}{|z-z_p|^2} = -\frac{Q}{2\pi}\frac{x-x_p}{(x-x_p)^2+(y-y_p)^2} \\ v_y = -\frac{Q}{2\pi}\frac{\Im(z-z_p)}{|z-z_p|^2} = -\frac{Q}{2\pi}\frac{y-y_p}{(x-x_p)^2+(y-y_p)^2} \end{cases} \tag{3.1b}$$

which have been partitioned into real and imaginary parts by multiplying numerator and denominator by $(z - z_p)$ with (2.76). The Analytic Element Method (AEM) frequently places elements in a background ***uniform vector field***, such as in Fig. 3.1c, and this complex potential and vector field is given by

$$\Omega(z) = -\overline{v_0}z + \Phi_0, \quad v(z) = v_0 = v_{x0} + iv_{y0} \tag{3.2}$$

where v_{x0} and v_{y0} are the real and imaginary components of the uniform vector field, and Φ_0 is a constant.

The stream function for a point-sink contains a ***branch cut*** in the stream function that is described using Fig. 3.1b. This branch cut occurs in the negative y-direction where the $\arg(z - z_s)$ jumps from $-\pi$ to $+\pi$, and Ψ jumps from $-Q/2$ to $+Q/2$. In each of the four quadrants, the difference in stream function is $Q/4$; e.g., $\Psi = Q/4$ along the positive y-axis and $\Psi = 0$ along the positive x-axis, and $\Delta Q = Q/4 - 0 = Q/4$ using (1.13). Thus, 1/4 of the flux to the point-sink occurs through each quadrant. Along the negative x-axis, the stream function jumps from $\Psi = -Q/2$ to $+Q/2$, giving $\Delta Q = Q$ directed away from the point-sink. This ***virtual flux along the branch cut*** serves as a conduit to transport the flux removed by the point-sink to $z = -\infty$ at an infinite velocity, and thus maintain zero divergence at the point-sink. A branch cuts in stream function are evident for the set of point-sinks in a uniform flow in Fig. 3.1c, and provide the virtual flux for each element.

A ***point-vortex***, or vortex filament (Helmholtz, 1858; Tait, 1867), generates circulation at a point with a curl perpendicular to the x–y plane as illustrated in Fig. 3.2a. The complex potential and vector field for a point-vortex with circulation Γ at z_v is

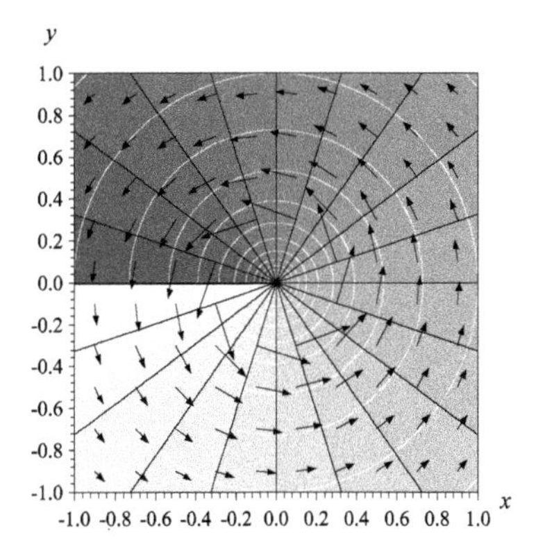

a. Complex potential and vector field

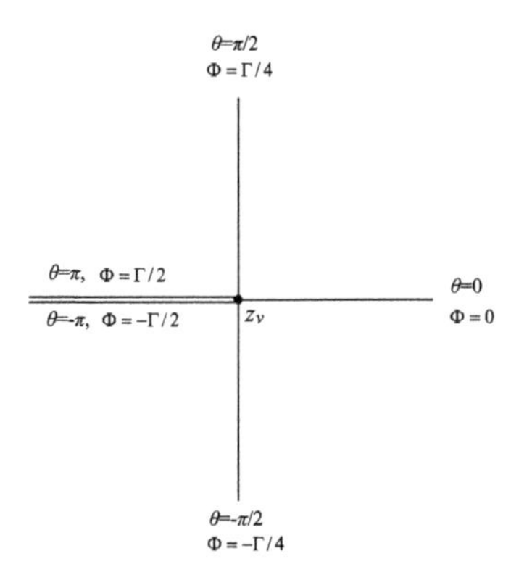

b. Branch cut with jump in potential

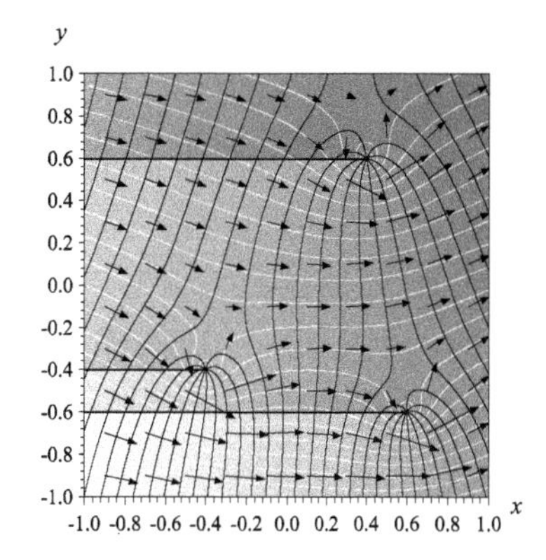

c. Point vortexes in a uniform vector field

Figure 3.2 *Point-vortex: circulation at a point.*

$$\Omega = \frac{\mathrm{i}\Gamma}{2\pi}\ln(z - z_v), \qquad v = \overline{\frac{\mathrm{i}\Gamma}{2\pi}\frac{1}{z - z_v}} \tag{3.3}$$

This element generates a ***branch cut*** in the potential function, which jumps from $-\Gamma/2$ to $+\Gamma/2$ across the negative x-axis as illustrated in Fig. 3.2b. A set of point-vortexes in a uniform vector field is illustrated in Fig. 3.2c, and a point-vortex will be used later to generate circulation about circle elements (in Fig. 3.5b).

A ***point-dipole*** generates a velocity that is infinite in magnitude at the point and oriented in the direction of its axis as illustrated in Fig. 3.3a. Its complex potential and vector field are

$$\Omega = \frac{|S|e^{\mathrm{i}\vartheta}}{2\pi}\frac{1}{z - z_d}, \qquad v = \overline{\frac{|S|e^{\mathrm{i}\vartheta}}{2\pi}\frac{1}{(z - z_d)^2}} \tag{3.4}$$

for an element centered at z_d with strength $|S|$ and orientation of the axis of the dipole ϑ. A point-dipole may be viewed as two point-sinks with equal and opposite strength become very close as illustrated in Fig. 3.3b. This limit occurs as the distance $\Delta\mathfrak{s}$ between point-sinks with magnitude $|S|/\Delta\mathfrak{s}$ approaches zero:

$$\Omega = \lim_{\Delta\mathfrak{s}\to 0}\frac{|S|}{\Delta\mathfrak{s}}\frac{1}{2\pi}\left\{-\ln\left[z - \left(z_d + \frac{\Delta\mathfrak{s}}{2}e^{\mathrm{i}\vartheta}\right)\right] + \ln\left[z - \left(z_d - \frac{\Delta\mathfrak{s}}{2}e^{\mathrm{i}\vartheta}\right)\right]\right\} = \frac{|S|e^{\mathrm{i}\vartheta}}{2\pi}\frac{1}{z - z_d} \tag{3.5a}$$

A point-dipole is also formed by two point-vortexes separated at a distance $\Delta\mathfrak{n}$ along a line at 90° from ϑ where the curl of the point-vortexes are opposite and of magnitude $|S|/\Delta\mathfrak{n}$, and evaluating the limit as the two vortexes approach z_d,

$$\Omega = \lim_{\Delta\mathfrak{n}\to 0}\frac{|S|}{\Delta\mathfrak{n}}\frac{\mathrm{i}}{2\pi}\left\{\ln\left[z - \left(z_d + \mathrm{i}\frac{\Delta\mathfrak{n}}{2}e^{\mathrm{i}\vartheta}\right)\right] - \ln\left[z - \left(z_d - \mathrm{i}\frac{\Delta\mathfrak{n}}{2}e^{\mathrm{i}\vartheta}\right)\right]\right\} = \frac{|S|e^{\mathrm{i}\vartheta}}{2\pi}\frac{1}{z - z_d} \tag{3.5b}$$

reproduces the complex potential of a point-dipole.

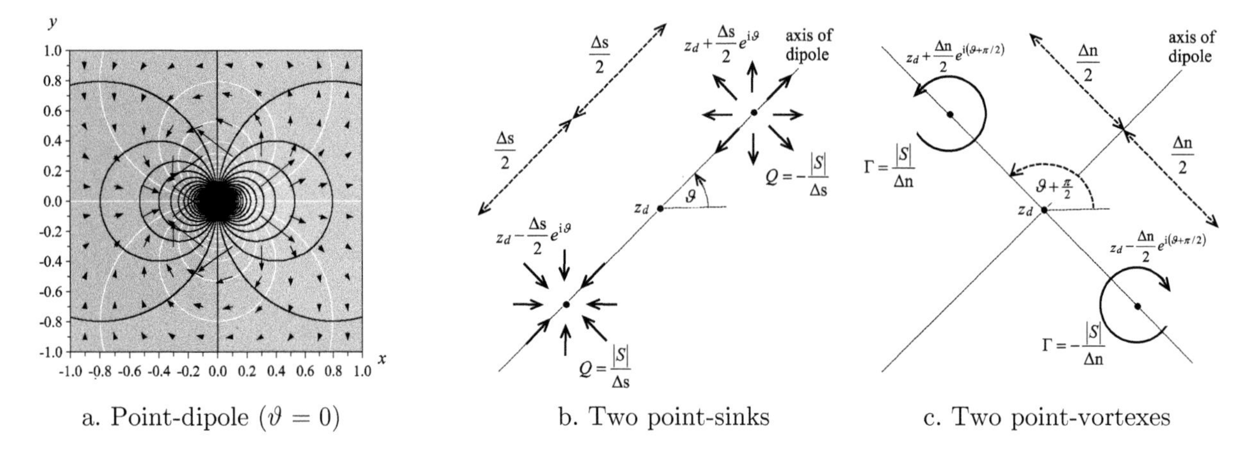

Figure 3.3 *A point-dipole at location z_d is formed in the limit as point-sinks or point-vortexes with infinite magnitude and opposite signs approach one another.*

Exact solutions for an analytic element with circular geometry may be obtained by placing these point elements in a uniform vector field of magnitude $|v_0|$ and direction ϑ, (3.2). For example, placing a point-dipole at z_c with an axis pointing in the same direction as the uniform vector field gives

$$\Omega = -\overline{v_0}z + \frac{v_0 r_0{}^2}{z - z_c}, \quad v = v_0 + \frac{\overline{v_0} r_0{}^2}{(z - z_c)^2} \tag{3.6}$$

which is illustrated in Fig. 3.4a. It may be shown that the potential is uniform along a boundary centered at z_c with radius r_0 by substituting $z = z_c + r_0 e^{i\theta}$ into this equation and solving for Φ. The potential for a point-sink centered at z_c, (3.1), is also uniform along this boundary as illustrated in Fig. 3.4b. Together the point-dipole and point-sink in uniform flow give the exact solution in Fig. 3.4c for a circle element with uniform boundary that removes a flux Q from the domain.

The exact solution for a circular element with an impermeable boundary in uniform flow is obtained by placing a point-dipole with the same magnitude but opposite direction in a uniform flow:

$$\Omega = -\overline{v_0}z - \frac{v_0 r_0{}^2}{z - z_c}, \quad v = v_0 - \frac{\overline{v_0} r_0{}^2}{(z - z_c)^2} \tag{3.7}$$

The components of the vector field normal and tangential to the circle, v_n and v_s, are obtained by rotating $v = v_x + iv_y$,

$$v_s + iv_n = ive^{-i\theta} \tag{3.8}$$

It may be shown that the element is impermeable by evaluating the vector field along the circular boundary at $z = z_0 + r_0 e^{i\theta}$ and showing that $v_n = 0$. A point-vortex centered at z_c, (3.1), is also impermeable as illustrated in Fig. 3.4b. Together the point-dipole and point-vortex in uniform flow give the exact solution in Fig. 3.4c for a circle element with $v_n = 0$ and circulation Γ.

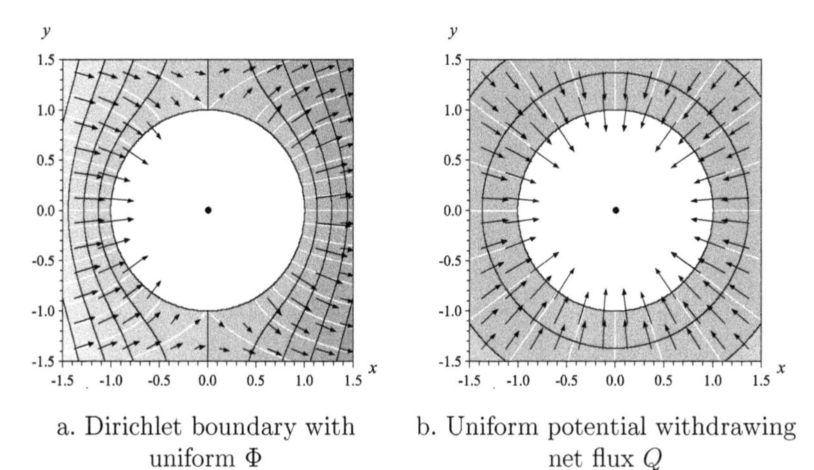

a. Dirichlet boundary with uniform Φ

b. Uniform potential withdrawing net flux Q

c. Dirichlet boundary with net flux

Figure 3.4 *Circle element in a uniform vector field with uniform potential and a net flux.*

Problem 3.1 Compute the potential, Φ, the stream function, Ψ, and the vector components, v_x and v_y, at the points $z_c + r_0$, $z_c + \mathrm{i}r_0$, $z_c - r_0$, and $z_c - \mathrm{i}r_0$, for the circle with uniform potential in Fig. 3.4a with $z_c = 0$ and:

A. $v_0 = 1.1, r_0 = 1$

B. $v_0 = 1.2, r_0 = 2$

C. $v_0 = 1.3, r_0 = 3$

D. $v_0 = 1.4, r_0 = 4$

E. $v_0 = 1.5, r_0 = 5$

F. $v_0 = 1.6, r_0 = 6$

G. $v_0 = 1.1\mathrm{i}, r_0 = 1$

H. $v_0 = 1.2\mathrm{i}, r_0 = 2$

I. $v_0 = 1.3\mathrm{i}, r_0 = 3$

J. $v_0 = 1.4\mathrm{i}, r_0 = 4$

K. $v_0 = 1.5\mathrm{i}, r_0 = 5$

L. $v_0 = 1.6\mathrm{i}, r_0 = 6$

Problem 3.2 Compute the potential, Φ, the stream function, Ψ, and the vector components, v_x and v_y, at the points $z_c + r_0$, $z_c + \mathrm{i}r_0$, $z_c - r_0$, and $z_c - \mathrm{i}r_0$, for the impermeable circle in Fig. 3.5a with $z_c = 1 + \mathrm{i}$, using the specified background vector field v_0 and radius r_0 in Problem 3.1.

Influence Function for Uniform Background

While the constants v_0 and Φ_0 of a uniform background are specified for some problems such as the circle elements just described, it will be convenient to solve for these variables to match prescribed conditions for other problems. This is achieved by separating the complex potential for the uniform background (3.2) from those for the additional elements, (2.7), with complex potential $\overset{\text{add}}{\Omega}$ as

$$\Omega = \Phi_0 + -\overline{v_0} z + \overset{\text{add}}{\Omega} \tag{3.9}$$

The values of Φ_0, v_{x0} and v_{y0} may be determined by setting the real part of Ω equal to specified values of the potential at a set of $M \geq 3$ control points $z_m = x_m + \mathrm{i}y_m$:

$$\Phi_m = \Phi_0 - v_{x0} x_m - v_{y0} y_m + \overset{\text{add}}{\Phi}(z_m) \qquad (m = 1, M) \tag{3.10}$$

This give a system of M equations that may be organized as

$$\mathbf{A}\mathbf{c} = \mathbf{b}v \tag{3.11a}$$

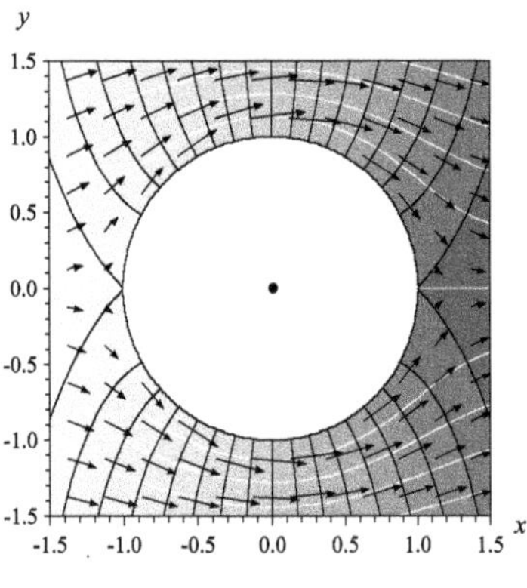

a. Neumann boundary with $v_n = 0$

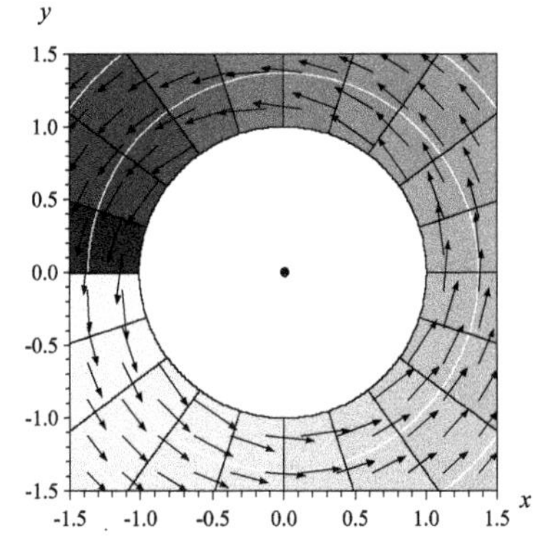

b. Uniform stream function with circulation Γ

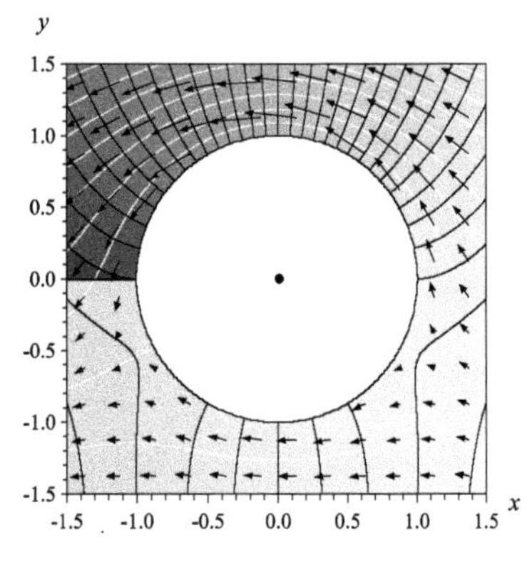

c. Neumann boundary with circulation

Figure 3.5 *Circle element in a uniform vector field with uniform stream function and circulation.*

with

$$\mathbf{A} = \begin{bmatrix} 1 & -x_1 & -y_1 \\ 1 & -x_2 & -y_2 \\ \cdots & \cdots & \cdots \\ 1 & -x_M & -y_M \end{bmatrix}, \quad \mathbf{c} = \begin{bmatrix} \Phi_0 \\ v_{x0} \\ v_{y0} \end{bmatrix}, \quad \mathbf{b} = \begin{bmatrix} \Phi_1 - \overset{\text{add}}{\Phi}(z_1) \\ \Phi_2 - \overset{\text{add}}{\Phi}(z_2) \\ \vdots \\ \Phi_M - \overset{\text{add}}{\Phi}(z_M) \end{bmatrix} \tag{3.11b}$$

This system of equations may be solved via the methods presented in Section 2.2.1 to obtain the variables for the background vector field

$$\mathbf{A}^{\mathrm{T}}\mathbf{A}\mathbf{c} = \mathbf{A}^{\mathrm{T}}\mathbf{b} \tag{3.12}$$

which is expounded in the next example and problem.

Example 3.1 This example illustrates how to obtain the solution for the vector field in Fig. 3.6 for a problem with uniform flow and the additional function set equal to zero. The potential was specified at three control points, with a system of equations given by

$$\begin{aligned} z_1 &= 10 + \mathrm{i}20, & \Phi_1 &= 20 \\ z_2 &= 40 + \mathrm{i}80, & \Phi_2 &= 14, \\ z_3 &= 90 + \mathrm{i}60, & \Phi_3 &= 10 \end{aligned} \qquad \begin{bmatrix} 1 & -10 & -20 \\ 1 & -40 & -80 \\ 1 & -90 & -60 \end{bmatrix} \begin{bmatrix} \Phi_0 \\ v_{x0} \\ v_{y0} \end{bmatrix} = \begin{bmatrix} 20 \\ 14 \\ 10 \end{bmatrix}$$

These equations may be solved to give

$$\Phi_0 = 22, \quad v_{x0} = 0.1, \quad v_{y0} = 0.05$$

Note that if more than three control points are specified, then the system of equations $\mathbf{A}\mathbf{c} = \mathbf{b}$ may be solved via least squares $\mathbf{A}^{\mathrm{T}}\mathbf{A}\mathbf{c} = \mathbf{A}^{\mathrm{T}}\mathbf{b}$ from Section 2.2.1. Once the solution is obtained, the background complex potential and vector field (3.2) may be evaluated, for example at the point $z = 50 + \mathrm{i}50$:

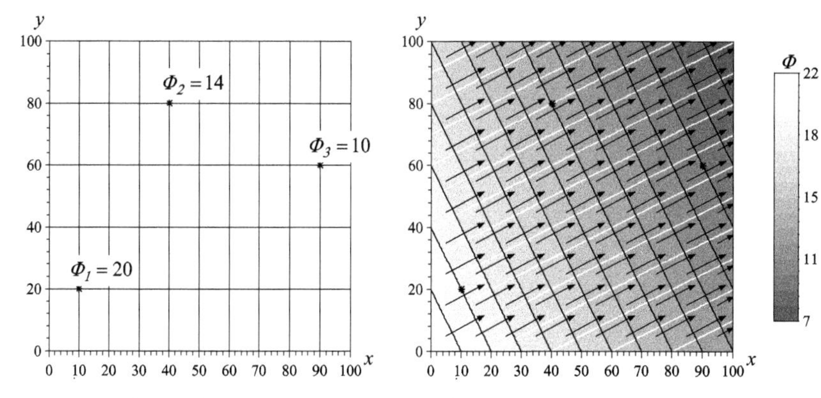

a. Control points   b. Complex potential and vector field

Figure 3.6 *Uniform vector field.*

$$\Omega(50 + i50) = -(0.1 - i0.05)(50 + i50) + 22 = 14.5 - i2.5$$
$$v(50 + i50) = v_0 = 0.1 + i0.05$$

Problem 3.3 Compute the values of Φ_0, v_0, and the complex potential at $z = 50 + i50$ for a uniform background vector field for the specified three or four control points with specified potential:

	A.	B.	C.	D.	E.	F.	G.
z_1	$10 + i20$	$10 + i20$	$10 + i20$	$10 + i10$	$10 + i10$	$10 + i10$	$10 + i10$
z_2	$40 + i80$	$40 + i80$	$40 + i80$	$90 + i10$	$90 + i10$	$90 + i10$	$90 + i10$
z_3	$90 + i60$	$90 + i60$	$90 + i60$	$90 + i90$	$90 + i90$	$90 + i90$	$90 + i90$
z_4				$10 + i90$	$10 + i90$	$10 + i90$	$10 + i90$
Φ_1	10	15	20	20	20	14	15
Φ_2	5	10	15	20	10	15	16
Φ_3	0	5	10	10	10	17	18
Φ_4				10	20	20	21

3.2 Domains with Circular Boundaries

The analytic elements introduced in the last section illustrate how Dirichlet and Neumann conditions may be satisfied along a single circular boundary using a simple mathematical function associated with singularities at points. This section extends these solutions to problems with many interacting elements with the geometry of circles. The mathematical development is easiest to formulate when the ith circle centered at $\underset{i}{z_c}$ with radius $\underset{i}{r_0}$ is translated and scaled to a standard local $\mathcal{Z}$-plane using

$$\mathcal{Z}\left(z, \underset{i}{z_c}, \underset{i}{r_0}\right) = \frac{z - \underset{i}{z_c}}{\underset{i}{r_0}} \tag{3.13}$$

The M control points where boundary conditions are to be applied are located at

$$\theta_m = \theta_1 + 2\pi\frac{m-1}{M}, \quad \mathcal{Z}_m = e^{i\theta_m}, \quad \underset{i}{z_m} = \underset{i}{z_c} + \underset{i}{r_0}e^{i\theta_m} \qquad (m = 1, M) \tag{3.14}$$

and illustrated in Fig. 3.7. Again, the notation convention used throughout this text, (2.13),

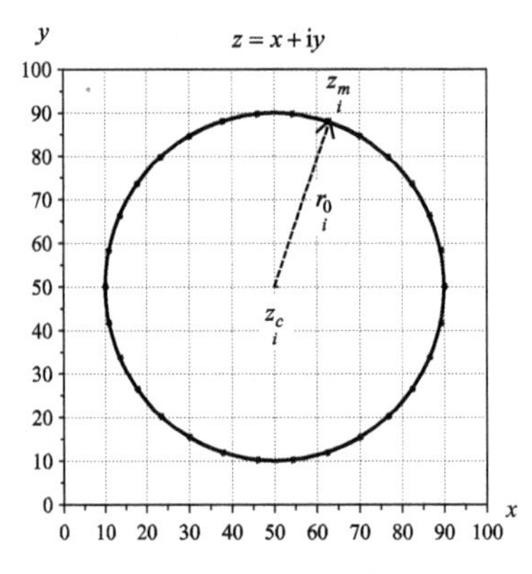

a. Physical z-coordinates

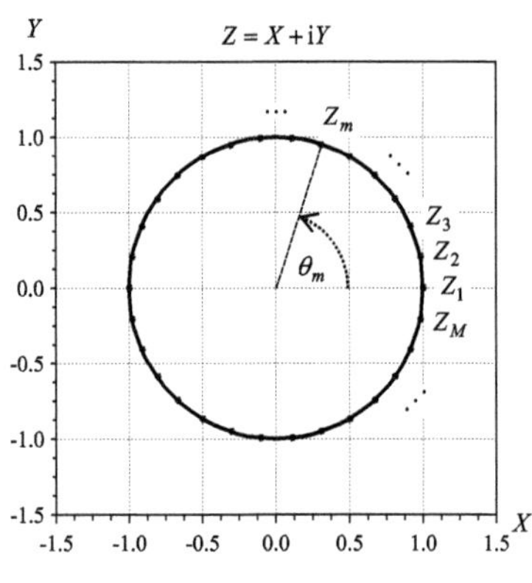

b. Local $\mathcal{Z}$-coordinates

Figure 3.7 *Coordinate systems used to delineate circular elements.*

drops this underscript i except when specifically needed to delineate solutions for problems with many I elements.

3.2.1 Laurent Series for Solutions outside Circles

The exact solutions just presented for circles utilized mathematical functions centered inside the element to achieve solutions to boundary value problems outside a circle. A more general formulation may be achieved for analytic elements with circular geometry using influence functions formed by a Laurent series expansion about its center. This complex potential and vector field is represented in the local $\mathcal{Z}$-plane as

$$\Omega = \sum_{n=1}^{N} \overset{\text{out}}{c}_n \mathcal{Z}^{-n}, \quad v = -\overline{\frac{d\Omega}{dz}} = \overline{\sum_{n=1}^{N} \overset{\text{out}}{c}_n \frac{n}{r_0} \mathcal{Z}^{-n-1}} \qquad (|\mathcal{Z}| \geq 1) \tag{3.15}$$

where summation occurs across the $n > 0$ terms, and the first few influence functions in this series are illustrated in Fig. 3.8. This series may be superimposed with a point-sink, (3.1), which is also shown in this figure; and the solution for circle elements may be represented as the sum of a point-sink that removes a net flux Q through the circular boundary, the Laurent series, and the contribution of all additional elements:

$$\Omega[\mathcal{Z}(z)] = Q\left(\frac{\ln \mathcal{Z}}{2\pi} - 1\right) + \sum_{n=1}^{N} \overset{\text{out}}{c}_n \mathcal{Z}^{-n} + \overset{\text{add}}{\Omega}(z) \tag{3.16}$$

This additional function contains the uniform vector field, (3.2), and the functions associated with the point-sink and Laurent series terms for all other elements. Note that 1 is subtracted from the logarithm term, which is physically intuitive since as the point-sink strength Q becomes larger, the value of the potential decreases along the circle.

A system of equations may be developed to satisfy Dirichlet conditions of prescribed potential, Φ_m, at locations z_m in (3.16) along the boundary, giving

$$\Re\Omega(z_m) = \Re\left[Q\left(\frac{\ln \mathcal{Z}_m}{2\pi} - 1\right) + \sum_{n=1}^{N} \overset{\text{out}}{c}_n (\mathcal{Z}_m)^{-n} + \overset{\text{add}}{\Omega}(z_m)\right] = \Phi_m \tag{3.17}$$

The Laurent series evaluated at these locations take on simpler forms using the exponential terms in (3.14) with Euler formulate (2.72), $\mathcal{Z}^{-n} = e^{-in\theta} = \cos n\theta - i \sin n\theta$, and separating the real and imaginary parts gives

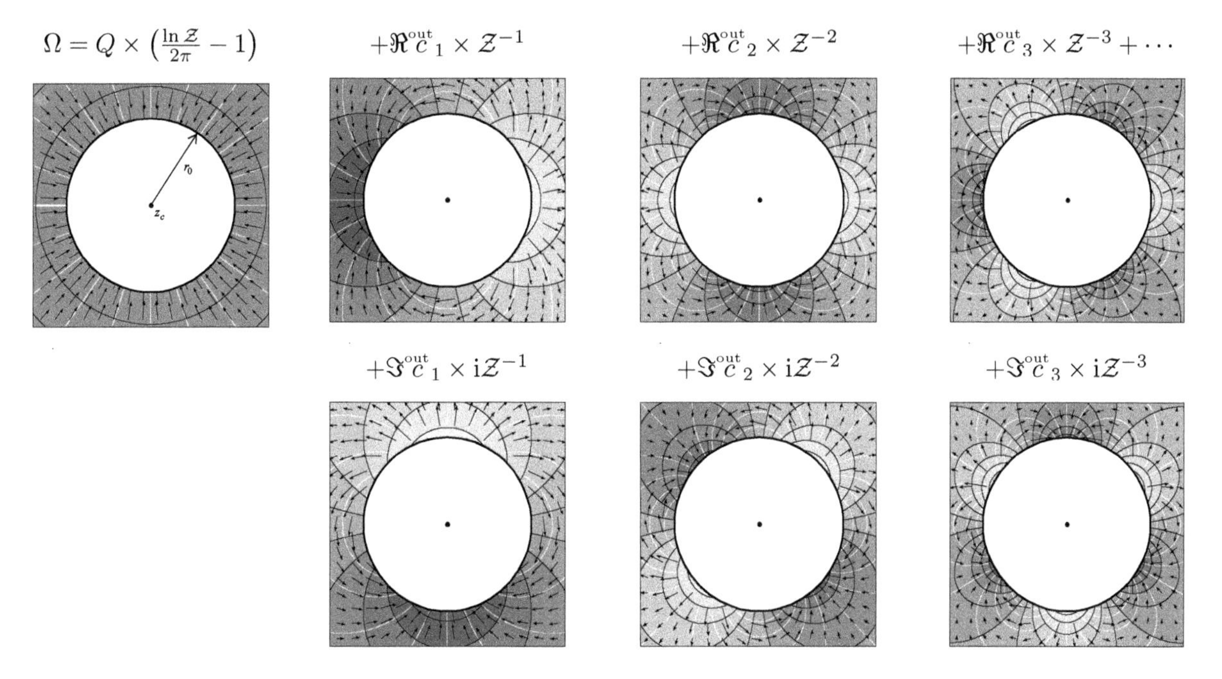

Figure 3.8 *Influence functions outside a circle, from (3.15).*

$$\begin{aligned}\overset{\text{out}}{c}_n(\mathcal{Z}_m)^{-n} &= \overset{\text{out}}{c}_n \mathrm{e}^{-\mathrm{i}n\theta_m} = \left(\Re\overset{\text{out}}{c}_n + \mathrm{i}\Im\overset{\text{out}}{c}_n\right)(\cos n\theta_m - \mathrm{i}\sin n\theta_m)\\ &= \left(\Re\overset{\text{out}}{c}_n \times \cos n\theta_m + \Im\overset{\text{out}}{c}_n \times \sin n\theta_m\right) + \mathrm{i}\left(-\Re\overset{\text{out}}{c}_n \times \sin n\theta_m + \Im\overset{\text{out}}{c}_n \times \cos n\theta_m\right)\end{aligned} \tag{3.18}$$

where the coefficients have been separated into real and imaginary parts, $c_n = \Re c_n + \mathrm{i}\Im c_n$, This gives an expression for the potential at control point m on the outside edge of the circle

$$\begin{aligned}\Phi_m^+ &= \Re\left[Q\left(\frac{\ln \mathcal{Z}_m}{2\pi} - 1\right) + \sum_{n=1}^{N} \overset{\text{out}}{c}_n \overset{\text{out}}{\Omega}_n(z_m) + \overset{\text{add}}{\Omega}(z_m)\right]\\ &= -Q + \sum_{n=1}^{N} \cos n\theta_m \Re\overset{\text{out}}{c}_n + \sin n\theta_m \Im\overset{\text{out}}{c}_n + \overset{\text{add}}{\Phi}(z_m)\end{aligned} \tag{3.19a}$$

and the value of the stream function at this location will be used later

$$\begin{aligned}\Psi_m^+ &= \Im\left[Q\left(\frac{\ln \mathcal{Z}_m}{2\pi} - 1\right) + \sum_{n=1}^{N} \overset{\text{out}}{c}_n \overset{\text{out}}{\Omega}_n(z_m) + \overset{\text{add}}{\Omega}(z_m)\right]\\ &= -Q\frac{\theta_m}{2\pi} + \sum_{n=1}^{N} -\sin n\theta_m \Re\overset{\text{out}}{c}_n + \cos n\theta_m \Im\overset{\text{out}}{c}_n + \overset{\text{add}}{\Psi}(z_m)\end{aligned} \tag{3.19b}$$

Solutions to the boundary condition for a circle with prescribed potential are obtained from the linear relation of coefficients, (3.19a), with the boundary conditions, (3.17)

$$-Q + \sum_{n=1}^{N} \cos n\theta_m \Re \overset{\text{out}}{c}_n + \sin n\theta_m \Im \overset{\text{out}}{c}_n + \overset{\text{add}}{\Phi}(z_m) = \Phi_m \tag{3.20}$$

This system of equations is organized in the matrices $\mathbf{Ac} = \mathbf{b}$ with

$$\mathbf{A} = \begin{bmatrix} 1 & \cos\theta_1 & \sin\theta_1 & \cdots & \cos N\theta_1 & \sin N\theta_1 \\ 1 & \cos\theta_2 & \sin\theta_2 & \cdots & \cos N\theta_2 & \sin N\theta_2 \\ \cdots & \cdots & \cdots & \cdots & \cdots & \cdots \\ 1 & \cos\theta_M & \sin\theta_M & \cdots & \cos N\theta_M & \sin N\theta_M \end{bmatrix}, \quad \mathbf{c} = \begin{bmatrix} -Q \\ \Re \overset{\text{out}}{c}_1 \\ \Im \overset{\text{out}}{c}_1 \\ \vdots \\ \Re \overset{\text{out}}{c}_N \\ \Im \overset{\text{out}}{c}_N \end{bmatrix}, \quad \mathbf{b} = \begin{bmatrix} \Phi_1^- - \overset{\text{add}}{\Phi}(z_1) \\ \Phi_2^- - \overset{\text{add}}{\Phi}(z_2) \\ \vdots \\ \Phi_M^- - \overset{\text{add}}{\Phi}(z_M) \end{bmatrix} \tag{3.21}$$

When the M points are evenly spaced around the circle using (3.14), the orthogonality of the Fourier series gives a solution similar to (2.47):

$$Q = -\frac{1}{M}\sum_{m=1}^{M}\left[\Phi_m^- - \overset{\text{add}}{\Phi}(z_m)\right], \quad \overset{\text{out}}{c}_n = \frac{2}{M}\sum_{m=1}^{M} e^{in\theta_m}\left[\Phi_m^- - \overset{\text{add}}{\Phi}(z_m)\right] \tag{3.22}$$

The solution obtained for a single element in Fig. 3.9a uses a single term in the Laurent series expansion with the same form as the point-dipole used for the exact solution in Fig. 3.4b, and will be explored in the next example and problem. A common problem is for multiple circular elements to be imbedded within a uniform background, as illustrated in Figs. 3.9b and 3.9c, where the complex potential is obtained by summing contributions for each of the I circle boundaries. These solutions were obtained using the iterative algorithm explained earlier with (2.9), where the additional function at the control points contains the contributions from the uniform vector field and all elements except the one being solved

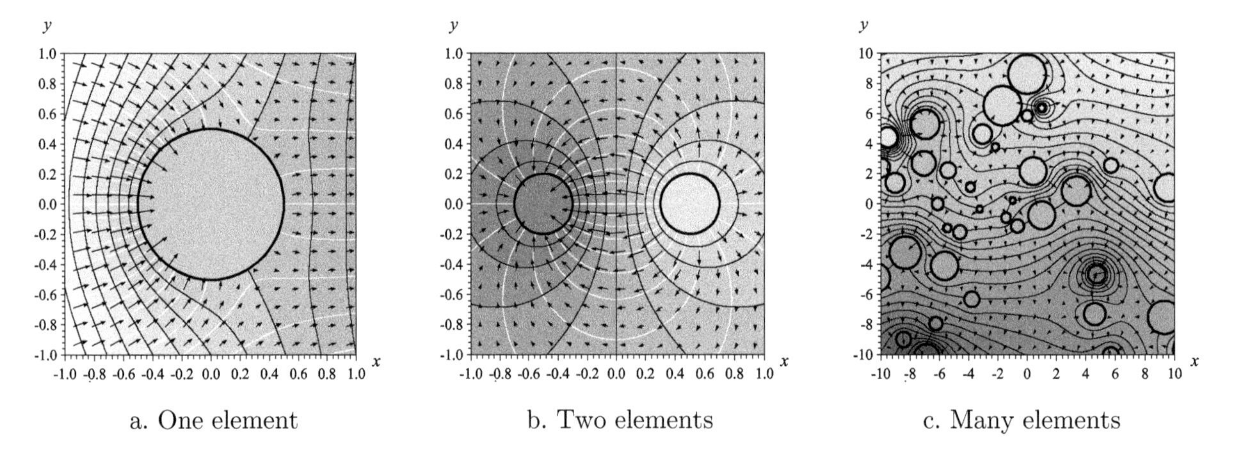

Figure 3.9 *Domain outside circles with specified potential.*

in (3.22). Solving progresses through all $i = 1, \cdots, I$ circles using Gauss–Seidel, and is repeated until the differences between successive iterates are small.

Example 3.2 The circle in Fig. 3.9a is centered at $z_c = 0$ and has radius $r_0 = 0.5$ and the potential at $M = 60$ control points is $\Phi_m = 1.2$. The uniform background is specified at the four corner points of this figure where $\Phi_1 = \Phi_2 = 2$ at $z_1 = -1 + \mathrm{i}$ and $z_2 = -1 - \mathrm{i}$, and $\Phi_3 = \Phi_4 = 1$ at $z_3 = 1 - \mathrm{i}$ and $z_4 = 1 + \mathrm{i}$. The solution is given by $\Phi_0 = 3.0129$ and $v_0 = 0.5714$ with $Q = 1.8129$, $\overset{\text{out}}{c}_1 = 0.2857$, and all other coefficients equal to zero.

Problem 3.4 Compute the coefficients Q and $\overset{\text{out}}{c}_1$ to satisfy a circular element with uniform potential control points $\Phi_m = 1$ in a uniform vector field for the additional function $\overset{\text{add}}{\Phi}(z) = -z$, where

A. $z_c = 0, r_0 = 1$	D. $z_c = 1, r_0 = 1$	G. $z_c = 2, r_0 = 1$	J. $z_c = 3, r_0 = 1$
B. $z_c = 0, r_0 = 2$	E. $z_c = 1, r_0 = 2$	H. $z_c = 2, r_0 = 2$	K. $z_c = 3, r_0 = 2$
C. $z_c = 0, r_0 = 3$	F. $z_c = 1, r_0 = 3$	I. $z_c = 2, r_0 = 3$	L. $z_c = 3, r_0 = 3$

Problem 3.5 Compute the coefficients Φ_0, v_0, Q, and $\overset{\text{out}}{c}_1$ to satisfy the following boundary conditions for the uniform vector field control points and circle geometry specified in Example 3.2.

	A.	B.	C.	D.	E.	F.	G.	H.	I.	J.	K.	L.	M.
Φ_1	3	4	5	6	2	3	4	5	6	3	4	5	6
Φ_2	3	4	5	6	1	1	1	1	1	2	3	4	5
Φ_3	1	1	1	1	1	1	1	1	1	1	2	3	4
Φ_4	1	1	1	1	2	3	4	5	6	2	3	4	5

Note that this solution requires an iterative procedure that independently solves for the uniform flow components and the element components, following (2.9).

3.2.2 Taylor Series for Solutions inside a Circle

An analytic element may also be formulated for problems inside a circle, where the potential is specified along the boundary of the domain. The influence functions for this element are obtained from a Taylor series expansion about the center of the circle, with complex potential and vector field given by

$$\Omega = \sum_{n=0}^{N} \overset{\text{in}}{c}_n Z^n, \quad v = -\overline{\sum_{n=1}^{N} \overset{\text{in}}{c}_n \frac{n}{r_0} Z^{n-1}} \qquad (|Z| < 1) \tag{3.23}$$

These influence functions remain finite inside the circle as shown in Fig. 3.10. These terms may be combined with additional functions specified inside the circle, and a solution is obtained by adjusting the coefficients to match a set of prescribed boundary conditions with specified potential Φ_m at control point m:

$$\Re\Omega(z_m) = \Re\left[\sum_{n=0}^{N} \overset{\text{in}}{c}_n (Z_m)^n + \overset{\text{add}}{\Omega}(z_m)\right] = \Phi_m \tag{3.24}$$

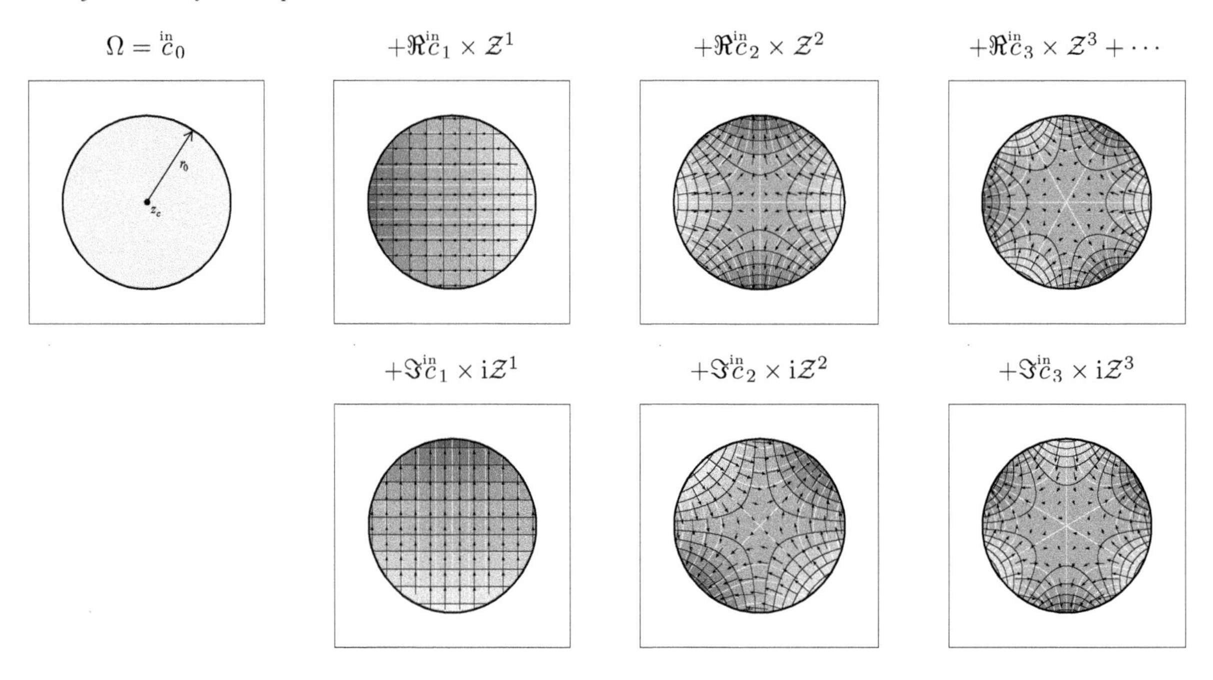

Figure 3.10 *Influence functions inside a circle, from (3.23).*

These Taylor series terms may be separated into real and imaginary parts

$$\begin{aligned}\overset{\text{in}}{c}_n(\mathcal{Z}_m)^n &= \overset{\text{in}}{c}_n e^{\mathrm{i}n\theta_m} = \left(\Re\overset{\text{in}}{c}_n + \mathrm{i}\Im\overset{\text{in}}{c}_n\right)(\cos n\theta_m + \mathrm{i}\sin n\theta_m)\\ &= \left(\Re\overset{\text{in}}{c}_n \times \cos n\theta_m - \Im\overset{\text{in}}{c}_n \times \sin n\theta_m\right) + \mathrm{i}\left(\Re\overset{\text{in}}{c}_n \times \sin n\theta_m + \Im\overset{\text{in}}{c}_n \times \cos n\theta_m\right)\end{aligned} \tag{3.25}$$

where use has been made of $\mathcal{Z}^n = e^{\mathrm{i}n\theta} = \cos n\theta + \mathrm{i}\sin n\theta$. This gives expressions at control points along the boundary inside the circle for the potential

$$\Phi_m^- = \Re\left[\sum_{n=0}^{N} \overset{\text{in}}{c}_n \overset{\text{in}}{\Omega}_n(z_m) + \overset{\text{add}}{\Omega}(z_m)\right] = \sum_{n=0}^{N} \cos n\theta_m \Re\overset{\text{in}}{c}_n - \sin n\theta_m \Im\overset{\text{in}}{c}_n + \overset{\text{add}}{\Phi}(z_m) \tag{3.26a}$$

and the stream function

$$\Psi_m^- = \Im\left[\sum_{n=0}^{N} \overset{\text{in}}{c}_n \overset{\text{in}}{\Omega}_n(z_m) + \overset{\text{add}}{\Omega}(z_m)\right] = \sum_{n=0}^{N} \sin n\theta_m \Re\overset{\text{in}}{c}_n + \cos n\theta_m \Im\overset{\text{in}}{c}_n + \overset{\text{add}}{\Psi}(z_m) \tag{3.26b}$$

Solutions inside a circle with prescribed potential are obtained from the linear relation of coefficients, (3.26a), with the boundary conditions, (3.24),

$$\sum_{n=0}^{N} \cos n\theta_m \Re\overset{\text{in}}{c}_n - \sin n\theta_m \Im\overset{\text{in}}{c}_n = \Phi_m - \overset{\text{add}}{\Phi}(z_m) \tag{3.27}$$

This provides a linear system of equations $\mathbf{Ac} = \mathbf{b}$ that may be written in matrix form as

$$\mathbf{A} = \begin{bmatrix} 1 & \cos\theta_1 & \sin\theta_1 & \cdots & \cos N\theta_1 & \sin N\theta_1 \\ 1 & \cos\theta_2 & \sin\theta_2 & \cdots & \cos N\theta_2 & \sin N\theta_2 \\ \cdots & \cdots & \cdots & \cdots & \cdots & \cdots \\ 1 & \cos\theta_M & \sin\theta_M & \cdots & \cos N\theta_M & \sin N\theta_M \end{bmatrix}, \quad \mathbf{c} = \begin{bmatrix} \Re\overset{\text{in}}{c}_0 \\ \Re\overset{\text{in}}{c}_1 \\ \Im\overset{\text{in}}{c}_1 \\ \vdots \\ \Re\overset{\text{in}}{c}_N \\ \Im\overset{\text{in}}{c}_N \end{bmatrix}, \quad \mathbf{b} = \begin{bmatrix} \Phi_1^- - \overset{\text{add}}{\Phi}(z_1) \\ \Phi_2^- - \overset{\text{add}}{\Phi}(z_2) \\ \vdots \\ \Phi_M^- - \overset{\text{add}}{\Phi}(z_M) \end{bmatrix} \tag{3.28}$$

When the M points are evenly spaced around the circle using (3.14), the orthogonality of the Fourier series (2.46) gives a least squares solution which may be expressed in summation form as

$$\Re\overset{\text{in}}{c}_0 = \frac{1}{M}\sum_{m=1}^{M}\left[\Phi_m^- - \overset{\text{add}}{\Phi}(z_m)\right] \quad \rightarrow \quad \overset{\text{in}}{c}_0 = \frac{1}{M}\sum_{m=1}^{M}\left[\Phi_m^- - \overset{\text{add}}{\Phi}(z_m)\right] \tag{3.29a}$$

and for $n > 0$

$$\left.\begin{aligned} \Re\overset{\text{in}}{c}_n &= \frac{2}{M}\sum_{m=1}^{M}\cos n\theta_m\left[\Phi_m^- - \overset{\text{add}}{\Phi}(z_m)\right] \\ \Im\overset{\text{in}}{c}_n &= \frac{2}{M}\sum_{m=1}^{M}\sin n\theta_m\left[\Phi_m^- - \overset{\text{add}}{\Phi}(z_m)\right] \end{aligned}\right\} \quad \rightarrow \quad \overset{\text{in}}{c}_n = \frac{2}{M}\sum_{m=1}^{M}\mathrm{e}^{\mathrm{i}n\theta_m}\left[\Phi_m^- - \overset{\text{add}}{\Phi}(z_m)\right] \tag{3.29b}$$

A solution for a boundary value problem with constant potential and point-sinks inside a domain is shown in Fig. 3.11. This figure also illustrates the boundary conditions and the functional approximation of these conditions, and the particular solution is presented next.

Example 3.3 A solution for a circular boundary with specified potential is shown in Fig. 3.11. This circle is centered at $z_c = 0$ with radius $r_0 = 4$ and contains $M = 60$

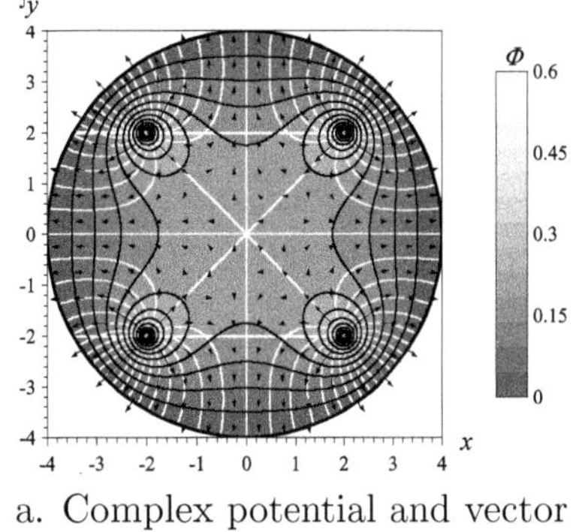

a. Complex potential and vector field

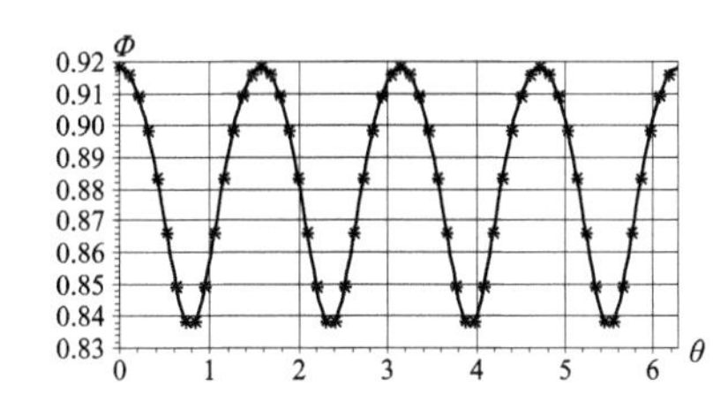

b. Potential $\left[\Phi_m^- - \overset{\text{add}}{\Phi}(z_m)\right]$ at control points θ_m, and function $\Re\left[\sum_{n=0}^{N}\overset{\text{in}}{c}_n\overset{\text{in}}{\Omega}_n(z)\right]$

Figure 3.11 *Circular domain inside a circle with point-sinks.*

control points with specified potential $\Phi_m = 0$. Four point-sinks, (3.1), are located in the domain with strength $Q = -1$, and are contained in

$$\overset{\text{add}}{\Omega} = \frac{Q}{2\pi}\ln[z-(2+i2)] + \frac{Q}{2\pi}\ln[z-(-2+i2)] + \frac{Q}{2\pi}\ln[z-(-2-i2)] + \frac{Q}{2\pi}\ln[z-(2-i2)]$$

The solution for N=20 is given by $\overset{\text{in}}{c}_0 = 0.8825$, $\overset{\text{in}}{c}_4 = 0.03979$, $\overset{\text{in}}{c}_8 = -0.004974$, $\overset{\text{in}}{c}_{12} = 0.0008289$, $\overset{\text{in}}{c}_{16} = -0.0001554$, and $\overset{\text{in}}{c}_{20} = 0.0000311$ with all other coefficients equal to zero. This solution gives a potential at $z = 0$ of $\Phi = 0.2206$.

Problem 3.6 Compute the coefficients and the value of the potential at the origin for the problem in Example 3.3 with the following changes:

A. $r_0 = 4.1$ C. $r_0 = 4.3$ E. $r_0 = 4.5$ G. $r_0 = 4.7$ I. $r_0 = 4.9$ K. $r_0 = 5.1$

B. $r_0 = 4.2$ D. $r_0 = 4.4$ F. $r_0 = 4.6$ H. $r_0 = 4.8$ J. $r_0 = 5.0$ L. $r_0 = 5.2$

3.2.3 Circular Interfaces with Continuity Conditions

Analytic elements with the geometry of circles may also be formulated as an interface problem where solutions are obtained both inside and outside each element. This problem is formulated here for heterogeneities with continuous partial derivative in the normal direction $\partial\Phi/\partial n$ across the interface, but a jump in the potential Φ occurs across each circle. The condition of continuous normal derivative is satisfied when the ***stream function is continuous across the interface*** and satisfies (2.65a):

$$\Psi^+ - \Psi^- = 0 \tag{3.30}$$

The series expansion for the stream function of a circular element at control points just outside and inside the interface, (3.19b) and (3.26b), gives

$$\begin{aligned}\Psi_m^+ &= \sum_{n=1}^{N} -\sin n\theta_m \Re \overset{\text{out}}{c}_n + \cos n\theta_m \Im \overset{\text{out}}{c}_n + \overset{\text{add}}{\Psi}(z_m)\\ \Psi_m^- &= \sum_{n=0}^{N} \sin n\theta_m \Re \overset{\text{in}}{c}_n + \cos n\theta_m \Im \overset{\text{in}}{c}_n + \overset{\text{add}}{\Psi}(z_m)\end{aligned} \tag{3.31}$$

when the additional function $\overset{\text{add}}{\Omega} = \overset{\text{add}}{\Phi} + i\overset{\text{add}}{\Psi}$ is evaluated both inside and outside the circle. The stream function is exactly a continuous condition when

$$\Im\overset{\text{in}}{c}_0 = 0, \qquad \begin{aligned}\Re\overset{\text{in}}{c}_n &= -\Re\overset{\text{out}}{c}_n\\ \Im\overset{\text{in}}{c}_n &= \Im\overset{\text{out}}{c}_n\end{aligned} \qquad (n > 0) \tag{3.32}$$

The system of equations will be formulated in terms of ***continuity coefficients***

$$\begin{aligned}\overset{\text{out}}{c}_n &= \overset{\text{cont}}{c}_n & (n = 1, N)\\ \overset{\text{in}}{c}_n &= -\Re\overset{\text{cont}}{c}_n + i\Im\overset{\text{cont}}{c}_n = -\overline{\overset{\text{cont}}{c}_n} & (n = 0, N)\end{aligned} \tag{3.33}$$

and the coefficients for the outside and inside expansions may be obtained from these after solving. The influence functions for continuity conditions are obtained when the influence

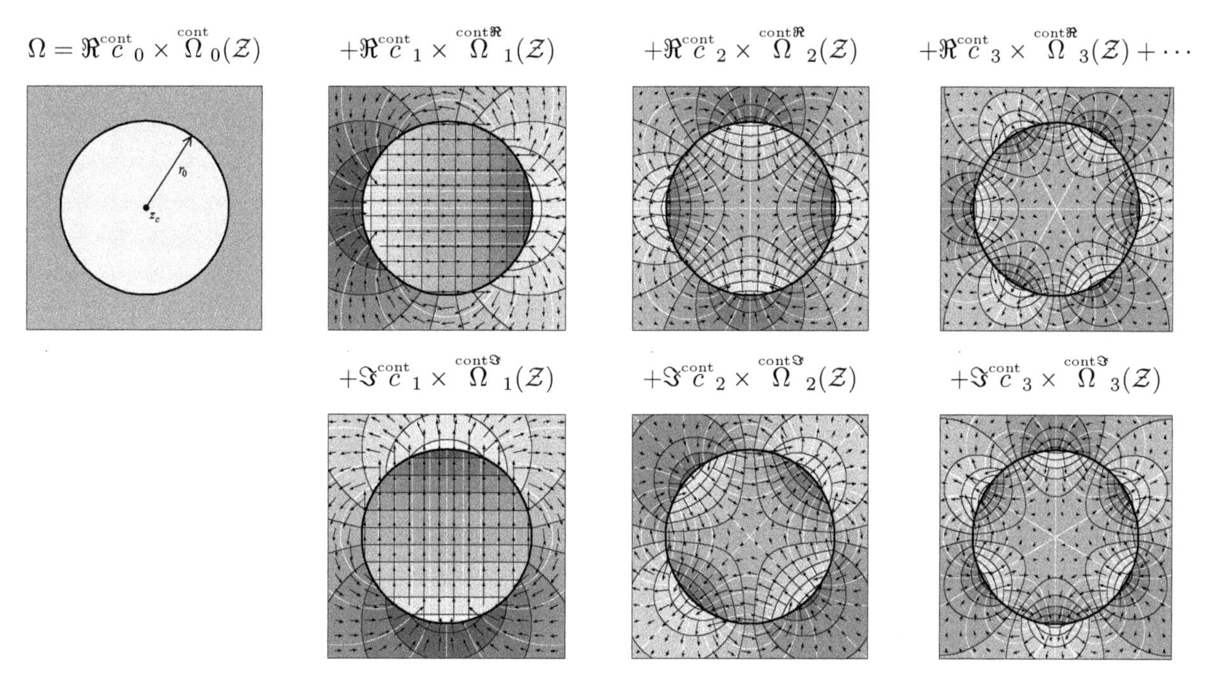

Figure 3.12 *Influence functions for continuity conditions with continuous stream function across a circle, from (3.33) with (3.15) and (3.23).*

functions outside a circle in Fig. 3.8 and inside a circle in Fig. 3.10 are multiplied by these factors. These influence functions are illustrated in Fig. 3.12, where the stream function and normal component of the vector field are continuous across the circular interface.

Circular heterogeneities with a continuous stream function also contain interface conditions for the potential from (2.64a):

$$\alpha^+\Phi^+ - \alpha^-\Phi^- = 0 \tag{3.34}$$

The series expansion for the potential of a circular element at control points just outside and inside the interface, (3.19a) and (3.26a), may be expressed in terms of the continuity coefficients (3.33) as

$$\begin{aligned}\Phi_m^+ &= \sum_{n=1}^{N} \cos n\theta_m \Re\overset{\text{cont}}{c}_n + \sin n\theta_m \Im\overset{\text{cont}}{c}_n + \overset{\text{add}}{\Phi}(z_m)\\ \Phi_m^- &= -\sum_{n=0}^{N} \cos n\theta_m \Re\overset{\text{cont}}{c}_n + \sin n\theta_m \Im\overset{\text{cont}}{c}_n + \overset{\text{add}}{\Phi}(z_m)\end{aligned} \tag{3.35}$$

Substituting these expressions for the potential into the continuity condition and rearranging gives

$$\begin{aligned}&\alpha^-\Re\overset{\text{cont}}{c}_0 + \sum_{n=1}^{N} \cos n\theta_m \left(\alpha^+ + \alpha^-\right)\Re\overset{\text{cont}}{c}_n\\ &\quad + \sin n\theta_m \left(\alpha^+ + \alpha^-\right)\Im\overset{\text{cont}}{c}_n = -(\alpha^+ - \alpha^-)\overset{\text{add}}{\Phi}(z_m)\end{aligned} \tag{3.36}$$

The continuity coefficients may be determined in a least square sense by applying this condition at a set of $M > 2N + 1$ control points on the boundary of the circle. This system of equations $\mathbf{Ac} = \mathbf{b}$ may be written in matrix form as

$$\mathbf{A} = \begin{bmatrix} 1 & \cos\theta_1 & \sin\theta_1 & \cdots & \cos N\theta_1 & \sin N\theta_1 \\ 1 & \cos\theta_2 & \sin\theta_2 & \cdots & \cos N\theta_2 & \sin N\theta_2 \\ \cdots & \cdots & \cdots & \cdots & \cdots & \cdots \\ 1 & \cos\theta_M & \sin\theta_M & \cdots & \cos N\theta_M & \sin N\theta_M \end{bmatrix} \tag{3.37a}$$

and

$$\mathbf{c} = \begin{bmatrix} \alpha^- \Re \overset{\text{cont}}{c}_0 \\ (\alpha^+ + \alpha^-)\Re \overset{\text{cont}}{c}_1 \\ (\alpha^+ + \alpha^-)\Im \overset{\text{cont}}{c}_1 \\ \vdots \\ (\alpha^+ + \alpha^-)\Re \overset{\text{cont}}{c}_N \\ (\alpha^+ + \alpha^-)\Im \overset{\text{cont}}{c}_N \end{bmatrix}, \quad \mathbf{b} = \begin{bmatrix} -(\alpha^+ - \alpha^-)\overset{\text{add}}{\Phi}(z_1) \\ -(\alpha^+ - \alpha^-)\overset{\text{add}}{\Phi}(z_2) \\ \vdots \\ -(\alpha^+ - \alpha^-)\overset{\text{add}}{\Phi}(z_M) \end{bmatrix} \tag{3.37b}$$

When the M points are evenly spaced around the circle using (3.14), the orthogonality of the Fourier series gives $\mathbf{A}^{\mathrm{T}}\mathbf{A} = \mathbf{D}$ where $\mathbf{D}$ is the diagonal matrix in (2.46), and the least squares solution $\mathbf{Dc} = \mathbf{A}^{\mathrm{T}}\mathbf{b}$ may be expressed in summation form as

$$\boxed{\overset{\text{cont}}{c}_0 = \Re \overset{\text{cont}}{c}_0 = -\frac{\alpha^+ - \alpha^-}{\alpha^-}\frac{1}{M}\sum_{m=1}^{M}\overset{\text{add}}{\Phi}(z_m)} \tag{3.38}$$

and

$$\begin{aligned} \Re \overset{\text{cont}}{c}_n &= -\frac{\alpha^+ - \alpha^-}{\alpha^+ + \alpha^-}\frac{2}{M}\sum_{m=1}^{M}\cos n\theta_m \overset{\text{add}}{\Phi}(z_m) \\ \Im \overset{\text{cont}}{c}_n &= -\frac{\alpha^+ - \alpha^-}{\alpha^+ + \alpha^-}\frac{2}{M}\sum_{m=1}^{M}\sin n\theta_m \overset{\text{add}}{\Phi}(z_m) \end{aligned} \qquad (n > 0) \tag{3.39}$$

Combining the real and imaginary parts of these coefficients gives the complex form (Barnes, and Janković, 1999)

$$\boxed{\begin{aligned} \overset{\text{cont}}{c}_n &= -\frac{\alpha^+ - \alpha^-}{\alpha^+ + \alpha^-}\frac{2}{M}\sum_{m=1}^{M}(\cos n\theta_m + \mathrm{i}\sin n\theta_m)\overset{\text{add}}{\Phi}(z_m) \\ &= -\frac{\alpha^+ - \alpha^-}{\alpha^+ + \alpha^-}\frac{2}{M}\sum_{m=1}^{M}\mathrm{e}^{\mathrm{i}n\theta_m}\overset{\text{add}}{\Phi}(z_m) \end{aligned} \qquad (n > 0)} \tag{3.40}$$

Once a solution for the coefficients $\overset{\text{cont}}{c}_n$ is obtained, the coefficients for the inside and outside expansions, $\overset{\text{in}}{c}_n$ and $\overset{\text{out}}{c}_n$, may be obtained using (3.33). Solutions across a range of conditions are illustrated in Fig. 3.13, for single isolated elements and for collections of elements.

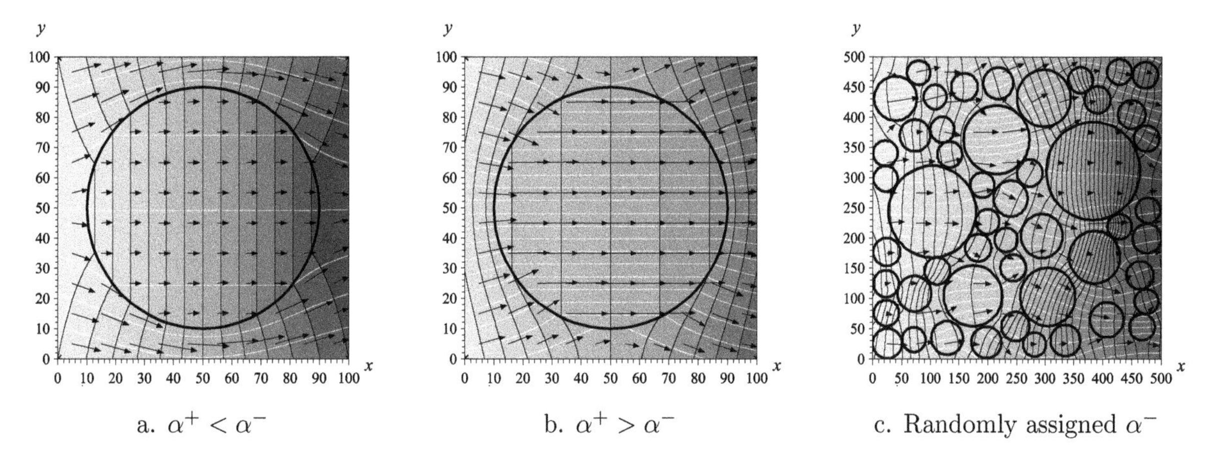

Figure 3.13 *Circle elements with distinct properties in a uniform background.*

Problem 3.7 Compute the coefficients and the value of the potential at $z = -1$, $z = 0$, and $z = 1$ for one inhomogeneity at $z_c = 0$ and radius $r_0 = 1$ in a uniform vector field with $v_0 = 1$ and $\Phi_0 = 0$ and

A. $\alpha^- = 1, \alpha^+ = 2$
B. $\alpha^- = 1, \alpha^+ = 3$
C. $\alpha^- = 1, \alpha^+ = 4$
D. $\alpha^- = 1, \alpha^+ = 5$
E. $\alpha^- = 1, \alpha^+ = 6$
F. $\alpha^- = 1, \alpha^+ = 7$
G. $\alpha^- = 2, \alpha^+ = 1$
H. $\alpha^- = 3, \alpha^+ = 1$
I. $\alpha^- = 4, \alpha^+ = 1$
J. $\alpha^- = 5, \alpha^+ = 1$
K. $\alpha^- = 6, \alpha^+ = 1$
L. $\alpha^- = 7, \alpha^+ = 1$

3.3 Ellipse Elements with Continuity Conditions

Analytic elements are formulated next for those with boundaries with the geometry of ellipses. This geometry is prescribed for ellipse i in Fig. 3.14a by specifying its center $\underset{i}{z_c}$ and orientation $\underset{i}{\vartheta}$, and the length of the major axis $2\underset{i}{L_1}$, and the minor axis $2\underset{i}{L_2}$. This ellipse has two foci, and the distance between them is given by the relation $\underset{i}{f}^2 = \underset{i}{L_1}^2 - \underset{i}{L_2}^2$. It is convenient to represent solutions in terms of the same local $\mathcal{Z}$-plane used for elements with circular geometry in the last section. This is achieved by first ***translating, rotating, and scaling*** locations in the physical z-plane using

$$\zeta = \frac{z - \underset{i}{z_c}}{\underset{i}{L_1} + \underset{i}{L_2}} \mathrm{e}^{-\mathrm{i}\underset{i}{\vartheta}}, \qquad z = \zeta \left(\underset{i}{L_1} + \underset{i}{L_2} \right) \mathrm{e}^{\mathrm{i}\underset{i}{\vartheta}} + \underset{i}{z_c} \tag{3.41}$$

where the element in the ζ-plane in Fig. 3.14b is centered at the origin and the major axis is aligned along the ξ-axis. The boundary of this ellipse is mapped to the unit circle in Fig. 3.14c using the ***conformal mapping***

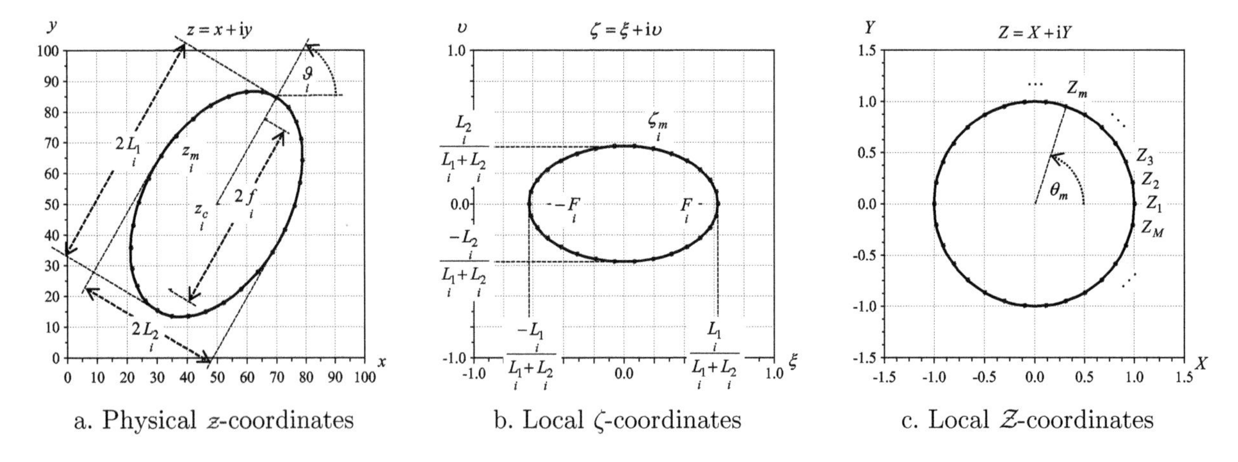

Figure 3.14 *Coordinate systems used to delineate elliptical elements.*

$$\mathcal{Z}=\zeta+\sqrt{\zeta+\underset{i}{F}}\sqrt{\zeta-\underset{i}{F}},\qquad \zeta=\frac{1}{2}\left[\mathcal{Z}+\frac{\underset{i}{F}^{2}}{\mathcal{Z}}\right],\qquad \underset{i}{F}=\frac{\underset{i}{f}}{\underset{i}{L_1}+\underset{i}{L_2}} \tag{3.42}$$

where $\pm\underset{i}{F}$ are the locations of the foci in the ζ-plane. As per the notation convention, the underscript i is dropped from the rest of the equations.

The ***influence functions*** for elliptical elements are formulated following Strack (1989, 2005). The complex potential takes on the same form as the ***Laurent series*** for a circle, (3.15), outside an ellipse

$$\begin{aligned}\Omega&=\sum_{n=1}^{N}\overset{\text{out}}{c}_n\mathcal{Z}^{-n}+\overset{\text{add}}{\Omega}\\ \upsilon&=\overline{\sum_{n=1}^{N}\overset{\text{out}}{c}_n\frac{n\mathrm{e}^{-\mathrm{i}\vartheta}}{L_1+L_2}\frac{\mathcal{Z}^{-n}}{\sqrt{\zeta+F}\sqrt{\zeta-F}}}-\overline{\frac{\mathrm{d}\overset{\text{add}}{\Omega}}{\mathrm{d}z}}\end{aligned}\qquad(|\mathcal{Z}|\geq 1) \tag{3.43}$$

where the derivatives associated with the vector field have been evaluated using the chain rule, $\frac{\mathrm{d}\mathcal{Z}}{\mathrm{d}z}=\frac{\mathrm{d}\mathcal{Z}}{\mathrm{d}\zeta}\frac{\mathrm{d}\zeta}{\mathrm{d}z}$, with derivatives of the conformal mappings in (3.42): $\frac{\mathrm{d}\mathcal{Z}}{\mathrm{d}\zeta}=\frac{\mathcal{Z}}{\sqrt{\zeta+F}\sqrt{\zeta-F}}$ and $\frac{\mathrm{d}\zeta}{\mathrm{d}z}=\frac{\mathrm{e}^{-\mathrm{i}\vartheta}}{L_1+L_2}$. The influence functions inside the ellipse are obtained by ***gathering terms in the Laurent and Taylor series***

$$\begin{aligned}\Omega&=\sum_{n=0}^{N}\overset{\text{in}}{c}_n\left[\mathcal{Z}^n+\left(\frac{F^2}{\mathcal{Z}}\right)^n\right]+\overset{\text{add}}{\Omega}\\ \upsilon&=-\overline{\sum_{n=1}^{N}\overset{\text{in}}{c}_n\frac{n\mathrm{e}^{-\mathrm{i}\vartheta}}{L_1+L_2}\frac{\mathcal{Z}^n-\left(\frac{F^2}{\mathcal{Z}}\right)^n}{\sqrt{\zeta+F}\sqrt{\zeta-F}}}+\overset{\text{add}}{\upsilon}\end{aligned}\qquad(|\mathcal{Z}|<1) \tag{3.44}$$

with common terms of $\mathcal{Z}^{\pm n}$ from the series ζ^n.

The coefficients in the influence functions for elliptical elements may be adjusted to solve a wide range of boundary conditions. This section illustrates such formulations by solving the same interface conditions used for circle elements in Section 3.2.3. These elements satisfy a condition of ***continuous normal component of the vector field***, which is achieved using expressions for the stream function just inside and outside the boundary of an ellipse where $\mathcal{Z}_m = \mathrm{e}^{\mathrm{i}\theta_m}$

$$\begin{aligned}
\Psi_m^+ &= \sum_{n=1}^{N} -\sin n\theta_m \Re \overset{\text{out}}{c}_n + \cos n\theta_m \Im \overset{\text{out}}{c}_n + \overset{\text{add}}{\Psi}(z_m) \\
\Psi_m^- &= \sum_{n=0}^{N} \sin n\theta_m (1 - F^{2n}) \Re \overset{\text{in}}{c}_n + \cos n\theta_m (1 + F^{2n}) \Im \overset{\text{in}}{c}_n + \overset{\text{add}}{\Psi}(z_m)
\end{aligned} \tag{3.45}$$

These expressions were obtained from the imaginary parts of (3.43) and (3.44) using $\mathcal{Z}^n + (F^2/\mathcal{Z})^n = \mathrm{e}^{\mathrm{i}n\theta_m} + F^{2n}\mathrm{e}^{-\mathrm{i}n\theta_m} = (1 + F^{2n}) \cos n\theta_m + \mathrm{i}(1 - F^{2n}) \sin n\theta_m$. The ***stream function is continuous across the boundary of the ellipse*** when

$$\Im \overset{\text{in}}{c}_0 = 0, \qquad \begin{aligned} \Re \overset{\text{in}}{c}_n &= -\frac{\Re \overset{\text{out}}{c}_n}{1 - F^{2n}} \\ \Im \overset{\text{in}}{c}_n &= \frac{\Im \overset{\text{out}}{c}_n}{1 + F^{2n}} \end{aligned} \qquad (n > 0) \tag{3.46}$$

These influence functions are illustrated in Fig. 3.15 and fulfill the condition of uniform stream function and continuous normal component of the vector field across the interface. The system of equations will be formulated in terms of ***continuity coefficients***

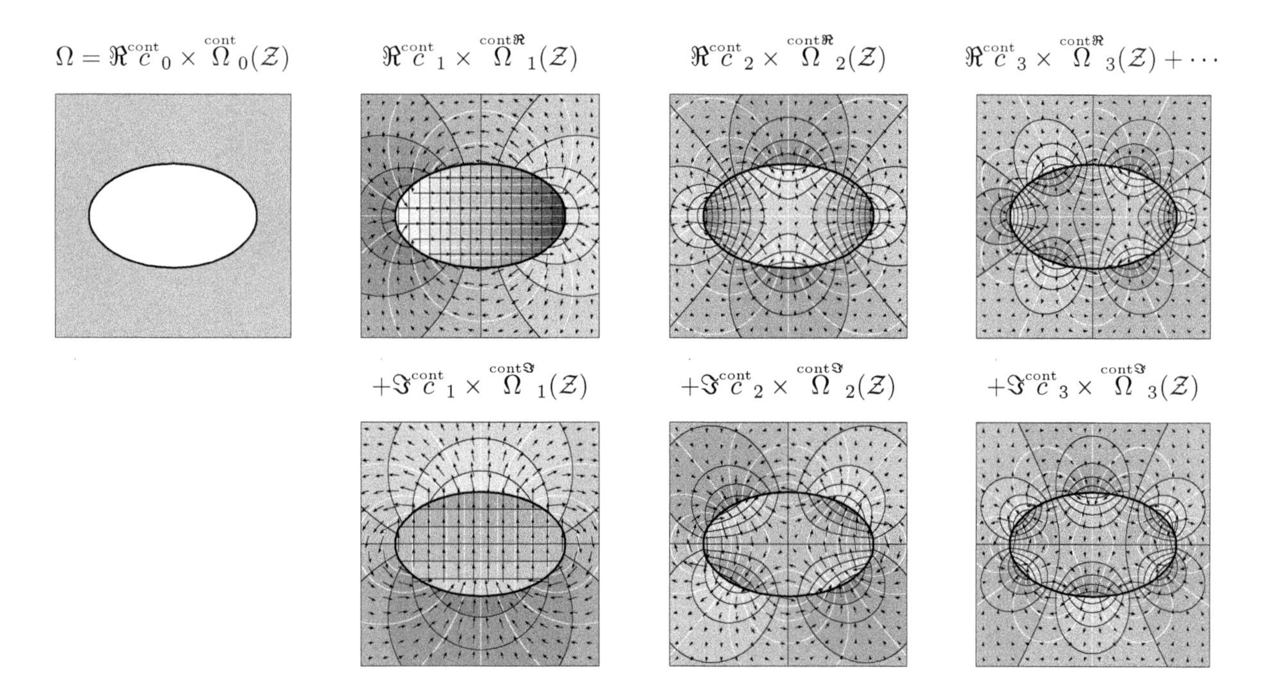

Figure 3.15 *Influence functions for continuity conditions across an ellipse, from (3.47) with (3.43) and (3.44).*

$$\overset{\text{out}}{c}_n = \overset{\text{cont}}{c}_n \quad \rightarrow \quad \begin{aligned} \overset{\text{in}}{c}_0 &= -\frac{1}{2}\Re\overset{\text{cont}}{c}_0 \\ \overset{\text{in}}{c}_n &= -\frac{\Re\overset{\text{cont}}{c}_n}{1-F^{2n}} + \mathrm{i}\frac{\Im\overset{\text{cont}}{c}_n}{1+F^{2n}} \end{aligned} \qquad (n = 1, N) \tag{3.47}$$

and the coefficients for the outside and inside expansions may be obtained from these after solving.

Heterogeneities also contain an interface condition imposed on the potential function, which is obtained from the real part of the complex potential, and is evaluated at control points just outside and inside the boundary to give

$$\begin{aligned} \Phi_m^+ &= \sum_{n=1}^{N} \cos n\theta_m \Re\overset{\text{out}}{c}_n + \sin n\theta_m \Im\overset{\text{out}}{c}_n + \overset{\text{add}}{\Phi}(z_m) \\ \Phi_m^- &= \sum_{n=0}^{N} \cos n\theta_m \left(1+F^{2n}\right)\Re\overset{\text{in}}{c}_n - \sin n\theta_m \left(1-F^{2n}\right)\Im\overset{\text{in}}{c}_n + \overset{\text{add}}{\Phi}(z_m) \end{aligned} \tag{3.48}$$

These expansions are substituted into the ***continuity condition across the elliptical interface***, $\alpha^+\Phi^+ - \alpha^-\Phi^- = 0$ from (2.64a), to obtain

$$\begin{aligned} &\alpha^+\left(\sum_{n=1}^{N} \cos n\theta_m \Re\overset{\text{out}}{c}_n + \sin n\theta_m \Im\overset{\text{out}}{c}_n\right) \\ &-\alpha^-\left[\sum_{n=0}^{N} \cos n\theta_m \left(1+F^{2n}\right)\Re\overset{\text{in}}{c}_n - \sin n\theta_m \left(1-F^{2n}\right)\Im\overset{\text{in}}{c}_n\right] = -(\alpha^+ - \alpha^-)\overset{\text{add}}{\Phi}(z_m) \end{aligned} \tag{3.49}$$

The inside and outside coefficients in this equation may be expressed in terms of ***continuity coefficients*** using the conditions imposed by a continuous stream function, (3.47), to give the system of equations with summations organized to gather the real and imaginary parts of these continuity coefficients

$$\begin{aligned} \alpha^-\Re\overset{\text{cont}}{c}_0 &+ \sum_{n=1}^{N} \cos n\theta_m \left(\alpha^+ + \frac{1+F^{2n}}{1-F^{2n}}\alpha^-\right)\Re\overset{\text{cont}}{c}_n \\ &+ \sin n\theta_m \left(\alpha^+ + \frac{1-F^{2n}}{1+F^{2n}}\alpha^-\right)\Im\overset{\text{cont}}{c}_n = -(\alpha^+ - \alpha^-)\overset{\text{add}}{\Phi}(z_m) \end{aligned} \tag{3.50}$$

A linear system of M equations with $2N+1$ unknowns may be represented in matrix format $\mathbf{Ac} = \mathbf{b}$ using the same matrix $\mathbf{A}$ and $\mathbf{b}$ as for the circles, (3.37), and the matrix $\mathbf{c}$ is

$$\mathbf{c} = \begin{bmatrix} \alpha^-\Re\overset{\text{cont}}{c}_0 \\ \left(\alpha^+ + \frac{1+F^2}{1-F^2}\alpha^-\right)\Re\overset{\text{cont}}{c}_1 \\ \left(\alpha^+ + \frac{1-F^2}{1+F^2}\alpha^-\right)\Im\overset{\text{cont}}{c}_1 \\ \vdots \\ \left(\alpha^+ + \frac{1+F^{2N}}{1-F^{2N}}\alpha^-\right)\Re\overset{\text{cont}}{c}_N \\ \left(\alpha^+ + \frac{1-F^{2N}}{1+F^{2N}}\alpha^-\right)\Im\overset{\text{cont}}{c}_N \end{bmatrix} = \begin{bmatrix} \alpha^-\Re\overset{\text{cont}}{c}_0 \\ \frac{(\alpha^++\alpha^-)-F^2(\alpha^+-\alpha^-)}{1-F^2}\Re\overset{\text{cont}}{c}_1 \\ \frac{(\alpha^++\alpha^-)+F^2(\alpha^+-\alpha^-)}{1+F^2}\Im\overset{\text{cont}}{c}_1 \\ \vdots \\ \frac{(\alpha^++\alpha^-)-F^{2N}(\alpha^+-\alpha^-)}{1-F^{2N}}\Re\overset{\text{cont}}{c}_N \\ \frac{(\alpha^++\alpha^-)+F^{2N}(\alpha^+-\alpha^-)}{1+F^{2N}}\Im\overset{\text{cont}}{c}_N \end{bmatrix} \tag{3.51}$$

When θ_m are equally spaced as in Eq. (3.14), the same orthogonality condition of the Fourier series that was used for the circle, (2.46), gives the coefficients for $n = 0$:

$$\overset{\text{cont}}{c}_0 = \Re \overset{\text{cont}}{c}_0 = -\frac{\alpha^+ - \alpha^+}{\alpha^-} \frac{1}{M} \sum_{m=1}^{M} \overset{\text{add}}{\Phi}(z_m) \tag{3.52}$$

and for higher-order ($n > 0$) terms

$$\begin{aligned}
\Re \overset{\text{cont}}{c}_n &= -\frac{(1 - F^{2n})(\alpha^+ - \alpha^-)}{(\alpha^+ + \alpha^-) - F^{2n}(\alpha^+ - \alpha^-)} \frac{2}{M} \sum_{m=1}^{M} \cos n\theta_m \overset{\text{add}}{\Phi}(z_m) \\
&= \frac{-(1 - F^{2n})(\alpha^{+2} - \alpha^{-2}) - (F^{2n} - F^{4n})(\alpha^+ - \alpha^-)^2}{(\alpha^+ + \alpha^-)^2 - F^{4n}(\alpha^+ - \alpha^-)^2} \frac{2}{M} \sum_{m=1}^{M} \cos n\theta_m \overset{\text{add}}{\Phi}(z_m) \\
&= \frac{-(\alpha^{+2} - \alpha^{-2}) + F^{2n} 2\alpha^-(\alpha^+ - \alpha^-) + F^{4n}(\alpha^+ - \alpha^-)^2}{(\alpha^+ + \alpha^-)^2 - F^{4n}(\alpha^+ - \alpha^-)^2} \frac{2}{M} \sum_{m=1}^{M} \cos n\theta_m \overset{\text{add}}{\Phi}(z_m) \\
\Im \overset{\text{cont}}{c}_n &= -\frac{(1 + F^{2n})(\alpha^+ - \alpha^-)}{(\alpha^+ + \alpha^-) + F^{2n}(\alpha^+ - \alpha^-)} \frac{2}{M} \sum_{m=1}^{M} \sin n\theta_m \overset{\text{add}}{\Phi}(z_m) \\
&= \frac{-(1 + F^{2n})(\alpha^{+2} - \alpha^{-2}) + (F^{2n} + F^{4n})(\alpha^+ - \alpha^-)^2}{(\alpha^+ + \alpha^-)^2 - F^{4n}(\alpha^+ - \alpha^-)^2} \frac{2}{M} \sum_{m=1}^{M} \sin n\theta_m \overset{\text{add}}{\Phi}(z_m) \\
&= \frac{-(\alpha^{+2} - \alpha^{-2}) - F^{2n} 2\alpha^-(\alpha^+ - \alpha^-) + F^{4n}(\alpha^+ - \alpha^-)^2}{(\alpha^+ + \alpha^-)^2 - F^{4n}(\alpha^+ - \alpha^-)^2} \frac{2}{M} \sum_{m=1}^{M} \sin n\theta_m \overset{\text{add}}{\Phi}(z_m)
\end{aligned} \tag{3.53}$$

where terms have been rearranged to have the same denominator for the real and imaginary parts, and then to gather terms with common powers of F in the numerator. These real and imaginary parts are added and gathered into positive and negative exponential terms using the Euler formula (2.72) to give a ***complex form for the continuity coefficients***:

$$\begin{aligned}
\overset{\text{cont}}{c}_n = &\frac{-(\alpha^{+2} - \alpha^{-2}) + F^{4n}(\alpha^+ - \alpha^-)^2}{(\alpha^+ + \alpha^-)^2 - F^{4n}(\alpha^+ - \alpha^-)^2} \frac{2}{M} \sum_{m=1}^{M} \mathrm{e}^{\mathrm{i}n\theta_m} \overset{\text{add}}{\Phi}(z_m) \\
&+ \frac{F^{2n} 2\alpha^-(\alpha^+ - \alpha^-)}{(\alpha^+ + \alpha^-)^2 - F^{4n}(\alpha^+ - \alpha^-)^2} \frac{2}{M} \sum_{m=1}^{M} \mathrm{e}^{-\mathrm{i}n\theta_m} \overset{\text{add}}{\Phi}(z_m) \qquad (n > 0)
\end{aligned} \tag{3.54}$$

Once a solution for the continuity coefficients $\overset{\text{cont}}{c}_n$ is obtained, the inside and outside coefficients may be obtained from (3.47). Note that these equations take on the same form as those for the coefficients of a circle (3.40) in the limit as $F \to 0$. The complex potential for elliptical elements is illustrated in Fig. 3.16 for isolated elements and for multiple elements with randomly assigned parameters in the interface condition.

Problem 3.8 Compute the coefficients and the value of the potential at $z = z_c - L_1 \mathrm{e}^{\mathrm{i}\vartheta}$ and $z = z_c + L_1 \mathrm{e}^{\mathrm{i}\vartheta}$ for one inhomogeneity with center $z_c = 0$ orientation, $\vartheta = \pi/3$, $L_1 = 2$, and $L_2 = 1$ in a uniform vector field with $v_0 = 1$ and $\Phi_0 = 0$ and:

A. $\alpha^- = 1, \alpha^+ = 2$	D. $\alpha^- = 1, \alpha^+ = 5$	G. $\alpha^- = 2, \alpha^+ = 1$	J. $\alpha^- = 5, \alpha^+ = 1$
B. $\alpha^- = 1, \alpha^+ = 3$	E. $\alpha^- = 1, \alpha^+ = 6$	H. $\alpha^- = 3, \alpha^+ = 1$	K. $\alpha^- = 6, \alpha^+ = 1$
C. $\alpha^- = 1, \alpha^+ = 4$	F. $\alpha^- = 1, \alpha^+ = 7$	I. $\alpha^- = 4, \alpha^+ = 1$	L. $\alpha^- = 7, \alpha^+ = 1$

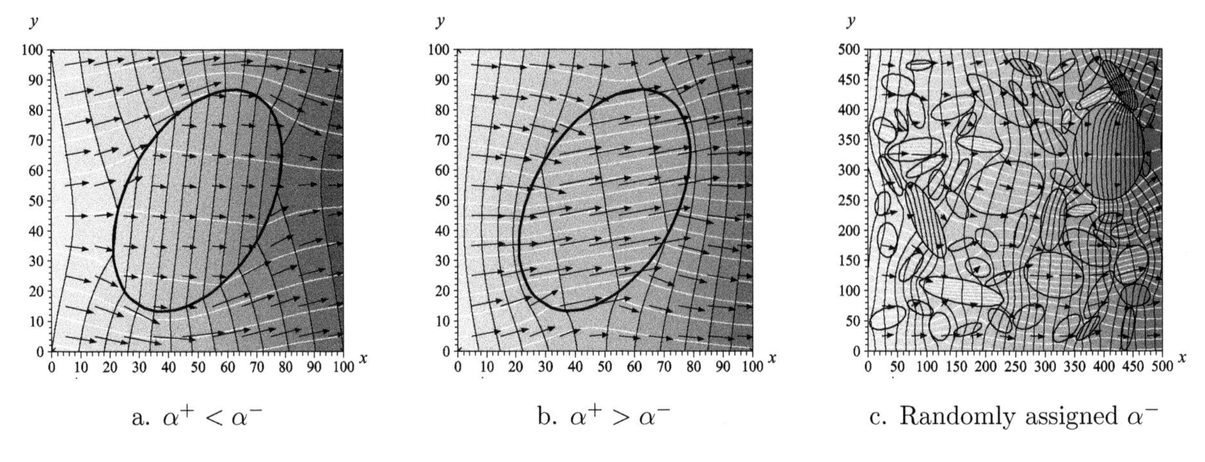

Figure 3.16 *Elliptical elements with distinct properties in a uniform background.*

3.4 Slit Element Formulation: Courant's Sewing Theorem with Circle Elements

This section extends complex functions, which have been formulated thus far to solve boundary conditions along circles and ellipses, to ***elements with the geometry of straight line segments***. In general, problems will be solved with multiple elements where the ith slit lies between the endpoints $\underset{i}{z}_{\min}$ and $\underset{i}{z}_{\max}$, and has length $2\underset{i}{L}$ and orientation $\underset{i}{\vartheta}$ as illustrated in Fig. 3.17a. Equations are formulated in a common reference system by first mapping locations in the physical z-plane to a slit in the ζ-plane between $\zeta = -1$ and $\zeta = 1$. This is accomplished through scaling, rotating, and translating using the following ***transformation and inverse mapping***

$$\zeta = \left(z - \frac{\underset{i}{z}_{\max} + \underset{i}{z}_{\min}}{2} \right) \frac{2}{\underset{i}{z}_{\max} - \underset{i}{z}_{\min}}, \qquad z = \zeta \frac{\underset{i}{z}_{\max} - \underset{i}{z}_{\min}}{2} + \frac{\underset{i}{z}_{\max} + \underset{i}{z}_{\min}}{2} \tag{3.55a}$$

The domain around this slit is then conformally mapped to the exterior of a unit circle in the $\mathcal{Z}$-plane using the ***Joukowsky transformation***

$$\mathcal{Z}\left[\zeta(z; \underset{i}{z}_{\min}, \underset{i}{z}_{\max}) \right] = \zeta + \sqrt{\zeta + 1}\sqrt{\zeta - 1}, \qquad \zeta(\mathcal{Z}) = \frac{1}{2}\left(\mathcal{Z} + \frac{1}{\mathcal{Z}} \right) \tag{3.55b}$$

This mapping may be viewed as the limiting case of an ellipse (3.55) with zero width and focal points occurring at its endpoints. The control points used later to solve boundary value problems, which are identified in Fig. 3.17, are evenly spaced about the circle in the $\mathcal{Z}$-plane using

$$\theta_m = \pi \frac{m - \frac{1}{2}}{M} \quad \rightarrow \quad \begin{aligned} \mathcal{Z}_m^+ &= \mathrm{e}^{\mathrm{i}\theta_m} \\ \mathcal{Z}_m^- &= \mathrm{e}^{-\mathrm{i}\theta_m} \end{aligned} \quad \rightarrow \quad \zeta_m = \cos\theta_m \tag{3.56}$$

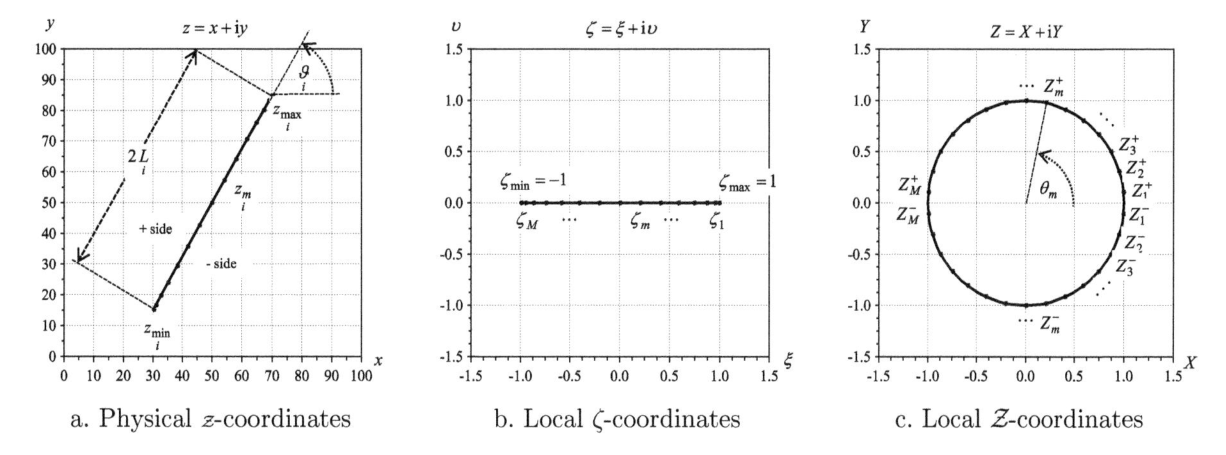

Figure 3.17 *Coordinate systems used to conformally map slit elements to the exterior of a unit circle. (Reprinted from Steward, 2015, Analysis of discontinuities across thin inhomogeneities, groundwater/surface water interactions in river networks, and circulation about slender bodies using slit elements in the Analytic Element Method,* Water Resources Research, *Vol. 51(11), Fig. 2, with permission from John Wiley and Sons. Copyright 2015 by the American Geophysical Union.)*

These locations lie on the upper and lower half of the element in the circular $\mathcal{Z}$-plane, yet they occur at the same location in physical space but on opposite sides of the slit element. This ***enables solutions to be stitched together across an element using the Courant sewing theorem*** (Courant, 1950).

The coordinate transformations and capacity of sew solutions together across a slit element are illustrated first for a set of exact solutions that lie along the x-axis, so $z = \zeta$. The complex potential for a slit-dipole element with ***uniform potential*** is obtained using the solution for the exterior of a unit circle centered at the origin in (3.6) and Fig. 3.4a

$$\begin{aligned}\Omega &= -\frac{\overline{v_0}}{2}\mathcal{Z} + \frac{v_0}{2}\frac{1}{\mathcal{Z}} = -\frac{\overline{v_0}}{2}\left(z + \sqrt{z+1}\sqrt{z-1}\right) + \frac{v_0}{2}\frac{1}{\left(z + \sqrt{z+1}\sqrt{z-1}\right)} + \Phi_0 \\ v &= \frac{v_0}{2}\frac{\overline{\left(z + \sqrt{z+1}\sqrt{z-1}\right)}}{\overline{\sqrt{z+1}\sqrt{z-1}}} + \frac{\overline{v_0}}{2}\frac{1}{\overline{\sqrt{z+1}\sqrt{z-1}\left(z + \sqrt{z+1}\sqrt{z-1}\right)}}\end{aligned} \tag{3.57}$$

This solution is illustrated in Fig. 3.18a, and the vector field is obtained using the derivatives

$$\frac{\mathrm{d}\mathcal{Z}}{\mathrm{d}z} = \frac{\mathcal{Z}}{\sqrt{z+1}\sqrt{z-1}}, \quad \frac{\mathrm{d}}{\mathrm{d}z}\frac{1}{\mathcal{Z}} = -\frac{1}{\mathcal{Z}\sqrt{z+1}\sqrt{z-1}} \tag{3.58}$$

Likewise, the vector field for a slit-doublet element with ***uniform stream function*** is obtained from the solution for the circle in Eq. (3.7) and Fig. 3.4b

$$\begin{aligned}\Omega &= -\frac{\overline{v_0}}{2}\left(z + \sqrt{z+1}\sqrt{z-1}\right) - \frac{v_0}{2}\frac{1}{\left(z + \sqrt{z+1}\sqrt{z-1}\right)} + \Phi_0 \\ v &= \frac{v_0}{2}\frac{\overline{\left(z + \sqrt{z+1}\sqrt{z-1}\right)}}{\overline{\sqrt{z+1}\sqrt{z-1}}} - \frac{\overline{v_0}}{2}\frac{1}{\overline{\sqrt{z+1}\sqrt{z-1}\left(z + \sqrt{z+1}\sqrt{z-1}\right)}}\end{aligned} \tag{3.59}$$

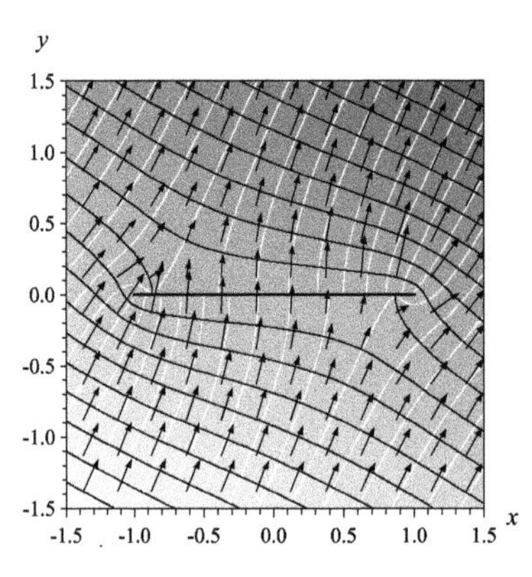

a. Slit-dipole with uniform Φ

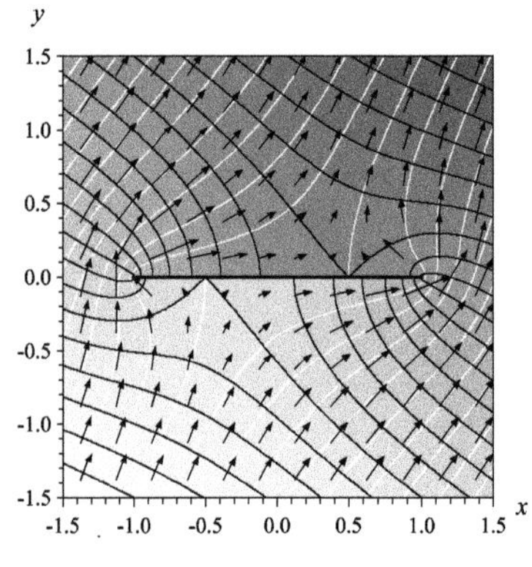

b. Slit-doublet with $v_n = 0$

Figure 3.18 *Slit-dipole and slit-doublet in a uniform vector field. (Reprinted from Steward, 2015, Analysis of discontinuities across thin inhomogeneities, groundwater/surface water interactions in river networks, and circulation about slender bodies using slit elements in the Analytic Element Method,* Water Resources Research, *Vol. 51(11), Fig. 1 with permission from John Wiley and Sons. Copyright 2015 by the American Geophysical Union.)*

to obtain the slit element shown in Fig. 3.18b. Note that the magnitude of the uniform vector field from the exterior of the unit circle is factored by 1/2, giving a regional vector field for the slit in the z-plane of magnitude $|v_0|$ and orientation $\arg(v_0)$. The exact solution for a ***slit-sink that withdraws a net flux*** Q is obtained by placing a point-sink, (3.1), at the center of the circle in the $\mathcal{Z}$-plane

$$\Omega = \frac{Q}{2\pi} \ln\left(z + \sqrt{z+1}\sqrt{z-1}\right), \quad v = -\frac{Q}{2\pi} \overline{\frac{1}{\sqrt{z+1}\sqrt{z-1}}} \tag{3.60}$$

and a ***slit-vortex with net circulation*** Γ is obtained using a point-vortex, (3.3), at this location:

$$\Omega = \frac{i\Gamma}{2\pi} \ln\left(z + \sqrt{z+1}\sqrt{z-1}\right), \quad v = \frac{i\Gamma}{2\pi} \overline{\frac{1}{\sqrt{z+1}\sqrt{z-1}}} \tag{3.61}$$

These exact solutions are illustrated in Fig. 3.19.

A general expression for the complex potential of a slit element is formulated using a ***Laurent series expansion*** about the element in the $\mathcal{Z}$-plane

$$\boxed{\Omega = \sum_{n=1}^{N} c_n \mathcal{Z}^{-n}} \tag{3.62}$$

similar to formulation of the ellipse elements, (3.62), following Steward (2015). This gives the complex vector field

$$\boxed{v = \sum_{n=1}^{N} \overline{c_n \frac{2n}{z_{\max} - z_{\min}} \frac{\mathcal{Z}^{-n}}{\sqrt{\zeta+1}\sqrt{\zeta-1}}} - \overline{\frac{\mathrm{d}\Omega}{\mathrm{d}z}}^{\mathrm{add}}} \tag{3.63}$$

where use has been made of the chain rule $\frac{\mathrm{d}\mathcal{Z}}{\mathrm{d}z} = \frac{\mathrm{d}\mathcal{Z}}{\mathrm{d}\zeta}\frac{\mathrm{d}\zeta}{\mathrm{d}z}$ and derivatives of the conformal mappings: $\frac{\mathrm{d}\mathcal{Z}}{\mathrm{d}\zeta} = \frac{\mathcal{Z}}{\sqrt{\zeta+1}\sqrt{\zeta-1}}$ and $\frac{\mathrm{d}\zeta}{\mathrm{d}z} = \frac{2}{z_{\max}-z_{\min}}$. Systems of equations will be developed

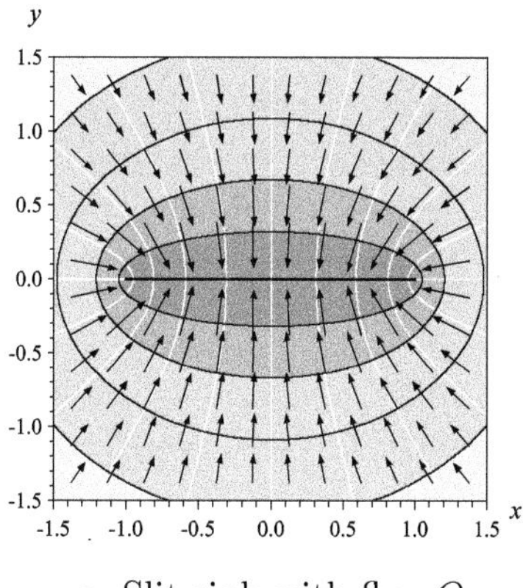

a. Slit-sink with flux Q

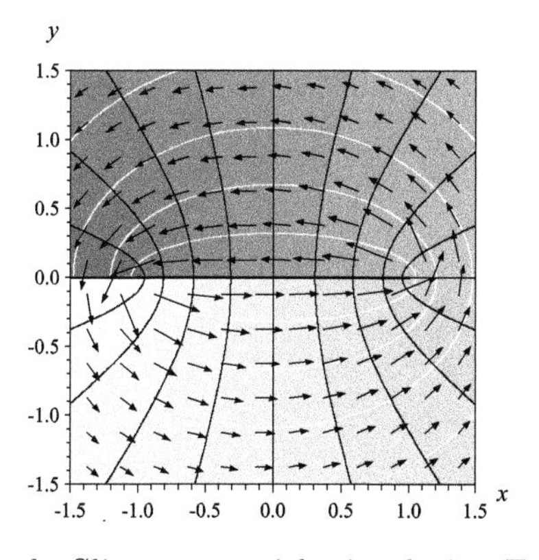

b. Slit-vortex with circulation Γ

Figure 3.19 *Slit-sink with a net flux removed by the element, and slit-vortex with a net circulation around the element. (Reprinted from Steward, 2015, Analysis of discontinuities across thin inhomogeneities, groundwater/surface water interactions in river networks, and circulation about slender bodies using slit elements in the Analytic Element Method,* Water Resources Research, *Vol. 51(11), Fig. 1 with permission from John Wiley and Sons. Copyright 2015 by the American Geophysical Union.)*

next to solve a wide range of interface conditions across slit elements by evaluating the complex potential (3.62) at control point m, where $\mathcal{Z}_m^+ = \mathrm{e}^{\mathrm{i}\theta_m}$ and $\mathcal{Z}_m^- = \mathrm{e}^{-\mathrm{i}\theta_m}$ at points on the sides just above and beneath the element. The real part gives the ***potential at points on the boundary***

$$
\begin{aligned}
\Phi_m^+ &= \sum_{n=1}^{N} \cos n\theta_m \Re c_n + \sin n\theta_m \Im c_n + \overset{\text{add}}{\Phi}(z_m) \\
\Phi_m^- &= \sum_{n=0}^{N} \cos n\theta_m \Re c_n - \sin n\theta_m \Im c_n + \overset{\text{add}}{\Phi}(z_m)
\end{aligned}
\tag{3.64}
$$

and the ***stream function along the boundary*** is

$$
\begin{aligned}
\Psi_m^+ &= \sum_{n=1}^{N} - \sin n\theta_m \Re c_n + \cos n\theta_m \Im c_n + \overset{\text{add}}{\Psi}(z_m) \\
\Psi_m^- &= \sum_{n=0}^{N} \sin n\theta_m \Re c_n + \cos n\theta_m \Im c_n + \overset{\text{add}}{\Psi}(z_m)
\end{aligned}
\tag{3.65}
$$

The complex vector field (3.63) may be expressed in terms of components tangential and normal to the slit element by multiplying v in (3.63) by $\mathrm{e}^{-\mathrm{i}\vartheta} = \frac{\mathcal{Z}_{\max}-\mathcal{Z}_{\min}}{|\mathcal{Z}_{\max}-\mathcal{Z}_{\min}|}$, giving

$$
v_{\mathrm{s}} + \mathrm{i}v_{\mathrm{n}} = \sum_{n=1}^{N} \overline{\frac{n}{L} c_n \frac{\mathcal{Z}^{-n}}{\sqrt{\zeta+1}\sqrt{\zeta-1}}} + \overset{\text{add}}{v}(z)\mathrm{e}^{-\mathrm{i}\vartheta}
\tag{3.66}
$$

where ϑ is the argument of the slit and L is half its length. These components of the vector field may be evaluated immediately above and beneath the slit with $\xi = \Re\zeta$ to give the ***tangential component of the vector field along the slit***

$$
\begin{aligned}
(v_{\mathrm{s}}^+)_m &= \sum_{n=1}^{N} \frac{n}{L} \frac{-\sin n\theta_m \Re c_n + \cos n\theta_m \Im c_n}{\sqrt{\xi_m+1}\sqrt{1-\xi_m}} + \Re\left[\overset{\text{add}}{v}(z_m)\mathrm{e}^{-\mathrm{i}\vartheta}\right] \\
(v_{\mathrm{s}}^-)_m &= \sum_{n=1}^{N} \frac{n}{L} \frac{-\sin n\theta_m \Re c_n - \cos n\theta_m \Im c_n}{\sqrt{\xi_m+1}\sqrt{1-\xi_m}} + \Re\left[\overset{\text{add}}{v}(z_m)\mathrm{e}^{-\mathrm{i}\vartheta}\right]
\end{aligned}
\tag{3.67a}
$$

and the ***normal component***

$$(v_n^+)_m = \sum_{n=1}^{N} \frac{n}{L} \frac{\cos n\theta_m \Re c_n + \sin n\theta_m \Im c_n}{\sqrt{\xi_m + 1}\sqrt{1 - \xi_m}} + \Im\left[\overset{\text{add}}{v}(z_m)e^{-i\vartheta}\right]$$
$$(v_n^-)_m = \sum_{n=1}^{N} \frac{n}{L} \frac{-\cos n\theta_m \Re c_n + \sin n\theta_m \Im c_n}{\sqrt{\xi_m + 1}\sqrt{1 - \xi_m}} + \Im\left[\overset{\text{add}}{v}(z_m)e^{-i\vartheta}\right] \tag{3.67b}$$

where $\frac{1}{\sqrt{\zeta+1}\sqrt{\zeta-1}}$ takes on values of $\frac{-i}{\sqrt{\xi+1}\sqrt{1-\xi}}$ at points $\zeta = \xi + i \lim_{\upsilon \to 0} \upsilon$ just above the slit and $\frac{i}{\sqrt{\xi+1}\sqrt{1-\xi}}$ at points $\zeta = \xi - i \lim_{\upsilon \to 0} \upsilon$ just beneath the slit.

Problem 3.9 Compute the potential, Φ, stream function, Ψ, and vector field $v = v_x + iv_y$ at the points $z = i$ and $z = \frac{1}{2}$ for a slit element with endpoints $z_{\min} = -1$ and $z_{\max} = +1$ for the influence function with the specified coefficient.

A. $c_1 = 1$
B. $c_1 = i$
C. $c_2 = 1$
D. $c_2 = i$
E. $c_3 = 1$
F. $c_3 = i$
G. $c_4 = 1$
H. $c_4 = i$
I. $c_5 = 1$
J. $c_5 = i$
K. $c_6 = 1$
L. $c_6 = i$

3.4.1 Slit-Dipole: Boundaries with Continuous Potential

The first set of boundary conditions formulated for a slit element will set the imaginary part of the coefficients equal to zero

$$\Im c_n = 0 \tag{3.68}$$

and is called a ***slit-dipole*** (Steward, 2015). This element has a continuous potential across the slit, since the non-zero terms in the potential along the boundary (3.64)

$$\Phi_m = \sum_{n=1}^{N} \cos n\theta_m \Re c_n + \overset{\text{add}}{\Phi}(z_m) = \Phi_U \tag{3.69}$$

have the same value on both sides of the slit. A slit-dipole has a discontinuous stream function across the element in (3.65) with a jump of

$$\Delta\Psi_m = \Psi_m^+ - \Psi_m^- = -2\sum_{n=1}^{N} \sin n\theta_m \Re c_n \tag{3.70}$$

The first few terms of the influence functions for a slit-dipole are shown in Fig. 3.20, and illustrate the ***continuous potential and discontinuous stream function***.

A boundary condition of ***uniform potential*** may be formulated by setting the potential in (3.69) equal to the unknown constant Φ_U at each control point. This gives a system of M equations with $N + 1$ unknowns that may be written in matrix form as $\mathbf{Ac} = \mathbf{b}$ with

$$\mathbf{A} = \begin{bmatrix} 1 & \cos\theta_1 & \cos 2\theta_1 & \cdots & \cos N\theta_1 \\ 1 & \cos\theta_2 & \cos 2\theta_2 & \cdots & \cos N\theta_2 \\ \cdots & \cdots & \cdots & \cdots & \cdots \\ 1 & \cos\theta_M & \cos 2\theta_M & \cdots & \cos N\theta_M \end{bmatrix}, \quad \mathbf{c} = \begin{bmatrix} -\Phi_U \\ \Re c_1 \\ \Re c_2 \\ \vdots \\ \Re c_N \end{bmatrix}, \quad \mathbf{b} = \begin{bmatrix} -\overset{\text{add}}{\Phi}(z_1) \\ -\overset{\text{add}}{\Phi}(z_2) \\ \vdots \\ -\overset{\text{add}}{\Phi}(z_M) \end{bmatrix} \tag{3.71}$$

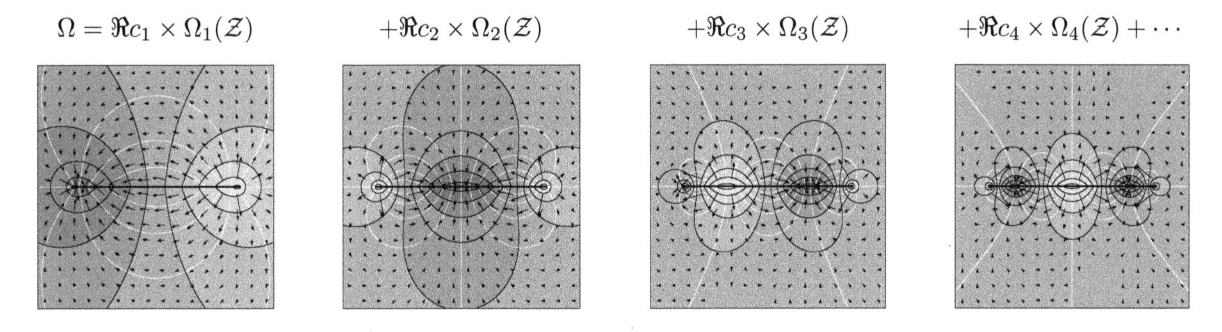

Figure 3.20 *Influence functions for slit-dipole with continuous potential across the element, from (3.62) and (3.63) with (3.68).*

The least squares solution $\mathbf{A}^{\mathrm{T}}\mathbf{A} = \mathbf{A}^{\mathrm{T}}\mathbf{b}$ takes on a simpler form for the special case where the $M > N + 1$ control points are evenly spaced around the circle in the $\mathcal{Z}$-plane using (3.56). In this case, the matrix $\mathbf{D} = \mathbf{A}^{\mathrm{T}}\mathbf{A}$ is the diagonal matrix in (2.46), and the least squares solution is given by

$$\mathbf{c} = \mathbf{D}^{-1}\mathbf{A}^{\mathrm{T}}\mathbf{b} \tag{3.72}$$

where the terms in the inverse matrix $\mathbf{D}^{-1}$ are $d_{11}^{-1} = 1/M$ for the first term and $2/M$ for all other diagonal terms. This solution may be expressed in summation form as

$$c_n = -\frac{2}{M}\sum_{m=1}^{M} \cos n\theta_m \overset{\text{add}}{\Phi}(z_m) \tag{3.73}$$

Examples are presented in Fig. 3.21 of an isolated slit with uniform potential in a uniform vector field, with the additional function set to the potential in (3.10) at each control point. This figure also illustrates the interactions of a collection of slit-dipoles, each with a uniform potential along the boundary that is distinct from the other elements, where the solution was obtained by iteratively solving for individual elements using Gauss–Seidel iteration where the additional function contains the uniform flow and contributions from all other elements.

A solution is formulated next for a slit-dipole with a ***Robin interface condition***, (2.66b), relating the tangential component of this vector field to a discontinuity of the stream function:

$$v_s = -\frac{1}{\delta}\left(\Psi^{+} - \Psi^{-}\right) \tag{3.74}$$

and δ is specified by the properties of the slit-dipole element and the surrounding domain. The series expansions for the stream function (3.65) and tangential component of the vector field (3.67a) at the control points are substituted to give an expression at the mth point on the boundary:

$$\sum_{n=1}^{N} \frac{n}{L} \frac{\sin n\theta_m \Re c_n}{\sqrt{\xi_m + 1}\sqrt{1 - \xi_m}} + \frac{2}{\delta}\sum_{n=1}^{N} \sin n\theta_m \Re c_n = \Re\left[\overset{\text{add}}{v}(z_m)\mathrm{e}^{-\mathrm{i}\vartheta}\right] \tag{3.75}$$

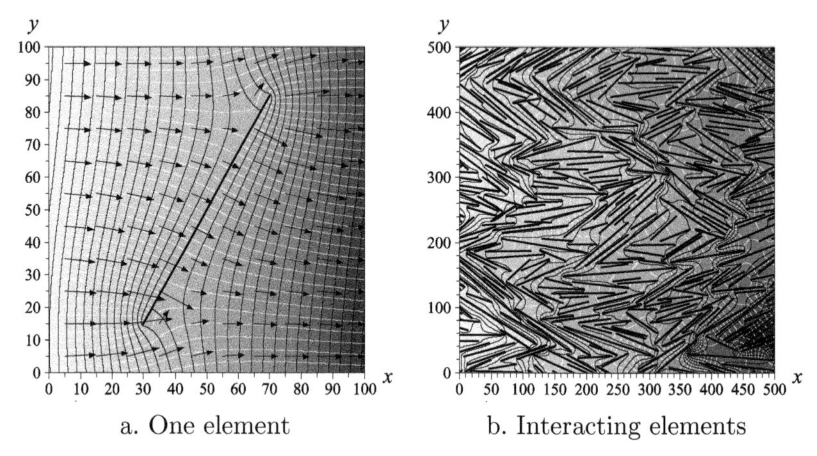

Figure 3.21 *Slit-dipoles with uniform potential along elements. (Reprinted from Steward, 2015, Analysis of discontinuities across thin inhomogeneities, groundwater/surface water interactions in river networks, and circulation about slender bodies using slit elements in the Analytic Element Method,* Water Resources Research, *Vol. 51(11), Fig. 4, with permission from John Wiley and Sons. Copyright 2015 by the American Geophysical Union.)*

with $\Im c_n = 0$, (3.64). A solution for the coefficients c_n is given by solving this set of equations at the M control points, which may be written in matrix form as

$$\left(\mathbf{BAC} + \frac{2}{\delta}\mathbf{A}\right)\mathbf{c} = \mathbf{b} \tag{3.76}$$

using

$$\mathbf{A} = \begin{bmatrix} \sin\theta_1 & \sin 2\theta_1 & \cdots & \sin N\theta_1 \\ \sin\theta_2 & \sin 2\theta_2 & \cdots & \sin N\theta_2 \\ \cdots & \cdots & \cdots & \cdots \\ \sin\theta_M & \sin 2\theta_M & \cdots & \sin N\theta_M \end{bmatrix}, \quad \mathbf{c} = \begin{bmatrix} \Re c_1 \\ \Re c_2 \\ \vdots \\ \Re c_N \end{bmatrix}, \quad \mathbf{b} = \begin{bmatrix} \Re\left[\overset{\text{add}}{v}(z_1)e^{-i\vartheta}\right] \\ \Re\left[\overset{\text{add}}{v}(z_2)e^{-i\vartheta}\right] \\ \vdots \\ \Re\left[\overset{\text{add}}{v}(z_M)e^{-i\vartheta}\right] \end{bmatrix} \tag{3.77a}$$

and the diagonal matrices

$$\mathbf{B} = \begin{bmatrix} \frac{1}{L\sqrt{\xi_1+1}\sqrt{1-\xi_1}} & 0 & \cdots & 0 \\ 0 & \frac{1}{L\sqrt{\xi_2+1}\sqrt{1-\xi_2}} & \cdots & 0 \\ \cdots & \cdots & \cdots & \cdots \\ 0 & 0 & \cdots & \frac{1}{L\sqrt{\xi_M+1}\sqrt{1-\xi_M}} \end{bmatrix}, \quad \mathbf{C} = \begin{bmatrix} 1 & 0 & \cdots & 0 \\ 0 & 2 & \cdots & 0 \\ \cdots & \cdots & \cdots & \cdots \\ 0 & 0 & \cdots & N \end{bmatrix} \tag{3.77b}$$

The least squares solution may be written as

$$\left(\mathbf{BAC} + \frac{2}{\delta}\mathbf{A}\right)^{\mathrm{T}}\left(\mathbf{BAC} + \frac{2}{\delta}\mathbf{A}\right)\mathbf{c} = \left(\mathbf{BAC} + \frac{2}{\delta}\mathbf{A}\right)^{\mathrm{T}}\mathbf{b} \tag{3.78}$$

Examples are presented in Fig. 3.22 to illustrate solutions for slit-dipoles that more readily transmit the flux associated with the vector field than the surrounding medium.

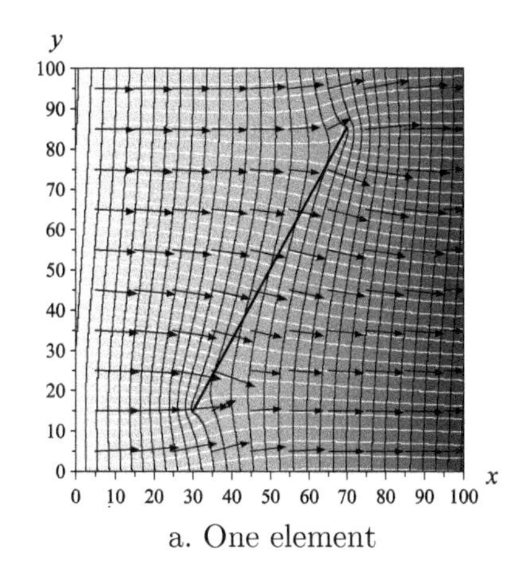

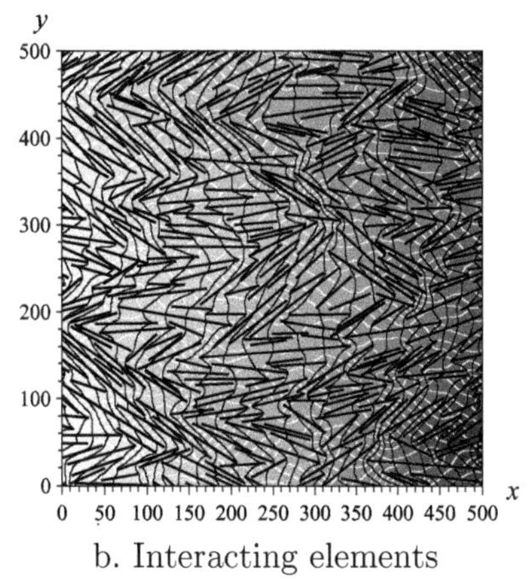

Figure 3.22 *Slit-dipoles with Robin interface condition for the tangential component* v_s. *(Reprinted from Steward, 2015, Analysis of discontinuities across thin inhomogeneities, groundwater/surface water interactions in river networks, and circulation about slender bodies using slit elements in the Analytic Element Method,* Water Resources Research, *Vol. 51(11), Fig. 4, with permission from John Wiley and Sons. Copyright 2015 by the American Geophysical Union.)*

Problem 3.10 Compute the coefficients of a slit-dipole with uniform potential for $N = 10$ and $M = 15$ in a uniform vector field with $v_0 = 1$ and $\Phi_0 = 0$, and compute the complex potential $\Omega = \Phi + \mathrm{i}\Psi$ and the vector field $v = v_x + \mathrm{i}v_y$ on each side of an element at its midpoint $z_c = (z_{\max} + z_{\min})/2$, for an element with specified endpoints:

A. $z_{\min} = 0, z_{\max} = 1 + 1\mathrm{i}$

B. $z_{\min} = 0, z_{\max} = 2 + 2\mathrm{i}$

C. $z_{\min} = 0, z_{\max} = 3 + 3\mathrm{i}$

D. $z_{\min} = 0, z_{\max} = 4 + 4\mathrm{i}$

E. $z_{\min} = 0, z_{\max} = 5 + 5\mathrm{i}$

F. $z_{\min} = 0, z_{\max} = 6 + 6\mathrm{i}$

G. $z_{\min} = 0, z_{\max} = 7 + 7\mathrm{i}$

H. $z_{\min} = 0, z_{\max} = 8 + 8\mathrm{i}$

I. $z_{\min} = 0, z_{\max} = 9 + 9\mathrm{i}$

J. $z_{\min} = 0, z_{\max} = 10 + 10\mathrm{i}$

K. $z_{\min} = 0, z_{\max} = 11 + 11\mathrm{i}$

L. $z_{\min} = 0, z_{\max} = 12 + 12\mathrm{i}$

Problem 3.11 Compute the coefficients of a slit-dipole with Robin conditions using $\delta = 10$, $N = 10$, and $M = 15$; and compute the complex potential Ω and the vector field v on each side of an element at its midpoint for specified endpoints and uniform vector field in Problem 3.10.

3.4.2 Slit-Doublet: Boundaries with Continuous Stream Function

The next set of slit elements will be formulated by setting the real part of the coefficients equal to zero

$$\Re c_n = 0 \tag{3.79}$$

and is called a ***slit-doublet*** (Steward, 2015). The non-zero terms in the stream function at control points along the boundary (3.65)

$$\Psi_m = \sum_{n=1}^{N} \cos n\theta_m \Im c_n + \overset{\text{add}}{\Psi}(z_m) = \Psi_U \tag{3.80}$$

are continuous across the element. A slit-doublet has a discontinuous potential across the element in (3.64) with a jump of

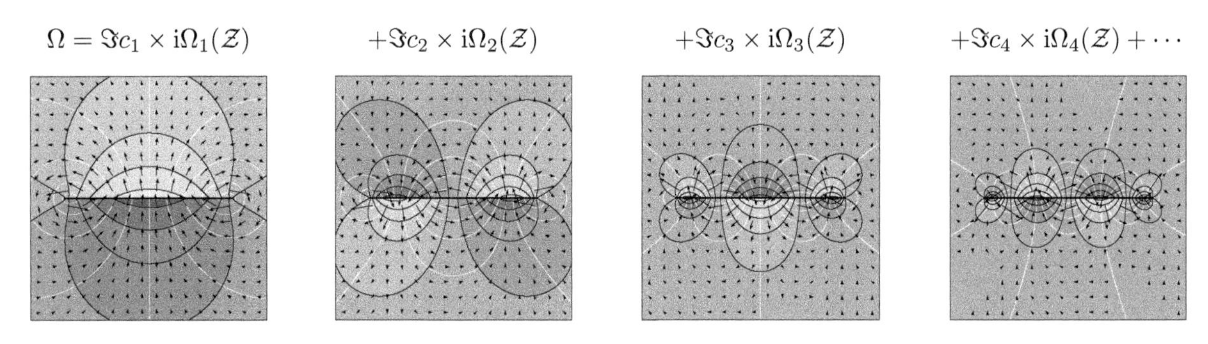

Figure 3.23 *Influence functions for slit-doublet with continuous stream function across the element from (3.62) and (3.63) with (3.79).*

$$\Delta\Phi = \Phi_m^+ - \Phi_m^+{}^- = 2\sum_{n=1}^{N} \sin n\theta_m \Im c_n \tag{3.81}$$

since the term $\sin n\theta_m$ in (3.62) jumps across the element. The influence functions in Fig. 3.23 illustrate the ***discontinuous potential and continuous stream function across a slit-dipole***.

A slit-doublet with ***uniform stream function*** along the element is obtained by setting the stream function at the control points, (3.80), equal to the unknown value of stream function Ψ_U. This provides a system of M equations with $N+1$ unknowns (including the unknown value of Ψ_U) in matrix form as $\mathbf{Ac} = \mathbf{b}$ where

$$\mathbf{A} = \begin{bmatrix} 1 & \cos\theta_1 & \cos 2\theta_1 & \cdots & \cos N\theta_1 \\ 1 & \cos\theta_2 & \cos 2\theta_2 & \cdots & \cos N\theta_2 \\ \cdots & \cdots & \cdots & \cdots & \cdots \\ 1 & \cos\theta_M & \cos 2\theta_M & \cdots & \cos N\theta_M \end{bmatrix}, \quad \mathbf{c} = \begin{bmatrix} -\Psi_U \\ \Im c_1 \\ \Im c_2 \\ \vdots \\ \Im c_N \end{bmatrix}, \quad \mathbf{b} = \begin{bmatrix} -\overset{\text{add}}{\Psi}(z_1) \\ -\overset{\text{add}}{\Psi}(z_2) \\ \vdots \\ -\overset{\text{add}}{\Psi}(z_M) \end{bmatrix} \tag{3.82}$$

This matrix $\mathbf{A}$ is identical to that for a uniform potential in (3.71), and the least squares solution is expressed in summation form as

$$c_n = -\mathrm{i}\frac{2}{M}\sum_{m=1}^{M} \cos n\theta_m \overset{\text{add}}{\Psi}(z_m) \tag{3.83}$$

Examples are presented in Fig. 3.24 of an isolated slit-doublet with uniform stream function in a uniform vector field, and a collection of slit-doublets, each with a uniform stream function along the boundary that is distinct from the other elements.

A slit-doublet may also be formulated with a ***Robin interface condition***, (2.66a), relating the normal component of the vector field, which is continuous across the element, to the jump in potential across the slit-doublet:

$$v_n = -\frac{1}{\delta}\left(\Phi^+ - \Phi^-\right) \tag{3.84}$$

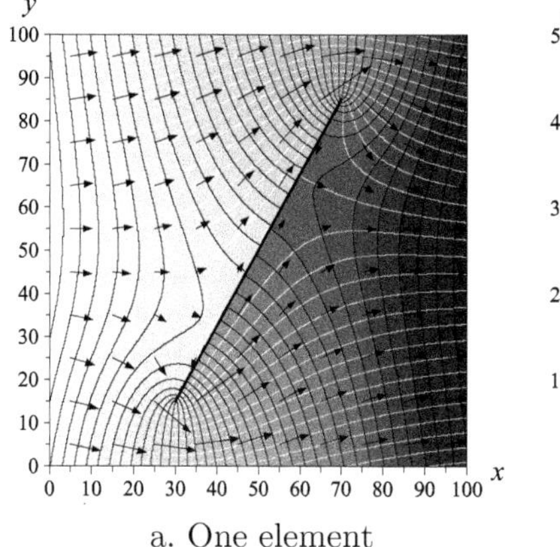

a. One element

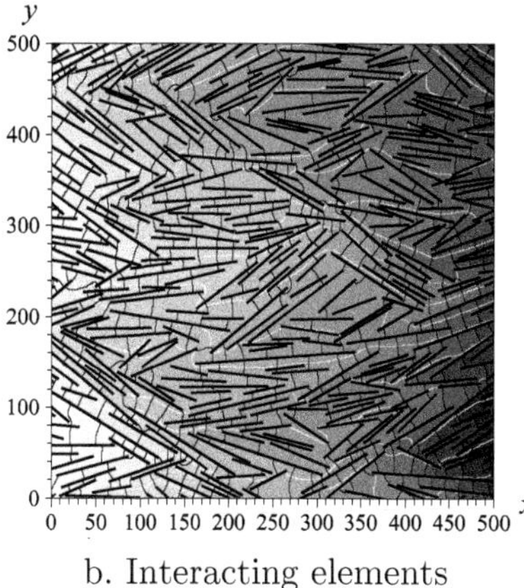

b. Interacting elements

Figure 3.24 *Slit-doublets with uniform stream function along elements. (Reprinted from Steward, 2015, Analysis of discontinuities across thin inhomogeneities, groundwater/surface water interactions in river networks, and circulation about slender bodies using slit elements in the Analytic Element Method,* Water Resources Research, *Vol. 51(11), Fig. 5, with permission from John Wiley and Sons. Copyright 2015 by the American Geophysical Union.)*

Expressions for the potential function (3.64) and the normal component of the vector field (3.67b) at the control points are substituted into this equation to give an expression at the mth point on the boundary:

$$\sum_{n=1}^{N} \frac{n}{L} \frac{\sin n\theta_m \Im c_n}{\sqrt{\xi_m + 1}\sqrt{1 - \xi_m}} + \frac{2}{\delta} \sum_{n=1}^{N} \sin n\theta_m \Im c_n = -\Im\left[\overset{\text{add}}{v}(z_m) e^{-i\vartheta}\right] \tag{3.85}$$

This element has a continuous stream function across the element, (2.65a), which is satisfied when $\Re c_n = 0$, (3.65). This set of equations at the M control points may be written in matrix form as

$$\left(\mathbf{BAC} + \frac{2}{\delta}\mathbf{A}\right)\mathbf{c} = \mathbf{b} \tag{3.86}$$

where **A**, **B**, and **C** are the same matrices in (3.77) and the other matrices are

$$\mathbf{c} = \begin{bmatrix} \Im c_1 \\ \Im c_2 \\ \vdots \\ \Im c_N \end{bmatrix}, \quad \mathbf{b} = \begin{bmatrix} -\Im\left[\overset{\text{add}}{v}(z_1) e^{-i\vartheta}\right] \\ -\Im\left[\overset{\text{add}}{v}(z_2) e^{-i\vartheta}\right] \\ \vdots \\ -\Im\left[\overset{\text{add}}{v}(z_M) e^{-i\vartheta}\right] \end{bmatrix} \tag{3.87}$$

This system of equations may be solved via least squares. An example is presented in Fig. 3.25 for slit-doublets with a lower flux associated with the vector field than the surrounding domain.

Problem 3.12 Compute the coefficients of a slit-doublet with uniform stream function, and compute the complex potential $\Omega = \Phi + i\Psi$ and the vector field $v = v_x + iv_y$ on each side of an element at its midpoint for specified endpoints and uniform vector field in Problem 3.10.

Problem 3.13 Compute the coefficients of a slit-doublet with Robin conditions using $\delta = 10$, $N = 10$, and $M = 15$; and compute the complex potential Ω and the vector field v on each side of an element at its midpoint for specified endpoints and uniform vector field in Problem 3.10.

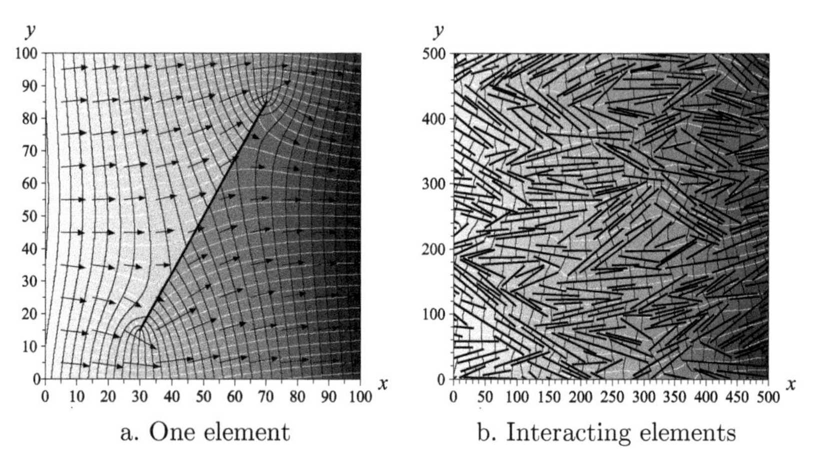

Figure 3.25 *Slit-doublets with Robin interface condition for the normal component* v_n. *(Reprinted from Steward, 2015, Analysis of discontinuities across thin inhomogeneities, groundwater/surface water interactions in river networks, and circulation about slender bodies using slit elements in the Analytic Element Method,* Water Resources Research, *Vol. 51(11), Fig. 5, with permission from John Wiley and Sons. Copyright 2015 by the American Geophysical Union.)*

3.4.3 Slit-Sink and Slit-Vortex: Divergence and Circulation

A ***slit-sink*** is formulated by adding a sink term, (3.60) to the Laurent series (3.62) for a slit-dipole

$$\Omega(z) = Q\left(\frac{\ln \mathcal{Z}}{2\pi} - 1\right) + \sum_{n=1}^{N} \Re c_n \mathcal{Z}^{-n} + \overset{\text{add}}{\Omega}(z) \tag{3.88}$$

where the imaginary part of the coefficients $\Im c_n$ is zero to maintain continuous potential across the slit-sink (Steward, 2015). This potential may be evaluated at the control points along the boundary, (3.64), similar to (3.73) to give

$$\boxed{\Phi_m = -Q + \sum_{n=1}^{N} \cos n\theta_m \Re c_n + \overset{\text{add}}{\Phi}(z_m)} \tag{3.89}$$

Note that inclusion of the sink term is necessary to be able to satisfy a condition of ***specified potential***, since this boundary condition can only be satisfied if a net flux is removed from the domain by a slit-sink.

This system of M equations with $N+1$ unknowns may be written in matrix form with Φ_m equal to the specified values of potential at each control point

$$\mathbf{A} = \begin{bmatrix} 1 & \cos\theta_1 & \cos 2\theta_1 & \cdots & \cos N\theta_1 \\ 1 & \cos\theta_2 & \cos 2\theta_2 & \cdots & \cos N\theta_2 \\ \cdots & \cdots & \cdots & \cdots & \cdots \\ 1 & \cos\theta_M & \cos 2\theta_M & \cdots & \cos N\theta_M \end{bmatrix}, \quad \mathbf{c} = \begin{bmatrix} -Q \\ \Re c_1 \\ \Re c_2 \\ \vdots \\ \Re c_N \end{bmatrix}, \quad \mathbf{b} = \begin{bmatrix} \Phi_1 - \overset{\text{add}}{\Phi}(z_1) \\ \Phi_2 - \overset{\text{add}}{\Phi}(z_2) \\ \vdots \\ \Phi_3 - \overset{\text{add}}{\Phi}(z_M) \end{bmatrix} \tag{3.90}$$

Since the matrix $\mathbf{A}$ is identical to that in (3.71), the orthogonality of the Fourier series for this control point spacing gives a series expansion for the coefficients as in (3.73):

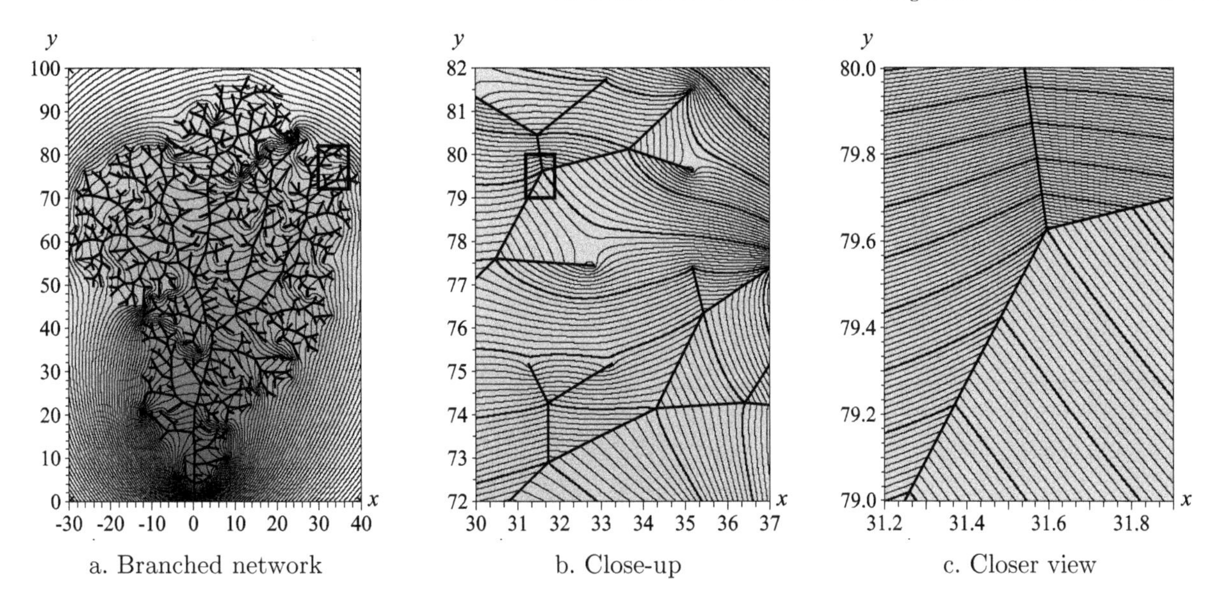

Figure 3.26 *Slit-sinks in a branched network with linear variation in Φ along elements. (Reprinted from Steward, 2015, Analysis of discontinuities across thin inhomogeneities, groundwater/surface water interactions in river networks, and circulation about slender bodies using slit elements in the Analytic Element Method,* Water Resources Research, *Vol. 51(11), Fig. 6, with permission from John Wiley and Sons. Copyright 2015 by the American Geophysical Union.)*

$$
\boxed{\begin{aligned} Q &= -\frac{1}{M}\sum_{m=1}^{M} \Phi_m - \overset{\text{add}}{\Phi}(z_m) \\ c_n &= \frac{2}{M}\sum_{m=1}^{M} \cos n\theta_m \left[\Phi_m - \overset{\text{add}}{\Phi}(z_m)\right] \end{aligned}} \tag{3.91}
$$

This solve process was implemented for a set of slit-sink elements with specified potential shown in Fig. 3.26. The control points for each element were specified so the potential varies linearly and increases in the magnitude of one of the displayed contour intervals. The additional function for each element contains the other slit-sinks and a uniform component that was solved using (3.11) to maintain the same value of potential at each corner of Fig. 3.26a. The close-up figures magnified a sequence of images by a factor of 10 to illustrate the capacity of the solution to nearly exactly match the specified boundary conditions along each element.

Problem 3.14 A slit element is placed in a uniform background $\Omega = -\overline{v_0} z + \Phi_0$ where $v_0 = 1$ and $\Phi_0 = 0$. Compute the coefficients Q, and c_n for $N = 10$ coefficients and $M = 15$ control points to satisfy the specified potential $\Phi_m = 1$ along the slit element with endpoints:

A. $z_{\text{min}} = 0$, $z_{\text{max}} = 1 + 1\text{i}$

B. $z_{\text{min}} = 0$, $z_{\text{max}} = 2 + 2\text{i}$

C. $z_{\text{min}} = 0$, $z_{\text{max}} = 3 + 3\text{i}$

D. $z_{\text{min}} = 0$, $z_{\text{max}} = 4 + 4\text{i}$

E. $z_{\text{min}} = 0$, $z_{\text{max}} = 5 + 5\text{i}$

F. $z_{\text{min}} = 0$, $z_{\text{max}} = 6 + 6\text{i}$

G. $z_{\text{min}} = 0$, $z_{\text{max}} = 7 + 7\text{i}$

H. $z_{\text{min}} = 0$, $z_{\text{max}} = 8 + 8\text{i}$

I. $z_{\text{min}} = 0$, $z_{\text{max}} = 9 + 9\text{i}$

J. $z_{\text{min}} = 0$, $z_{\text{max}} = 10 + 10\text{i}$

K. $z_{\text{min}} = 0$, $z_{\text{max}} = 11 + 11\text{i}$

L. $z_{\text{min}} = 0$, $z_{\text{max}} = 12 + 12\text{i}$

Problem 3.15 A slit element is placed in a uniform background $\Omega = -\overline{v_0}z + \Phi_0$ with $N = 10$ coefficients and $M = 15$ control points with specified potential $\Phi_m = 1$ and endpoints $z_{\min} = -1 - \mathrm{i}$ and $z_{\max} = 1 + \mathrm{i}$. Compute the coefficients Q, and c_n to satisfy the specified conditions along the slit element while solving for Φ_0, v_0 to satisfy the specified potential at $z_1 = -2 + 2\mathrm{i}$, $z_2 = 2 + 2\mathrm{i}$, and $z_3 = 2 - 2\mathrm{i}$.

	A.	B.	C.	D.	E.	F.	G.	H.	I.	J.	K.	L.
Φ_1	2	3	4	5	1	1	1	1	1	1	1	1
Φ_2	1	1	1	1	2	3	4	5	1	1	1	1
Φ_3	1	1	1	1	1	1	1	1	2	3	4	5

Note that this solution requires an iterative procedure that independently solves for the uniform flow components and the element components, following (2.9).

A ***slit-vortex*** is obtained by adding a vortex term, (3.61), to the Laurent series for a slit-doublet

$$\Omega(z) = \Gamma \mathrm{i}\left(\frac{\ln \mathcal{Z}}{2\pi} - 1\right) + \sum_{n=1}^{N} \Im c_n \mathcal{Z}^{-n} + \overset{\text{add}}{\Omega}(z) \tag{3.92}$$

where the real part $\Re c_n = 0$ to maintain continuous stream function across the slit-vortex (3.79). A slit-vortex may be solved to satisfy a boundary condition of zero normal component of the vector field by adjusting the coefficients in the Laurent series of the slit-doublet terms using (3.82). The vortex term enables a slit-vortex to satisfy the Kutta condition, since this boundary condition requires a net circulation about the slit-vortex.

The ***Kutta condition*** requires the vector field to remain finite at the trailing edge of an element, which is satisfied by adjusting Γ as follows. The complex vector field in the $\mathcal{Z}$-plane is related to the partial derivative of the complex potential, (3.92), with respect to $\mathcal{Z}$

$$\frac{\partial \Omega}{\partial \mathcal{Z}} = \Gamma \frac{\mathrm{i}}{2\pi}\frac{1}{\mathcal{Z}} + \sum_{n=1}^{N} \mathrm{i}\Im c_n(-n)\mathcal{Z}^{-n-1} + \frac{\partial \overset{\text{add}}{\Omega}}{\partial \mathcal{Z}} \tag{3.93}$$

The derivative of the complex potential for the additional functions with respect to $\mathcal{Z}$ maybe be evaluated using

$$\frac{\partial \overset{\text{add}}{\Omega}(z)}{\partial \mathcal{Z}} = \frac{\partial \overset{\text{add}}{\Omega}}{\partial z}\frac{\partial z}{\partial \zeta}\frac{\partial \zeta}{\partial \mathcal{Z}} = \frac{\partial \overset{\text{add}}{\Omega}}{\partial z}\left(\frac{z_{\max} - z_{\min}}{2}\right)\left(\frac{\sqrt{\zeta+1}\sqrt{\zeta-1}}{\mathcal{Z}}\right) = 0 \qquad (\zeta = \pm 1) \tag{3.94}$$

and this term is zero at the endpoints. The Kutta condition is satisfied when $V = -\overline{\partial\Omega/\partial\mathcal{Z}} = 0$ at one of the endpoints, which gives

$$\Gamma = \begin{cases} 2\pi \sum_{n=1}^{N} n(-1)^n \Im c_n & (z = z_{\min},\ \mathcal{Z} = -1) \\ 2\pi \sum_{n=1}^{N} n \Im c_n & (z = z_{\max},\ \mathcal{Z} = 1) \end{cases} \tag{3.95}$$

where the Kutta condition is applied at the endpoint along the trailing edge. The example in Fig. 3.27a illustrates a slit-vortex with zero normal component of the vector field and a Kutta condition on its left endpoint.

A string of slit-vortex elements may be aligned with contiguous endpoints to satisfy the Kutta condition along more general geometries. For example, the first element in Fig. 3.27b

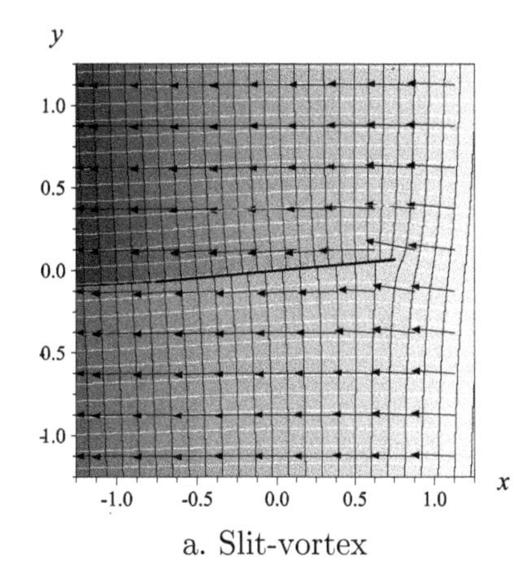

a. Slit-vortex

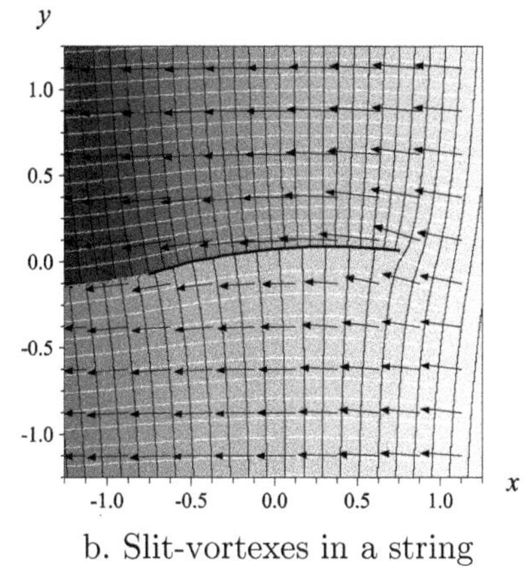

b. Slit-vortexes in a string

Figure 3.27 *Slit-vortex elements with uniform stream function and Kutta condition at the left endpoint. (Reprinted from Steward, 2015, Analysis of discontinuities across thin inhomogeneities, groundwater/surface water interactions in river networks, and circulation about slender bodies using slit elements in the Analytic Element Method,* Water Resources Research, *Vol. 51(11), Fig. 7 with permission from John Wiley and Sons. Copyright 2015 by the American Geophysical Union.)*

satisfies the Kutta condition along its left endpoint, and the other elements lie along the same wing. This is achieved by setting the stream function for these elements to have the same value as the left-most element. Satisfying a boundary condition of ***specified stream function*** at control points along the element is achieved using the imaginary part of the complex potential for the slit-vortex, (3.92)

$$\Psi_m = -\Gamma + \sum_{n=1}^{N} \cos n\theta_m \Im c_n + \overset{\text{add}}{\Psi}(z_m) \tag{3.96}$$

This leads to the same set of equations as a slit-sink with specified potential (3.90). The solution is given by (3.91)

$$\begin{aligned} \Gamma &= -\frac{1}{M}\sum_{m=1}^{M} \Psi_m - \overset{\text{add}}{\Psi}(z_m) \\ c_n &= \mathrm{i}\frac{2}{M}\sum_{m=1}^{M} \cos n\theta_m \left[\Psi_m - \overset{\text{add}}{\Psi}(z_m)\right] \end{aligned} \tag{3.97}$$

The remaining terms in the stream function along the boundary (3.65) may be evaluated at the control points, similar to the slit-doublet in (3.80), to give

$$\Psi_m + \Gamma = \sum_{n=1}^{N} \cos n\theta_m \Im c_n + \overset{\text{add}}{\Psi}(z_m) = \Psi_U \tag{3.98}$$

Since the vortex term Γ does not contribute to the value of the stream function along the slit, this system of M equations with $N+1$ unknowns has the same form as (3.80) and the values of these unknown coefficients are given by (3.83):

$$c_n = -\mathrm{i}\frac{2}{M}\sum_{m=1}^{M} \cos n\theta_m \overset{\text{add}}{\Psi}(z_m) \tag{3.99}$$

The solve process is partitioned into the coefficients associated with each ith slit-vortex, similar to what has been done before, where the additional function contains the specified uniform vector field v_{x0} and all other elements. This is solved iteratively, by solving for the element at the end with the Kutta condition, determining its stream function, setting the stream function of the other elements equal to this, and continuing until small differences between iterates occurs.

Problem 3.16 Compute Γ and the coefficients of a slit-vortex with $N = 10$ and $M = 15$ that satisfies a Kutta condition at end z_{max} in a uniform vector field with $v_0 = 1$ and $\Phi_0 = 0$, for an element with specified endpoints:

A. $z_{\min} = 0, z_{\max} = 10 - 1.1\text{i}$
B. $z_{\min} = 0, z_{\max} = 10 - 1.2\text{i}$
C. $z_{\min} = 0, z_{\max} = 10 - 1.3\text{i}$
D. $z_{\min} = 0, z_{\max} = 10 - 1.4\text{i}$
E. $z_{\min} = 0, z_{\max} = 10 - 1.5\text{i}$
F. $z_{\min} = 0, z_{\max} = 10 - 1.6\text{i}$
G. $z_{\min} = 0, z_{\max} = 10 - 1.7\text{i}$
H. $z_{\min} = 0, z_{\max} = 10 - 1.8\text{i}$
I. $z_{\min} = 0, z_{\max} = 10 - 1.9\text{i}$
J. $z_{\min} = 0, z_{\max} = 10 - 2.0\text{i}$
K. $z_{\min} = 0, z_{\max} = 10 - 2.1\text{i}$
L. $z_{\min} = 0, z_{\max} = 10 - 2.2\text{i}$

3.5 Circular Arcs and Joukowsky's Wing

The conformal mappings used to formulate slit elements in the last section may be applied to study analytic elements with shapes associated with potential flow around wings. The ***Joukowsky transformation*** was used to map the exterior of a unit circle in the $\mathcal{Z}$-plane to a slit in the z-plane in (3.55), and is repeated here for a slit of length $2L$:

$$\frac{z}{L} = \frac{1}{2}\left(\mathcal{Z} + \frac{1}{\mathcal{Z}}\right) \tag{3.100}$$

This conformal mapping may also be used to map other geometries. For example, the thicker circle centered at $\mathcal{Z}_c = \text{i}\tan(\vartheta_c/2)$ with radius $r_0 = 1/\cos(\vartheta_c/2)$ in the $\mathcal{Z}$-plane in Fig. 3.28a maps to a ***circular arc*** in the z-plane shown in Fig. 3.28b. This arc forms an angle ϑ_c between the line tangent to the arc and the chord along the x-axis at the endpoints where $\mathcal{Z} = \pm 1$ map to $z = \pm L$. The camber of this thin wing, $\frac{L}{\sin\vartheta_c} - \frac{L}{\tan\vartheta_c}$, is the maximum distance it lies above the x-axis, which occurs at its center.

The Joukowsky transformation forms a ***two-sheeted Riemann surface*** where every point in the z-plane has two corresponding locations in the $\mathcal{Z}$-plane. For example

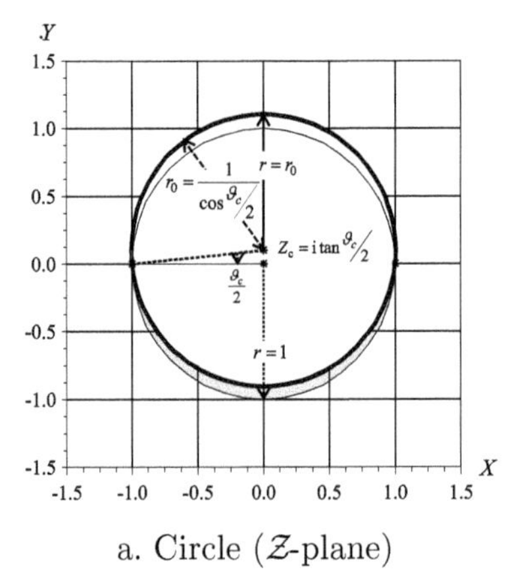

a. Circle ($\mathcal{Z}$-plane)

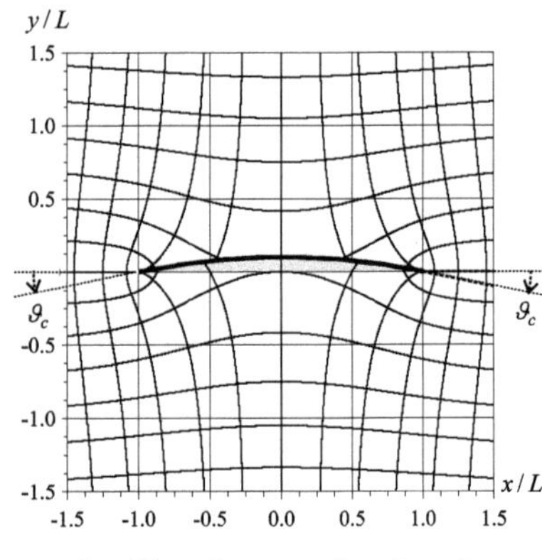

b. Circular arc (z-plane)

Figure 3.28 *Conformally mapping the outside of a circle to a circular arc, showing lines of constant X and Y.*

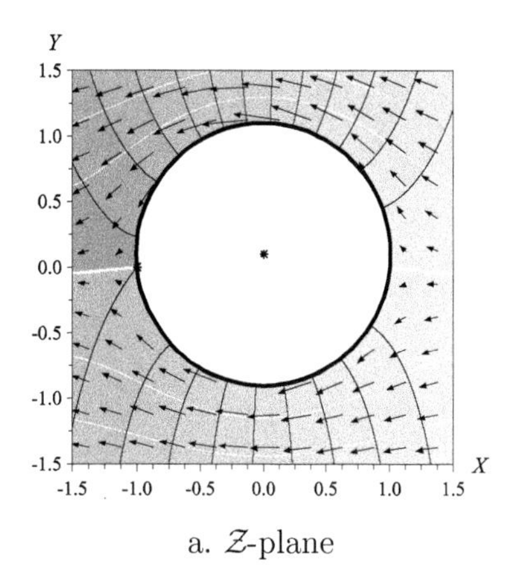

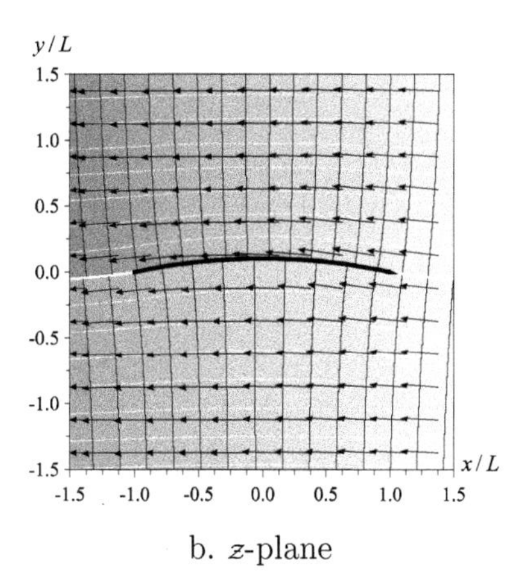

Figure 3.29 *Vector field about a circular arc with Kutta condition.*

$\mathcal{Z} = 2 - \sqrt{3}$ lies inside the unit circle, $\mathcal{Z} = 2 + \sqrt{3}$ is outside, and they both map to $z = 2L$. The particular sheet of the Riemann surface that is utilized in the conformal mapping is specified by the choice of the inverse mapping $z(\mathcal{Z})$. A mapping that is continuous in the physical z-plane is achieved when the Riemann sheet outside the circle ($|\mathcal{Z}| > 1$) is used everywhere except the shaded area above the y-axis and below the circular arc. This is given by

$$\mathcal{Z} = \begin{cases} \frac{z}{L} + \sqrt{\frac{z}{L} + 1}\sqrt{\frac{z}{L} - 1} & \left(y < 0 \text{ OR } \left|\frac{z}{L} + \mathrm{i}\frac{1}{\tan\vartheta_c}\right| > \frac{1}{\sin\vartheta_c}\right) \\ \frac{z}{L} - \sqrt{\frac{z}{L} + 1}\sqrt{\frac{z}{L} - 1} & \left(y \geq 0 \text{ AND } \left|\frac{z}{L} + \mathrm{i}\frac{1}{\tan\vartheta_c}\right| \leq \frac{1}{\sin\vartheta_c}\right) \end{cases} \tag{3.101}$$

and continuity of the mapping is illustrated by the lines of constant X and Y across the interface on the x-axis.

The boundary of a wing satisfies a Neumann condition with $v_n = 0$ and generates circulation about the element. These conditions are satisfied for a wing in a uniform vector field v_0 by the complex potential in the $\mathcal{Z}$-plane of the superposition of a point-dipole in uniform flow, (3.3), with a point-vortex, (3.7):

$$\Omega = -\frac{\overline{v_0}}{2}\mathcal{Z} - \frac{v_0 r_0{}^2}{2}\frac{1}{\mathcal{Z} - \mathcal{Z}_c} + \frac{\mathrm{i}\Gamma}{2\pi}\ln\frac{\mathcal{Z} - \mathcal{Z}_c}{r_0} \tag{3.102}$$

Note that v_0 needs to be factored by 1/2 in the $\mathcal{Z}$-plane to give the correct far-field vector field in the physical plane, and Γ needs to not be factored in the $\mathcal{Z}$-plane, to give the correct circulation, since the net circulation is the same in both planes. The vector field in the z-plane is obtained using the chain rule

$$v = -\overline{\frac{\mathrm{d}\Omega}{\mathrm{d}z}} = -\overline{\frac{\mathrm{d}\Omega}{\mathrm{d}\mathcal{Z}}\frac{\mathrm{d}\mathcal{Z}}{\mathrm{d}z}} = \overline{\frac{\mathcal{Z}^2}{\mathcal{Z}^2 - 1}}\left[v_0 - \overline{v_0} r_0{}^2\overline{\frac{1}{(\mathcal{Z} - \mathcal{Z}_c)^2}} + \frac{\mathrm{i}\Gamma}{\pi}\overline{\frac{1}{\mathcal{Z} - \mathcal{Z}_c}}\right] \tag{3.103}$$

The Kutta condition requires that flow remain finite at the sharp tip at the tail endpoint of the wing. This is satisfied when the stagnation point in the $\mathcal{Z}$-plane lies at this location,

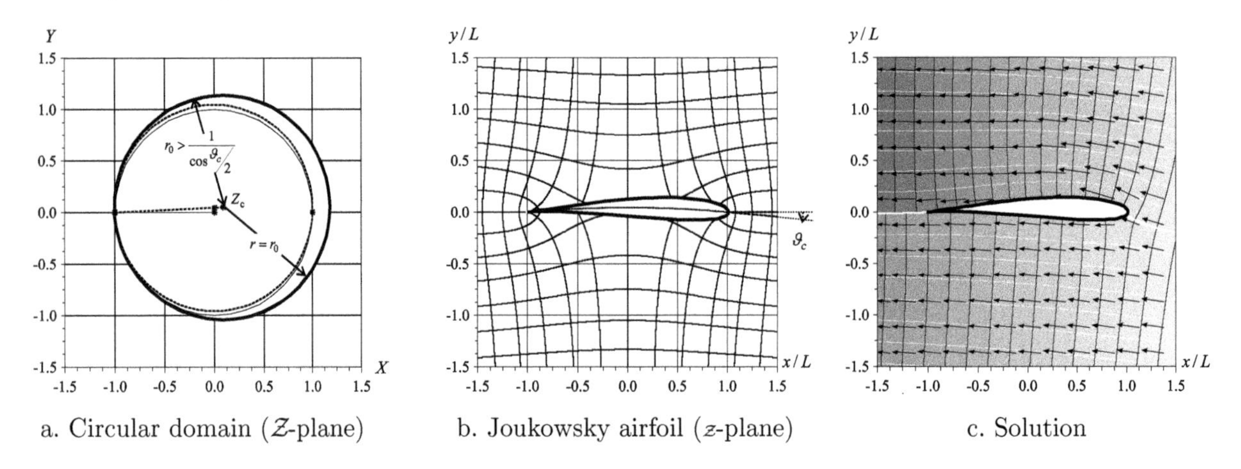

Figure 3.30 *Exact solutions for two-dimensional wings from the Joukowsky transformation.*

which occurs when the circulation of the point-vortex is

$$\Gamma = 2\pi |v_0| r_0 \sin\left(\arg v_0 - \frac{\vartheta_c}{2}\right) \tag{3.104}$$

similar to (Abbott, 1959, chap.3). The complex potential and vector field are illustrated in Fig. 3.29 both in the circular $\mathcal{Z}$-plane and in the physical z-plane about the ***thin wing with the geometry of a circular arc***.

These methods may be extended to wings with finite thickness, and there are two parameters that control the shape of the ***Joukowsky airfoil*** illustrated in Fig. 3.30. In this case, the Joukowsky transformation is used to map a circle passing through $\mathcal{Z} = -1$ and centered at $\mathcal{Z}_c = X_c + \mathrm{i}Y_c$ onto the z-plane, with $X_c \geq 0$ and $Y_c \geq 0$. The angle ϑ_c formed between the ξ-axis and the line passing through $\mathcal{Z} = -1$ and $\mathcal{Z} = \mathcal{Z}_c$ controls the camber (convex upward curvature). This is illustrated by the thick dotted line that passes through $\mathcal{Z} = -1$ and $\mathcal{Z} = +1$ and is centered at $\mathcal{Z} = \mathrm{i}\tan\vartheta_c$. This line maps to the dotted circular arc in the z-plane that passes through $z = -1$ and $z = 1$ and has arguments of $\pm 2\vartheta_c$ at these points, and this multiplicity in angle follows directly from the Taylor series and Fig. 2.14. The second parameter r_0 controls with thickness of the airfoil. In the limit that $r_0 = 1/\cos(\vartheta_c)$ and $\mathcal{Z}_c$ lies on the Y-axis, the airfoil has a thickness of zero and takes on the shape of the dotted circular arc in Fig. 3.29a. As r_0 increases, the airfoil becomes thicker, as illustrated by the solution for ***flow about a Joukowsky airfoil with Kutta condition at its trailing sharp edge***.

Problem 3.17 Compute the ratio of length $2L$ to camber, the circulation Γ, and the velocity at the top and bottom interfaces of a circular arc at its middle where $x = 0$, for the following configuration in a background velocity $v_0 = -1$.

A. L=1, ϑ_c=8.1°
B. L=1, ϑ_c=8.2°
C. L=1, ϑ_c=8.3°
D. L=1, ϑ_c=8.4°
E. L=1, ϑ_c=8.5°
F. L=1, ϑ_c=8.6°
G. L=1, ϑ_c=8.7°
H. L=1, ϑ_c=8.8°
I. L=1, ϑ_c=8.9°
J. L=1, ϑ_c=9.0°
K. L=1, ϑ_c=9.1°
L. L=1, ϑ_c=9.2°

3.6 Complex Vector Fields with Divergence and Curl

Complex functions can be extended to ***two-dimensional rotational or divergent vector fields by expressing the complex function*** Ω ***as functions of both*** z ***and*** $\overline{z}$. This provides a set of relations between the complex function Ω and the potential function Φ and stream function Ψ:

$$\begin{aligned} \Omega(z,\overline{z}) &= \Phi + \mathrm{i}\Psi \\ \overline{\Omega(z,\overline{z})} &= \Phi - \mathrm{i}\Psi \end{aligned} \quad \rightarrow \quad \Phi = \frac{\Omega + \overline{\Omega}}{2}, \quad \Psi = \frac{\Omega - \overline{\Omega}}{2\mathrm{i}} \tag{3.105}$$

Likewise, the real and complex variables may be written as

$$\begin{aligned} z &= x + \mathrm{i}y = r\mathrm{e}^{\mathrm{i}\theta} \\ \overline{z} &= x - \mathrm{i}y = r\mathrm{e}^{-\mathrm{i}\theta} \end{aligned} \quad \rightarrow \quad x = \frac{z + \overline{z}}{2}, \quad y = \frac{z - \overline{z}}{2\mathrm{i}} \tag{3.106}$$

A complex vector field $v = v_x + \mathrm{i}v_y$ is related to the first derivatives of Ω, and partial differentiation is organized using the chain rule with the derivatives of the last equation, which give

$$\begin{aligned} \frac{\partial \Omega}{\partial z} &= \frac{\partial x}{\partial z}\frac{\partial \Omega}{\partial x} + \frac{\partial y}{\partial z}\frac{\partial \Omega}{\partial y} = \frac{1}{2}\left(\frac{\partial \Omega}{\partial x} - \mathrm{i}\frac{\partial \Omega}{\partial y}\right) \\ \frac{\partial \Omega}{\partial \overline{z}} &= \frac{\partial x}{\partial \overline{z}}\frac{\partial \Omega}{\partial x} + \frac{\partial y}{\partial \overline{z}}\frac{\partial \Omega}{\partial y} = \frac{1}{2}\left(\frac{\partial \Omega}{\partial x} + \mathrm{i}\frac{\partial \Omega}{\partial y}\right) \end{aligned} \quad \rightarrow \quad \begin{aligned} \frac{\partial \Omega}{\partial x} &= \frac{\partial \Omega}{\partial z} + \frac{\partial \Omega}{\partial \overline{z}} \\ \frac{\partial \Omega}{\partial y} &= \mathrm{i}\left(\frac{\partial \Omega}{\partial z} - \frac{\partial \Omega}{\partial \overline{z}}\right) \end{aligned} \tag{3.107}$$

The ***irrotational, divergent component of a vector field*** may be represented using a scalar potential by Helmholtz theorem, (1.6), and this vector field is given by

$$v = v_x + \mathrm{i}v_y = -\frac{\partial \Phi}{\partial x} - \mathrm{i}\frac{\partial \Phi}{\partial y} = -\left(\frac{\partial \Phi}{\partial z} + \frac{\partial \Phi}{\partial \overline{z}}\right) + \left(\frac{\partial \Phi}{\partial z} - \frac{\partial \Phi}{\partial \overline{z}}\right) = -2\frac{\partial \Phi}{\partial \overline{z}} \tag{3.108a}$$

Likewise, the ***divergence-free, rotational component of a vector field*** may be represented using a stream function for two-dimensional flow, (1.11), giving

$$v = v_x + \mathrm{i}v_y = -\frac{\partial \Psi}{\partial y} + \mathrm{i}\frac{\partial \Psi}{\partial x} = -\mathrm{i}\left(\frac{\partial \Psi}{\partial z} - \frac{\partial \Psi}{\partial \overline{z}}\right) + \mathrm{i}\left(\frac{\partial \Psi}{\partial z} + \frac{\partial \Psi}{\partial \overline{z}}\right) = 2\mathrm{i}\frac{\partial \Psi}{\partial \overline{z}} \tag{3.108b}$$

Together, these two expressions enable computation of a vector field from Ω or its complex conjugate:

$$\boxed{v = v_x + \mathrm{i}v_y = -2\frac{\partial \overline{\Omega}}{\partial \overline{z}} = -2\overline{\frac{\partial \Omega}{\partial z}}} \tag{3.109}$$

for general two-dimensional vector fields containing divergent and/or rotational components.

The governing equations are formulated using the Laplacian operator, (2.1), which needs the second derivatives of a function, given by

$$\begin{aligned} \frac{\partial^2\Omega}{\partial z^2} &= \frac{1}{4}\left(\frac{\partial^2\Omega}{\partial x^2} - 2\mathrm{i}\frac{\partial^2\Omega}{\partial x\partial y} - \frac{\partial^2\Omega}{\partial y^2}\right) \\ \frac{\partial^2\Omega}{\partial \overline{z}^2} &= \frac{1}{4}\left(\frac{\partial^2\Omega}{\partial x^2} + 2\mathrm{i}\frac{\partial^2\Omega}{\partial x\partial y} - \frac{\partial^2\Omega}{\partial y^2}\right) \\ \frac{\partial^2\Omega}{\partial z\partial \overline{z}} &= \frac{1}{4}\left(\frac{\partial^2\Omega}{\partial x^2} + \frac{\partial^2\Omega}{\partial y^2}\right) \end{aligned} \quad \Rightarrow \quad \begin{aligned} \frac{\partial^2\Omega}{\partial x^2} &= \frac{\partial^2\Omega}{\partial z^2} + 2\frac{\partial^2\Omega}{\partial z\partial \overline{z}} + \frac{\partial^2\Omega}{\partial \overline{z}^2} \\ \frac{\partial^2\Omega}{\partial y^2} &= -\frac{\partial^2\Omega}{\partial z^2} + 2\frac{\partial^2\Omega}{\partial z\partial \overline{z}} - \frac{\partial^2\Omega}{\partial \overline{z}^2} \\ \frac{\partial^2\Omega}{\partial x\partial y} &= \mathrm{i}\left(\frac{\partial^2\Omega}{\partial z^2} - \frac{\partial^2\Omega}{\partial \overline{z}^2}\right) \end{aligned} \tag{3.110}$$

This gives a complex form for the ***Laplacian*** of Ω, and the fourth derivative terms in the ***biharmonic equation*** follow directly:

$$\boxed{\nabla^2\Omega = \frac{\partial^2\Omega}{\partial x^2} + \frac{\partial^2\Omega}{\partial y^2} = 4\frac{\partial^2\Omega}{\partial z\partial \overline{z}}}, \quad \boxed{\nabla^4\Omega = \frac{\partial^4\Omega}{\partial x^4} + 2\frac{\partial^4\Omega}{\partial x^2\partial y^2} + \frac{\partial^4\Omega}{\partial y^4} = 16\frac{\partial^4\Omega}{\partial z^2\partial \overline{z}^2}} \tag{3.111}$$

The real and imaginary components of the Laplacian may be separated as follows. The scalar potential exists for the component of a vector field that is irrotational, (1.6), and the divergence, (1.3), is

$$\mathbf{div} = \nabla \cdot \mathbf{v} = \frac{\partial v_x}{\partial x} + \frac{\partial v_y}{\partial y} = -\frac{\partial^2\Phi}{\partial x^2} - \frac{\partial^2\Phi}{\partial y^2} = -4\frac{\partial^2\Phi}{\partial z\partial \overline{z}} \tag{3.112a}$$

Likewise, the stream function exists for a vector field that is divergence-free, (1.10), and the component of curl in the vertical direction, (1.4), is given by

$$\mathbf{curl}_3 = \nabla_3 \times \mathbf{v} = \frac{\partial v_y}{\partial x} - \frac{\partial v_x}{\partial y} = \frac{\partial^2\Psi}{\partial x^2} + \frac{\partial^2\Psi}{\partial y^2} = 4\frac{\partial^2\Psi}{\partial z\partial \overline{z}} \tag{3.112b}$$

Adding minus the divergence and i times the vertical component of the curl gives the Laplacian in complex form for vector fields with divergent and/or rotational components,

$$\boxed{\frac{\partial^2\Omega}{\partial x^2} + \frac{\partial^2\Omega}{\partial y^2} = 4\frac{\partial^2\Omega}{\partial z\partial \overline{z}} = -\mathbf{div}(z,\overline{z}) + \mathrm{i}\,\mathbf{curl}_3(z,\overline{z})} \tag{3.113}$$

Thus, ***the divergence and curl of a two-dimensional vector field are given by the Laplacian of its complex function*** $\Omega(z,\overline{z})$.

Example 3.4 The following potential and stream functions are represented in complex form, and expressions for the vector field, (3.109), and Laplacian, (3.111), are evaluated:

$$\Phi = -\frac{r^2}{4} \rightarrow \begin{cases} \Omega & = -\frac{z\overline{z}}{4} \\ v & = \frac{z}{2} \\ \nabla^2\Omega = -1 & \end{cases} \rightarrow \quad \mathbf{div} = 1 \tag{3.114a}$$

$$\Psi = \frac{r^2}{4} \rightarrow \begin{cases} \Omega & = \mathrm{i}\frac{z\overline{z}}{4} \\ v & = \mathrm{i}\frac{z}{2} \\ \nabla^2\Omega & = \mathrm{i} \end{cases} \rightarrow \quad \mathbf{curl}_3 = 1 \tag{3.114b}$$

$$\Psi = -\frac{1}{r} \rightarrow \begin{cases} \Omega & = -\frac{\mathrm{i}}{(z\overline{z})^{1/2}} \\ v & = \frac{\mathrm{i}z}{(z\overline{z})^{3/2}} \\ \nabla^2\Omega & = -\frac{\mathrm{i}}{(z\overline{z})^{3/2}} \end{cases} \rightarrow \quad \mathbf{curl}_3 = -\frac{1}{r^3} \tag{3.114c}$$

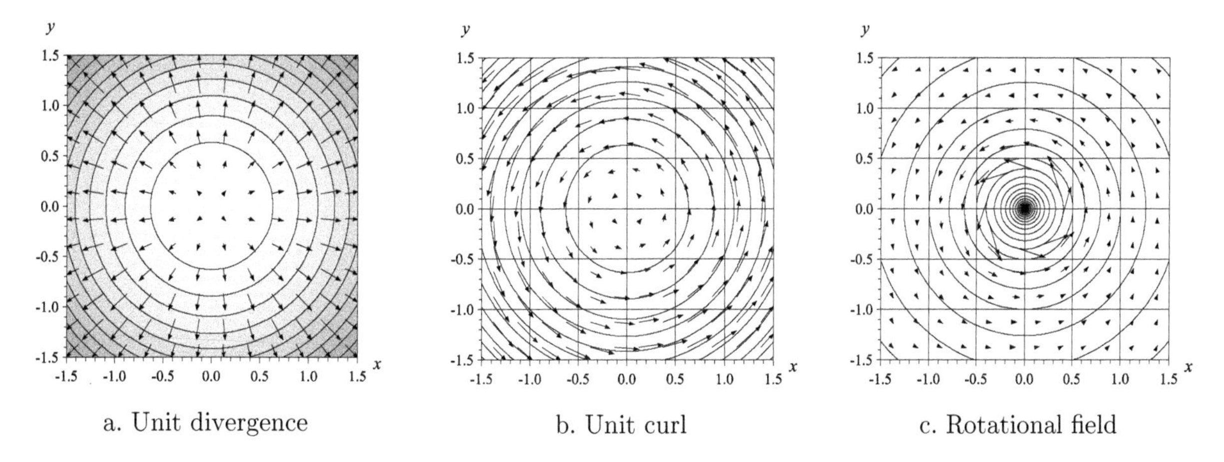

Figure 3.31 *Complex vector fields with divergence and rotation.*

The potential in (3.114a) generates a vector field with a uniform divergence and is illustrated in Fig. 3.31a. The stream function (3.114b) generates rigid body motion with uniform rotation about the origin in Fig. 3.31b, and the stream function (3.114c) generates the vortex in Fig. 3.31c.

Problem 3.18 Develop expressions for the divergence and curl associated with the following complex functions. Compute the value of the complex function, the complex vector field, and the Laplacian at $z = 1 + \mathrm{i}$.

A. $\Omega = x$
B. $\Omega = x^2$
C. $\Omega = x^3$
D. $\Omega = x^4$
E. $\Omega = \mathrm{i}y$
F. $\Omega = \mathrm{i}y^2$
G. $\Omega = \mathrm{i}y^3$
H. $\Omega = \mathrm{i}y^4$
I. $\Omega = r$
J. $\Omega = r^2$
K. $\Omega = r^3$
L. $\Omega = r^4$

Formulation of a function of both z and its complex conjugate $\overline{z}$ is useful to for ***obtaining the complex potential for two-dimensional vector fields with specified divergence and vertical components of curl through direct integration***. The partial differential equation in (3.113) may be integrated with respect to z and $\overline{z}$ to provide

$$\Omega(z,\overline{z}) = \frac{1}{4}\int\int\left[-\mathbf{div}(z,\overline{z}) + \mathrm{i}\mathbf{curl}_3(z,\overline{z})\right]\mathrm{d}\overline{z}\mathrm{d}z \tag{3.115}$$

It is desirable to also obtain the vector field, which may be obtained by performing each integration separately, with the first integration of

$$\boxed{v = -\frac{1}{2}\int\overline{-\mathbf{div}(z,\overline{z}) + \mathrm{i}\mathbf{curl}_3(z,\overline{z})}\mathrm{d}z = -\frac{1}{2}\int\overline{\nabla^2\Omega}\mathrm{d}z} \tag{3.116}$$

which follows directly by differentiating the expression (3.115) using (3.109). The complex potential $\Omega(z,\overline{z})$ may be directly from this vector field using the second integration:

$$\boxed{\Omega = -\frac{1}{2}\int\overline{v}\mathrm{d}z} \tag{3.117}$$

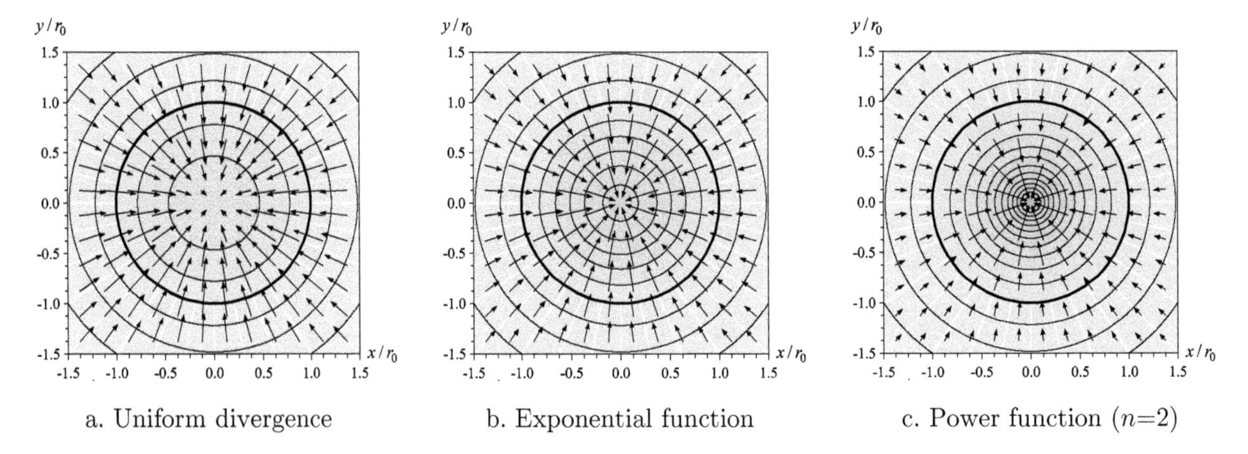

Figure 3.32 *Complex vector fields with specified divergence inside $r < r_0$. (Reprinted from Steward and Ahring, 2009, An analytic solution for groundwater uptake by phreatophytes spanning spatial scales from plant to field to regional,* Journal of Engineering Mathematics, *Vol. 64(2), Fig. 2, used with permission from Springer Nature; distributed under Creative Commons license (Attribution-Noncommercial), creativecommons.org.)*

The constants of integration provide a uniform complex potential Ω_0 and a uniform vector field $v_0 = v_{x0} + iv_{y0}$ that may be chosen to match the constraints of a particular problem.

Example 3.5 A specified divergences was used by Steward and Ahring (2009) to study the root water uptake of groundwater by plants (phreatophytes) using mathematical approximations for this divergence that are easily integrable to provide the vector field and potential function. Each functional form removes a flux Q inside a circle of radius r_0, with the divergences varying from uniform to focused beneath the center of the plant at z_c.

A uniform divergence and the vector field, (3.116), and complex potential function, (3.117), obtained through integration are given by

$$\nabla^2\Omega = \frac{Q}{\pi {r_0}^2}, \quad v = -\frac{Q}{2\pi {r_0}^2}(z - z_c), \quad \Omega = \frac{Q}{4\pi {r_0}^2}\left(r^2 - {r_0}^2\right) \qquad (r < r_0) \tag{3.118}$$

where $r^2 = (z - z_c)\overline{(z - z_c)}$, and the constant of integration is chosen so $\Omega = 0$ on the circle. Functions with a divergence that varies as an exponential function of r are given by

$$\begin{aligned} \nabla^2\Omega &= \frac{Qa}{\pi {r_0}^2} \mathrm{e}^{-a(r/r_0)^2}, \quad v = \frac{Q}{2\pi} \frac{1}{\overline{z - z_c}} \left(\mathrm{e}^{-a(r/r_0)^2} - 1\right) \\ \Omega &= \frac{Q}{4\pi}\left[E_1\left(a\frac{r^2}{{r_0}^2}\right) - E_1(a) + \ln\frac{r^2}{{r_0}^2}\right], \quad a = \ln 100 \end{aligned} \tag{3.119}$$

where $E_1(x)$ is the exponential integral. Functions with divergence that varies as an integer power $n \geq 2$ of a function with r^2 terms are given by

$$\begin{aligned} \nabla^2\Omega &= \frac{Q(n-1)b}{\pi {r_0}^2}\left(\frac{{r_0}^2}{br^2 + {r_0}^2}\right)^n, \quad v = \frac{Q}{2\pi}\frac{1}{\overline{z - z_c}}\left[\left(\frac{{r_0}^2}{br^2 + {r_0}^2}\right)^{n-1} - 1\right] \\ \Omega &= \frac{Q}{4\pi}\left\{\ln\frac{br^2 + {r_0}^2}{b{r_0}^2 + {r_0}^2} - \sum_{l=1}^{n-2}\frac{1}{l}\left[\left(\frac{{r_0}^2}{br^2 + {r_0}^2}\right)^l - \left(\frac{1}{b+1}\right)^l\right]\right\}, \quad b = 100^{\frac{1}{n-1}} - 1 \end{aligned} \tag{3.120}$$

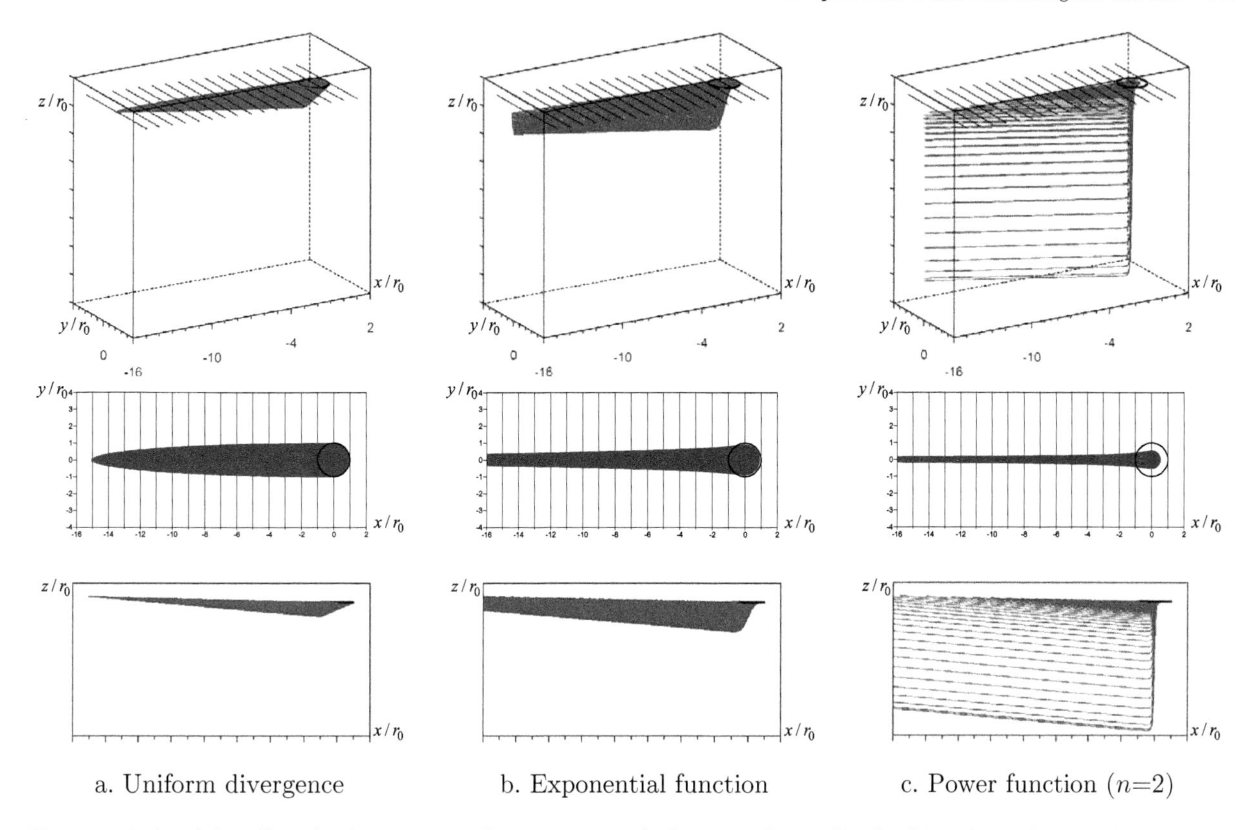

Figure 3.33 *Quasi three-dimensional capture zones for root water uptake from groundwater. (Reprinted from Steward and Ahring, 2009, An analytic solution for groundwater uptake by phreatophytes spanning spatial scales from plant to field to regional,* Journal of Engineering Mathematics, *Vol. 64(2), Fig. 6, used with permission from Springer Nature; distributed under Creative Commons license (Attribution-Noncommercial), creativecommons.org.)*

These functions match the potential function and vector field along the interface at $r = r_0$ when a point-sink, (3.1), at z_c with flux Q is used outside the circle:

$$\nabla^2\Omega = 0, \quad v = -\frac{Q}{2\pi}\frac{1}{\overline{z - z_c}}, \quad \Omega = \frac{Q}{2\pi}\ln\frac{z - z_c}{r_0} \qquad (r \geq r_0) \tag{3.121}$$

The potential function and vector field for these functions are shown in Fig. 3.32, and additional details may be found in Steward and Ahring (2009). These complex functions were used to develop three-dimensional depictions of the capture zones for phreatophytes, or trees that capture groundwater. This is illustrated in Fig. 3.33, where the divergence in Fig. 3.32 is placed in a uniform vector field v_{x0} with recharge, and particles are traced backwards from the boundary of the tree until they intersect the top of the flow domain, and the quasi three-dimensional approximation, (1.32), is used to compute the vertical component of the vector field.

Problem 3.19 Develop expressions for the vector fields and complex potential functions for the following Laplacians. Chose the constants of integration such that $\Omega = 0$ and $v = 0$ at the origin, and compute the value of the complex function, the complex vector field, and the Laplacian at $z = 1 + \mathrm{i}$.

A. $\nabla^2\Omega = r^2$
B. $\nabla^2\Omega = r^4$
C. $\nabla^2\Omega = r^6$
D. $\nabla^2\Omega = r^8$
E. $\nabla^2\Omega = r^{10}$
F. $\nabla^2\Omega = r^{12}$
G. $\nabla^2\Omega = \mathrm{i}r^2$
H. $\nabla^2\Omega = \mathrm{i}r^4$
I. $\nabla^2\Omega = \mathrm{i}r^6$
J. $\nabla^2\Omega = \mathrm{i}r^8$
K. $\nabla^2\Omega = \mathrm{i}r^{10}$
L. $\nabla^2\Omega = \mathrm{i}r^{12}$

3.7 Biharmonic Equation and the Kolosov Formulas

A ***complex formulation of the biharmonic equation*** is presented next using functions of z and $\overline{z}$, as introduced in the previous section. Solutions were presented in terms of the Airy (1863) stress function in (1.107), where its partial derivatives provide the normal stresses in the x- and y-directions, σ_x and σ_y, and the shear stress τ_{xy}:

$$\begin{aligned} &\nabla^4\mathcal{F} = \frac{\partial^4\mathcal{F}}{\partial x^4} + 2\frac{\partial^4\mathcal{F}}{\partial x^2\partial y^2} + \frac{\partial^4\mathcal{F}}{\partial y^4} = 0 \\ &\rightarrow \quad \sigma_x = \frac{\partial^2\mathcal{F}}{\partial y^2}, \quad \sigma_y = \frac{\partial^2\mathcal{F}}{\partial x^2}, \quad \tau_{xy} = -\frac{\partial^2\mathcal{F}}{\partial x\partial y} \end{aligned} \tag{3.122}$$

The complex formulation of the biharmonic equation, (3.111), may be integrated with respect to z and $\overline{z}$ to give

$$\begin{aligned} &\nabla^4\mathcal{F} = 16\frac{\partial^4\mathcal{F}}{\partial z^2\partial\overline{z}^2} = 0 \\ &\rightarrow \quad \mathcal{F} = \frac{1}{2}\left[\overline{z}\phi(z) + z\overline{\phi(z)} + \chi(z) + \overline{\chi(z)}\right] = \Re\left(\overline{z}\phi + \chi\right) \end{aligned} \tag{3.123}$$

where ϕ and χ are complex functions, and $\overline{\phi}$ and $\overline{\chi}$ are their complex conjugates. The last two equations lead to the ***Kolosov–Muskhelishvili formulas*** where components of stress are expressed in terms of the complex functions ϕ and ψ, (1.112):

$$\begin{aligned} &\frac{\sigma_x + \sigma_y}{2} = 2\Re\phi' & & \sigma_x = \Re\left(2\phi' - \overline{z}\phi'' - \psi'\right) \\ &\frac{\sigma_y - \sigma_x}{2} + \mathrm{i}\tau_{xy} = \overline{z}\phi'' + \psi' & \Rightarrow \quad & \sigma_y = \Re\left(2\phi' + \overline{z}\phi'' + \psi'\right) \\ &u = u_x + \mathrm{i}u_y = \frac{1}{2\mu}\left(\kappa\phi - z\overline{\phi'} - \overline{\psi}\right) & & \tau_{xy} = \Im\left(\overline{z}\phi'' + \psi'\right) \end{aligned} \tag{3.124}$$

with $\psi = \chi'$. This equation also presents the real and imaginary components of a complex displacement vector, u_x and u_y, which are also directly related to the Kolosov functions ϕ and ψ, as shown by Muskhelishvili (1953*b*, eqn 41.1) The coefficients used to compute displacement are provided for μ in (1.99) and κ in (1.113).

Example 3.6 The Kolosov functions provide a background of uniform stress into which analytic elements may be inserted with components σ_{x0}, σ_{y0}, and τ_0, where (Jaeger, 1969, p.129); (Jaeger and Cook, 1976, p.247)

$$\begin{aligned} \phi &= \frac{\sigma_{x0} + \sigma_{y0}}{4}z + \frac{2\mu}{\kappa}u_0 & & \sigma_x = \sigma_{x0} \\ & & \rightarrow \quad & \sigma_y = \sigma_{y0} \quad , \quad u_x + \mathrm{i}u_y = u_0 \\ \psi &= \frac{-\sigma_{x0} + \sigma_{y0}}{2}z + \mathrm{i}\tau_0 z & & \tau_{xy} = \tau_0 \end{aligned} \tag{3.125}$$

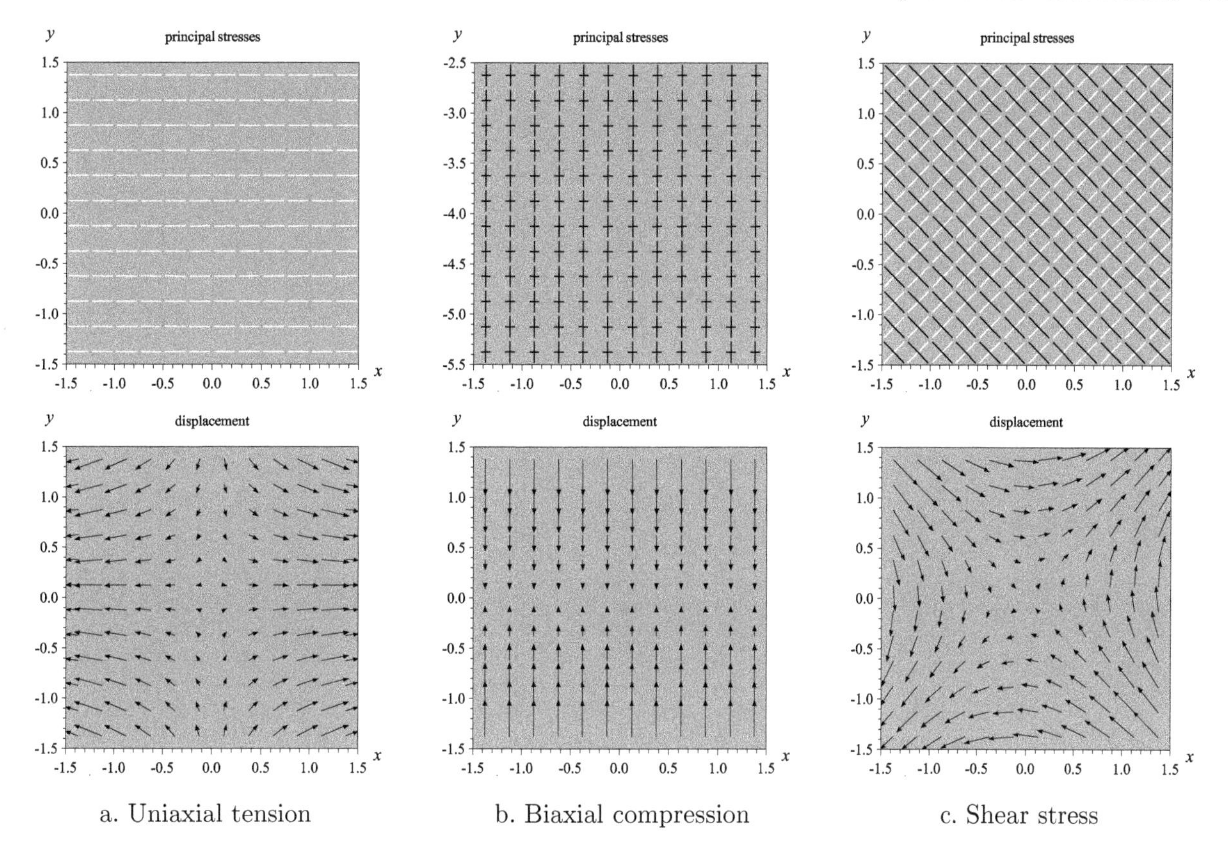

Figure 3.34 *Uniform stress.*

Examples in Fig. 3.34 illustrate uniaxial tension, biaxial compression (occurring at a specified elevation due to gravity body forces as in Fig. 1.56), and a pure shear with equal and opposite principal stress components. The complex coefficient u_0 is chosen so displacement is zero at the origin.

Solutions are visualized using the principal stress components, σ_1 and σ_2 in (1.119), and the maximum shear stress, $\tau_{\max}$ in (1.120). These functions are related to the Kolosov formula as follows:

$$\begin{aligned}\sigma_1 &= \frac{\sigma_x+\sigma_y}{2} + \sqrt{\left(\frac{\sigma_x-\sigma_y}{2}\right)^2 + {\tau_{xy}}^2} \\ \sigma_2 &= \frac{\sigma_x+\sigma_y}{2} - \sqrt{\left(\frac{\sigma_x-\sigma_y}{2}\right)^2 + {\tau_{xy}}^2} \\ \tau_{\max} &= \sqrt{\left(\frac{\sigma_x-\sigma_y}{2}\right)^2 + {\tau_{xy}}^2}\end{aligned} \quad \Rightarrow \quad \boxed{\begin{aligned}\sigma_1 &= 2\Re\phi' + |\overline{z}\phi'' + \psi'| \\ \sigma_2 &= 2\Re\phi' - |\overline{z}\phi'' + \psi'| \\ \tau_{\max} &= |\overline{z}\phi'' + \psi'|\end{aligned}} \tag{3.126}$$

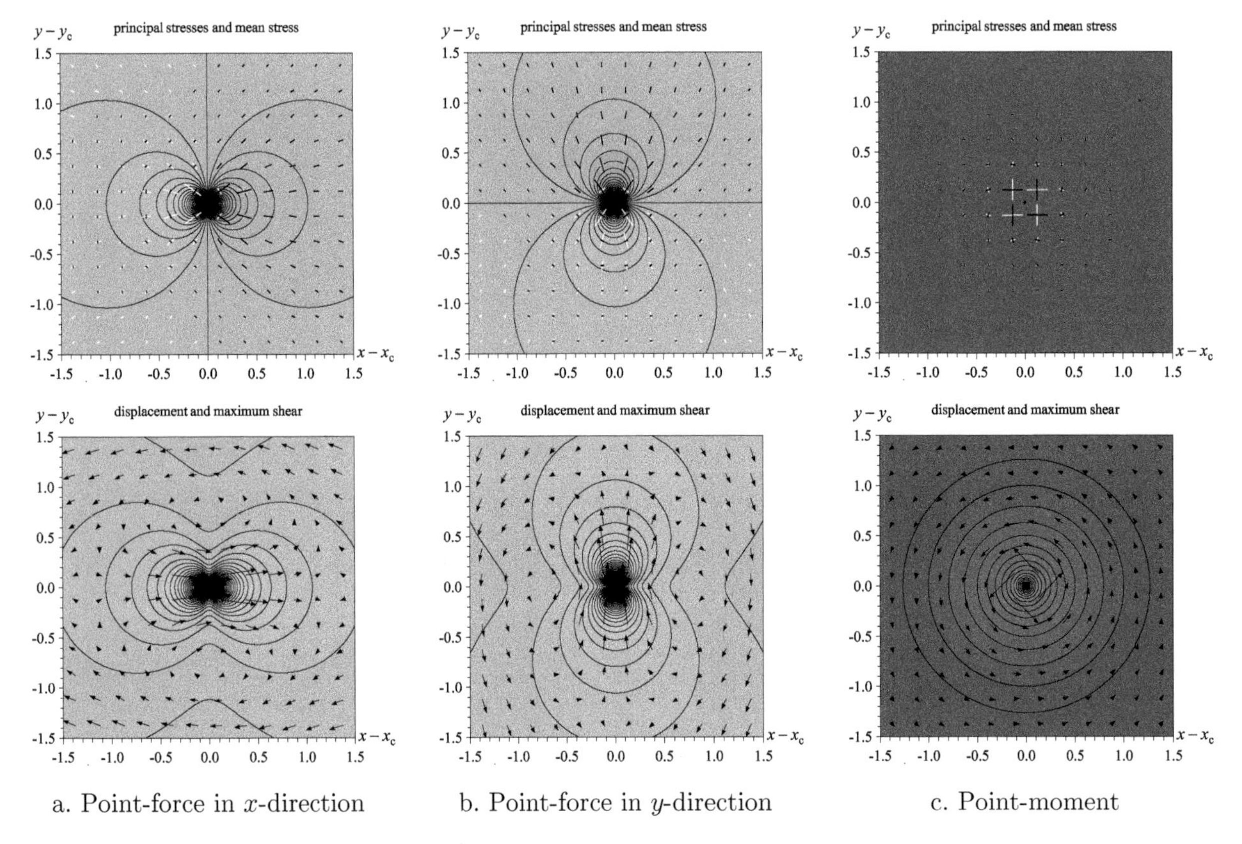

Figure 3.35 *Biharmonic equation solutions for a point force and moment.*

which follows directly from (3.124). The direction of the principal stresses, illustrated in Fig. 3.35, lie in the direction of the eigenvectors of the Cauchy stress tensor, (1.118). A set of ***isostress*** contour lines with uniform mean stress, $(\sigma_1 + \sigma_2)/2$, is also illustrated, and identify regions of compression directly in front of the point-force and tension behind it.

Example 3.7 The stress field generated by a point-force is illustrated in Fig. 3.35. This field is generated by force with components $F = F_x + \mathrm{i}F_y$ located at z_c by (Muskhelishvili, 1953*b*, eqn 57.3)

$$\begin{aligned} \phi &= -\frac{F}{2\pi(\kappa+1)}\ln(z-z_c) \\ \psi &= \frac{\kappa\overline{F}}{2\pi(\kappa+1)}\ln(z-z_c) + \frac{F\overline{z_c}}{2\pi(\kappa+1)}\frac{1}{z-z_c} \end{aligned} \tag{3.127}$$

Note that each of these logarithm terms individually generates a dislocation with a discontinuous displacement across a branch cut (Jaeger and Cook, 1976, p.249). However, the factors by which ϕ and ψ are scaled causes these dislocations to cancel, and the displacement is a continuous function. These factors also generate stresses where the integral of their normal and shear components around the point element is equal to and opposite the force.

The Kolosov formulas for a point-moment with couple M at z_c (Green and Zerna, 1968, p.256) is given by

$$\begin{aligned} &\phi = 0 \\ &\psi = \frac{iM}{2\pi}\frac{1}{z - z_c} \end{aligned} \quad \rightarrow \quad \begin{aligned} \sigma_r - i\tau_{r\theta} &= \frac{iM}{2\pi r^2} \\ u_r + iu_\theta &= \frac{1}{2\mu}\frac{iM}{2\pi r} \end{aligned} \quad \rightarrow \quad \begin{aligned} \sigma_r &= 0,\ \tau_{r\theta} = -\frac{M}{2\pi r^2} \\ u_r &= 0,\ u_\theta = \frac{1}{2\mu}\frac{M}{2\pi r} \end{aligned} \tag{3.128}$$

using $z - z_c = re^{i\theta}$ with stress from (3.134) and displacement from (3.133). This solution is illustrated in Fig. 3.35c.

Problem 3.20 Compute the stress components $\sigma_x, \sigma_y, \tau_{xy}, \sigma_1, \sigma_2$, and $\tau_{\max}$ at the specified location z [m] for the given forces components F [N/m] and location $z_c = 0$.

A. $F = 1.2 \times 10^6$, $z = 0.1 - 0.1i$
B. $F = 1.4 \times 10^6$, $z = 0.1 + 0.1i$
C. $F = 1.6 \times 10^6$, $z = -0.1 + 0.1i$
D. $F = 1.8 \times 10^6$, $z = -0.1 - 0.1i$
E. $F = i1.2 \times 10^6$, $z = 0.1 - 0.1i$
F. $F = i1.4 \times 10^6$, $z = 0.1 + 0.1i$
G. $F = i1.6 \times 10^6$, $z = -0.1 + 0.1i$
H. $F = i1.8 \times 10^6$, $z = -0.1 - 0.1i$
I. $F = 1.2 \times (10^6 + i10^6)$, $z = 0.1$
J. $F = 1.4 \times (10^6 + i10^6)$, $z = 0.1i$
K. $F = 1.6 \times (10^6 + i10^6)$, $z = -0.1$
L. $F = 1.8 \times (10^6 + i10^6)$, $z = -0.1i$

Problem 3.21 Compute the displacement components u_x and u_y at the specified location z, force F, and location $z_c = 0$ from Problem 3.20, for a steel beam with coefficient of elasticity $E = 200 * 10^9 \text{N/m}^2$, Poisson ratio $\nu = 0.3$, and coefficient κ for plane stress, from (1.113).

3.7.1 Biharmonic Solutions for Domains outside Circles

Solutions to problems with boundaries of circular geometry are presented in terms of the same local $\mathcal{Z}$-plane presented earlier in Eq. (3.13) and Fig. 3.7, with the element centered at $\mathcal{Z} = 0$ with unit radius. The series expansions for the functions used in the Kolosov formulas outside a circle element are given by

$$\boxed{\begin{aligned} \phi &= \overset{\text{out}}{c}_0 r_0 \ln(r_0\mathcal{Z}) - \sum_{n=1}^{N} \overset{\phi\text{out}}{c}_n \frac{r_0}{n}\mathcal{Z}^{-n} \\ \phi' &= \frac{\overset{\text{out}}{c}_0}{\mathcal{Z}} + \sum_{n=1}^{N} \overset{\phi\text{out}}{c}_n \mathcal{Z}^{-n-1} \\ \phi'' &= -\frac{\overset{\text{out}}{c}_0}{r_0\mathcal{Z}^2} - \sum_{n=1}^{N} \overset{\phi\text{out}}{c}_n \frac{n+1}{r_0}\mathcal{Z}^{-n-2} \\ \psi &= -\overline{\overset{\text{out}}{c}_0}\kappa r_0 \ln(r_0\mathcal{Z}) - \frac{\overset{\text{out}}{c}_0\overline{z_c}}{\mathcal{Z}} - \sum_{n=1}^{N} \overset{\psi\text{out}}{c}_n \frac{r_0}{n}\mathcal{Z}^{-n} \\ \psi' &= -\frac{\overline{\overset{\text{out}}{c}_0}\kappa}{\mathcal{Z}} + \frac{\overset{\text{out}}{c}_0\overline{z_c}}{r_0\mathcal{Z}^2} + \sum_{n=1}^{N} \overset{\psi\text{out}}{c}_n \mathcal{Z}^{-n-1} \end{aligned}}, \quad F = -2\pi(\kappa+1)r_0\overset{\text{out}}{c}_0 \tag{3.129}$$

in terms of the local variable $\mathcal{Z} = (z - z_c)/r_0$, (3.13), which is translated to the center of the circle z_c and scaled by its radius r_0. This choice of c_0 terms removes the singularity in logarithm terms of the ϕ and ψ functions to generate a continuous displacement, and this term is related to the force $\boldsymbol{F}$ generated at z_c, as in (3.127). These Kolosov functions are substituted into (3.124) to give the stress components

$$
\begin{aligned}
\sigma_x &= \Re\left[\frac{\overset{\text{out}}{c}_0 2 + \overline{\overset{\text{out}}{c}_0}\kappa}{\mathcal{Z}} + \frac{\overset{\text{out}}{c}_0\overline{\mathcal{Z}}}{\mathcal{Z}^2} + \sum_{n=1}^{N} \overset{\phi\text{out}}{c}_n 2\mathcal{Z}^{-n-1} \right.\\
&\left. + \sum_{n=1}^{N} \overset{\phi\text{out}}{c}_n \frac{\overline{z}}{r_0}(n+1)\mathcal{Z}^{-n-2} - \sum_{n=1}^{N} \overset{\psi\text{out}}{c}_n \mathcal{Z}^{-n-1}\right] + \overset{\text{add}}{\sigma_x}\\
\sigma_y &= \Re\left[\frac{\overset{\text{out}}{c}_0 2 - \overline{\overset{\text{out}}{c}_0}\kappa}{\mathcal{Z}} - \frac{\overset{\text{out}}{c}_0\overline{\mathcal{Z}}}{\mathcal{Z}^2} + \sum_{n=1}^{N} \overset{\phi\text{out}}{c}_n 2\mathcal{Z}^{-n-1} \right.\\
&\left. - \sum_{n=1}^{N} \overset{\phi\text{out}}{c}_n \frac{\overline{z}}{r_0}(n+1)\mathcal{Z}^{-n-2} + \sum_{n=1}^{N} \overset{\psi\text{out}}{c}_n \mathcal{Z}^{-n-1}\right] + \overset{\text{add}}{\sigma_y}\\
\tau_{xy} &= \Im\left[-\frac{\overline{\overset{\text{out}}{c}_0}\kappa}{\mathcal{Z}} - \frac{\overset{\text{out}}{c}_0\overline{\mathcal{Z}}}{\mathcal{Z}^2} - \sum_{n=1}^{N} \overset{\phi\text{out}}{c}_n \frac{\overline{z}}{r_0}(n+1)\mathcal{Z}^{-n-2} + \sum_{n=1}^{N} \overset{\psi\text{out}}{c}_n \mathcal{Z}^{-n-1}\right] + \overset{\text{add}}{\tau_{xy}}
\end{aligned}
\tag{3.130}
$$

and the complex displacement

$$
\begin{aligned}
u &= \frac{1}{2\mu}\left[2\overset{\text{out}}{c}_0\kappa r_0 \ln(r_0|\mathcal{Z}|) - \overline{\overset{\text{out}}{c}_0} r_0 \frac{\mathcal{Z}}{\overline{\mathcal{Z}}}\right.\\
&\left. - \sum_{n=1}^{N} \overset{\phi\text{out}}{c}_n \kappa r_0 \frac{\mathcal{Z}^{-n}}{n} - \sum_{n=1}^{N} \overline{\overset{\phi\text{out}}{c}_n z \mathcal{Z}^{-n-1}} + \sum_{n=1}^{N} \overline{\overset{\psi\text{out}}{c}_n r_0 \frac{\mathcal{Z}^{-n}}{n}}\right] + \overset{\text{add}}{u}
\end{aligned}
\tag{3.131}
$$

where the additional functions contain the contribution from the setting and all other elements. The stress and displacement are illustrated in Fig. 3.36 for these first few influence functions in these series.

The ***components of stress and displacement in cylindrical*** (r, θ) ***coordinates*** are obtained from Mohr's representation, (1.122), that may be reorganized with the Kolosov formula, (1.124), giving (Jaeger, 1969 p.197; Muskhelishvilim 1953*b* p.138)

$$
\begin{aligned}
\sigma_r &= \frac{\sigma_x + \sigma_y}{2} + \frac{\sigma_x - \sigma_y}{2}\cos 2\theta + \tau_{xy}\sin 2\theta\\
\sigma_\theta &= \frac{\sigma_x + \sigma_y}{2} - \frac{\sigma_x - \sigma_y}{2}\cos 2\theta - \tau_{xy}\sin 2\theta\\
\tau_{r\theta} &= -\frac{\sigma_x - \sigma_y}{2}\sin 2\theta + \tau_{xy}\cos 2\theta
\end{aligned}
\quad\Rightarrow\quad
\boxed{\begin{aligned}\frac{\sigma_r + \sigma_\theta}{2} &= 2\Re\phi'\\ \frac{\sigma_\theta - \sigma_r}{2} + \mathrm{i}\tau_{r\theta} &= \mathrm{e}^{2\mathrm{i}\theta}\left(\overline{z}\phi'' + \psi'\right)\end{aligned}}
\tag{3.132}
$$

and (Jaeger, 1969, p.204)

$$
u_r + \mathrm{i}u_\theta = \mathrm{e}^{-\mathrm{i}\theta}\left(u_x + \mathrm{i}u_y\right) \quad\Rightarrow\quad \boxed{u_r + \mathrm{i}u_\theta = \frac{\mathrm{e}^{-\mathrm{i}\theta}}{2\mu}\left(\kappa\phi - z\overline{\phi'} - \overline{\psi}\right)}
\tag{3.133}
$$

where the radial and θ components of displacement follow from (1.123) and Jaeger (1969, p.204). Expressions for the radial and tangential stress components along a circular boundary follow directly, and provide a complex form for these components (Muskhelishvili, 1953*b*, p.147; Jaeger, 1969, p.195):

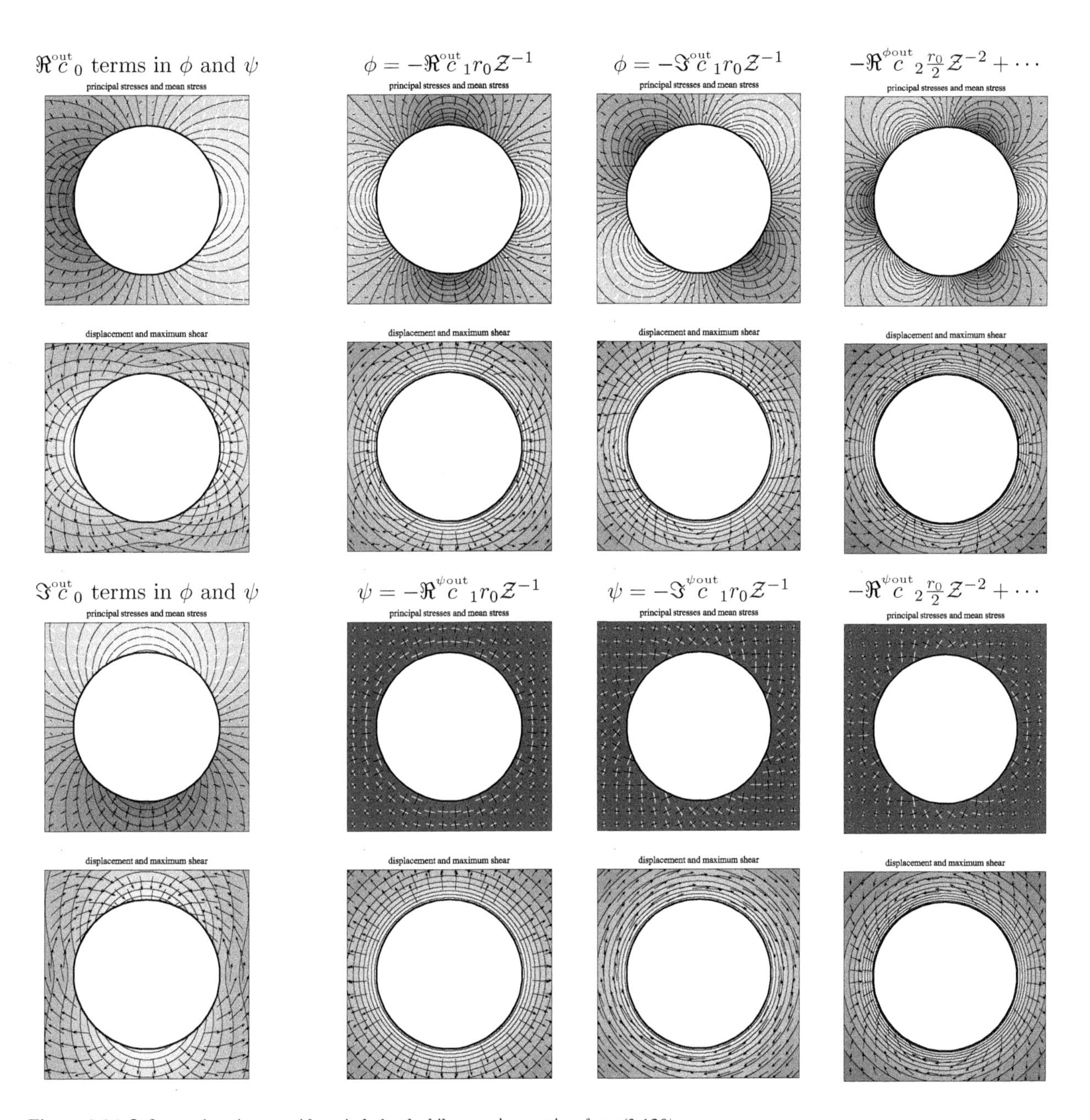

Figure 3.36 *Influence functions outside a circle for the biharmonic equation, from (3.129).*

$$\begin{aligned} \sigma_r &= 2\Re\phi' - \Re\left[\mathrm{e}^{2\mathrm{i}\theta}\left(\overline{z}\phi'' + \psi'\right)\right] \\ \sigma_\theta &= 2\Re\phi' + \Re\left[\mathrm{e}^{2\mathrm{i}\theta}\left(\overline{z}\phi'' + \psi'\right)\right] \quad \Rightarrow \quad \sigma_r - \mathrm{i}\tau_{r\theta} = 2\Re\phi' - \mathrm{e}^{2\mathrm{i}\theta}\left(\overline{z}\phi'' + \psi'\right) \\ \tau_{r\theta} &= \Im\left[\mathrm{e}^{2\mathrm{i}\theta}\left(\overline{z}\phi'' + \psi'\right)\right] \end{aligned} \tag{3.134}$$

The Kolosov–Muskhelishvili formulas provide a framework for developing solutions for analytic elements, such as for a ***circle with zero traction along its boundary in a field of uniform stress***. The solution for a circle centered at $z_c = 0$ with radius r_0 is given by superposition of the uniform stress in (3.125) with the series for a circle in (3.129)

$$\begin{aligned} \phi &= \frac{\sigma_{x0} + \sigma_{y0}}{4} z - \sum_{n=1}^{N} {}^{\phi\text{out}}c_n \frac{r_0}{n} Z^{-n} \\ \psi &= \frac{-\sigma_{x0} + \sigma_{y0}}{2} z + \mathrm{i}\tau_0 z - \sum_{n=1}^{N} {}^{\psi\text{out}}c_n \frac{r_0}{n} Z^{-n} \end{aligned} \tag{3.135a}$$

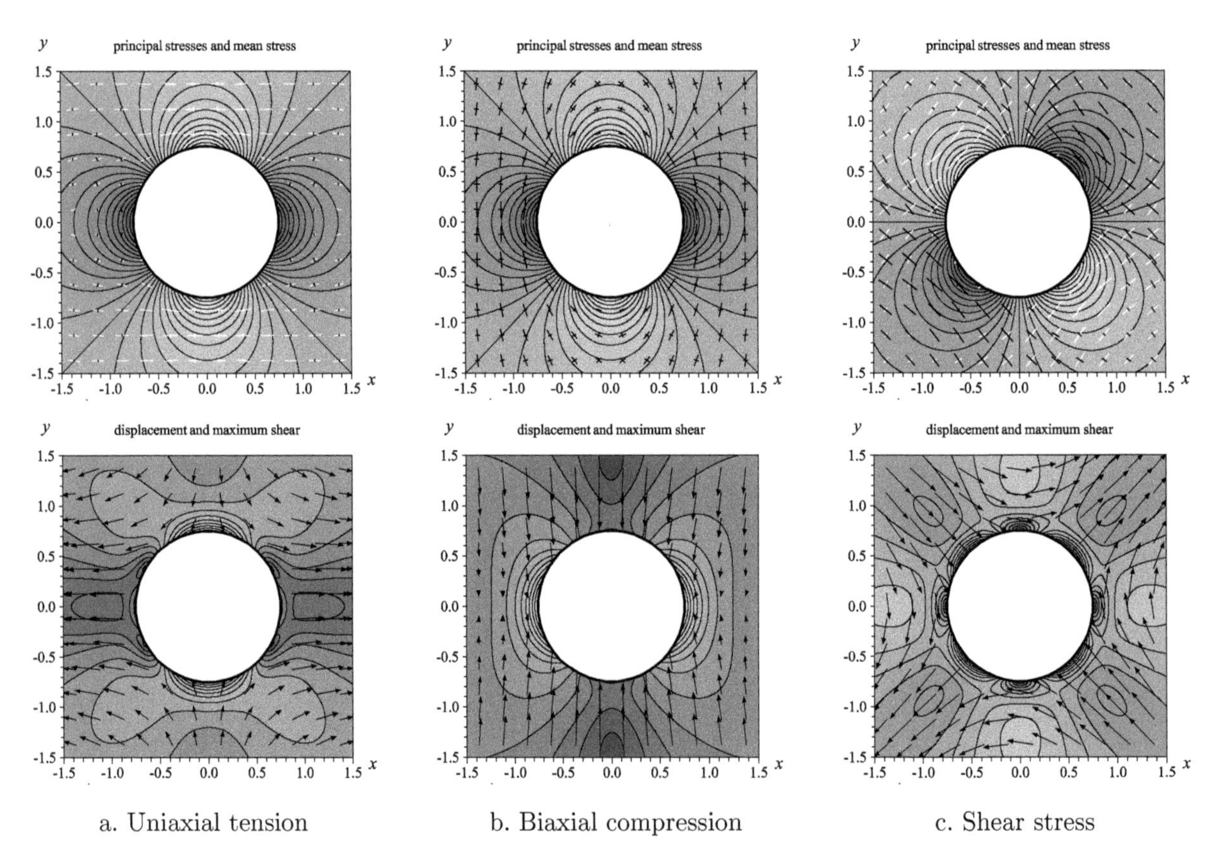

Figure 3.37 *Circle with zero traction in uniform stress.*

with coefficients

$$\begin{aligned}
&\overset{\phi\text{out}}{c}_1 = -\frac{\sigma_{x0} - \sigma_{y0}}{2} - \mathrm{i}\tau_0 \\
&\overset{\psi\text{out}}{c}_1 = \frac{\sigma_{x0} + \sigma_{y0}}{2}, \quad \overset{\psi\text{out}}{c}_3 = -3\frac{\sigma_{x0} - \sigma_{y0}}{2} - 3\mathrm{i}\tau_0 \\
&\overset{\phi\text{out}}{c}_n = 0\ (n \neq 1), \quad \overset{\psi\text{out}}{c}_n = 0\ (n \neq 1, 3)
\end{aligned} \tag{3.135b}$$

where the expression for the coefficients is adapted from the uniaxial solution by Jaeger (1969, p.197). It may be shown using (3.134) that this solution satisfies a traction condition with $\sigma_r = \tau_{r\theta} = 0$ along the circle $|z| = r_0$, and the Laurent series reproduces the background uniform stress as $|z| \to \infty$. This exact solution is illustrated in Fig. 3.37.

Problem 3.22 Compute the stress components σ_x, σ_y, and τ_{xy} at $z = r_0$, $z = r_0 e^{\mathrm{i}\pi/4}$, and $z = \mathrm{i}r_0$ for a circle with $N = 20$ for a elastic media in plane strain with coefficient of elasticity $E = 80 * 10^6 \mathrm{N/m^2}$ and Poisson ratio $\nu = 0.3$, for the specified uniform stress and radius r_0:

A. $\sigma_{x0} = 1.2 \times 10^6 \mathrm{N/m^2}, r_0 = 0.2$ m
B. $\sigma_{x0} = 1.4 \times 10^6 \mathrm{N/m^2}, r_0 = 0.4$ m
C. $\sigma_{x0} = 1.6 \times 10^6 \mathrm{N/m^2}, r_0 = 0.6$ m
D. $\sigma_{x0} = 1.8 \times 10^6 \mathrm{N/m^2}, r_0 = 0.8$ m
E. $\sigma_{y0} = 1.2 \times 10^6 \mathrm{N/m^2}, r_0 = 0.2$ m
F. $\sigma_{y0} = 1.4 \times 10^6 \mathrm{N/m^2}, r_0 = 0.4$ m
G. $\sigma_{y0} = 1.6 \times 10^6 \mathrm{N/m^2}, r_0 = 0.6$ m
H. $\sigma_{y0} = 1.8 \times 10^6 \mathrm{N/m^2}, r_0 = 0.8$ m
I. $\tau_0 = 1.2 \times 10^6 \mathrm{N/m^2}, r_0 = 0.2$ m
J. $\tau_0 = 1.4 \times 10^6 \mathrm{N/m^2}, r_0 = 0.4$ m
K. $\tau_0 = 1.6 \times 10^6 \mathrm{N/m^2}, r_0 = 0.6$ m
L. $\tau_0 = 1.8 \times 10^6 \mathrm{N/m^2}, r_0 = 0.8$ m

Methods for satisfying boundary conditions with specified stress along a circle element are formulated at a set of M equally spaced control points (3.14)

$$\theta_m = \theta_1 + 2\pi \frac{m-1}{M}, \quad Z_m = e^{\mathrm{i}\theta_m}, \quad z_m = z_c + r_0 e^{\mathrm{i}\theta_m} \quad (m = 1, M) \tag{3.136}$$

The components of stress in (r,θ) coordinates are obtained using (3.134) and evaluated at these locations along the boundary, giving

$$\begin{aligned}
\sigma_r^+(z_m) - \mathrm{i}\tau_{r\theta}^+(z_m) = 2\Re &\left[\sum_{n=1}^{N} \overset{\phi\text{out}}{c}_n e^{-\mathrm{i}(n+1)\theta_m} \right] + \sum_{n=1}^{N} \overset{\phi\text{out}}{c}_n (n+1) \frac{\overline{z_m}}{r_0} e^{-\mathrm{i}n\theta_m} \\
&- \sum_{n=1}^{N} \overset{\psi\text{out}}{c}_n e^{-\mathrm{i}(n-1)\theta_m} + \overset{\text{add}}{\sigma_r}(z_m) - \mathrm{i}\overset{\text{add}}{\tau_{r\theta}}(z_m)
\end{aligned} \tag{3.137}$$

where the radial and θ components for the additional functions may be obtained from their Cartesian components using (3.132). Note that the terms associated with $n = 0$ are not presented, since this solution is being developed for an element that generates no net force. These M equations are rearranged to gather terms with unknown coefficients on the left-hand side of the equation, and to separate the complex coefficients into real and imaginary parts:

$$\begin{aligned}
&\sum_{n=1}^{N} \left(\Re \overset{\phi\text{out}}{c}_n \right) 2\cos(n+1)\theta_m + \left(\Im \overset{\phi\text{out}}{c}_n \right) 2\sin(n+1)\theta_m \\
&\quad + \sum_{n=1}^{N} \left(\Re \overset{\phi\text{out}}{c}_n + \mathrm{i}\Im \overset{\phi\text{out}}{c}_n \right) (n+1) \frac{\overline{z_m}}{r_0} e^{-\mathrm{i}n\theta_m} - \sum_{n=1}^{N} \left(\Re \overset{\psi\text{out}}{c}_n + \mathrm{i}\Im \overset{\psi\text{out}}{c}_n \right) e^{-\mathrm{i}(n-1)\theta_m} \\
&\quad = \left[\sigma_{r\,m}^+ - \overset{\text{add}}{\sigma_r}(z_m) \right] - \mathrm{i}\left[\tau_{r\theta\,m}^+ - \overset{\text{add}}{\tau_{r\theta}}(z_m) \right]
\end{aligned} \tag{3.138}$$

The real terms in this system of equations relate coefficients to the radial component of stress and the imaginary terms relate shear, and both sets of conditions are organized within the following matrices:

$$\begin{bmatrix} \overset{\Re\phi\sigma}{\mathbf{A}} & \overset{\Im\phi\sigma}{\mathbf{A}} & \overset{\Re\psi\sigma}{\mathbf{A}} & \overset{\Im\psi\sigma}{\mathbf{A}} \\ \overset{\Re\phi\tau}{\mathbf{A}} & \overset{\Im\phi\tau}{\mathbf{A}} & \overset{\Re\psi\tau}{\mathbf{A}} & \overset{\Im\psi\tau}{\mathbf{A}} \end{bmatrix} \begin{bmatrix} \overset{\Re\phi}{\mathbf{c}} \\ \overset{\Im\phi}{\mathbf{c}} \\ \overset{\Re\psi}{\mathbf{c}} \\ \overset{\Im\psi}{\mathbf{c}} \end{bmatrix} = \begin{bmatrix} \boldsymbol{\sigma} \\ -\boldsymbol{\tau} \end{bmatrix} \tag{3.139}$$

where the unknown coefficients are gathered in a set of column vectors

$$\overset{\Re\phi}{\mathbf{c}} = \begin{bmatrix} \vdots \\ \Re \overset{\phi\text{out}}{c}_n \\ \vdots \end{bmatrix}, \quad \overset{\Im\phi}{\mathbf{c}} = \begin{bmatrix} \vdots \\ \Im \overset{\phi\text{out}}{c}_n \\ \vdots \end{bmatrix}, \quad \overset{\Re\psi}{\mathbf{c}} = \begin{bmatrix} \vdots \\ \Re \overset{\psi\text{out}}{c}_n \\ \vdots \end{bmatrix}, \quad \overset{\Im\psi}{\mathbf{c}} = \begin{bmatrix} \vdots \\ \Im \overset{\psi\text{out}}{c}_n \\ \vdots \end{bmatrix} \tag{3.140}$$

the specified conditions minus the additional functions evaluated at the control points are gathered in column vectors

$$\boldsymbol{\sigma} = \begin{bmatrix} \vdots \\ \sigma^+_{r\,m} - \overset{\text{add}}{\sigma_r}(z_m) \\ \vdots \end{bmatrix}, \quad \boldsymbol{\tau} = \begin{bmatrix} \vdots \\ \tau^+_{r\theta\,m} - \overset{\text{add}}{\tau_{r\theta}}(z_m) \\ \vdots \end{bmatrix} \tag{3.141}$$

and the terms within the coefficient matrices are obtained by evaluating

$$\begin{aligned} \overset{\Re\phi\sigma}{\mathbf{A}}_{mn} &= 2\cos(n+1)\theta_m + \Re\left[(n+1)\frac{\overline{z_m}}{r_0}\mathrm{e}^{-\mathrm{i}n\theta_m}\right] & \overset{\Re\psi\sigma}{\mathbf{A}}_{mn} &= \Re\left[-\mathrm{e}^{-\mathrm{i}(n-1)\theta_m}\right] \\ \overset{\Im\phi\sigma}{\mathbf{A}}_{mn} &= 2\sin(n+1)\theta_m + \Re\left[\mathrm{i}(n+1)\frac{\overline{z_m}}{r_0}\mathrm{e}^{-\mathrm{i}n\theta_m}\right] & \overset{\Im\psi\sigma}{\mathbf{A}}_{mn} &= \Re\left[-\mathrm{i}\mathrm{e}^{-\mathrm{i}(n-1)\theta_m}\right] \\ \overset{\Re\phi\tau}{\mathbf{A}}_{mn} &= \Im\left[(n+1)\frac{\overline{z_m}}{r_0}\mathrm{e}^{-\mathrm{i}n\theta_m}\right] & \overset{\Re\psi\tau}{\mathbf{A}}_{mn} &= \Im\left[-\mathrm{e}^{-\mathrm{i}(n-1)\theta_m}\right] \\ \overset{\Im\phi\tau}{\mathbf{A}}_{mn} &= \Im\left[\mathrm{i}(n+1)\frac{\overline{z_m}}{r_0}\mathrm{e}^{-\mathrm{i}n\theta_m}\right] & \overset{\Im\psi\tau}{\mathbf{A}}_{mn} &= \Im\left[-\mathrm{i}\mathrm{e}^{-\mathrm{i}(n-1)\theta_m}\right] \end{aligned} \tag{3.142}$$

This system of $2M$ equations with $4N$ real coefficients is solved via least squares to provide a solution for stress-specified boundary conditions along circular boundaries. This solution is illustrated in Fig. 3.38 for an array of circles with boundaries of zero traction placed in a field of uniform stress. For problems where the center of an element z_c is located a large distance from the origin, the system of equations may be scaled so terms in the **A** matrices vary between ± 1.

Circle elements with boundary conditions of ***specified displacement*** in (r, θ) coordinates (3.133) may be related to the coefficients in the series expansion (3.131):

$$\begin{aligned} u_r + \mathrm{i}u_\theta = \frac{\mathrm{e}^{-\mathrm{i}\theta}}{2\mu} \Bigg[&\overset{\text{out}}{c}_0 2\kappa r_0 \ln(r_0|Z|) - \overline{\overset{\text{out}}{c}}_0 r_0 \frac{Z}{\overline{Z}} - \sum_{n=1}^{N} \overset{\phi\text{out}}{c}_n \kappa r_0 \frac{Z^{-n}}{n} \\ &- \sum_{n=1}^{N} \overline{\overset{\phi\text{out}}{c}_n} z \overline{Z^{-n-1}} + \sum_{n=1}^{N} \overline{\overset{\psi\text{out}}{c}_n} r_0 \frac{\overline{Z^{-n}}}{n} \Bigg] + \overset{\text{add}}{u}\mathrm{e}^{-\mathrm{i}\theta} \end{aligned} \tag{3.143}$$

This is evaluated at the control points, where $z_m = \mathrm{e}^{\mathrm{i}\theta_m}$ from (3.136), to provide a system of M complex equations:

$$\begin{aligned} u_r^+(z_m) + \mathrm{i}u_\theta^+(z_m) = \frac{1}{2\mu}\Bigg[& \overset{\text{out}}{c}_0 2\kappa r_0 \ln(r_0)\mathrm{e}^{-\mathrm{i}\theta_m} - \overline{\overset{\text{out}}{c}_0} r_0 \mathrm{e}^{\mathrm{i}\theta_m} \\ & - \sum_{n=1}^{N} \overset{\phi\text{out}}{c}_n \kappa r_0 \frac{\mathrm{e}^{-\mathrm{i}(n+1)\theta_m}}{n} - \sum_{n=1}^{N} \overline{\overset{\phi\text{out}}{c}_n} z_m \mathrm{e}^{\mathrm{i}n\theta_m} \\ & + \sum_{n=1}^{N} \overline{\overset{\psi\text{out}}{c}_n} r_0 \frac{\mathrm{e}^{\mathrm{i}(n-1)\theta_m}}{n} \Bigg] + \overset{\text{add}}{u_r}(z_m) + \mathrm{i}\overset{\text{add}}{u_\theta}(z_m) \end{aligned} \tag{3.144}$$

The terms with unknown coefficients are gathered on the left-hand side of the equation,

$$\begin{aligned} & \left(\Re\overset{\text{out}}{c}_0 + \mathrm{i}\Im\overset{\text{out}}{c}_0\right) 2\kappa \ln(r_0)\mathrm{e}^{-\mathrm{i}\theta_m} - \left(\Re\overset{\text{out}}{c}_0 - \mathrm{i}\Im\overset{\text{out}}{c}_0\right)\mathrm{e}^{\mathrm{i}\theta_m} \\ & - \sum_{n=1}^{N}\left(\Re\overset{\phi\text{out}}{c}_n + \mathrm{i}\Im\overset{\phi\text{out}}{c}_n\right)\kappa\frac{\mathrm{e}^{-\mathrm{i}(n+1)\theta_m}}{n} - \sum_{n=1}^{N}\left(\Re\overset{\phi\text{out}}{c}_n - \mathrm{i}\Im\overset{\phi\text{out}}{c}_n\right)\frac{z_m}{r_0}\mathrm{e}^{\mathrm{i}n\theta_m} \\ & + \sum_{n=1}^{N}\left(\Re\overset{\psi\text{out}}{c}_n - \mathrm{i}\Im\overset{\psi\text{out}}{c}_n\right)\frac{\mathrm{e}^{\mathrm{i}(n-1)\theta_m}}{n} = \frac{2\mu}{r_0}\left[u_{r\,m}^+ - \overset{\text{add}}{u_r}(z_m)\right] + \mathrm{i}\frac{2\mu}{r_0}\left[u_{\theta\,m}^+ - \overset{\text{add}}{u_\theta}(z_m)\right] \end{aligned} \tag{3.145}$$

Example 3.8 The exact solution for a circle centered at $z_c = 0$ with radius r_0 is given by the series in (3.135) with coefficients

$$\begin{aligned} & \overset{\phi\text{out}}{c}_1 = \frac{\overset{\psi\text{out}}{c}_3}{3} = \frac{1}{\kappa}\left(\frac{\sigma_{x0} - \sigma_{y0}}{2} + \mathrm{i}\tau_0\right), && \overset{\phi\text{out}}{c}_n = 0 \ (n \neq 1) \\ & \overset{\psi\text{out}}{c}_1 = (1-\kappa)\frac{\sigma_{x0} + \sigma_{y0}}{4}, && \overset{\psi\text{out}}{c}_n = 0 \ (n \neq 1, 3) \end{aligned} \tag{3.146}$$

This exact solution is illustrated in Fig. 3.39.

This system of equations, (3.146), provides $2M$ conditions for $4N + 2$ unknowns (including $\overset{\phi\text{out}}{c}_0$ terms), represented in matrix form as

$$\begin{bmatrix} {}^{0u_r}\mathbf{A} & {}^{\Re\phi u_r}\mathbf{A} & {}^{\Im\phi u_r}\mathbf{A} & {}^{\Re\psi u_r}\mathbf{A} & {}^{\Im\psi u_r}\mathbf{A} \\ {}^{0u_\theta}\mathbf{A} & {}^{\Re\phi u_\theta}\mathbf{A} & {}^{\Im\phi u_\theta}\mathbf{A} & {}^{\Re\psi u_\theta}\mathbf{A} & {}^{\Im\psi u_\theta}\mathbf{A} \end{bmatrix} \begin{bmatrix} \overset{0}{\mathbf{c}} \\ \overset{\Re\phi}{\mathbf{c}} \\ \overset{\Im\phi}{\mathbf{c}} \\ \overset{\Re\psi}{\mathbf{c}} \\ \overset{\Im\psi}{\mathbf{c}} \end{bmatrix} = \begin{bmatrix} \boldsymbol{u_r} \\ \boldsymbol{u_\theta} \end{bmatrix}, \quad \overset{0}{\mathbf{c}} = \begin{bmatrix} \Re\overset{\text{out}}{c}_0 \\ \Im\overset{\text{out}}{c}_0 \end{bmatrix} \tag{3.147}$$

where the specified conditions minus the additional functions are gathered in

$$\boldsymbol{u_r} = \frac{2\mu}{r_0}\begin{bmatrix} \vdots \\ u_{r\,m}^+ - \overset{\text{add}}{u_r}(z_m) \\ \vdots \end{bmatrix}, \quad \boldsymbol{u_\theta} = \frac{2\mu}{r_0}\begin{bmatrix} \vdots \\ u_{\theta\,m}^+ - \overset{\text{add}}{u_\theta}(z_m) \\ \vdots \end{bmatrix} \tag{3.148}$$

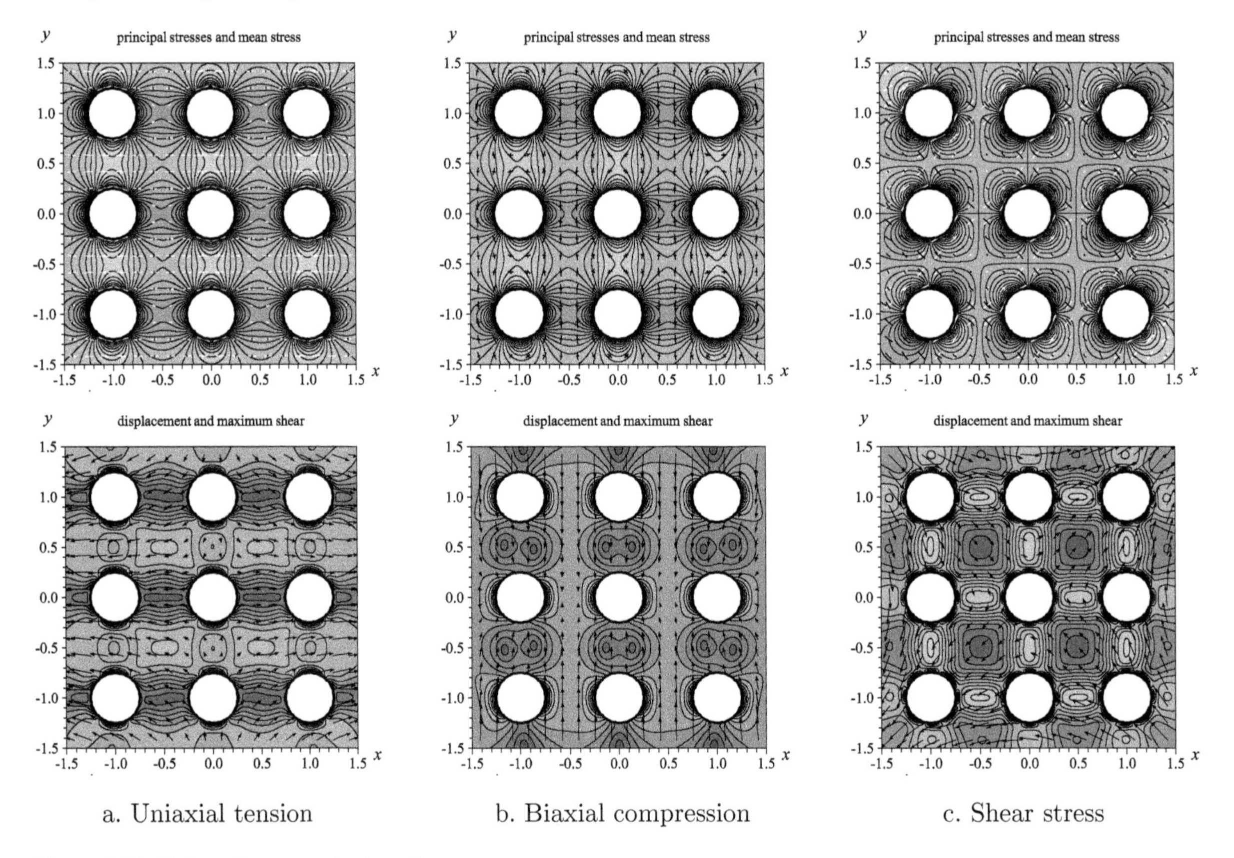

Figure 3.38 *Circles with zero traction in uniform stress.*

and the terms relating coefficients to boundary conditions are obtained by evaluating

$$\begin{aligned}
\overset{\Re\phi u_r}{\mathbf{A}}_{mn} &= \Re\left[-\kappa\frac{\mathrm{e}^{-\mathrm{i}(n+1)\theta_m}}{n} - \frac{z_m}{r_0}\mathrm{e}^{\mathrm{i}n\theta_m}\right], & \overset{\Re\psi u_r}{\mathbf{A}}_{mn} &= \Re\left[\frac{\mathrm{e}^{\mathrm{i}(n-1)\theta_m}}{n}\right]\\
\overset{\Im\phi u_r}{\mathbf{A}}_{mn} &= \Re\left[-\mathrm{i}\kappa\frac{\mathrm{e}^{-\mathrm{i}(n+1)\theta_m}}{n} + \mathrm{i}\frac{z_m}{r_0}\mathrm{e}^{\mathrm{i}n\theta_m}\right], & \overset{\Im\psi u_r}{\mathbf{A}}_{mn} &= \Re\left[-\mathrm{i}\frac{\mathrm{e}^{\mathrm{i}(n-1)\theta_m}}{n}\right]\\
\overset{\Re\phi u_\theta}{\mathbf{A}}_{mn} &= \Im\left[-\kappa\frac{\mathrm{e}^{-\mathrm{i}(n+1)\theta_m}}{n} - \frac{z_m}{r_0}\mathrm{e}^{\mathrm{i}n\theta_m}\right], & \overset{\Re\psi u_\theta}{\mathbf{A}}_{mn} &= \Im\left[\frac{\mathrm{e}^{\mathrm{i}(n-1)\theta_m}}{n}\right]\\
\overset{\Im\phi u_\theta}{\mathbf{A}}_{mn} &= \Im\left[-\mathrm{i}\kappa\frac{\mathrm{e}^{-\mathrm{i}(n+1)\theta_m}}{n} + \mathrm{i}\frac{z_m}{r_0}\mathrm{e}^{\mathrm{i}n\theta_m}\right], & \overset{\Im\psi u_\theta}{\mathbf{A}}_{mn} &= \Im\left[-\mathrm{i}\frac{\mathrm{e}^{\mathrm{i}(n-1)\theta_m}}{n}\right]
\end{aligned} \tag{3.149}$$

and for the terms related to $\overset{\text{out}}{c}_0$,

$$\begin{aligned}
\overset{0u_r}{\mathbf{A}}_{m1} &= \Re\left[2\kappa\ln(r_0)\mathrm{e}^{-\mathrm{i}\theta_m} - \mathrm{e}^{\mathrm{i}\theta_m}\right], & \overset{0u_r}{\mathbf{A}}_{m2} &= \Re\left[\mathrm{i}2\kappa\ln(r_0)\mathrm{e}^{-\mathrm{i}\theta_m} + \mathrm{i}\mathrm{e}^{\mathrm{i}\theta_m}\right]\\
\overset{0u_\theta}{\mathbf{A}}_{m1} &= \Im\left[2\kappa\ln(r_0)\mathrm{e}^{-\mathrm{i}\theta_m} - \mathrm{e}^{\mathrm{i}\theta_m}\right], & \overset{0u_\theta}{\mathbf{A}}_{m2} &= \Im\left[\mathrm{i}2\kappa\ln(r_0)\mathrm{e}^{-\mathrm{i}\theta_m} + \mathrm{i}\mathrm{e}^{\mathrm{i}\theta_m}\right]
\end{aligned} \tag{3.150}$$

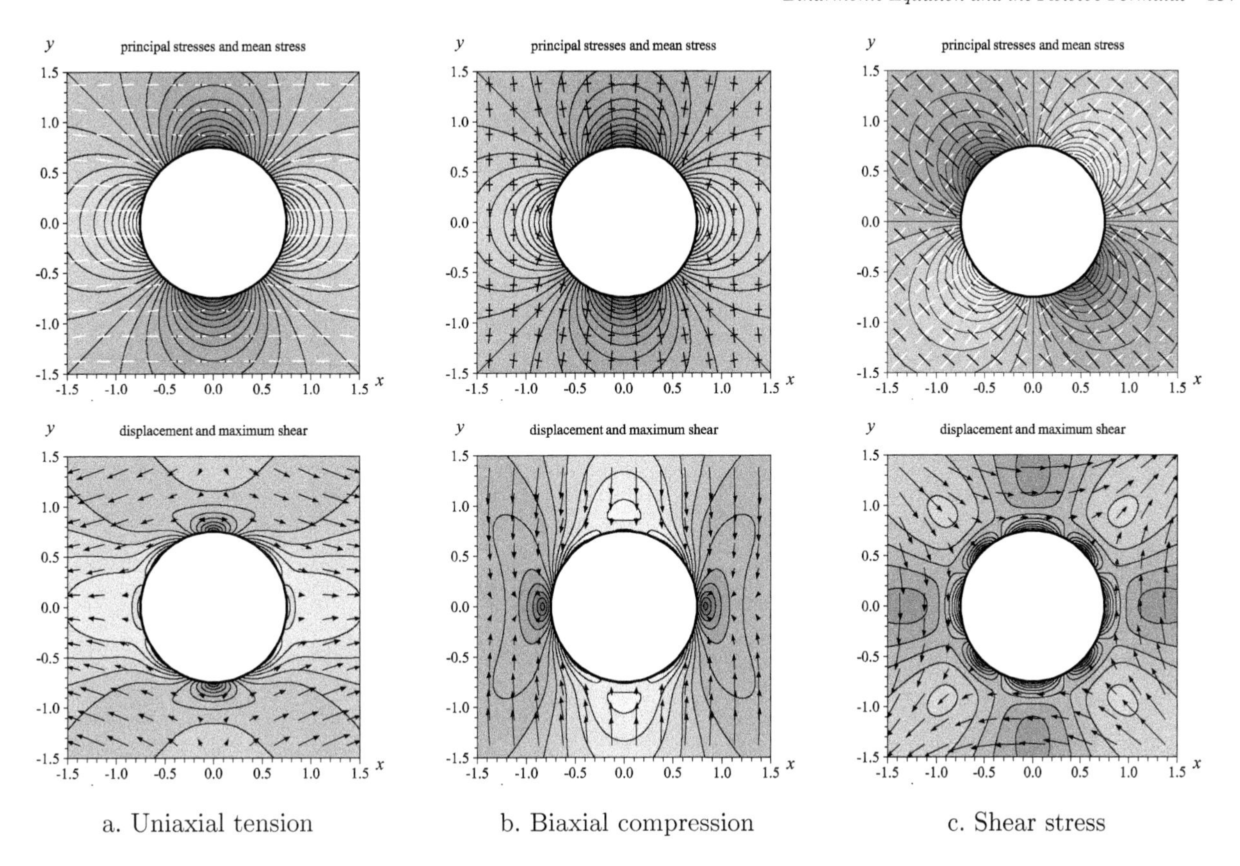

Figure 3.39 *Circle with zero displacement in uniform stress.*

Two additional steps were taken to achieve the solutions illustrated in Fig. 3.40. First the constant term u_0 for the background field in (3.125) was solved, using a condition that $u_x = 0$ along the y-axis and $u_y = 0$ along the x-axis for uniaxial and biaxial stress, and these components are zero along the opposite axes for a background field of pure shear. These conditions were solved for iteratively at the end of each sequence of solving all circles. Secondly, the solution for displacement-specified conditions tends to overshoot the correct value with many elements. This was solved using Successive Over Relaxation where the coefficient for each circle used 0.75 times the previous solution and 0.25 times the next iterates matrix solution for coefficients.

Example 3.9 A force is generated by each circle element to hold its position. This force is obtained directly from the $n = 0$ coefficients by equating this to that in the force example, giving ${}^{\phi\text{out}}c_0 = -\frac{F}{2\pi(\kappa+1)r0}$ and ${}^{\psi\text{out}}c_0 = \frac{\kappa*\overline{F}}{2\pi(\kappa+1)r0}$, where use has been made of ${}^{\psi\text{out}}c_0 = -\kappa\,\overline{{}^{\phi\text{out}}c_0}$ in (3.129) so the discontinuities in displacement cancel.

Problem 3.23 Compute the stress components σ_x, σ_y, and τ_{xy}, and σ_r, σ_θ, and $\tau_{r\theta}$ at $z = r_0$, $z = r_0 e^{i\pi/4}$, and $z = ir_0$ for a circle with $N = 20$ in the specified uniform stress and radius r_0 from Problem 3.22.

3.7.2 Biharmonic Elements for Circular Interfaces and Slits

The Kolosov formula for the domain inside a circle is provided by the terms in a Taylor series that remain finite within the circle (Muskhelishvili, 1953*b*, eqn 54.6)

$$\begin{aligned} \phi &= \sum_{n=1}^{N} \overset{\phi\text{in}}{c}_n \frac{r_0}{n} \mathcal{Z}^n \quad , \quad \psi = \overset{\psi\text{in}}{c}_0 r_0 + \sum_{n=1}^{N} \overset{\psi\text{in}}{c}_n \frac{r_0}{n} \mathcal{Z}^n \\ \phi' &= \sum_{n=1}^{N} \overset{\phi\text{in}}{c}_n \mathcal{Z}^{n-1} \quad , \quad \psi' = \sum_{n=1}^{N} \overset{\psi\text{in}}{c}_n \mathcal{Z}^{n-1} \\ \phi'' &= \sum_{n=2}^{N} \overset{\phi\text{in}}{c}_n \frac{n-1}{r_0} \mathcal{Z}^{n-2} \end{aligned} \tag{3.151}$$

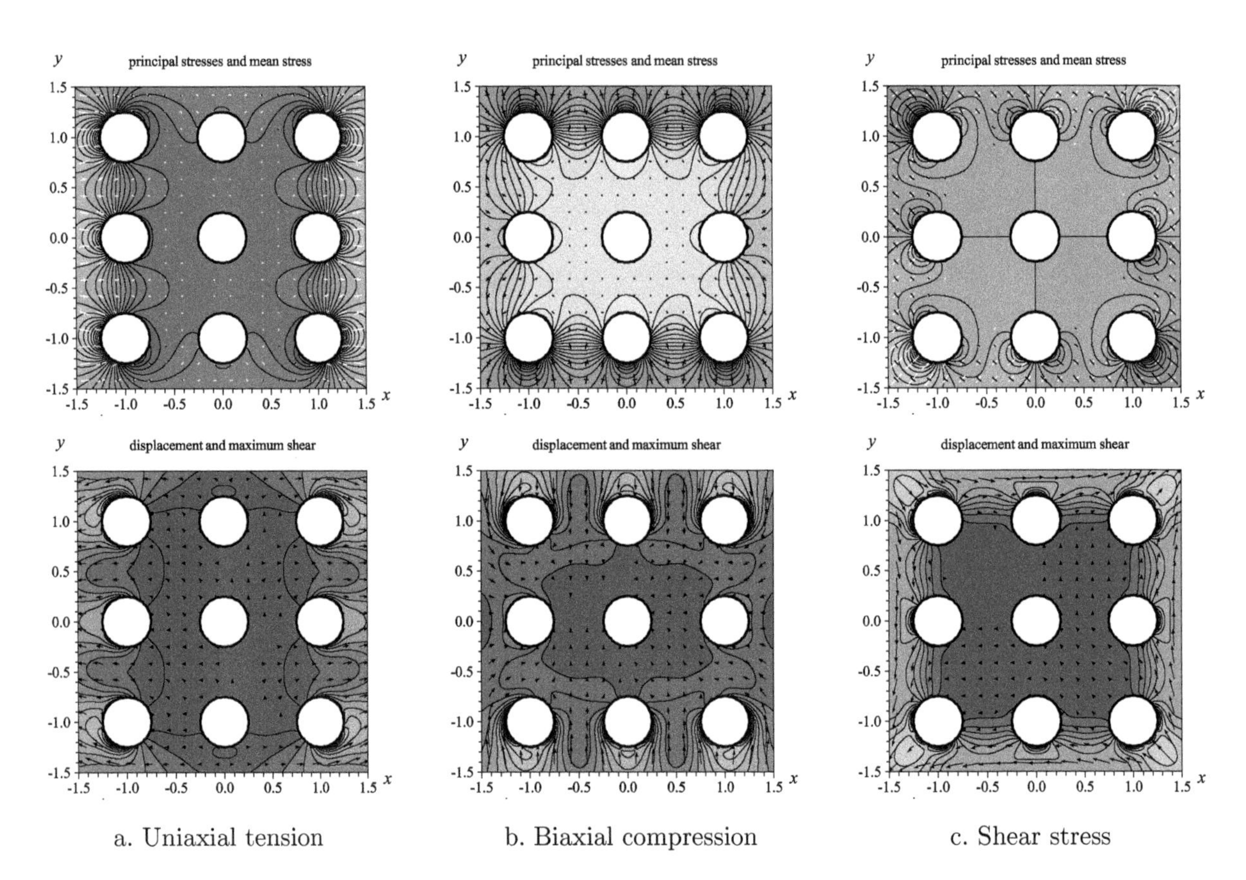

Figure 3.40 *Circles with zero displacement in uniform stress.*

Linear superposition of these influence functions is illustrated in Fig. 3.41. Expressions for the normal and shear components of stress along the boundary follow directly from the methods utilized outside a circle, (3.137),

$$
\begin{aligned}
\sigma_r^-(z_m) - \mathrm{i}\tau_{r\theta}^-(z_m) = 2\Re & \left[\sum_{n=1}^{N} \overset{\phi\mathrm{in}}{c}_n \mathrm{e}^{\mathrm{i}(n-1)\theta_m}\right] \\
& - \left(\frac{\overline{z_c}}{r_0} + \mathrm{e}^{-\mathrm{i}\theta_m}\right) \sum_{n=1}^{N} \overset{\phi\mathrm{in}}{c}_n (n-1)\mathrm{e}^{\mathrm{i}n\theta_m} \\
& - \sum_{n=1}^{N} \overset{\psi\mathrm{in}}{c}_n \mathrm{e}^{\mathrm{i}(n+1)\theta_m} + \overset{\mathrm{add}}{\sigma_r}(z_m) - \mathrm{i}\overset{\mathrm{add}}{\tau_{r\theta}}(z_m)
\end{aligned}
\tag{3.152}
$$

Likewise, displacement along the boundary follows from (3.144)

$$
\begin{aligned}
u_r^-(z_m) + \mathrm{i}u_\theta^-(z_m) = \frac{1}{2\mu} & \left[\kappa r_0 \sum_{n=1}^{N} \overset{\phi\mathrm{in}}{c}_n \frac{\mathrm{e}^{\mathrm{i}(n-1)\theta_m}}{n}\right. \\
& - \left(z_c + r_0\mathrm{e}^{\mathrm{i}\theta_m}\right) \sum_{n=1}^{N} \overline{\overset{\phi\mathrm{in}}{c}_n} \mathrm{e}^{-\mathrm{i}n\theta_m} - \overline{\overset{\psi\mathrm{in}}{c}_0} r_0 \mathrm{e}^{-\mathrm{i}\theta_m} \\
& \left. - r_0 \sum_{n=1}^{N} \overline{\overset{\psi\mathrm{in}}{c}_n} \frac{\mathrm{e}^{-\mathrm{i}(n+1)\theta_m}}{n}\right] + \overset{\mathrm{add}}{u_r}(z_m) + \mathrm{i}\overset{\mathrm{add}}{u_\theta}(z_m)
\end{aligned}
\tag{3.153}
$$

Following the development of solutions to stress-specified conditions in (3.138), the terms may be gather terms with common exponential powers for stress (3.152)

$$
\begin{aligned}
& 2\Re\left[\sum_{n=0}^{N-1} \overset{\phi\mathrm{in}}{c}_{n+1} \mathrm{e}^{\mathrm{i}n\theta_m}\right] - \frac{\overline{z_c}}{r_0} \sum_{n=1}^{N} \overset{\phi\mathrm{in}}{c}_n (n-1)\mathrm{e}^{\mathrm{i}n\theta_m} - \sum_{n=0}^{N-1} \overset{\phi\mathrm{in}}{c}_{n+1} n \mathrm{e}^{\mathrm{i}n\theta_m} \\
& - \sum_{n=2}^{N+1} \overset{\psi\mathrm{in}}{c}_{n-1} \mathrm{e}^{\mathrm{i}n\theta_m} = \sigma_{r\,m}^- - \mathrm{i}\tau_{r\theta\,m}^- - \left(\overset{\mathrm{add}}{\sigma}_{r\,m} - \mathrm{i}\overset{\mathrm{add}}{\tau}_{r\theta\,m}\right)
\end{aligned}
\tag{3.154}
$$

Likewise, displacement-specified conditions are facilitated by gathering their common terms, as done in (3.145) outside the circle:

$$
\begin{aligned}
& \frac{r_0}{2\mu}\left[\kappa \sum_{n=0}^{N-1} \frac{\overset{\phi\mathrm{in}}{c}_{n+1}}{n+1} \mathrm{e}^{\mathrm{i}n\theta_m} - \frac{z_c}{r_0} \sum_{n=1}^{N} \overline{\overset{\phi\mathrm{in}}{c}_n} \mathrm{e}^{-\mathrm{i}n\theta_m} - \sum_{n=0}^{N-1} \overline{\overset{\phi\mathrm{in}}{c}_{n+1}} \mathrm{e}^{-\mathrm{i}n\theta_m}\right. \\
& \left. - \overline{\overset{\psi\mathrm{in}}{c}_0} r_0 \mathrm{e}^{-\mathrm{i}\theta_m} - \sum_{n=2}^{N+1} \frac{\overline{\overset{\psi\mathrm{in}}{c}_{n-1}}}{n-1} \mathrm{e}^{-\mathrm{i}n\theta_m}\right] = u_{r\,m}^- + \mathrm{i}u_{\theta\,m}^- - \left(\overset{\mathrm{add}}{u}_{r\,m} + \mathrm{i}\overset{\mathrm{add}}{u}_{\theta\,m}\right)
\end{aligned}
\tag{3.155}
$$

The system of equations to solve for a domain inside a circle with a boundary of specified stress or displacement follows directly from the formulation of these conditions for domains outside a circle. Furthermore, interface conditions that apply continuity conditions across the interface of circles may be applied using the formulation of heterogeneities in Section 3.2.3. In this case, continuity of the radial and tangential components of stress may be applied, and conditions of continuity of the normal and tangential displacement may be applied for domains with distinct differences in elasticity properties (E, μ, and ν) across the interface (Green and Zerna, 1968, p.278; Jaeger and Cook, 1976, p.251).

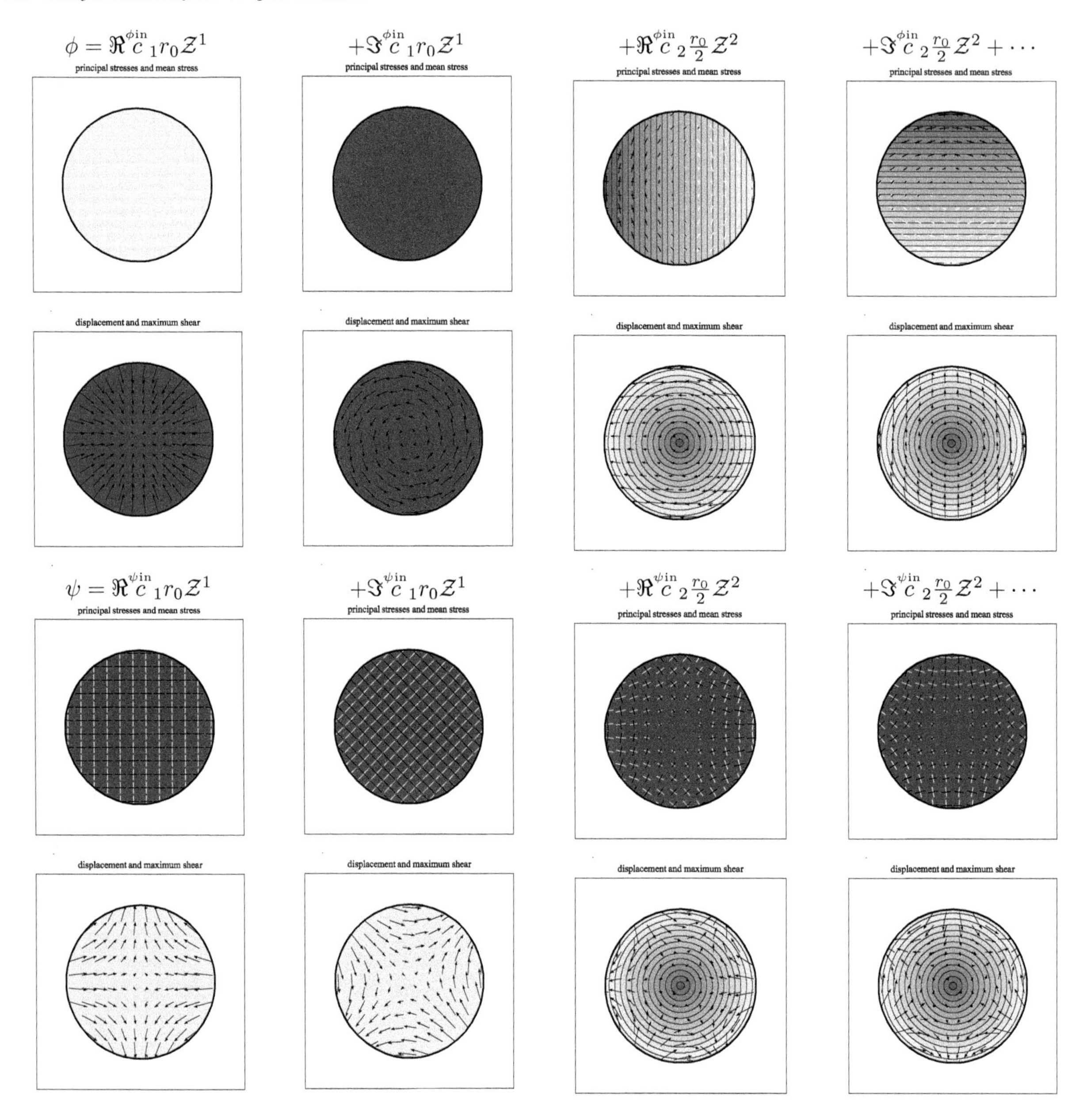

Figure 3.41 *Influence functions inside a circle for the biharmonic equation, from (3.151).*

Solutions to the biharmonic equation may be extended to ***slit elements*** using the conformal mapping illustrated earlier in Fig. 3.17. Expressions for these coordinate transformations are given by (3.55):

$$\zeta = \left(\frac{z - z_{\max} + z_{\min}}{2}\right)\frac{2}{z_{\max} - z_{\min}}, \qquad z = \zeta\frac{z_{\max} - z_{\min}}{2} + \frac{z_{\max} + z_{\min}}{2} \tag{3.156}$$

and

$$\mathcal{Z} = \zeta + \sqrt{\zeta+1}\sqrt{\zeta-1}, \qquad \zeta = \frac{1}{2}\left(\mathcal{Z} + \frac{1}{\mathcal{Z}}\right) \tag{3.157}$$

The ***Kolosov formulas for a slit element*** may be represented as a series expansion of influences functions with a similar form as the complex functions for those outside a circle element, (3.129):

$$\boxed{\begin{aligned}
&\phi = -\sum_{n=1}^{N} \overset{\phi}{c}_n \frac{Le^{i\vartheta}}{n}\mathcal{Z}^{-n}, \qquad \psi = -\sum_{n=1}^{N} \overset{\psi}{c}_n \frac{Le^{i\vartheta}}{n}\mathcal{Z}^{-n} \\
&\phi' = \sum_{n=1}^{N} \overset{\phi}{c}_n \frac{\mathcal{Z}^{-n}}{\sqrt{\zeta+1}\sqrt{\zeta-1}}, \qquad \psi' = \sum_{n=1}^{N} \overset{\psi}{c}_n \frac{\mathcal{Z}^{-n}}{\sqrt{\zeta+1}\sqrt{\zeta-1}} \\
&\phi'' = -\sum_{n=1}^{N} \overset{\phi}{c}_n \frac{1}{Le^{i\vartheta}}\left[\frac{(n+1)\mathcal{Z}^{-n}}{(\zeta+1)(\zeta-1)} + \frac{\mathcal{Z}^{-n-1}}{(\zeta+1)^{3/2}(\zeta-1)^{3/2}}\right]
\end{aligned}} \tag{3.158}$$

These functions are substituted into the expressions for stress components (3.124) to give

$$\begin{aligned}
\sigma_x &= \Re\left\{\sum_{n=1}^{N} \overset{\phi}{c}_n \frac{2\mathcal{Z}^{-n}}{\sqrt{\zeta+1}\sqrt{\zeta-1}}\right. \\
&\quad + \sum_{n=1}^{N} \overset{\phi}{c}_n \frac{\overline{z}}{Le^{i\vartheta}}\left[\frac{(n+1)\mathcal{Z}^{-n}}{(\zeta+1)(\zeta-1)} + \frac{\mathcal{Z}^{-n-1}}{(\zeta+1)^{3/2}(\zeta-1)^{3/2}}\right] \\
&\quad \left. - \sum_{n=1}^{N} \overset{\psi}{c}_n \frac{\mathcal{Z}^{-n}}{\sqrt{\zeta+1}\sqrt{\zeta-1}}\right\} + \overset{\text{add}}{\sigma_x} \\
\sigma_y &= \Re\left\{\sum_{n=1}^{N} \overset{\phi}{c}_n \frac{2\mathcal{Z}^{-n}}{\sqrt{\zeta+1}\sqrt{\zeta-1}}\right. \\
&\quad - \sum_{n=1}^{N} \overset{\phi}{c}_n \frac{\overline{z}}{Le^{i\vartheta}}\left[\frac{(n+1)\mathcal{Z}^{-n}}{(\zeta+1)(\zeta-1)} + \frac{\mathcal{Z}^{-n-1}}{(\zeta+1)^{3/2}(\zeta-1)^{3/2}}\right] \\
&\quad \left. + \sum_{n=1}^{N} \overset{\psi}{c}_n \frac{\mathcal{Z}^{-n}}{\sqrt{\zeta+1}\sqrt{\zeta-1}}\right\} + \overset{\text{add}}{\sigma_y} \\
\tau_{xy} &= \Im\left\{-\sum_{n=1}^{N} \overset{\phi}{c}_n \frac{\overline{z}}{Le^{i\vartheta}}\left[\frac{(n+1)\mathcal{Z}^{-n}}{(\zeta+1)(\zeta-1)} + \frac{\mathcal{Z}^{-n-1}}{(\zeta+1)^{3/2}(\zeta-1)^{3/2}}\right]\right. \\
&\quad \left. + \sum_{n=1}^{N} \overset{\psi}{c}_n \frac{\mathcal{Z}^{-n}}{\sqrt{\zeta+1}\sqrt{\zeta-1}}\right\} + \overset{\text{add}}{\tau_{xy}}
\end{aligned} \tag{3.159}$$

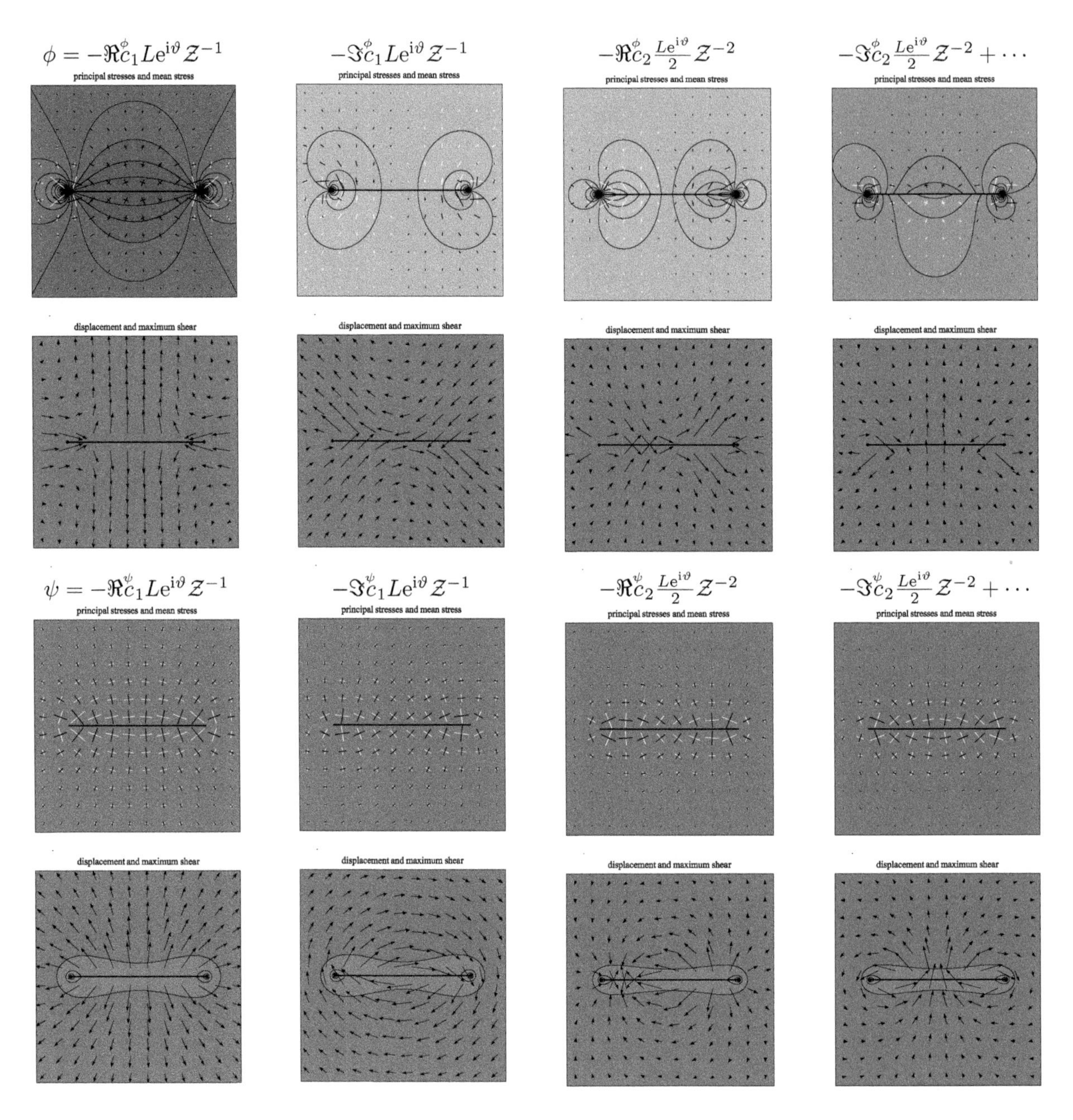

Figure 3.42 *Influence functions for a slit element with the biharmonic equation, from (3.158).*

Likewise, the complex displacement, (3.124), for slit elements is given by

$$u = \frac{1}{2\mu}\left(-\sum_{n=1}^{N} \overset{\phi}{c}_n \frac{\kappa L e^{i\vartheta}}{n} \mathcal{Z}^{-n} - \sum_{n=1}^{N} z \overline{\overset{\phi}{c}_n \frac{\mathcal{Z}^{-n}}{\sqrt{\zeta+1}\sqrt{\zeta-1}}} + \sum_{n=1}^{N} \overline{\overset{\psi}{c}_n \frac{L e^{i\vartheta}}{n} \mathcal{Z}^{-n}}\right) + \overset{\text{add}}{u} \quad (3.160)$$

This representation as the summation of influence functions for stress and displacement is visualized in Fig. 3.42 for a slit of unit length lying along the x-axis.

Development of boundary conditions requires expressions of ***stress and displacement along a slit element***. The normal σ_{n} and shear $\tau_{\mathfrak{s}\mathrm{n}}$ components of the stress are (Muskhelishvili, 1953*b*, p.147; Jaeger, 1969, p.195)

$$(\sigma_{\mathrm{n}} - i\tau_{\mathfrak{s}\mathrm{n}}) = 2\Re\left[\phi'(s)\right] - \left[\overline{z}\phi''(s) + \psi'(s)\right] e^{i2\vartheta} \quad (3.161)$$

where ϑ is the orientation of the slit element with respect to the x-axis The displacement components along the slit follow from the rotated coordinates in (1.123) (Jaeger, 1969, p.204)

$$u_{\mathfrak{s}} + i u_{\mathrm{n}} = e^{-i\vartheta}\left(u_x + i u_y\right) = \frac{e^{-i\vartheta}}{2\mu}\left(\kappa\phi - z\overline{\phi'} - \overline{\psi}\right) \quad (3.162)$$

Substitution of the Kolosov formulas, (3.158), provides expressions for the values on the +side of a slit, where $\theta > \pi$ and the angle associated with $\arg(\zeta - 1) = \pi$, and on the −side, which follows directly from the slit element formulation in Section 3.4. Note that a broad range of problems may be solved with this linear expression relating coefficients to interface conditions, utilizing combinations of normal and tangential components of stress and displacement across the element (Green and Zerna, 1968, p.276; Jaeger and Cook, 1976, p.273) Formulation of the matrices **A** and **b** grouping common terms of the cosine and sine functions follows directly to provide linear systems of equations to solve for the coefficients.

Further Reading

Sections 3.1, 3.2, and 3.3, Point, Circle, and Ellipse Elements

- Helmholtz (1858)
- Bakker and Nieber (2004*b*)
- Barnes and Janković (1999)
- Janković, Fiori, and Dagan (2003)
- Steward, Peterson, Yang, Bulatewicz, Herrera-Rodriguez, Mao and Hendricks (2009)
- Strack (1989)
- Strack (2005)
- Suribhatla, Bakker, Bandilla and Janković (2004)
- Tait (1867)

Sections 3.4 and 3.5, Slit Elements, Circular Arcs, and Joukowsky's Wing

- Abbott (1959)
- Bakker (2008)
- Courant (1950)
- Hess and Smith (1967)
- Joukowsky (1910)
- Joukowsky (1912)
- McCormick (1995)
- Steward (2015)
- Strack (1989)

- Sunada (1997)

Section 3.6, Complex Vector Fields with Divergence and Curl

- Muskhelishvili (1953*b*)
- Remmert (1991)
- Steward and Ahring (2009)
- Strack (2009*b*)
- Strack (2009*a*)
- Strack and Namazi (2014)
- Wirtinger (1927)

Section 3.7, Biharmonic Equation and the Kolosov Formulas

- Airy (1863)
- Boresi (1965)
- Green and Zerna (1968)
- Goursat (1898)
- Goursat (1904)
- Jaeger (1969)
- Jaeger and Cook (1976)
- Muskhelishvili (1953*b*)
- Sanford (2003)
- Westergaard (1952)

Analytic Elements from Separation of Variables

4

- Separation of variables provides influence functions for analytic elements, which extend the solutions available with complex functions to problems involving the Helmholtz and modified Helmholtz equations.
- Methods are introduced for one-dimensional problems that provide the background vector field for many problems, and these solutions are extended to finite domains with interconnected rectangle elements in Section 4.3.
- Circular elements are developed in Section 4.4 using series of Bessel and Fourier functions to model wave propagation around and through collections of elements, and vadose zone solutions are extended to solve the non-linear interface conditions occurring along circles.
- Methods are extended to three-dimensional problems for spheres (Section 4.5), and prolate and oblate spheroids in Section 4.6.

4.1 Overview

Analytic elements may be formulated using the separable conditions that exist when problems are formulated in ***orthogonal coordinate systems*** (Moon and Spencer, 1961*a*,*b*). Examples of coordinate systems with mutually orthogonal axes are illustrated in Fig. 4.1 for cases where one of the coordinate axes is normal to planes of uniform z. Specifically, the Cartesian coordinate system in Fig. 4.1a utilizes the x-, y-, and z-axes; circular cylindrical coordinates in Fig. 4.1b use the distance r measured from the z-axis and the angle θ measured about the z-axis; and the elliptic cylindrical coordinates in Fig. 4.1c contain ellipses with constant η and hyperbolas with constant ξ. The ***geometry of analytic elements emerges naturally using surfaces of uniform position along a coordinate axis***. For example, circle elements (with constant r in Fig. 4.1b) are formulated later using a separation of variables to extend the solutions developed in Section 3.2 using complex variables. Likewise, the separable coordinate system in Fig. 4.1c may be utilized to extend the solution methods previously presented for ellipse elements in Section 3.3 and slit elements in Section 3.4.

Analytic elements are developed in this chapter to solve problems that satisfy the Laplace equation, (2.2a), the Poisson equation, (2.2b), the Helmholtz equation, (2.2c), and the modified Helmholtz equation, (2.2d). These developments utilize the ***Laplacian*** and the ***gradient***, which may be expressed in terms of the variables in each of the coordinate systems. This gives the following forms for ***Cartesian coordinates***

Analytic Element Method: Complex Interactions of Boundaries and Interfaces. David R. Steward, Oxford University Press (2020). © David R. Steward.
DOI: 10.1093/oso/9780198856788.001.0001

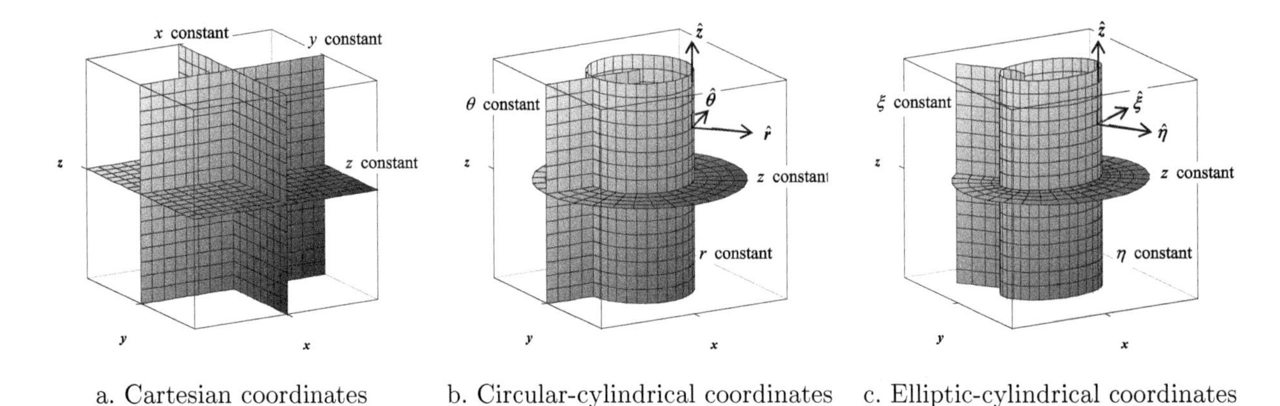

Figure 4.1 *Orthogonal coordinate systems containing planes of constant z.*

$$\nabla^2\Phi = \frac{\partial^2\Phi}{\partial x^2} + \frac{\partial^2\Phi}{\partial y^2} + \frac{\partial^2\Phi}{\partial z^2}, \quad \nabla\Phi = \frac{\partial\Phi}{\partial x}\hat{\mathbf{x}} + \frac{\partial\Phi}{\partial y}\hat{\mathbf{y}} + \frac{\partial\Phi}{\partial z}\hat{\mathbf{z}} \tag{4.1a}$$

for ***circular coordinates***

$$\nabla^2\Phi = \frac{\partial^2\Phi}{\partial r^2} + \frac{1}{r}\frac{\partial\Phi}{\partial r} + \frac{1}{r^2}\frac{\partial^2\Phi}{\partial\theta^2} + \frac{\partial^2\Phi}{\partial z^2}, \quad \nabla\Phi = \frac{\partial\Phi}{\partial r}\hat{\mathbf{r}} + \frac{1}{r}\frac{\partial\Phi}{\partial\theta}\hat{\boldsymbol{\theta}} + \frac{\partial\Phi}{\partial z}\hat{\mathbf{z}} \tag{4.1b}$$

and for ***elliptic coordinates***

$$\begin{aligned}\nabla^2\Phi &= \frac{1}{f^2\left(\cosh^2\eta - \cos^2\xi\right)}\left(\frac{\partial^2\Phi}{\partial\eta^2} + \frac{\partial^2\Phi}{\partial\xi^2}\right) + \frac{\partial^2\Phi}{\partial z^2} \\ \nabla\Phi &= \frac{1}{f\sqrt{\cosh^2\eta - \cos^2\xi}}\left(\frac{\partial\Phi}{\partial\eta}\hat{\eta} + \frac{\partial\Phi}{\partial\xi}\hat{\boldsymbol{\xi}}\right) + \frac{\partial\Phi}{\partial z}\hat{\mathbf{z}}\end{aligned} \tag{4.1c}$$

where f is a constant related to the foci of the element.

Orthogonal coordinate systems may also be formulated by rotation about an axis of symmetry in the z-direction, as illustrated in Fig. 4.2. The spherical coordinate system in Fig. 4.2a contains spheres with constant r and cones with uniform ξ; Fig. 4.2b contains prolate spheroids with constant η and hyperboloids of revolution with constant ξ; and Fig. 4.2c contains oblate spheroids with constant η and hyperboloids with constant ξ. And, each coordinate system contains half-planes with uniform θ. The Laplacian and gradient may be written in terms of the coordinate directions, giving the following expressions for ***spherical coordinates***

$$\begin{aligned}\nabla^2\Phi &= \frac{\partial^2\Phi}{\partial r^2} + \frac{2}{r}\frac{\partial\Phi}{\partial r} + \frac{1}{r^2}\frac{\partial^2\Phi}{\partial\xi^2} + \frac{\cot\xi}{r^2}\frac{\partial\Phi}{\partial\xi} + \frac{1}{r^2\sin^2\xi}\frac{\partial^2\Phi}{\partial\theta^2} \\ \nabla\Phi &= \frac{\partial\Phi}{\partial r}\hat{\mathbf{r}} + \frac{1}{r}\frac{\partial\Phi}{\partial\xi}\hat{\boldsymbol{\xi}} + \frac{1}{r\sin\xi}\frac{\partial\Phi}{\partial\theta}\hat{\boldsymbol{\theta}}\end{aligned} \tag{4.2a}$$

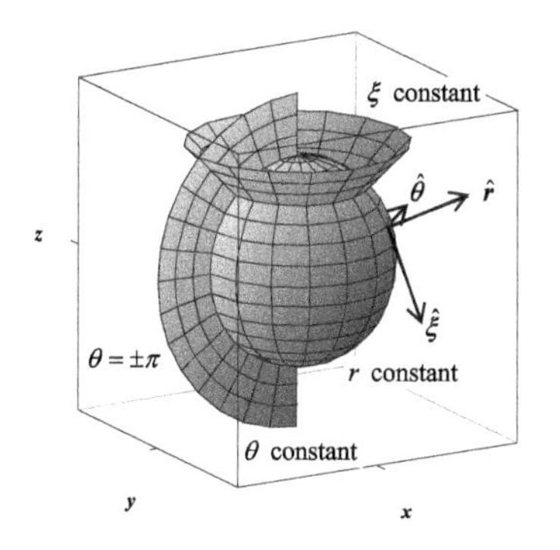

a. Spherical coordinates

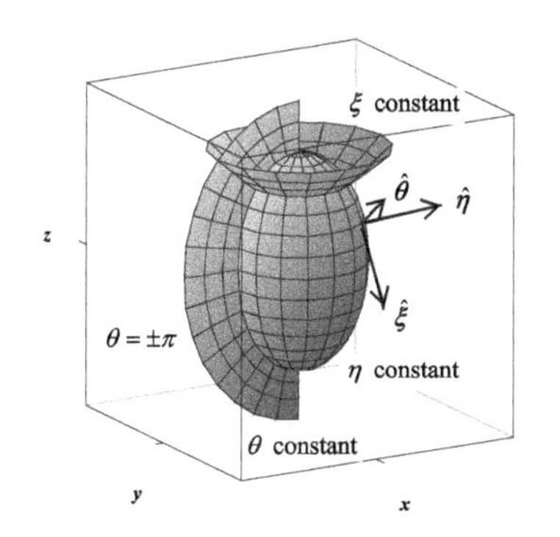

b. Prolates pheroidal coordinates

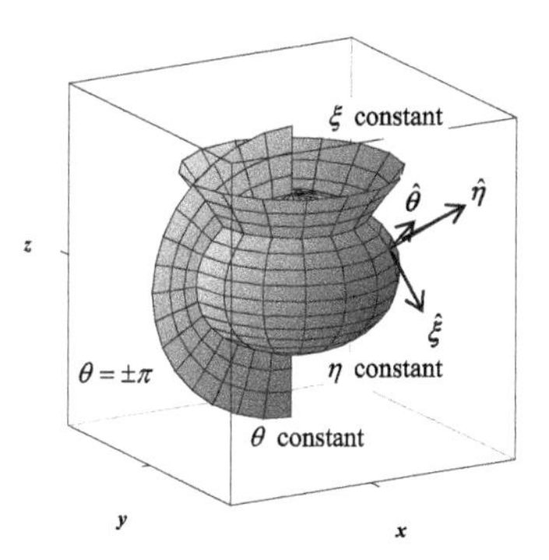

c. Oblate spheroidal coordinates

Figure 4.2 *Orthogonal coordinate systems with axisymmetry.*

for ***prolate spheroid coordinates***

$$
\begin{aligned}
\nabla^2\Phi &= \frac{1}{f^2\left(\sinh^2\eta + \sin^2\xi\right)}\left(\frac{\partial^2\Phi}{\partial\eta^2} + \coth\eta\frac{\partial\Phi}{\partial\eta} + \frac{\partial^2\Phi}{\partial\xi^2} + \cot\xi\frac{\partial\Phi}{\partial\xi}\right) + \frac{1}{f^2\sinh^2\eta\sin^2\xi}\frac{\partial^2\Phi}{\partial\theta^2} \\
\nabla\Phi &= \frac{1}{f\sqrt{\sinh^2\eta + \sin^2\xi}}\left(\frac{\partial\Phi}{\partial\eta}\hat{\boldsymbol{\eta}} + \frac{\partial\Phi}{\partial\xi}\hat{\boldsymbol{\xi}}\right) + \frac{1}{f\sinh\eta\sin\xi}\frac{\partial\Phi}{\partial\theta}\hat{\boldsymbol{\theta}}
\end{aligned}
\tag{4.2b}
$$

and for ***oblate spheroid coordinates***

$$
\begin{aligned}
\nabla^2\Phi &= \frac{1}{f^2\left[\cosh^2\eta - \sin^2\xi\right]}\left(\frac{\partial^2\Phi}{\partial\eta^2} + \tanh\eta\frac{\partial\Phi}{\partial\eta} + \frac{\partial^2\Phi}{\partial\xi^2} + \cot\xi\frac{\partial\Phi}{\partial\xi}\right) + \frac{1}{f^2\cosh^2\eta\sin^2\xi}\frac{\partial^2\Phi}{\partial\theta^2} \\
\nabla\Phi &= \frac{1}{f\sqrt{\cosh^2\eta - \sin^2\xi}}\left(\frac{\partial\Phi}{\partial\eta}\hat{\boldsymbol{\eta}} + \frac{\partial\Phi}{\partial\xi}\hat{\boldsymbol{\xi}}\right) + \frac{1}{f\cosh\eta\sin\xi}\frac{\partial\Phi}{\partial\theta}\hat{\boldsymbol{\theta}}
\end{aligned}
\tag{4.2c}
$$

Thus, the separation of variables provides solutions for ***analytic elements with the geometry of circles, spheres, prolate, and oblate spheroids***.

This chapter formulates analytic elements using the separable conditions of orthogonal coordinate systems, whereby solutions may be partitioned into the multiplication of functions of only one coordinate direction. These separable forms are found in Table 4.1, along with the solutions to the ordinary differential equations that emerge from the separation of variables. These solutions are formulated next for analytic elements to develop their influence functions and their systems of equations that solve a wide range of important problems. In particular, ***the separation of variables extends analytic element solutions to problems governed by the Helmholtz equation and the modified Helmholtz equation***.

Table 4.1 *Separation of variables into orthogonal coordinate systems*

Coordinates	Separable form	Solutions to ordinary differential equations
Cartesian	$\Phi(x, y, z) = X(x)Y(y)Z(z)$	Fourier series, exponential functions
circular	$\Phi(r, \theta, z) = R(r)\Theta(\theta)Z(z)$	Bessel functions, Fourier series
elliptic	$\Phi(\eta, \xi, z) = H(\eta)\Xi(\xi)Z(z)$	Mathieu functions
spherical	$\Phi(r, \xi, \theta) = R(r)\Xi(\xi)\Theta(\theta)$	power and Fourier series, associated Legendre functions
prolate spheroidal	$\Phi(\eta, \xi, \theta) = H(\eta)\Xi(\xi)\Theta(\theta)$	associated Legendre functions, Fourier series
oblate spheroidal	$\Phi(\eta, \xi, \theta) = H(\eta)\Xi(\xi)\Theta(\theta)$	complex associated Legendre functions, Fourier series

Problem 4.1 Compute the outward unit vector normal to the surface of analytic elements with the following geometries at the specified locations.

Circle element with specified radius r_0, at $\theta = 0, \pi/6, \pi/4, \pi/3$, and $\pi/2$:

A. $r_0 = 1$ B. $r_0 = 2$ C. $r_0 = 3$ D. $r_0 = 4$

Sphere element with specified radius r_0 and θ, at $\xi = 0, \pi/4, \pi/2, 3\pi/4$, and π:

E. $r_0 = 1, \theta = \pi/6$ F. $r_0 = 2, \theta = \pi/4$ G. $r_0 = 3, \theta = \pi/3$ H. $r_0 = 4, \theta = \pi/2$

Prolate spheroid with specified coordinate η_0, θ, and $f = 0.5$, at $\xi = 0, \pi/4, \pi/2, 3\pi/4$, and π:

I. $\eta_0 = 1, \theta = \pi/6$ J. $\eta_0 = 2, \theta = \pi/4$ K. $\eta_0 = 3, \theta = \pi/3$ L. $\eta_0 = 4, \theta = \pi/2$

Oblate spheroid with specified coordinate η_0, θ, and $f = 0.5$, at $\xi = 0, \pi/4, \pi/2, 3\pi/4$, and π:

M. $\eta_0 = 1, \theta = \pi/6$ N. $\eta_0 = 2, \theta = \pi/4$ O. $\eta_0 = 3, \theta = \pi/3$ P. $\eta_0 = 4, \theta = \pi/2$

(Hint: The definitions of coordinates are found later in this chapter.)

4.2 Separation for One-Dimensional Problems

The method of separation of variables is applied first to problems where a function varies only in one direction. For problems governed by the ***Laplace equation*** (2.2a), this equation takes on the form of an ordinary differential equation with general solution given by

$$\frac{d^2\Phi}{dx^2} = 0 \quad \rightarrow \quad \Phi = c_1 + c_2 x \tag{4.3}$$

where c_1 and c_2 are constants. These constants may be chosen, for example to satisfy conditions such that the potential takes on the value of Φ_0 at $x = x_0$, and minus the gradient is equal to the vector component v_{x0}, giving

$$\Phi(x_0) = \Phi_0, \; v_{x0} = -\frac{\partial \Phi}{\partial x} \quad \rightarrow \quad \Phi(x) = -v_{x0}(x - x_0) + \Phi_0 \tag{4.4}$$

This solution may be extended to a uniform vector field in a three-dimensional domain:

$$\boxed{\Phi(x, y, z) = -v_{x0}(x - x_0) - v_{y0}(y - y_0) - v_{z0}(z - z_0) + \Phi_0} \tag{4.5}$$

with uniform vector components v_{x0}, v_{y0}, and v_{z0} in the coordinate directions, and the potential is equal to Φ_0 at (x_0, y_0, z_0). This provides a background potential for many of the examples found later in this chapter and in the three-dimensional vector fields of Section 5.7.

4.2.1 One-Dimensional Helmholtz Equation for Waves

The separation of variables provides a method for developing solutions for the Helmholtz equation, (2.2c), which also reduces to an ordinary differential equation for one-dimensional problems. The general solution for the complex wave function φ for fields varying in the x-direction is given by

$$\frac{\mathrm{d}^2\varphi}{\mathrm{d}x^2} + k^2\varphi = 0 \quad \rightarrow \quad \varphi = c_1\mathrm{e}^{-\mathrm{i}kx} + c_2\mathrm{e}^{\mathrm{i}kx} \tag{4.6}$$

A particular solution with coefficient c_2 set equal to the complex constant φ_0 provides a function φ, (1.73), that satisfies the wave equation, (1.72):

$$\begin{matrix} c_1 = 0 \\ c_2 = \varphi_0 \end{matrix} \quad \rightarrow \quad \varphi = \varphi_0\mathrm{e}^{\mathrm{i}kx} \quad \rightarrow \quad \Phi = \Re\left(\varphi\mathrm{e}^{-\mathrm{i}\omega t}\right) = \Re\left[\varphi_0\mathrm{e}^{\mathrm{i}(kx-\omega t)}\right] \tag{4.7}$$

where ω is the angular frequency of the wave field. This function may be rearranged using the complex form $\varphi_0 = |\varphi_0|\mathrm{e}^{\mathrm{i}\arg\varphi_0}$ with magnitude $|\varphi_0|$ and phase $\arg\varphi_0$ to provide

$$\Phi = \Re\left[|\varphi_0|\,\mathrm{e}^{\mathrm{i}(kx-\omega t+\arg\varphi_0)}\right] = |\varphi_0|\cos(kx - \omega t + \arg\varphi_0) \tag{4.8}$$

Thus, the wave field has uniform amplitude $|\varphi_0|$, traveling in the x-direction with wave velocity $C = \omega/k = L/T$, (1.75), where L is the wave length and T is the wave period. This solution may be reoriented for ***plane waves*** traveling in the θ_0-direction:

$$\boxed{\varphi = \varphi_0\mathrm{e}^{\mathrm{i}k(x\cos\theta_0 + y\sin\theta_0)}} \tag{4.9}$$

where the coordinate axes in which waves propagate were rotated by angle θ_0 using

$$\Re\left[(x + \mathrm{i}y)\mathrm{e}^{-\mathrm{i}\theta_0}\right] = \Re\left[(x + \mathrm{i}y)(\cos\theta_0 - \mathrm{i}\sin\theta_0)\right] = x\cos\theta_0 + y\sin\theta_0 \tag{4.10}$$

This solution in Fig. 4.3a illustrates the uniform amplitude, and the phase corresponds with monochromatic waves traveling in the $\theta_0 = 30°$ direction.

This solution may be adapted to study the wave field generated by two sets of plane waves traveling in different directions. For example, ***wave reflection*** was introduced in Fig. 1.43a for a set of incident plane waves that approaches a straight boundary at angle γ measured normal to the boundary with amplitude φ_0. A second set of reflected plane waves travels

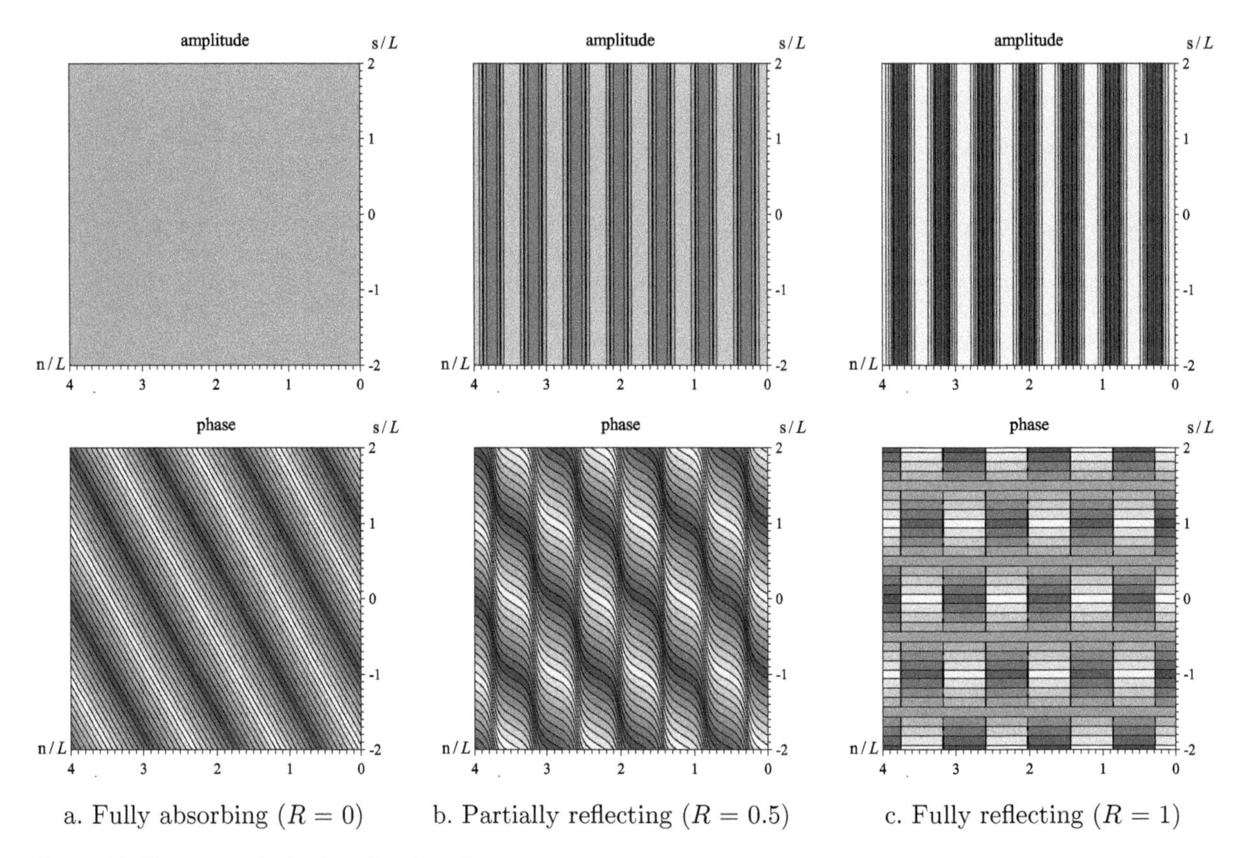

Figure 4.3 *Plane wave reflection by a planar boundary.*

away from the boundary at angle $-\gamma$ with amplitude $R\varphi_0$, where the reflection coefficient R takes on values from 0 for an absorbing boundary to 1 for full reflection of the incident waves. Superposition of the incident and reflected waves

$$\varphi = \varphi_0 \left[e^{ik(\mathfrak{s} \sin\gamma - \mathfrak{n} \cos\gamma)} + R e^{ik(\mathfrak{s} \sin\gamma + \mathfrak{n} \cos\gamma)} \right] \tag{4.11}$$

is represented in terms of the tangent $\mathfrak{s}$ and normal $\mathfrak{n}$ directions, as in (1.84). This solution reproduces a ***Robin boundary condition***, (1.86):

$$\boxed{\frac{\partial\varphi}{\partial\mathfrak{n}} + \alpha\varphi = 0, \quad \alpha = ik\cos\gamma\frac{1-R}{1+R} \qquad (\mathfrak{n}=0)} \tag{4.12}$$

and reflection of plane waves by a planar boundary is illustrated in Fig. 4.3 for boundaries varying from fully absorbing to partially absorbing to fully reflecting.

Superposition of sets of plane waves may also be adapted to study the transmission of incident waves between regions with different wave number, k. ***Wave refraction*** was introduced in Fig. 1.43c as sets of incident and reflected waves occurring in a region where $\mathfrak{n} > 0$ with wave number k_1 and transmitted waves where $\mathfrak{n} < 0$ with k_2:

$$\varphi = \begin{cases} \varphi_0 \left[\mathrm{e}^{\mathrm{i}k_1(\mathfrak{s}\sin\gamma_1 - \mathfrak{n}\cos\gamma_1)} + R\mathrm{e}^{\mathrm{i}k_1(\mathfrak{s}\sin\gamma_1 + \mathfrak{n}\cos\gamma_1)} \right] & (\mathfrak{n} > 0) \\ \varphi_0 \tau \mathrm{e}^{\mathrm{i}k_2(\mathfrak{s}\sin\gamma_2 - \mathfrak{n}\cos\gamma_2)} & (\mathfrak{n} < 0) \end{cases} \tag{4.13}$$

Here, the transmission coefficient τ is a ratio of the amplitude of transmitted waves to incident waves, and the change in direction of waves across the interface is given by ***Snell's law*** (1.79)

$$k_1 \sin\gamma_1 = k_2 \sin\gamma_2 \tag{4.14}$$

At the two sides of the interface, the complex function and its partial derivative in the direction normal to the interface take on the values of

$$\begin{matrix} \varphi^+ = \varphi_0(1+R)\mathrm{e}^{\mathrm{i}\mathfrak{s}k_1\sin\gamma_1} \\ \varphi^- = \varphi_0\tau\mathrm{e}^{\mathrm{i}\mathfrak{s}k_2\sin\gamma_2} \end{matrix}, \quad \begin{matrix} \dfrac{\partial\varphi^+}{\partial\mathfrak{n}} = -\varphi_0\mathrm{i}k_1\cos\gamma_1(1-R)\mathrm{e}^{\mathrm{i}\mathfrak{s}k_1\sin\gamma_1} \\ \dfrac{\partial\varphi^-}{\partial\mathfrak{n}} = -\varphi_0\mathrm{i}k_2\cos\gamma_2\tau\mathrm{e}^{\mathrm{i}\mathfrak{s}k_2\sin\gamma_2} \end{matrix} \quad (\mathfrak{n} = 0) \tag{4.15}$$

The reflection and transmission coefficients may be obtained from interface conditions for the complex function and the normal derivative (1.78):

$$\begin{matrix} \varphi^+ = \varphi^- \\ \beta^+\dfrac{\partial\varphi^+}{\partial\mathfrak{n}} = \beta^-\dfrac{\partial\varphi^-}{\partial\mathfrak{n}} \end{matrix} \quad \rightarrow \quad \begin{matrix} 1+R = \tau \\ (1-R)\beta^+\tan\gamma_2 = \tau\beta^-\tan\gamma_1 \end{matrix} \tag{4.16}$$

The solution to these two equations provides expressions for the reflection and transmission coefficients:

$$R = \frac{\beta^+\tan\gamma_2 - \beta^-\tan\gamma_1}{\beta^+\tan\gamma_2 + \beta^-\tan\gamma_1}, \quad \tau = \frac{2\beta^+\tan\gamma_2}{\beta^+\tan\gamma_2 + \beta^-\tan\gamma_1} \tag{4.17}$$

The coefficients for gravity waves, $\beta^+ = \frac{\sinh k_1 h_1}{k_1 \cosh k_1 h_1}$ and $\beta^- = \frac{\sinh k_2 h_2}{k_2 \cosh k_2 h_2}$ in (1.91), are related to the water depth (h_1 and h_2) on the two sides of the interface, and the solutions in Fig. 4.4 illustrate how changes in the ratio of $\beta = \beta^+/\beta^-$ impact the wave field.

Problem 4.2 For wave refraction with $\varphi_0 = 1, \gamma = 30°$, and specified angular frequency ω traveling between water depths h_1 and h_2: compute the wave numbers k_1 and k_2 using the dispersion relation for shallow water waves in Table 1.2, compute the reflection and transmission coefficients, R and τ, and compute φ at $(x, y) = (-10\text{m}, 0), (0, 0)$, and $(10\text{ m}, 0)$.

A. $\omega = 2\pi/(8\text{ sec}), h_1 = 0.5\text{ m}, h_2 = 0.4\text{ m}$

B. $\omega = 2\pi/(8\text{ sec}), h_1 = 0.5\text{ m}, h_2 = 0.3\text{ m}$

C. $\omega = 2\pi/(8\text{ sec}), h_1 = 0.5\text{ m}, h_2 = 0.2\text{ m}$

D. $\omega = 2\pi/(10\text{ sec}), h_1 = 0.5\text{ m}, h_2 = 0.4\text{ m}$

E. $\omega = 2\pi/(10\text{ sec}), h_1 = 0.5\text{ m}, h_2 = 0.3\text{ m}$

F. $\omega = 2\pi/(10\text{ sec}), h_1 = 0.5\text{ m}, h_2 = 0.2\text{ m}$

G. $\omega = 2\pi/(12\text{ sec}), h_1 = 0.5\text{ m}, h_2 = 0.4\text{ m}$

H. $\omega = 2\pi/(12\text{ sec}), h_1 = 0.5\text{ m}, h_2 = 0.3\text{ m}$

I. $\omega = 2\pi/(12\text{ sec}), h_1 = 0.5\text{ m}, h_2 = 0.2\text{ m}$

J. $\omega = 2\pi/(14\text{ sec}), h_1 = 0.5\text{ m}, h_2 = 0.4\text{ m}$

K. $\omega = 2\pi/(14\text{ sec}), h_1 = 0.5\text{ m}, h_2 = 0.3\text{ m}$

L. $\omega = 2\pi/(14\text{ sec}), h_1 = 0.5\text{ m}, h_2 = 0.2\text{ m}$

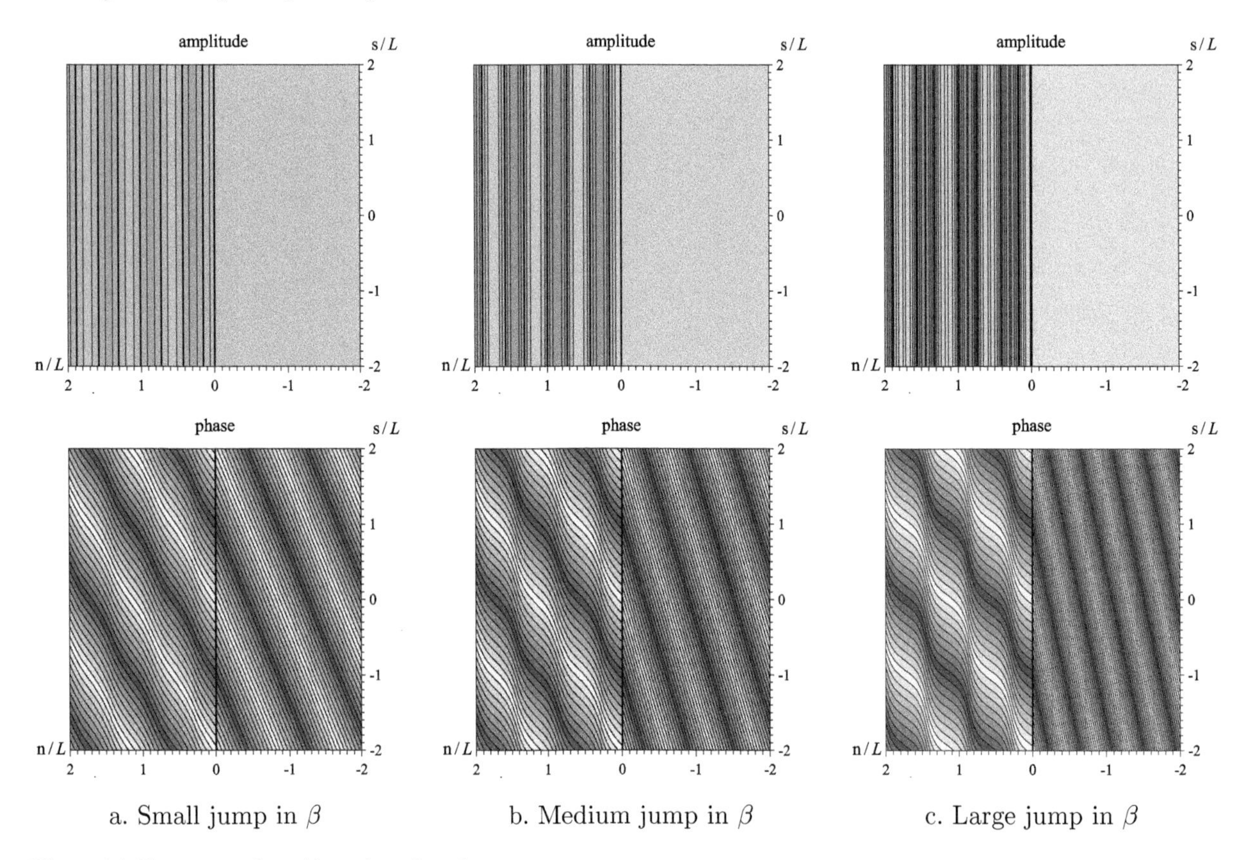

Figure 4.4 *Plane wave refracted by a planar boundary.*

4.2.2 One-Dimensional Modified Helmholtz Equation

The ***modified Helmholtz equation*** (2.2d) also provides an ordinary differential equation for one-dimensional problems that is written here in the z-direction, with general solution

$$\frac{\mathrm{d}^2\Phi}{\mathrm{d}z^2} - k^2\Phi = 0 \quad \rightarrow \quad \boxed{\Phi = c_1\mathrm{e}^{-kz} + c_2\mathrm{e}^{kz}} \tag{4.18}$$

Solutions for Φ may be used to compute the pressure head, p in (1.41), and specific discharge, q_z in (1.42), for seepage problems in the vadose zone, using

$$p = \frac{1}{2k}\ln\frac{\Phi\mathrm{e}^{-kz}}{F_s} = \frac{1}{2k}\ln\frac{c_1\mathrm{e}^{-2kz} + c_2}{F_s}, \quad q_z = -e^{-kz}\left(\frac{\partial\Phi}{\partial z} + k\Phi\right) = -2kc_2 \tag{4.19}$$

The coefficients may be set by matching boundary conditions, for example by setting the pressure head to p_0 at elevation z_0, and setting the specific discharge to minus the recharge rate R, giving

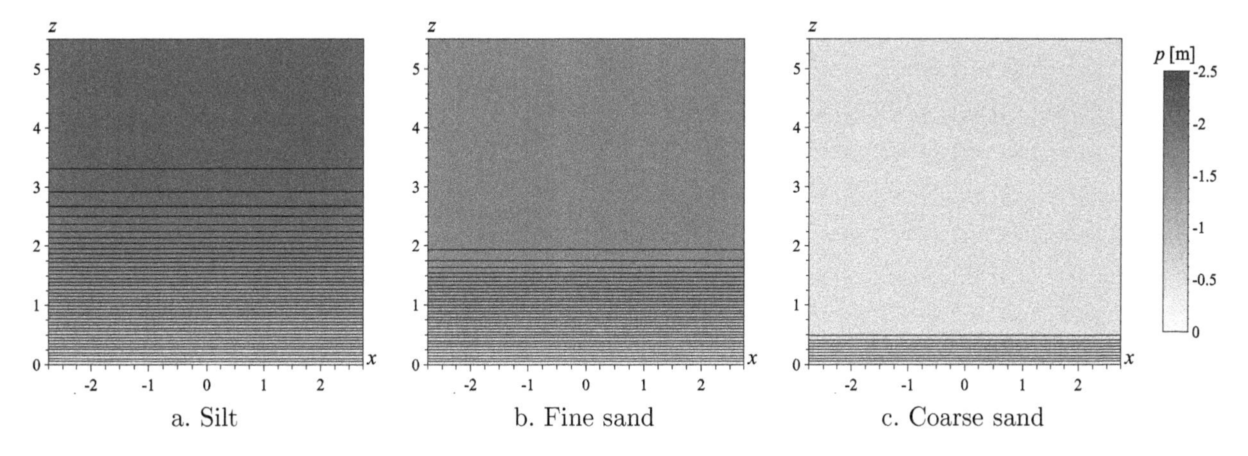

Figure 4.5 *One-dimensional solution for modified Helmholtz equation.*

$$\begin{aligned} p(z_0) &= p_0 \\ q_z(z_0) &= -R \end{aligned} \quad \rightarrow \quad c_1 = \left(F_s e^{2kp_0} - \frac{R}{2k} \right) e^{2kz_0}, \quad c_2 = \frac{R}{2k} \tag{4.20}$$

Examples are shown in Fig. 4.5 for the pressure head distribution for problems with a uniform q_z and pressure head $p = 0$ occurring at elevation $z_0 = 0$ in a uniform soil horizon using the representative soil parameters in Table 1.1.

Problem 4.3 Compute the coefficients c_1 and c_2, and the pressure head p at $z = 2$ m above an interface where $p = 0$ and $z_0 = 0$ for the following values of R and soil properties from Table 1.1:

A. silt, $R = 0.025$ m/yr

B. silt, $R = 0.05$ m/yr

C. silt, $R = 0.1$ m/yr

D. silt, $R = 0.2$ m/yr

E. fine sand, $R = 0.025$ m/yr

F. fine sand, $R = 0.05$ m/yr

G. fine sand, $R = 0.1$ m/yr

H. fine sand, $R = 0.2$ m/yr

I. coarse sand, $R = 0.025$ m/yr

J. coarse sand, $R = 0.05$ m/yr

K. coarse sand, $R = 0.1$ m/yr

L. coarse sand, $R = 0.2$ m/yr

4.3 Separation in Cartesian Coordinates

4.3.1 Rectangle Elements for Laplace and Poisson Equations

The method of separtion of variables is developed next for problems in a rectangular domain in the x–y plane as illustrated in Fig. 4.6b. A potential that satisfies the Laplace equation (2.2a) for a two-dimensional vector field is written in separated form as the product of two functions, one of which varies as a function of only x and the other varies as a function of only y:

$$\boxed{\Phi(x,y) = X(x)Y(y)}, \quad \frac{\partial^2 \Phi}{\partial x^2} + \frac{\partial^2 \Phi}{\partial y^2} = 0 \tag{4.21}$$

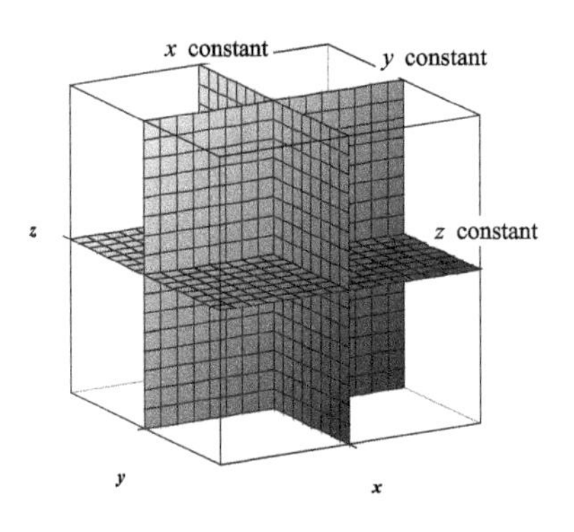

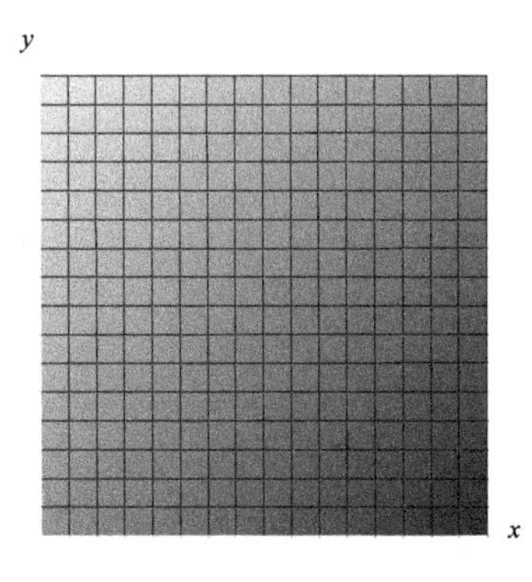

a. Three-dimensional domain

b. Plane with constant z

Figure 4.6 *Cartesian coordinates* (x, y, z)*: with planes (*x*,* y*, or* z *constant).*

Substituting the separated functions of $\Phi = XY$ into the Laplacian, evaluating the partial derivatives, and dividing both sides of the equation by Φ gives

$$\frac{1}{XY}\left(Y\frac{d^2X}{dx^2} + X\frac{d^2Y}{dy^2}\right) = \frac{1}{XY}(0) \quad \rightarrow \quad \frac{1}{X}\frac{d^2X}{dx^2} + \frac{1}{Y}\frac{d^2Y}{dy^2} = 0 \tag{4.22}$$

This equation may be rearranged to place all terms containing x on the left side and those containing y on the right side, and both sides must be equal to a constant here set to be a negative number and called $-n^2$:

$$\frac{1}{X}\frac{d^2X}{dx^2} = -\frac{1}{Y}\frac{d^2Y}{dy^2} = -n^2 \tag{4.23}$$

This provides two ordinary differential equations with particular solutions given by

$$\begin{aligned} \frac{d^2X}{dx^2} + n^2X = 0 \quad &\rightarrow \quad X = \left\{\cos nx, \ \sin nx\right. \\ \frac{d^2Y}{dy^2} - n^2Y = 0 \quad &\rightarrow \quad Y = \left\{\cosh ny, \ \sinh ny\right. \end{aligned} \tag{4.24a}$$

Alternately, the constant in (4.23) could be chosen to be positive by using n^2 instead of $-n^2$, giving particular solutions to the two ordinary differential equations of

$$\begin{aligned} \frac{d^2X}{dx^2} - n^2X = 0 \quad &\rightarrow \quad X = \left\{\cosh nx, \ \sinh nx\right. \\ \frac{d^2Y}{dy^2} + n^2Y = 0 \quad &\rightarrow \quad Y = \left\{\cos ny, \ \sin ny\right. \end{aligned} \tag{4.24b}$$

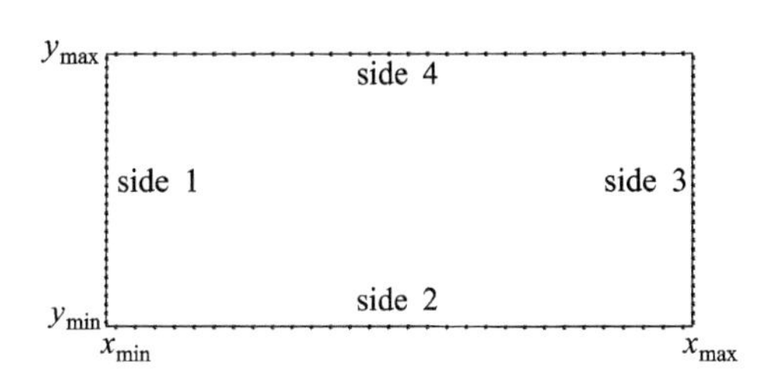

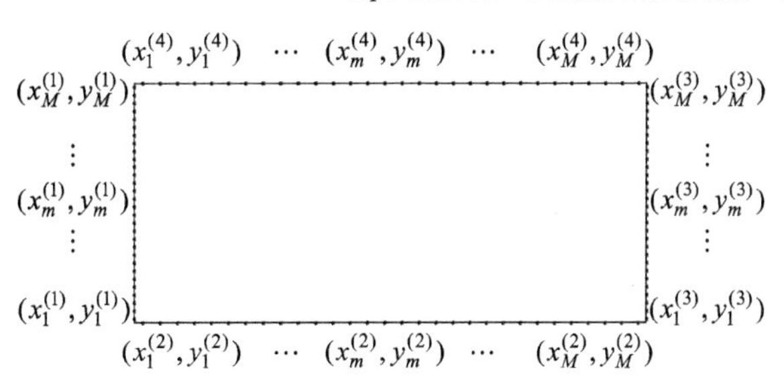

Figure 4.7 *A rectangular element with sides S=1,2,3,4 and the location of control points. (Reprinted from Steward and Allen, 2013, The Analytic Element Method for rectangular gridded domains, benchmark comparisons and application to the High Plains Aquifer,* Advances in Water Resources, *Vol. 60, Fig. 1, with permission from Elsevier.)*

Or, if the constant is zero, and this gives

$$\begin{aligned} \frac{d^2X}{dx^2} &= 0 \quad \rightarrow \quad X = \Big\{1,\ x \\ \frac{d^2Y}{dy^2} &= 0 \quad \rightarrow \quad Y = \Big\{1,\ y \end{aligned} \tag{4.24c}$$

Together, the particular solutions to these ordinary differential equations may be multiplied in (4.21) to provide a set of functional forms that satisfy the two-dimensional Laplace equation:

$$\Phi = XY = \begin{cases} 1,\ x,\ y,\ xy, \\ \cos nx \cosh ny,\ \cos nx \sinh ny,\ \sin nx \cosh ny,\ \sin nx \sinh ny, \\ \cosh nx \cos ny,\ \cosh nx \sin ny,\ \sinh nx \cos ny,\ \sinh nx \sin ny \end{cases} \tag{4.25}$$

These equations follow from the Fourier series developed by Fourier (1878, chap.3) with the method for separating variables presented in Moon and Spencer (1961*b*, p.92–3).

These functions provide the mathematical basis for an analytic element with the geometry of a rectangle. Its geometry is specified in Fig. 4.7, with edges of constant value x_{min} on side 1, y_{min} on side 2, x_{max} on side 3, and y_{max} on side 4. The following influence functions contain terms that are the constant, linear, or quadratic function of x and y:

$$\begin{aligned} &\overset{0}{\Phi} = 1, \quad \overset{x}{\Phi} = \frac{2x - (x_{max} + x_{min})}{x_{max} - x_{min}}, \quad \overset{y}{\Phi} = \frac{2y - (y_{max} + y_{min})}{y_{max} - y_{min}}, \\ &\overset{xy}{\Phi} = \left[\frac{2x - (x_{max} + x_{min})}{x_{max} - x_{min}}\right]\left[\frac{2y - (y_{max} + y_{min})}{y_{max} - y_{min}}\right], \\ &\overset{x2y2}{\Phi} = \left[\frac{2x - (x_{max} + x_{min})}{x_{max} - x_{min}}\right]^2 - \left[\frac{2y - (y_{max} + y_{min})}{x_{max} - x_{min}}\right]^2 \end{aligned} \tag{4.26a}$$

Note that each function has been scaled to vary between ± 1 across the element, as lower-order terms in a local power series, (2.35). The influence functions containing cos and sin terms are scaled to vary between ± 1 on one of the sides, to repeat n times on this side, and to be zero on the opposite side (Steward and Allen, 2013):

$$
\begin{aligned}
\overset{1\cos}{\Phi}_n &= \frac{\sinh 2\pi n \dfrac{x_{\max}-x}{y_{\max}-y_{\min}}}{\sinh 2\pi n \dfrac{x_{\max}-x_{\min}}{y_{\max}-y_{\min}}} \cos 2\pi n \frac{y-y_{\min}}{y_{\max}-y_{\min}} \\
\overset{1\sin}{\Phi}_n &= \frac{\sinh 2\pi n \dfrac{x_{\max}-x}{y_{\max}-y_{\min}}}{\sinh 2\pi n \dfrac{x_{\max}-x_{\min}}{y_{\max}-y_{\min}}} \sin 2\pi n \frac{y-y_{\min}}{y_{\max}-y_{\min}} \\
\overset{2\cos}{\Phi}_n &= \cos 2\pi n \frac{x-x_{\min}}{x_{\max}-x_{\min}} \frac{\sinh 2\pi n \dfrac{y_{\max}-y}{x_{\max}-x_{\min}}}{\sinh 2\pi n \dfrac{y_{\max}-y_{\min}}{x_{\max}-x_{\min}}} \\
\overset{2\sin}{\Phi}_n &= \sin 2\pi n \frac{x-x_{\min}}{x_{\max}-x_{\min}} \frac{\sinh 2\pi n \dfrac{y_{\max}-y}{x_{\max}-x_{\min}}}{\sinh 2\pi n \dfrac{y_{\max}-y_{\min}}{x_{\max}-x_{\min}}} \\
\overset{3\cos}{\Phi}_n &= \frac{\sinh 2\pi n \dfrac{x-x_{\min}}{y_{\max}-y_{\min}}}{\sinh 2\pi n \dfrac{x_{\max}-x_{\min}}{y_{\max}-y_{\min}}} \cos 2\pi n \frac{y-y_{\min}}{y_{\max}-y_{\min}} \\
\overset{3\sin}{\Phi}_n &= \frac{\sinh 2\pi n \dfrac{x-x_{\min}}{y_{\max}-y_{\min}}}{\sinh 2\pi n \dfrac{x_{\max}-x_{\min}}{y_{\max}-y_{\min}}} \sin 2\pi n \frac{y-y_{\min}}{y_{\max}-y_{\min}} \\
\overset{4\cos}{\Phi}_n &= \cos 2\pi n \frac{x-x_{\min}}{x_{\max}-x_{\min}} \frac{\sinh 2\pi n \dfrac{y-y_{\min}}{x_{\max}-x_{\min}}}{\sinh 2\pi n \dfrac{y_{\max}-y_{\min}}{x_{\max}-x_{\min}}} \\
\overset{4\sin}{\Phi}_n &= \sin 2\pi n \frac{x-x_{\min}}{x_{\max}-x_{\min}} \frac{\sinh 2\pi n \dfrac{y-y_{\min}}{x_{\max}-x_{\min}}}{\sinh 2\pi n \dfrac{y_{\max}-y_{\min}}{x_{\max}-x_{\min}}}
\end{aligned}
\tag{4.26b}
$$

This notation utilizes the first term in the over-script $S = 1, 2, 3, 4$ to denote the four sides, and the cos or sin term represents the Fourier series variation along this side. The potential and vector field associated with the influence functions are shown in Fig. 4.8.

A solution to the Laplace equation is obtained by taking linear combinations of these influence functions where the coefficients are gathered in the matrix $\mathbf{c}$:

$$
\begin{aligned}
\Phi(x,y,\mathbf{c}) &= \overset{0}{c}\overset{0}{\Phi} + \overset{x}{c}\overset{x}{\Phi}(x) + \overset{y}{c}\overset{y}{\Phi}(y) + \overset{xy}{c}\overset{xy}{\Phi}(x,y) + \overset{x2y2}{c}\overset{x2y2}{\Phi}(x,y) \\
&+ \left[\sum_{n=1}^{N}\sum_{S=1}^{4} \overset{S\cos}{c}_n \overset{S\cos}{\Phi}_n(x,y) + \overset{S\sin}{c}_n \overset{S\sin}{\Phi}_n(x,y)\right] + \overset{\text{add}}{\Phi}(x,y) \\
v_x(x,y,\mathbf{c}) &= \overset{0}{c}\overset{0}{v}_x + \overset{x}{c}\overset{x}{v}_x(x) + \overset{y}{c}\overset{y}{v}_x(y) + \overset{xy}{c}\overset{xy}{v}_x(x,y) + \overset{x2y2}{c}\overset{x2y2}{v}_x(x,y) \\
&+ \left[\sum_{n=1}^{N}\sum_{S=1}^{4} \overset{S\cos}{c}_n \overset{S\cos}{v}_{x\,n}(x,y) + \overset{S\sin}{c}_n \overset{S\sin}{v}_{x\,n}(x,y)\right] + \overset{\text{add}}{v}_x(x,y), \\
v_y(x,y,\mathbf{c}) &= \overset{0}{c}\overset{0}{v}_y + \overset{x}{c}\overset{x}{v}_y(x) + \overset{y}{c}\overset{y}{v}_y(y) + \overset{xy}{c}\overset{xy}{v}_y(x,y) + \overset{x2y2}{c}\overset{x2y2}{v}_y(x,y) \\
&+ \left[\sum_{n=1}^{N}\sum_{S=1}^{4} \overset{S\cos}{c}_n \overset{S\cos}{v}_{y\,n}(x,y) + \overset{S\sin}{c}_n \overset{S\sin}{v}_{y\,n}(x,y)\right] + \overset{\text{add}}{v}_y(x,y)
\end{aligned}
\qquad
\mathbf{c} = \begin{bmatrix} \overset{0}{c} \\ \overset{x}{c} \\ \overset{y}{c} \\ \overset{xy}{c} \\ \overset{x2y2}{c} \\ \vdots \\ \overset{S\cos}{c}_n \\ \overset{S\sin}{c}_n \\ \vdots \end{bmatrix}
\tag{4.27}
$$

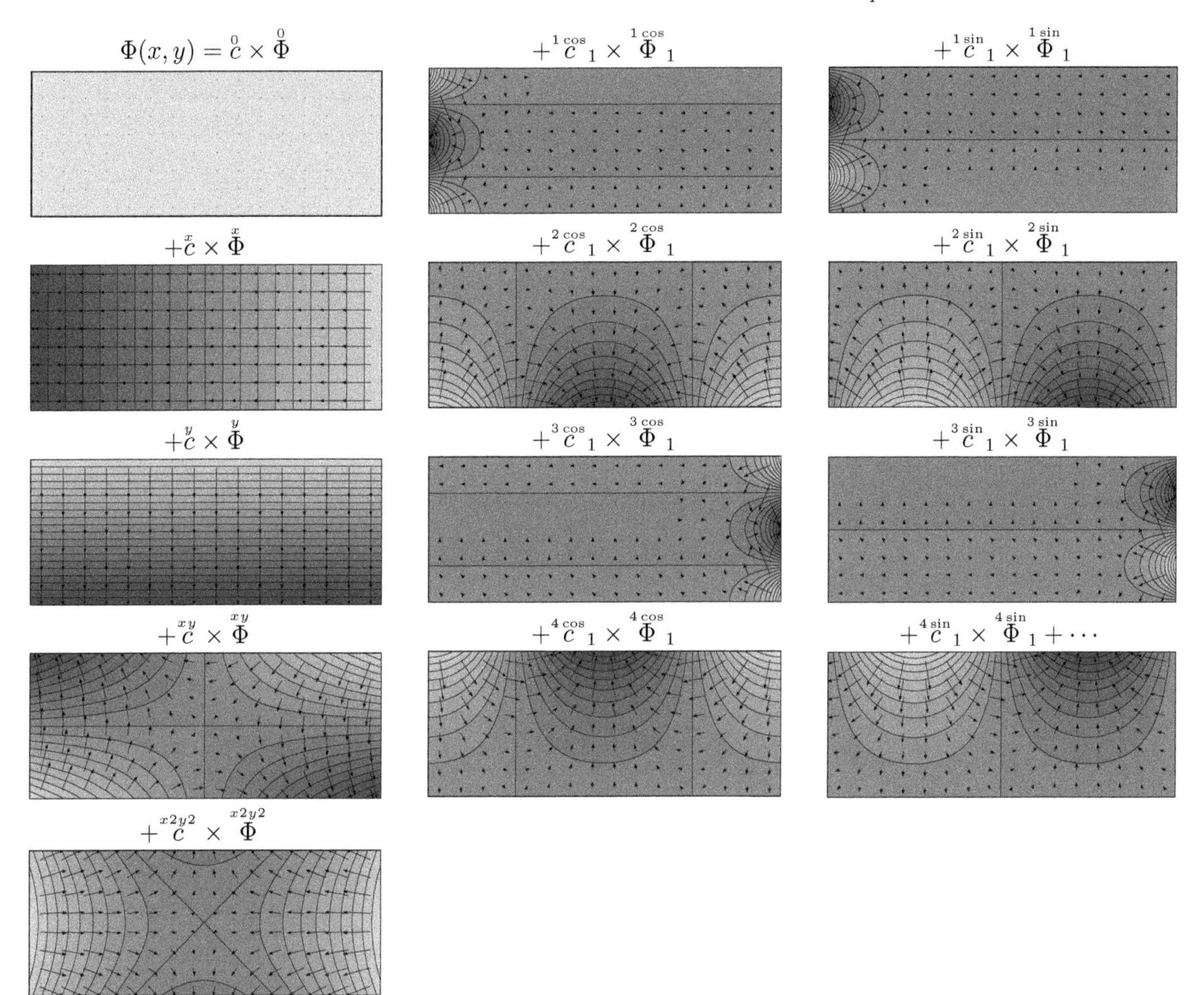

Figure 4.8 *Influence functions for rectangular domains with the Laplace equation, from (4.26) and (4.33).*

and the Cartesian components of the vector field are obtained from linear combinations of the derivatives $v_x = -\partial\Phi/\partial x$ and $v_y = -\partial\Phi/\partial y$ of the influence functions, (4.26). The function $\overset{\text{add}}{\Phi}$ and its derivatives are used to incorporate additional functions in the rectangle element.

Boundary conditions may be applied at a set of M equally spaced control points on the four sides of the model domain identified in Fig. 4.7:

$$
\begin{aligned}
x_m^{(1)} &= x_{\min} & y_m^{(1)} &= y_{\min} + (y_{\max} - y_{\min})\frac{m - \frac{1}{2}}{M} \\
x_m^{(2)} &= x_{\min} + (x_{\max} - x_{\min})\frac{m - \frac{1}{2}}{M} & y_m^{(2)} &= y_{\min} \\
x_m^{(3)} &= x_{\max} & y_m^{(3)} &= y_{\min} + (y_{\max} - y_{\min})\frac{m - \frac{1}{2}}{M} \\
x_m^{(4)} &= x_{\min} + (x_{\max} - x_{\min})\frac{m - \frac{1}{2}}{M} & y_m^{(4)} &= y_{\max}
\end{aligned}
\tag{4.28}
$$

For a side $\mathcal{S}$ with a Dirichlet boundary condition where the value of Φ is specified at each control point, a set of M equations is obtained by evaluating the influence functions and the additional function in (4.27) at each control point on this side, giving conditions for $m = 1, \cdots, M$ organized as rows in the matrices

$$\overset{\Phi}{\mathbf{A}}{}^{(\mathcal{S})}\mathbf{c} = \overset{\Phi}{\mathbf{b}}{}^{(\mathcal{S})}, \quad \overset{\Phi}{\mathbf{b}}{}^{(\mathcal{S})} = \begin{bmatrix} \vdots \\ \Phi_m^{(\mathcal{S})} - \overset{\text{add}}{\Phi}(x_m^{(\mathcal{S})}, y_m^{(\mathcal{S})}) \\ \vdots \end{bmatrix}$$
$$\overset{\Phi}{\mathbf{A}}{}^{(\mathcal{S})} = \begin{bmatrix} \vdots & \vdots & \vdots & \vdots & \vdots & & \vdots & \vdots & \\ 1 & \overset{x}{\Phi}(x_m^{(\mathcal{S})}) & \overset{y}{\Phi}(y_m^{(\mathcal{S})}) & \overset{xy}{\Phi}(x_m^{(\mathcal{S})}, y_m^{(\mathcal{S})}) & \overset{x2y2}{\Phi}(x_m^{(\mathcal{S})}, y_m^{(\mathcal{S})}) & \cdots & \overset{\mathcal{S}\cos}{\Phi}{}_n(x_m^{(\mathcal{S})}, y_m^{(\mathcal{S})}) & \overset{\mathcal{S}\sin}{\Phi}{}_n(x_m^{(\mathcal{S})}, y_m^{(\mathcal{S})}) & \cdots \\ \vdots & \vdots & \vdots & \vdots & \vdots & & \vdots & \vdots & \end{bmatrix} \tag{4.29}$$

Likewise, the matrices that relate the unknown coefficients to Neumann boundary conditions where the normal components of the vector field v_{n} are specified at the mth control points by

$$\overset{v}{\mathbf{A}}{}^{(\mathcal{S})}\mathbf{c} = \overset{v}{\mathbf{b}}{}^{(\mathcal{S})}, \quad \overset{v}{\mathbf{b}}{}^{(\mathcal{S})} = \begin{bmatrix} \vdots \\ (v_{\text{n}})_m^{(\mathcal{S})} - \overset{\text{add}}{v_{\text{n}}}(x_m^{(\mathcal{S})}, y_m^{(\mathcal{S})}) \\ \vdots \end{bmatrix}$$
$$\overset{v}{\mathbf{A}}{}^{(\mathcal{S})} = \begin{bmatrix} \vdots & \vdots & \vdots & \vdots & \vdots & & \vdots & \vdots & \\ 0 & \overset{x}{v_{\text{n}}}(x_m^{(\mathcal{S})}) & \overset{y}{v_{\text{n}}}(y_m^{(\mathcal{S})}) & \overset{xy}{v_{\text{n}}}(x_m^{(\mathcal{S})}, y_m^{(\mathcal{S})}) & \overset{x2y2}{v_{\text{n}}}(x_m^{(\mathcal{S})}, y_m^{(\mathcal{S})}) & \cdots & \overset{\mathcal{S}\cos}{v_{\text{n}}}{}_n(x_m^{(\mathcal{S})}, y_m^{(\mathcal{S})}) & \overset{\mathcal{S}\sin}{v_{\text{n}}}{}_n(x_m^{(\mathcal{S})}, y_m^{(\mathcal{S})}) & \cdots \\ \vdots & \vdots & \vdots & \vdots & \vdots & & \vdots & \vdots & \end{bmatrix} \tag{4.30}$$

where the normal component v_{n} is equal to v_x for side 1, v_y for side 2, $-v_x$ for side 3, and $-v_y$ for side 4. Together, the four sides of a rectangle provide a set of $4M$ equations to solve for the $5 + 8N$ unknown coefficients

$$\begin{bmatrix} \mathbf{A}^{(1)} \\ \mathbf{A}^{(2)} \\ \mathbf{A}^{(3)} \\ \mathbf{A}^{(4)} \end{bmatrix} \mathbf{c} = \begin{bmatrix} \mathbf{b}^{(1)} \\ \mathbf{b}^{(2)} \\ \mathbf{b}^{(3)} \\ \mathbf{b}^{(4)} \end{bmatrix} \quad \text{with} \quad \begin{cases} \mathbf{A}^{(\mathcal{S})} = \overset{\Phi}{\mathbf{A}}{}^{(\mathcal{S})}, \ \mathbf{b}^{(\mathcal{S})} = \overset{\Phi}{\mathbf{b}}{}^{(\mathcal{S})} & \text{Dirichlet condition for side } \mathcal{S} \\ \mathbf{A}^{(\mathcal{S})} = \overset{v}{\mathbf{A}}{}^{(\mathcal{S})}, \ \mathbf{b}^{(\mathcal{S})} = \overset{v}{\mathbf{b}}{}^{(\mathcal{S})} & \text{Neumann condition for side } \mathcal{S} \end{cases} \tag{4.31}$$

Note that use of the linear and quadratic influence functions provides functional forms that allow trends in the value of a function to occur across the edge of an element. Attempting to reproduce such variation using only the Fourier series terms would result in the Gibbs effect, with inaccuracies near the corners of the rectangle. It was found that the analytic element as formulated above accurately matches boundary conditions with a relatively small number of Fourier terms, and examples illustrate results with N=5 and $M = 15$.

Example 4.1 Solutions are presented in Fig. 4.9 for rectangle elements with Dirichlet, (4.29), and Neumann conditions, (4.30). The boundary in Fig. 4.9a has uniform potential, and contains an additional function that solves the ***Poisson equation*** (2.2b) for a one-dimensional vector field

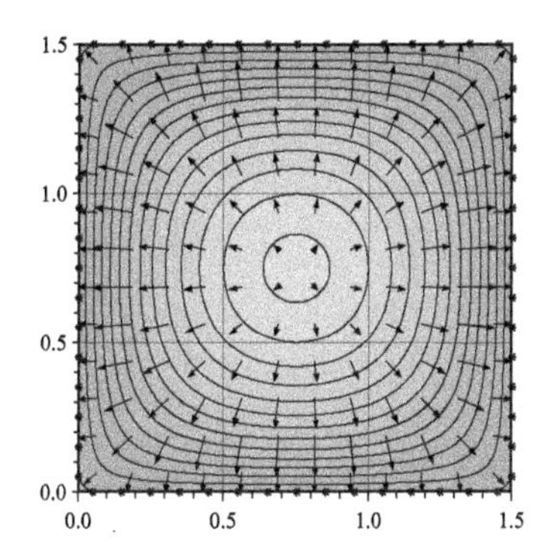

a. Uniform potential boundary with a divergent vector field

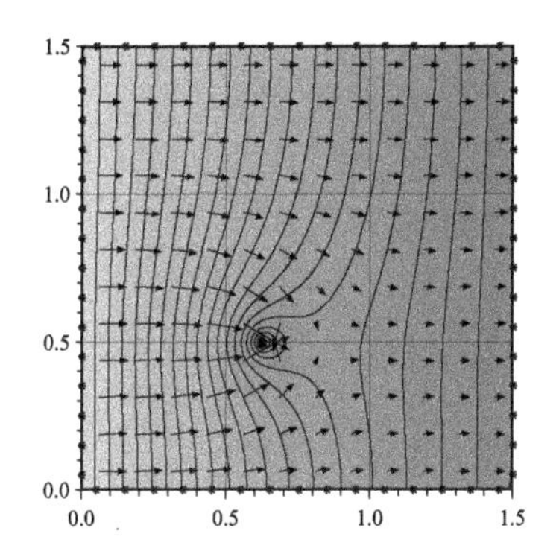

b. Uniform potential and zero normal component witha point-sink

Figure 4.9 *Separation of variable solutions with interior vector fields.*

$$\frac{d^2 \overset{\text{add}}{\Phi}}{dx^2} = -d_0 \quad \rightarrow \quad \boxed{\overset{\text{add}}{\Phi} = -\int\int d_0 dx dx = -\frac{d_0}{2}x^2} \tag{4.32}$$

Thus, this element contains a uniform divergence d_0. The solution in Fig. 4.9b satisfies boundaries containing mixed conditions with uniform potential Φ_1 along the left and $\Phi_3 < \Phi_1$ along the right, and Neumann conditions of $v_y = 0$ along the bottom and top. Its additional function represents a point-sink, (3.1).

Problem 4.4 For a rectangular element bounded by $x_{\min} = -10, x_{\max} = 10, y_{\min} = -10, y_{\max} = 10$, compute the potential Φ and the vector field components $v_x = -\partial\Phi/\partial x$ and $v_y = -\partial\Phi/\partial y$ at the four corners of the rectangle and at the four midpoints of the rectangle for the following coefficients, with all other coefficients set to zero

A. $\overset{1\cos}{c}_1 = 1$
B. $\overset{1\sin}{c}_1 = 1$
C. $\overset{1\cos}{c}_2 = 1$
D. $\overset{1\sin}{c}_2 = 1$
E. $\overset{2\cos}{c}_1 = 1$
F. $\overset{2\sin}{c}_1 = 1$
G. $\overset{2\cos}{c}_2 = 1$
H. $\overset{2\sin}{c}_2 = 1$
I. $\overset{3\cos}{c}_1 = 1$
J. $\overset{3\sin}{c}_1 = 1$
K. $\overset{3\cos}{c}_2 = 1$
L. $\overset{3\sin}{c}_2 = 1$
M. $\overset{4\cos}{c}_1 = 1$
N. $\overset{4\sin}{c}_1 = 1$
O. $\overset{4\cos}{c}_2 = 1$
P. $\overset{4\sin}{c}_2 = 1$

Problem 4.5 Develop a solution using the separation of variables for a rectangle with $N = 5, M = 15$ for a Dirichlet problem with $\Phi_1 = 10$ on side 1 at $x_{\min} = 0$, $\Phi_2 = 10$ on side 2 at $y_{\min} = 0$, $\Phi_3 = 10$ on side 3 at $x_{\max} = 1$, $\Phi_4 = 10$ on side 4 at $y_{\max} = 1$, and the specified additional function. Turn in the coefficients $\overset{0}{c}, \overset{x}{c}, \overset{y}{c}, \overset{xy}{c}$, and $\overset{x2y2}{c}$ and the value of Φ at the center of the rectangle where $(x, y) = (0.5, 0.5)$.

A. $\overset{\text{add}}{\Phi} = -x^2$
B. $\overset{\text{add}}{\Phi} = -2x^2$
C. $\overset{\text{add}}{\Phi} = -3x^2$
D. $\overset{\text{add}}{\Phi} = -4x^2$
E. $\overset{\text{add}}{\Phi} = -y^2$
F. $\overset{\text{add}}{\Phi} = -2y^2$
G. $\overset{\text{add}}{\Phi} = -3y^2$
H. $\overset{\text{add}}{\Phi} = -4y^2$
I. $\overset{\text{add}}{\Phi} = -x^2 - y^2$
J. $\overset{\text{add}}{\Phi} = -2x^2 - 2y^2$
K. $\overset{\text{add}}{\Phi} = -3x^2 - 3y^2$
L. $\overset{\text{add}}{\Phi} = -4x^2 - 4y^2$

Problems that solve ***boundaries with Neumann conditions*** require the normal component of the vector field. Equations that enable implementation of the vector components in (4.27) are obtained from minus the gradient of of the influence functions, (4.26), giving

$$
\begin{aligned}
&\overset{0}{v}_x = 0, \quad \overset{x}{v}_x = -\frac{2}{x_{\max} - x_{\min}}, \quad \overset{y}{v}_x = 0, \\
&\overset{xy}{v}_x = -\frac{2}{x_{\max} - x_{\min}} \left[\frac{2y - (y_{\max} + y_{\min})}{y_{\max} - y_{\min}} \right], \quad \overset{x2y2}{v}_x = -4\frac{2x - (x_{\max} + x_{\min})}{(x_{\max} - x_{\min})^2} \\
&\overset{0}{v}_y = 0, \quad \overset{x}{v}_y = 0, \quad \overset{y}{v}_y = -\frac{2}{y_{\max} - y_{\min}}, \\
&\overset{xy}{v}_y = -\left[\frac{2x - (x_{\max} + x_{\min})}{x_{\max} - x_{\min}} \right] \frac{2}{y_{\max} - y_{\min}}, \quad \overset{x2y2}{v}_y = 4\frac{2y - (y_{\max} + y_{\min})}{(x_{\max} - x_{\min})^2}
\end{aligned}
\tag{4.33a}
$$

and

$$
\begin{aligned}
\overset{1\cos}{v}_{x\,n} &= +\frac{2\pi n}{y_{\max} - y_{\min}} \frac{\cosh 2\pi n \dfrac{x_{\max} - x}{y_{\max} - y_{\min}}}{\sinh 2\pi n \dfrac{x_{\max} - x_{\min}}{y_{\max} - y_{\min}}} \cos 2\pi n \frac{y - y_{\min}}{y_{\max} - y_{\min}} \\
\overset{1\sin}{v}_{x\,n} &= +\frac{2\pi n}{y_{\max} - y_{\min}} \frac{\cosh 2\pi n \dfrac{x_{\max} - x}{y_{\max} - y_{\min}}}{\sinh 2\pi n \dfrac{x_{\max} - x_{\min}}{y_{\max} - y_{\min}}} \sin 2\pi n \frac{y - y_{\min}}{y_{\max} - y_{\min}} \\
\overset{2\cos}{v}_{x\,n} &= +\frac{2\pi n}{x_{\max} - x_{\min}} \sin 2\pi n \frac{x - x_{\min}}{x_{\max} - x_{\min}} \frac{\sinh 2\pi n \dfrac{y_{\max} - y}{x_{\max} - x_{\min}}}{\sinh 2\pi n \dfrac{y_{\max} - y_{\min}}{x_{\max} - x_{\min}}} \\
\overset{2\sin}{v}_{x\,n} &= -\frac{2\pi n}{x_{\max} - x_{\min}} \cos 2\pi n \frac{x - x_{\min}}{x_{\max} - x_{\min}} \frac{\sinh 2\pi n \dfrac{y_{\max} - y}{x_{\max} - x_{\min}}}{\sinh 2\pi n \dfrac{y_{\max} - y_{\min}}{x_{\max} - x_{\min}}} \\
\overset{3\cos}{v}_{x\,n} &= -\frac{2\pi n}{y_{\max} - y_{\min}} \frac{\cosh 2\pi n \dfrac{x - x_{\min}}{y_{\max} - y_{\min}}}{\sinh 2\pi n \dfrac{x_{\max} - x_{\min}}{y_{\max} - y_{\min}}} \cos 2\pi n \frac{y - y_{\min}}{y_{\max} - y_{\min}} \\
\overset{3\sin}{v}_{x\,n} &= -\frac{2\pi n}{y_{\max} - y_{\min}} \frac{\cosh 2\pi n \dfrac{x - x_{\min}}{y_{\max} - y_{\min}}}{\sinh 2\pi n \dfrac{x_{\max} - x_{\min}}{y_{\max} - y_{\min}}} \sin 2\pi n \frac{y - y_{\min}}{y_{\max} - y_{\min}} \\
\overset{4\cos}{v}_{x\,n} &= +\frac{2\pi n}{x_{\max} - x_{\min}} \sin 2\pi n \frac{x - x_{\min}}{x_{\max} - x_{\min}} \frac{\sinh 2\pi n \dfrac{y - y_{\min}}{x_{\max} - x_{\min}}}{\sinh 2\pi n \dfrac{y_{\max} - y_{\min}}{x_{\max} - x_{\min}}} \\
\overset{4\sin}{v}_{x\,n} &= -\frac{2\pi n}{x_{\max} - x_{\min}} \cos 2\pi n \frac{x - x_{\min}}{x_{\max} - x_{\min}} \frac{\sinh 2\pi n \dfrac{y - y_{\min}}{x_{\max} - x_{\min}}}{\sinh 2\pi n \dfrac{y_{\max} - y_{\min}}{x_{\max} - x_{\min}}}
\end{aligned}
\tag{4.33b}
$$

and

$$
\begin{aligned}
\overset{1\cos}{v}_{y\,n} &= +\frac{2\pi n}{y_{\max}-y_{\min}} \frac{\sinh 2\pi n \dfrac{x_{\max}-x}{y_{\max}-y_{\min}}}{\sinh 2\pi n \dfrac{x_{\max}-x_{\min}}{y_{\max}-y_{\min}}} \sin 2\pi n \frac{y-y_{\min}}{y_{\max}-y_{\min}} \\
\overset{1\sin}{v}_{y\,n} &= -\frac{2\pi n}{y_{\max}-y_{\min}} \frac{\sinh 2\pi n \dfrac{x_{\max}-x}{y_{\max}-y_{\min}}}{\sinh 2\pi n \dfrac{x_{\max}-x_{\min}}{y_{\max}-y_{\min}}} \cos 2\pi n \frac{y-y_{\min}}{y_{\max}-y_{\min}} \\
\overset{2\cos}{v}_{y\,n} &= +\frac{2\pi n}{x_{\max}-x_{\min}} \cos 2\pi n \frac{x-x_{\min}}{x_{\max}-x_{\min}} \frac{\cosh 2\pi n \dfrac{y_{\max}-y}{x_{\max}-x_{\min}}}{\sinh 2\pi n \dfrac{y_{\max}-y_{\min}}{x_{\max}-x_{\min}}} \\
\overset{2\sin}{v}_{y\,n} &= +\frac{2\pi n}{x_{\max}-x_{\min}} \sin 2\pi n \frac{x-x_{\min}}{x_{\max}-x_{\min}} \frac{\cosh 2\pi n \dfrac{y_{\max}-y}{x_{\max}-x_{\min}}}{\sinh 2\pi n \dfrac{y_{\max}-y_{\min}}{x_{\max}-x_{\min}}} \\
\overset{3\cos}{v}_{y\,n} &= +\frac{2\pi n}{y_{\max}-y_{\min}} \frac{\sinh 2\pi n \dfrac{x-x_{\min}}{y_{\max}-y_{\min}}}{\sinh 2\pi n \dfrac{x_{\max}-x_{\min}}{y_{\max}-y_{\min}}} \sin 2\pi n \frac{y-y_{\min}}{y_{\max}-y_{\min}} \\
\overset{3\sin}{v}_{y\,n} &= -\frac{2\pi n}{y_{\max}-y_{\min}} \frac{\sinh 2\pi n \dfrac{x-x_{\min}}{y_{\max}-y_{\min}}}{\sinh 2\pi n \dfrac{x_{\max}-x_{\min}}{y_{\max}-y_{\min}}} \cos 2\pi n \frac{y-y_{\min}}{y_{\max}-y_{\min}} \\
\overset{4\cos}{v}_{y\,n} &= -\frac{2\pi n}{x_{\max}-x_{\min}} \cos 2\pi n \frac{x-x_{\min}}{x_{\max}-x_{\min}} \frac{\cosh 2\pi n \dfrac{y-y_{\min}}{x_{\max}-x_{\min}}}{\sinh 2\pi n \dfrac{y_{\max}-y_{\min}}{x_{\max}-x_{\min}}} \\
\overset{4\sin}{v}_{y\,n} &= -\frac{2\pi n}{x_{\max}-x_{\min}} \sin 2\pi n \frac{x-x_{\min}}{x_{\max}-x_{\min}} \frac{\cosh 2\pi n \dfrac{y-y_{\min}}{x_{\max}-x_{\min}}}{\sinh 2\pi n \dfrac{y_{\max}-y_{\min}}{x_{\max}-x_{\min}}}
\end{aligned}
\tag{4.33c}
$$

Problem 4.6 Develop a solution for a rectangle with $N = 5, M = 15$ for the mixed boundary value problem with $v_x = 1$ on side 1 at $x_{\min} = 0$, $\Phi_2 = 10$ on side 2 at $y_{\min} = 0$, $v_x = 1$ on side 3 at $x_{\max} = 2$, Φ_4 on side 4 at $y_{\max} = 1$, where the value of Φ_4 is specified below. Turn in the coefficients $\overset{0}{c}, \overset{x}{c}, \overset{y}{c}, \overset{xy}{c}$, and $\overset{x2y2}{c}$ and the value of Φ, v_x, and v_y at the center of the rectangle where $(x, y) = (1, 0.5)$.

A. $\Phi_4 = -5$	C. $\Phi_4 = -3$	E. $\Phi_4 = -1$	G. $\Phi_4 = 1$	I. $\Phi_4 = 3$	K. $\Phi_4 = 5$
B. $\Phi_4 = -4$	D. $\Phi_4 = -2$	F. $\Phi_4 = 0$	H. $\Phi_4 = 2$	J. $\Phi_4 = 4$	L. $\Phi_4 = 6$

4.3.2 Interconnected Rectangular Domains

The analytic elements with the geometry of a rectangle may be extended to problems with domains consisting of interconnected rectangles. Their interface conditions are solved first for ***adjacent elements that connect along the right and left sides of an interface***, as illustrated in Fig. 4.10a, where side 3 of the left element and side 1 of the right element share a common edge. This problem may be formulated at $m = 1 \cdots M$ conditions along the interface that satisfy continuity conditions for the potential, (2.64a), and the normal component of the vector field, (2.64b), using

$$
\begin{aligned}
&\alpha^- \Phi(x_m^{(3)}, y_m^{(3)}, \underset{\text{left}}{\mathbf{c}}) - \alpha^+ \Phi(x_m^{(1)}, y_m^{(1)}, \underset{\text{right}}{\mathbf{c}}) = \gamma_m \\
&\beta^- v_x(x_m^{(3)}, y_m^{(3)}, \underset{\text{left}}{\mathbf{c}}) - \beta^+ v_x(x_m^{(1)}, y_m^{(1)}, \underset{\text{right}}{\mathbf{c}}) = 0
\end{aligned}
\tag{4.34}
$$

where $\underset{\text{left}}{\mathbf{c}}$ and $\underset{\text{right}}{\mathbf{c}}$ are the coefficients of the elements on the left and right sides of the interface. These equations may be solved along with those specified along the other sides of the adjoining elements using the system of equations

$$
\begin{bmatrix}
\mathbf{A}^{(1)} & 0 \\
\mathbf{A}^{(2)} & 0 \\
\mathbf{A}^{(4)} & 0 \\
\alpha^- \overset{\Phi}{\mathbf{A}}{}^{(3)} & -\alpha^+ \overset{\Phi}{\mathbf{A}}{}^{(1)} \\
\beta^- \overset{v}{\mathbf{A}}{}^{(3)} & -\beta^+ \overset{v}{\mathbf{A}}{}^{(1)} \\
0 & \mathbf{A}^{(2)} \\
0 & \mathbf{A}^{(3)} \\
0 & \mathbf{A}^{(4)}
\end{bmatrix}
\begin{bmatrix}
\underset{\text{left}}{\mathbf{c}} \\
\underset{\text{right}}{\mathbf{c}}
\end{bmatrix}
=
\begin{bmatrix}
\mathbf{b}^{(1)} \\
\mathbf{b}^{(2)} \\
\mathbf{b}^{(4)} \\
\gamma_m - \alpha^- \overset{\Phi}{\mathbf{b}}{}^{(3)} + \alpha^+ \overset{\Phi}{\mathbf{b}}{}^{(1)} \\
-\beta^- \overset{v}{\mathbf{b}}{}^{(3)} + \beta^+ \overset{v}{\mathbf{b}}{}^{(1)} \\
\mathbf{b}^{(2)} \\
\mathbf{b}^{(3)} \\
\mathbf{b}^{(4)}
\end{bmatrix}
\tag{4.35}
$$

Here, the matrices $\mathbf{A}^{(S)}$ and $\mathbf{b}^{(S)}$ for the first three sets of rows are filled with terms necessary for solving Dirichlet or Neumann conditions along the sides of the left element, and the last three sets of rows are filled for the right element, using (4.31). For example, the four sides of the left element in Fig. 4.10a have constant Φ_1, $v_y = 0$, interface conditions and $v_y = 0$; and the four sides of the right element have interface conditions, $v_y = 0$, $v_x = 0$, and constant Φ_4. Note that the matrices $\mathbf{b}^{(3)}$ and $\mathbf{b}^{(1)}$ for the continuity conditions are only filled with terms associated with the additional functions in (4.29) and (4.30), and so are zero for this example.

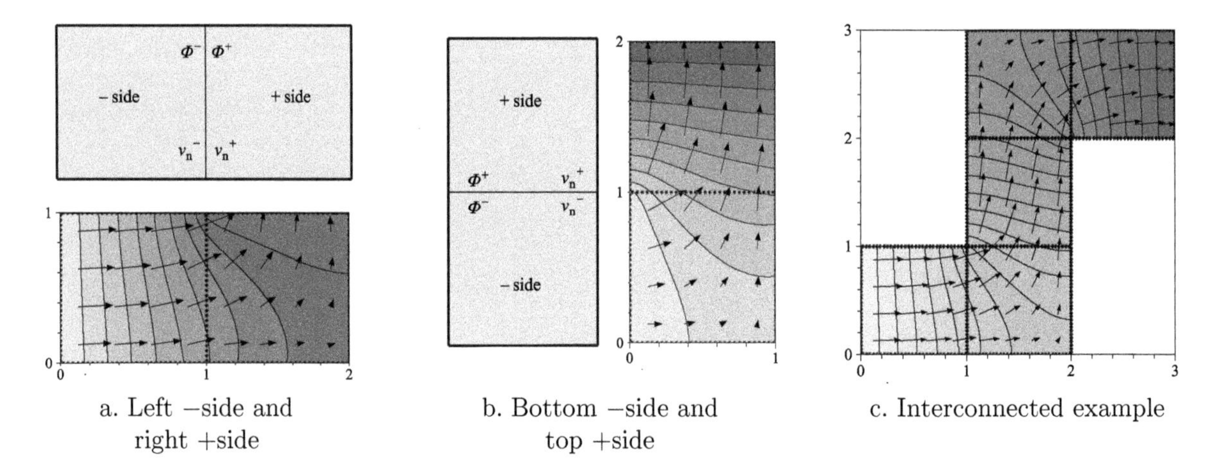

Figure 4.10 *Interface conditions for adjacent rectangle elements.*

Interface conditions may also be formulated for ***adjacent elements that connect along the top and bottom of an interface*** as illustrated in Fig. 4.10b. These conditions are given by

$$\begin{aligned} \alpha^-\Phi(x_m^{(4)}, y_m^{(4)}, \underset{\text{bottom}}{\mathbf{c}}) - \alpha^+\Phi(x_m^{(2)}, y_m^{(2)}, \underset{\text{top}}{\mathbf{c}}) &= \gamma_m \\ \beta^- v_y(x_m^{(4)}, y_m^{(4)}, \underset{\text{bottom}}{\mathbf{c}}) - \beta^+ v_y(x_m^{(2)}, y_m^{(2)}, \underset{\text{top}}{\mathbf{c}}) &= 0 \end{aligned} \tag{4.36}$$

where $\underset{\text{bottom}}{\mathbf{c}}$ and $\underset{\text{top}}{\mathbf{c}}$ are the coefficients of the elements on the bottom and top sides. These equations may be solved along with those specified along the other sides of the adjoining matrices using the system of equations

$$\begin{bmatrix} \mathbf{A}^{(1)} & 0 \\ \mathbf{A}^{(2)} & 0 \\ \mathbf{A}^{(3)} & 0 \\ \alpha^- \overset{\Phi}{\mathbf{A}}{}^{(4)} & -\alpha^+ \overset{\Phi}{\mathbf{A}}{}^{(2)} \\ \beta^- \overset{v}{\mathbf{A}}{}^{(4)} & -\beta^+ \overset{v}{\mathbf{A}}{}^{(2)} \\ 0 & \mathbf{A}^{(1)} \\ 0 & \mathbf{A}^{(3)} \\ 0 & \mathbf{A}^{(4)} \end{bmatrix} \begin{bmatrix} \underset{\text{bottom}}{\mathbf{c}} \\ \underset{\text{top}}{\mathbf{c}} \end{bmatrix} = \begin{bmatrix} \mathbf{b}^{(1)} \\ \mathbf{b}^{(2)} \\ \mathbf{b}^{(3)} \\ \gamma_m - \alpha^- \overset{\Phi}{\mathbf{b}}{}^{(4)} + \alpha^+ \overset{\Phi}{\mathbf{b}}{}^{(2)} \\ -\beta^- \overset{v}{\mathbf{b}}{}^{(4)} + \beta^+ \overset{v}{\mathbf{b}}{}^{(2)} \\ \mathbf{b}^{(1)} \\ \mathbf{b}^{(3)} \\ \mathbf{b}^{(4)} \end{bmatrix} \tag{4.37}$$

Here, the matrices $\mathbf{A}^{(S)}$ and $\mathbf{b}^{(S)}$ for the first three sets of rows are filled for the bottom element, and the last three sets of rows are filled for the right element. For example, the four sides of the bottom element in Fig. 4.10b have constant Φ_1, $v_y = 0$, $v_x = 0$, and interface conditions; and the four sides of the top element have $v_x = 0$, interface conditions, $v_x = 0$, and constant Φ_4.

Problems with ***multiple interconnected rectangles***, as illustrated in Fig. 4.10c, may be solved by sequentially solving for the coefficients of adjacent elements for all interfaces existing in the problem domain. After the coefficients are determined to satisfy the continuity conditions across an interface, the value of the potential may be computed for the rectangles on each side of the interface, and subsequently used in a Dirichlet condition to solve interface conditions that may exist across other sides of a rectangle, following Steward and Allen (2013). Examples illustrate the solution for interfaces across which $\alpha^- \neq \alpha^+$, $\beta^- = \beta^+$, and γ_m=0 in Fig. 4.11a, and a solution with $\gamma_m \neq 0$ is shown in Fig. 4.11b.

Problem 4.7 Solve for the coefficients for two rectangle elements in Fig. 4.10a with sides of length 1, each with $N = 10$ and $M = 30$. For each rectangle, turn in the coefficients $\overset{0}{c}$, $\overset{x}{c}$, $\overset{y}{c}$, $\overset{xy}{c}$, and $\overset{x2y2}{c}$ and the value of Φ, v_x, and v_y at the center of each rectangle. The potential on side 4 of the right element is $\Phi_4 = 0$ and the potential on side 1 of the left element is

A. $\Phi_1 = 1$	C. $\Phi_1 = 3$	E. $\Phi_1 = 5$	G. $\Phi_1 = 7$	I. $\Phi_1 = 9$	K. $\Phi_1 = 11$
B. $\Phi_1 = 2$	D. $\Phi_1 = 4$	F. $\Phi_1 = 6$	H. $\Phi_1 = 8$	J. $\Phi_1 = 10$	L. $\Phi_1 = 12$

Problem 4.8 Solve for the coefficients for two rectangle elements in Fig. 4.10b with sides of length 1, each with $N = 10$ and $M = 30$. The potential for the top element is $\Phi_4 = 0$ and the potential Φ_1 on side 1 of the bottom element is from Problem 4.7. For each rectangle, turn in the coefficients $\overset{0}{c}$, $\overset{x}{c}$, $\overset{y}{c}$, $\overset{xy}{c}$, and $\overset{x2y2}{c}$ and the value of Φ, v_x, and v_y at the center of each rectangle.

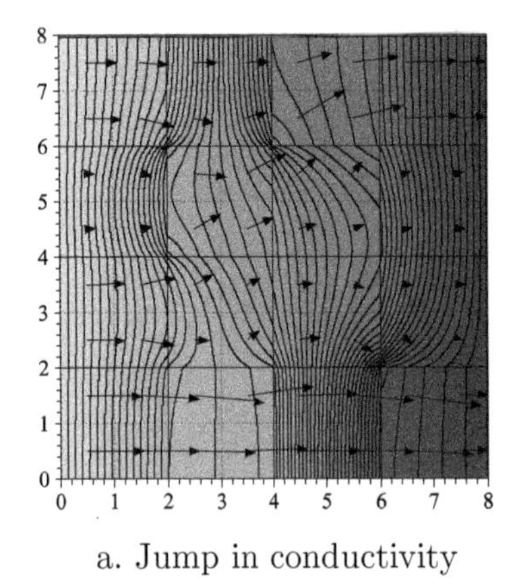

a. Jump in conductivity

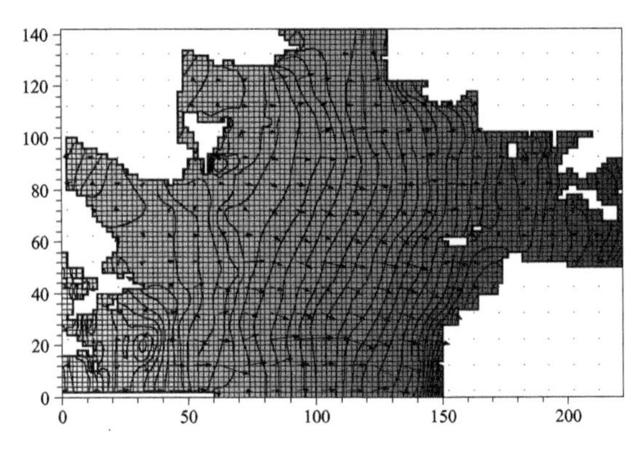

b. Regional ground water model

Figure 4.11 *The separation of variables solution for adjacent rectangles with discontinuities in parameters. (Reprinted from Steward and Allen, 2013, The Analytic Element Method for rectangular gridded domains, benchmark comparisons and application to the High Plains Aquifer,* Advances in Water Resources, *Vol. 60, Figs. 3,5, with permission from Elsevier.)*

4.4 Separation in Circular–Cylindrical Coordinates

Analytic elements are developed next using the separation of variables in the circular–cylindrical coordinates shown in Fig. 4.12, where coordinate surfaces occur along circles with constant r, half-planes with constant θ, and planes with constant z. ***Coordinate transformations*** are specified such that z has the same position in both systems, and the relationship between (r, θ) and Cartesian (x, y) coordinates is

$$\left.\begin{array}{l} r = \sqrt{x^2 + y^2} \\ \theta = \arctan \dfrac{y}{x} \end{array}\right\} \quad \leftrightarrow \quad \left\{\begin{array}{l} x = r\cos\theta \\ y = r\sin\theta \end{array}\right. \tag{4.38}$$

The ***coordinate unit vectors*** in Fig. 4.12 are directed towards the gradient of coordinate surfaces, and may be expressed in terms of their Cartesian components:

$$\hat{\mathbf{r}} = \frac{\nabla r}{|\nabla r|} = \cos\theta\hat{\mathbf{x}} + \sin\theta\hat{\mathbf{y}}, \quad \hat{\boldsymbol{\theta}} = \frac{\nabla\theta}{|\nabla\theta|} = -\sin\theta\hat{\mathbf{x}} + \cos\theta\hat{\mathbf{y}} \tag{4.39}$$

Problems are formulated using the ***Laplacian and gradient in circular–cylindrical coordinates***, which are repeated here from Eq. (4.1b):

$$\boxed{\nabla^2\Phi = \frac{\partial^2\Phi}{\partial r^2} + \frac{1}{r}\frac{\partial f}{\partial r} + \frac{1}{r^2}\frac{\partial^2\Phi}{\partial\theta^2} + \frac{\partial^2\Phi}{\partial z^2}}, \quad \boxed{\nabla\Phi = \frac{\partial\Phi}{\partial r}\hat{\mathbf{r}} + \frac{1}{r}\frac{\partial\Phi}{\partial\theta}\hat{\boldsymbol{\theta}} + \frac{\partial\Phi}{\partial z}\hat{\mathbf{z}}} \tag{4.40}$$

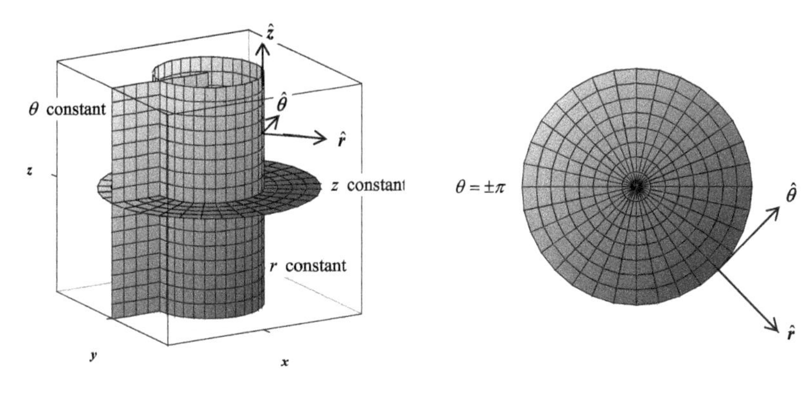

Figure 4.12 *Circular–cylindrical coordinates (r, θ, z): with circular cylinders (r constant), half-planes (θ constant), and planes (z constant).*

Analytic elements are developed next to solve boundary value problems in circular coordinates. While methods are applicable to problems governed by the Laplace equation, such problems have already been effectively covered using complex variables in Section 3.2. So, the rest of this section focuses on problems governed by the ***Helmholtz and modified Helmholtz equations***.

4.4.1 Radial Waves Emanating from a Point

The use of separation of variables to solve wave problems in circular coordinates is illustrated for waves emanating from a point. This problem is governed by the Helmholtz equation with the complex wave function φ, which is an ordinary differential equation in the r-direction with a general solution given by

$$\frac{\mathrm{d}^2\varphi}{\mathrm{d}r^2} + \frac{1}{r}\frac{\mathrm{d}\varphi}{\mathrm{d}r} + k^2\varphi = 0 \quad \rightarrow \quad \varphi = c_1 H_0^{(1)}(kr) + c_2 H_0^{(2)}(kr) \tag{4.41}$$

Note that $H_0^{(1)}$ and $H_0^{(2)}$ are Hankel functions of zero order that are described shortly, beginning with Eq. (4.52). The particular solution given by

$$\varphi = \frac{S}{2\pi} H_0^{(1)}(kr) \tag{4.42}$$

represents an outgoing wave emanating from a ***point-source*** at the origin, and the wave field is illustrated in Fig. 4.13a. This wave's amplitude decreases as waves propagate away from the point-source, and the phase illustrates wave crests and troughs spaced at wavelength $L = 2\pi/k$. This wave field is easily expanded to waves fully reflected by a boundary at $x = 0$,

$$\varphi = \frac{S}{2\pi} H_0^{(1)}\left(k\sqrt{(x-d)^2 + y^2}\right) + \frac{S}{2\pi} H_0^{(1)}\left(k\sqrt{(x+d)^2 + y^2}\right) \quad \rightarrow \quad \frac{\partial\varphi}{\partial n} = 0 \quad (x = 0) \tag{4.43}$$

which is illustrated in Fig. 4.13b. This solution utilizes the ***method of images*** to superimpose two point-sources, which are mirror images across the boundary, to exactly satisfy the fully reflecting boundary condition.

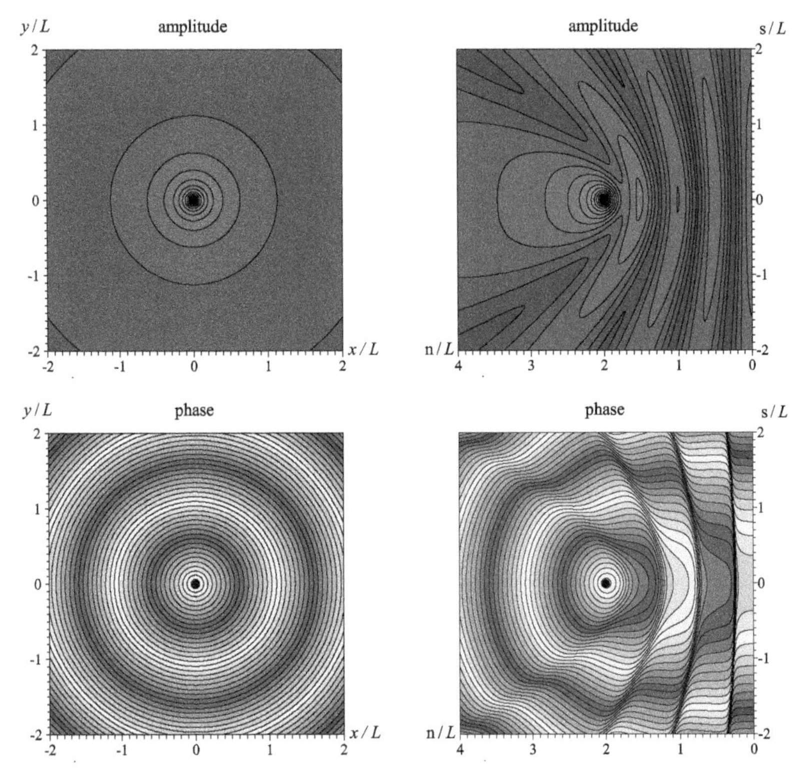

Figure 4.13 *Radial waves in circular coordinates.*

Problem 4.9 Compute the amplitude, $|\varphi|$, and phase, $\arg\varphi$, from the complex wave function $\varphi = |\varphi|e^{i\arg\varphi}$ for the wave field of a point-source with $S = 1$ and the specified wave number k at locations r=1, 2, 5, and 10.

A. $k = \frac{2\pi}{1}$
B. $k = \frac{2\pi}{2}$
C. $k = \frac{2\pi}{3}$
D. $k = \frac{2\pi}{4}$
E. $k = \frac{2\pi}{5}$
F. $k = \frac{2\pi}{6}$
G. $k = \frac{2\pi}{7}$
H. $k = \frac{2\pi}{8}$
I. $k = \frac{2\pi}{9}$
J. $k = \frac{2\pi}{10}$
K. $k = \frac{2\pi}{11}$
L. $k = \frac{2\pi}{12}$

4.4.2 Waves around Circle Elements

The separation of variables is formulated next for analytic elements with circular geometry. A potential Φ that satisfies the Helmholtz equation (2.2c) for a two-dimensional vector field is written in separated form as the product of two function, one of which varies as a function of only r and the other varies as a function of only θ:

$$\boxed{\Phi(r,\theta) = R(r)\Theta(\theta)}, \quad \nabla^2\Phi = \frac{\partial^2\Phi}{\partial r^2} + \frac{1}{r}\frac{\partial\Phi}{\partial r} + \frac{1}{r^2}\frac{\partial^2\Phi}{\partial\theta^2} = -k^2\Phi \tag{4.44}$$

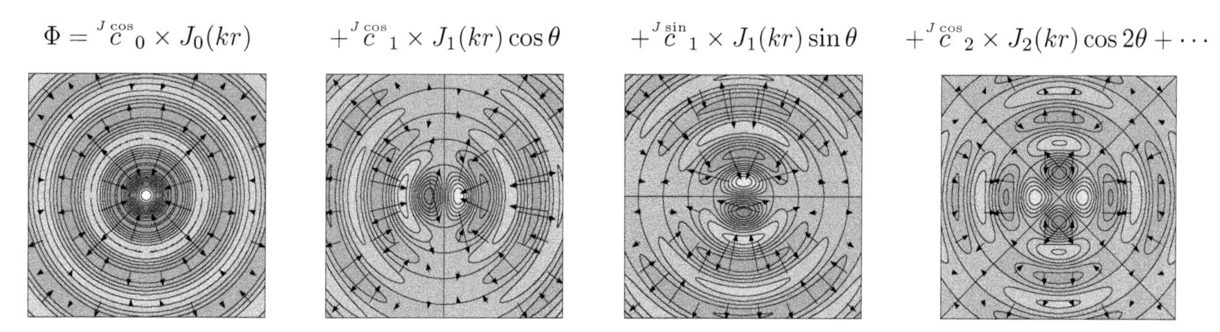

Figure 4.14 *Influence functions for circular domains using Bessel functions of the first kind, from (4.48).*

The separated form $\Phi = R\Theta$ is substituted into the Helmholtz equation and the equation is organized by multiplying both sides by r^2/Φ, gathering terms of r and θ on opposite sides of the equation, and then setting each of these equal to a constant designated as n^2:

$$\begin{aligned} &\frac{r^2}{R\Theta}\left(\Theta\frac{\mathrm{d}^2R}{\mathrm{d}r^2} + \Theta\frac{1}{r}\frac{\mathrm{d}R}{\mathrm{d}r} + R\frac{1}{r^2}\frac{\mathrm{d}^2\Theta}{\mathrm{d}\theta^2}\right) = \frac{r^2}{R\Theta}\left(-k^2R\Theta\right) \\ \rightarrow\quad & r^2\left[\frac{1}{R}\left(\frac{\mathrm{d}^2R}{\mathrm{d}r^2} + \frac{1}{r}\frac{\mathrm{d}R}{\mathrm{d}r}\right) + k^2\right] = -\frac{1}{\Theta}\frac{\mathrm{d}^2\Theta}{\mathrm{d}\theta^2} = n^2 \end{aligned} \tag{4.45}$$

This provides two ordinary differential equations with particular solutions obtained following Moon and Spencer (1961*b*, p.16) for $n = 0$

$$\begin{aligned} \frac{\mathrm{d}^2R}{\mathrm{d}r^2} + \frac{1}{r}\frac{\mathrm{d}R}{\mathrm{d}r} + k^2R = 0 \quad &\rightarrow \quad R = \left\{\mathcal{J}_0(kr),\ Y_0(kr)\right. \\ \frac{\mathrm{d}^2\Theta}{\mathrm{d}\theta^2} = 0 \quad &\rightarrow \quad \Theta = \left\{1,\ \theta\right. \end{aligned} \tag{4.46}$$

and for $n^2 > 0$

$$\begin{aligned} \frac{\mathrm{d}^2R}{\mathrm{d}r^2} + \frac{1}{r}\frac{\mathrm{d}R}{\mathrm{d}r} + \left(k^2 - \frac{n^2}{r^2}\right)R = 0 \quad &\rightarrow \quad R = \left\{\mathcal{J}_n(kr),\ Y_n(kr)\right. \\ \frac{\mathrm{d}^2\Theta}{\mathrm{d}\theta^2} + n^2\Theta = 0 \quad &\rightarrow \quad \Theta = \left\{\cos n\theta,\ \sin n\theta\right. \end{aligned} \tag{4.47}$$

These particular solutions are represented in terms of ***Bessel functions*** of the first kind $\mathcal{J}_n$ and the second kind Y_n, each of order n. Computational methods useful for evaluating these functions are presented in Abramowitz and Stegun (1972, chap.9).

The influence functions for an analytic element are obtained through the linear superposition of these separable solutions times coefficients. This gives the following for ***Bessel function of the first kind***

$$\boxed{\Phi(r,\theta) = \sum_{n=0}^{N} {}^{\mathcal{J}\cos}c_n\,\mathcal{J}_n(kr)\cos n\theta + \sum_{n=1}^{N} {}^{\mathcal{J}\sin}c_n\,\mathcal{J}_n(kr)\sin n\theta} \tag{4.48}$$

where the over-script denotes that the coefficient is associated with the Bessel function $\mathcal{J}$ and the cos or sin function. These influence functions are illustrated in Fig. 4.14, and they

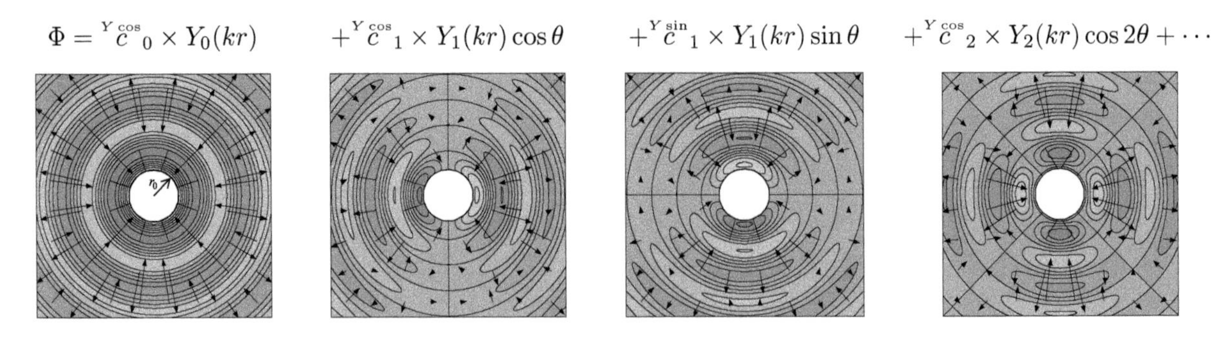

Figure 4.15 *Influence functions for circular domains using Bessel functions of the second kind, from (4.49).*

remain finite everywhere. The influence functions associated with ***Bessel functions of the second kind*** are given by

$$\Phi(r,\theta) = \sum_{n=0}^{N} {}^{Y\cos}c_n Y_n(kr)\cos n\theta + \sum_{n=1}^{N} {}^{Y\sin}c_n Y_n(kr)\sin n\theta \qquad (r \geq r_0) \tag{4.49}$$

These functions are singular at $r = 0$ and so are evaluated outside an element of radius r_0 as illustrated in Fig. 4.15. For elements with radius r_0, the influence functions contain cos and sin terms that repeat n times on its boundary, where these functions take on values associated with the Bessel functions evaluated at kr_0.

The ***vector field associated with minus the gradient of*** Φ is illustrated for influence functions in Figs. 4.14 and 4.15. Its components in the r and θ directions are given by these partial derivatives in (4.40), and the components in Cartesian components are related to these from the components in coordinate directions from the unit vectors (4.39), giving

$$\begin{aligned} v_r &= -\frac{\partial\Phi}{\partial r} & v_x &= -\cos\theta\frac{\partial\Phi}{\partial r} + \frac{\sin\theta}{r}\frac{\partial\Phi}{\partial\theta} \\ v_\theta &= -\frac{1}{r}\frac{\partial\Phi}{\partial\theta}, & v_y &= -\sin\theta\frac{\partial\Phi}{\partial r} - \frac{\cos\theta}{r}\frac{\partial\Phi}{\partial\theta} \end{aligned} \tag{4.50}$$

Derivatives of influence functions with respect to r may be evaluated using

$$\mathcal{J}_n'(kr) = \frac{n}{r}\mathcal{J}_n(kr) - k\mathcal{J}_{n+1}(kr), \qquad Y_n'(kr) = \frac{n}{r}Y_n(kr) - kY_{n+1}(kr) \tag{4.51}$$

which was adapted from Moon and Spencer (1961*b*, p.191) and Abramowitz and Stegun (1972, p.361).

Problem 4.10 Compute the potential Φ and the components of the vector field v_x, v_y, v_r, and v_θ, at (x,y)=(1,0), (0,1), and (1,1) for a circle element in a domain with $k = 1$ with the following coefficient and all other coefficients set to zero:

A. ${}^{\mathcal{J}\cos}c_0 = 1$
B. ${}^{\mathcal{J}\cos}c_1 = 1$
C. ${}^{\mathcal{J}\cos}c_2 = 1$
D. ${}^{\mathcal{J}\sin}c_1 = 1$
E. ${}^{\mathcal{J}\sin}c_2 = 1$
F. ${}^{\mathcal{J}\sin}c_3 = 1$
G. ${}^{Y\cos}c_0 = 1$
H. ${}^{Y\cos}c_1 = 1$
I. ${}^{Y\cos}c_2 = 1$
J. ${}^{Y\sin}c_1 = 1$
K. ${}^{Y\sin}c_2 = 1$
L. ${}^{Y\sin}c_3 = 1$

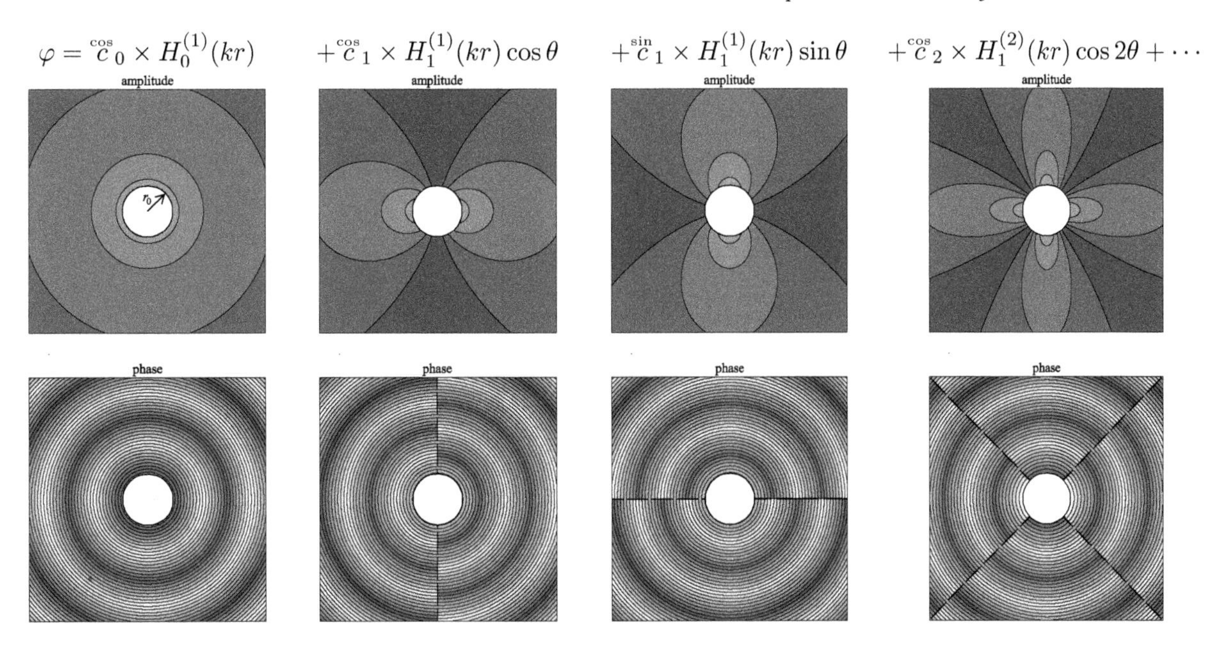

Figure 4.16 *Influence functions for circular domains using Hankel functions, from (4.53), illustrating the modulus and cosine of the phase for the influence functions. (Reprinted from Steward, 2018, Wave resonance and dissipation in collections of partially reflecting vertical cylinders,* Journal of Waterways, Ports, Coastal, and Ocean Engineering, *Vol. 144(4), Fig. 2, with permission from ASCE.)*

Complex Wave Functions

Solutions to the Helmholtz equation for complex wave functions are obtained by gathering the Bessel functions $\mathcal{J}_n$ and Y_n in terms of ***Bessel functions of the third kind (Hankel functions)***:

$$
\begin{aligned}
H_n^{(1)}(r) &= \mathcal{J}_n(r) + \mathrm{i}Y_n(r) \\
H_n^{(2)}(r) &= \mathcal{J}_n(r) - \mathrm{i}Y_n(r)
\end{aligned}
\tag{4.52}
$$

The complex function for a wave field is obtained by gathering these functions for a circle element with radius r_0, as linear superposition

$$
\boxed{\varphi(r,\theta) = \sum_{n=0}^{N} \overset{\cos}{c}_n H_n^{(1)}(kr)\cos n\theta + \sum_{n=1}^{N} \overset{\sin}{c}_n H_n^{(1)}(kr)\sin n\theta \qquad (r \geq r_0)}
\tag{4.53}
$$

where the amplitude and phase associated with these influence functions are illustrated in Fig. 4.16. Note that this solution contain only the functions $H_n^{(1)}$, since $H_n^{(2)}$ do not satisfy the Sommerfeld radiation condition, (1.77). The ***gradient of the complex wave function*** is needed later to formulate boundary conditions and is given by

$$
\begin{aligned}
\frac{\partial\varphi}{\partial x} &= \cos\theta\frac{\partial\varphi}{\partial r} - \frac{\sin\theta}{r}\frac{\partial\varphi}{\partial\theta} \\
\frac{\partial\varphi}{\partial y} &= \sin\theta\frac{\partial\varphi}{\partial r} + \frac{\cos\theta}{r}\frac{\partial\varphi}{\partial\theta}
\end{aligned}
\tag{4.54a}
$$

with

$$\frac{\partial \varphi}{\partial r} = \sum_{n=0}^{N} \overset{\cos}{c}_n H_n^{'(1)}(kr) \cos n\theta + \sum_{n=1}^{N} \overset{\sin}{c}_n H_n^{'(1)}(kr) \sin n\theta$$
$$\frac{\partial \varphi}{\partial \theta} = -\sum_{n=0}^{N} \overset{\cos}{c}_n H_n^{(1)}(kr) n \sin n\theta + \sum_{n=1}^{N} \overset{\sin}{c}_n H_n^{(1)}(kr) n \cos n\theta \tag{4.54b}$$

where the partial derivative of influence functions with respect to r may be evaluated using

$$H_n^{'(1)}(kr) = \frac{n}{r} H_n^{(1)}(kr) - k H_{n+1}^{(1)}(kr) \tag{4.55}$$

similar to (4.51) for the real Bessel functions.

Boundary and interface conditions along circular elements are prescribed in terms of the complex function φ, (4.54), and its ***solution requires separating terms into real and imaginary parts***. This is accomplished for the circular element with complex coefficients ($\overset{\cos}{c}_n = \Re \overset{\cos}{c}_n + i\Im \overset{\cos}{c}_n$ and $\overset{\sin}{c}_n = \Re \overset{\sin}{c}_n + i\Im \overset{\sin}{c}_n$) using the real and imaginary parts of the Hankel functions in (4.52) to obtain

$$\begin{aligned}\varphi = \sum_{n=0}^{N} & \left[\mathcal{J}_n(kr) \Re \overset{\cos}{c}_n - Y_n(kr) \Im \overset{\cos}{c}_n \right] \cos n\theta \\ & + i \left[Y_n(kr) \Re \overset{\cos}{c}_n + \mathcal{J}_n(kr) \Im \overset{\cos}{c}_n \right] \cos n\theta \\ + \sum_{n=1}^{N} & \left[\mathcal{J}_n(kr) \Re \overset{\sin}{c}_n - Y_n(kr) \Im \overset{\sin}{c}_n \right] \sin n\theta \\ & + i \left[Y_n(kr) \Re \overset{\sin}{c}_n + \mathcal{J}_n(kr) \Im \overset{\sin}{c}_n \right] \sin n\theta + \overset{\text{add}}{\varphi}(r, \theta)\end{aligned} \tag{4.56}$$

The partial derivative of this function in the r-direction normal to the boundary is also required for the solution and given by

$$\begin{aligned}\frac{\partial \varphi}{\partial r} = \sum_{n=0}^{N} & \left[\mathcal{J}'_n(kr) \Re \overset{\cos}{c}_n - Y'_n(kr) \Im \overset{\cos}{c}_n \right] \cos n\theta \\ & + i \left[Y'_n(kr) \Re \overset{\cos}{c}_n + \mathcal{J}'_n(kr) \Im \overset{\cos}{c}_n \right] \cos n\theta \\ + \sum_{n=1}^{N} & \left[\mathcal{J}'_n(kr) \Re \overset{\sin}{c}_n - Y'_n(kr) \Im \overset{\sin}{c}_n \right] \sin n\theta \\ & + i \left[Y'_n(kr) \Re \overset{\sin}{c}_n + \mathcal{J}'_n(kr) \Im \overset{\sin}{c}_n \right] \sin n\theta + \frac{\partial \overset{\text{add}}{\varphi}}{\partial r}(r, \theta)\end{aligned} \tag{4.57}$$

Note that the additional function $\overset{\text{add}}{\varphi}$ contains the contributions to the complex wave function for the background wave field and all other elements, and its partial derivative in the r-direction (4.39) is given by

$$\frac{\partial \overset{\text{add}}{\varphi}}{\partial r} = \cos\theta \frac{\partial \overset{\text{add}}{\varphi}}{\partial x} + \sin\theta \frac{\partial \overset{\text{add}}{\varphi}}{\partial y} \tag{4.58}$$

A circle element with a ***partially reflecting boundary*** satisfies a Robin condition given by $\partial\varphi/\partial r + \alpha\varphi = 0$, (2.61), where $\alpha = ik(1-R)/(1+R)$ with wave number k and reflection coefficient R. The separated form of φ in (4.56) with derivatives from (4.51) are substituted into this condition and evaluated at control point m where $r = r_0$ and $\theta = \theta_m$ to give

$$\begin{aligned}&\sum_{n=0}^{N} \cos n\theta_m \left(\mathcal{J}_n \Re \overset{\cos}{c}_n - \mathcal{Y}_n \Im \overset{\cos}{c}_n \right) \\ &\quad + \sum_{n=1}^{N} \sin n\theta_m \left(\mathcal{J}_n \Re \overset{\sin}{c}_n - \mathcal{Y}_n \Im \overset{\sin}{c}_n \right) = -\Re \left(\frac{\partial \overset{\text{add}}{\varphi}_m}{\partial r} + \alpha \overset{\text{add}}{\varphi}_m \right)\end{aligned} \tag{4.59a}$$

and

$$\begin{aligned}&\sum_{n=0}^{N} \cos n\theta_m \left(\mathcal{Y}_n \Re \overset{\cos}{c}_n + \mathcal{J}_n \Im \overset{\cos}{c}_n \right) \\ &\quad + \sum_{n=1}^{N} \sin n\theta_m \left(\mathcal{Y}_n \Re \overset{\sin}{c}_n + \mathcal{J}_n \Im \overset{\sin}{c}_n \right) = -\Im \left(\frac{\partial \overset{\text{add}}{\varphi}_m}{\partial r} + \alpha \overset{\text{add}}{\varphi}_m \right)\end{aligned} \tag{4.59b}$$

using a condensed notation from Steward (2018)

$$\begin{aligned}\mathcal{J}_n &= J'_n(kr_0) - (\Im\alpha)\, Y_n(kr_0) \\ \mathcal{Y}_n &= Y'_n(kr_0) + (\Im\alpha)\, J_n(kr_0)\end{aligned} \tag{4.60}$$

These $m = 1 \cdots M$ conditions may be organized as ***two systems of equations*** in the matrices

$$\begin{aligned}\mathbf{A}\overset{\Re}{\mathbf{c}} &= \overset{\Re}{\mathbf{b}} \\ \mathbf{A}\overset{\Im}{\mathbf{c}} &= \overset{\Im}{\mathbf{b}}\end{aligned}, \quad \mathbf{A} = \begin{bmatrix} 1 & \cos\theta_1 & \sin\theta_1 & \cdots & \cos N\theta_1 & \sin N\theta_1 \\ 1 & \cos\theta_2 & \sin\theta_2 & \cdots & \cos N\theta_2 & \sin N\theta_2 \\ \cdots & \cdots & \cdots & \cdots & \cdots & \cdots \\ 1 & \cos\theta_M & \sin\theta_M & \cdots & \cos N\theta_M & \sin N\theta_M \end{bmatrix} \tag{4.61}$$

with

$$\overset{\Re}{\mathbf{c}} = \begin{bmatrix} \mathcal{J}_0 \Re \overset{\cos}{c}_0 - \mathcal{Y}_0 \Im \overset{\cos}{c}_0 \\ \mathcal{J}_1 \Re \overset{\cos}{c}_1 - \mathcal{Y}_1 \Im \overset{\cos}{c}_1 \\ \mathcal{J}_1 \Re \overset{\sin}{c}_1 - \mathcal{Y}_1 \Im \overset{\sin}{c}_1 \\ \vdots \\ \mathcal{J}_N \Re \overset{\cos}{c}_N - \mathcal{Y}_N \Im \overset{\cos}{c}_N \\ \mathcal{J}_N \Re \overset{\sin}{c}_N - \mathcal{Y}_N \Im \overset{\sin}{c}_N \end{bmatrix}, \quad \overset{\Re}{\mathbf{b}} = \begin{bmatrix} -\Re\left(\frac{\partial \overset{\text{add}}{\varphi}_1}{\partial r} + \alpha \overset{\text{add}}{\varphi}_1 \right) \\ -\Re\left(\frac{\partial \overset{\text{add}}{\varphi}_2}{\partial r} + \alpha \overset{\text{add}}{\varphi}_2 \right) \\ \vdots \\ -\Re\left(\frac{\partial \overset{\text{add}}{\varphi}_M}{\partial r} + \alpha \overset{\text{add}}{\varphi}_M \right) \end{bmatrix} \tag{4.62a}$$

and

$$\overset{\Im}{\mathbf{c}} = \begin{bmatrix} \mathcal{Y}_0 \Re \overset{\cos}{c}_0 + \mathcal{J}_0 \Im \overset{\cos}{c}_0 \\ \mathcal{Y}_1 \Re \overset{\cos}{c}_1 + \mathcal{J}_1 \Im \overset{\cos}{c}_1 \\ \mathcal{Y}_1 \Re \overset{\sin}{c}_1 + \mathcal{J}_1 \Im \overset{\sin}{c}_1 \\ \vdots \\ \mathcal{Y}_N \Re \overset{\cos}{c}_N + \mathcal{J}_N \Im \overset{\cos}{c}_N \\ \mathcal{Y}_N \Re \overset{\sin}{c}_N + \mathcal{J}_N \Im \overset{\sin}{c}_N \end{bmatrix}, \quad \overset{\Im}{\mathbf{b}} = \begin{bmatrix} -\Im\left(\frac{\partial \overset{\text{add}}{\varphi}_1}{\partial r} + \alpha \overset{\text{add}}{\varphi}_1 \right) \\ -\Im\left(\frac{\partial \overset{\text{add}}{\varphi}_2}{\partial r} + \alpha \overset{\text{add}}{\varphi}_2 \right) \\ \vdots \\ -\Im\left(\frac{\partial \overset{\text{add}}{\varphi}_M}{\partial r} + \alpha \overset{\text{add}}{\varphi}_M \right) \end{bmatrix} \tag{4.62b}$$

The least squares solution for equally spaced control points (2.47) gives

$$\begin{aligned}
\mathcal{J}_0 \Re \overset{\cos}{c}_0 - \mathcal{Y}_0 \Im \overset{\cos}{c}_0 &= -\frac{1}{M}\sum_{m=1}^{M} \Re\left(\frac{\partial \overset{\text{add}}{\varphi}_m}{\partial r} + \alpha \overset{\text{add}}{\varphi}_m\right) \\
\mathcal{J}_n \Re \overset{\cos}{c}_n - \mathcal{Y}_n \Im \overset{\cos}{c}_n &= -\frac{2}{M}\sum_{m=1}^{M} \Re\left(\frac{\partial \overset{\text{add}}{\varphi}_m}{\partial r} + \alpha \overset{\text{add}}{\varphi}_m\right)\cos n\theta_m \\
\mathcal{J}_n \Re \overset{\sin}{c}_n - \mathcal{Y}_n \Im \overset{\sin}{c}_n &= -\frac{2}{M}\sum_{m=1}^{M} \Re\left(\frac{\partial \overset{\text{add}}{\varphi}_m}{\partial r} + \alpha \overset{\text{add}}{\varphi}_m\right)\sin n\theta_m
\end{aligned} \tag{4.63a}$$

and

$$\begin{aligned}
\mathcal{Y}_0 \Re \overset{\cos}{c}_0 + \mathcal{J}_0 \Im \overset{\cos}{c}_0 &= -\frac{1}{M}\sum_{m=1}^{M} \Im\left(\frac{\partial \overset{\text{add}}{\varphi}_m}{\partial r} + \alpha \overset{\text{add}}{\varphi}_m\right) \\
\mathcal{Y}_n \Re \overset{\cos}{c}_n + \mathcal{J}_n \Im \overset{\cos}{c}_n &= -\frac{2}{M}\sum_{m=1}^{M} \Im\left(\frac{\partial \overset{\text{add}}{\varphi}_m}{\partial r} + \alpha \overset{\text{add}}{\varphi}_m\right)\cos n\theta_m \\
\mathcal{Y}_n \Re \overset{\sin}{c}_n + \mathcal{J}_n \Im \overset{\sin}{c}_n &= -\frac{2}{M}\sum_{m=1}^{M} \Im\left(\frac{\partial \overset{\text{add}}{\varphi}_m}{\partial r} + \alpha \overset{\text{add}}{\varphi}_m\right)\sin n\theta_m
\end{aligned} \tag{4.63b}$$

The coefficients are rearranged by solving the set of two equations with two unknowns for each n to give the real and imaginary components,

$$\begin{aligned}
\Re \overset{\cos}{c}_0 &= \frac{1}{M}\sum_{m=1}^{M}\left[-\frac{\mathcal{J}_0}{\mathcal{J}_0^2+\mathcal{Y}_0^2}\Re\left(\frac{\partial \overset{\text{add}}{\varphi}_m}{\partial r} + \alpha \overset{\text{add}}{\varphi}_m\right) - \frac{\mathcal{Y}_0}{\mathcal{J}_0^2+\mathcal{Y}_0^2}\Im\left(\frac{\partial \overset{\text{add}}{\varphi}_m}{\partial r} + \alpha \overset{\text{add}}{\varphi}_m\right)\right] \\
\Im \overset{\cos}{c}_0 &= \frac{1}{M}\sum_{m=1}^{M}\left[-\frac{\mathcal{J}_0}{\mathcal{J}_0^2+\mathcal{Y}_0^2}\Im\left(\frac{\partial \overset{\text{add}}{\varphi}_m}{\partial r} + \alpha \overset{\text{add}}{\varphi}_m\right) + \frac{\mathcal{Y}_0}{\mathcal{J}_0^2+\mathcal{Y}_0^2}\Re\left(\frac{\partial \overset{\text{add}}{\varphi}_m}{\partial r} + \alpha \overset{\text{add}}{\varphi}_m\right)\right]
\end{aligned} \tag{4.64a}$$

$$\begin{aligned}
\Re \overset{\cos}{c}_n &= \frac{2}{M}\sum_{m=1}^{M}\left[-\frac{\mathcal{J}_n}{\mathcal{J}_n^2+\mathcal{Y}_n^2}\Re\left(\frac{\partial \overset{\text{add}}{\varphi}_m}{\partial r} + \alpha \overset{\text{add}}{\varphi}_m\right) - \frac{\mathcal{Y}_n}{\mathcal{J}_n^2+\mathcal{Y}_n^2}\Im\left(\frac{\partial \overset{\text{add}}{\varphi}_m}{\partial r} + \alpha \overset{\text{add}}{\varphi}_m\right)\right]\cos n\theta_m \\
\Im \overset{\cos}{c}_n &= \frac{2}{M}\sum_{m=1}^{M}\left[-\frac{\mathcal{J}_n}{\mathcal{J}_n^2+\mathcal{Y}_n^2}\Im\left(\frac{\partial \overset{\text{add}}{\varphi}_m}{\partial r} + \alpha \overset{\text{add}}{\varphi}_m\right) + \frac{\mathcal{Y}_n}{\mathcal{J}_n^2+\mathcal{Y}_n^2}\Re\left(\frac{\partial \overset{\text{add}}{\varphi}_m}{\partial r} + \alpha \overset{\text{add}}{\varphi}_m\right)\right]\cos n\theta_m
\end{aligned} \tag{4.64b}$$

$$\begin{aligned}
\Re \overset{\sin}{c}_n &= \frac{2}{M}\sum_{m=1}^{M}\left[-\frac{\mathcal{J}_n}{\mathcal{J}_n^2+\mathcal{Y}_n^2}\Re\left(\frac{\partial \overset{\text{add}}{\varphi}_m}{\partial r} + \alpha \overset{\text{add}}{\varphi}_m\right) - \frac{\mathcal{Y}_n}{\mathcal{J}_n^2+\mathcal{Y}_n^2}\Im\left(\frac{\partial \overset{\text{add}}{\varphi}_m}{\partial r} + \alpha \overset{\text{add}}{\varphi}_m\right)\right]\sin n\theta_m \\
\Im \overset{\sin}{c}_n &= \frac{2}{M}\sum_{m=1}^{M}\left[-\frac{\mathcal{J}_n}{\mathcal{J}_n^2+\mathcal{Y}_n^2}\Im\left(\frac{\partial \overset{\text{add}}{\varphi}_m}{\partial r} + \alpha \overset{\text{add}}{\varphi}_m\right) + \frac{\mathcal{Y}_n}{\mathcal{J}_n^2+\mathcal{Y}_n^2}\Re\left(\frac{\partial \overset{\text{add}}{\varphi}_m}{\partial r} + \alpha \overset{\text{add}}{\varphi}_m\right)\right]\sin n\theta_m
\end{aligned} \tag{4.64c}$$

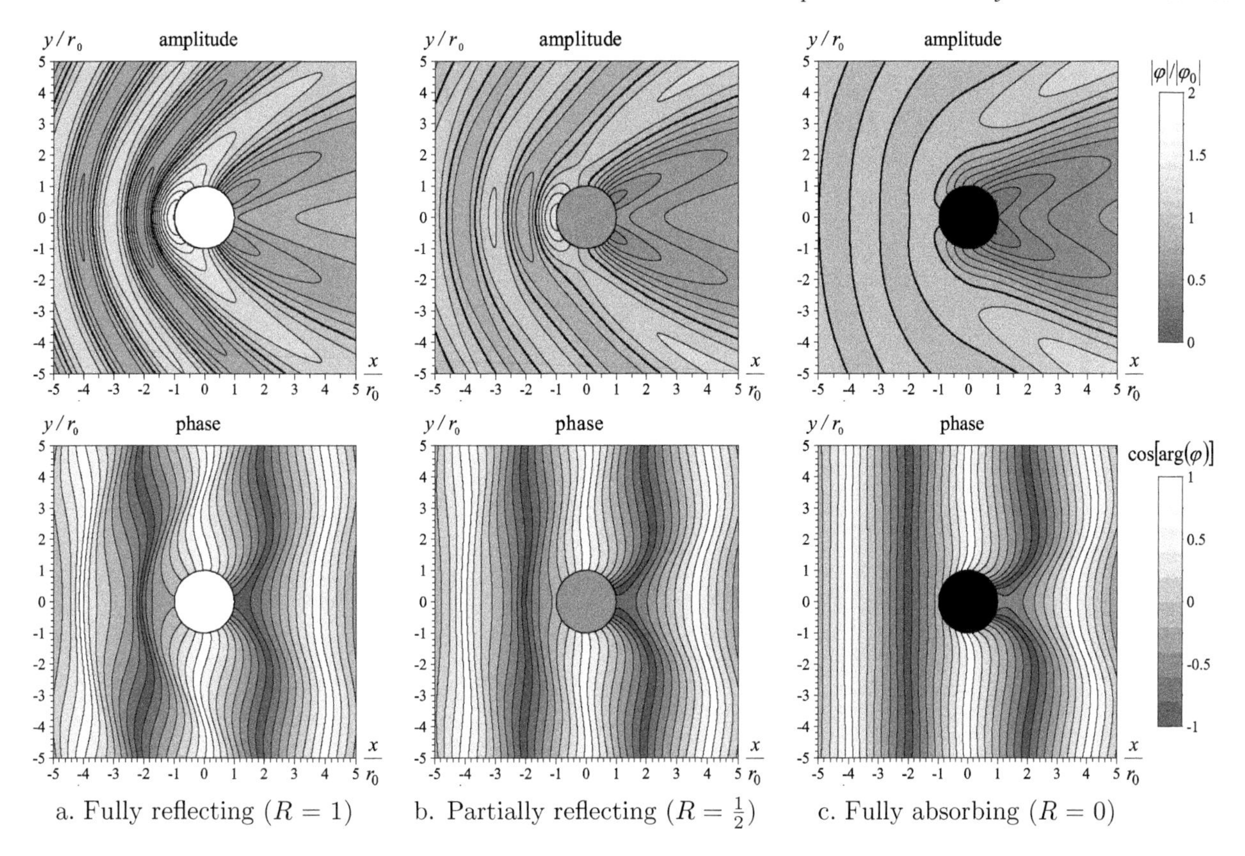

Figure 4.17 *Circle element in plane waves with a partially reflecting boundary. (Reprinted from Steward, 2018, Wave resonance and dissipation in collections of partially reflecting vertical cylinders,* Journal of Waterways, Ports, Coastal, and Ocean Engineering, *Vol. 144(4), Fig. 1, with permission from ASCE.)*

which are added to give the ***complex coefficients***:

$$
\begin{aligned}
\overset{\cos}{c}_0 &= \Re\overset{\cos}{c}_0 + \mathrm{i}\Im\overset{\cos}{c}_0 = \frac{1}{M}\sum_{m=1}^{M} \frac{-\mathcal{J}_0 + \mathrm{i}\mathcal{Y}_0}{\mathcal{J}_0{}^2 + \mathcal{Y}_0{}^2} \left(\frac{\partial \overset{\text{add}}{\varphi}_m}{\partial r} + \alpha \overset{\text{add}}{\varphi}_m \right) \\
\overset{\cos}{c}_n &= \frac{2}{M}\sum_{m=1}^{M} \frac{-\mathcal{J}_n + \mathrm{i}\mathcal{Y}_n}{\mathcal{J}_n{}^2 + \mathcal{Y}_n{}^2} \left(\frac{\partial \overset{\text{add}}{\varphi}_m}{\partial r} + \alpha \overset{\text{add}}{\varphi}_m \right) \cos n\theta_m \\
\overset{\sin}{c}_n &= \frac{2}{M}\sum_{m=1}^{M} \frac{-\mathcal{J}_n + \mathrm{i}\mathcal{Y}_n}{\mathcal{J}_n{}^2 + \mathcal{Y}_n{}^2} \left(\frac{\partial \overset{\text{add}}{\varphi}_m}{\partial r} + \alpha \overset{\text{add}}{\varphi}_m \right) \sin n\theta_m
\end{aligned}
\qquad (4.65)
$$

The solution in Fig. 4.17 illustrates a single circle element placed in a background uniform wave field obtained by setting the additional function equal to that wave field in (4.9). The solution in Fig. 4.18 illustrates the wave field in randomly distributed circle elements with partially and fully reflecting boundaries.

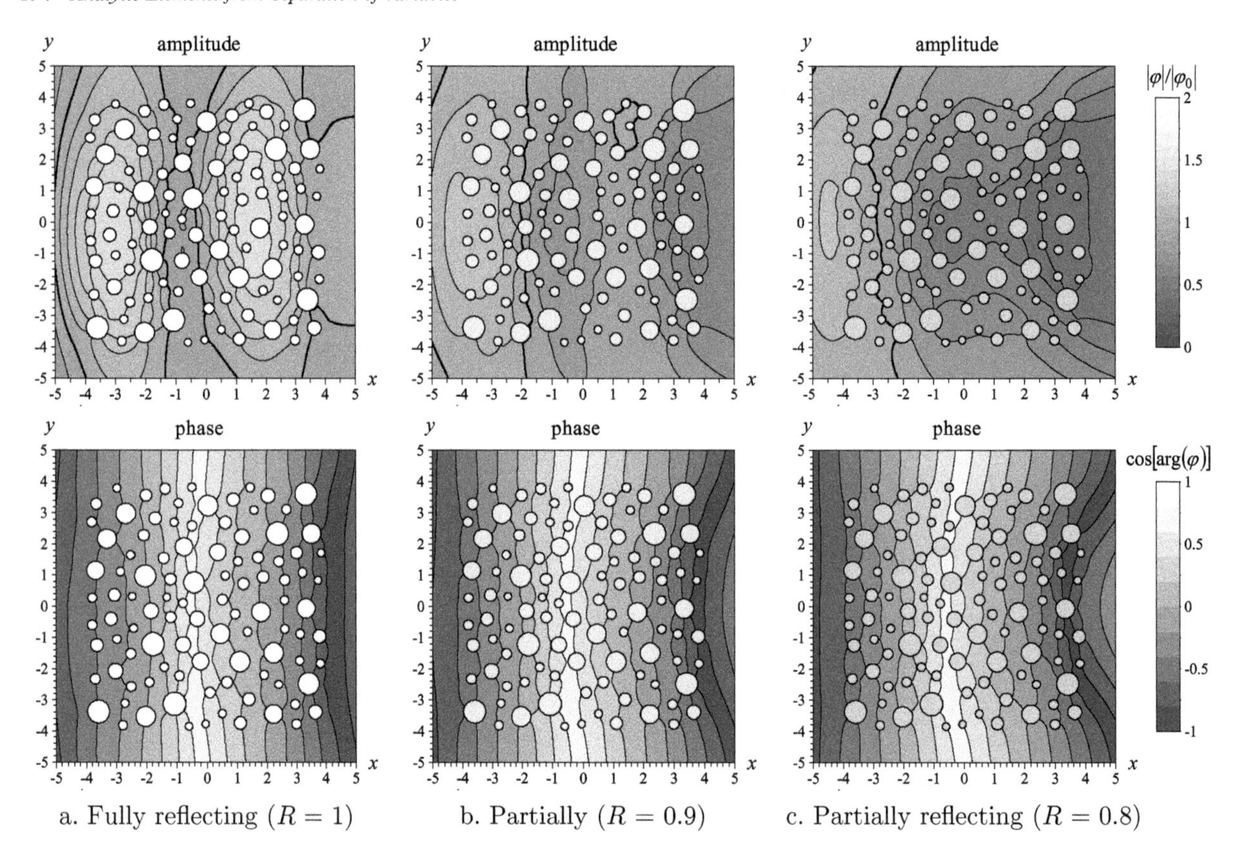

Figure 4.18 *Randomly distributed circle elements in plane waves with fully and partially reflecting boundaries. (Reprinted from Steward, 2018, Wave resonance and dissipation in collections of partially reflecting vertical cylinders,* Journal of Waterways, Ports, Coastal, and Ocean Engineering, *Vol. 144(4), Fig. 9, with permission from ASCE.)*

Example 4.2 The solution for partial reflection of plane waves by a circle element in Fig. 4.17 contains coefficients obtained from (4.65) with the additional function set equal to the plane waves in (4.9). This reproduces the exact solution given by

$$\begin{aligned}\overset{\cos}{c}_0 &= -\varphi_0 \frac{\mathcal{J}_0'(kr_0) + \mathrm{i}k\frac{1-R}{1+R}\mathcal{J}_0(kr_0)}{H_0'^{(1)}(kr_0) + \mathrm{i}k\frac{1-R}{1+R}H_0^{(1)}(kr_0)} \\ \overset{\cos}{c}_n &= -2\mathrm{i}^n\varphi_0 \frac{\mathcal{J}_n'(kr_0) + \mathrm{i}k\frac{1-R}{1+R}\mathcal{J}_n(kr_0)}{H_n'^{(1)}(kr_0) + \mathrm{i}k\frac{1-R}{1+R}H_n^{(1)}(kr_0)}, \qquad \overset{\sin}{c}_n = 0 \qquad (n > 0)\end{aligned} \tag{4.66}$$

which may be derived with use of the Jacobi–Anger expansion for plane waves

$$\varphi = \varphi_0 \mathrm{e}^{\mathrm{i}kr\cos(\theta-\theta_0)} = \varphi_0 \left[\mathcal{J}_0(kr) + 2\sum_{n=1}^{\infty} \mathrm{i}^n \mathcal{J}_n(kr)\cos n(\theta-\theta_0)\right] \tag{4.67}$$

in the Robin boundary condition.

Problem 4.11 Solve for coefficients $\overset{\cos}{c}_n$ for $n = 0 \cdots 5$ for a fully reflective circle ($R = 1$) with specified radius r_0 in a background uniform wave field in the x-direction with $\varphi_0 = 1$ and specified wave number k.

A. $r_0 = 0.5, k = \pi$
B. $r_0 = 0.5, k = \pi/2$
C. $r_0 = 0.5, k = \pi/4$
D. $r_0 = 1, k = \pi$
E. $r_0 = 1, k = \pi/2$
F. $r_0 = 1, k = \pi/4$
G. $r_0 = 2, k = \pi$
H. $r_0 = 2, k = \pi/2$
I. $r_0 = 2, k = \pi/4$
J. $r_0 = 4, k = \pi$
K. $r_0 = 4, k = \pi/2$
L. $r_0 = 4, k = \pi/4$

Problem 4.12 Solve for coefficients $\overset{\cos}{c}_n$ for $n = 0 \cdots 5$ for a partially reflective circle ($R = 0.5$) with the specified radius r_0, background uniform wave field, and wave number from Problem 4.11.

Problem 4.13 Solve for coefficients $\overset{\cos}{c}_n$ for $n = 0 \cdots 5$ for a fully absorbing circle ($R = 0$) with the specified radius r_0, background uniform wave field, and wave number from Problem 4.11.

4.4.3 Waves through Circle Elements

Analytic elements are developed next to study the propagation of waves through elements. This problem is formulated as a ***circular interface across which the medium has a different capacity for transmitting waves***, where the wave number has a value k^+ outside an element and k^- inside the element. Its solution utilizes the same expansion of Hankel functions previously used outside a circle, (4.53), with an inner expansion of complex coefficients times the Bessel functions $\mathcal{J}_n$, (4.48), which remain finite inside a circle, giving (Sommerfeld, 1972, p.108)

$$\begin{aligned}\varphi(r,\theta) &= \sum_{n=0}^{N} \overset{H\cos}{c}_n H_n^{(1)}(k^+ r)\cos n\theta \qquad (r \geq r_0)\\ &\quad + \sum_{n=1}^{N} \overset{H\sin}{c}_n H_n^{(1)}(k^+ r)\sin n\theta + \overset{\text{add}}{\varphi}\\ \varphi(r,\theta) &= \sum_{n=0}^{N} \overset{\mathcal{J}\cos}{c}_n \mathcal{J}_n(k^- r)\cos n\theta + \sum_{n=1}^{N} \overset{\mathcal{J}\sin}{c}_n \mathcal{J}_n(k^- r)\sin n\theta \quad (r < r_0)\end{aligned} \tag{4.68}$$

Note that the additional contributions $\overset{\text{add}}{\varphi}$ from the regional wave field and all other elements are only evaluated outside the element where the wave number is k^+.

Development of a solution to this problem utilizes expressions for the ***complex wave function and its partial derivative at each side of the interface*** at control point m. Evaluating the complex wave function at these points gives

$$\begin{aligned}\varphi_m^+ &= \sum_{n=0}^{N} \overset{H\cos}{c}_n H_n^{(1)}(k^+ r_0)\cos n\theta_m + \sum_{n=1}^{N} \overset{H\sin}{c}_n H_n^{(1)}(k^+ r_0)\sin n\theta_m + \overset{\text{add}}{\varphi}_m\\ \varphi_m^- &= \sum_{n=0}^{N} \overset{\mathcal{J}\cos}{c}_n \mathcal{J}_n(k^- r_0)\cos n\theta_m + \sum_{n=1}^{N} \overset{\mathcal{J}\sin}{c}_n \mathcal{J}_n(k^- r_0)\sin n\theta_m\end{aligned} \tag{4.69}$$

These functions may be separated into real and imaginary parts, similar to what was done for waves outside a circle in (4.56), to give

$$
\begin{aligned}
\varphi_m^+ = \sum_{n=0}^{N} & \left[\mathcal{J}_n(k^+ r_0) \Re \overset{H\cos}{c}_n - Y_n(k^+ r_0) \Im \overset{H\cos}{c}_n \right] \cos n\theta_m \\
& + \mathrm{i} \left[Y_n(k^+ r_0) \Re \overset{H\cos}{c}_n + \mathcal{J}_n(k^+ r_0) \Im \overset{H\cos}{c}_n \right] \cos n\theta_m \\
+ \sum_{n=1}^{N} & \left[\mathcal{J}_n(k^+ r_0) \Re \overset{H\sin}{c}_n - Y_n(k^+ r_0) \Im \overset{H\sin}{c}_n \right] \sin n\theta_m \\
& + \mathrm{i} \left[Y_n(k^+ r_0) \Re \overset{H\sin}{c}_n + \mathcal{J}_n(k^+ r_0) \Im \overset{H\sin}{c}_n \right] \sin n\theta_m + \overset{\text{add}}{\varphi}_m
\end{aligned}
\tag{4.70a}
$$

and

$$
\begin{aligned}
\varphi_m^- = & \sum_{n=0}^{N} \left[\mathcal{J}_n(k^- r_0) \Re \overset{\mathcal{J}\cos}{c}_n \right] \cos n\theta_m + \mathrm{i} \left[\mathcal{J}_n(k^- r_0) \Im \overset{\mathcal{J}\cos}{c}_n \right] \cos n\theta_m \\
& + \sum_{n=1}^{N} \left[\mathcal{J}_n(k^- r_0) \Re \overset{\mathcal{J}\sin}{c}_n \right] \sin n\theta_m + \mathrm{i} \left[\mathcal{J}_n(k^- r_0) \Im \overset{\mathcal{J}\sin}{c}_n \right] \sin n\theta_m
\end{aligned}
\tag{4.70b}
$$

The partial derivatives of these functions in the r-direction normal to the interface may be evaluated using (4.54) with (4.51) to give

$$
\begin{aligned}
\frac{\partial \varphi_m^+}{\partial r} &= \sum_{n=0}^{N} \overset{H\cos}{c}_n H_n^{\prime(1)}(k^+ r_0) \cos n\theta_m + \sum_{n=1}^{N} \overset{H\sin}{c}_n H_n^{\prime(1)}(k^+ r_0) \sin n\theta_m + \frac{\partial \overset{\text{add}}{\varphi}_m}{\partial r} \\
\frac{\partial \varphi_m^-}{\partial r} &= \sum_{n=0}^{N} \overset{\mathcal{J}\cos}{c}_n \mathcal{J}_n^{\prime}(k^- r_0) \cos n\theta_m + \sum_{n=1}^{N} \overset{\mathcal{J}\sin}{c}_n \mathcal{J}_n^{\prime}(k^- r_0) \sin n\theta_m
\end{aligned}
\tag{4.71}
$$

with the derivative of the additional function in (4.58). These complex expressions are also easily separated into real and imaginary parts, as in (4.57).

The coefficients in these influence functions are determined for wave refraction by satisfying two interface conditions for waves moving between mediums with different wave numbers. ***Conservation of energy***, (1.88), requires that a wave's amplitude and phase be continuous across the interface, which gives the ***first interface condition*** for control point m:

$$
\varphi^+ = \varphi^- \quad \rightarrow \quad \boxed{f_m^{(1)} = \varphi_m^+ - \varphi_m^- = 0}
\tag{4.72}
$$

Conservation of mass, (1.91), provides the ***second interface condition*** for control point m:

$$
\beta^+ \frac{\partial \varphi^+}{\partial r} = \beta^- \frac{\partial \varphi^-}{\partial r} \quad \rightarrow \quad \boxed{f_m^{(2)} = \frac{\partial \varphi_m^+}{\partial r} - \beta \frac{\partial \varphi_m^-}{\partial r} = 0, \quad \beta = \frac{\beta^-}{\beta^+}}
\tag{4.73}
$$

The coefficient β is related to the wave numbers, k^+ and k^-; for example, gravity waves have $\beta^+ = \frac{\sinh k^+ h^+}{k^+ \cosh k^+ h^+}$ and $\beta^- = \frac{\sinh k^- h^-}{k^- \cosh k^- h^-}$, with water depth h^+ and h^- corresponding to the two sides of the interface. Solutions for background plane waves traveling through a circle with a different wave number than the background are illustrated in Fig. 4.19.

The ***systems of equations*** that arise from evaluating these interface conditions at control points may be ***organized in matrices*** as follows. Substituting the real and imaginary parts of the complex wave function (4.70) and its normal derivative (4.71) into the conditions in (4.72) and (4.73) gives

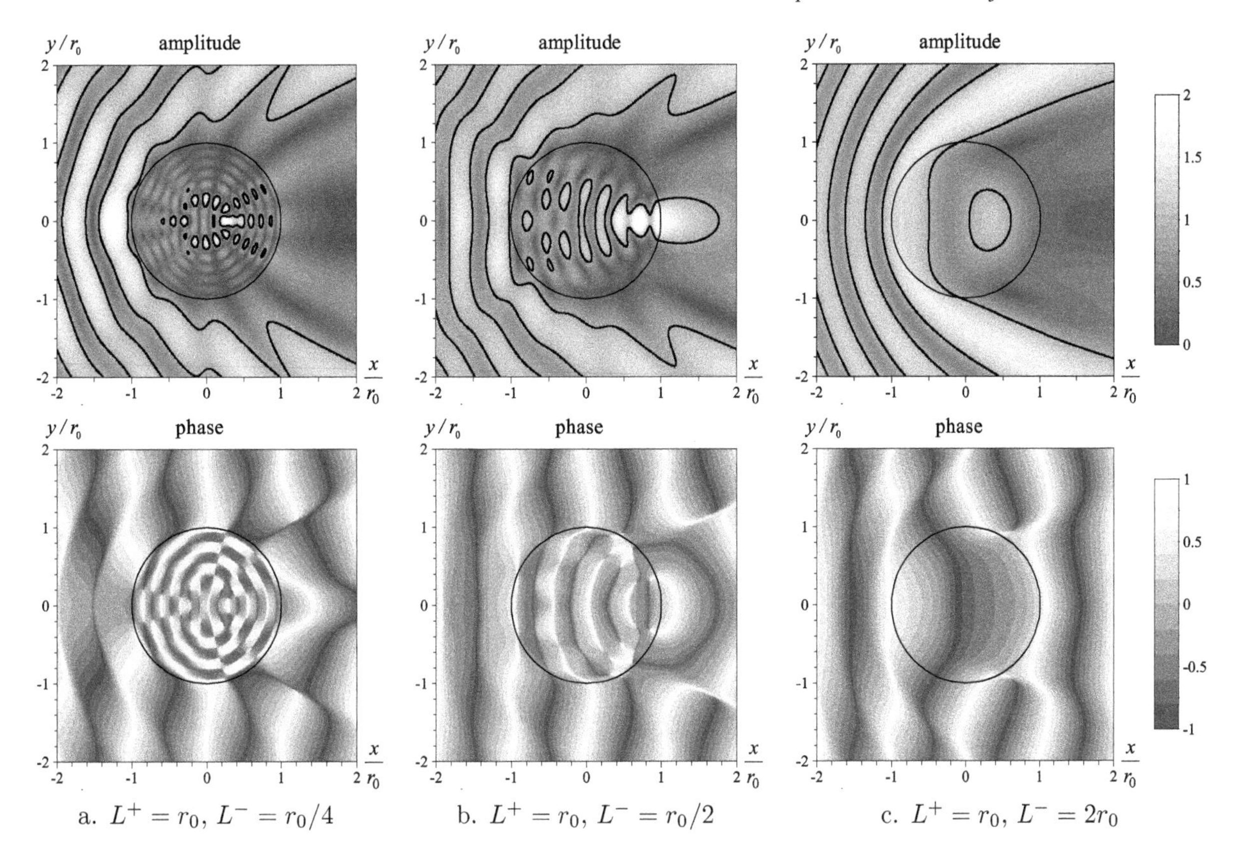

Figure 4.19 *Plane waves traveling through a circle element. (Reprinted from Steward, 2020, Waves in collections of circular shoals and bathymetric depressions,* Journal of Waterways, Ports, Coastal, and Ocean Engineering, *Vol. 146, Figs. 2, preproduction version DOI 10.1061/(ASCE)WW.1943-5460.0000570, with permission from ASCE.)*

$$\begin{matrix}\mathbf{A}\overset{\Re}{\mathbf{c}}^{(1)} = \overset{\Re}{\mathbf{b}}^{(1)} \\ \mathbf{A}\overset{\Im}{\mathbf{c}}^{(1)} = \overset{\Im}{\mathbf{b}}^{(1)}\end{matrix}, \quad \begin{matrix}\mathbf{A}\overset{\Re}{\mathbf{c}}^{(2)} = \overset{\Re}{\mathbf{b}}^{(2)} \\ \mathbf{A}\overset{\Im}{\mathbf{c}}^{(2)} = \overset{\Im}{\mathbf{b}}^{(2)}\end{matrix}, \quad \mathbf{A} = \begin{bmatrix} 1 & \cos\theta_1 & \sin\theta_1 & \cdots & \cos N\theta_1 & \sin N\theta_1 \\ 1 & \cos\theta_2 & \sin\theta_2 & \cdots & \cos N\theta_2 & \sin N\theta_2 \\ \cdots & \cdots & \cdots & \cdots & \cdots & \cdots \\ 1 & \cos\theta_M & \sin\theta_M & \cdots & \cos N\theta_M & \sin N\theta_M \end{bmatrix} \tag{4.74}$$

The coefficient and known vectors for the first condition, $f_m^{(1)}$, contain the following values for the real part

$$\overset{\Re}{\mathbf{c}}^{(1)} = \begin{bmatrix} \mathcal{J}_0(k^+r_0)\Re \overset{H\cos}{c}_0 - Y_0(k^+r_0)\Im \overset{H\cos}{c}_0 - \mathcal{J}_0(k^-r_0)\Re \overset{\mathcal{J}\cos}{c}_0 \\ \mathcal{J}_1(k^+r_0)\Re \overset{H\cos}{c}_1 - Y_1(k^+r_0)\Im \overset{H\cos}{c}_1 - \mathcal{J}_1(k^-r_0)\Re \overset{\mathcal{J}\cos}{c}_1 \\ \mathcal{J}_1(k^+r_0)\Re \overset{H\sin}{c}_1 - Y_1(k^+r_0)\Im \overset{H\sin}{c}_1 - \mathcal{J}_1(k^-r_0)\Re \overset{\mathcal{J}\sin}{c}_1 \\ \vdots \\ \mathcal{J}_N(k^+r_0)\Re \overset{H\cos}{c}_N - Y_N(k^+r_0)\Im \overset{H\cos}{c}_N - \mathcal{J}_N(k^-r_0)\Re \overset{\mathcal{J}\cos}{c}_N \\ \mathcal{J}_N(k^+r_0)\Re \overset{H\sin}{c}_N - Y_N(k^+r_0)\Im \overset{H\sin}{c}_N - \mathcal{J}_N(k^-r_0)\Re \overset{\mathcal{J}\sin}{c}_N \end{bmatrix}, \quad \overset{\Re}{\mathbf{b}}^{(1)} = \begin{bmatrix} -\Re \overset{\text{add}}{\varphi}_1 \\ -\Re \overset{\text{add}}{\varphi}_2 \\ \vdots \\ -\Re \overset{\text{add}}{\varphi}_M \end{bmatrix} \tag{4.75a}$$

and the imaginary parts

$$\overset{\Im}{\mathbf{c}}^{(1)} = \begin{bmatrix} Y_0(k^+r_0)\Re\overset{H\cos}{c}_0 + \mathcal{J}_0(k^+r_0)\Im\overset{H\cos}{c}_0 - \mathcal{J}_0(k^-r_0)\Im\overset{\mathcal{J}\cos}{c}_0 \\ Y_1(k^+r_0)\Re\overset{H\cos}{c}_1 + \mathcal{J}_1(k^+r_0)\Im\overset{H\cos}{c}_1 - \mathcal{J}_1(k^-r_0)\Im\overset{\mathcal{J}\cos}{c}_1 \\ Y_1(k^+r_0)\Re\overset{H\sin}{c}_1 + \mathcal{J}_1(k^+r_0)\Im\overset{H\sin}{c}_1 - \mathcal{J}_1(k^-r_0)\Im\overset{\mathcal{J}\sin}{c}_1 \\ \vdots \\ Y_N(k^+r_0)\Re\overset{H\cos}{c}_N + \mathcal{J}_N(k^+r_0)\Im\overset{H\cos}{c}_N - \mathcal{J}_N(k^-r_0)\Im\overset{\mathcal{J}\cos}{c}_N \\ Y_N(k^+r_0)\Re\overset{H\sin}{c}_N + \mathcal{J}_N(k^+r_0)\Im\overset{H\sin}{c}_N - \mathcal{J}_N(k^-r_0)\Im\overset{\mathcal{J}\sin}{c}_N \end{bmatrix}, \quad \overset{\Im}{\mathbf{b}}^{(1)} = \begin{bmatrix} -\Im\overset{\text{add}}{\varphi}_1 \\ -\Im\overset{\text{add}}{\varphi}_2 \\ \vdots \\ -\Im\overset{\text{add}}{\varphi}_M \end{bmatrix} \tag{4.75b}$$

Likewise, the real part of the second condition, $f_m^{(2)}$, is contained in the vectors

$$\overset{\Re}{\mathbf{c}}^{(2)} = \begin{bmatrix} \mathcal{J}_0'(k^+r_0)\Re\overset{H\cos}{c}_0 - Y_0'(k^+r_0)\Im\overset{H\cos}{c}_0 - \beta\mathcal{J}_0'(k^-r_0)\Re\overset{\mathcal{J}\cos}{c}_0 \\ \mathcal{J}_1'(k^+r_0)\Re\overset{H\cos}{c}_1 - Y_1'(k^+r_0)\Im\overset{H\cos}{c}_1 - \beta\mathcal{J}_1'(k^-r_0)\Re\overset{\mathcal{J}\cos}{c}_1 \\ \mathcal{J}_1'(k^+r_0)\Re\overset{H\sin}{c}_1 - Y_1'(k^+r_0)\Im\overset{H\sin}{c}_1 - \beta\mathcal{J}_1'(k^-r_0)\Re\overset{\mathcal{J}\sin}{c}_1 \\ \vdots \\ \mathcal{J}_N'(k^+r_0)\Re\overset{H\cos}{c}_N - Y_N'(k^+r_0)\Im\overset{H\cos}{c}_N - \beta\mathcal{J}_N'(k^-r_0)\Re\overset{\mathcal{J}\cos}{c}_N \\ \mathcal{J}_N'(k^+r_0)\Re\overset{H\sin}{c}_N - Y_N'(k^+r_0)\Im\overset{H\sin}{c}_N - \beta\mathcal{J}_N'(k^-r_0)\Re\overset{\mathcal{J}\sin}{c}_N \end{bmatrix}, \quad \overset{\Re}{\mathbf{b}}^{(2)} = \begin{bmatrix} -\Re\frac{\partial\overset{\text{add}}{\varphi}_1}{\partial r} \\ -\Re\frac{\partial\overset{\text{add}}{\varphi}_2}{\partial r} \\ \vdots \\ -\Re\frac{\partial\overset{\text{add}}{\varphi}_M}{\partial r} \end{bmatrix} \tag{4.75c}$$

and the imaginary parts are in

$$\overset{\Im}{\mathbf{c}}^{(2)} = \begin{bmatrix} Y_0'(k^+r_0)\Re\overset{H\cos}{c}_0 + \mathcal{J}_0'(k^+r_0)\Im\overset{H\cos}{c}_0 - \beta\mathcal{J}_0'(k^-r_0)\Im\overset{\mathcal{J}\cos}{c}_0 \\ Y_1'(k^+r_0)\Re\overset{H\cos}{c}_1 + \mathcal{J}_1'(k^+r_0)\Im\overset{H\cos}{c}_1 - \beta\mathcal{J}_1'(k^-r_0)\Im\overset{\mathcal{J}\cos}{c}_1 \\ Y_1'(k^+r_0)\Re\overset{H\sin}{c}_1 + \mathcal{J}_1'(k^+r_0)\Im\overset{H\sin}{c}_1 - \beta\mathcal{J}_1'(k^-r_0)\Im\overset{\mathcal{J}\sin}{c}_1 \\ \vdots \\ Y_N'(k^+r_0)\Re\overset{H\cos}{c}_N + \mathcal{J}_N'(k^+r_0)\Im\overset{H\cos}{c}_N - \beta\mathcal{J}_N'(k^-r_0)\Im\overset{\mathcal{J}\cos}{c}_N \\ Y_N'(k^+r_0)\Re\overset{H\sin}{c}_N + \mathcal{J}_N'(k^+r_0)\Im\overset{H\sin}{c}_N - \beta\mathcal{J}_N'(k^-r_0)\Im\overset{\mathcal{J}\sin}{c}_N \end{bmatrix}, \quad \overset{\Im}{\mathbf{b}}^{(2)} = \begin{bmatrix} -\Im\frac{\partial\overset{\text{add}}{\varphi}_1}{\partial r} \\ -\Im\frac{\partial\overset{\text{add}}{\varphi}_2}{\partial r} \\ \vdots \\ -\Im\frac{\partial\overset{\text{add}}{\varphi}_M}{\partial r} \end{bmatrix} \tag{4.75d}$$

This four systems of equations may be easily ***solved using least squares*** for equally spaced control points using the ***Fourier series*** methods developed in (2.47). This provides four sets of related coefficients associated with the system of equations in matrices $\mathbf{A}\overset{\Re}{\mathbf{c}}^{(1)} = \overset{\Re}{\mathbf{b}}^{(1)}$

$$\begin{aligned} \mathcal{J}_0(k^+r_0)\Re\overset{H\cos}{c}_0 - Y_0(k^+r_0)\Im\overset{H\cos}{c}_0 - \mathcal{J}_0(k^-r_0)\Re\overset{\mathcal{J}\cos}{c}_0 &= -\frac{1}{M}\sum_{m=1}^{M}\Re\left(\overset{\text{add}}{\varphi}_m\right) \\ \mathcal{J}_n(k^+r_0)\Re\overset{H\cos}{c}_n - Y_n(k^+r_0)\Im\overset{H\cos}{c}_n - \mathcal{J}_n(k^-r_0)\Re\overset{\mathcal{J}\cos}{c}_n &= -\frac{2}{M}\sum_{m=1}^{M}\Re\left(\overset{\text{add}}{\varphi}_m\right)\cos n\theta_m \\ \mathcal{J}_n(k^+r_0)\Re\overset{H\sin}{c}_n - Y_n(k^+r_0)\Im\overset{H\sin}{c}_n - \mathcal{J}_n(k^-r_0)\Re\overset{\mathcal{J}\sin}{c}_n &= -\frac{2}{M}\sum_{m=1}^{M}\Re\left(\overset{\text{add}}{\varphi}_m\right)\sin n\theta_m \end{aligned} \tag{4.76a}$$

$\mathbf{A}\overset{\Im}{\mathbf{c}}^{(1)} = \overset{\Im}{\mathbf{b}}^{(1)}$

$$
\begin{aligned}
Y_0(k^+r_0)\Re\overset{H\cos}{c}_0 + \mathcal{J}_0(k^+r_0)\Im\overset{H\cos}{c}_0 - \mathcal{J}_0(k^-r_0)\Im\overset{\mathcal{J}\cos}{c}_0 &= -\frac{1}{M}\sum_{m=1}^{M}\Im\left(\overset{\text{add}}{\varphi}_m\right)\\
Y_n(k^+r_0)\Re\overset{H\cos}{c}_n + \mathcal{J}_n(k^+r_0)\Im\overset{H\cos}{c}_n - \mathcal{J}_n(k^-r_0)\Im\overset{\mathcal{J}\cos}{c}_n &= -\frac{2}{M}\sum_{m=1}^{M}\Im\left(\overset{\text{add}}{\varphi}_m\right)\cos n\theta_m \qquad (4.76\text{b})\\
Y_n(k^+r_0)\Re\overset{H\sin}{c}_n + \mathcal{J}_n(k^+r_0)\Im\overset{H\sin}{c}_n - \mathcal{J}_n(k^-r_0)\Im\overset{\mathcal{J}\sin}{c}_n &= -\frac{2}{M}\sum_{m=1}^{M}\Im\left(\overset{\text{add}}{\varphi}_m\right)\sin n\theta_m
\end{aligned}
$$

$\mathbf{A}\overset{\Re}{\mathbf{c}}^{(2)} = \overset{\Re}{\mathbf{b}}^{(2)}$

$$
\begin{aligned}
\mathcal{J}_0'(k^+r_0)\Re\overset{H\cos}{c}_0 - Y_0'(k^+r_0)\Im\overset{H\cos}{c}_0 - \beta\mathcal{J}_0'(k^-r_0)\Re\overset{\mathcal{J}\cos}{c}_0 &= -\frac{1}{M}\sum_{m=1}^{M}\Re\left(\frac{\partial\overset{\text{add}}{\varphi}_m}{\partial r}\right)\\
\mathcal{J}_n'(k^+r_0)\Re\overset{H\cos}{c}_n - Y_n'(k^+r_0)\Im\overset{H\cos}{c}_n - \beta\mathcal{J}_n'(k^-r_0)\Re\overset{\mathcal{J}\cos}{c}_n &= -\frac{2}{M}\sum_{m=1}^{M}\Re\left(\frac{\partial\overset{\text{add}}{\varphi}_m}{\partial r}\right)\cos n\theta_m\\
\mathcal{J}_n'(k^+r_0)\Re\overset{H\sin}{c}_n - Y_n'(k^+r_0)\Im\overset{H\sin}{c}_n - \beta\mathcal{J}_n'(k^-r_0)\Re\overset{\mathcal{J}\sin}{c}_n &= -\frac{2}{M}\sum_{m=1}^{M}\Re\left(\frac{\partial\overset{\text{add}}{\varphi}_m}{\partial r}\right)\sin n\theta_m
\end{aligned}
\qquad (4.76\text{c})
$$

and $\mathbf{A}\overset{\Im}{\mathbf{c}}^{(2)} = \overset{\Im}{\mathbf{b}}^{(2)}$

$$
\begin{aligned}
Y_0'(k^+r_0)\Re\overset{H\cos}{c}_0 + \mathcal{J}_0'(k^+r_0)\Im\overset{H\cos}{c}_0 - \beta\mathcal{J}_0'(k^-r_0)\Im\overset{\mathcal{J}\cos}{c}_0 &= -\frac{1}{M}\sum_{m=1}^{M}\Im\left(\frac{\partial\overset{\text{add}}{\varphi}_m}{\partial r}\right)\\
Y_n'(k^+r_0)\Re\overset{H\cos}{c}_n + \mathcal{J}_n'(k^+r_0)\Im\overset{H\cos}{c}_n - \beta\mathcal{J}_n'(k^-r_0)\Im\overset{\mathcal{J}\cos}{c}_n &= -\frac{2}{M}\sum_{m=1}^{M}\Im\left(\frac{\partial\overset{\text{add}}{\varphi}_m}{\partial r}\right)\cos n\theta_m\\
Y_n'(k^+r_0)\Re\overset{H\sin}{c}_n + \mathcal{J}_n'(k^+r_0)\Im\overset{H\sin}{c}_n - \beta\mathcal{J}_n'(k^-r_0)\Im\overset{\mathcal{J}\sin}{c}_n &= -\frac{2}{M}\sum_{m=1}^{M}\Im\left(\frac{\partial\overset{\text{add}}{\varphi}_m}{\partial r}\right)\sin n\theta_m
\end{aligned}
\qquad (4.76\text{d})
$$

These systems of equations may be rearranged to remove the Bessel $\mathcal{J}$ terms for those equations, giving the following equations for those with a right-hand side containing real terms

$$
\begin{aligned}
\mathcal{T}_0\Re\overset{H\cos}{c}_0 - \mathcal{Y}_0\Im\overset{H\cos}{c}_0 &= -\frac{1}{M}\sum_{m=1}^{M}\Re\left[\mathcal{J}_0(k^-r_0)\frac{\partial\overset{\text{add}}{\varphi}_m}{\partial r} - \beta\mathcal{J}_0'(k^-r_0)\overset{\text{add}}{\varphi}_m\right]\\
\mathcal{T}_n\Re\overset{H\cos}{c}_n - \mathcal{Y}_n\Im\overset{H\cos}{c}_n &= -\frac{2}{M}\sum_{m=1}^{M}\Re\left[\mathcal{J}_n(k^-r_0)\frac{\partial\overset{\text{add}}{\varphi}_m}{\partial r} - \beta\mathcal{J}_n'(k^-r_0)\overset{\text{add}}{\varphi}_m\right]\cos n\theta_m \qquad (4.77\text{a})\\
\mathcal{T}_n\Re\overset{H\sin}{c}_n - \mathcal{Y}_n\Im\overset{H\sin}{c}_n &= -\frac{2}{M}\sum_{m=1}^{M}\Re\left[\mathcal{J}_n(k^-r_0)\frac{\partial\overset{\text{add}}{\varphi}_m}{\partial r} - \beta\mathcal{J}_n'(k^-r_0)\overset{\text{add}}{\varphi}_m\right]\sin n\theta_m
\end{aligned}
$$

and those containing imaginary terms

$$
\begin{aligned}
\mathcal{Y}_0 \Re \overset{H\cos}{c}_0 + \mathcal{J}_0 \Im \overset{H\cos}{c}_0 &= -\frac{1}{M}\sum_{m=1}^{M} \Im\left[\mathscr{J}_0(k^- r_0)\frac{\partial \overset{\text{add}}{\varphi}_m}{\partial r} - \beta \mathscr{J}_0'(k^- r_0)\overset{\text{add}}{\varphi}_m\right] \\
\mathcal{Y}_n \Re \overset{H\cos}{c}_n + \mathcal{J}_n \Im \overset{H\cos}{c}_n &= -\frac{2}{M}\sum_{m=1}^{M} \Im\left[\mathscr{J}_n(k^- r_0)\frac{\partial \overset{\text{add}}{\varphi}_m}{\partial r} - \beta \mathscr{J}_n'(k^- r_0)\overset{\text{add}}{\varphi}_m\right]\cos n\theta_m \\
\mathcal{Y}_n \Re \overset{H\sin}{c}_n + \mathcal{J}_n \Im \overset{H\sin}{c}_n &= -\frac{2}{M}\sum_{m=1}^{M} \Im\left[\mathscr{J}_n(k^- r_0)\frac{\partial \overset{\text{add}}{\varphi}_m}{\partial r} - \beta \mathscr{J}_n'(k^- r_0)\overset{\text{add}}{\varphi}_m\right]\sin n\theta_m
\end{aligned}
\tag{4.77b}
$$

which uses a condensed notation with

$$
\begin{aligned}
\mathcal{J}_n &= \mathscr{J}_n'(k^+ r_0)\mathscr{J}_n(k^- r_0) - \beta \mathscr{J}_n(k^+ r_0)\mathscr{J}_n'(k^- r_0) \\
\mathcal{Y}_n &= Y_n'(k^+ r_0)\mathscr{J}_n(k^- r_0) - \beta Y_n(k^+ r_0)\mathscr{J}_n'(k^- r_0)
\end{aligned}
\tag{4.78}
$$

This system of two equations with two unknowns for each n may be solved for the real and imaginary components, which are then combined to give the ***complex coefficients for the exterior influence functions***:

$$
\begin{aligned}
\overset{H\cos}{c}_0 &= \frac{1}{M}\sum_{m=1}^{M} \frac{-\mathcal{J}_0 + \mathrm{i}\mathcal{Y}_0}{\mathcal{J}_0^2 + \mathcal{Y}_0^2}\left[\mathscr{J}_0(k^- r_0)\frac{\partial \overset{\text{add}}{\varphi}_m}{\partial r} - \beta \mathscr{J}_0'(k^- r_0)\overset{\text{add}}{\varphi}_m\right] \\
\overset{H\cos}{c}_n &= \frac{2}{M}\sum_{m=1}^{M} \frac{-\mathcal{J}_n + \mathrm{i}\mathcal{Y}_n}{\mathcal{J}_n^2 + \mathcal{Y}_n^2}\left[\mathscr{J}_n(k^- r_0)\frac{\partial \overset{\text{add}}{\varphi}_m}{\partial r} - \beta \mathscr{J}_n'(k^- r_0)\overset{\text{add}}{\varphi}_m\right]\cos n\theta_m \\
\overset{H\sin}{c}_n &= \frac{2}{M}\sum_{m=1}^{M} \frac{-\mathcal{J}_n + \mathrm{i}\mathcal{Y}_n}{\mathcal{J}_n^2 + \mathcal{Y}_n^2}\left[\mathscr{J}_n(k^- r_0)\frac{\partial \overset{\text{add}}{\varphi}_m}{\partial r} - \beta \mathscr{J}_n'(k^- r_0)\overset{\text{add}}{\varphi}_m\right]\sin n\theta_m
\end{aligned}
\tag{4.79}
$$

The ***coefficients for the influence functions inside an element*** may be obtained directly from the exterior Hankel coefficients by combining the real with i times the imaginary parts of the $f^{(1)}$ conditions, (4.72) to give

$$
\begin{aligned}
\mathscr{J}_0(k^- r_0)\overset{\mathscr{J}\cos}{c}_0 &= \left[\mathscr{J}_0(k^+ r_0) + \mathrm{i}Y_0(k^+ r_0)\right]\overset{H\cos}{c}_0 + \frac{1}{M}\sum_{m=1}^{M}\overset{\text{add}}{\varphi}_m \\
\mathscr{J}_n(k^- r_0)\overset{\mathscr{J}\cos}{c}_n &= \left[\mathscr{J}_n(k^+ r_0) + \mathrm{i}Y_n(k^+ r_0)\right]\overset{H\cos}{c}_n + \frac{2}{M}\sum_{m=1}^{M}\overset{\text{add}}{\varphi}_m \cos n\theta_m \\
\mathscr{J}_n(k^- r_0)\overset{\mathscr{J}\sin}{c}_n &= \left[\mathscr{J}_n(k^+ r_0) + \mathrm{i}Y_n(k^+ r_0)\right]\overset{H\sin}{c}_n + \frac{2}{M}\sum_{m=1}^{M}\overset{\text{add}}{\varphi}_m \sin n\theta_m
\end{aligned}
\tag{4.80}
$$

Solutions that illustrate the ***wave interactions as plane waves travel through a collection of regularly spaced elements*** with different wave number than the background are shown in Fig. 4.20. The plane waves in Fig. 4.21 travel through a ***random collection of circle elements*** with a different values of internal wave number k^- than the exterior domain k^+.

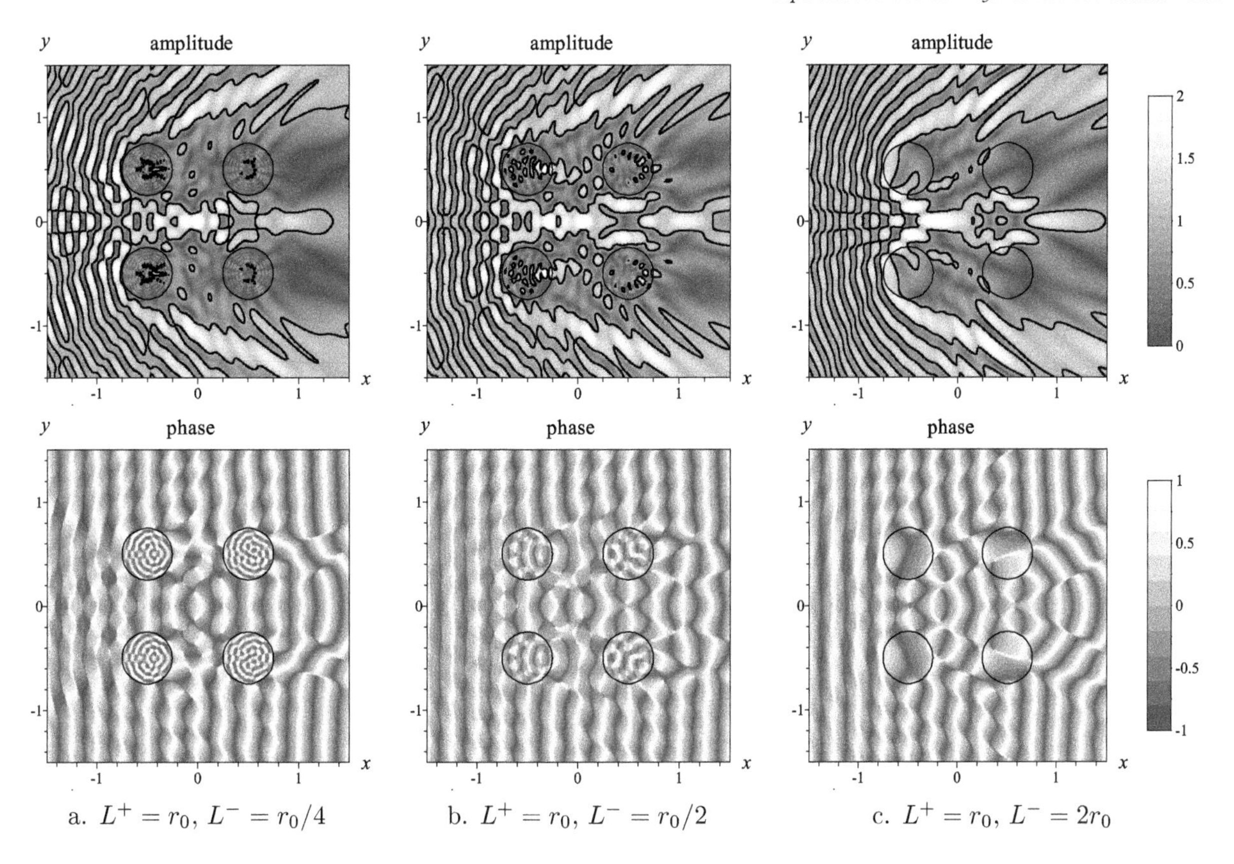

Figure 4.20 *Plane waves traveling through circle elements. (Reprinted from Steward, 2020, Waves in collections of circular shoals and bathymetric depressions,* Journal of Waterways, Ports, Coastal, and Ocean Engineering, *Vol. 146, Figs. 3, preproduction version DOI 10.1061/(ASCE)WW.1943-5460.0000570, with permission from ASCE.)*

Problem 4.14 Solve for coefficients ${}^{H}c^{\cos}_{n}$ for $n = 0 \cdots 5$ for a wave through a circle with $\beta = 1$ and specified radius r_0 in a background uniform wave field in the x-direction with $\varphi_0 = 1$ and specified wave numbers k^+ and k^-.

A. $r_0 = 0.5, k^+ = \pi/2, k^- = \pi/8$

B. $r_0 = 0.5, k^+ = \pi/2, k^- = \pi/4$

C. $r_0 = 0.5, k^+ = \pi/2, k^- = \pi$

D. $r_0 = 1, k^+ = \pi/2, k^- = \pi/8$

E. $r_0 = 1, k^+ = \pi/2, k^- = \pi/4$

F. $r_0 = 1, k^+ = \pi/2, k^- = \pi$

G. $r_0 = 2, k^+ = \pi/2, k^- = \pi/8$

H. $r_0 = 2, k^+ = \pi/2, k^- = \pi/4$

I. $r_0 = 2, k^+ = \pi/2, k^- = \pi$

J. $r_0 = 4, k^+ = \pi/2, k^- = \pi/8$

K. $r_0 = 4, k^+ = \pi/2, k^- = \pi/4$

L. $r_0 = 4, k^+ = \pi/2, k^- = \pi$

Problem 4.15 Solve for coefficients ${}^{J}c^{\cos}_{n}$ for $n = 0 \cdots 5$ for the circle element specified in Problem 4.14.

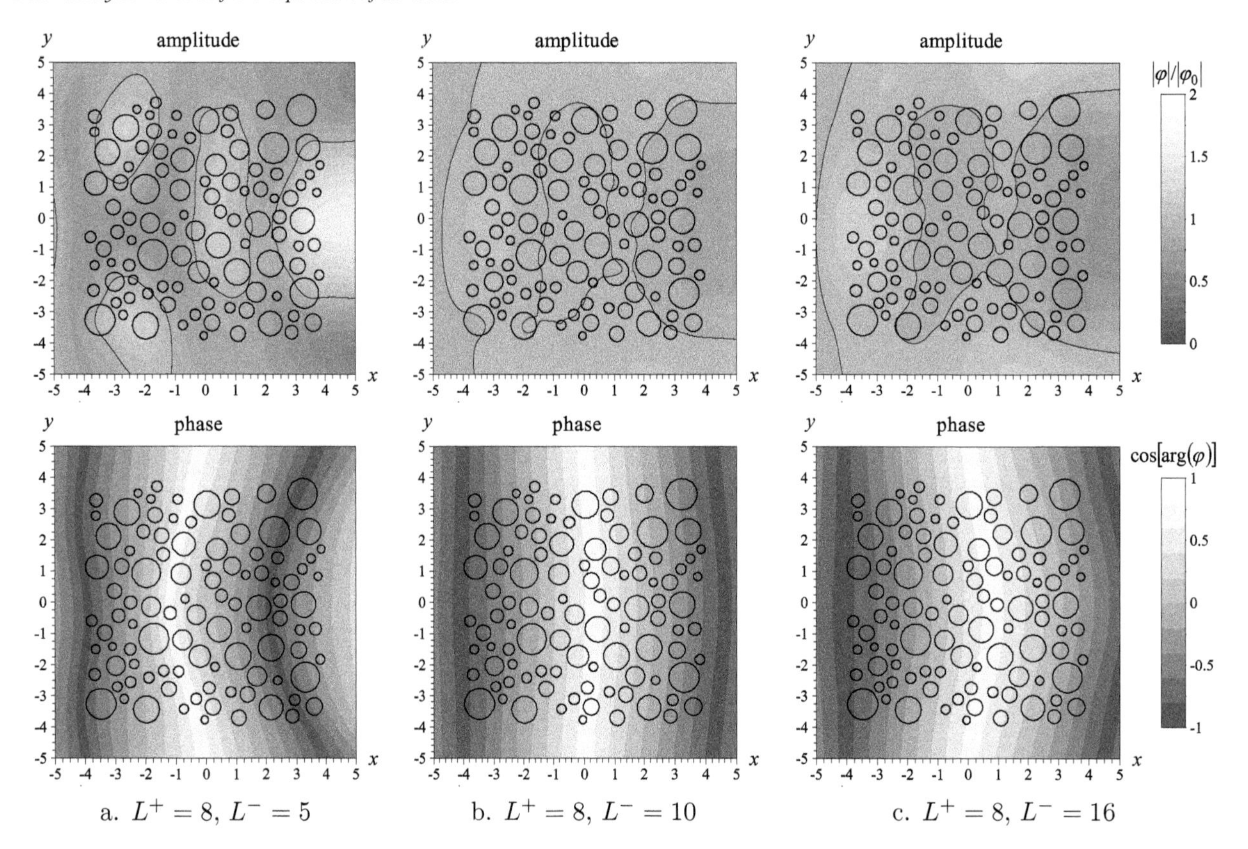

Figure 4.21 *Plane waves traveling through a random collection of circle elements across variations in wavelength L^- inside elements. (Reprinted from Steward, 2020, Waves in collections of circular shoals and bathymetric depressions,* Journal of Waterways, Ports, Coastal, and Ocean Engineering, *Vol. 146, Figs. 4, preproduction version DOI 10.1061/(ASCE)WW.1943-5460.0000570, with permission from ASCE.)*

4.4.4 Circle Elements with Modified Helmholtz Equation

The separation of variables is applied next to analytic elements that satisfy the modified Helmholtz equation, (2.2d). A potential Φ that satisfies this equation for a two-dimensional vector field is written in separated form as the product of a function of only r and a function of only θ:

$$\boxed{\Phi(r,\theta) = R(r)\Theta(\theta)}, \quad \nabla^2\Phi = \frac{\partial^2\Phi}{\partial r^2} + \frac{1}{r}\frac{\partial\Phi}{\partial r} + \frac{1}{r^2}\frac{\partial^2\Phi}{\partial\theta^2} = +k^2\Phi \tag{4.81}$$

Similar to the organization of the Helmholtz equation in (4.45), the potential function $\Phi = R\Theta$ is substituted into the modified Helmholtz equation, and terms of r and θ are separated,

$$r^2\left[\frac{1}{R}\left(\frac{\mathrm{d}^2R}{\mathrm{d}r^2} + \frac{1}{r}\frac{\mathrm{d}R}{\mathrm{d}r}\right) - k^2\right] = -\frac{1}{\Theta}\frac{\mathrm{d}^2\Theta}{\mathrm{d}\theta^2} = n^2 \tag{4.82}$$

This provides the following ordinary differential equations with particular solutions for $n = 0$

$$\begin{aligned} \frac{\mathrm{d}^2R}{\mathrm{d}r^2} + \frac{1}{r}\frac{\mathrm{d}R}{\mathrm{d}r} - k^2R = 0 \quad &\rightarrow \quad R = \left\{ I_0(kr),\ K_0(kr) \right. \\ \frac{\mathrm{d}^2\Theta}{\mathrm{d}\theta^2} = 0 \quad &\rightarrow \quad \Theta = \left\{ 1,\ \theta \right. \end{aligned} \tag{4.83}$$

and for $n^2 > 0$

$$\begin{aligned} \frac{\mathrm{d}^2R}{\mathrm{d}r^2} + \frac{1}{r}\frac{\mathrm{d}R}{\mathrm{d}r} - \left(k^2 + \frac{n^2}{r^2}\right)R = 0 \quad &\rightarrow \quad R = \left\{ I_n(kr),\ K_n(kr) \right. \\ \frac{\mathrm{d}^2\Theta}{\mathrm{d}\theta^2} + n^2\Theta = 0 \quad &\rightarrow \quad \Theta = \left\{ \cos n\theta,\ \sin n\theta \right. \end{aligned} \tag{4.84}$$

(Moon and Spencer, 1961*b*, p.16). These particular solutions are represented in terms of the ***modified Bessel functions*** of the first kind I_n and the second kind K_n, each of order n (Abramowitz and Stegun, 1972, chap.9).

The influence functions for analytic elements are formulated using linear combinations of the separated solutions, where the modified Bessel functions are evaluated in the domain (outside or inside a circle of radius r_0) where they remain finite:

$$\Phi(r,\theta) = \begin{cases} \sum_{n=0}^{N} {}^{K\cos}_{\ \ c}{}_n K_n(kr)\cos n\theta + \sum_{n=1}^{N} {}^{K\sin}_{\ \ c}{}_n K_n(kr)\sin n\theta & (r \geq r_0) \\ \sum_{n=0}^{N} {}^{I\cos}_{\ \ c}{}_n I_n(kr)\cos n\theta + \sum_{n=1}^{N} {}^{I\sin}_{\ \ c}{}_n I_n(kr)\sin n\theta & (r < r_0) \end{cases} \tag{4.85}$$

These functions are illustrated in Fig. 4.22 along with a vector equal to minus the gradient of Φ. The vector field components in the r and θ directions, (4.50), may be evaluated using

$$\begin{aligned} \frac{\partial\Phi}{\partial r} &= \begin{cases} \sum_{n=0}^{N} {}^{K\cos}_{\ \ c}{}_n K_n'(kr)\cos n\theta + \sum_{n=1}^{N} {}^{K\sin}_{\ \ c}{}_n K_n'(kr)\sin n\theta & (r \geq r_0) \\ \sum_{n=0}^{N} {}^{I\cos}_{\ \ c}{}_n I_n'(kr)\cos n\theta + \sum_{n=1}^{N} {}^{I\sin}_{\ \ c}{}_n I_n'(kr)\sin n\theta & (r < r_0) \end{cases} \\ \frac{\partial\Phi}{\partial\theta} &= \begin{cases} -\sum_{n=0}^{N} {}^{K\cos}_{\ \ c}{}_n K_n(kr) n\sin n\theta + \sum_{n=1}^{N} {}^{K\sin}_{\ \ c}{}_n K_n(kr) n\cos n\theta & (r \geq r_0) \\ -\sum_{n=0}^{N} {}^{I\cos}_{\ \ c}{}_n I_n(kr) n\sin n\theta + \sum_{n=1}^{N} {}^{I\sin}_{\ \ c}{}_n I_n(kr) n\cos n\theta & (r < r_0) \end{cases} \end{aligned} \tag{4.86}$$

where the partial derivatives with respect to r are given by

$$I_n'(kr) = \frac{n}{r}I_n(kr) + kI_{n+1}(kr), \qquad K_n'(kr) = \frac{n}{r}K_n(kr) - kK_{n+1}(kr) \tag{4.87}$$

a similar recursive relation for Bessel functions in (4.51).

Non-linear Solutions for Vadose Zone Studies

The methods necessary for solving a non-linear interface condition are developed for the problem of ***seepage through heterogeneity in the vadose zone***, which is governed by the modified Helmholtz equation. This problem was illustrated in Fig. 1.19, where an inhomogeneity of radius r_0 is placed in the x–z plane with the z-axis oriented against gravity.

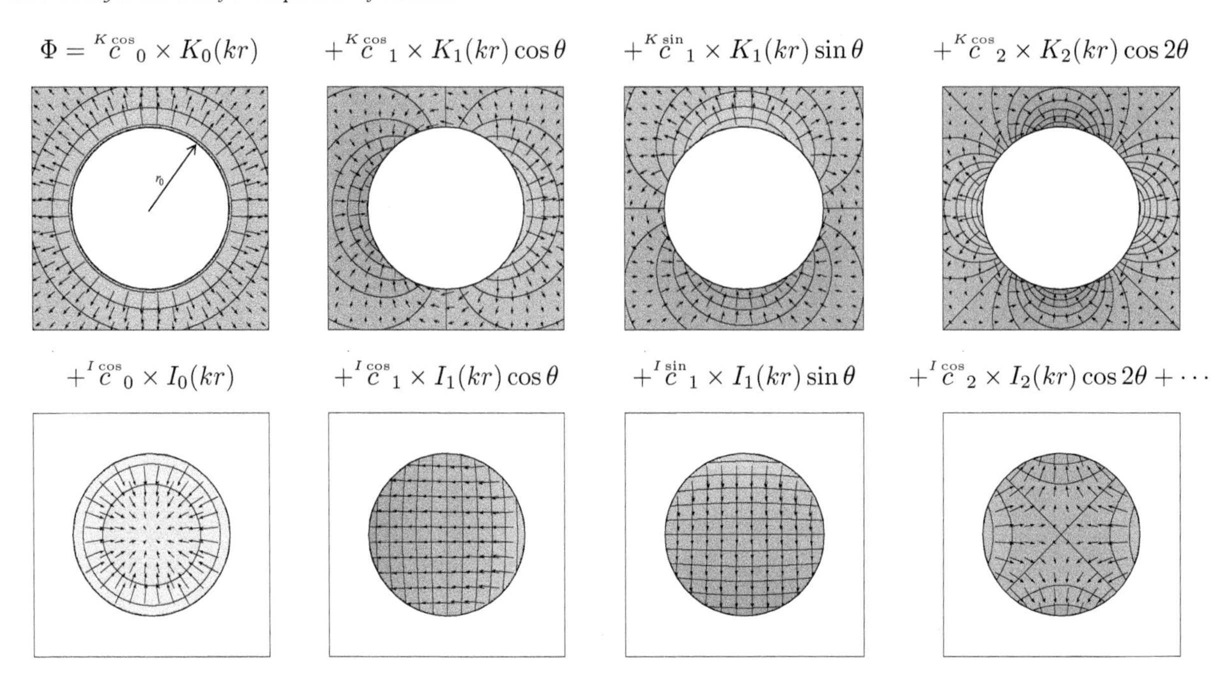

Figure 4.22 *Influence functions for circular domains using modified Bessel functions, from (4.85). (Reprinted from Steward, 2016, Analysis of vadose zone inhomogeneity toward distinguishing recharge rates: Solving the nonlinear interface problem with Newton method,* Water Resources Research, *Vol. 52(11), Fig. 2, with permission from John Wiley and Sons. Copyright 2016 by the American Geophysical Union.)*

The coefficient k in the modified Helmholtz equation has a value of k^+ outside the element and k^- inside the element, and the ***influence functions for a circular element*** from (4.85) takes on the form of

$$\begin{aligned}
\Phi &= \sum_{n=0}^{N} {}^{K\cos}c_n K_n(k^+ r)\cos n\theta + \sum_{n=1}^{N} {}^{K\sin}c_n K_n(k^+ r)\sin n\theta + \overset{\text{add}}{\Phi} \qquad (r \geq r_0) \\
\Phi &= \sum_{n=0}^{N} {}^{I\cos}c_n I_n(k^- r)\cos n\theta + \sum_{n=1}^{N} {}^{I\sin}c_n I_n(k^- r)\sin n\theta + {}^{1D}c_1 e^{-k^- z} + {}^{1D}c_2 e^{k^- z} \qquad (r < r_0)
\end{aligned} \tag{4.88}$$

Note that the additional contributions $\overset{\text{add}}{\Phi}$ contain the regional background, (4.18), with all other elements, and it is only evaluated outside the element where the Helmholtz coefficient is k^+. These influence functions also contain terms for steady seepage, (4.18), within the element, following Steward (2016).

Two interface conditions exist for heterogeneity in the vadose zone. ***Conservation of energy*** requires a continuous pressure across the interface, (1.43):

$$\begin{aligned}
&p^+ = p^-, \quad p^+ = \frac{1}{2k^+}\ln\frac{e^{-k^+ z}\Phi^+}{F_s^{\,+}}, \quad p^- = \frac{1}{2k^-}\ln\frac{e^{-k^- z}\Phi^-}{F_s^{\,-}} \\
&\rightarrow \quad \left(\frac{e^{-k^+ z}}{F_s^{\,+}}\Phi^+\right)^{\frac{1}{2k^+}} = \left(\frac{e^{-k^- z}}{F_s^{\,-}}\Phi^-\right)^{\frac{1}{2k^-}}
\end{aligned} \tag{4.89}$$

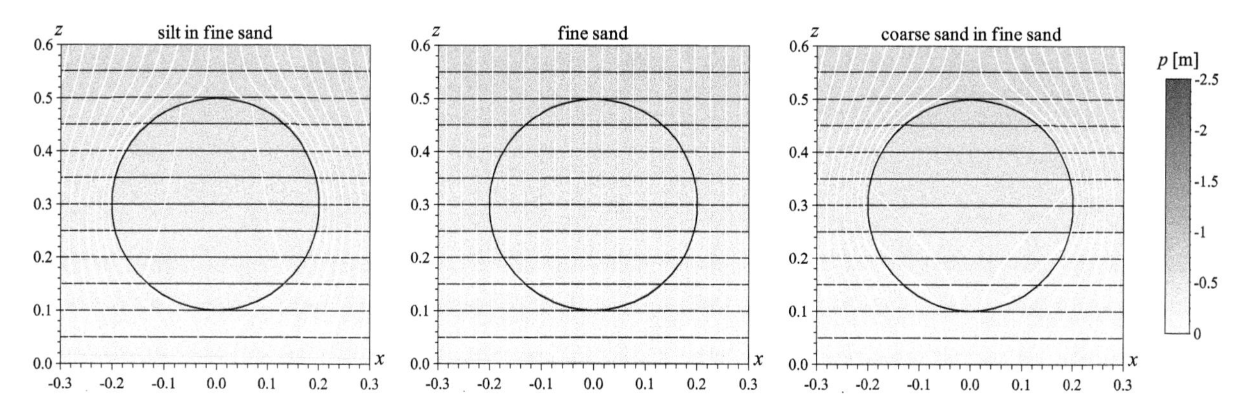

Figure 4.23 *Modified Bessel functions used for a circular heterogeneity. (Reprinted from Steward, 2016, Analysis of vadose zone inhomogeneity toward distinguishing recharge rates: Solving the nonlinear interface problem with Newton method,* Water Resources Research, *Vol. 52(11), Fig. 4, with permission from John Wiley and Sons. Copyright 2016 by the American Geophysical Union.)*

This results in a ***non-linear interface condition in terms of the potential*** at control point m:

$$\boxed{f_m^{(1)} = \left(\frac{e^{-k^+ z_m}}{F_s^{\,+}}\Phi_m^+\right)^{\gamma^+} - \left(\frac{e^{-k^- z_m}}{F_s^{\,-}}\Phi_m^-\right)^{\gamma^-} = 0} \qquad \begin{cases}\gamma^+ = \frac{k^-}{k^+},\ \gamma^- = 1 & (k^- \le k^+)\\ \gamma^+ = 1,\ \gamma^- = \frac{k^+}{k^-} & (k^- > k^+)\end{cases} \tag{4.90}$$

and the constants γ_+ and γ^- are set to have values less than one. ***Conservation of mass*** requires a continuous normal component of the specific discharge vector, (1.44), across the circular interface:

$$q_r^{\,+} = q_r^{\,-}, \qquad \begin{aligned} q_r^{\,+} &= -e^{-k^+ z}\left[(k^+ \sin\theta)\,\Phi^+ + \frac{\partial \Phi^+}{\partial r}\right] \\ q_r^{\,-} &= -e^{-k^- z}\left[(k^- \sin\theta)\,\Phi^- + \frac{\partial \Phi^-}{\partial r}\right] \end{aligned} \tag{4.91}$$

which gives a ***Robin continuity condition***, (2.64d), at each control point

$$\boxed{f_m^{(2)} = -e^{-k^+ z_m}\left[(k^+ \sin\theta_m)\,\Phi_m^+ + \frac{\partial \Phi_m^+}{\partial r}\right] + e^{-k^- z_m}\left[(k^- \sin\theta_m)\,\Phi_m^- + \frac{\partial \Phi_m^-}{\partial r}\right] = 0} \tag{4.92}$$

Solving these two interface conditions requires evaluation of the potential, (4.88), at control point m along the circular boundary outside and inside the element:

$$\begin{aligned} \Phi_m^+ &= \sum_{n=0}^{N} {}^{K\cos}\!c_n K_n(k^+ r_0)\cos n\theta_m + \sum_{n=1}^{N} {}^{K\sin}\!c_n K_n(k^+ r_0)\sin n\theta_m + \overset{\text{add}}{\Phi}_m \\ \Phi_m^- &= \sum_{n=0}^{N} {}^{I\cos}\!c_n I_n(k^- r_0)\cos n\theta_m + \sum_{n=1}^{N} {}^{I\sin}\!c_n I_n(k^- r_0)\sin n\theta_m + {}^{1D}\!c_1 e^{-k^- z_m} + {}^{1D}\!c_2 e^{k^- z_m} \end{aligned} \tag{4.93}$$

The normal derivative at the control points may be obtained by evaluating

$$\frac{\partial \Phi_m^+}{\partial r} = \sum_{n=0}^{N} {}^{K\cos}_{\;\;c\;\;n} K_n'(k^+ r_0) \cos n\theta_m + \sum_{n=1}^{N} {}^{K\sin}_{\;\;c\;\;n} K_n'(k^+ r_0) \sin n\theta_m + \frac{\partial \overset{\text{add}}{\Phi}_m}{\partial r}$$
$$\frac{\partial \Phi_m^-}{\partial r} = \sum_{n=0}^{N} {}^{I\cos}_{\;\;c\;\;n} I_n'(k^- r_0) \cos n\theta_m + \sum_{n=1}^{N} {}^{I\sin}_{\;\;c\;\;n} I_n'(k^- r_0) \sin n\theta_m - {}^{1D}_{\;c\;1} k^- \sin\theta_m e^{-k^- z_m} + {}^{1D}_{\;c\;2} k^- \sin\theta_m e^{k^- z_m} \tag{4.94}$$

and relations between components of the gradient in x–z and r–θ components are articulated in (4.86).

Newton's method, (2.56), may be used to ***solve the non-linear condition*** $f_m^{(1)}$, (4.90), with $f_m^{(2)}$, (4.92), for each analytic element. This iterative method is organized in matrices associated with the Jacobian $\mathbf{J}$, coefficients $\mathbf{c}$, and functions $\mathbf{f}$ for iterate l:

$$\mathbf{J}|_l \, (\Delta \mathbf{c}) \,|_l = -\mathbf{f}|_l \tag{4.95}$$

with

$$\mathbf{J} = \begin{bmatrix} \frac{\partial f_m^{(1)}}{\partial {}^{K\cos}_{\;c\;n}} & \frac{\partial f_m^{(1)}}{\partial {}^{K\sin}_{\;c\;n}} & \frac{\partial f_m^{(1)}}{\partial {}^{I\cos}_{\;c\;n}} & \frac{\partial f_m^{(1)}}{\partial {}^{I\sin}_{\;c\;n}} & \frac{\partial f_m^{(1)}}{\partial {}^{1D}_{\;c\;1}} & \frac{\partial f_m^{(1)}}{\partial {}^{1D}_{\;c\;2}} \\ \frac{\partial f_m^{(2)}}{\partial {}^{K\cos}_{\;c\;n}} & \frac{\partial f_m^{(2)}}{\partial {}^{K\sin}_{\;c\;n}} & \frac{\partial f_m^{(2)}}{\partial {}^{I\cos}_{\;c\;n}} & \frac{\partial f_m^{(2)}}{\partial {}^{I\sin}_{\;c\;n}} & \frac{\partial f_m^{(2)}}{\partial {}^{1D}_{\;c\;1}} & \frac{\partial f_m^{(2)}}{\partial {}^{1D}_{\;c\;2}} \end{bmatrix}, \quad \mathbf{c} = \begin{bmatrix} {}^{K\cos}_{\;c\;n} \\ {}^{K\sin}_{\;c\;n} \\ {}^{I\cos}_{\;c\;n} \\ {}^{I\sin}_{\;c\;n} \\ {}^{1D}_{\;c\;1} \\ {}^{1D}_{\;c\;2} \end{bmatrix}, \quad \mathbf{f} = \begin{bmatrix} f_m^{(1)} \\ f_m^{(2)} \end{bmatrix} \tag{4.96}$$

This system of $2M$ equations with $4N + 2$ unknowns is solved for each iterate using least squares. The ***first iterate*** $l = 1$ utilizes initial estimates for the coefficients given by

$$\left. {}^{K\cos}_{\;c\;n} \right|_1 = \left. {}^{K\sin}_{\;c\;n} \right|_1 = \left. {}^{I\cos}_{\;c\;n} \right|_1 = \left. {}^{I\sin}_{\;c\;n} \right|_1 = 0$$
$$\begin{bmatrix} e^{-k^- z_m} & e^{k^- z_m} \end{bmatrix} \begin{bmatrix} \left. {}^{1D}_{\;c\;1} \right|_1 \\ \left. {}^{1D}_{\;c\;2} \right|_1 \end{bmatrix} = \begin{bmatrix} (F_s^-)^{\frac{1}{2k^-}} \left(\frac{\overset{\text{add}}{\Phi}_m}{F_s^+} \right)^{\frac{1}{2k^+}} \end{bmatrix} \tag{4.97}$$

The ***method of successive over relaxation*** is utilized to ensure that the pressure head remains within physically realistic limits during the iteration process. This is achieved using

$$\mathbf{c}|_{l+1} = \mathbf{c}|_l + SOR \, \Delta \mathbf{c}|_l \tag{4.98}$$

with $SOR = 0.1$.

The partial derivatives in the Jacobian matrix, (4.96), are necessary to achieve an iterative solution

$$
\begin{aligned}
\frac{\partial f_m^{(1)}}{\partial\, {}^{K\cos}c_n} &= \left(\frac{\mathrm{e}^{-k^+ z_m}}{F_s^+}\right)^{\gamma^+} \frac{\gamma^+ K_n(k^+ r_0)\cos n\theta_m}{(\Phi_m^+)^{1-\gamma^+}} \\
\frac{\partial f_m^{(2)}}{\partial\, {}^{K\cos}c_n} &= -\mathrm{e}^{-k^+ z_m}\left[k^+ \sin\theta_m K_n(k^+ r_0) + K_n'(k^+ r_0)\right]\cos n\theta_m \\
\frac{\partial f_m^{(1)}}{\partial\, {}^{K\sin}c_n} &= \left(\frac{\mathrm{e}^{-k^+ z_m}}{F_s^+}\right)^{\gamma^+} \frac{\gamma^+ K_n(k^+ r_0)\sin n\theta_m}{(\Phi_m^+)^{1-\gamma^+}} \\
\frac{\partial f_m^{(2)}}{\partial\, {}^{K\sin}c_n} &= -\mathrm{e}^{-k^+ z_m}\left[k^+ \sin\theta_m K_n(k^+ r_0) + K_n'(k^+ r_0)\right]\sin n\theta_m \\
\frac{\partial f_m^{(1)}}{\partial\, {}^{I\cos}c_n} &= -\left(\frac{\mathrm{e}^{-k^- z_m}}{F_s^-}\right)^{\gamma^-} \frac{\gamma^- I_n(k^- r_0)\cos n\theta_m}{(\Phi_m^-)^{1-\gamma^-}} \\
\frac{\partial f_m^{(2)}}{\partial\, {}^{I\cos}c_n} &= \mathrm{e}^{-k^- z_m}\left[k^- \sin\theta_m I_n(k^- r_0) + I_n'(k^- r_0)\right]\cos n\theta_m \\
\frac{\partial f_m^{(1)}}{\partial\, {}^{I\sin}c_n} &= -\left(\frac{\mathrm{e}^{-k^- z_m}}{F_s^-}\right)^{\gamma^-} \frac{\gamma^- I_n(k^- r_0)\sin n\theta_m}{(\Phi_m^-)^{1-\gamma^-}} \\
\frac{\partial f_m^{(2)}}{\partial\, {}^{I\sin}c_n} &= \mathrm{e}^{-k^- z_m}\left[k^- \sin\theta_m I_n(k^- r_0) + I_n'(k^- r_0)\right]\sin n\theta_m \\
\frac{\partial f_m^{(1)}}{\partial\, {}^{1D}c_1} &= -\left(\frac{\mathrm{e}^{-k^- z_m}}{F_s^-}\right)^{\gamma^-} \frac{\gamma^- \mathrm{e}^{-k^- z_m}}{(\Phi_m^-)^{1-\gamma^-}}, \quad \frac{\partial f_m^{(2)}}{\partial\, {}^{1D}c_1} = 0 \\
\frac{\partial f_m^{(1)}}{\partial\, {}^{1D}c_2} &= -\left(\frac{\mathrm{e}^{-k^- z_m}}{F_s^-}\right)^{\gamma^-} \frac{\gamma^- \mathrm{e}^{k^- z_m}}{(\Phi_m^-)^{1-\gamma^-}}, \quad \frac{\partial f_m^{(2)}}{\partial\, {}^{1D}c_2} = 2k^- \sin\theta_m
\end{aligned}
\tag{4.99}
$$

The values of the potential inside and outside the element, Φ_m^+ and Φ_m^-, in these functions are adjusted to lie within the following bounds:

$$
p \in (-\infty, 0] \quad \rightarrow \quad \begin{cases} \Phi_m^+ \in (0, F_s^+ \mathrm{e}^{k^+ z_m}] \\ \Phi_m^- \in (0, F_s^- \mathrm{e}^{k^- z_m}] \end{cases} \tag{4.100}
$$

The solutions in Fig. 4.23 illustrate vadose zone seepage through an analytic element, utilizing soil properties from Section 1.2.2.

Problem 4.16 Solve for coefficients ${}^{1D}c_1$, ${}^{1D}c_2$, ${}^{K\cos}c_n$, and ${}^{I\cos}c_n$ for $n = 0 \cdots 5$, and ${}^{K\sin}c_n$ and ${}^{I\sin}c_n$ for $n = 1 \cdots 5$ using the soil properties for the vadose zone in Table 1.1. Each element has radius $r_0 = 0.2m$ and located elevation $z = 0$. It is placed in a background uniform seepage R through fine sand with potential, (4.18), and background coefficients, c_1 and c_2 in (4.20) using $p_0 = 0$ and $z_0 = -1m$.

A. silt, $R = 0.025$ m/yr
B. silt, $R = 0.05$ m/yr
C. silt, $R = 0.075$ m/yr
D. silt, $R = 0.1$ m/yr
E. silt, $R = 0.125$ m/yr
F. silt, $R = 0.15$ m/yr
G. coarse sand, $R = 0.025$ m/yr
H. coarse sand, $R = 0.05$ m/yr
I. coarse sand, $R = 0.075$ m/yr
J. coarse sand, $R = 0.1$ m/yr
K. coarse sand, $R = 0.125$ m/yr
L. coarse sand, $R = 0.15$ m/yr

Additional details are necessary to achieve solutions for the complicated interactions of many elements, such as shown in Fig. 4.24. Methods utilized by Steward (2016, App.A3,A4)

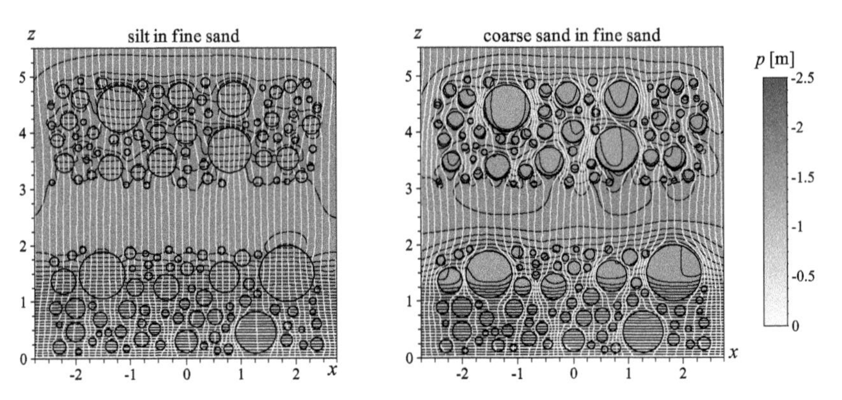

Figure 4.24 *Modified Bessel functions used for circular heterogeneities. (Reprinted from Steward, 2016, Analysis of vadose zone inhomogeneity toward distinguishing recharge rates: Solving the nonlinear interface problem with Newton method,* Water Resources Research, *Vol. 52(11), Fig. 5, with permission from John Wiley and Sons. Copyright 2016 by the American Geophysical Union.)*

for solving such problems include weighted least squares, preconditioning the Jacobian to normalize the system of equation, and precomputing influence functions at control points, (2.12).

4.5 Separation in Spherical Coordinates

Separation of variables is applied next for three-dimensional problems associated with the spherical coordinate system in Fig. 4.25. ***Coordinate transformation*** are specified such that

$$\left.\begin{aligned} r^2 &= x^2+y^2+z^2 \\ \xi &= \arctan\frac{\sqrt{x^2+y^2}}{z} \\ \theta &= \arctan\frac{y}{x} \end{aligned}\right\} \quad \leftrightarrow \quad \left\{\begin{aligned} x &= r\sin\xi\cos\theta \\ y &= r\sin\xi\sin\theta \\ z &= r\cos\xi \end{aligned}\right. \tag{4.101}$$

The ***coordinate unit vectors*** in Fig. 4.25 are directed towards the gradient of coordinate surfaces, with Cartesian components

$$\hat{\mathbf{r}} = \frac{\nabla r}{|\nabla r|}, \qquad \nabla r = \frac{x}{r}\hat{\mathbf{x}} + \frac{y}{r}\hat{\mathbf{y}} + \frac{z}{r}\hat{\mathbf{z}} \tag{4.102a}$$

$$\hat{\xi} = \frac{\nabla\xi}{|\nabla\xi|}, \qquad \nabla\xi = \frac{xz}{r^2\sqrt{x^2+y^2}}\hat{\mathbf{x}} + \frac{yz}{r^2\sqrt{x^2+y^2}}\hat{\mathbf{y}} - \frac{\sqrt{x^2+y^2}}{r^2}\hat{\mathbf{z}} \tag{4.102b}$$

$$\hat{\theta} = \frac{\nabla\theta}{|\nabla\theta|}, \qquad \nabla\theta = \frac{-y}{x^2+y^2}\hat{\mathbf{x}} + \frac{x}{x^2+y^2}\hat{\mathbf{y}} \tag{4.102c}$$

Problems are formulated using the ***Laplacian and gradient in spherical coordinates*** in terms of spherical coordinate directions, (4.2a),

$$\boxed{\nabla^2\Phi = \frac{\partial^2\Phi}{\partial r^2} + \frac{2}{r}\frac{\partial\Phi}{\partial r} + \frac{1}{r^2}\frac{\partial^2\Phi}{\partial\xi^2} + \frac{\cot\xi}{r^2}\frac{\partial\Phi}{\partial\xi} + \frac{1}{r^2\sin^2\xi}\frac{\partial^2\Phi}{\partial\theta^2}}, \qquad \boxed{\nabla\Phi = \frac{\partial\Phi}{\partial r}\hat{\mathbf{r}} + \frac{1}{r}\frac{\partial\Phi}{\partial\xi}\hat{\boldsymbol{\xi}} + \frac{1}{r\sin\xi}\frac{\partial\Phi}{\partial\theta}\hat{\boldsymbol{\theta}}} \tag{4.103}$$

Solutions are developed next for three-dimensional waves and for three-dimensional heterogeneity.

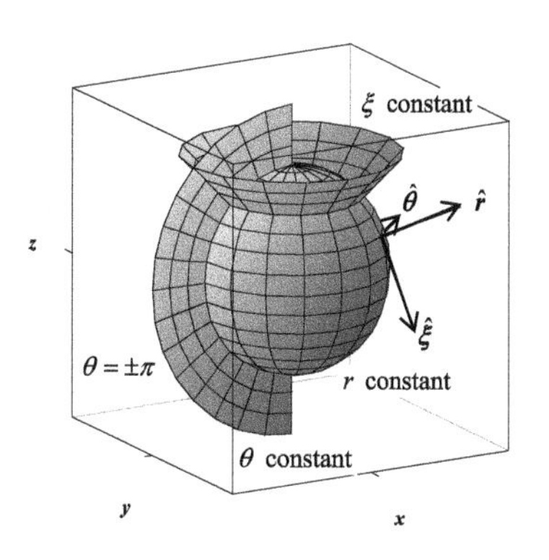

a. Three-dimensional domain

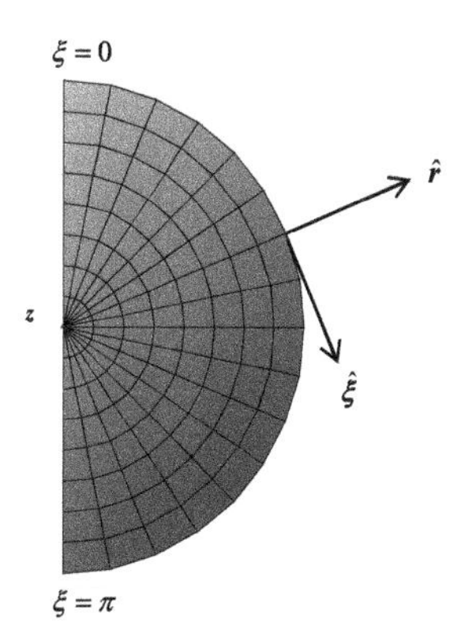

b. Half-plane with constant θ

Figure 4.25 *Spherical coordinates (r, ξ, θ): with spheres (r constant), circular cones (ξ constant), and half-planes (θ constant).*

4.5.1 Radial Waves Emanating from a Point

Methods are applied first to three-dimensional waves that satisfy the ***Helmholtz equation***. A separable solution for radial waves emanating from a point is represented in terms of spherical Bessel functions of the first and second kind (Moon and Spencer, 1961*b*, p.27):

$$\frac{\mathrm{d}^2R}{\mathrm{d}r^2} + \frac{2}{r}\frac{\mathrm{d}R}{\mathrm{d}r} + k^2R = 0 \quad \rightarrow \quad R = \left\{ j_0(kr),\ y_0(kr) \right. \tag{4.104}$$

These particular solutions may be gathered in a ***spherical Bessel function of the third kind*** that satisfies the Sommerfeld radiation condition

$$\varphi = \frac{S}{4\pi} h_0^{(1)}(kr) \tag{4.105}$$

Note that the spherical Bessel function is presented in more detail later, and may be obtained directly from the circular Bessel functions using (4.111). This represents a ***three-dimensional point-source***, and the amplitude and phase are illustrated in Fig. 4.26a. This solution is easily adapted to wave reflection of a point-source near a fully reflective boundary and $x = 0$ using the ***method of images***

$$\begin{aligned} \varphi &= \frac{S}{2\pi} h_0^{(1)}\left(k\sqrt{(x-d)^2 + y^2}\right) + \frac{S}{2\pi} h_0^{(1)}\left(k\sqrt{(x+d)^2 + y^2}\right) \\ &\rightarrow \quad \frac{\partial \varphi}{\partial \mathrm{n}} = 0 \quad (x = 0) \end{aligned} \tag{4.106}$$

The solution in Fig. 4.26b illustrates the partially standing waves occurring between the point-source and the ***fully reflecting boundary***.

Problem 4.17 Compute the amplitude, $|\varphi|$, and phase, $\arg\varphi$, from the complex wave function $\varphi = |\varphi|\mathrm{e}^{\mathrm{i}\arg\varphi}$ for the wave field of a point-source with $S = 1$ and the specified wave number k at locations: r=1, 2, 5, and 10.

A. $k = \frac{2\pi}{1}$ C. $k = \frac{2\pi}{3}$ E. $k = \frac{2\pi}{5}$ G. $k = \frac{2\pi}{7}$ I. $k = \frac{2\pi}{9}$ K. $k = \frac{2\pi}{11}$

B. $k = \frac{2\pi}{2}$ D. $k = \frac{2\pi}{4}$ F. $k = \frac{2\pi}{6}$ H. $k = \frac{2\pi}{8}$ J. $k = \frac{2\pi}{10}$ L. $k = \frac{2\pi}{12}$

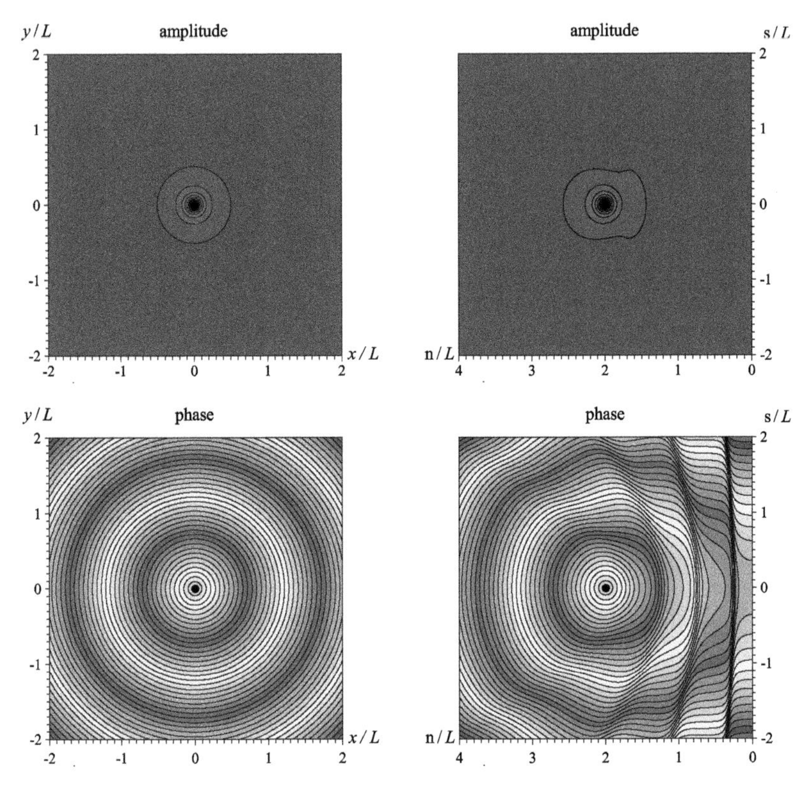

Figure 4.26 *Radial waves in spherical coordinates.*

4.5.2 Three-Dimensional Solutions for Spherical Objects

Separation of variables with $\Phi(r,\xi,\theta) = R(r)\Xi(\xi)\Theta(\theta)$ leads to ordinary differential equations for the ***Laplace equation*** with particular solutions (Moon and Spencer, 1961*b*, p. 26):

$$\begin{aligned} \frac{d^2R}{dr^2} + \frac{2}{r}\frac{dR}{dr} - \frac{n_1(n_1+1)}{r^2}R = 0 \quad &\rightarrow \quad R = \left\{ r^{-(n_1+1)},\ r^{n_1} \right. \\ \frac{d^2\Xi}{d\xi^2} + \cot\xi\frac{d\Xi}{d\xi} + \left[n_1(n_1+1) - \frac{(n_2)^2}{\sin^2\xi} \right]\Xi = 0 \quad &\rightarrow \quad \Xi = \left\{ P_{n_1}^{n_2}(\cos\xi),\ Q_{n_1}^{n_2}(\cos\xi) \right. \\ \frac{d^2\Theta}{d\theta^2} + (n_2)^2\Theta = 0 \quad &\rightarrow \quad \Theta = \left\{ \cos n_2\theta,\ \sin n_2\theta \right. \end{aligned} \tag{4.107}$$

where P_n^m and Q_n^m are the ***associated Legendre functions*** of integer order m and degree n. The ***influence functions for analytic elements of spherical geometry*** are obtained using linear superposition of the functions that remain finite outside the sphere

$$\boxed{\begin{aligned}\Phi(r,\xi,\theta) = &\sum_{n_1=1}^{N_1}\sum_{n_2=0}^{n_1} \overset{\text{out}}{\underset{\cos}{c}}_{n_1 n_2} \left(\frac{r}{r_0}\right)^{-(n_1+1)} P_{n_1}^{n_2}(\cos\xi)\cos n_2\theta \\ &+ \sum_{n_1=1}^{N_1}\sum_{n_2=1}^{n_1} \overset{\text{out}}{\underset{\sin}{c}}_{n_1 n_2} \left(\frac{r}{r_0}\right)^{-(n_1+1)} P_{n_1}^{n_2}(\cos\xi)\sin n_2\theta + \overset{\text{add}}{\Phi}(r,\xi,\theta)\end{aligned} \quad (r \ge r_0)} \tag{4.108a}$$

and the functions that remain finite inside the sphere (Moon and Spencer, 1961*a*, p.232)

$$\boxed{\begin{aligned}\Phi(r,\xi,\theta) = \overset{\text{in}}{c}_0 + &\sum_{n_1=1}^{N_1}\sum_{n_2=0}^{n_1} \overset{\text{in}}{\underset{\cos}{c}}_{n_1 n_2} \left(\frac{r}{r_0}\right)^{n_1} P_{n_1}^{n_2}(\cos\xi)\cos n_2\theta \\ &+ \sum_{n_1=1}^{N_1}\sum_{n_2=1}^{n_1} \overset{\text{in}}{\underset{\sin}{c}}_{n_1 n_2} \left(\frac{r}{r_0}\right)^{n_1} P_{n_1}^{n_2}(\cos\xi)\sin n_2\theta + \overset{\text{add}}{\Phi}(r,\xi,\theta)\end{aligned} \quad (r < r_0)} \tag{4.108b}$$

Note that the Q functions are singular at $\theta = 0$ and π along the z-axis, and so these terms are not included (Moon and Spencer, 1961*a*, chap.8). Expressions that enable computation of the associated Legendre polynomials are presented in Abramowitz and Stegun (1972, chap.8). The derivative of the Legendre function, which is necessary for evaluating the gradient, (4.103), may be evaluated using Abramowitz and Stegun (1972, eqn 8.5.4)

$$P'^{m}_{n}(z) = \frac{nzP_n^m(z) - (n+m)P_{n-1}^m(z)}{z^2-1}, \quad Q'^{m}_{n}(z) = \frac{nzQ_n^m(z) - (n+m)Q_{n-1}^m(z)}{z^2-1} \tag{4.109}$$

The influence functions for elements and the vector field associated with minus the gradient of Φ are illustrated in Fig. 4.27.

Separation of variables for problems with the ***Helmholtz equation*** in spherical coordinates leads to the following particular solutions (Moon and Spencer, 1961*b*, p.27)

$$\begin{aligned} \frac{\mathrm{d}^2 R}{\mathrm{d}r^2} + \frac{2}{r}\frac{\mathrm{d}R}{\mathrm{d}r} + \left[k^2 - \frac{n_1(n_1+1)}{r^2}\right] R = 0 \quad &\rightarrow \quad R = \left\{ j_{n_1}(kr),\ y_{n_1}(kr) \right. \\ \frac{\mathrm{d}^2\Xi}{\mathrm{d}\xi^2} + \cot\xi\,\frac{\mathrm{d}\Xi}{\mathrm{d}\xi} + \left[n_1(n_1+1) - \frac{(n_2)^2}{\sin^2\xi}\right]\Xi = 0 \quad &\rightarrow \quad \Xi = \left\{ P_{n_1}^{n_2}(\cos\xi),\ Q_{n_1}^{n_2}(\cos\xi) \right. \\ \frac{\mathrm{d}^2\Theta}{\mathrm{d}\theta^2} + (n_2)^2\Theta = 0 \quad &\rightarrow \quad \Theta = \left\{ \cos n_2\theta,\ \sin n_2\theta \right. \end{aligned} \tag{4.110}$$

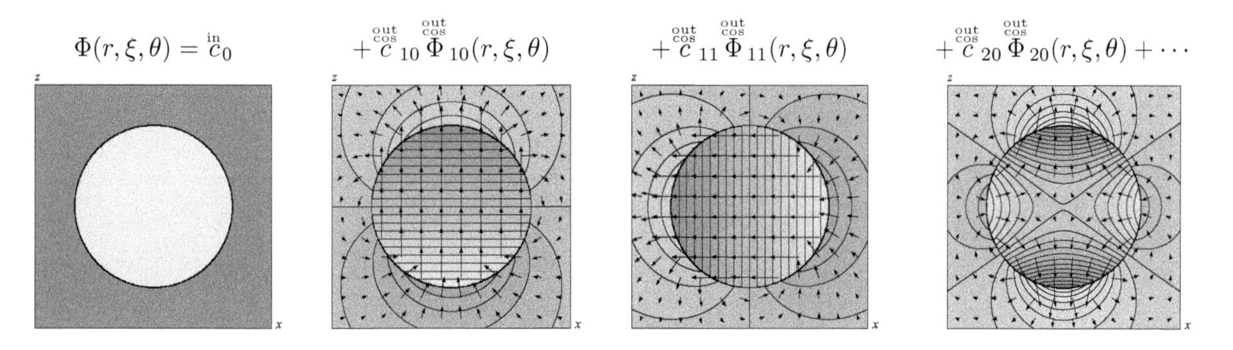

Figure 4.27 *Influence functions of the Laplace equation for spherical interface conditions, from (4.108), shown in plane with $y = 0$.*

The spherical Bessel functions of the first kind and the second kind (Abramowitz and Stegun, 1972, eqn 10.1.1) are related to the Bessel function solutions for circular coordinates,

$$j_n(z) = \sqrt{\frac{\pi}{2z}} \mathcal{J}_{n+\frac{1}{2}}(z), \quad y_n(z) = \sqrt{\frac{\pi}{2z}} Y_{n+\frac{1}{2}}(z) \tag{4.111a}$$

Waves may be formulating using the complex ***spherical Bessel functions of the third kind***

$$\begin{aligned} h_n^{(1)}(z) &= j_n(z) + \mathrm{i} y_n(z) = \sqrt{\frac{\pi}{2z}} H^{(1)}_{n+\frac{1}{2}}(z) \\ h_n^{(2)}(z) &= j_n(z) - \mathrm{i} y_n(z) = \sqrt{\frac{\pi}{2z}} H^{(2)}_{n+\frac{1}{2}}(z) \end{aligned} \tag{4.111b}$$

which are related to the circular Bessel functions of the third kind, (4.52). Waves may be modeled using linear superposition outside the sphere

$$\begin{aligned} \varphi(r,\xi,\theta) = &\sum_{n_1=0}^{N_1} \sum_{n_2=0}^{N_2} \overset{h\cos}{c}_{n_1 n_2} h_{n_1}^{(1)}(k^+ r) P_{n_1}^{n_2}(\cos\xi)\cos n_2\theta \\ &+ \sum_{n_1=0}^{N_1} \sum_{n_2=1}^{N_2} \overset{h\sin}{c}_{n_1 n_2} h_{n_1}^{(1)}(k^+ r) P_{n_1}^{n_2}(\cos\xi)\sin n_2\theta + \overset{\text{add}}{\varphi}(r,\xi,\theta) \end{aligned} \quad (r \geq r_0) \tag{4.112a}$$

and inside the sphere

$$\begin{aligned} \varphi(r,\xi,\theta) = &\sum_{n_1=0}^{N_1} \sum_{n_2=0}^{N_2} \overset{j\cos}{c}_{n_1 n_2} j_{n_1}(k^- r) P_{n_1}^{n_2}(\cos\xi)\cos n_2\theta \\ &+ \sum_{n_1=0}^{N_1} \sum_{n_2=1}^{N_2} \overset{j\sin}{c}_{n_1 n_2} j_{n_1}(k^- r) P_{n_1}^{n_2}(\cos\xi)\sin n_2\theta \end{aligned} \quad (r < r_0) \tag{4.112b}$$

Separation of variables for problems with the ***modified Helmholtz Equation*** in spherical coordinates leads to linear superposition of the particular solutions for those functions that remain finite outside the sphere

$$\begin{aligned} \Phi(r,\xi,\theta) = &\sum_{n_1=0}^{N_1} \sum_{n_2=0}^{N_2} \overset{k\cos}{c}_{n_1 n_2} k_{n_1}(k^+ r) P_{n_1}^{n_2}(\cos\xi)\cos n_2\theta \\ &+ \sum_{n_1=0}^{N_1} \sum_{n_2=1}^{N_2} \overset{k\sin}{c}_{n_1 n_2} k_{n_1}(k^+ r) P_{n_1}^{n_2}(\cos\xi)\sin n_2\theta + \overset{\text{add}}{\Phi}(r,\xi,\theta) \end{aligned} \quad (r \geq r_0) \tag{4.113a}$$

and inside the sphere

$$\begin{aligned} \Phi(r,\xi,\theta) = &\sum_{n_1=0}^{N_1} \sum_{n_2=0}^{N_2} \overset{i\cos}{c}_{n_1 n_2} i_{n_1}(k^- r) P_{n_1}^{n_2}(\cos\xi)\cos n_2\theta \\ &+ \sum_{n_1=0}^{N_1} \sum_{n_2=1}^{N_2} \overset{i\sin}{c}_{n_1 n_2} i_{n_1}(k^- r) P_{n_1}^{n_2}(\cos\xi)\sin n_2\theta \end{aligned} \quad (r < r_0) \tag{4.113b}$$

These influence functions are formulated in terms of the ***modified spherical Bessel functions of the first kind*** (Abramowitz and Stegun, 1972, eqn 10.2.2)

$$\sqrt{\frac{\pi}{2z}} I_{n+\frac{1}{2}}(z) = \begin{cases} \mathrm{e}^{-\mathrm{i}n\pi/2} i_n(z\mathrm{e}^{\mathrm{i}\pi/2}) & (-\pi < \arg z \le \frac{\pi}{2}) \\ \mathrm{e}^{\mathrm{i}3n\pi/2} i_n(z\mathrm{e}^{-\mathrm{i}3\pi/2}) & (\frac{\pi}{2} < \arg z \le \pi) \end{cases} \tag{4.114}$$

and the ***modified spherical Bessel functions of the third kind*** (Abramowitz and Stegun, 1972, eqn 10.2.4)

$$k_n(z) = \sqrt{\frac{\pi}{2z}} K_{n+\frac{1}{2}}(z) = \frac{\pi}{2}(-1)^{n+1}\sqrt{\frac{\pi}{2z}}\left[I_{n+\frac{1}{2}}(z) - I_{-n-\frac{1}{2}}(z)\right] \tag{4.115}$$

See the references Knight et al. (1989) and Warrick and Knight (2004) for examples of spherical inclusions in vadose zone problems.

Spheres with Continuity Conditions for the Laplace Equation

Solution methods are presented for the Laplace equation for a sphere that satisfies continuity conditions across its interface. This problem is formulated to have a ***continuous normal component*** of the vector field, (2.65b), which provides a relation between the coefficients inside and outside the sphere:

$$\frac{\partial \Phi^+}{\partial r} = \frac{\partial \Phi^-}{\partial r} \quad \rightarrow \quad \begin{cases} \overset{\text{in}}{\underset{\cos}{c}}{}_{n_1 n_2} = -\frac{n_1+1}{n_1} \overset{\text{out}}{\underset{\cos}{c}}{}_{n_1 n_2} \\ \overset{\text{in}}{\underset{\sin}{c}}{}_{n_1 n_2} = -\frac{n_1+1}{n_1} \overset{\text{out}}{\underset{\sin}{c}}{}_{n_1 n_2} \end{cases} \tag{4.116}$$

A second condition is provided by ***continuity condition for the potential***, (2.64a), at the interface of a heterogeneity, which gives

$$\alpha^+\Phi^+ = \alpha^-\Phi^- \quad \rightarrow \quad \begin{aligned} &-\alpha^- \overset{\text{in}}{c}_0 + \sum_{n_1=1}^{N_1}\sum_{n_2=0}^{N_2}\left(\alpha^+ + \frac{n_1+1}{n_1}\alpha^-\right) \overset{\text{out}}{\underset{\cos}{c}}{}_{n_1 n_2} P_{n_1}^{n_2}(\cos\xi_m)\cos n_2\theta_m \\ &+ \sum_{n_1=1}^{N_1}\sum_{n_2=1}^{N_2}\left(\alpha^+ + \frac{n_1+1}{n_1}\alpha^-\right) \overset{\text{out}}{\underset{\sin}{c}}{}_{n_1 n_2} P_{n_1}^{n_2}(\cos\xi_m)\sin n_2\theta_m = \left(\alpha^- - \alpha^+\right) \overset{\text{add}}{\Phi}_m \end{aligned} \tag{4.117}$$

These conditions are applied at control points laying at evenly spaced locations across the sphere with ξ between 0 and π, and θ between $-\pi$ and π, chosen sequentially from

$$\xi_m \in \pi\frac{\tilde{m} - \frac{1}{2}}{M}, \quad \theta_m \in -\pi + 2\pi\frac{\tilde{m} - 1}{M}, \quad \tilde{m} \in [1, 2, \cdots, M] \tag{4.118}$$

This gives a system of $2M^2$ equations to solve for the $N = (N_1 + 1)(2N_2 + 1)$ unknown coefficients that may be gathered in the matrices $\mathbf{Ac} = \mathbf{b}$ with

$$\mathbf{A} = \begin{bmatrix} 1 & \cdots & P_{n_1}^{n_2}(\cos\xi_1)\cos n_2\theta_1 & \cdots & P_{n_1}^{n_2}(\cos\xi_1)\sin n_2\theta_1 & \cdots \\ 1 & \cdots & P_{n_1}^{n_2}(\cos\xi_2)\cos n_2\theta_2 & \cdots & P_{n_1}^{n_2}(\cos\xi_2)\sin n_2\theta_2 & \cdots \\ \cdots & \cdots & \cdots & \cdots & \cdots & \cdots \\ 1 & \cdots & P_{n_1}^{n_2}(\cos\xi_{(2M^2)})\cos n_2\theta_{(2M^2)} & \cdots & P_{n_1}^{n_2}(\cos\xi_{(2M^2)})\sin n_2\theta_{(2M^2)} & \cdots \end{bmatrix} \tag{4.119a}$$

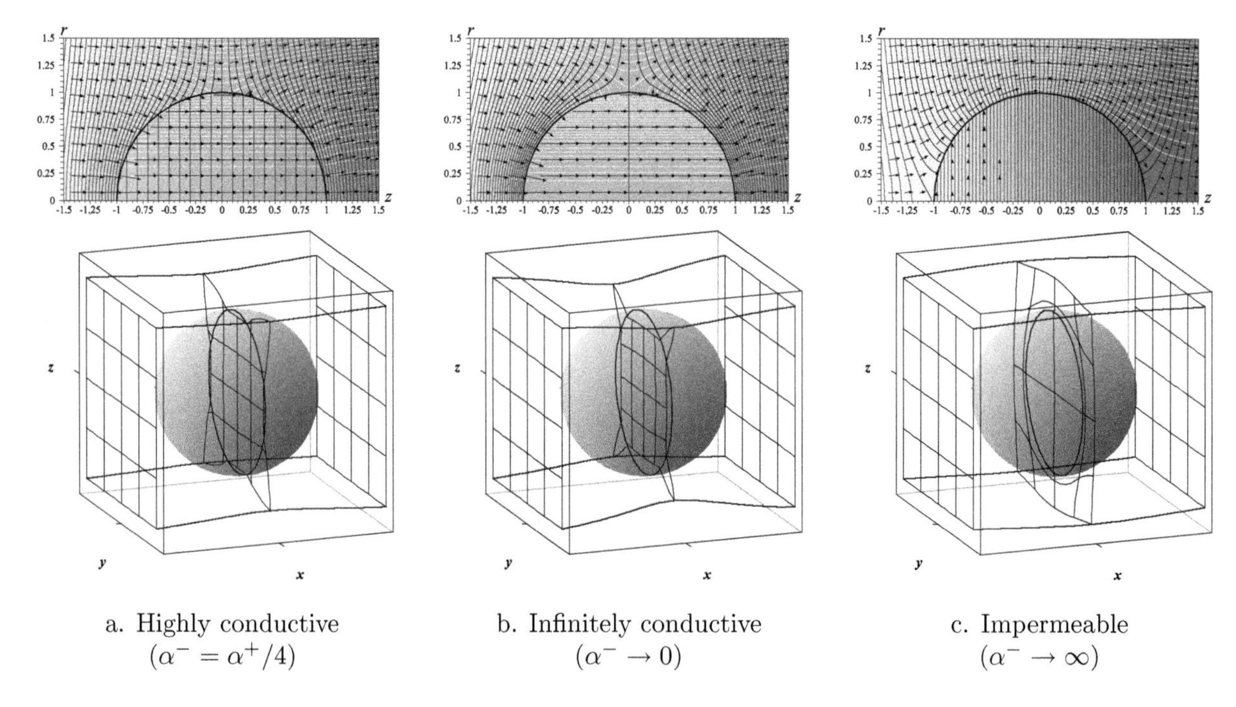

Figure 4.28 *Streamlines in half-plane with constant θ, and three-dimensional breakthrough planes near spherical heterogeneity.*

and

$$\mathbf{c} = \begin{bmatrix} -\alpha^- \overset{\text{in}}{c}_0 \\ \vdots \\ \left(\alpha^+ + \frac{n_1+1}{n_1}\alpha^-\right) \overset{\text{out}}{\underset{\text{cos}}{c}}_{n_1 n_2} \\ \vdots \\ \left(\alpha^+ + \frac{n_1+1}{n_1}\alpha^-\right) \overset{\text{out}}{\underset{\text{sin}}{c}}_{n_1 n_2} \\ \vdots \end{bmatrix}, \quad \mathbf{b} = \begin{bmatrix} \left(\alpha^- - \alpha^+\right) \overset{\text{add}}{\Phi}_1 \\ \left(\alpha^- - \alpha^+\right) \overset{\text{add}}{\Phi}_2 \\ \vdots \\ \left(\alpha^- - \alpha^+\right) \overset{\text{add}}{\Phi}_{(2M^2)} \end{bmatrix} \tag{4.119b}$$

The case of a single sphere in a uniform vector field v_{x0} directed along the x-axis is shown in Fig. 4.28. This example is reproduced using the potential for the background vector field in (4.5) when all coefficients are set to zero except for

$$\overset{\text{add}}{\Phi} = -v_{x0}x \;\rightarrow\; \overset{\text{out}}{\underset{\text{cos}}{c}}_{11} = \frac{\alpha^- - \alpha^+}{2\alpha^- + \alpha^+} v_{x0} r_0, \; \overset{\text{in}}{\underset{\text{cos}}{c}}_{11} = \frac{-2\alpha^- + 2\alpha^+}{2\alpha^- + \alpha^+} v_{x0} r_0 \tag{4.120}$$

for a sphere centered at the origin with radius r_0. This follows directly from (4.117). Solutions with a set of spherical objects that collectively interact to shape the vector field are illustrated in Fig. 4.29.

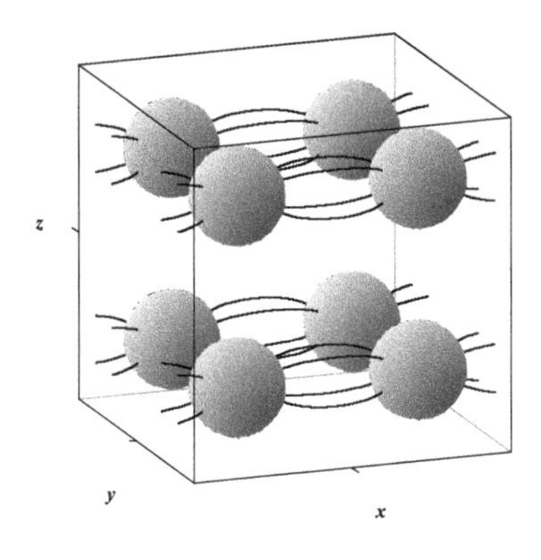

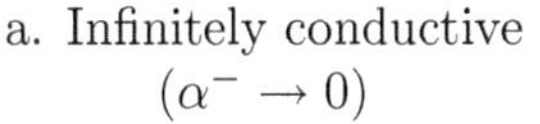

a. Infinitely conductive ($\alpha^- \to 0$)

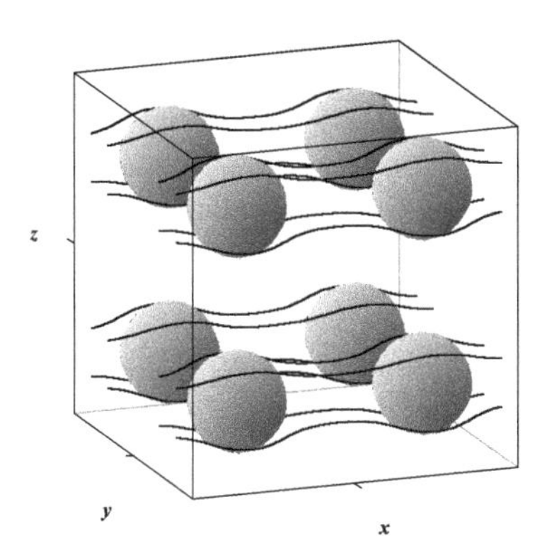

b. Impermeable ($\alpha^- \to \infty$)

Figure 4.29 *Spherical heterogeneity in a uniform background.*

Problem 4.18 Compute the coefficients for a sphere and the potential Φ and velocity component v_x at locations $(x, y, z) = (-r_0, 0, 0)$ and $(r_0, 0, 0)$ for a sphere of radius $r_0 = 1$ in a uniform field of $v_{x0} = 1$, using the following coefficients.

A. $\alpha^+ = 1, \alpha^- = 0.01$
B. $\alpha^+ = 1, \alpha^- = 0.2$
C. $\alpha^+ = 1, \alpha^- = 0.4$
D. $\alpha^+ = 1, \alpha^- = 0.6$
E. $\alpha^+ = 1, \alpha^- = 0.8$
F. $\alpha^+ = 1, \alpha^- = 1.0$
G. $\alpha^+ = 1, \alpha^- = 1.2$
H. $\alpha^+ = 1, \alpha^- = 1.4$
I. $\alpha^+ = 1, \alpha^- = 1.6$
J. $\alpha^+ = 1, \alpha^- = 1.8$
K. $\alpha^+ = 1, \alpha^- = 2.0$
L. $\alpha^+ = 1, \alpha^- = 100.0$

4.6 Separation in Spheroidal Coordinates

4.6.1 Three-Dimensional Solutions for Prolate Spheroids

The geometry of a prolate spheroid and its coordinate system is illustrated in Fig. 4.30. Coordinate transformation between (x, y, z) and (η, ξ, θ) axes is given by

$$\left.\begin{aligned} &\frac{x^2}{f^2\sinh^2\eta} + \frac{y^2}{f^2\sinh^2\eta} + \frac{z^2}{f^2\cosh^2\eta} = 1 \\ &\frac{x^2}{f^2\sin^2\xi} + \frac{y^2}{f^2\sin^2\xi} - \frac{z^2}{f^2\cos^2\xi} = -1 \\ &\theta = \arctan\frac{y}{x} \end{aligned}\right\} \leftrightarrow \left\{\begin{aligned} x &= f\sinh\eta\sin\xi\cos\theta \\ y &= f\sinh\eta\sin\xi\sin\theta \\ z &= f\cosh\eta\cos\xi \end{aligned}\right. \tag{4.121}$$

A set of ***inverse transformations*** useful for calculations is given by Morse and Feshbach (1953, page 661),

$$\cosh\eta = \frac{d_1 + d_2}{2f}, \quad \cos\xi = \frac{d_1 - d_2}{2f} \tag{4.122a}$$

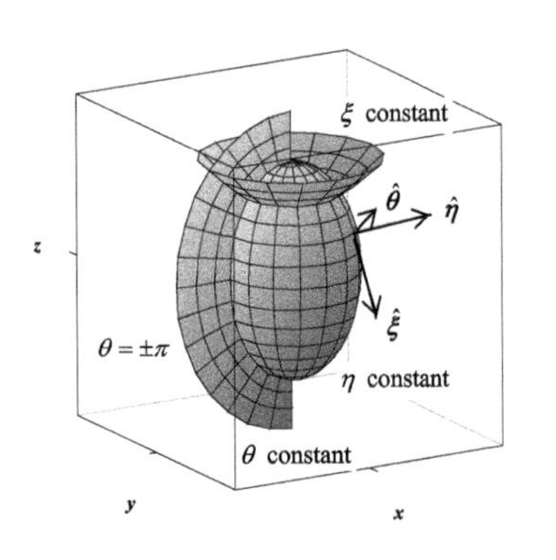

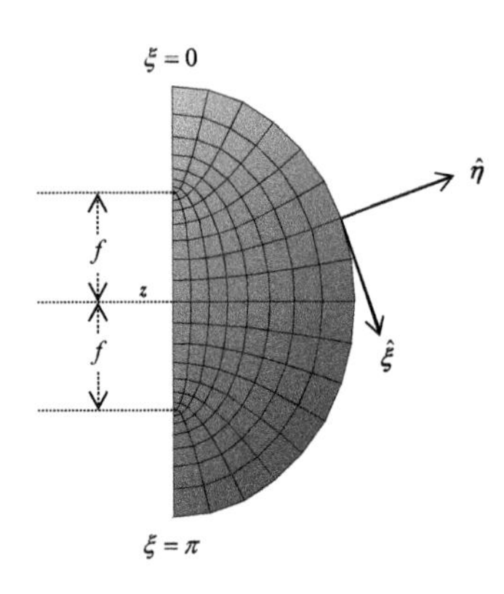

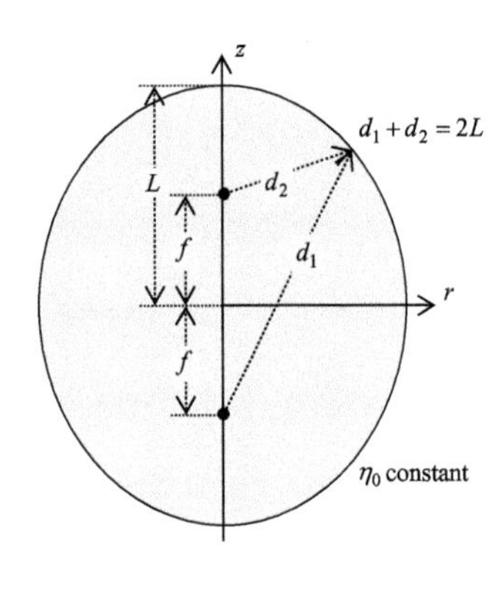

a. Three-dimensional domain b. Half-plane with constant θ c. Spheroid geometry

Figure 4.30 *Prolate spheroidal coordinates* (η, ξ, θ)*: with prolate spheroids (*η *constant), hyperboloids of revolution (*ξ *constant), and half-planes (*θ *constant).*

where

$$d_1 = \sqrt{x^2 + y^2 + (z + f)^2}, \quad d_2 = \sqrt{x^2 + y^2 + (z - f)^2} \tag{4.122b}$$

are the distances to the foci. The prolate spheroid is formed by the revolution of an ellipse about its major z-axis, as shown in Fig. 4.30c, and the value of $\cosh \eta_0 = L/f$ is constant along its boundary. Inside the spheroid $d_1 + d_2$ is less than $2L$, and outside it is greater than $2L$. Coordinate unit vectors

$$\hat{\eta} = \frac{\nabla \eta}{|\nabla \eta|}, \quad \hat{\xi} = \frac{\nabla \xi}{|\nabla \xi|}, \quad \hat{\theta} = \frac{\nabla \theta}{|\nabla \theta|} \tag{4.123}$$

are directed towards the gradient of coordinate surfaces, with Cartesian components

$$\begin{aligned}
\nabla \eta &= \frac{1}{2f \sinh \eta}\left(\frac{x}{d_1} + \frac{x}{d_2}\right)\hat{\mathbf{x}} + \frac{1}{2f \sinh \eta}\left(\frac{y}{d_1} + \frac{y}{d_2}\right)\hat{\mathbf{y}} + \frac{1}{2f \sinh \eta}\left(\frac{z+f}{d_1} + \frac{z-f}{d_2}\right)\hat{\mathbf{z}} \\
\nabla \xi &= -\frac{1}{2f \sin \xi}\left(\frac{x}{d_1} - \frac{x}{d_2}\right)\hat{\mathbf{x}} - \frac{1}{2f \sin \xi}\left(\frac{y}{d_1} - \frac{y}{d_2}\right)\hat{\mathbf{y}} - \frac{1}{2f \sin \xi}\left(\frac{z+f}{d_1} - \frac{z-f}{d_2}\right)\hat{\mathbf{z}} \\
\nabla \theta &= \frac{-y}{x^2 + y^2}\hat{\mathbf{x}} + \frac{x}{x^2 + y^2}\hat{\mathbf{y}}
\end{aligned} \tag{4.124}$$

Problems are formulated in terms of the ***Laplacian and gradient*** from (4.2b)

$$\begin{aligned}
\nabla^2 \Phi &= \frac{1}{f^2\left(\sinh^2 \eta + \sin^2 \xi\right)}\left(\frac{\partial^2 \Phi}{\partial \eta^2} + \coth \eta \frac{\partial \Phi}{\partial \eta} + \frac{\partial^2 \Phi}{\partial \xi^2} + \cot \xi \frac{\partial \Phi}{\partial \xi}\right) + \frac{1}{f^2 \sinh^2 \eta \sin^2 \xi}\frac{\partial^2 \Phi}{\partial \theta^2}, \\
\nabla \Phi &= \frac{1}{f\sqrt{\sinh^2 \eta + \sin^2 \xi}}\left(\frac{\partial \Phi}{\partial \eta}\hat{\eta} + \frac{\partial \Phi}{\partial \xi}\hat{\xi}\right) + \frac{1}{f \sinh \eta \sin \xi}\frac{\partial \Phi}{\partial \theta}\hat{\theta}
\end{aligned} \tag{4.125}$$

The separation of variables using $\Phi(\eta,\xi,\theta) = H(\eta)\Xi(\xi)\Theta(\theta)$ leads to ordinary differential equations with particular solutions (Moon and Spencer, 1961*b*, p.29)

$$\begin{aligned}
\frac{d^2H}{d\eta^2} + \coth\eta\frac{dH}{d\eta} - \left[n_1(n_1+1) + \frac{(n_2)^2}{\sinh^2\eta}\right]H = 0 \quad &\rightarrow \quad H = \left\{P_{n_1}^{n_2}(\cosh\eta),\ Q_{n_1}^{n_2}(\cosh\eta)\right. \\
\frac{d^2\Xi}{d\xi^2} + \cot\xi\frac{d\Xi}{d\xi} + \left[n_1(n_1+1) - \frac{(n_2)^2}{\sin^2\xi}\right]\Xi = 0 \quad &\rightarrow \quad \Xi = \left\{P_{n_1}^{n_2}(\cos\xi),\ Q_{n_1}^{n_2}(\cos\xi)\right. \\
\frac{d^2\Theta}{d\theta^2} + (n_2)^2\Theta = 0 \quad &\rightarrow \quad \Theta = \left\{\cos n_2\theta,\ \sin n_2\theta\right.
\end{aligned} \tag{4.126}$$

A solution may be obtained through the linear superposition of these ***influence functions***, which gives the following equations for those functions that remain finite outside

$$\begin{aligned}
\Phi(r,\xi,\theta) = &\sum_{n_1=1}^{N_1}\sum_{n_2=0}^{N_2} \overset{Q\cos}{c}_{n_1n_2} Q_{n_1}^{n_2}(\cosh\eta)P_{n_1}^{n_2}(\cos\xi)\cos n_2\theta \\
&+ \sum_{n_1=1}^{N_1}\sum_{n_2=1}^{N_2} \overset{Q\sin}{c}_{n_1n_2} Q_{n_1}^{n_2}(\cosh\eta)P_{n_1}^{n_2}(\cos\xi)\sin n_2\theta + \overset{\text{add}}{\Phi}(r,\xi,\theta)
\end{aligned} \quad (d_1+d_2 \geq 2L) \tag{4.127a}$$

and inside the prolate spheroid

$$\begin{aligned}
\Phi(r,\xi,\theta) = \overset{\text{in}}{c}_0 + &\sum_{n_1=1}^{N_1}\sum_{n_2=0}^{N_2} \overset{P\cos}{c}_{n_1n_2} P_{n_1}^{n_2}(\cosh\eta)P_{n_1}^{n_2}(\cos\xi)\cos n_2\theta \\
&+ \sum_{n_1=1}^{N_1}\sum_{n_2=1}^{N_2} \overset{P\sin}{c}_{n_1n_2} P_{n_1}^{n_2}(\cosh\eta)P_{n_1}^{n_2}(\cos\xi)\sin n_2\theta
\end{aligned} \quad (d_1+d_2 < 2L) \tag{4.127b}$$

following Moon and Spencer (1961*a*, p.232) and MacRobert (1967, chap.11).

These methods are applied to the problem of an isolated ***prolate spheroid with a background vector field directed along the major axis***, where the potential satisfies a jump condition across its interface. This problem may be modeled with the additional function associated with a uniform vector field in the z-direction and choosing the coefficients of the spheroid as

$$\overset{\text{add}}{\Phi} = -v_{z0}z, \quad \overset{Q\cos}{c}_{10} = -v_{z0}c_1f, \quad \overset{P\cos}{c}_{10} = -v_{z0}c_2f \tag{4.128}$$

with all other coefficients in (4.127) equal to zero, and c_1 and c_2 are constants that will be adjusted to match interface conditions. This gives (Steward and Janković, 2001, eqn 14)

$$\Phi = \begin{cases} -v_{z0}c_1f\left[\frac{\cosh\eta}{2}\ln\frac{\cosh\eta+1}{\cosh\eta-1} - 1\right]\cos\xi - v_{z0}z & \eta \geq \eta_0 \\ -v_{z0}c_2f\cosh\eta\cos\xi & \eta < \eta_0 \end{cases} \tag{4.129}$$

It is convenient to rearrange this function in terms of cylindrical coordinates (r,θ,z) using the inverse transformations (4.122) with $d_1 = [r^2+(z+f)^2]^{1/2}$ and $d_2 = [r^2+(z-f)^2]^{1/2}$, along with the identities

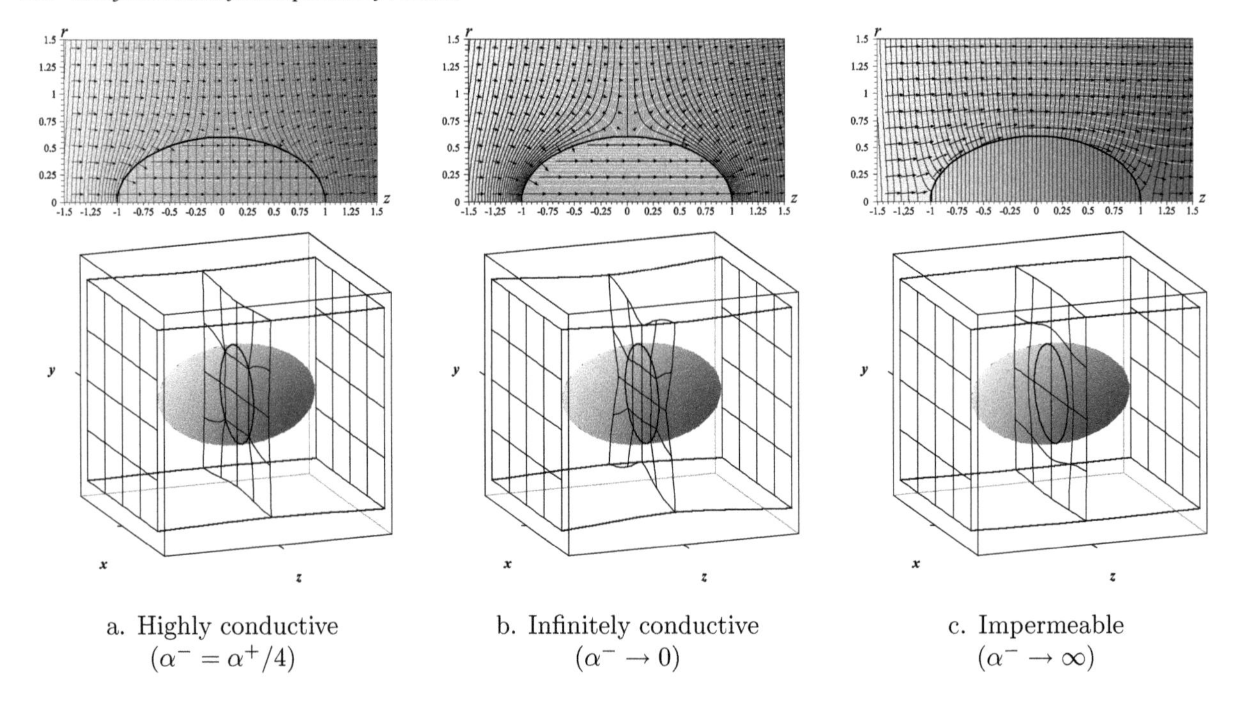

Figure 4.31 *Streamlines in half-plane with constant θ, and three-dimensional breakthrough planes near prolate spheroidal heterogeneity.*

$$\ln\frac{d_1+d_2+2f}{d_1+d_2-2f} = -\ln\frac{d_1-(z+f)}{d_2-(z-f)}, \quad \frac{2f}{d_1+d_2} = \frac{d_1-d_2}{2z} \tag{4.130}$$

to obtain

$$\Phi = \begin{cases} -v_{z0}z\left\{1-\frac{c_1}{2}\left[\ln\left(\frac{d_1-(z+f)}{d_2-(z-f)}\right)+\frac{d_1-d_2}{z}\right]\right\} & d_1+d_2 \geq 2L \\ -v_{z0}z\, c_2 & d_1+d_2 < 2L \end{cases} \tag{4.131}$$

This solution is illustrated in Fig. 4.31.

The potential function Φ is related to the components of the vector field in the r- and z-directions for this axisymmetric flow with flow symmetric about the z-axis, and the ***Stokes stream function***, Ψ (Stokes, 1842), using

$$v_r = -\frac{\partial\Phi}{\partial r} = -\frac{1}{r}\frac{\partial\Psi}{\partial z}, \quad v_\theta = 0, \quad v_z = -\frac{\partial\Phi}{\partial z} = \frac{1}{r}\frac{\partial\Psi}{\partial r} \tag{4.132}$$

This gives the vector field

$$v_r = \begin{cases} -v_{z0}\frac{c_1}{2}\left[\frac{z}{r}\left(\frac{z+f}{d_1}-\frac{z-f}{d_2}\right)+\frac{r}{d_1}-\frac{r}{d_2}\right] & d_1+d_2 \geq 2L \\ 0 & d_1+d_2 < 2L \end{cases} \tag{4.133}$$

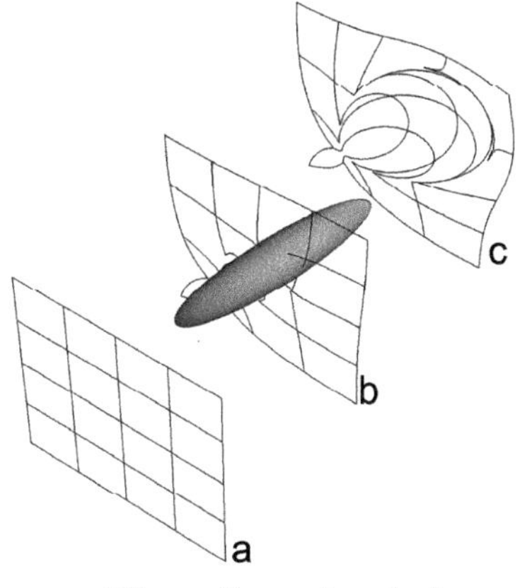

a. Three-dimensional view

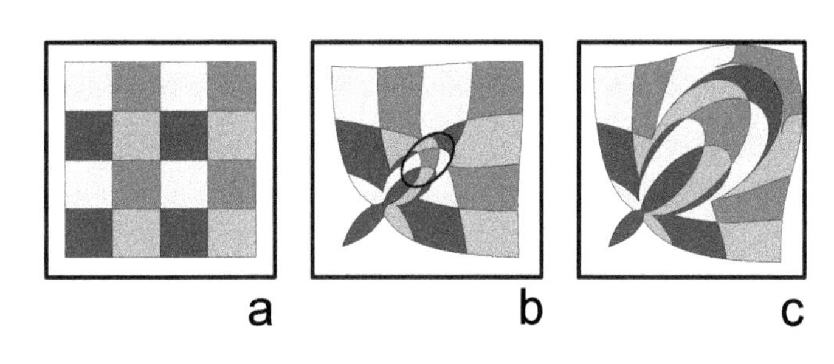

b. Intersections with breakthrough planes

Figure 4.32 *Three-dimensional stream surfaces for a prolate spheroid. (Reprinted from Janković, Steward, Barnes, and Dagan, 2009, Is Transverse Macrodispersivity in Three-dimensional Transport Through Isotropic Heterogeneous Formation Equal to Zero? A Counterexample,* Water Resources Research, *Vol. 45(8), Fig. 2, with permission from John Wiley and Sons. Copyright 2009 by the American Geophysical Union.)*

and

$$v_z = \begin{cases} v_{z0}\left\{1 - \frac{c_1}{2}\left[\ln\left(\frac{d_1-(z+f)}{d_2-(z-f)}\right) + \frac{f}{d_1} + \frac{f}{d_2}\right]\right\} & d_1 + d_2 \geq 2L \\ v_{z0}c_2 & d_1 + d_2 < 2L \end{cases} \tag{4.134}$$

And integration of the vector field yields the stream function (Steward and Janković, 2001, eqn 21)

$$\Psi = \begin{cases} v_{z0}\left\{\frac{r^2}{2} + \frac{c_1}{4}\left[-r^2\ln\left(\frac{d_1-(z+f)}{d_2-(z-f)}\right) + (z-f)\,d_1 - (z+f)\,d_2\right]\right\} & d_1 + d_2 \geq 2L \\ v_{z0}\frac{r^2}{2}c_2 & d_1 + d_2 < 2L \end{cases} \tag{4.135}$$

The coefficients c_1 and c_2 may be obtained by satisfying interface conditions on the normal component of the vector field and the potential

$$v_\eta{}^+ = v_\eta{}^-, \quad \alpha^+\Phi^+ = \alpha^-\Phi^- \tag{4.136}$$

across the interface where $\eta = \eta_0$. This gives coefficients (Fitts, 1990)

$$\begin{aligned} c_1 &= \frac{-\alpha^+ + \alpha^-}{\alpha^+\left(\frac{1}{2}\ln\frac{\cosh\eta_0+1}{\cosh\eta_0-1} - \frac{1}{\cosh\eta_0}\right) - \alpha^-\left(\frac{1}{2}\ln\frac{\cosh\eta_0+1}{\cosh\eta_0-1} - \frac{\cosh\eta_0}{\sinh^2\eta_0}\right)} \\ c_2 &= \frac{\alpha^+\left(-\frac{1}{\cosh\eta_0} + \frac{\cosh\eta_0}{\sinh^2\eta_0}\right)}{\alpha^+\left(\frac{1}{2}\ln\frac{\cosh\eta_0+1}{\cosh\eta_0-1} - \frac{1}{\cosh\eta_0}\right) - \alpha^-\left(\frac{1}{2}\ln\frac{\cosh\eta_0+1}{\cosh\eta_0-1} - \frac{\cosh\eta_0}{\sinh^2\eta_0}\right)} \end{aligned} \tag{4.137}$$

Similar development related to axisymmetric vector fields and the Stokes stream function are found in Maxwell (1881, pp.62–8), Carslaw and Jaeger (1959, pp.426-8), and Moon and Spencer (1961*b*, p.8). The three-dimensional stream surfaces for a prolate spheroid at oblique orientation to the regional vector field are illustrated in Fig. 4.32.

Problem 4.19 Compute the coefficients for a prolate spheroid with major axis aligned with a uniform background with $v_{z0} = 1$, and compute the potential Φ and velocity component v_z at locations $(x, y, z) = (0, 0, -L)$ and $(0, 0, L)$ for a spheroid with geometry:

A. $f = 0.1, L = 1, \alpha^+ = 1, \alpha^- = 0.01$
B. $f = 0.1, L = 1, \alpha^+ = 1, \alpha^- = 100$
C. $f = 0.2, L = 1, \alpha^+ = 1, \alpha^- = 0.01$
D. $f = 0.2, L = 1, \alpha^+ = 1, \alpha^- = 100$
E. $f = 0.4, L = 1, \alpha^+ = 1, \alpha^- = 0.01$
F. $f = 0.4, L = 1, \alpha^+ = 1, \alpha^- = 100$
G. $f = 0.6, L = 1, \alpha^+ = 1, \alpha^- = 0.01$
H. $f = 0.6, L = 1, \alpha^+ = 1, \alpha^- = 100$
I. $f = 0.8, L = 1, \alpha^+ = 1, \alpha^- = 0.01$
J. $f = 0.8, L = 1, \alpha^+ = 1, \alpha^- = 100$
K. $f = 0.9, L = 1, \alpha^+ = 1, \alpha^- = 0.01$
L. $f = 0.9, L = 1, \alpha^+ = 1, \alpha^- = 100$

4.6.2 Three-Dimensional Solutions for Oblate Spheroids

The geometry of an oblate spheroid and its coordinate system is illustrated in Fig. 4.33. Coordinate transformation between (x, y, z) and (η, ξ, θ) axes is given by

$$\left.\begin{array}{l} \dfrac{x^2}{f^2\cosh^2\eta} + \dfrac{y^2}{f^2\cosh^2\eta} + \dfrac{z^2}{f^2\sinh^2\eta} = 1 \\ \dfrac{x^2}{f^2\sin^2\xi} + \dfrac{y^2}{f^2\sin^2\xi} - \dfrac{z^2}{f^2\cos^2\xi} = 1 \\ \theta = \arctan\dfrac{y}{x} \end{array}\right\} \leftrightarrow \begin{cases} x = f\cosh\eta\sin\xi\cos\theta \\ y = f\cosh\eta\sin\xi\sin\theta \\ z = f\sinh\eta\cos\xi \end{cases} \tag{4.138}$$

Inverse transformations are given by (Steward and Janković, 2001)

$$\sinh\eta = \frac{d_1 + d_2}{2f}, \quad \cos\xi = \frac{d_1 - d_2}{i2f} \tag{4.139a}$$

where

$$d_1 = \sqrt{x^2 + y^2 + (z + if)^2}, \quad d_2 = \sqrt{x^2 + y^2 + (z - if)^2} \tag{4.139b}$$

Coordinate unit vectors

$$\hat{\eta} = \frac{\nabla\eta}{|\nabla\eta|}, \quad \hat{\xi} = \frac{\nabla\xi}{|\nabla\xi|}, \quad \hat{\theta} = \frac{\nabla\theta}{|\nabla\theta|} \tag{4.140}$$

are directed towards the gradient of coordinate surfaces, with Cartesian components

$$\begin{aligned} \nabla\eta &= \frac{1}{2f\cosh\eta}\left(\frac{x}{d_1} + \frac{x}{d_2}\right)\hat{\mathbf{x}} + \frac{1}{2f\cosh\eta}\left(\frac{y}{d_1} + \frac{y}{d_2}\right)\hat{\mathbf{y}} + \frac{1}{2f\cosh\eta}\left(\frac{z+if}{d_1} + \frac{z-if}{d_2}\right)\hat{\mathbf{z}} \\ \nabla\xi &= -\frac{1}{2if\sin\xi}\left(\frac{x}{d_1} - \frac{x}{d_2}\right)\hat{\mathbf{x}} - \frac{1}{2if\sin\xi}\left(\frac{y}{d_1} - \frac{y}{d_2}\right)\hat{\mathbf{y}} - \frac{1}{2if\sin\xi}\left(\frac{z+if}{d_1} - \frac{z-if}{d_2}\right)\hat{\mathbf{z}} \\ \nabla\theta &= \frac{-y}{x^2+y^2}\hat{\mathbf{x}} + \frac{x}{x^2+y^2}\hat{\mathbf{y}} \end{aligned} \tag{4.141}$$

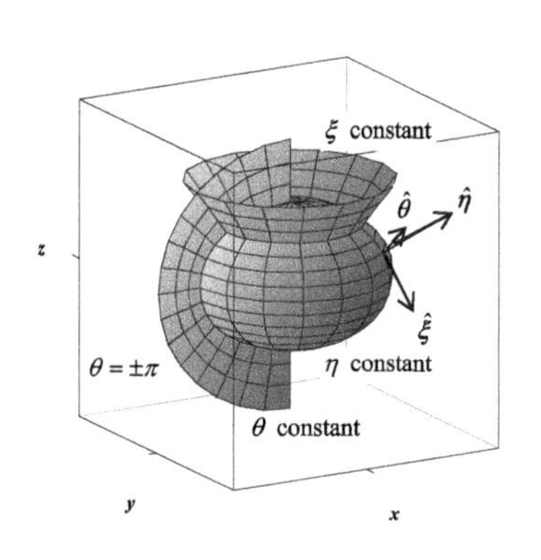

a. Three-dimensional domain

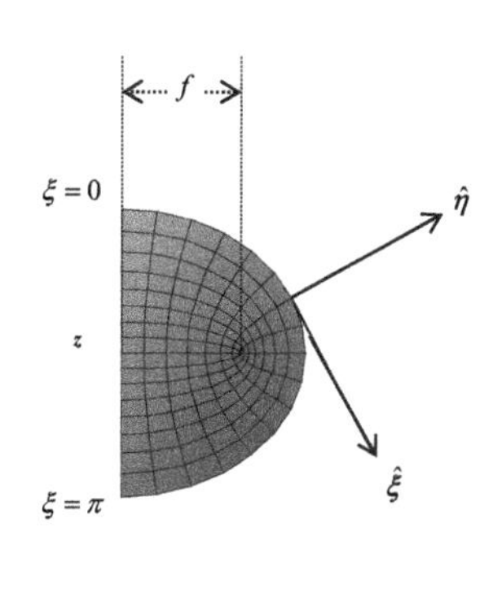

b. Half-plane with constant θ

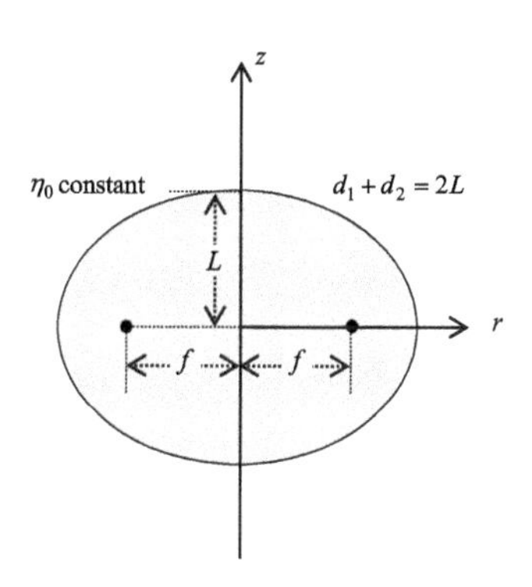

c. Spheroid geometry

Figure 4.33 *Oblate spheroidal coordinates* (η, ξ, θ)*: with oblate spheroids (*η *constant), hyperboloids of revolution (*ξ *constant), and half-planes (*θ *constant).*

Problems are formulated in terms of the ***Laplacian and gradient*** from (4.2c)

$$\boxed{\nabla^2\Phi = \frac{1}{f^2\left[\cosh^2\eta - \sin^2\xi\right]}\left(\frac{\partial^2\Phi}{\partial\eta^2} + \tanh\eta\,\frac{\partial\Phi}{\partial\eta} + \frac{\partial^2\Phi}{\partial\xi^2} + \cot\xi\,\frac{\partial\Phi}{\partial\xi}\right) + \frac{1}{f^2\cosh^2\eta\sin^2\xi}\frac{\partial^2\Phi}{\partial\theta^2}},$$
$$\boxed{\nabla\Phi = \frac{1}{f\sqrt{\cosh^2\eta - \sin^2\xi}}\left(\frac{\partial\Phi}{\partial\eta}\hat{\boldsymbol{\eta}} + \frac{\partial\Phi}{\partial\xi}\hat{\boldsymbol{\xi}}\right) + \frac{1}{f\cosh\eta\sin\xi}\frac{\partial\Phi}{\partial\theta}\hat{\boldsymbol{\theta}}} \tag{4.142}$$

The separation of variables with $\Phi(\eta,\xi,\theta) = H(\eta)\Xi(\xi)\Theta(\theta)$ leads to ordinary differential equations with particular solutions (Moon and Spencer, 1961*b*, p.32)

$$\begin{aligned}
\frac{\mathrm{d}^2H}{\mathrm{d}\eta^2} + \tan\eta\frac{\mathrm{d}H}{\mathrm{d}\eta} + \left[-n_1(n_1+1) + \frac{(n_2)^2}{\cosh^2\eta}\right]H = 0 \quad &\rightarrow \quad H = \left\{P_{n_1}^{n_2}(\mathrm{i}\sinh\eta),\ Q_{n_1}^{n_2}(\mathrm{i}\sinh\eta)\right. \\
\frac{\mathrm{d}^2\Xi}{\mathrm{d}\xi^2} + \cot\xi\frac{\mathrm{d}\Xi}{\mathrm{d}\xi} + \left[n_1(n_1+1) - \frac{(n_2)^2}{\sin^2\xi}\right]\Xi = 0 \quad &\rightarrow \quad \Xi = \left\{P_{n_1}^{n_2}(\cos\xi),\ Q_{n_1}^{n_2}(\cos\xi)\right. \\
\frac{\mathrm{d}^2\Theta}{\mathrm{d}\theta^2} + (n_2)^2\Theta = 0 \quad &\rightarrow \quad \Theta = \left\{\cos n_2\theta,\ \sin n_2\theta\right.
\end{aligned} \tag{4.143}$$

A solution may be obtained through linear superposition of these functions, which gives the following equations for the ***influence functions*** that remain finite outside and inside an oblate spheroid (Moon and Spencer, 1961*a*, p.232):

$$\boxed{\begin{aligned}\Phi(r,\xi,\theta) = &\sum_{n_1=1}^{N_1}\sum_{n_2=1}^{N_2} \overset{Q\cos}{c}_{n_1n_2}\overset{Q\cos}{\Phi}_{n_1n_2}(r,\theta) \\ &+ \sum_{n_1=1}^{N_1}\sum_{n_2=1}^{N_2} \overset{Q\sin}{c}_{n_1n_2}\overset{Q\sin}{\Phi}_{n_1n_2}(r,\theta) + \overset{\text{add}}{\Phi}(r,\xi,\theta)\end{aligned} \quad (d_1 + d_2 \geq 2L)} \tag{4.144a}$$

and

$$\Phi(r,\xi,\theta) = \overset{\text{in}}{c}_0 + \sum_{n_1=1}^{N_1}\sum_{n_2=0}^{N_2} \overset{P\cos}{c}_{n_1 n_2} \overset{P\cos}{\Phi}_{n_1 n_2}(r,\xi,\theta) + \sum_{n_1=1}^{N_1}\sum_{n_2=1}^{N_2} \overset{P\sin}{c}_{n_1 n_2} \overset{P\sin}{\Phi}_{n_1 n_2}(r,\xi,\theta) \qquad (d_1+d_2<2L) \tag{4.144b}$$

The influence functions are developed to remove singularities of P and Q functions with complex arguments (Moon and Spencer, 1961*a*, p.274). This gives the following influence functions where the set of functions are non-singular in domains inside and outside the spheroid following Janković (1997, p.30):

$$\begin{aligned}
\overset{Q\cos}{\Phi}_{n_1 n_2} &= \begin{cases} \Im\left[Q_{n_1}^{n_2}(\mathrm{i}\sinh\eta)\right] P_{n_1}^{n_2}(\cos\xi)\cos n_2\theta & n_1 \text{ even} \\ \Re\left[Q_{n_1}^{n_2}(\mathrm{i}\sinh\eta)\right] P_{n_1}^{n_2}(\cos\xi)\cos n_2\theta & n_1 \text{ odd} \end{cases} \\
\overset{Q\sin}{\Phi}_{n_1 n_2} &= \begin{cases} \Im\left[Q_{n_1}^{n_2}(\mathrm{i}\sinh\eta)\right] P_{n_1}^{n_2}(\cos\xi)\sin n_2\theta & n_1 \text{ even} \\ \Re\left[Q_{n_1}^{n_2}(\mathrm{i}\sinh\eta)\right] P_{n_1}^{n_2}(\cos\xi)\sin n_2\theta & n_1 \text{ odd} \end{cases} \\
\overset{P\cos}{\Phi}_{n_1 n_2} &= \begin{cases} \Re\left[P_{n_1}^{n_2}(\mathrm{i}\sinh\eta)\right] P_{n_1}^{n_2}(\cos\xi)\cos n_2\theta & n_1 \text{ even} \\ \Im\left[P_{n_1}^{n_2}(\mathrm{i}\sinh\eta)\right] P_{n_1}^{n_2}(\cos\xi)\cos n_2\theta & n_1 \text{ odd} \end{cases} \\
\overset{P\sin}{\Phi}_{n_1 n_2} &= \begin{cases} \Re\left[P_{n_1}^{n_2}(\mathrm{i}\sinh\eta)\right] P_{n_1}^{n_2}(\cos\xi)\sin n_2\theta & n_1 \text{ even} \\ \Im\left[P_{n_1}^{n_2}(\mathrm{i}\sinh\eta)\right] P_{n_1}^{n_2}(\cos\xi)\sin n_2\theta & n_1 \text{ odd} \end{cases}
\end{aligned} \tag{4.145}$$

These methods are applied to an isolated ***oblate spheroid with a vector field directed along the minor axis*** and a jump in potential across its interface, using the additional function associated with a uniform vector field in the z-direction and choosing the coefficients of the spheroid as

$$\overset{\text{add}}{\Phi} = -v_{z0}z, \qquad \overset{Q\cos}{c}_{10} = -v_{z0}c_1 f, \qquad \overset{P\cos}{c}_{10} = -v_{z0}c_2 f \tag{4.146}$$

with all other coefficients in (4.144) equal to zero. This gives (Steward and Janković, 2001, eqn 22)

$$\Phi = \begin{cases} -v_{z0}c_1 f\left[\frac{\mathrm{i}\sinh\eta}{2}\ln\frac{\mathrm{i}\sinh\eta+1}{\mathrm{i}\sinh\eta-1} - 1\right]\cos\xi - v_{z0}z & \eta \geq \eta_0 \\ -v_{z0}c_2 f\sinh\eta\cos\xi & \eta < \eta_0 \end{cases} \tag{4.147}$$

It is convenient to rearrange this function in terms of cylindrical coordinates (r,θ,z) using the inverse transformations (4.139) with $d_1 = [r^2+(z+\mathrm{i}f)^2]^{1/2}$ and $d_2 = [r^2+(z-\mathrm{i}f)^2]^{1/2}$, along with the identities

$$\ln\frac{d_1+d_2-\mathrm{i}2f}{d_1+d_2+\mathrm{i}2f} = \ln\frac{d_1-(z+\mathrm{i}f)}{d_2-(z-\mathrm{i}f)}, \qquad \frac{2f}{d_1+d_2} = -\mathrm{i}\frac{d_1-d_2}{2z} \tag{4.148}$$

This gives (Steward and Janković, 2001, eqn 26)

$$\Phi = \begin{cases} -v_{z0}z\left\{1+\mathrm{i}\frac{c_1}{2}\left[\ln\left(\frac{d_1-(z+\mathrm{i}f)}{d_2-(z-\mathrm{i}f)}\right)+\frac{d_1-d_2}{z}\right]\right\} & d_1+d_2 \geq 2L \\ -v_{z0}zc_2 & d_1+d_2 < 2L \end{cases} \tag{4.149}$$

The ***Stokes stream function*** is obtained by first taking derivatives of Φ, (4.132), to obtain

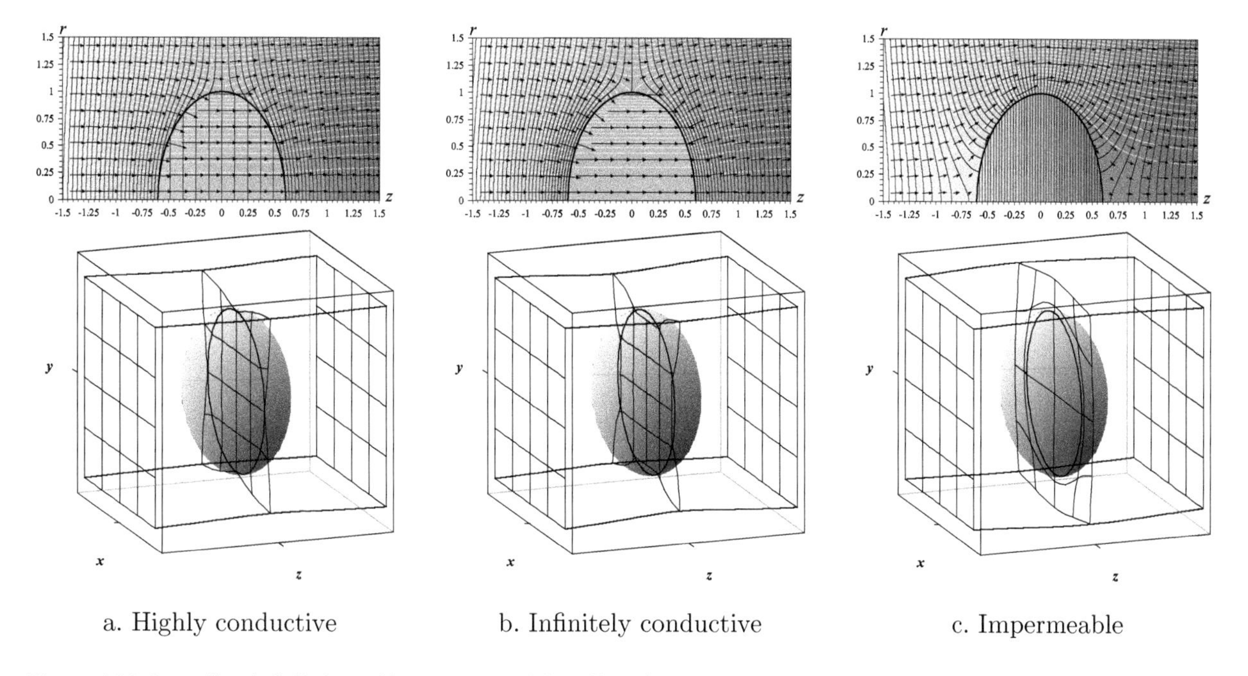

Figure 4.34 *Streamlines in half-plane with constant θ, and three-dimensional breakthrough planes near oblate spheroidal heterogeneity.*

$$v_r = \begin{cases} +v_{z0}\mathrm{i}\frac{c_1}{2}\left[\frac{z}{r}\left(\frac{z+\mathrm{i}f}{d_1} - \frac{z-\mathrm{i}f}{d_2}\right) + \frac{r}{d_1} - \frac{r}{d_2}\right] & d_1 + d_2 \geq 2L \\ 0 & d_1 + d_2 < 2L \end{cases} \tag{4.150}$$

and

$$v_z = \begin{cases} v_{z0}\left\{1 + \mathrm{i}\frac{c_1}{2}\left[\ln\left(\frac{d_1-(z+\mathrm{i}f)}{d_2-(z-\mathrm{i}f)}\right) + \frac{\mathrm{i}f}{d_1} + \frac{\mathrm{i}f}{d_2}\right]\right\} & d_1 + d_2 \geq 2L \\ v_{z0}c_2 & d_1 + d_2 < 2L \end{cases} \tag{4.151}$$

and integrating to obtain (Steward and Janković, 2001, eqn 28)

$$\Psi = \begin{cases} v_{z0}\left\{\frac{r^2}{2} - \mathrm{i}\frac{c_1}{4}\left[-r^2\ln\left(\frac{d_1-(z+\mathrm{i}f)}{d_2-(z-\mathrm{i}f)}\right) + (z-\mathrm{i}f)\,d_1 - (z+\mathrm{i}f)\,d_2\right]\right\} & d_1 + d_2 \geq 2L \\ v_{z0}\frac{r^2}{2}c_2 & d_1 + d_2 < 2L \end{cases} \tag{4.152}$$

Satisfying the interface conditions, (4.136), across the interface where $\eta = \eta_0$ gives coefficients

$$c_1 = \frac{-\alpha^+ + \alpha^-}{\alpha^+\left(\frac{\mathrm{i}}{2}\ln\frac{\mathrm{i}\sinh\eta_0+1}{\mathrm{i}\sinh\eta_0-1} - \frac{1}{\sinh\eta_0}\right) - \alpha^-\left(\frac{\mathrm{i}}{2}\ln\frac{\mathrm{i}\sinh\eta_0+1}{\mathrm{i}\sinh\eta_0-1} - \frac{\sinh\eta_0}{\cosh^2\eta_0}\right)} \tag{4.153a}$$

and

$$c_2 = \frac{\alpha^+\left(-\frac{1}{\sinh\eta_0} + \frac{\sinh\eta_0}{\cosh^2\eta_0}\right)}{\alpha^+\left(\frac{\mathrm{i}}{2}\ln\frac{\mathrm{i}\sinh\eta_0+1}{\mathrm{i}\sinh\eta_0-1} - \frac{1}{\sinh\eta_0}\right) - \alpha^-\left(\frac{\mathrm{i}}{2}\ln\frac{\mathrm{i}\sinh\eta_0+1}{\mathrm{i}\sinh\eta_0-1} - \frac{\sinh\eta_0}{\cosh^2\eta_0}\right)} \tag{4.153b}$$

This solution is illustrated in Fig. 4.34. The three-dimensional stream surfaces near an oblate spheroid placed obliquely to the background vector field are illustrated in Fig. 4.35.

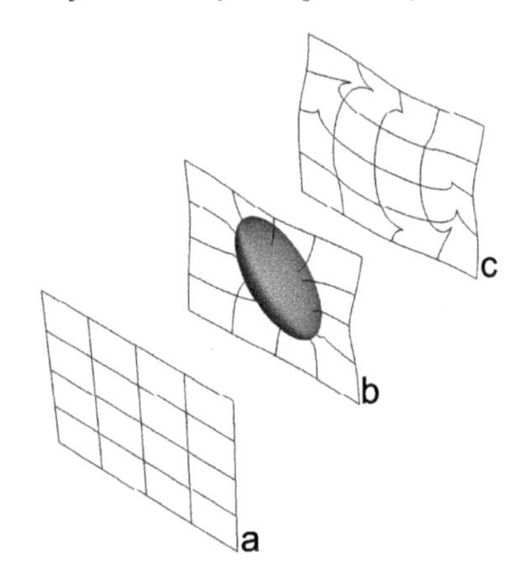

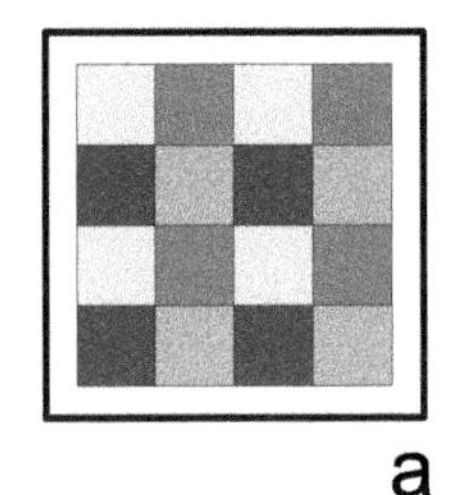

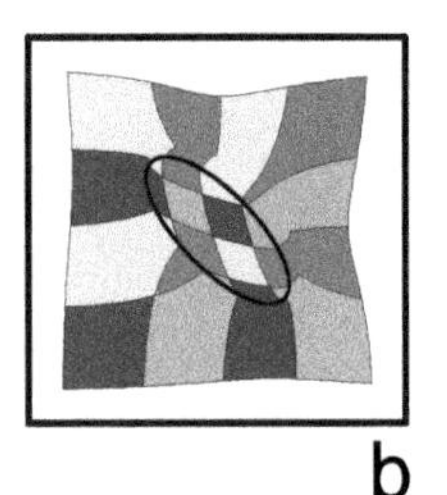

a. Three-dimensional view

b. Intersections with breakthrough planes

Figure 4.35 *Three-dimensional stream surfaces for an oblate spheroid. (Reprinted from Janković, Steward, Barnes, and Dagan, 2009, Is Transverse Macrodispersivity in Three-dimensional Transport Through Isotropic Heterogeneous Formation Equal to Zero? A Counterexample,* Water Resources Research, *Vol. 45(8), Fig. 2, with permission from John Wiley and Sons. Copyright 2009 by the American Geophysical Union.)*

Problem 4.20 Compute the coefficients for an oblate spheroid with minor axis aligned with a uniform background with $v_{z0} = 1$, and compute the potential Φ and velocity component v_z at locations $(x, y, z) = (0, 0, -L)$ and $(0, 0, L)$ for a spheroid with geometry:

A. $L = 0.1, f = 1, \alpha^+ = 1, \alpha^- = 0.01$

B. $L = 0.1, f = 1, \alpha^+ = 1, \alpha^- = 100$

C. $L = 0.2, f = 1, \alpha^+ = 1, \alpha^- = 0.01$

D. $L = 0.2, f = 1, \alpha^+ = 1, \alpha^- = 100$

E. $L = 0.4, f = 1, \alpha^+ = 1, \alpha^- = 0.01$

F. $L = 0.4, f = 1, \alpha^+ = 1, \alpha^- = 100$

G. $L = 0.6, f = 1, \alpha^+ = 1, \alpha^- = 0.01$

H. $L = 0.6, f = 1, \alpha^+ = 1, \alpha^- = 100$

I. $L = 0.8, f = 1, \alpha^+ = 1, \alpha^- = 0.01$

J. $L = 0.8, f = 1, \alpha^+ = 1, \alpha^- = 100$

K. $L = 0.9, f = 1, \alpha^+ = 1, \alpha^- = 0.01$

L. $L = 0.9, f = 1, \alpha^+ = 1, \alpha^- = 100$

Further Reading

Section 4.3, Cartesian Coordinates

- Abate and Valkó (2004)
- Braester (1973)
- Carslaw and Jaeger (1959)
- Davies (1978)
- Fourier (1878)
- Fourier (1808)
- Janković and Barnes (1999*a*)
- Moon and Spencer (1961*a*)
- Moon and Spencer (1961*b*)
- Oberhettinger (1973)
- Selvadurai (2000)

- Stehfest (1970)
- Steward and Allen (2013)
- Warrick (1974)

Section 4.4, Circular-Cylindrical Coordinates

- Abramowitz and Stegun (1972)
- Bakker and Nieber (2004*a*)
- Berkhoff (1976)
- Dagan, Fiori, and Janković (2004)
- Levenberg (1944)
- MacRobert (1967)
- Marquardt (1963)
- Moon and Spencer (1961*a*)
- Moon and Spencer (1961*b*)
- Philip, Knight, and Waechter (1989)
- Sommerfeld (1972)
- Steward (2016)
- Steward (2018)
- Steward (2020)
- Warrick and Knight (2002)

Section 4.5, Spherical Coordinates

- Abramowitz and Stegun (1972)
- Byerly (1893)
- Dagan, Fiori, and Janković (2003)
- Janković (1997)
- Janković, Fiori, and Dagan (2006)
- Kellogg (1929)
- Knight, Philip, and Waechter (1989)
- MacRobert (1967)
- Moon and Spencer (1961*a*)
- Moon and Spencer (1961*b*)
- Warrick and Knight (2004)

Section 4.6, Spheroidal Coordinates

- Abramowitz and Stegun (1972)
- Barnes and Janković (1999)
- Carslaw and Jaeger (1959)
- Fitts (1990)
- Fitts (1991)
- Janković (1997)
- Janković and Barnes (1999*b*)
- Janković, Steward, Barnes, and Dagan (2009)
- Kellogg (1929)
- Knight, Philip, and Waechter (1989)
- Legendre (1806)
- MacRobert (1967)
- Moon and Spencer (1961*a*)
- Moon and Spencer (1961*b*)
- Morse and Feshbach (1953)
- Maxwell (1865)
- Maxwell (1881)
- Steward and Janković (2001)
- Stokes (1842)

Analytic Elements from Singular Integral Equations

5

- Solutions to interface problems may be developed using analytic elements with mathematical solutions to the Laplace equation developed by singular integral equations.
- This formulation leads to solutions with discontinuities occurring across line segments, where the potential or stream function is discontinuous across double-layer elements in Section 5.2, and the normal or tangential component of the vector field is discontinuous across single-layer elements in Section 5.3. Examples illustrate a broad range of solutions to interface conditions possible with these elements.
- Series expansions are used to represent the far-field at larger distances from elements in Section 5.4, which leads to higher-order elements with nearly exact solutions and also provides a simpler representation for contiguous strings of adjacent elements.
- Such strings of elements are used with polygon elements in 5.5 to solve conditions along the interfaces of heterogeneities, and to provide a common series expansion to represent the far-field for a group of neighboring elements.
- Methods are extended to analytic elements with curvilinear geometry using conformal mappings (Section 5.6) and to three-dimensional fields in Section 5.7.

5.1 Formulation of Singular Integral Equations

Analytic elements may be formulated from a ***singular integral equation***, which is an integral with an integrand that is singular along the path of integration and the integrand is expressed as an unknown function that must be solved to achieve a solution. For example, the ***Cauchy integral formula*** provides the complex potential at the point z within a closed domain D with a simple, smooth boundary ∂D using

$$\Omega(z) = \frac{\mathrm{i}}{2\pi} \oint_{\partial D} \frac{\Omega(\tilde{z})}{z - \tilde{z}} \mathrm{d}\tilde{z} \tag{5.1}$$

as illustrated in Fig. 5.1a. This formula utilizes the values of the function $\Omega(\tilde{z}) = \Phi(\tilde{z}) + \mathrm{i}\Psi(\tilde{z})$ at points $\tilde{z}$ along the path of integration on the boundary of the closed domain. However, most problems specify boundary conditions in terms of either Φ or Ψ. For example, the domain in Fig. 5.1b specifies the potential Φ along part of the boundary, where $\Psi(\tilde{z})$ is

Analytic Element Method: Complex Interactions of Boundaries and Interfaces. David R. Steward, Oxford University Press (2020). © David R. Steward.
DOI: 10.1093/oso/9780198856788.001.0001

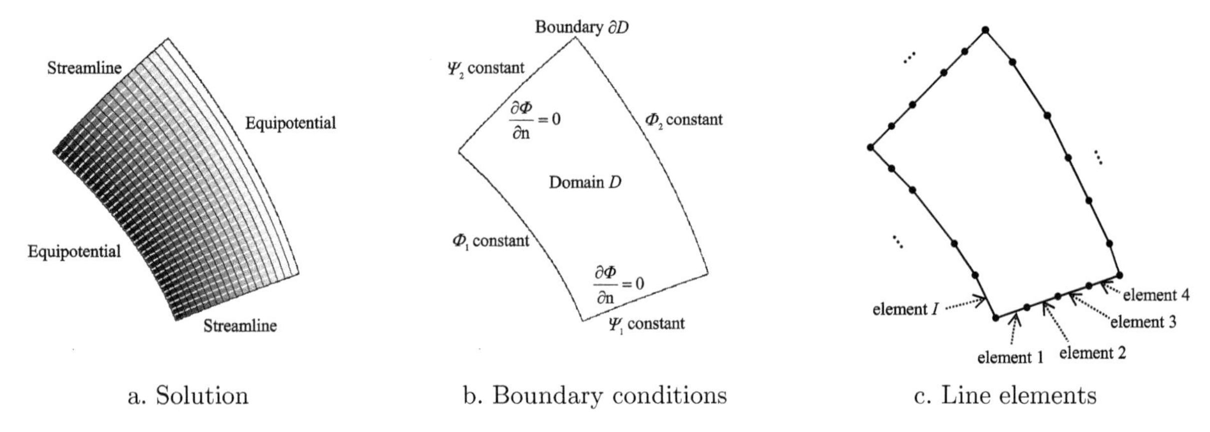

Figure 5.1 *Partitioning a boundary value problem for a simply connected, closed domain into a family of line elements.*

unknown, and the stream function Ψ is specified on the rest, where $\Phi(\tilde{z})$ is unknown along this part. Thus, the Cauchy integral formula is a singular integral equation, with a Cauchy kernel $1/(z-\tilde{z})$ that is singular along the boundary, and with unknown functions that must be calculated to achieve a solution for a boundary value problem.

The integrals observed in the Cauchy integral formula may be formulated as analytic elements. This is achieved by first partitioning the integral equation with a Cauchy kernel into a set of elements connecting endpoints z_{min} and z_{max}, as shown in Fig. 5.1c, where each element is a ***double layer*** (Muskhelishvili, 1953*a*):

$$\Omega(z) = \frac{\mathrm{i}}{2\pi} \int_{z_{\text{min}}}^{z_{\text{max}}} \frac{\Omega(\tilde{z})}{z-\tilde{z}} \mathrm{d}\tilde{z} \tag{5.2}$$

Formulation is aided by separating the real and imaginary parts of the continuous function $\Omega(\tilde{z}) = \mu(\tilde{z}) - \mathrm{i}\nu(\tilde{z})$ along the line segment to provide two integrals, where the first is called a ***line-dipole*** and the second is a ***line-doublet*** (Strack, 1989):

$$\Omega(z) = \boxed{\frac{1}{2\pi} \int_{z_{\text{min}}}^{z_{\text{max}}} \frac{\nu(\tilde{z})}{z-\tilde{z}} \mathrm{d}\tilde{z}} + \boxed{\frac{1}{2\pi} \int_{z_{\text{min}}}^{z_{\text{max}}} \frac{\mathrm{i}\mu(\tilde{z})}{z-\tilde{z}} \mathrm{d}\tilde{z}} \tag{5.3}$$

Methods are developed in the next section to adjust the unknown functions ν and μ to solve boundary conditions associated with the simple closed domain in Fig. 5.1, and to extend these functions to more general problems.

There exists two more singular integral equations, which will be used to solve an even wider range of problems. These elements may be obtained by applying integration by parts to the integral equations in the double layer, (5.2), giving

$$\frac{1}{2\pi}\int_{z_{\min}}^{z_{\max}} \frac{\nu(\tilde{z})+\mathrm{i}\mu(\tilde{z})}{z-\tilde{z}}\mathrm{d}\tilde{z} = \frac{1}{2\pi}\int_{z_{\min}}^{z_{\max}} \left(\frac{\mathrm{d}\nu}{\mathrm{d}\tilde{z}}+\mathrm{i}\frac{\mathrm{d}\mu}{\mathrm{d}\tilde{z}}\right)\ln(z-\tilde{z})\mathrm{d}\tilde{z} - \frac{\nu(z_{\max})+\mathrm{i}\mu(z_{\max})}{2\pi}\ln(z-z_{\max}) + \frac{\nu(z_{\min})+\mathrm{i}\mu(z_{\min})}{2\pi}\ln(z-z_{\min}) \tag{5.4}$$

The integral on the right-hand side is called a ***single layer*** (Burton and Miller, 1971) or a simple layer (Muskhelishvili, 1953*a*). A single layer may be separated into two singular integral equations using a change of variable

$$\sigma = \frac{\mathrm{d}\nu}{\mathrm{d}\tilde{z}}, \quad \gamma = \frac{\mathrm{d}\mu}{\mathrm{d}\tilde{z}} \tag{5.5}$$

to give a ***line-sink*** (Strack, 1989) and a ***line-vortex*** (Saffman, 1992):

$$\Omega(z) = \boxed{\frac{1}{2\pi}\int_{z_{\min}}^{z_{\max}} \sigma(\tilde{z})\ln(z-\tilde{z})\mathrm{d}s} + \boxed{\frac{1}{2\pi}\int_{z_{\min}}^{z_{\max}} \mathrm{i}\gamma(\tilde{z})\ln(z-\tilde{z})\mathrm{d}s} \tag{5.6}$$

Note that the remaining terms in (5.4) are dropped, for now, to give the Fredholm integral equations of the first kind. For readers familiar with Green's third identity, formulation of these four types of singular integral equations follow directly through use of the free space Green's function for the Laplace equation.

Mathematical development of the singular integral equations takes on a simpler form when expressed in terms of the ***local coordinate system*** illustrated in Fig. 5.2. Each element is located along a straight segment in the physical $z = x + \mathrm{i}y$ plane in Fig. 5.2a with endpoints $z_{\min}$ and $z_{\max}$, length $2L$, and orientation ϑ:

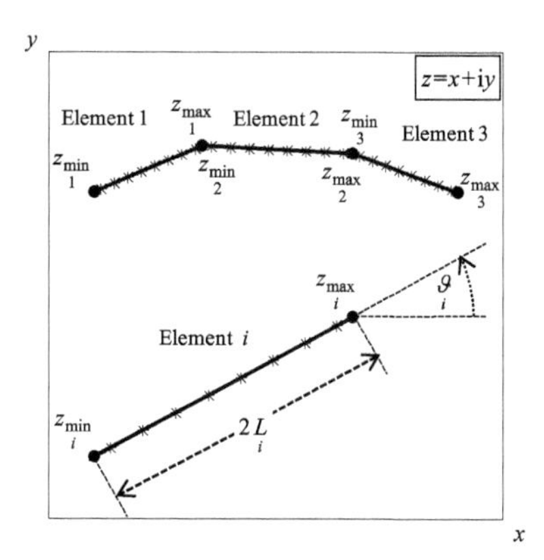

a. Line elements in physical z-coordinates

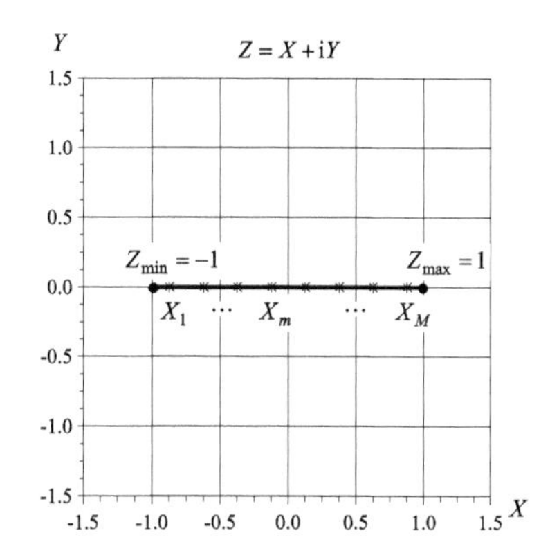

b. Each element is mapped to a local $\mathcal{Z}$-plane

Figure 5.2 *Coordinate system used to formulate singular line integrals.*

$$L = \left| \frac{z_{\max} + z_{\min}}{2} \right| , \qquad \vartheta = \arg(z_{\max} - z_{\min}) \tag{5.7}$$

A common coordinate system is provided by translation, rotation, and scaling each element to the local $\mathcal{Z} = X + \mathrm{i}Y$ plane with endpoints at $\mathcal{Z} = -1$ and $\mathcal{Z} = 1$ using (Strack, 1989, p.284)

$$\mathcal{Z} = \left(z - \frac{z_{\max} + z_{\min}}{2} \right) \frac{2}{z_{\max} - z_{\min}} , \qquad z = \mathcal{Z} \frac{z_{\max} - z_{\min}}{2} + \frac{z_{\max} + z_{\min}}{2} \tag{5.8}$$

Figure 5.2 also identifies the M control points where boundary conditions will be applied later, which are evenly spaced along the straight line segment at

$$\mathcal{Z}_m = -1 + \frac{2m-1}{M} , \quad z_m = z_{\min} + \frac{m - \frac{1}{2}}{M} (z_{\max} - z_{\min}) \quad (m = 1, M) \tag{5.9}$$

The complex potential is formulated such that $\Omega(z) = \Omega(\mathcal{Z}(z))$ takes on the same value at the same relative positions in the z and $\mathcal{Z}$ planes (Strack, 1989). This gives integrals in local $\mathcal{Z}$-coordinates on the boundary along the X-axis for the line-dipole and line-doublet (5.3) (Steward et al., 2008, eqn 2.7):

$$\Omega(\mathcal{Z}) = \boxed{\frac{1}{2\pi} \int_{-1}^{1} \frac{\nu(\tilde{X})}{\mathcal{Z} - \tilde{X}} \mathrm{d}\tilde{X}} + \boxed{\frac{\mathrm{i}}{2\pi} \int_{-1}^{1} \frac{\mu(\tilde{X})}{\mathcal{Z} - \tilde{X}} \mathrm{d}\tilde{X}} \tag{5.10}$$

and for the line-sink and line-vortex (5.6) (Steward et al., 2008, eqn 2.7):

$$\Omega(\mathcal{Z}) = \boxed{\frac{L}{2\pi} \int_{-1}^{1} \sigma(\tilde{X}) \ln(\mathcal{Z} - \tilde{X}) \mathrm{d}\tilde{X}} + \boxed{\frac{L}{2\pi} \int_{-1}^{1} \mathrm{i}\gamma(\tilde{X}) \ln(\mathcal{Z} - \tilde{X}) \mathrm{d}\tilde{X}} \tag{5.11}$$

The vector field is obtained by differentiating these complex potential functions with the change of variables, (5.8), to give

$$v = v_x + \mathrm{i}v_y = -\overline{\frac{\mathrm{d}\Omega}{\mathrm{d}z}} = -\overline{\frac{\mathrm{d}\Omega}{\mathrm{d}\mathcal{Z}} \frac{\mathrm{d}\mathcal{Z}}{\mathrm{d}z}} = -\frac{e^{\mathrm{i}\vartheta}}{L} \overline{\frac{\mathrm{d}\Omega}{\mathrm{d}\mathcal{Z}}} \tag{5.12a}$$

The components tangential and normal to the line element are given by rotating this v:

$$v_s + \mathrm{i}v_n = v e^{-\mathrm{i}\vartheta} \quad \rightarrow \quad \begin{cases} v_s = \Re(e^{-\mathrm{i}\vartheta} v) \\ v_n = \Im(e^{-\mathrm{i}\vartheta} v) \end{cases} \tag{5.12b}$$

These functions are developed next for singular integral equations along with their solution methods.

Problem 5.1 Compute the local position $\mathcal{Z}$ of point $z = 1 + \mathrm{i}$, and compute the components v_s and v_n of the vector $v = 1 + \mathrm{i}$ for a line element with endpoints:

A. $z_{\min} = -1 + \mathrm{i}1$, $z_{\max} = 1 - \mathrm{i}1$

B. $z_{\min} = -1 + \mathrm{i}2$, $z_{\max} = 1 - \mathrm{i}2$

C. $z_{\min} = -1 + \mathrm{i}3$, $z_{\max} = 1 - \mathrm{i}3$

D. $z_{\min} = -1 + \mathrm{i}4$, $z_{\max} = 1 - \mathrm{i}4$

E. $z_{\min} = -1 + \mathrm{i}5$, $z_{\max} = 1 - \mathrm{i}5$

F. $z_{\min} = -1 + \mathrm{i}6$, $z_{\max} = 1 - \mathrm{i}6$

G. $z_{\min} = 1 - \mathrm{i}1$, $z_{\max} = -1 + \mathrm{i}1$

H. $z_{\min} = 1 - \mathrm{i}2$, $z_{\max} = -1 + \mathrm{i}2$

I. $z_{\min} = 1 - \mathrm{i}3$, $z_{\max} = -1 + \mathrm{i}3$

J. $z_{\min} = 1 - \mathrm{i}4$, $z_{\max} = -1 + \mathrm{i}4$

K. $z_{\min} = 1 - \mathrm{i}5$, $z_{\max} = -1 + \mathrm{i}5$

L. $z_{\min} = 1 - \mathrm{i}6$, $z_{\max} = -1 + \mathrm{i}6$

5.2 Double-Layer Elements

This section develops the mathematical expressions necessary to compute the complex potential and vector field for double-layer elements. These singular integral equations are formulated in terms of the variable X that is real along an element and varies between ± 1, as shown in Fig. 5.2. It is convenient to approximate the variations occurring along an element in (5.3) as a ***local power series***, as in (2.36):

$$\nu(X) + \mathrm{i}\mu(X) = \sum_{n=0}^{N} \overset{\mathrm{dl}}{c}_n X^n \quad \rightarrow \quad \begin{aligned} \nu(X) &= \sum_{n=0}^{N} \Re \overset{\mathrm{dl}}{c}_n X^n \\ \mu(X) &= \sum_{n=0}^{N} \Im \overset{\mathrm{dl}}{c}_n X^n \end{aligned} \tag{5.13}$$

Since X is real, the line-dipole (ν) and line-doublet (μ) functions are provided directly in terms of either the real or the imaginary part of the complex coefficients $\overset{\mathrm{dl}}{c}_n$ for the double layer. Substituting the power series into (5.10) gives

$$\Omega = \frac{1}{2\pi} \int_{-1}^{1} \frac{\nu(\tilde{X}) + \mathrm{i}\mu(\tilde{X})}{\mathcal{Z} - \tilde{X}} \mathrm{d}\tilde{X} = \sum_{n=0}^{N} \frac{\overset{\mathrm{dl}}{c}_n}{2\pi} \int_{-1}^{1} \frac{\tilde{X}^n}{\mathcal{Z} - \tilde{X}} \mathrm{d}\tilde{X} \tag{5.14}$$

thereby reducing the double layer into linear superposition of coefficients times influence functions.

Closed-form expressions are obtained using the integration provided by Gröbner and Hofreiter (1975, p.7), substituting the limits of integration, and reorganizing the summation to skip terms where $1 - (-1)^{l-1}$ cancel:

$$\begin{aligned} \int_{-1}^{1} \frac{\tilde{X}^n}{\mathcal{Z} - \tilde{X}} \mathrm{d}\tilde{X} &= -\mathcal{Z}^n \ln(\mathcal{Z} - \tilde{X}) - \sum_{l=2}^{n+1} \frac{\mathcal{Z}^{n+1-l}}{l-1} \tilde{X}^{l-1} \Bigg|_{\tilde{X}=-1}^{1} \\ &= -\mathcal{Z}^n \ln(\mathcal{Z} - 1) + \mathcal{Z}^n \ln(\mathcal{Z} + 1) - \sum_{l=2}^{n+1} \frac{\mathcal{Z}^{n+1-l}}{l-1} \left[1 - (-1)^{l-1}\right] \\ &= \mathcal{Z}^n \ln \frac{\mathcal{Z} + 1}{\mathcal{Z} - 1} - \sum_{l=1}^{\frac{n+1}{2}} \frac{2}{2l-1} \mathcal{Z}^{n+1-2l} \qquad (n = 0, 1, \cdots) \end{aligned} \tag{5.15}$$

Note that the upper limit of summation $\frac{n+1}{2}$ is rounded down to the nearest integral, a notation convention utilized throughout this chapter. Together, the last two equations provide the ***complex potential for a double layer*** (Steward et al., 2008, eqn 2.12):

$$\Omega = \sum_{n=0}^{N} \overset{\text{dl}}{c}_n \overset{\text{dl}}{\Omega}_n(\mathcal{Z}) = \left[\sum_{n=0}^{N} \Re \overset{\text{dl}}{c}_n \times \overset{\text{dl}}{\Omega}_n(\mathcal{Z})\right] + \left[\sum_{n=0}^{N} \Im \overset{\text{dl}}{c}_n \times \mathrm{i}\overset{\text{dl}}{\Omega}_n(\mathcal{Z})\right],$$
$$\boxed{\overset{\text{dl}}{\Omega}_n[\mathcal{Z}(z)] = \frac{1}{2\pi}\left(\mathcal{Z}^n \ln\frac{\mathcal{Z}+1}{\mathcal{Z}-1} - \sum_{l=1}^{\frac{n+1}{2}} \frac{2}{2l-1}\mathcal{Z}^{n+1-2l}\right)} \tag{5.16}$$

The terms associated with real coefficients form ***line-dipole*** elements, and the imaginary coefficients form ***line-doublet*** elements.

An expression for the ***vector field of a double layer*** is obtained by differentiating, (5.12a), the singular integral for the complex potential of a double layer in (5.14):

$$v = -\frac{e^{\mathrm{i}\vartheta}}{L}\overline{\frac{\mathrm{d}\Omega}{\mathrm{d}\mathcal{Z}}} = \sum_{n=0}^{N} \overline{\overset{\text{dl}}{c}_n \frac{e^{\mathrm{i}\vartheta}}{2\pi L}\int_{-1}^{1} \frac{\tilde{X}^n}{\left(\mathcal{Z}-\tilde{X}\right)^2}\mathrm{d}\tilde{X}} \tag{5.17}$$

A closed-form expression may be obtained by first rearranging terms using integration by parts and substituting the limits of integration

$$\begin{aligned} \int_{-1}^{1} \frac{\tilde{X}^n}{(\mathcal{Z}-\tilde{X})^2}\mathrm{d}\tilde{X} &= \left.\frac{\tilde{X}^n}{\mathcal{Z}-\tilde{X}}\right|_{\tilde{X}=-1}^{1} - n\int_{-1}^{1} \frac{\tilde{X}^{n-1}}{\mathcal{Z}-\tilde{X}}\mathrm{d}\tilde{X} \\ &= \frac{1}{\mathcal{Z}-1} - \frac{(-1)^n}{\mathcal{Z}+1} - n\int_{-1}^{1} \frac{\tilde{X}^{n-1}}{\mathcal{Z}-\tilde{X}}\mathrm{d}\tilde{X} \end{aligned} \tag{5.18}$$

and then applying (5.15) to the remaining integral to give the ***complex vector*** of a double layer (Steward et al., 2008, eqn 2.15):

$$v = \sum_{n=0}^{N} \overline{\overset{\text{dl}}{c}_n \overset{\text{dl}}{v}_n[\mathcal{Z}(z)]} = \left[\sum_{n=0}^{N} \Re \overset{\text{dl}}{c}_n \times \overset{\text{dl}}{v}_n(\mathcal{Z})\right] - \left[\sum_{n=0}^{N} \Im \overset{\text{dl}}{c}_n \times \mathrm{i}\overset{\text{dl}}{v}_n(\mathcal{Z})\right],$$
$$\boxed{\overset{\text{dl}}{v}_n = -\frac{e^{\mathrm{i}\vartheta}}{2\pi L}\overline{\left[\frac{(-1)^n}{\mathcal{Z}+1} - \frac{1}{\mathcal{Z}-1} + n\mathcal{Z}^{n-1}\ln\frac{\mathcal{Z}+1}{\mathcal{Z}-1} - n\sum_{l=1}^{\frac{n}{2}} \frac{2}{2l-1}\mathcal{Z}^{n-2l}\right]}} \tag{5.19}$$

These complex potential and vector field functions will be illustrated shortly for double layers. First, however, problems are provided to learn how to compute the values of the first few influence functions:

$$
\begin{aligned}
\overset{\text{dl}}{\Omega}_0 &= \frac{1}{2\pi}\ln\frac{\mathcal{Z}+1}{\mathcal{Z}-1} &\quad \overset{\text{dl}}{v}_0 &= \overline{\frac{e^{i\vartheta}}{2\pi L}\left(-\frac{1}{\mathcal{Z}+1}+\frac{1}{\mathcal{Z}-1}\right)}\\
\overset{\text{dl}}{\Omega}_1 &= \frac{1}{2\pi}\left(\mathcal{Z}\ln\frac{\mathcal{Z}+1}{\mathcal{Z}-1}-2\right) &\quad \overset{\text{dl}}{v}_1 &= \overline{\frac{e^{i\vartheta}}{2\pi L}\left(\frac{1}{\mathcal{Z}+1}+\frac{1}{\mathcal{Z}-1}-\ln\frac{\mathcal{Z}+1}{\mathcal{Z}-1}\right)}\\
\overset{\text{dl}}{\Omega}_2 &= \frac{1}{2\pi}\left(\mathcal{Z}^2\ln\frac{\mathcal{Z}+1}{\mathcal{Z}-1}-2\mathcal{Z}\right) &\quad \overset{\text{dl}}{v}_2 &= \overline{\frac{e^{i\vartheta}}{2\pi L}\left(-\frac{1}{\mathcal{Z}+1}+\frac{1}{\mathcal{Z}-1}-2\mathcal{Z}\ln\frac{\mathcal{Z}+1}{\mathcal{Z}-1}+4\right)}\\
\overset{\text{dl}}{\Omega}_3 &= \frac{1}{2\pi}\left(\mathcal{Z}^3\ln\frac{\mathcal{Z}+1}{\mathcal{Z}-1}-2\mathcal{Z}^2-\frac{2}{3}\right) &\quad \overset{\text{dl}}{v}_3 &= \overline{\frac{e^{i\vartheta}}{2\pi L}\left(\frac{1}{\mathcal{Z}+1}+\frac{1}{\mathcal{Z}-1}-3\mathcal{Z}^2\ln\frac{\mathcal{Z}+1}{\mathcal{Z}-1}+6\mathcal{Z}\right)}
\end{aligned}
\tag{5.20}
$$

Note that the position where functions are to be evaluated in physical coordinates z must be transformed in to local $\mathcal{Z}$ coordinates with (5.8) to evaluate these functions.

Problem 5.2 Compute the influence functions $\overset{\text{dl}}{\Omega}_0, \overset{\text{dl}}{\Omega}_1, \overset{\text{dl}}{\Omega}_2, \overset{\text{dl}}{\Omega}_3, \overset{\text{dl}}{v}_0, \overset{\text{dl}}{v}_1, \overset{\text{dl}}{v}_2$, and $\overset{\text{dl}}{v}_3$ at the specified point z for a double layer lying between $z_{\min} = -1$ and $z_{\max} = 1$:

A. $z = -1 + i2$	D. $z = -1 + i$	G. $z = -1 - i$	J. $z = -1 - i2$
B. $z = i2$	E. $z = i$	H. $z = -i$	K. $z = -i2$
C. $z = 1 + i2$	F. $z = 1 + i$	I. $z = 1 - i$	L. $z = 1 - i2$

5.2.1 Line-Dipole: Boundaries with Discontinuous Stream Function

A line-dipole is formed by a double layer, (5.10), with real coefficients, (5.13). Its integral represents a distribution of point-dipoles, (3.4), with the axis of dipole oriented in the X-direction tangential to the element, and its complex potential and vector field are expressed as linear combinations of coefficients times influence functions:

$$
\begin{aligned}
\Omega &= \frac{1}{2\pi}\int_{-1}^{1}\frac{\nu(\tilde{X})}{\mathcal{Z}-\tilde{X}}\,d\tilde{X} \\
\nu &= \sum_{n=0}^{N}\Re\overset{\text{dl}}{c}_n X^n
\end{aligned}
\quad\rightarrow\quad
\boxed{
\begin{aligned}
\Omega &= \sum_{n=0}^{N}\Re\overset{\text{dl}}{c}_n \times \overset{\text{dl}}{\Omega}_n(\mathcal{Z})\\
v &= \sum_{n=0}^{N}\Re\overset{\text{dl}}{c}_n \times \overset{\text{dl}}{v}_n(\mathcal{Z})
\end{aligned}
}
\tag{5.21}
$$

The closed-form expressions for the influences functions in (5.16) and (5.19) are substituted into these equations to provide the potential and stream functions, and the vector field may be rotated using (5.12b) to provide the components tangential and normal to the element:

$$
\Phi + i\Psi = \sum_{n=0}^{N}\frac{\Re\overset{\text{dl}}{c}_n}{2\pi}\left(\mathcal{Z}^n\ln\frac{\mathcal{Z}+1}{\mathcal{Z}-1}-\sum_{l=1}^{\frac{n+1}{2}}\frac{2}{2l-1}\mathcal{Z}^{n+1-2l}\right)
$$

$$
v_s + iv_n = -\sum_{n=0}^{N}\frac{\Re\overset{\text{dl}}{c}_n}{2\pi L}\overline{\left[\frac{(-1)^n}{\mathcal{Z}+1}-\frac{1}{\mathcal{Z}-1}+n\mathcal{Z}^{n-1}\ln\frac{\mathcal{Z}+1}{\mathcal{Z}-1}-n\sum_{l=1}^{\frac{n}{2}}\frac{2}{2l-1}\mathcal{Z}^{n-2l}\right]}
\tag{5.22}
$$

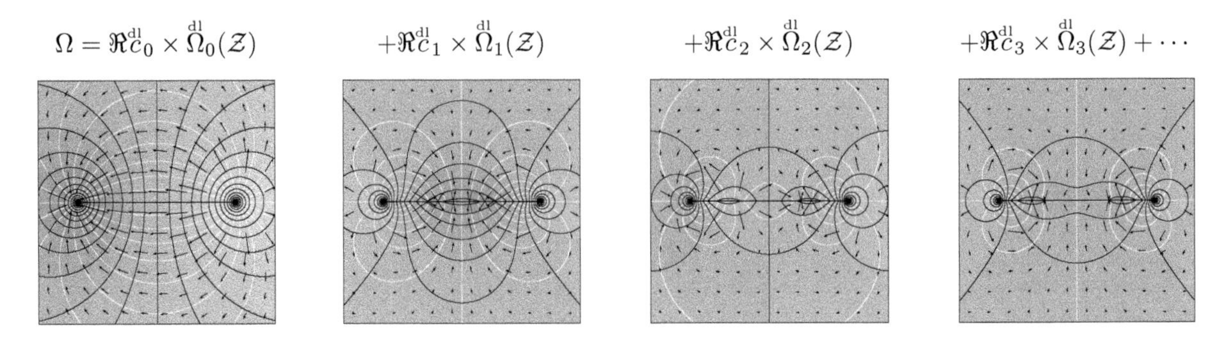

Figure 5.3 *Influence functions for a line-dipole, from (5.21) with (5.16) and (5.19).*

The first few influence functions are illustrated in Fig. 5.3, and it is observed that ***the potential function and the normal component of the vector field are continuous across a line-dipole, while the stream function and the normal component of the vector field jump across it***.

The conditions across the interface of a line-dipole become elucidated by examining the logarithm terms occurring in the complex potential and vector functions. These logarithm functions may be separated into real and imaginary parts, as done earlier for a point-sink in (3.1):

$$\begin{aligned} \ln(\mathcal{Z}+1) &= \ln|\mathcal{Z}+1| + \mathrm{i}\theta_{\min} \\ \ln(\mathcal{Z}-1) &= \ln|\mathcal{Z}-1| + \mathrm{i}\theta_{\max} \end{aligned} \quad \rightarrow \quad \ln\frac{\mathcal{Z}+1}{\mathcal{Z}-1} = \ln\left|\frac{\mathcal{Z}+1}{\mathcal{Z}-1}\right| + \mathrm{i}\,(\theta_{\min}-\theta_{\max}) \tag{5.23}$$

The imaginary part of these logarithm functions are illustrated in Fig. 5.4, where the arguments $\theta_{\min} = \arg(\mathcal{Z}+1)$ and $\theta_{\max} = \arg(\mathcal{Z}-1)$ represent the angles between the location $\mathcal{Z}$ and the ends of the element where $\mathcal{Z} = -1$ and $\mathcal{Z} = +1$. For a point $\mathcal{Z}$ lying along the X-axis, these angles are equal to 0 and $\pm\pi$ as illustrated in Fig. 5.4. Specifically, these angles take on the following values on the +side immediately above the X-axis and the −side immediately below the X-axis (Steward et al., 2008, fig.2):

$$\begin{aligned} &\theta^+_{\min} = \begin{cases} \pi & (X < -1) \\ 0 & (-1 < X) \end{cases}, \quad \theta^-_{\min} = \begin{cases} -\pi & (X < -1) \\ 0 & (-1 < X) \end{cases} \\ &\rightarrow \quad \Delta\theta_{\min} = \theta^+_{\min} - \theta^-_{\min} = \begin{cases} 2\pi & (X < -1) \\ 0 & (-1 < X) \end{cases} \\ &\theta^+_{\max} = \begin{cases} \pi & (X < 1) \\ 0 & (1 < X) \end{cases}, \quad \theta^-_{\max} = \begin{cases} -\pi & (X < 1) \\ 0 & (1 < X) \end{cases} \\ &\rightarrow \quad \Delta\theta_{\max} = \theta^+_{\max} - \theta^-_{\max} = \begin{cases} 2\pi & (X < 1) \\ 0 & (1 < X) \end{cases} \end{aligned} \tag{5.24a}$$

and the summation of these differences is

$$\Delta(\theta_{\min} - \theta_{\max}) = \begin{cases} 0 & (X < -1) \\ -2\pi & (-1 < X < 1) \\ 0 & (1 < X) \end{cases} \tag{5.24b}$$

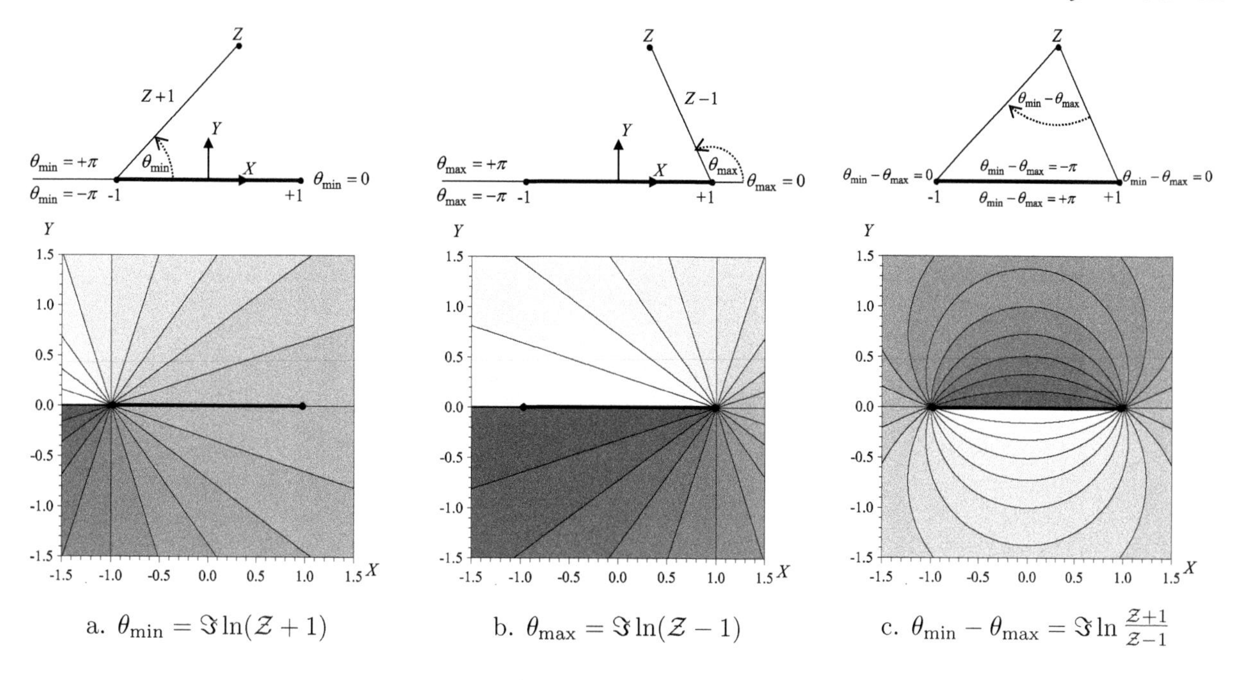

Figure 5.4 *Branch cuts in the imaginary part of logarithm terms along the X-axis.*

Thus, the imaginary parts of the logarithm functions in (5.23) jump across the X-axis due to the branch cuts formed by discontinuities in the arctan functions, while the real part of these logarithm terms are continuous across the X-axis.

The complex potential and the vector field for the line-dipole, (5.22), may be evaluated at points on the +side and the −side of the element using this knowledge of the complex logarithm. The differences of these functions across the element are quantified using the simplifications associated with the arguments of the complex logarithm as

$$\begin{aligned}(\Phi^+ + \mathrm{i}\Psi^+) - (\Phi^- + \mathrm{i}\Psi^-) &= \sum_{n=0}^{N} \Re \overset{\mathrm{dl}}{c}_n X^n \mathrm{i} \frac{\Delta\left(\theta_{\min} - \theta_{\max}\right)}{2\pi} \\ (v_s^+ + \mathrm{i}v_n^+) - (v_s^- + \mathrm{i}v_n^-) &= \sum_{n=0}^{N} \frac{\Re \overset{\mathrm{dl}}{c}_n}{L} n X^{n-1} \mathrm{i} \frac{\Delta\left(\theta_{\min} - \theta_{\max}\right)}{2\pi}\end{aligned} \tag{5.25}$$

Thus, ***a line-dipole has a continuous potential and a jump in stream function of −v, minus the strength of the distribution of dipoles along the element, from points immediately above to immediately below the element*** (Steward and Ahring, 2009, eqn 32):

$$\boxed{\Phi^+ = \Phi^-}, \quad \boxed{\Delta\Psi = -\sum_{n=0}^{N} \Re \overset{\mathrm{dl}}{c}_n X^n = -\nu(X)} \quad \begin{pmatrix} -1 < X < 1 \\ Y = 0 \end{pmatrix} \tag{5.26a}$$

Likewise, ***a line-dipole has a continuous tangential component of the vector field, and generates a jump in the normal component related to minus the derivative of v along the element***:

$$\boxed{v_s^{+} = v_s^{-}}, \quad \boxed{\Delta v_n = -\sum_{n=0}^{N} \Re \overset{\text{dl}}{c}_n \frac{nX^{n-1}}{L} = -\frac{1}{L}\frac{\mathrm{d}v}{\mathrm{d}X}} \quad \begin{pmatrix} -1 < X < 1 \\ Y = 0 \end{pmatrix} \tag{5.26b}$$

Problem 5.3 Compute Φ, Ψ, v_s and v_n at points $z = -0.5$, and $z = 0.5$, for a double layer lying between $z_{\min} = -1$ and $z_{\max} = 1$ with the following coefficients:

A. $\overset{\text{dl}}{c}_0 = 1$
B. $\overset{\text{dl}}{c}_1 = 1$
C. $\overset{\text{dl}}{c}_2 = 1$
D. $\overset{\text{dl}}{c}_3 = 1$
E. $\overset{\text{dl}}{c}_4 = 1$
F. $\overset{\text{dl}}{c}_5 = 1$
G. $\overset{\text{dl}}{c}_6 = 1$
H. $\overset{\text{dl}}{c}_7 = 1$
I. $\overset{\text{dl}}{c}_8 = 1$
J. $\overset{\text{dl}}{c}_9 = 1$
K. $\overset{\text{dl}}{c}_{10} = 1$
L. $\overset{\text{dl}}{c}_{11} = 1$

A line-dipole element has a potential that is continuous across the element, (5.26a), and its coefficients may be adjusted to satisfy a ***boundary condition of uniform potential*** along the element, (2.59b). This solution is achieved by evaluating the potential at control point m, which is the real part of the complex potential in (5.21), and settings the value of all control points equal to an unknown constant Φ_U (Steward, 2015, eqn 16):

$$\Phi_m = \sum_{n=0}^{N} \Re \overset{\text{dl}}{c}_n \Re\left[\overset{\text{dl}}{\Omega}_n(\mathcal{Z}_m)\right] + \overset{\text{add}}{\Phi}(z_m) = \Phi_U \tag{5.27}$$

This system of M equations with $N + 2$ unknowns may be expressed in matrix form as $\mathbf{Ac} = \mathbf{b}$ with

$$\mathbf{A} = \begin{bmatrix} 1 & \Phi_{10} & \Phi_{11} & \cdots & \Phi_{1N} \\ 1 & \Phi_{20} & \Phi_{21} & \cdots & \Phi_{2N} \\ \cdots & \cdots & \cdots & \cdots & \cdots \\ 1 & \Phi_{M0} & \Phi_{M1} & \cdots & \Phi_{MN} \end{bmatrix}, \quad \mathbf{c} = \begin{bmatrix} -\Phi_U \\ \Re \overset{\text{dl}}{c}_0 \\ \Re \overset{\text{dl}}{c}_1 \\ \vdots \\ \Re \overset{\text{dl}}{c}_N \end{bmatrix}, \quad \mathbf{b} = \begin{bmatrix} -\overset{\text{add}}{\Phi}(z_1) \\ -\overset{\text{add}}{\Phi}(z_2) \\ \vdots \\ -\overset{\text{add}}{\Phi}(z_M) \end{bmatrix} \tag{5.28}$$

and the coefficients $\Phi_{mn} = \Re[\overset{\text{dl}}{\Omega}_n(\mathcal{Z}_m)]$ are the real part of Ω at control point m for coefficient n of a double layer. The least squares solution to this problem is shown in Fig. 5.5, for a single element and a set of adjacent elements with uniform potential.

Discontinuities occur in both the stream function (5.26a) and the normal component of the vector field (5.26b), which were computed for the single element, and are also shown in the figure. Note that the jump $\Delta\Psi$ goes to zero at the endpoints of a single element. However, this discontinuity in stream function is non-zero across the intersection of adjacent elements in the string of elements. While line-dipoles can generate a non-zero jump in stream function at the endpoints, (5.26a), the jump associated with a slit-dipole goes to zero at the endpoints, (3.65). Thus, ***line-dipole elements can simulate uniform potential along a contiguous string of adjacent elements, while slit-dipole elements cannot be interconnected at their endpoints***.

A line-dipole may also be formulated to satisfy a ***Robin interface condition relating the tangential component of the vector field to the discontinuity of the stream function***, (2.66b), at control point m

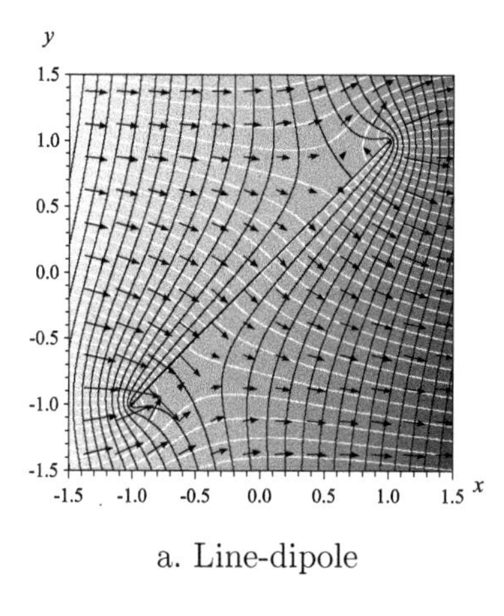

a. Line-dipole

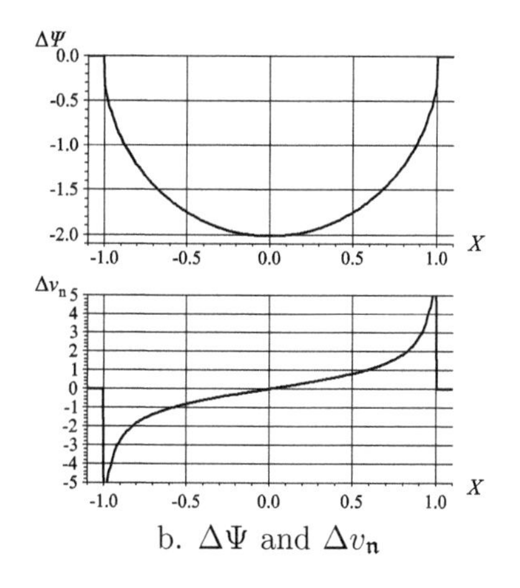

b. $\Delta\Psi$ and $\Delta v_{\mathfrak{n}}$

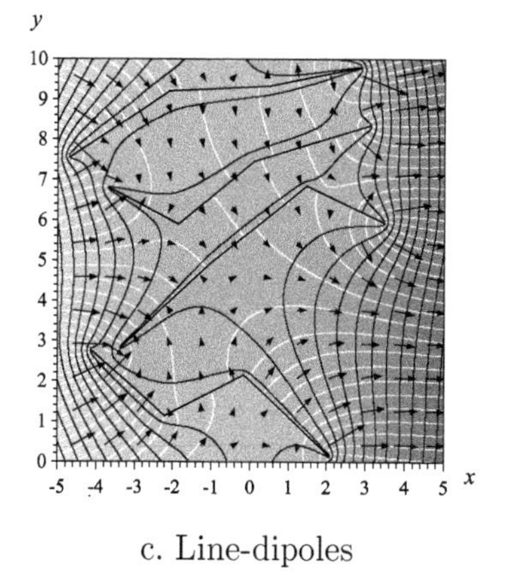

c. Line-dipoles

Figure 5.5 *Line-dipole with uniform potential.*

$$(v_{\mathfrak{s}})_m = -\frac{\Psi_m^+ - \Psi_m^-}{\delta} \tag{5.29}$$

where δ is a property of the line-dipole and its setting (Steward, 2015, eqn 20). This equation may be expanded using the series for the vector field in (5.21) with the tangential component of the vector field for the additional functions, and the series for the discontinuity in the stream function in (5.26a) to give

$$\sum_{n=0}^{N} \Re \overset{\text{dl}}{c}_n \Re\left[\overset{\text{dl}}{v}_n(\mathcal{Z}_m)\mathrm{e}^{-\mathrm{i}\vartheta}\right] + \Re\left[\overset{\text{add}}{v}(z_m)\mathrm{e}^{-\mathrm{i}\vartheta}\right] = \frac{1}{\delta}\sum_{n=0}^{N} \Re \overset{\text{dl}}{c}_n (X_m)^n \tag{5.30}$$

This system of equations at the M control points may be rearranged with the unknown coefficients on the left-hand side, which is expressed in matrix form as

$$\mathbf{A} = \begin{bmatrix} v_{\mathfrak{s}10} - \frac{(X_1)^0}{\delta} & v_{\mathfrak{s}11} - \frac{(X_1)^1}{\delta} & \cdots & v_{\mathfrak{s}1N} - \frac{(X_1)^N}{\delta} \\ v_{\mathfrak{s}20} - \frac{(X_2)^0}{\delta} & v_{\mathfrak{s}21} - \frac{(X_2)^1}{\delta} & \cdots & v_{\mathfrak{s}2N} - \frac{(X_2)^N}{\delta} \\ \cdots & \cdots & \cdots & \cdots \\ v_{\mathfrak{s}M0} - \frac{(X_M)^0}{\delta} & v_{\mathfrak{s}M1} - \frac{(X_M)^1}{\delta} & \cdots & v_{\mathfrak{s}MN} - \frac{(X_M)^N}{\delta} \end{bmatrix}, \quad \mathbf{c} = \begin{bmatrix} \Re \overset{\text{dl}}{c}_0 \\ \Re \overset{\text{dl}}{c}_1 \\ \vdots \\ \Re \overset{\text{dl}}{c}_N \end{bmatrix}, \quad \mathbf{b} = \begin{bmatrix} -\Re\left[\overset{\text{add}}{v}(z_1)\mathrm{e}^{-\mathrm{i}\vartheta}\right] \\ -\Re\left[\overset{\text{add}}{v}(z_2)\mathrm{e}^{-\mathrm{i}\vartheta}\right] \\ \vdots \\ -\Re\left[\overset{\text{add}}{v}(z_M)\mathrm{e}^{-\mathrm{i}\vartheta}\right] \end{bmatrix} \tag{5.31}$$

where $v_{\mathfrak{s}mn} = \Re[\overset{\text{dl}}{v}_n(\mathcal{Z}_m)\mathrm{e}^{-\mathrm{i}\vartheta}]$ represents the tangential component of influence function n at the element's control point m. The example in Fig. 5.6 illustrates this solution for an isolated line-dipole and strings of adjacent elements. In the limit as $\delta \to \infty$, the matrix $\mathbf{A}$ in the system of Eqs (5.31) reduces to

$$\mathbf{A} = \begin{bmatrix} v_{\mathfrak{s}10} & v_{\mathfrak{s}11} & \cdots & v_{\mathfrak{s}1N} \\ v_{\mathfrak{s}20} & v_{\mathfrak{s}21} & \cdots & v_{\mathfrak{s}2N} \\ \cdots & \cdots & \cdots & \cdots \\ v_{\mathfrak{s}M0} & v_{\mathfrak{s}M1} & \cdots & v_{\mathfrak{s}MN} \end{bmatrix} \tag{5.32}$$

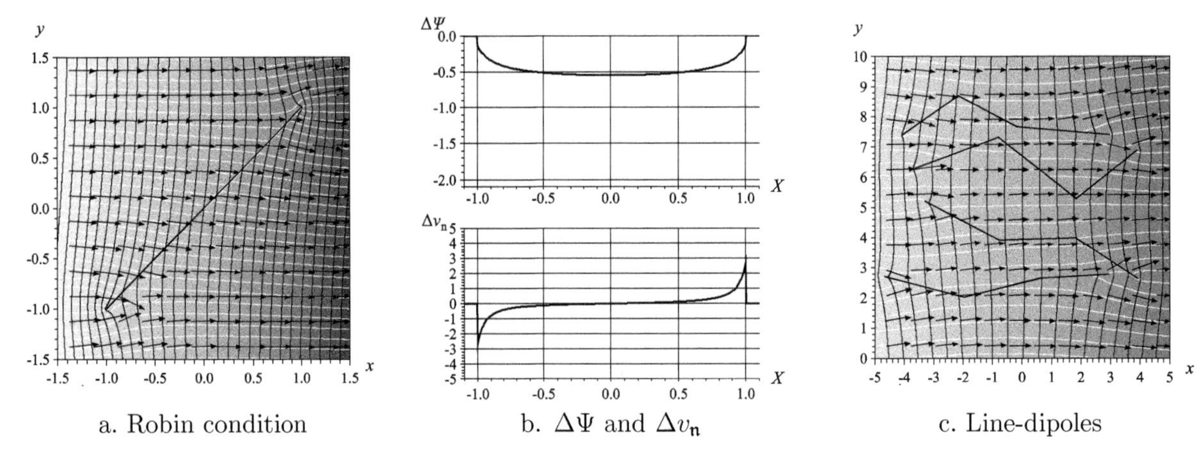

Figure 5.6 *Line-dipole with Robin condition.*

This solution reproduces the condition of uniform potential, (5.27), by setting the tangential component of the vector field equal to zero at each control point.

Problem 5.4 Compute the coefficients of a line-dipole with uniform potential for $N = 10$ and $M = 15$ in a background uniform vector field, (3.2), with $v_0 = 1$ and $\Phi_0 = 0$, and compute the complex potential $\Omega = \Phi + \mathrm{i}\Psi$ on each side of an element at its midpoint $z_c = (z_{\max} + z_{\min})/2$, for an element with specified endpoints:

A. $z_{\min} = 0,\ z_{\max} = 1 + 1\mathrm{i}$

B. $z_{\min} = 0,\ z_{\max} = 2 + 2\mathrm{i}$

C. $z_{\min} = 0,\ z_{\max} = 3 + 3\mathrm{i}$

D. $z_{\min} = 0,\ z_{\max} = 4 + 4\mathrm{i}$

E. $z_{\min} = 0,\ z_{\max} = 5 + 5\mathrm{i}$

F. $z_{\min} = 0,\ z_{\max} = 6 + 6\mathrm{i}$

G. $z_{\min} = 0,\ z_{\max} = 7 + 7\mathrm{i}$

H. $z_{\min} = 0,\ z_{\max} = 8 + 8\mathrm{i}$

I. $z_{\min} = 0,\ z_{\max} = 9 + 9\mathrm{i}$

J. $z_{\min} = 0,\ z_{\max} = 10 + 10\mathrm{i}$

K. $z_{\min} = 0,\ z_{\max} = 11 + 11\mathrm{i}$

L. $z_{\min} = 0,\ z_{\max} = 12 + 12\mathrm{i}$

Problem 5.5 Compute the coefficients of a line-dipole with Robin conditions using $\delta = 10$, $N = 10$, and $M = 15$; and compute the complex potential Ω on each side of an element at its midpoint for specified endpoints and uniform vector field in Problem 5.4.

5.2.2 Line-Doublet: Boundaries with Discontinuous Potential

A line-doublet is formed by a double layer, (5.10), with imaginary coefficients, (5.13). Its integral represents a distribution of point-dipoles, (3.4), with the axis of dipole normal to the element, and its complex potential and vector field are

$$\begin{aligned}\Omega &= \frac{\mathrm{i}}{2\pi}\int_{-1}^{1}\frac{\mu(\tilde{X})}{\mathcal{Z}-\tilde{X}}\mathrm{d}\tilde{X} \\ \mu &= \sum_{n=0}^{N}\Im\overset{\text{dl}}{c}_n X^n\end{aligned} \quad\rightarrow\quad \boxed{\begin{aligned}\Omega &= \sum_{n=0}^{N}\Im\overset{\text{dl}}{c}_n \times \mathrm{i}\overset{\text{dl}}{\Omega}_n(\mathcal{Z}) \\ v &= \sum_{n=0}^{N}\Im\overset{\text{dl}}{c}_n \times [-\mathrm{i}\overset{\text{dl}}{v}_n(\mathcal{Z})]\end{aligned}} \tag{5.33}$$

The closed-form expressions for the influences functions in (5.16) provide the potential and stream functions, and the vector field in (5.19) provides the components tangential and normal to the element:

$$\Phi + \mathrm{i}\Psi = \sum_{n=0}^{N} \frac{\mathrm{i}\Im\overset{\mathrm{dl}}{c}_n}{2\pi} \left(\mathcal{Z}^n \ln \frac{\mathcal{Z}+1}{\mathcal{Z}-1} - \sum_{l=1}^{\frac{n+1}{2}} \frac{2}{2l-1} \mathcal{Z}^{n+1-2l} \right)$$

$$v_{\mathrm{s}} + \mathrm{i}v_{\mathrm{n}} = \sum_{n=0}^{N} \frac{\mathrm{i}\Im\overset{\mathrm{dl}}{c}_n}{2\pi L} \left[\frac{(-1)^n}{\mathcal{Z}+1} - \frac{1}{\mathcal{Z}-1} + n\mathcal{Z}^{n-1} \ln \frac{\mathcal{Z}+1}{\mathcal{Z}-1} - n \sum_{l=1}^{\frac{n}{2}} \frac{2}{2l-1} \mathcal{Z}^{n-2l} \right] \quad (5.34)$$

The first few influence functions in Fig. 5.7 illustrate that ***the stream function and the normal component of the vector field are continuous across a line-dipole, while the potential function and the normal component of the vector field jump across it***.

These expressions for complex potential and the vector field of a line-doublet may be evaluated on the +side and the −side of the element, and their differences in the complex logarithm (5.23) may be expressed in terms of (5.24b) to give

$$\left(\Phi^+ + \mathrm{i}\Psi^+\right) - \left(\Phi^- + \mathrm{i}\Psi^-\right) = -\sum_{n=0}^{N} \Im\overset{\mathrm{dl}}{c}_n X^n \frac{\Delta\left(\theta_{\min} - \theta_{\max}\right)}{2\pi}$$

$$\left(v_{\mathrm{s}}^+ + \mathrm{i}v_{\mathrm{n}}^+\right) - \left(v_{\mathrm{s}}^- + \mathrm{i}v_{\mathrm{n}}^-\right) = \sum_{n=0}^{N} \frac{\Im\overset{\mathrm{dl}}{c}_n}{L} n X^{n-1} \frac{\Delta\left(\theta_{\min} - \theta_{\max}\right)}{2\pi} \quad (5.35)$$

Thus, ***a line-doublet creates a jump in potential of μ, the strength of the distribution of doublets along the element, from points immediately above to immediately below the element*** (Steward and Ahring, 2009, eqn 29):

$$\boxed{\Delta\Phi = \sum_{n=0}^{N} \Im\overset{\mathrm{dl}}{c}_n X^n = \mu(X)}, \quad \boxed{\Psi^+ = \Psi^-} \quad \begin{pmatrix} -1 < X < 1 \\ Y = 0 \end{pmatrix} \quad (5.36a)$$

Likewise, ***a line-doublet generates a jump in the tangential component of the vector field related to minus the derivative of μ along the element***:

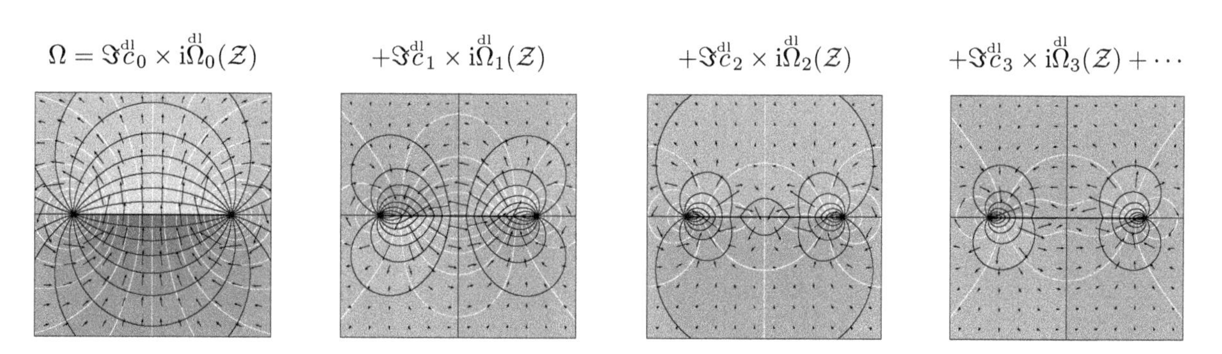

Figure 5.7 *Influence functions for a line-doublet, from (5.33) with (5.16) and (5.19).*

$$\boxed{\Delta v_{\mathrm{s}} = -\sum_{n=0}^{N} \Im \overset{\mathrm{dl}}{c}_n \frac{nX^{n-1}}{L} = -\frac{1}{L}\frac{\mathrm{d}\mu}{\mathrm{d}X}}, \quad \boxed{v_{\mathrm{n}}^{+} = v_{\mathrm{n}}^{-}} \quad \begin{pmatrix} -1 < X < 1 \\ Y = 0 \end{pmatrix} \tag{5.36b}$$

Problem 5.6 Compute Φ, Ψ, v_{s}, and v_{n} at points $z = -0.5$ and $z = 0.5$, for a double layer lying between $z_{\min} = -1$ and $z_{\max} = 1$ with the following coefficients:

A. $\overset{\mathrm{dl}}{c}_0 = \mathrm{i}$ B. $\overset{\mathrm{dl}}{c}_1 = \mathrm{i}$ C. $\overset{\mathrm{dl}}{c}_2 = \mathrm{i}$ D. $\overset{\mathrm{dl}}{c}_3 = \mathrm{i}$ E. $\overset{\mathrm{dl}}{c}_4 = \mathrm{i}$ F. $\overset{\mathrm{dl}}{c}_5 = \mathrm{i}$ G. $\overset{\mathrm{dl}}{c}_6 = \mathrm{i}$ H. $\overset{\mathrm{dl}}{c}_7 = \mathrm{i}$ I. $\overset{\mathrm{dl}}{c}_8 = \mathrm{i}$ J. $\overset{\mathrm{dl}}{c}_9 = \mathrm{i}$ K. $\overset{\mathrm{dl}}{c}_{10} = \mathrm{i}$ L. $\overset{\mathrm{dl}}{c}_{11} = \mathrm{i}$

The line-doublet element has a continuous stream function across the element, (5.36a), and its coefficients may be adjusted to satisfy a ***boundary condition of uniform stream function*** along the element, (2.59b). This solution is achieved by evaluating the potential at control point m, which is the real part of the complex potential in (5.33), and setting the value of all control points equal to an unknown constant Ψ_U (Steward, 2015, eqn 24):

$$\Psi_m = \sum_{n=0}^{N} \Im \overset{\mathrm{dl}}{c}_n \Im\left[\mathrm{i}\overset{\mathrm{dl}}{\Omega}_n(\mathcal{Z}_m)\right] + \overset{\mathrm{add}}{\Psi}(z_m) = \Psi_U \tag{5.37}$$

This provides a system of M equations with $N+2$ unknowns

$$\mathbf{A} = \begin{bmatrix} 1 & \Psi_{10} & \Psi_{11} & \cdots & \Psi_{1N} \\ 1 & \Psi_{20} & \Psi_{21} & \cdots & \Psi_{2N} \\ \cdots & \cdots & \cdots & \cdots & \cdots \\ 1 & \Psi_{M0} & \Psi_{M1} & \cdots & \Psi_{MN} \end{bmatrix}, \quad \mathbf{c} = \begin{bmatrix} -\Psi_U \\ \Im \overset{\mathrm{dl}}{c}_0 \\ \Im \overset{\mathrm{dl}}{c}_1 \\ \vdots \\ \Im \overset{\mathrm{dl}}{c}_N \end{bmatrix}, \quad \mathbf{b} = \begin{bmatrix} -\overset{\mathrm{add}}{\Psi}(z_1) \\ -\overset{\mathrm{add}}{\Psi}(z_2) \\ \vdots \\ -\overset{\mathrm{add}}{\Psi}(z_M) \end{bmatrix} \tag{5.38}$$

with $\Psi_{mn} = \Im[\mathrm{i}\overset{\mathrm{dl}}{\Omega}_n(\mathcal{Z}_m)]$. The least squares solution to this problem is shown in Fig. 5.8. Discontinuity occurs in both the potential (5.36a) and the tangential component of the vector field (5.36b), which is also show in the figure. While line-doublets can generate a non-zero jump in potential at the endpoints, the jump associated with a slit-doublet goes to zero at the endpoints, (3.81), and so is not appropriate by itself for strings of elements with discontinuities in potential. Thus, ***line-doublet elements can simulate uniform stream function along a contiguous string of adjacent elements, while slit-doublet elements cannot be interconnected at their endpoints***.

A line-doublet may be formulated to satisfy the ***Robin interface condition*** (2.66a), where the normal component of the vector field is continuous across the element and related to the jump in potential across the element at the mth control point by

$$(v_n)_m = -\frac{\Phi_m^{+} - \Phi_m^{-}}{\delta} \tag{5.39}$$

(Steward, 2015, eqn 27). This equation may be expanded using the series for the vector field in (5.33) with the normal component of the vector field for the additional functions, and the series for the discontinuity in the potential in (5.36a) to give

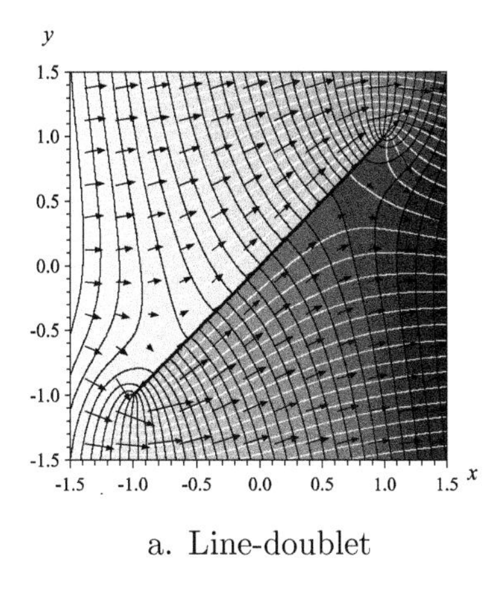

a. Line-doublet

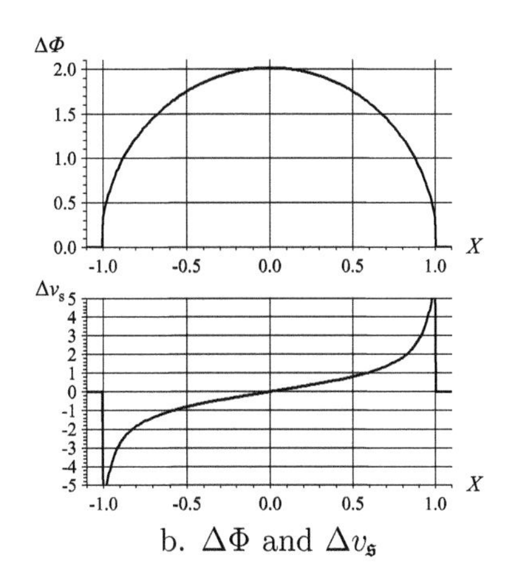

b. $\Delta\Phi$ and $\Delta v_{\mathfrak{s}}$

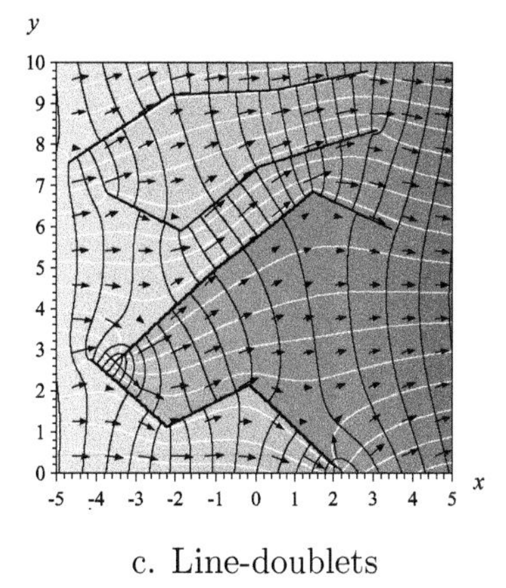

c. Line-doublets

Figure 5.8 *Line-doublet with uniform stream function.*

$$\sum_{n=0}^{N} \Im \overset{\text{dl}}{c}_n \Im\left[-\mathrm{i}\overset{\text{dl}}{v}_n(\mathcal{Z}_m)\mathrm{e}^{-\mathrm{i}\vartheta}\right] + \Im\left[\overset{\text{add}}{v}(z_m)\mathrm{e}^{-\mathrm{i}\vartheta}\right] = -\frac{1}{\delta}\sum_{n=0}^{N} \Im \overset{\text{dl}}{c}_n (X_m)^n \tag{5.40}$$

A solution for the coefficients is given by solving this set of equations at the M control points, which is expressed in matrix form using

$$\mathbf{A} = \begin{bmatrix} v_{\mathrm{n}\,10} + \frac{(X_1)^0}{\delta} & v_{\mathrm{n}\,11} + \frac{(X_1)^1}{\delta} & \cdots & v_{\mathrm{n}\,1N} + \frac{(X_1)^N}{\delta} \\ v_{\mathrm{n}\,20} + \frac{(X_2)^0}{\delta} & v_{\mathrm{n}\,21} + \frac{(X_2)^1}{\delta} & \cdots & v_{\mathrm{n}\,2N} + \frac{(X_2)^N}{\delta} \\ \cdots & \cdots & \cdots & \cdots \\ v_{\mathrm{n}\,M0} + \frac{(X_M)^0}{\delta} & v_{\mathrm{n}\,M1} + \frac{(X_M)^1}{\delta} & \cdots & v_{\mathrm{n}\,MN} + \frac{(X_M)^N}{\delta} \end{bmatrix}, \quad \mathbf{c} = \begin{bmatrix} \Im\overset{\text{dl}}{c}_0 \\ \Im\overset{\text{dl}}{c}_1 \\ \vdots \\ \Im\overset{\text{dl}}{c}_N \end{bmatrix}, \quad \mathbf{b} = \begin{bmatrix} -\Im\left[\overset{\text{add}}{v}(z_1)\mathrm{e}^{-\mathrm{i}\vartheta}\right] \\ -\Im\left[\overset{\text{add}}{v}(z_2)\mathrm{e}^{-\mathrm{i}\vartheta}\right] \\ \vdots \\ -\Im\left[\overset{\text{add}}{v}(z_M)\mathrm{e}^{-\mathrm{i}\vartheta}\right] \end{bmatrix} \tag{5.41}$$

where $v_{\mathrm{n}\,mn} = \Im[-\mathrm{i}\overset{\text{dl}}{v}_n(\mathcal{Z}_m)\mathrm{e}^{-\mathrm{i}\vartheta}]$ represents the normal component of influence function n at the element's control point m. An example is presented in Fig. 5.9 for a line-doublet with lower conductivity than the surrounding medium. In the limit as $\delta \to \infty$, the matrix $\mathbf{A}$ in the system of Eqs (5.41) reduces to

$$\mathbf{A} = \begin{bmatrix} v_{\mathrm{n}\,10} & v_{\mathrm{n}\,11} & \cdots & v_{\mathrm{n}\,1N} \\ v_{\mathrm{n}\,20} & v_{\mathrm{n}\,21} & \cdots & v_{\mathrm{n}\,2N} \\ \cdots & \cdots & \cdots & \cdots \\ v_{\mathrm{n}\,M0} & v_{\mathrm{n}\,M1} & \cdots & v_{\mathrm{n}\,MN} \end{bmatrix} \tag{5.42}$$

This solution reproduces the condition of uniform stream function, (5.37), by setting the normal component of the vector field equal to zero at each control point (Steward et al., 2008, eqn 4.7).

Problem 5.7 Compute the coefficients for a line-doublet with uniform stream function with $N = 10$ and $M = 15$; and compute the complex potential Ω on each side of an element at its midpoint for specified endpoints and uniform vector field in Problem 5.4.

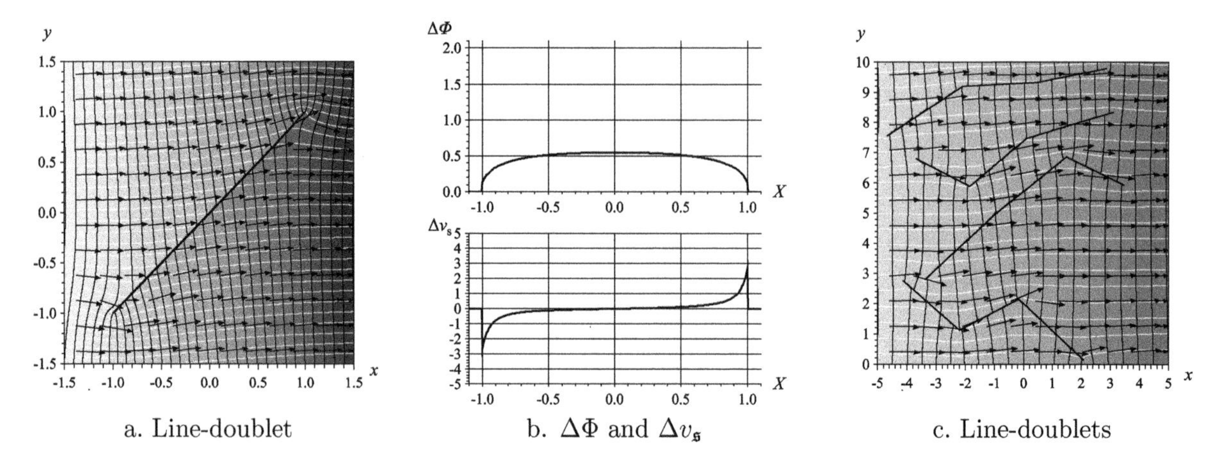

Figure 5.9 *Line-doublet with Robin condition.*

Problem 5.8 Compute the coefficients for a line-doublet with Robin conditions using $\delta = 10, N = 10$, and $M = 15$; and compute the complex potential Ω on each side of an element at its midpoint for specified endpoints and uniform vector field in Problem 5.4.

5.3 Single-Layer Elements

The single-layer elements presented in (5.4) are developed in this section to solve a wider range of interface conditions than those available with double layers. These singular integral equations are formulated using the same local power series approximation in terms of X used for the double layer in (5.13):

$$\sigma(X) + \mathrm{i}\gamma(X) = \frac{1}{L}\sum_{n=0}^{N} \overset{\mathrm{sl}}{c}_n X^n \quad \to \quad \begin{array}{l} \sigma(X) = \dfrac{1}{L}\displaystyle\sum_{n=0}^{N} \Re \overset{\mathrm{sl}}{c}_n X^n \\ \gamma(X) = \dfrac{1}{L}\displaystyle\sum_{n=0}^{N} \Im \overset{\mathrm{sl}}{c}_n X^n \end{array} \tag{5.43}$$

Since X is real along the element, the strengths of a line-sink (σ) and line-vortex (γ) follow directly in terms of either the real or imaginary part of the complex coefficient $\overset{\mathrm{sl}}{c}_n$. This polynomial approximation is substituted into the integral in (5.11) to obtain

$$\Omega = \frac{L}{2\pi}\int_{-1}^{1}\left[\sigma(\tilde{X}) + \mathrm{i}\gamma(\tilde{X})\right]\ln(\mathcal{Z} - \tilde{X})\mathrm{d}\tilde{X} = \sum_{n=0}^{N}\frac{\overset{\mathrm{sl}}{c}_n}{2\pi}\int_{-1}^{1}\tilde{X}^n \ln\left(\mathcal{Z} - \tilde{X}\right)\mathrm{d}\tilde{X} \tag{5.44}$$

Closed-form expressions are obtained using the integral from Gröbner and Hofreiter (1975, p.113), which is simplified by substituting the limits of integration and reorganizing the summation to skip terms where $1 - (-1)^l$ cancel:

$$\int_{-1}^{1} \tilde{X}^n \ln(\mathcal{Z} - \tilde{X}) \mathrm{d}\tilde{X} = \frac{1}{n+1} \left[\left(-\mathcal{Z}^{n+1} + \tilde{X}^{n+1} \right) \ln(\mathcal{Z} - \tilde{X}) - \sum_{l=1}^{n+1} \frac{1}{l} \tilde{X}^l \mathcal{Z}^{n+1-l} \right] \Bigg|_{\tilde{X}=-1}^{1}$$
$$= \frac{1}{n+1} \left[\mathcal{Z}^{n+1} \ln \frac{\mathcal{Z}+1}{\mathcal{Z}-1} + (-1)^n \ln(\mathcal{Z}+1) + \ln(\mathcal{Z}-1) - \sum_{l=1}^{\frac{n+2}{2}} \frac{2}{2l-1} \mathcal{Z}^{n+2-2l} \right] \tag{5.45}$$

This gives the ***complex potential of a single layer*** (Steward et al., 2008, eqn 2.12):

$$\Omega = \sum_{n=0}^{N} \overset{\text{sl}}{c}_n \overset{\text{sl}}{\Omega}_n(\mathcal{Z}) = \left[\sum_{n=0}^{N} \Re \overset{\text{sl}}{c}_n \times \overset{\text{sl}}{\Omega}_n(\mathcal{Z}) \right] + \left[\sum_{n=0}^{N} \Im \overset{\text{sl}}{c}_n \times \mathrm{i} \overset{\text{sl}}{\Omega}_n(\mathcal{Z}) \right],$$
$$\boxed{\overset{\text{sl}}{\Omega}_n[\mathcal{Z}(z)] = \frac{1}{2\pi} \frac{1}{n+1} \left[\mathcal{Z}^{n+1} \ln \frac{\mathcal{Z}+1}{\mathcal{Z}-1} + (-1)^n \ln(\mathcal{Z}+1) + \ln(\mathcal{Z}-1) - \sum_{l=1}^{\frac{n+2}{2}} \frac{2}{2l-1} \mathcal{Z}^{n+2-2l} \right]} \tag{5.46}$$

The summation on the first line with real coefficients provides terms necessary to achieve solutions for ***line-sink*** elements and those with the imaginary part of the coefficients provide terms for ***line-vortex*** elements.

The vector field for a single layer is obtained by differentiating (5.12a) the summation in (5.44) and taking the complex conjugate to give

$$v = -\frac{e^{\mathrm{i}\vartheta}}{L} \overline{\frac{\mathrm{d}\Omega}{\mathrm{d}\mathcal{Z}}} = -\sum_{n=0}^{N} \overline{\overset{\text{sl}}{c}_n} \overline{\frac{e^{\mathrm{i}\vartheta}}{2\pi L} \int_{-1}^{1} \frac{\tilde{X}^n}{\mathcal{Z} - \tilde{X}} \mathrm{d}\tilde{X}} \tag{5.47}$$

Integration is performed using (5.15) to obtain a closed-form expression for the ***complex vector of a single layer*** (Steward et al., 2008, eqn 2.15):

$$v = \sum_{n=0}^{N} \overline{\overset{\text{sl}}{c}_n} \overset{\text{sl}}{v}_n(\mathcal{Z}) = \left[\sum_{n=0}^{N} \Re \overset{\text{sl}}{c}_n \times \overset{\text{sl}}{v}_n(\mathcal{Z}) \right] - \left[\sum_{n=0}^{N} \Im \overset{\text{sl}}{c}_n \times \mathrm{i} \overset{\text{sl}}{v}_n(\mathcal{Z}) \right],$$
$$\boxed{\overset{\text{sl}}{v}_n[\mathcal{Z}(z)] = -\frac{e^{\mathrm{i}\vartheta}}{2\pi L} \overline{\left(\mathcal{Z}^n \ln \frac{\mathcal{Z}+1}{\mathcal{Z}-1} - \sum_{l=1}^{\frac{n+1}{2}} \frac{2}{2l-1} \mathcal{Z}^{n+1-2l} \right)}} \tag{5.48}$$

The first few influence functions for the complex potential are given by

$$\overset{\text{sink}}{\Omega}_0 = \frac{1}{2\pi} \left[\mathcal{Z} \ln \frac{\mathcal{Z}+1}{\mathcal{Z}-1} + \ln(\mathcal{Z}+1) + \ln(\mathcal{Z}-1) - 2 \right]$$
$$\overset{\text{sink}}{\Omega}_1 = \frac{1}{2\pi} \frac{1}{2} \left[\mathcal{Z}^2 \ln \frac{\mathcal{Z}+1}{\mathcal{Z}-1} - \ln(\mathcal{Z}+1) + \ln(\mathcal{Z}-1) - 2\mathcal{Z} \right]$$
$$\overset{\text{sink}}{\Omega}_2 = \frac{1}{2\pi} \frac{1}{3} \left[\mathcal{Z}^3 \ln \frac{\mathcal{Z}+1}{\mathcal{Z}-1} + \ln(\mathcal{Z}+1) + \ln(\mathcal{Z}-1) - 2\mathcal{Z}^2 - \frac{2}{3} \right]$$
$$\overset{\text{sink}}{\Omega}_3 = \frac{1}{2\pi} \frac{1}{4} \left[\mathcal{Z}^4 \ln \frac{\mathcal{Z}+1}{\mathcal{Z}-1} - \ln(\mathcal{Z}+1) + \ln(\mathcal{Z}-1) - 2\mathcal{Z}^3 - \frac{2}{3} \mathcal{Z} \right] \tag{5.49}$$

The influence functions for the vector field in (5.48) take on the same form as those for the complex potential of the double layer in (5.20).

Problem 5.9 Compute the influence functions $\overset{\text{sl}}{\Omega}_0$, $\overset{\text{sl}}{\Omega}_1$, $\overset{\text{sl}}{\Omega}_2$, $\overset{\text{sl}}{\Omega}_3$, $\overset{\text{sl}}{v}_0$, $\overset{\text{sl}}{v}_1$, $\overset{\text{sl}}{v}_2$, and $\overset{\text{sl}}{v}_3$ at the specified point z for a single layer lying between $z_{\min} = -1$ and $z_{\max} = 1$:

A. $z = -1 + \mathrm{i}2$ D. $z = -1 + \mathrm{i}$ G. $z = -1 - \mathrm{i}$ J. $z = -1 - \mathrm{i}2$

B. $z = \mathrm{i}2$ E. $z = \mathrm{i}$ H. $z = -\mathrm{i}$ K. $z = -\mathrm{i}2$

C. $z = 1 + \mathrm{i}2$ F. $z = 1 + \mathrm{i}$ I. $z = 1 - \mathrm{i}$ L. $z = 1 - \mathrm{i}2$

5.3.1 Line-Sink: Boundaries with Divergence and Discontinuous v_n

A line-sink is formed by a single layer, (5.11), with real coefficients, (5.43). Its integral represents a distribution of point-sinks, (3.1), that removes a flux per unit length σ along the element, and its complex potential and vector field are expressed as linear combinations of coefficients times influence functions:

$$\left.\begin{aligned} \Omega &= \frac{L}{2\pi}\int_{-1}^{1} \sigma(\tilde{X})\ln(\mathcal{Z} - \tilde{X})\mathrm{d}\tilde{X} \\ \sigma &= \frac{1}{L}\sum_{n=0}^{N} \Re\overset{\text{sl}}{c}_n X^n \end{aligned}\right. \quad \rightarrow \quad \boxed{\begin{aligned} \Omega &= \sum_{n=0}^{N} \Re\overset{\text{sl}}{c}_n \times \overset{\text{sl}}{\Omega}_n(\mathcal{Z}) \\ v &= \sum_{n=0}^{N} \Re\overset{\text{sl}}{c}_n \times \overset{\text{sl}}{v}_n(\mathcal{Z}) \end{aligned}} \tag{5.50}$$

using the representation for the complex potential from (5.46), and vector field from (5.48) Closed-form expressions are obtained using these integrations to provide the potential and stream function, and the vector field is rotated, (5.12b), to provide the components tangential and normal to the element:

$$\begin{aligned} \Phi + \mathrm{i}\Psi &= \sum_{n=0}^{N} \frac{\Re\overset{\text{sl}}{c}_n}{2\pi(n+1)} \left[\mathcal{Z}^{n+1} \ln\frac{\mathcal{Z}+1}{\mathcal{Z}-1} + (-1)^n \ln(\mathcal{Z}+1) + \ln(\mathcal{Z}-1) - \sum_{l=1}^{\frac{n+2}{2}} \frac{2}{2l-1}\mathcal{Z}^{n+2-2l} \right] \\ v_s + \mathrm{i}v_n &= -\sum_{n=0}^{N} \frac{\Re\overset{\text{sl}}{c}_n}{2\pi L} \overline{\left(\mathcal{Z}^n \ln\frac{\mathcal{Z}+1}{\mathcal{Z}-1} - \sum_{l=1}^{\frac{n+1}{2}} \frac{2}{2l-1}\mathcal{Z}^{n+1-2l} \right)} \end{aligned} \tag{5.51}$$

The influence functions in Fig. 5.10 illustrate that ***the potential and tangential component of the vector field are continuous across a line-sink, while the stream function and normal components jump across it,*** similar to the line-dipole in Fig. 5.3.

These expressions for complex potential and the vector field of a line-sink may be evaluated on the +side and the −side of the element. The branch cuts in the logarithm terms (5.23) that are discontinuous across the element give

$$\begin{aligned} (\Phi^+ + \mathrm{i}\Psi^+) - (\Phi^- + \mathrm{i}\Psi^-) &= \sum_{n=0}^{N} \frac{\Re\overset{\text{sl}}{c}_n}{(n+1)} \mathrm{i} \left[X^{n+1}\frac{\Delta(\theta_{\min} - \theta_{\max})}{2\pi} + (-1)^n \frac{\Delta\theta_{\min}}{2\pi} + \frac{\Delta\theta_{\max}}{2\pi} \right] \\ (v_s^+ + \mathrm{i}v_n^+) - (v_s^- + \mathrm{i}v_n^-) &= \sum_{n=0}^{N} \frac{\Re\overset{\text{sl}}{c}_n}{L} X^n \mathrm{i}\frac{\Delta(\theta_{\min} - \theta_{\max})}{2\pi} \end{aligned} \tag{5.52}$$

$$\Omega = \Re\overset{\text{sl}}{c}_0 \times \overset{\text{sl}}{\Omega}_0(\mathcal{Z}) \qquad + \Re\overset{\text{sl}}{c}_1 \times \overset{\text{sl}}{\Omega}_1(\mathcal{Z}) \qquad + \Re\overset{\text{sl}}{c}_2 \times \overset{\text{sl}}{\Omega}_2(\mathcal{Z}) \qquad + \Re\overset{\text{sl}}{c}_3 \times \overset{\text{sl}}{\Omega}_3(\mathcal{Z}) + \cdots$$

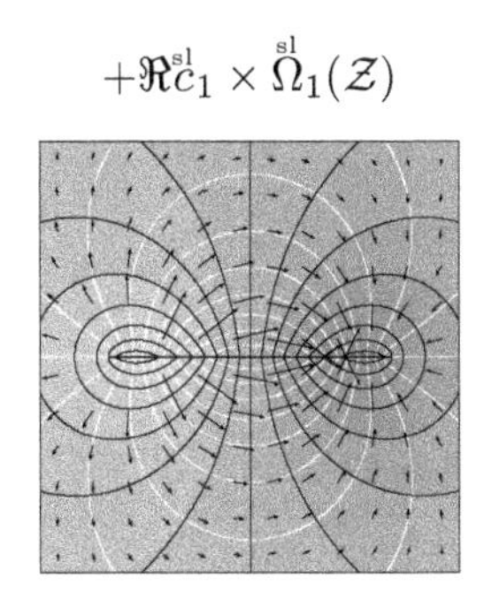

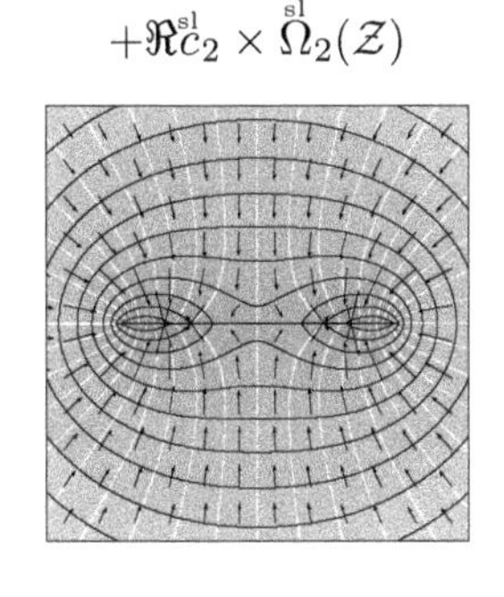

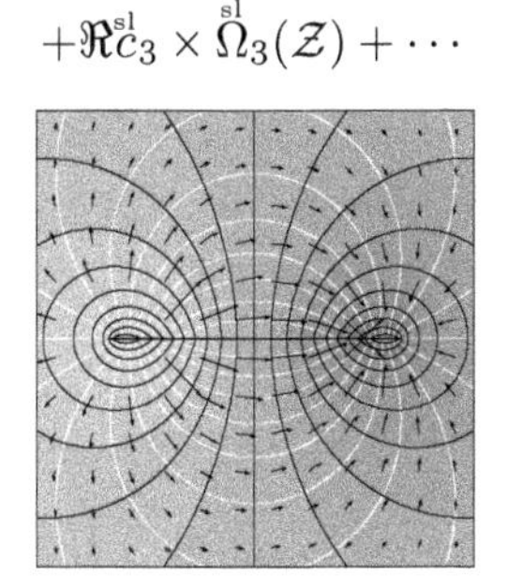

Figure 5.10 *Influence functions for a line-sink, from (5.50) with (5.46) and (5.48).*

where the difference in these angles across the element is given by (5.24b). Thus, ***a line-sink creates a jump in stream function related to the integral of σ, the strength of the distribution of sinks along the element***:

$$\boxed{\Phi^+ = \Phi^-} \;, \quad \boxed{\Delta\Psi = \begin{cases} 0 & (1 < X \,,\; Y = 0) \\ \sum\limits_{n=0}^{N} \Re\overset{\text{sl}}{c}_n \frac{1-X^{n+1}}{n+1} = L \int\limits_X^1 \sigma(\tilde{X})\mathrm{d}\tilde{X} & (-1 \le X \le 1 \,,\; Y = 0) \\ \sum\limits_{n=0}^{N} \Re\overset{\text{sl}}{c}_n \frac{1+(-1)^n}{n+1} = L \int\limits_{-1}^1 \sigma(\tilde{X})\mathrm{d}\tilde{X} & (X < -1 \,,\; Y = 0) \end{cases}} \tag{5.53a}$$

The discontinuity in $\Delta\Psi$ also exists along the negative X-axis and is equal to the integral of σL along the element, which is the net flux removed from the vector field by the line-sink. Thus, ***a line-sink generates a jump in the normal component of the vector field of $-\sigma$, which is equal to the flux per unit length removed from the vector field by a line-sink from points immediately above to immediately below the element***:

$$\boxed{v_{\mathfrak{s}}^+ = v_{\mathfrak{s}}^-} \;, \quad \boxed{\Delta v_{\mathfrak{n}} = -\sum_{n=0}^{N} \frac{\Re\overset{\text{sl}}{c}_n X^n}{L} = -\sigma(X)} \quad \begin{pmatrix} -1 < X < 1 \\ Y = 0 \end{pmatrix} \tag{5.53b}$$

The net flux removed by the line-sink between the end at $X = 1$ and a point at location $\mathfrak{s}$ on the element is given by

$$\Delta\Psi = -\int\limits_{\mathfrak{s}}^{L} \Delta v_{\mathfrak{n}}(\mathfrak{s})\mathrm{d}\mathfrak{s} = -L\int\limits_X^1 \Delta v_{\mathfrak{n}}(\tilde{X})\mathrm{d}\tilde{X} = \int\limits_X^1 \sum_{n=0}^{N} \overset{\text{sink}}{c}_n \tilde{X}^n \mathrm{d}\tilde{X} = \sum_{n=0}^{N} \frac{\overset{\text{sink}}{c}_n}{n+1}\left(1 - X^{n+1}\right) \tag{5.54}$$

Thus, ***the jump in stream function, (5.53a), represents the net flux removed between a point on the element and the end at $X = 1$***.

Problem 5.10 Compute Φ, Ψ, $v_{\mathfrak{s}}$, and $v_{\mathfrak{n}}$ at points $z = -0.5$ and $z = 0.5$, for a single layer lying between $z_{\min} = -1$ and $z_{\max} = 1$ with the following coefficients:

A. $\overset{\text{sl}}{c}_0 = 1$ C. $\overset{\text{sl}}{c}_2 = 1$ E. $\overset{\text{sl}}{c}_4 = 1$ G. $\overset{\text{sl}}{c}_6 = 1$ I. $\overset{\text{sl}}{c}_8 = 1$ K. $\overset{\text{sl}}{c}_{10} = 1$

B. $\overset{\text{sl}}{c}_1 = 1$ D. $\overset{\text{sl}}{c}_3 = 1$ F. $\overset{\text{sl}}{c}_5 = 1$ H. $\overset{\text{sl}}{c}_7 = 1$ J. $\overset{\text{sl}}{c}_9 = 1$ L. $\overset{\text{sl}}{c}_{11} = 1$

The line-sink element with its continuous potential across the element, (5.53a), may be formulated to satisfy a boundary condition of ***specified potential*** along the element, (2.59a). The real part of the complex potential in (5.50) is set equal to a specified value of the potential Φ_m at control point m (Steward, 2015, eqn 31):

$$\Phi_m = \sum_{n=0}^{N} \Re \overset{\text{sl}}{c}_n \Re \left[\overset{\text{sl}}{\Omega}_n(\mathcal{Z}_m) \right] + \overset{\text{add}}{\Phi}(z_m) \tag{5.55}$$

The terms associated with the unknown coefficients for the line-sink are separated on the left-hand side and organized as a system of M equations with $N+1$ unknowns in matrix form

$$\mathbf{A} = \begin{bmatrix} \Phi_{10} & \Phi_{11} & \cdots & \Phi_{1N} \\ \Phi_{20} & \Phi_{21} & \cdots & \Phi_{2N} \\ \cdots & \cdots & \cdots & \cdots \\ \Phi_{M0} & \Phi_{M1} & \cdots & \Phi_{MN} \end{bmatrix}, \quad \mathbf{c} = \begin{bmatrix} \Re \overset{\text{sl}}{c}_0 \\ \Re \overset{\text{sl}}{c}_1 \\ \vdots \\ \Re \overset{\text{sl}}{c}_N \end{bmatrix}, \quad \mathbf{b} = \begin{bmatrix} \Phi_1 - \overset{\text{add}}{\Phi}(z_1) \\ \Phi_2 - \overset{\text{add}}{\Phi}(z_2) \\ \vdots \\ \Phi_m - \overset{\text{add}}{\Phi}(z_M) \end{bmatrix} \tag{5.56}$$

with $\Phi_{mn} = \Re[\overset{\text{sl}}{\Omega}_n(\mathcal{Z}_m)]$. The least squares solution is shown in Fig. 5.11 for a single element with constant potential and for a set of elements with linearly varying potential along each element. A discontinuity exists in both the stream function and the normal component of the vector field, which is also shown in the figure. Since a net flux is removed by these line-sinks, the branch cut in the stream function extends to infinity in Fig. 5.11a.

A line-sink may be formulated to solve boundary conditions for ***specified resistance*** where the potential of the element Φ_m is specified at the control points but is separated from the potential $\Phi(z_m)$ at this point by a zone with higher resistance. This boundary satisfies a ***Robin interface condition*** (2.66a), where a net input to the domain along the line-sink is equal to the difference in potential divided by δ:

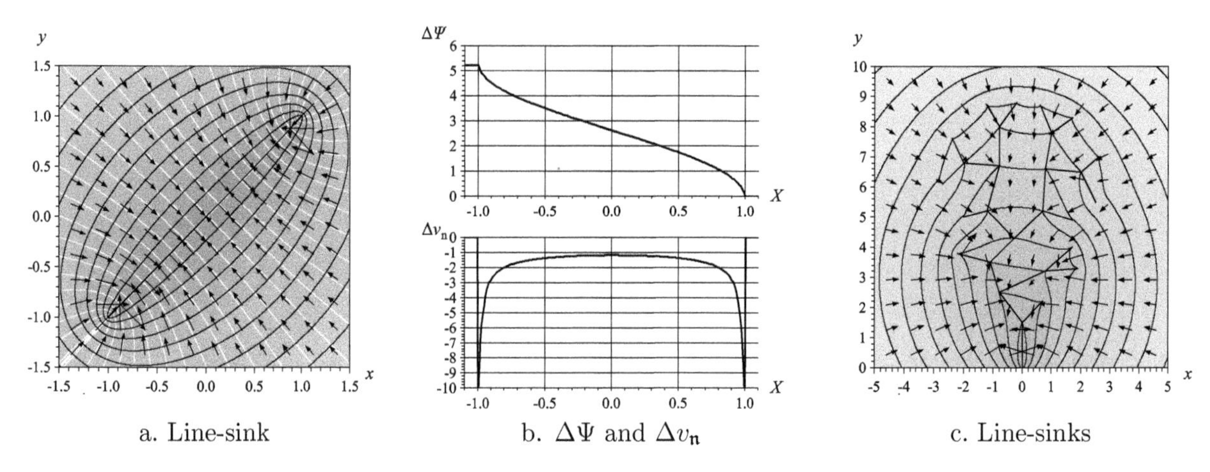

Figure 5.11 *Line-sink with specified potential.*

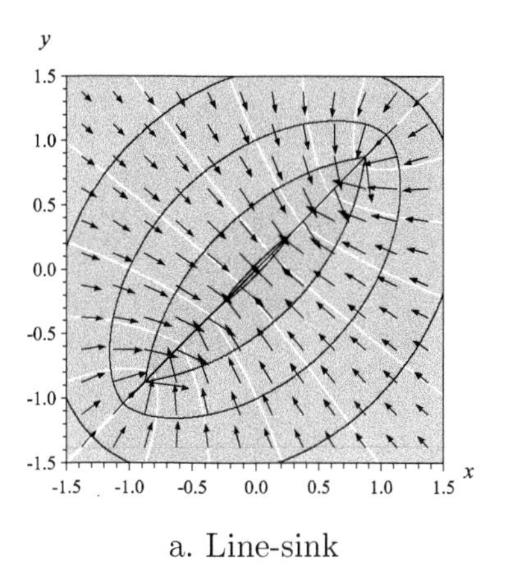

a. Line-sink

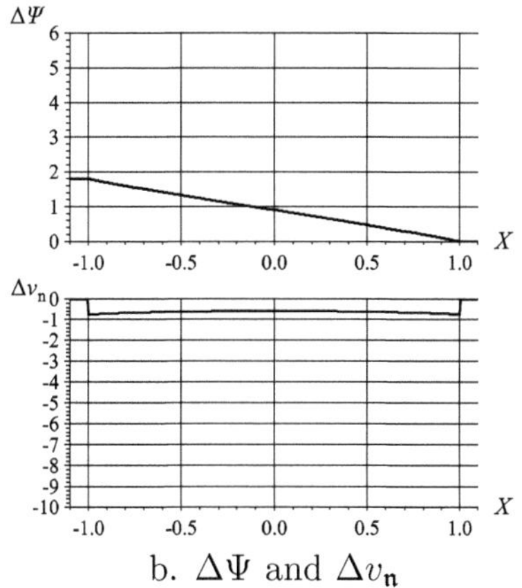

b. $\Delta\Psi$ and Δv_n

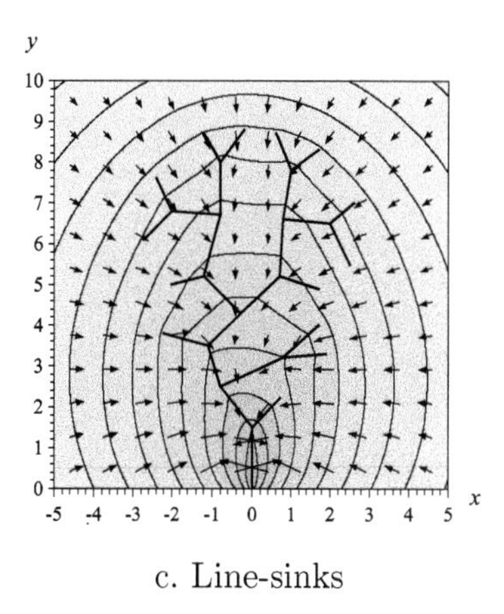

c. Line-sinks

Figure 5.12 *Line-sink with specified resistance.*

$$(\Delta v_n)_m = \frac{\Phi_m - \Phi(z_m)}{\delta} \tag{5.57}$$

This equation may be expanded using the series for the potential in (5.50) with the additional function, and the series for the discontinuity in the normal component of the vector field in (5.53b) to give

$$-\delta \sum_{n=0}^{N} \frac{\Re \overset{\text{sl}}{c}_n X^n}{L} = \Phi_m - \sum_{n=0}^{N} \Re \overset{\text{sl}}{c}_n \Re\left[\overset{\text{sl}}{\Omega}_n(Z_m)\right] - \overset{\text{add}}{\Phi}(z_m) \tag{5.58}$$

This provides the same system of equations as (5.56), but with additional terms in the coefficient matrix to account for resistance:

$$\mathbf{A} = \begin{bmatrix} \Phi_{10} - \frac{\delta (X_1)^0}{L} & \Phi_{11} - \frac{\delta (X_1)^1}{L} & \cdots & \Phi_{1N} - \frac{\delta (X_1)^N}{L} \\ \Phi_{20} - \frac{\delta (X_2)^0}{L} & \Phi_{21} - \frac{\delta (X_2)^1}{L} & \cdots & \Phi_{2N} - \frac{\delta (X_2)^N}{L} \\ \cdots & \cdots & \cdots & \cdots \\ \Phi_{M0} - \frac{\delta (X_M)^0}{L} & \Phi_{M1} - \frac{\delta (X_M)^1}{L} & \cdots & \Phi_{MN} - \frac{\delta (X_M)^N}{L} \end{bmatrix} \tag{5.59}$$

An example is presented in Fig. 5.12 for a line-sink with resistance.

Problem 5.11 A line-sink is placed in a uniform background $\Omega = -\overline{v_0} z + \Phi_0$ where $v_0 = 1$ and $\Phi_0 = 0$. Compute Φ, Ψ, v_s, and v_n on each side of an element at its midpoint $z_c = (z_{\max} + z_{\min})/2$, and compute the coefficients $\overset{\text{sl}}{c}_n$ for $N = 10$ coefficients and $M = 15$ control points to satisfy the specified potential $\Phi_m = 1$ along the line-sink with endpoints:

A. $z_{\min} = 0$, $z_{\max} = 1 + 1\text{i}$
B. $z_{\min} = 0$, $z_{\max} = 2 + 2\text{i}$
C. $z_{\min} = 0$, $z_{\max} = 3 + 3\text{i}$
D. $z_{\min} = 0$, $z_{\max} = 4 + 4\text{i}$
E. $z_{\min} = 0$, $z_{\max} = 5 + 5\text{i}$
F. $z_{\min} = 0$, $z_{\max} = 6 + 6\text{i}$
G. $z_{\min} = 0$, $z_{\max} = 7 + 7\text{i}$
H. $z_{\min} = 0$, $z_{\max} = 8 + 8\text{i}$
I. $z_{\min} = 0$, $z_{\max} = 9 + 9\text{i}$
J. $z_{\min} = 0$, $z_{\max} = 10 + 10\text{i}$
K. $z_{\min} = 0$, $z_{\max} = 11 + 11\text{i}$
L. $z_{\min} = 0$, $z_{\max} = 12 + 12\text{i}$

Problem 5.12 Repeat the calculations from Problem 5.11 for a resistance-specified line-sink with $\delta = 1, N = 10$, and $M = 15$.

Problem 5.13 A line-sink is placed in a uniform background $\Omega = -\overline{v_0}z + \Phi_0$ with $N = 10$ coefficients and $M = 15$ control points with specified potential $\Phi_m = 1$ and endpoints $z_{\text{min}} = -1 - \text{i}$ and $z_{\text{max}} = 1 + \text{i}$. Compute the coefficients $\overset{\text{sl}}{c}_n$ to satisfy the specified conditions along the line-sink while solving for Φ_0, v_0 to satisfy the specified potential at $z_1 = -2 + 2\text{i}$, $z_2 = 2 + 2\text{i}$, and $z_3 = 2 - 2\text{i}$.

	A.	B.	C.	D.	E.	F.	G.	H.	I.	J.	K.	L.
Φ_1	2	3	4	5	1	1	1	1	1	1	1	1
Φ_2	1	1	1	1	2	3	4	5	1	1	1	1
Φ_3	1	1	1	1	1	1	1	1	2	3	4	5

Note that this solution requires an iterative procedure that independently solves for the uniform flow components and the element components, following (2.9). This was solved using Successive Over-Relaxation where the coefficient for the line used 0.75 times the previous solution and 0.25 times the next iterates a matrix solution for the coefficients.

5.3.2 Line-Vortex: Boundaries with Circulation and Discontinuous v_s

A line-vortex is formed by a single layer, (5.11), with imaginary coefficients, (5.43). Its integral represents a distribution of point-vortexes or vortex filaments (Helmholtz, 1858; Tait, 1867) along a line with circulation γ from (3.3),

$$\begin{aligned} \Omega &= \frac{L}{2\pi}\int_{-1}^{1} \text{i}\gamma(\tilde{X})\ln(\mathcal{Z}-\tilde{X})\,\text{d}\tilde{X} \\ \gamma &= \frac{1}{L}\sum_{n=0}^{N} \Im\overset{\text{sl}}{c}_n X^n \end{aligned} \quad \rightarrow \quad \boxed{\begin{aligned} \Omega &= \sum_{n=0}^{N} \Im\overset{\text{sl}}{c}_n \times \text{i}\overset{\text{sl}}{\Omega}_n(\mathcal{Z}) \\ v &= \sum_{n=0}^{N} \Im\overset{\text{sl}}{c}_n \times [-\text{i}\overset{\text{sl}}{v}_n(\mathcal{Z})] \end{aligned}} \tag{5.60}$$

where the influence functions for the complex potential and vector field are found in (5.46) and (5.48). Integration with closed-form expressions from (5.60) provide expressions for the potential and stream function, and the components of the vector field rotated, (5.12b), to provide the components tangential and normal to the element:

$$\begin{aligned} \Phi + \text{i}\Psi &= \sum_{n=0}^{N} \frac{\text{i}\Im\overset{\text{sl}}{c}_n}{2\pi(n+1)} \left[\mathcal{Z}^{n+1}\ln\frac{\mathcal{Z}+1}{\mathcal{Z}-1} + (-1)^n\ln(\mathcal{Z}+1) + \ln(\mathcal{Z}-1) - \sum_{l=1}^{\frac{n+2}{2}} \frac{2}{2l-1}\mathcal{Z}^{n+2-2l} \right] \\ v_s + \text{i}v_n &= \overline{\sum_{n=0}^{N} \frac{\text{i}\Im\overset{\text{sl}}{c}_n}{2\pi L} \left(\mathcal{Z}^n\ln\frac{\mathcal{Z}+1}{\mathcal{Z}-1} - \sum_{l=1}^{\frac{n+1}{2}} \frac{2}{2l-1}\mathcal{Z}^{n+1-2l} \right)} \end{aligned} \tag{5.61}$$

The complex potential and vector field for the influence functions of a line-vortex are illustrated in Fig. 5.13, and show that ***the stream function and normal component of the vector field are continuous across a line-vortex, while the potential and tangential components jump across it***, similar to the line-doublet in Fig. 5.7.

These expressions for complex potential and the vector field of a line-vortex may be evaluated on the +side and the −side of the element, with the branch cuts in the logarithm terms (5.23) to obtain

$\Omega = \Im\overset{\text{sl}}{c}_0 \times \mathrm{i}\overset{\text{sl}}{\Omega}_0(\mathcal{Z})$ $+\Im\overset{\text{sl}}{c}_1 \times \mathrm{i}\overset{\text{sl}}{\Omega}_1(\mathcal{Z})$ $+\Im\overset{\text{sl}}{c}_2 \times \mathrm{i}\overset{\text{sl}}{\Omega}_2(\mathcal{Z})$ $+\Im\overset{\text{sl}}{c}_3 \times \mathrm{i}\overset{\text{sl}}{\Omega}_3(\mathcal{Z}) + \cdots$

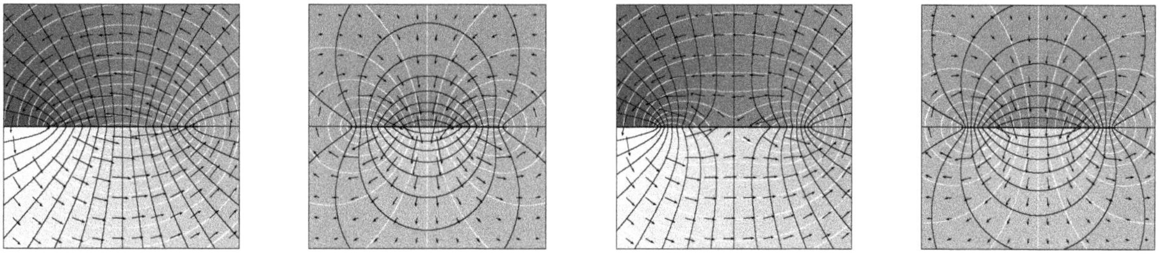

Figure 5.13 *Influence functions for a line-vortex, from (5.60) with (5.46) and (5.48).*

$$
\begin{aligned}
\left(\Phi^{+} + \mathrm{i}\Psi^{+}\right) - \left(\Phi^{-} + \mathrm{i}\Psi^{-}\right) &= -\sum_{n=0}^{N} \frac{\Im\overset{\text{sl}}{c}_n}{(n+1)} \left[X^{n+1} \frac{\Delta\left(\theta_{\min} - \theta_{\max}\right)}{2\pi} + (-1)^n \frac{\Delta\theta_{\min}}{2\pi} + \frac{\Delta\theta_{\max}}{2\pi} \right] \\
\left(v_{\mathfrak{s}}^{+} + \mathrm{i}v_{\mathrm{n}}^{+}\right) - \left(v_{\mathfrak{s}}^{-} + \mathrm{i}v_{\mathrm{n}}^{-}\right) &= \sum_{n=0}^{N} \frac{\Im\overset{\text{sl}}{c}_n}{L} X^n \frac{\Delta\left(\theta_{\min} - \theta_{\max}\right)}{2\pi}
\end{aligned}
\tag{5.62}
$$

where the difference in these angles across the element is given by (5.24). Thus, ***a line-vortex creates a jump in potential related to the integral of γ, the strength of the distribution of vortexes along the element***:

$$
\boxed{\Delta\Phi = \begin{cases} 0 & (1 < X\,,\; Y = 0) \\ -\sum\limits_{n=0}^{N} \Im\overset{\text{sl}}{c}_n \frac{1 - X^{n+1}}{n+1} = -L\int\limits_{X}^{1} \gamma(\tilde{X})\mathrm{d}\tilde{X} & (-1 \le X \le 1\,,\; Y = 0) \\ -\sum\limits_{n=0}^{N} \Im\overset{\text{sl}}{c}_n \frac{1 + (-1)^n}{n+1} = -L\int\limits_{-1}^{1} \gamma(\tilde{X})\mathrm{d}\tilde{X} & (X < -1\,,\; Y = 0) \end{cases}}\,, \quad \boxed{\Psi^{+} = \Psi^{-}}
\tag{5.63a}
$$

A line-vortex generates a jump in the tangential component of the vector field of $-\gamma$, from points immediately above to immediately below the element:

$$
\boxed{\Delta v_{\mathfrak{s}} = -\sum_{n=0}^{N} \frac{\Im\overset{\text{sl}}{c}_n X^n}{L} = -\gamma(X)}\,, \quad \boxed{v_{\mathrm{n}}^{+} = v_{\mathrm{n}}^{-}} \quad \begin{pmatrix} -1 < X < 1 \\ Y = 0 \end{pmatrix}
\tag{5.63b}
$$

The net circulation generated by a line-vortex between the end at $X = 1$ and a point at location $\mathfrak{s}$ on the element is given by

$$
-\Delta\Phi = -\int_{\mathfrak{s}}^{L} \Delta v_{\mathfrak{s}}(\mathfrak{s})\mathrm{d}\mathfrak{s} = -L\int_{X}^{1} \Delta v_{\mathfrak{s}}(\tilde{X})\mathrm{d}\tilde{X} = \int_{X}^{1} \sum_{n=0}^{N} \Im\overset{\text{vort}}{c}_n \tilde{X}^n \mathrm{d}\tilde{X} = \sum_{n=0}^{N} \frac{\Im\overset{\text{vort}}{c}_n}{n+1}\left(1 - X^{n+1}\right)
\tag{5.64}
$$

Thus, ***the jump in potential, (5.63a), is equal to minus the net circulation generated between a point on the element and the end at $X = 1$.***

Problem 5.14 Compute Φ, Ψ, $v_{\mathfrak{s}}$, and $v_{\mathfrak{n}}$ at points $z = -0.5$ and $z = 0.5$, for a single layer lying between $z_{\min} = -1$ and $z_{\max} = 1$ with the following coefficients:

A. $\overset{\text{sl}}{c}_0 = \mathrm{i}$
B. $\overset{\text{sl}}{c}_1 = \mathrm{i}$
C. $\overset{\text{sl}}{c}_2 = \mathrm{i}$
D. $\overset{\text{sl}}{c}_3 = \mathrm{i}$
E. $\overset{\text{sl}}{c}_4 = \mathrm{i}$
F. $\overset{\text{sl}}{c}_5 = \mathrm{i}$
G. $\overset{\text{sl}}{c}_6 = \mathrm{i}$
H. $\overset{\text{sl}}{c}_7 = \mathrm{i}$
I. $\overset{\text{sl}}{c}_8 = \mathrm{i}$
J. $\overset{\text{sl}}{c}_9 = \mathrm{i}$
K. $\overset{\text{sl}}{c}_{10} = \mathrm{i}$
L. $\overset{\text{sl}}{c}_{11} = \mathrm{i}$

The line-vortex element with its continuous stream function across the element, (5.63a), may be formulated to satisfy a boundary condition of ***uniform potential***, where the stream function at control point m, in (5.60), is set equal to an unknown constant Ψ (Steward, 2015, eqn 36):

$$\Psi_m = \sum_{n=0}^{N} \Im \overset{\text{sl}}{c}_n \Im \left[\mathrm{i} \overset{\text{sl}}{\Omega}_n(\mathcal{Z}_m) \right] + \overset{\text{add}}{\Psi}(z_m) = \Psi_U \tag{5.65}$$

An additional ***Kutta condition*** removes the singularity at the left end of the element by setting the jump in tangential component in (5.63b) equal to zero at $X = -1$, giving

$$\Delta v_{\mathfrak{s}} = -\sum_{n=0}^{N} \frac{\Im \overset{\text{sl}}{c}_n (-1)^n}{L} = 0 \tag{5.66}$$

This provides a system of $M + 1$ equations with $N + 2$ unknowns that is written in matrix form as $\mathbf{Ac} = \mathbf{b}$ with

$$\mathbf{A} = \begin{bmatrix} 1 & \Psi_{10} & \Psi_{11} & \cdots & \Psi_{1N} \\ 1 & \Psi_{20} & \Psi_{21} & \cdots & \Psi_{2N} \\ \cdots & \cdots & \cdots & \cdots & \cdots \\ 1 & \Psi_{M0} & \Psi_{M1} & \cdots & \Psi_{MN} \\ 0 & \frac{1}{L} & -\frac{1}{L} & \cdots & \frac{(-1)^N}{L} \end{bmatrix}, \quad \mathbf{c} = \begin{bmatrix} -\Psi_U \\ \Im \overset{\text{sl}}{c}_0 \\ \Im \overset{\text{sl}}{c}_1 \\ \vdots \\ \Im \overset{\text{sl}}{c}_N \end{bmatrix}, \quad \mathbf{b} = \begin{bmatrix} -\overset{\text{add}}{\Psi}(z_1) \\ -\overset{\text{add}}{\Psi}(z_2) \\ \vdots \\ -\overset{\text{add}}{\Psi}(z_M) \\ 0 \end{bmatrix} \tag{5.67}$$

and the coefficients $\Psi_{mn} = \Im[\mathrm{i}\overset{\text{sl}}{\Omega}_n(\mathcal{Z}_m)]$ at control point m for coefficient n of a single layer. The vector fields in Fig. 5.14 illustrate a single element that satisfies a Kutta condition, and two strings of elements where the left-most element of each string satisfies this condition. The net circulation in these solutions results in a branch cut in the potential function that extends to infinity.

The line-vortex element may be formulated to satisfy a Dirichlet boundary condition of ***specified stream function*** along the element, (2.59a). The imaginary part of the complex potential in (5.60), is set equal to a specified value of the potential Ψ_m at control point m (Steward, 2015, eqn 39):

$$\Psi_m = \sum_{n=0}^{N} \Im \overset{\text{sl}}{c}_n \Im \left[\mathrm{i} \overset{\text{sl}}{\Omega}_n(\mathcal{Z}_m) \right] + \overset{\text{add}}{\Psi}(z_m) \tag{5.68}$$

The terms associated with the unknown coefficients for the line-vortex are separated on the left-hand side and organized as a system of M equations with $N + 1$ unknowns in matrix form

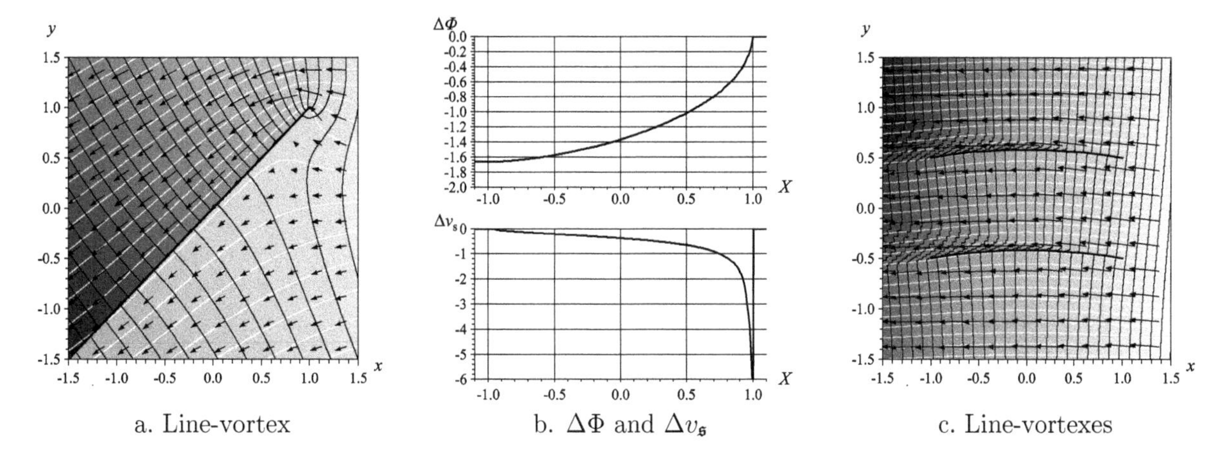

Figure 5.14 *Line-vortex with Kutta condition.*

$$
\mathbf{A} = \begin{bmatrix} \Psi_{10} & \Psi_{11} & \cdots & \Psi_{1N} \\ \Psi_{20} & \Psi_{21} & \cdots & \Psi_{2N} \\ \cdots & \cdots & \cdots & \cdots \\ \Psi_{M0} & \Psi_{M1} & \cdots & \Psi_{MN} \end{bmatrix}, \quad \mathbf{c} = \begin{bmatrix} \Im \overset{\text{sl}}{c}_0 \\ \Im \overset{\text{sl}}{c}_1 \\ \vdots \\ \Im \overset{\text{sl}}{c}_N \end{bmatrix}, \quad \mathbf{b} = \begin{bmatrix} \Psi_1 - \overset{\text{add}}{\Psi}(z_1) \\ \Psi_2 - \overset{\text{add}}{\Psi}(z_2) \\ \vdots \\ \Psi_m - \overset{\text{add}}{\Psi}(z_M) \end{bmatrix} \tag{5.69}
$$

with $\Psi_{mn} = \Im[\mathrm{i}\overset{\text{sl}}{\Omega}_n(\mathcal{Z}_m)]$. Such elements are illustrated in Fig. 5.14c, where the left-most element in each string is first solved to satisfy the Kutta condition in (5.67), and all other elements are set to the same value of stream function using (5.69).

Problem 5.15 Compute the coefficients with $N = 10$ and $M = 15$ for a line-vortex that satisfies a Kutta condition at end z_{max} and uniform vector field $v_0 = 1$ for an element with specified endpoints:

A. $z_{\min} = 0, z_{\max} = 10 - 1.1\mathrm{i}$

B. $z_{\min} = 0, z_{\max} = 10 - 1.2\mathrm{i}$

C. $z_{\min} = 0, z_{\max} = 10 - 1.3\mathrm{i}$

D. $z_{\min} = 0, z_{\max} = 10 - 1.4\mathrm{i}$

E. $z_{\min} = 0, z_{\max} = 10 - 1.5\mathrm{i}$

F. $z_{\min} = 0, z_{\max} = 10 - 1.6\mathrm{i}$

G. $z_{\min} = 0, z_{\max} = 10 - 1.7\mathrm{i}$

H. $z_{\min} = 0, z_{\max} = 10 - 1.8\mathrm{i}$

I. $z_{\min} = 0, z_{\max} = 10 - 1.9\mathrm{i}$

J. $z_{\min} = 0, z_{\max} = 10 - 2.0\mathrm{i}$

K. $z_{\min} = 0, z_{\max} = 10 - 2.1\mathrm{i}$

L. $z_{\min} = 0, z_{\max} = 10 - 2.2\mathrm{i}$

5.4 Simpler Far-Field Representation

5.4.1 Far-Field Expansions for Double- and Single-Layer Elements

The double- and single-layer elements may be formulated using a simpler far-field representation that enables computations to be performed more accurately and efficiently outside the neighborhood of an element. The complex potential (5.16) and the complex vector field (5.19) for these elements are summarized here:

$$\Omega = \sum_{n=0}^{N} \overset{\text{dl}}{c}_n \overset{\text{dl}}{\Omega}_n[\mathcal{Z}(z)]\,, \quad v = \sum_{n=0}^{N} \overline{\overset{\text{dl}}{c}_n \overset{\text{dl}}{v}_n[\mathcal{Z}(z)]}$$

$$\overset{\text{dl}}{\Omega}_n = \frac{1}{2\pi}\left(\mathcal{Z}^n \ln\frac{\mathcal{Z}+1}{\mathcal{Z}-1} - \sum_{l=1}^{\frac{n+1}{2}} \frac{2}{2l-1}\mathcal{Z}^{n+1-2l}\right) \qquad (|\mathcal{Z}| \le R_{\text{far}}) \tag{5.70}$$

$$\overset{\text{dl}}{v}_n = -\frac{e^{i\vartheta}}{2\pi L}\overline{\left[\frac{(-1)^n}{\mathcal{Z}+1} - \frac{1}{\mathcal{Z}-1} + n\mathcal{Z}^{n-1}\ln\frac{\mathcal{Z}+1}{\mathcal{Z}-1} - n\sum_{l=1}^{\frac{n}{2}}\frac{2}{2l-1}\mathcal{Z}^{n-2l}\right]}$$

These equations have numerical difficulty achieving accurate computations that become evident with increasing order N at large distance from the element. This is observed by expanding the logarithm terms as a Laurent series using the series expansion (Abramowitz and Stegun, 1972)

$$\ln\frac{\mathcal{Z}+1}{\mathcal{Z}-1} = 2\left(\frac{1}{\mathcal{Z}} + \frac{1}{3\mathcal{Z}^3} + \frac{1}{5\mathcal{Z}^5} + \frac{1}{7\mathcal{Z}^7} + \cdots\right) = \sum_{l=1}^{\infty}\frac{2}{2l-1}\mathcal{Z}^{1-2l} \qquad (|\mathcal{Z}| > 1) \tag{5.71}$$

which is valid in the far-field neighborhood excluding a circle about the element. This series is multiplied by powers of $\mathcal{Z}^n$, and the remaining series in the nth influence function of the complex potential is also expanded to give

$$\begin{aligned} \mathcal{Z}^n \ln\frac{\mathcal{Z}+1}{\mathcal{Z}-1} &= 2\mathcal{Z}^{n-1} + \frac{2\mathcal{Z}^{n-3}}{3} + \cdots + \frac{2}{n} + \cdots + \frac{2}{2l-1}\mathcal{Z}^{1-2l} + \cdots \\ -\sum_{l=1}^{\frac{n+1}{2}}\frac{2}{2l-1}\mathcal{Z}^{n+1-2l} &= -2\mathcal{Z}^{n-1} - \frac{2\mathcal{Z}^{n-3}}{3} - \cdots - \frac{2}{n} \end{aligned} \tag{5.72}$$

The terms in these series with positive powers of $\mathcal{Z}$ are equal and opposite in sign, and, while they cancel, they become quite large with increased distance from the element.

This numerical problem is resolved by explicitly canceling the higher-order terms for computations at far-field locations outside the neighborhood of the element. This gives the complex potential as a far-field Laurent series expansion (Steward et al., 2008, eqn 2.14, 2.15):

$$\overset{\text{db}}{\Omega}_n = \frac{1}{2\pi}\sum_{l=\frac{n+3}{2}}^{\frac{n+3}{2}+N_{\text{far}}}\frac{2}{2l-1}\mathcal{Z}^{n+1-2l} \qquad (|\mathcal{Z}| > R_{\text{far}}) \tag{5.73}$$

$$\overset{\text{db}}{v}_n = -\frac{e^{i\vartheta}}{L}\overline{\frac{d\Omega}{d\mathcal{Z}}} = -\frac{e^{i\vartheta}}{2\pi L}\overline{\sum_{l=\frac{n+3}{2}}^{\frac{n+3}{2}+N_{\text{far}}}\frac{2(n+1-2l)}{2l-1}\mathcal{Z}^{n-2l}}$$

where Ω behaves like $1/\mathcal{Z}$ or lower orders of $\mathcal{Z}$ as $\mathcal{Z} \to \infty$ (Strack, 1989, 2003). The far-field expansion for the vector field in the previous equation may be obtained directly by evaluating the derivative of Ω with respect to z, as in (5.12a). In practice, the near-field forms (5.70) provide the correct logarithm singularity near the element, and the far-field forms (5.73) are applied outside radius R_{far}, where N_{far}=11 and $R_{\text{far}} = 1.5$ for the figures presented here. ***This formulation removes the need to compute logarithm terms in the far-field for double layers, which provides faster computational speed and increased numerical accuracy.***

The complex potential (5.46) and vector field (5.48) for a single layer, which are summarized next, also contain terms that cancel at large distances from the element:

$$\Omega = \sum_{n=0}^{N} \overset{\text{sl}}{c}_n \overset{\text{sl}}{\Omega}_n[\mathcal{Z}(z)]\,, \quad v = \sum_{n=0}^{N} \overset{\text{sl}}{c}_n \overline{\overset{\text{sl}}{v}_n[\mathcal{Z}(z)]}$$

$$\overset{\text{sl}}{\Omega}_n = \frac{1}{2\pi}\frac{1}{n+1}\left[\mathcal{Z}^{n+1}\ln\frac{\mathcal{Z}+1}{\mathcal{Z}-1} + (-1)^n\ln(\mathcal{Z}+1) + \ln(\mathcal{Z}-1) - \sum_{l=1}^{\frac{n+2}{2}}\frac{2}{2l-1}\mathcal{Z}^{n+2-2l}\right] \tag{5.74}$$

$$\overset{\text{sl}}{v}_n = -\frac{e^{i\vartheta}}{2\pi L}\overline{\left(\mathcal{Z}^n\ln\frac{\mathcal{Z}+1}{\mathcal{Z}-1} - \sum_{l=1}^{\frac{n+1}{2}}\frac{2}{2l-1}\mathcal{Z}^{n+1-2l}\right)} \qquad (|\mathcal{Z}| \le R_{\text{far}})$$

Expanding the logarithm as a series and canceling terms leads to the same formulation in the far-field for the vector field as the complex potential of a double layer:

$$\overset{\text{sl}}{\Omega}_n = \frac{1}{2\pi}\frac{1}{n+1}\left[(-1)^n\ln(\mathcal{Z}+1) + \ln(\mathcal{Z}-1) + \sum_{l=\frac{n+4}{2}}^{\frac{n+4}{2}+N_{\text{far}}}\frac{2}{2l-1}\mathcal{Z}^{n+2-2l}\right]$$

$$\overset{\text{sl}}{v}_n = -\frac{e^{i\vartheta}}{2\pi L}\overline{\sum_{l=\frac{n+3}{2}}^{\frac{n+3}{2}+N_{\text{far}}}\frac{2}{2l-1}\mathcal{Z}^{n+1-2l}} \qquad (|\mathcal{Z}| > R_{\text{far}}) \tag{5.75}$$

where the details necessary for organizing the logarithm terms for the complex potential may be found at Steward et al. (2008, eqn 2.14, 2.15). Note that ***this complex potential of single layers contains a far-field with logarithm terms centered at the endpoints of each element, while the far-field expansion of a double layer, (5.73), contains only a Laurent series***.

Problem 5.16 Compute the value of the complex potential and vector field for the specified coefficient of a double layer with $z_{\text{min}} = -1$ and $z_{\text{max}} = 1$ at three locations $\mathcal{Z} = (1.1e^{i\pi/4}, 1.5e^{i\pi/4}, 5e^{i\pi/4})$ using the expressions for the near-field (5.70) and the far-field (5.73) with $N_{\text{far}} = 12$ terms.

A. $\overset{\text{dl}}{c}_{10} = 1$
B. $\overset{\text{dl}}{c}_{11} = 1$
C. $\overset{\text{dl}}{c}_{12} = 1$
D. $\overset{\text{dl}}{c}_{13} = 1$
E. $\overset{\text{dl}}{c}_{14} = 1$
F. $\overset{\text{dl}}{c}_{15} = 1$
G. $\overset{\text{dl}}{c}_{16} = 1$
H. $\overset{\text{dl}}{c}_{17} = 1$
I. $\overset{\text{dl}}{c}_{18} = 1$
J. $\overset{\text{dl}}{c}_{19} = 1$
K. $\overset{\text{dl}}{c}_{20} = 1$
L. $\overset{\text{dl}}{c}_{21} = 1$

Problem 5.17 Compute the value of the complex potential and vector field for the specified coefficient of a single layer with $z_{\text{min}} = -1$ and $z_{\text{max}} = 1$ at three locations $\mathcal{Z} = (1.1e^{i\pi/4}, 1.5e^{i\pi/4}, 5e^{i\pi/4})$ using the expressions for the near-field (5.74) and the far-field (5.75) with $N_{\text{far}} = 12$ terms.

A. $\overset{\text{sl}}{c}_{10} = 1$
B. $\overset{\text{sl}}{c}_{11} = 1$
C. $\overset{\text{sl}}{c}_{12} = 1$
D. $\overset{\text{sl}}{c}_{13} = 1$
E. $\overset{\text{sl}}{c}_{14} = 1$
F. $\overset{\text{sl}}{c}_{15} = 1$
G. $\overset{\text{sl}}{c}_{16} = 1$
H. $\overset{\text{sl}}{c}_{17} = 1$
I. $\overset{\text{sl}}{c}_{18} = 1$
J. $\overset{\text{sl}}{c}_{19} = 1$
K. $\overset{\text{sl}}{c}_{20} = 1$
L. $\overset{\text{sl}}{c}_{21} = 1$

5.4.2 Single Layers versus Double Layers with a Point Element

A natural relation exists between single- and double-layer elements that is utilized in this section to enable the equations of a double layer to help compute the complex potential and vector field of a single layer. This is computationally attractive, since it was just shown that the logarithm functions in the far-field of a single layer do not exist in double layers. Commonality is established by analyzing the discontinuities in the real or imaginary part across these elements, which are summarized in Table 5.1.

The coefficients of these elements may be organized so single layers and double layers provide the same values in discontinuity along an element. Specifically, the branch cuts for a line-dipole (5.26) and line-sink (5.53) along the element where $-1 < X < 1$ and $Y = 0$ are given by

$$\Delta\Psi = \begin{cases} -\sum\limits_{n=0}^{N+1} \Re \overset{\text{dl}}{c}_n X^n = -\nu(X) \\ \sum\limits_{n=0}^{N} \Re \overset{\text{sl}}{c}_n \frac{1-X^{n+1}}{n+1} = L\int\limits_X^1 \sigma(\tilde{X})\mathrm{d}\tilde{X} \end{cases}, \quad \Delta v_{\mathfrak{n}} = \begin{cases} -\sum\limits_{n=0}^{N+1} \Re \overset{\text{dl}}{c}_n \frac{nX^{n-1}}{L} = -\frac{1}{L}\frac{\mathrm{d}\nu}{\mathrm{d}X} \\ -\sum\limits_{n=0}^{N} \frac{\Re \overset{\text{sl}}{c}_n X^n}{L} = -\sigma(X) \end{cases} \tag{5.76}$$

Thus, these elements give the same discontinuities across the element when the real part of the coefficients are related by

$$\Re \overset{\text{dl}}{c}_0 = -\sum_{n=0}^{N} \frac{1}{n+1} \Re \overset{\text{sl}}{c}_n\,, \quad \Re \overset{\text{dl}}{c}_{n+1} = \frac{\Re \overset{\text{sl}}{c}_n}{n+1} \quad (n = 0, \cdots, N) \tag{5.77}$$

Likewise, the branch cuts associated with a line-doublet (5.36) and line-vortex (5.63) take on the following values along the element:

$$\Delta\Phi = \begin{cases} \sum\limits_{n=0}^{N+1} \Im \overset{\text{dl}}{c}_n X^n = \mu(X) \\ -\sum\limits_{n=0}^{N} \Im \overset{\text{sl}}{c}_n \frac{1-X^{n+1}}{n+1} = -L\int\limits_X^1 \gamma(\tilde{X})\mathrm{d}\tilde{X} \end{cases}, \quad \Delta v_{\mathfrak{s}} = \begin{cases} -\sum\limits_{n=0}^{N+1} \Im \overset{\text{dl}}{c}_n \frac{nX^{n-1}}{L} = -\frac{1}{L}\frac{\mathrm{d}\mu}{\mathrm{d}X} \\ -\sum\limits_{n=0}^{N} \frac{\Im \overset{\text{sl}}{c}_n X^n}{L} = -\gamma(X) \end{cases} \tag{5.78}$$

and these elements give the same discontinuities across the element when the imaginary part of the coefficients is related by

Table 5.1 *Discontinuities across singular integral equations*

Line type	Potential	Stream function	Tangential	Normal
	Φ	Ψ	$v_{\mathfrak{s}}$	$v_{\mathfrak{n}}$
line-dipole, (5.26)	continuous	jump	continuous	jump
line-doublet, (5.36)	jump	continuous	jump	continuous
line-sink, (5.53)	continuous	jump	continuous	jump
line-vortex, (5.63)	jump	continuous	jump	continuous

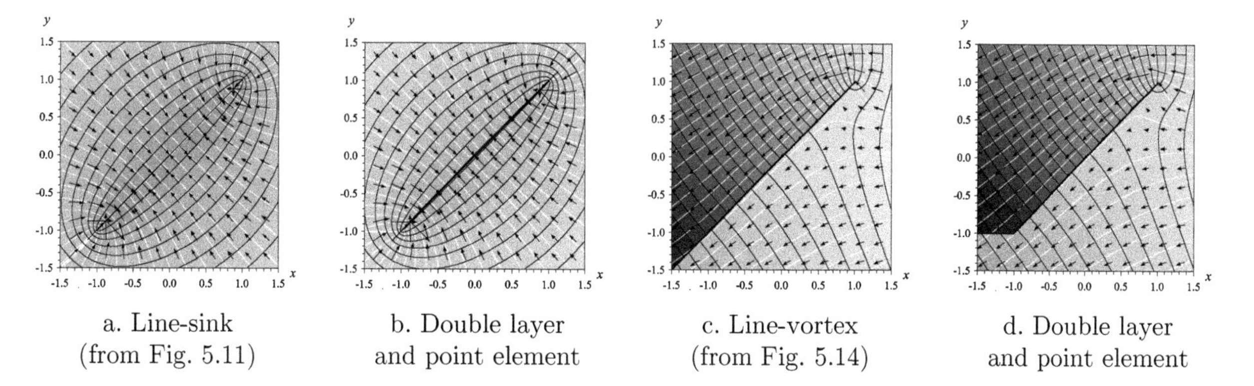

Figure 5.15 *Single layer vs. double layer plus a point element.*

$$\Im \overset{\text{dl}}{c}_0 = -\sum_{n=0}^{N} \frac{1}{n+1} \Im \overset{\text{sl}}{c}_n \,, \quad \Im \overset{\text{dl}}{c}_{n+1} = \frac{\Im \overset{\text{sl}}{c}_n}{n+1} \quad (n = 0, \cdots, N) \tag{5.79}$$

Together, ***the discontinuity of a double layer element has the same value as a single-layer element when the complex coefficients take on values where***

$$\boxed{\overset{\text{dl}}{c}_0 = -\sum_{n=0}^{N} \frac{1}{n+1} \overset{\text{sl}}{c}_n \quad , \quad \overset{\text{dl}}{c}_{n+1} = \frac{1}{n+1} \overset{\text{sl}}{c}_n \quad (n = 0, \cdots, N)} \tag{5.80}$$

However, single-layer elements also contain a discontinuity that extends along the ray where $X < -1$ and $Y = 0$. The discontinuity in the stream function of a line-sink in Fig. 5.15a is generated by the net flux of the element, (5.53a),

$$Q = L \int_{-1}^{1} \sigma(\tilde{X}) \mathrm{d}\tilde{X} = \Delta\Psi|_{X=-1} = \sum_{n=0}^{N} \Re \overset{\text{sl}}{c}_n \frac{1 + (-1)^n}{n+1} \tag{5.81}$$

Likewise, the discontinuity in potential of a line-vortex in Fig. 5.15c is generated by the net circulation of this element, (5.63a)

$$\Gamma = L \int_{-1}^{1} \gamma(\tilde{X}) \mathrm{d}\tilde{X} = -\Delta\Phi|_{X=-1} = \sum_{n=0}^{N} \Im \overset{\text{sl}}{c}_n \frac{1 + (-1)^n}{n+1} \tag{5.82}$$

This same value of these discontinuities may alternately be generated by a point-sink with flux Q and a point-vortex with circulation Γ placed at the endpoint z_{min} (Steward and Ahring, 2009, eqn 37):

$$\boxed{\Omega = \frac{Q + \mathrm{i}\Gamma}{2\pi} \ln(z - z_{\text{min}}) \quad , \quad Q + \mathrm{i}\Gamma = \sum_{n=0}^{N} \overset{\text{sl}}{c}_n \frac{1 + (-1)^n}{n+1}} \tag{5.83}$$

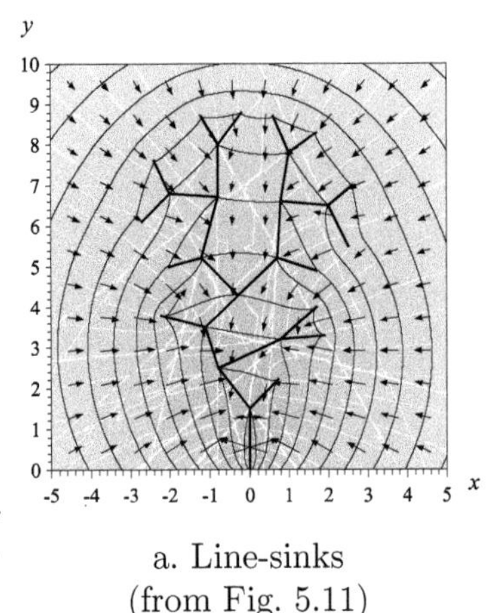

a. Line-sinks (from Fig. 5.11)

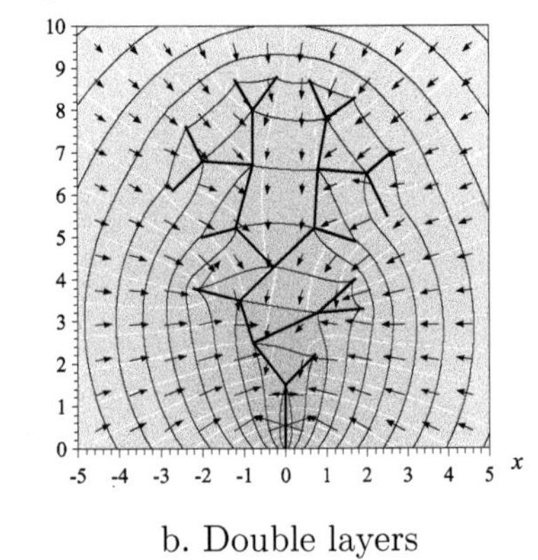

b. Double layers and point element

Figure 5.16 *Aligning the branch cuts of strings of single layers along double layers plus a point element for line-sinks.*

Thus, ***the vector field for a single layer may be generated by a double layer with coefficients computed using (5.80) with a point element at the endpoint*** $z_{\min}$ ***with strengths obtained from (5.83)***. These comparisons are illustrated in Fig. 5.15.

This facilitates the organization of strings of adjacent single layers to be represented as a string of double layers with the same geometry, but with all logarithm terms for the strings gathered at a single point. It is observed that the constant term in the double-layer element effectively transports $Q + \mathrm{i}\Gamma$ through the element and then the point element conveys this net flux or circulation along the branch cut. This notion may be extended to transport the branch cuts of adjacent double layers through the elements. This may be accomplished by first computing the net flux and circulation generated by each single-layer element from their coefficients, as in (5.83):

$$\underset{i}{Q} + \mathrm{i}\underset{i}{\Gamma} = \sum_{n=0}^{N} \underset{i}{\overset{\mathrm{sl}}{c}}_n \frac{1+(-1)^n}{n+1} \tag{5.84}$$

Flux may be summed across the j elements with branch cuts directed to the end of the element at $z_{\max}$, and the branch cut at the endpoint $z_{\min}$ is this plus that occurring due to the element:

$$\underset{i}{Q}_{\max} + \mathrm{i}\underset{i}{\Gamma}_{\max} = \sum_{j \in \underset{i}{z}_{\max}} \left(\underset{j}{Q} + \mathrm{i}\underset{j}{\Gamma} \right), \quad \underset{i}{Q}_{\min} + \mathrm{i}\underset{i}{\Gamma}_{\min} = \left(\underset{i}{Q}_{\max} + \mathrm{i}\underset{i}{\Gamma}_{\max} \right) + \left(\underset{i}{Q} + \mathrm{i}\underset{i}{\Gamma} \right) \tag{5.85}$$

The coefficients for double-layer elements are computed as before (5.80); however, the constant term is adjusted so that the branch cut of all elements attached to its end $z_{\max}$ is equal to the jump in $\Delta\Phi$ and $\Delta\Psi$ occurring there, occurring when

$$\underset{i}{\overset{\mathrm{dl}}{c}}_0 = -\sum_{n=0}^{N} \frac{1}{n+1} \underset{i}{\overset{\mathrm{sl}}{c}}_n - \left(\underset{i}{Q}_{\max} + \mathrm{i}\underset{i}{\Gamma}_{\max} \right), \quad \underset{i}{\overset{\mathrm{dl}}{c}}_{n+1} = \frac{1}{n+1} \underset{i}{\overset{\mathrm{sl}}{c}}_n \qquad (n = 0, \cdots, N) \tag{5.86}$$

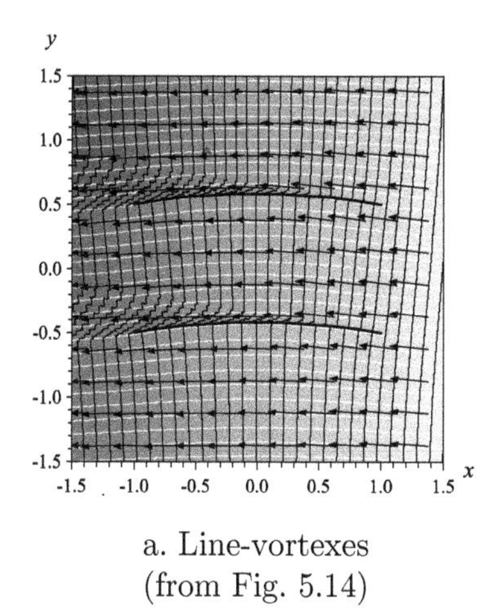

a. Line-vortexes (from Fig. 5.14)

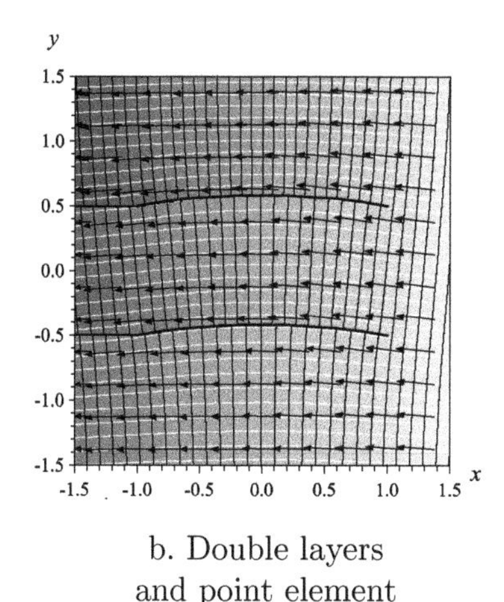

b. Double layers and point element

Figure 5.17 *Aligning the branch cuts of strings of single layers along double layers plus a point element for line-vortexes.*

where summation occurs over those j elements with branch cut directed towards $z_{\max}$. Then a point element must be added to $z_{\min}$ at the first element i of the string with contributions from all adjacent elements contributing to this branch cut (Steward and Ahring, 2009, eqn 36)

$$\Omega = \frac{\underset{i}{Q}_{\min} + \mathrm{i}\underset{i}{\Gamma}_{\min}}{2\pi} \ln\left(z - \underset{i}{z}_{\min}\right) \tag{5.87}$$

and summation j occurs over all single-layer elements with branch cut directed to $z_{\min}$.

Thus, ***a string of adjacent single-layer elements may be reformulated to provide the same vector field using a string of double-layer elements with one point element at the end of the string that gathers the branchcuts of all elements.*** This is illustrated in Fig. 5.16a containing the branch cuts in the stream function for all line-sinks in Fig. 5.11. The figure in 5.16b results when the single-layer elements are replaced with double-layer elements, and the branch cuts lie along the elements except for that occurring at the end of all strings. Note that, while the branch cuts are preserved between the use of line-sinks and line-doublets, there is a difference in Φ between the two solutions equal to an additive constant, which requires adjusting the uniform coefficient, Φ_0 in (3.2), so the values of potential are equal in both figures. Likewise, a branch cut in the potential function occurs for the line-vortexes in Fig. 5.17a, and Fig. 5.17b illustrates the field occurring when these are aligned along the elements using double layers with a point-sink. ***This provides a more computationally efficient far-field calculations for single layers using the Laurent series of double layers with a single logarithm term.***

Problem 5.18 Compute the coefficients for a double layer and point element that will reproduce the solution for the line-sink with endpoints $z_{\min} = -1 - \mathrm{i}$ and $z_{\max} = 1 + \mathrm{i}$, and with coefficients equal to one ($\overset{\text{sl}}{c}_0 = 1, \overset{\text{sl}}{c}_1 = 1, \cdots, \overset{\text{sl}}{c}_N = 1$) for:

A. $N = 3$ B. $N = 4$ C. $N = 5$ D. $N = 6$ E. $N = 7$ F. $N = 8$ G. $N = 9$ H. $N = 10$ I. $N = 11$ J. $N = 12$ K. $N = 13$ L. $N = 14$

Problem 5.19 Compute the coefficients for a double layer and point element that will reproduce the solution for the line-vortex with endpoints $z_{\min} = -1 - \mathrm{i}$ and $z_{\max} = 1 + \mathrm{i}$, and with coefficients equal to i ($\overset{\text{sl}}{c}_0 = \mathrm{i}, \overset{\text{sl}}{c}_1 = \mathrm{i}, \cdots, \overset{\text{sl}}{c}_N = \mathrm{i}$) for:

A. $N = 3$
B. $N = 4$
C. $N = 5$
D. $N = 6$
E. $N = 7$
F. $N = 8$
G. $N = 9$
H. $N = 10$
I. $N = 11$
J. $N = 12$
K. $N = 13$
L. $N = 14$

5.5 Polygon Elements

Features with the geometry of polygons may be modeled using double and single layers placed along their boundaries. This is illustrated in Fig. 5.18a, where the boundary of a polygon is composed of elements located along straight segments connecting adjacent vertices. Systems of equations are formulated next to match interface conditions occurring across the edges of a polygon.

5.5.1 Heterogeneities

A heterogeneity with the geometry of a polygon satisfies conditions associated with a jump in domain properties across its edges. The example in Fig. 5.18a contains a discontinuity in values of α with distinct values outside α^+ and inside α^- the heterogeneity. This particular problem is formulated using these parameters to satisfy a continuity condition for the potential (2.64a)

$$\alpha^+\Phi^+ - \alpha^-\Phi^- = 0 \quad \rightarrow \quad \boxed{\alpha^-\left(\Phi^+ - \Phi^-\right) = -\left(\alpha^+ - \alpha^-\right)\Phi^+} \tag{5.88a}$$

which has been rearranged to separate the jump in potential across the interface from the value of potential outside the polygon. This problem also has a continuous of the normal component of the vector field across the interface (2.65b),

$$v_n^+ = v_n^- \tag{5.88b}$$

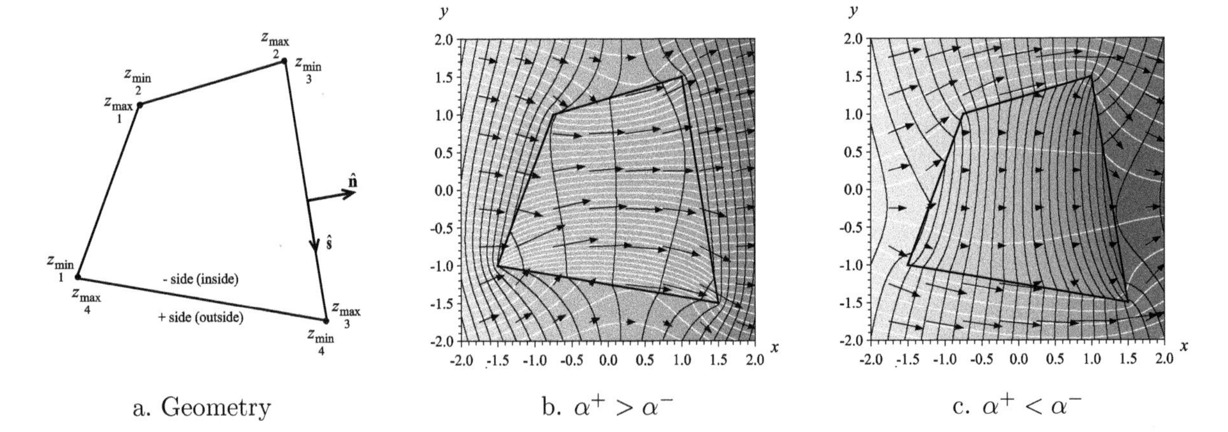

Figure 5.18 *Heterogeneity with the geometry of a polygon.*

Together, these conditions may be satisfied by a line-doublet across which the potential jumps and the normal component of the vector field is continuous, (5.36). Note that the solution in Fig. 5.18 plots $\alpha^+\Phi$ outside and $\alpha^-\Phi$ inside the polygon.

A system of equations may be formulated to solve for the coefficients of each line-doublet by substituting the jump in potential, (5.36a), and the complex potential, (5.33) into the interface condition (5.88a),

$$\alpha^- \sum_{n=0}^{N} \Im \overset{\text{dl}}{c}_n X^n = -\left(\alpha^+ - \alpha^-\right) \left\{ \sum_{n=0}^{N} \Im \overset{\text{dl}}{c}_n \Re \left[\mathrm{i} \overset{\text{dl}}{\Omega}_n \left(\mathcal{Z}^+ \right) \right] + \overset{\text{add}}{\Phi} (z) \right\} \tag{5.89}$$

where the contributions to the complex potential for all other elements are gathered in the additional function. Note that the complex potential for a line-doublet (5.34) may be evaluated at control point m using the complex logarithm (5.23) at points along the outer edge of the heterogeneity (5.24b):

$$\Re \left[\mathrm{i} \overset{\text{dl}}{\Omega}_n \left(\mathcal{Z}_m^+ \right) \right] = \sum_{n=0}^{N} \Im \overset{\text{dl}}{c}_n \frac{(X_m)^n}{2} \tag{5.90}$$

The last two equations may be rearranged to gather terms with unknown coefficients on the left-hand side to provide the conditions at control point m:

$$\boxed{\sum_{n=0}^{N} \Im \overset{\text{dl}}{c}_n (X_m)^n = -2 \frac{\alpha^+ - \alpha^-}{\alpha^+ + \alpha^-} \overset{\text{add}}{\Phi} (z_m)} \tag{5.91}$$

This system of equations may be gathered in the following matrices for each element

$$\mathbf{A} = \begin{bmatrix} (X_1)^0 & (X_1)^1 & \cdots & (X_1)^N \\ (X_2)^0 & (X_2)^1 & \cdots & (X_2)^N \\ \cdots & \cdots & \cdots & \cdots \\ (X_M)^0 & (X_M)^1 & \cdots & (X_M)^N \end{bmatrix}, \quad \mathbf{c} = \begin{bmatrix} \Im \overset{\text{dl}}{c}_0 \\ \Im \overset{\text{dl}}{c}_1 \\ \vdots \\ \Im \overset{\text{dl}}{c}_N \end{bmatrix}, \quad \mathbf{b} = \begin{bmatrix} -2 \frac{\alpha^+ - \alpha^-}{\alpha^+ + \alpha^-} \overset{\text{add}}{\Phi} (z_1) \\ -2 \frac{\alpha^+ - \alpha^-}{\alpha^+ + \alpha^-} \overset{\text{add}}{\Phi} (z_2) \\ \vdots \\ -2 \frac{\alpha^+ - \alpha^-}{\alpha^+ + \alpha^-} \overset{\text{add}}{\Phi} (z_M) \end{bmatrix} \tag{5.92}$$

The elements along the boundary are solved iteratively, and solutions are shown in Fig. 5.18. This illustrates conditions when the inside parameter α^- is either higher or lower than the exterior, and it is visualized by plotting contours of Φ times the value of α in the appropriate domain to illustrate achievement of the interface condition, (5.88a).

Problem 5.20 Compute the value of the potential at $z = 1$, $z = 2 + \mathrm{i}$, and $z = 3$, for a square inhomogeneity with endpoints $\underset{1}{z}_{\min} = \underset{4}{z}_{\max} = 1 - \mathrm{i}$, $\underset{2}{z}_{\min} = \underset{1}{z}_{\max} = 1 + \mathrm{i}$, $\underset{3}{z}_{\min} = \underset{2}{z}_{\max} = 3 + \mathrm{i}$, $\underset{4}{z}_{\min} = \underset{3}{z}_{\max} = 3 - \mathrm{i}$ in a uniform vector field with $v_0 = 1$ and $\Phi_0 = 0$ and:

A. $\alpha^- = 1, \alpha^+ = 2$
B. $\alpha^- = 1, \alpha^+ = 3$
C. $\alpha^- = 1, \alpha^+ = 4$
D. $\alpha^- = 1, \alpha^+ = 5$
E. $\alpha^- = 1, \alpha^+ = 6$
F. $\alpha^- = 1, \alpha^+ = 7$
G. $\alpha^- = 2, \alpha^+ = 1$
H. $\alpha^- = 3, \alpha^+ = 1$
I. $\alpha^- = 4, \alpha^+ = 1$
J. $\alpha^- = 5, \alpha^+ = 1$
K. $\alpha^- = 6, \alpha^+ = 1$
L. $\alpha^- = 7, \alpha^+ = 1$

Note that this solution requires an iterative procedure that independently solves for the element components, following (2.9).

5.5.2 Simpler Far-Fields for Functions Gathered within Polygons

Polygons may also be used to gather a collection of functions associated with analytic elements within a region, and provide a simpler representation and more numerically efficient computations outside the polygon. The methods used to achieve such a solution are developed for a specified function of potential inside a polygon, where these functions are not used in computations outside the polygon:

$$\Phi = \begin{cases} \overset{\text{in}}{\Phi}(x,y) & z \in \text{polygon} \\ 0 & z \notin \text{polygon} \end{cases} \tag{5.93}$$

following Strack (1989). The domain over which this function is computed is illustrated for a square in Fig. 5.19a. A complex vector function also exists inside the polygon, which may be computed from the derivatives of the potential, (3.108)

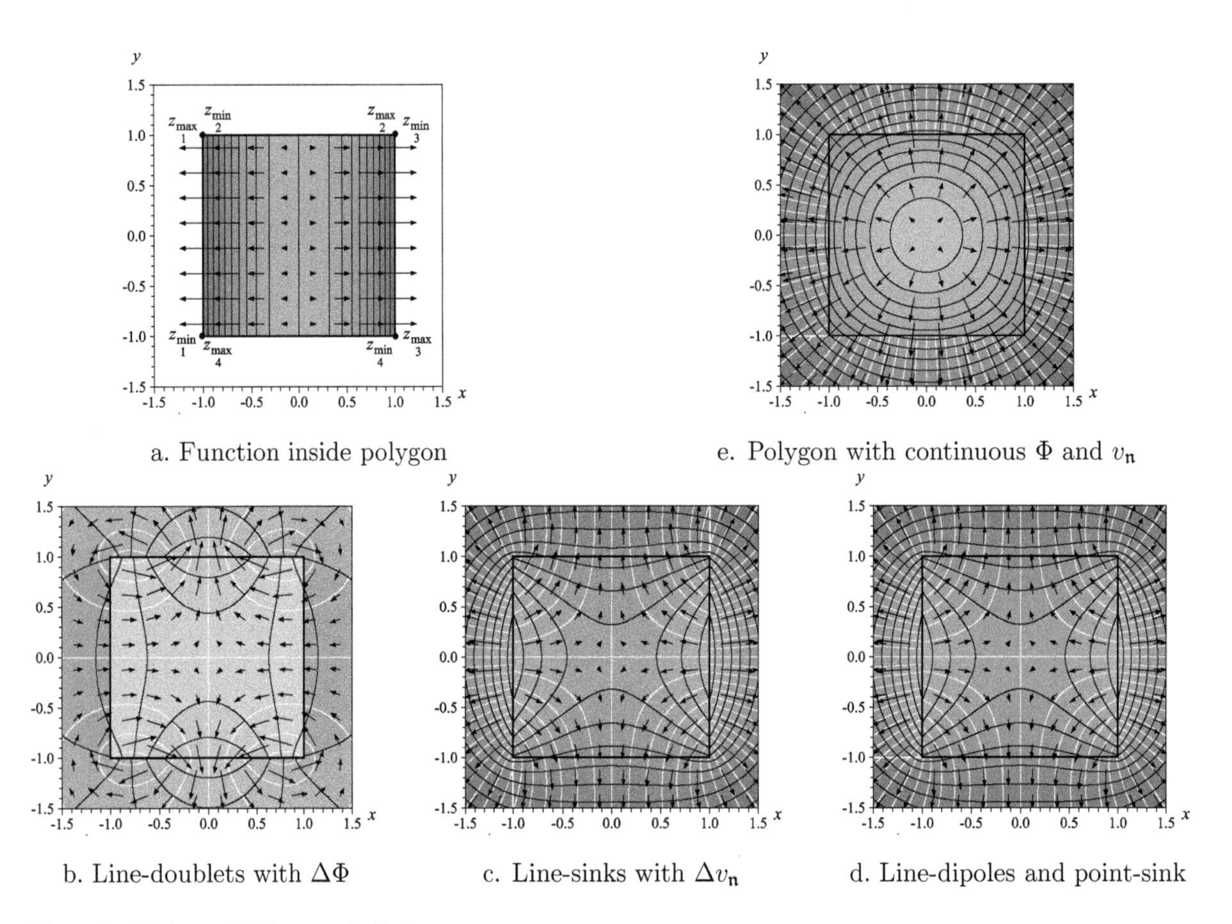

Figure 5.19 *Polygon with divergence inside element.*

$$\overset{\text{in}}{v} = -\frac{\partial \overset{\text{in}}{\Phi}}{\partial x} - \mathrm{i}\frac{\partial \overset{\text{in}}{\Phi}}{\partial y} \quad , \quad \overset{\text{in}}{v}_{\mathrm{n}} = \Im\left(\mathrm{e}^{-\mathrm{i}\vartheta}\overset{\text{in}}{v}\right) \tag{5.94}$$

where the normal component to each line segment is given by (5.12b). Clearly, the potential and the normal component of the vector field are both discontinuous across edges of this polygon.

A system of equations may be formulated to ***cancel the jump in potential across the interface*** from zero outside to $\overset{\text{in}}{\Phi}$ inside the polygon. This is accomplished by placing a set of line-doublets along the interface with coefficients, (5.36a), related to the value of the function at control points

$$\Phi^{+} - \Phi^{-} = \overset{\text{in}}{\Phi} \quad \rightarrow \quad \boxed{\sum_{n=0}^{N} \Im\overset{\text{dl}}{c}_n \left(X_m\right)^n = \overset{\text{in}}{\Phi}(z_m)} \tag{5.95}$$

similar to how the jump in potential was just canceled for heterogeneities. This system of equations may be organized in matrices as (Steward and Ahring, 2009, eqn 31)

$$\mathbf{A} = \begin{bmatrix} (X_1)^0 & (X_1)^1 & \cdots & (X_1)^N \\ (X_2)^0 & (X_2)^1 & \cdots & (X_2)^N \\ \cdots & \cdots & \cdots & \cdots \\ (X_M)^0 & (X_M)^1 & \cdots & (X_M)^N \end{bmatrix}, \quad \mathbf{c} = \begin{bmatrix} \Im\overset{\text{dl}}{c}_0 \\ \Im\overset{\text{dl}}{c}_1 \\ \vdots \\ \Im\overset{\text{dl}}{c}_N \end{bmatrix}, \quad \mathbf{b} = \begin{bmatrix} \overset{\text{in}}{\Phi}(z_1) \\ \overset{\text{in}}{\Phi}(z_2) \\ \vdots \\ \overset{\text{in}}{\Phi}(z_M) \end{bmatrix} \tag{5.96}$$

and the solution is illustrated in Fig. 5.19b.

A system of equations may be formulated to ***satisfied continuity of the normal component of the vector field using a line-sink***, (5.53b), with coefficients related to the value of the function at control points

$$v_{\mathrm{n}}^{+} - v_{\mathrm{n}}^{-} = \overset{\text{in}}{v}_{\mathrm{n}} \quad \rightarrow \quad \boxed{\sum_{n=0}^{N} \frac{\Re\overset{\text{sl}}{c}_n \left(X_m\right)^n}{L} = -\overset{\text{in}}{v}_{\mathrm{n}}(z_m)} \tag{5.97}$$

This system of equations may be also be organized in matrices as (Steward and Ahring, 2009, eqn 34)

$$\mathbf{A} = \begin{bmatrix} (X_1)^0 & (X_1)^1 & \cdots & (X_1)^N \\ (X_2)^0 & (X_2)^1 & \cdots & (X_2)^N \\ \cdots & \cdots & \cdots & \cdots \\ (X_M)^0 & (X_M)^1 & \cdots & (X_M)^N \end{bmatrix}, \quad \mathbf{c} = \begin{bmatrix} \Re\overset{\text{sl}}{c}_0 \\ \Re\overset{\text{sl}}{c}_1 \\ \vdots \\ \Re\overset{\text{sl}}{c}_N \end{bmatrix}, \quad \mathbf{b} = -L \begin{bmatrix} \overset{\text{in}}{v}_{\mathrm{n}}(z_1) \\ \overset{\text{in}}{v}_{\mathrm{n}}(z_2) \\ \vdots \\ \overset{\text{in}}{v}_{\mathrm{n}}(z_M) \end{bmatrix} \tag{5.98}$$

and the set of line-sinks that cancel discontinuities in the normal component of the vector field for all sides of the polygon are shown in Fig. 5.19c. Note that ***this collection of single layers may be replaced with a collection of double layers and one point-sink***, following the methods of Section 5.4.2, and these sets of elements are shown in Fig. 5.19d.

Collectively, the double layers for the line-doublets and the single layer for the line-sink or the equivalent double layers with a point element cancel the discontinuities associated with the potential function evaluated inside the element (5.93). The solution obtained with all these functions is shown in Fig. 5.19e. Clearly, the final solution satisfies continuity of potential and the normal component of the vector field. Note that ***these coefficients computed for this solution only depend upon the function specified inside the polygon, (5.93), and so, they may be computed independent of the additional functions outside the polygon***.

Example 5.1 Figure 5.1 was constructed with endpoints $\underset{1}{z}_{\min} = \underset{4}{z}_{\max} = -1 - \mathrm{i}$, $\underset{2}{z}_{\min} = \underset{1}{z}_{\max} = -1 + \mathrm{i}$, $\underset{3}{z}_{\min} = \underset{2}{z}_{\max} = 1 + \mathrm{i}$, $\underset{4}{z}_{\min} = \underset{3}{z}_{\max} = 1 - \mathrm{i}$, and interior function $\overset{\text{in}}{\Phi} = -x^2$. The conditions for the line-doublets, (5.95), and the line-sink conditions, (5.97), were matched with coefficients

$$\underset{1}{\overset{\text{dl}}{c}}_0 = -\mathrm{i}\,, \quad \underset{2}{\overset{\text{dl}}{c}}_2 = -\mathrm{i}\,, \quad \underset{3}{\overset{\text{dl}}{c}}_0 = -\mathrm{i}\,, \quad \underset{4}{\overset{\text{dl}}{c}}_2 = -\mathrm{i}\,, \quad \underset{1}{\overset{\text{sl}}{c}}_0 = -2\,, \quad \underset{3}{\overset{\text{sl}}{c}}_0 = -2$$

with all other coefficients equal to zero.

This method of isolating functions inside a polygon may be extended to develop simpler far-field representations for other problems, by simply using those functions inside a polygon, and using double layers with a point-sink to provide the vector field outside the polygon. In general, the complex potential for all elements inside the polygon and its complex vector field from (3.109) are evaluated inside the polygon, but not outside this domain:

$$\Omega = \begin{cases} \overset{\text{in}}{\Omega}(z, \overline{z})) & z \in \text{polygon} \\ 0 & z \notin \text{polygon} \end{cases} \quad , \quad \overset{\text{in}}{v} = -2\overline{\frac{\partial \overset{\text{in}}{\Omega}}{\partial z}} \tag{5.99}$$

For example, the field of phreatophytes from Steward and Ahring (2009) in Fig. 5.20a was obtained by superimposing the complex functions for these elements developed in Example 3.5. This figure also illustrates the double layers used to cancel the jump in potential, and the double layer and point element used to cancel the jump in the normal component of the vector field. Collectively the solution satisfies continuity of the potential

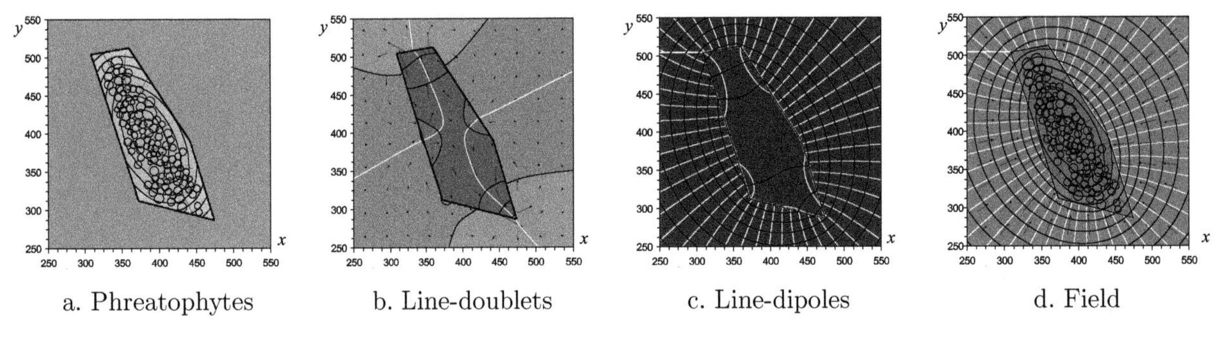

Figure 5.20 *Polygon with root water uptake from many plants in a field bounded by a polygon. (Reprinted from Steward and Ahring, 2009, An analytic solution for groundwater uptake by phreatophytes spanning spatial scales from plant to field to regional,* Journal of Engineering Mathematics, *Vol. 64(2), Figs. 4,5, used with permission from Springer Nature; distributed under Creative Commons license (Attribution-Noncommercial), creativecommons.org.)*

and vector field as shown in Fig. 5.20d. Note that ***the point-sink used in the far-field outside the polygon represents the net flux removed by all elements in the polygon, and thus provides a convenient method for computing this value for a collection of elements. This also significantly reduces computations, particularly when a large number of analytic elements are gathered within the polygon.***

Problem 5.21 Compute the coefficients required to satisfy boundary conditions for continuity of Φ and v_n for a polygon with geometry in Example 5.1 and interior function:

A. $\overset{\text{in}}{\Phi} = -x^3$ C. $\overset{\text{in}}{\Phi} = -x^5$ E. $\overset{\text{in}}{\Phi} = -y^2$ G. $\overset{\text{in}}{\Phi} = -y^4$ I. $\overset{\text{in}}{\Phi} = -x^2 - y^2$ K. $\overset{\text{in}}{\Phi} = -x^4 - y^4$

B. $\overset{\text{in}}{\Phi} = -x^4$ D. $\overset{\text{in}}{\Phi} = -x^6$ F. $\overset{\text{in}}{\Phi} = -y^3$ H. $\overset{\text{in}}{\Phi} = -y^5$ J. $\overset{\text{in}}{\Phi} = -x^3 - y^3$ L. $\overset{\text{in}}{\Phi} = -x^5 - y^5$

5.6 Curvilinear Elements

5.6.1 Formulation of Curvilinear Elements

Analytic elements may be formulated for curvilinear geometries to extend the previous development of singular integral equations for straight segments using conformal mappings. This is illustrated in Fig. 5.21a for an element i located along a curvilinear boundary in physical z-coordinates with endpoints at z_{min} and z_{max}. This curvilinear geometry is preserved when each element is translated, rotated, and scaled to the local $\mathcal{Z} = X + \text{i}Y$ plane in Fig. 5.21b with endpoints $\mathcal{Z} = -1$ and $\mathcal{Z} = 1$ using

$$\mathcal{Z} = \left(z - \frac{z_{\text{max}} + z_{\text{min}}}{2} \right) \frac{2}{z_{\text{max}} - z_{\text{min}}}, \quad z = \mathcal{Z} \frac{z_{\text{max}} - z_{\text{min}}}{2} + \frac{z_{\text{max}} + z_{\text{min}}}{2} \tag{5.100}$$

as done previously for straight line elements with (5.8). A conformal mapping will be developed shortly for a range of geometries that then maps this local curvilinear geometry to the real axis in the ζ-plane

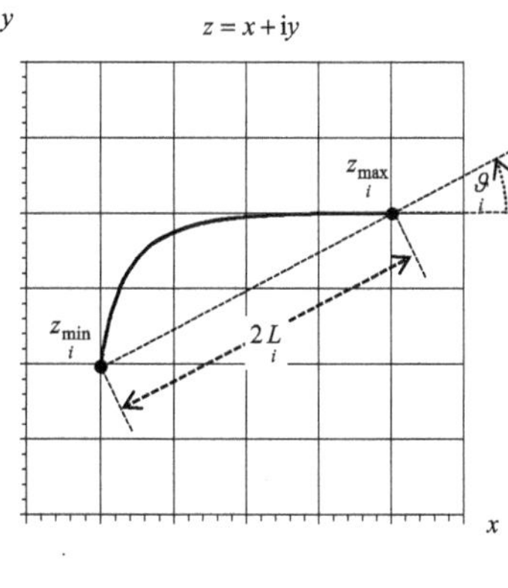

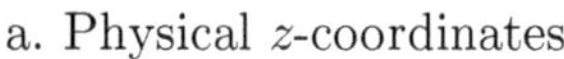
a. Physical z-coordinates

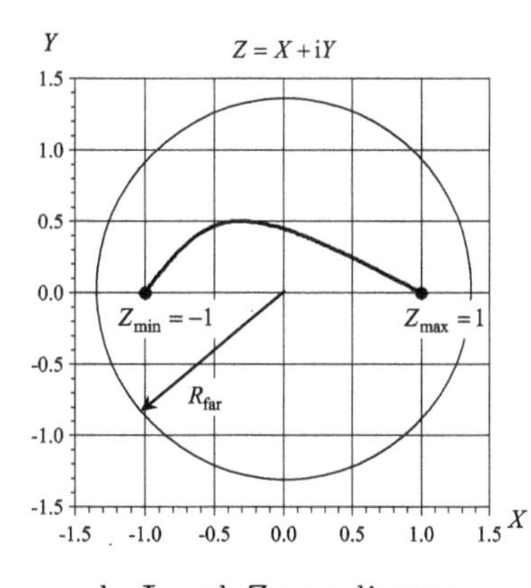

b. Local Z-coordinates

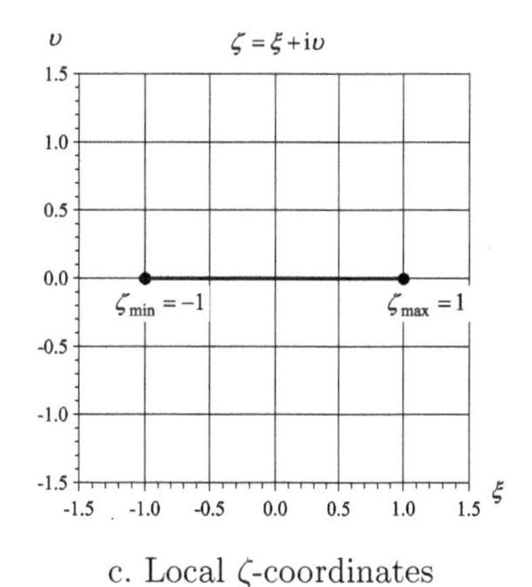

c. Local ζ-coordinates

Figure 5.21 *Conformal mapping and coordinate transformations for curvilinear elements.*

$$\tilde{\varsigma} = \tilde{\varsigma}(z)\,, \quad \tilde{\varsigma}_{\min} = \tilde{\varsigma}(z_{\min})\,, \quad \tilde{\varsigma}_{\max} = \tilde{\varsigma}(z_{\max}) \tag{5.101}$$

where the endpoints map to $\tilde{\varsigma}_{\min}$ and $\tilde{\varsigma}_{\max}$. Finally, this segment is translated and scaled to the standardized straight segment in the ζ-plane between $\zeta_{\min} = -1$ and $\zeta_{\max} = 1$ shown in Fig. 5.21c by

$$\zeta = \left[\tilde{\varsigma}(z) - \frac{\tilde{\varsigma}_{\max} + \tilde{\varsigma}_{\min}}{2}\right] \frac{2}{\tilde{\varsigma}_{\max} - \tilde{\varsigma}_{\min}} \tag{5.102}$$

Thus, every curvilinear element is conformally mapped to the same unit real axis used to formulate straight elements.

The complex potential for curvilinear elements are formulated using the singular integral equations developed for straight elements as double layers (5.10)

$$\Omega(\mathcal{Z}) = \frac{1}{2\pi} \int_{-1}^{1} \frac{\nu(\tilde{\mathcal{Z}})}{\mathcal{Z} - \tilde{\mathcal{Z}}} \mathrm{d}\tilde{\mathcal{Z}} + \frac{\mathrm{i}}{2\pi} \int_{-1}^{1} \frac{\mu(\tilde{\mathcal{Z}})}{\mathcal{Z} - \tilde{\mathcal{Z}}} \mathrm{d}\tilde{\mathcal{Z}} \tag{5.103}$$

with line-dipole ν and line-doublet μ functions, and single layers (5.11)

$$\Omega(\mathcal{Z}) = \frac{L}{2\pi} \int_{-1}^{1} \sigma(\tilde{\mathcal{Z}}) \ln(\mathcal{Z} - \tilde{\mathcal{Z}}) \mathrm{d}\tilde{\mathcal{Z}} + \frac{L}{2\pi} \int_{-1}^{1} \mathrm{i}\gamma(\tilde{\mathcal{Z}}) \ln(\mathcal{Z} - \tilde{\mathcal{Z}}) \mathrm{d}\tilde{\mathcal{Z}} \tag{5.104}$$

with line-sink σ and line-vortex γ functions. Note that these functions are specified as functions of $\mathcal{Z}$, which is real along a straight element where $\mathcal{Z} = X$; however, $\mathcal{Z} = X + \mathrm{i}Y$ has both real and imaginary components along a curvilinear element. Therefore, the complex potential and vector fields for curvilinear elements are formulated as with strengths specified as functions of ζ:

$$\begin{aligned} \boxed{\nu + \mathrm{i}\mu = \sum_{n=0}^{N} \overset{\mathrm{dl}}{c}_n \zeta^n} \quad &\rightarrow \quad \nu = \sum_{n=0}^{N} \Re \overset{\mathrm{dl}}{c}_n \zeta^n\,, \quad \mu = \sum_{n=0}^{N} \Im \overset{\mathrm{dl}}{c}_n \zeta^n \\ \boxed{\sigma + \mathrm{i}\gamma = \frac{1}{L} \sum_{n=0}^{N} \overset{\mathrm{sl}}{c}_n \zeta^n} \quad &\rightarrow \quad \sigma = \frac{1}{L} \sum_{n=0}^{N} \Re \overset{\mathrm{sl}}{c}_n \zeta^n\,, \quad \gamma = \frac{1}{L} \sum_{n=0}^{N} \Im \overset{\mathrm{sl}}{c}_n \zeta^n \end{aligned} \tag{5.105}$$

This enables the strengths of the singular integral equations to be expressed as summations of either the real or imaginary part of the coefficients along the element times powers of $\zeta = \xi$, which is real on the element.

The formulation of curvilinear elements is facilitated through use of a ***Taylor series approximation to represent the conformal mapping between ζ and $\mathcal{Z}$ coordinates***. This gives

$$\zeta = \sum_{p=0}^{P} a_p \mathcal{Z}^p\,, \quad a_p = \frac{1}{p!} \left.\frac{\mathrm{d}^p \zeta}{\mathrm{d}\mathcal{Z}^p}\right|_{\mathcal{Z}=0} \tag{5.106}$$

where the coefficients are obtained from the derivatives of the coordinate transformations and conformal mapping

$$a_p = \begin{cases} \dfrac{2}{\tilde{\varsigma}_{\max}-\tilde{\varsigma}_{\min}} \left[\tilde{\varsigma}(z) - \dfrac{\tilde{\varsigma}_{\max}+\tilde{\varsigma}_{\min}}{2}\right]\Bigg|_{z=\frac{z_{\max}+z_{\min}}{2}} & (p=0) \\ \dfrac{2}{\tilde{\varsigma}_{\max}-\tilde{\varsigma}_{\min}} \dfrac{1}{p!} \left(\dfrac{z_{\max}-z_{\min}}{2}\right)^p \dfrac{\mathrm{d}^p \tilde{\varsigma}(z)}{\mathrm{d}z^p}\Bigg|_{z=\frac{z_{\max}+z_{\min}}{2}} & (p>0) \end{cases} \tag{5.107}$$

evaluated at the center of the element. These power series are substituted into (5.106) to representation the strength of double and single layers in terms of $\mathcal{Z}$:

$$\nu + \mathrm{i}\mu = \sum_{n=0}^{N} \overset{\mathrm{dl}}{c}_n \left(\sum_{p=0}^{P} a_p \mathcal{Z}^p\right)^n, \quad \sigma + \mathrm{i}\gamma = \frac{1}{L}\sum_{n=0}^{N} \overset{\mathrm{sl}}{c}_n \left(\sum_{p=0}^{P} a_p \mathcal{Z}^p\right)^n \tag{5.108}$$

The powers of these series may be expanded to gather terms with common factors of $\mathcal{Z}$ using

$$\left(\sum_{p=0}^{P} a_p \mathcal{Z}^p\right)^n = \sum_{q=0}^{Pn} \mathbf{d}_{nq} \mathcal{Z}^q \quad \rightarrow \quad \boxed{\nu + \mathrm{i}\mu = \sum_{n=0}^{N} \overset{\mathrm{dl}}{c}_n \sum_{q=0}^{Pn} \mathbf{d}_{nq} \mathcal{Z}^q} \quad \boxed{\sigma + \mathrm{i}\gamma = \frac{1}{L}\sum_{n=0}^{N} \overset{\mathrm{sl}}{c}_n \sum_{q=0}^{Pn} \mathbf{d}_{nq} \mathcal{Z}^q} \tag{5.109}$$

where formulation of the coefficients $\mathbf{d}_{nq}$ is deferred to (5.128).

The complex potential for curvilinear element is obtained by substituting this relation between strengths into the integral equation for a double layer, (5.10), and integrating

$$\Omega(\mathcal{Z}) = \sum_{n=0}^{N} \overset{\mathrm{dl}}{c}_n \sum_{q=0}^{Pn} \mathbf{d}_{nq} \frac{1}{2\pi} \int_{-1}^{1} \frac{\tilde{\mathcal{Z}}^q}{\mathcal{Z} - \tilde{\mathcal{Z}}} \mathrm{d}\tilde{\mathcal{Z}}, \quad \boxed{\Omega = \sum_{n=0}^{N} \overset{\mathrm{dl}}{c}_n \sum_{q=0}^{Pn} \mathbf{d}_{nq} \overset{\mathrm{dl}}{\Omega}_q(\mathcal{Z})} \tag{5.110}$$

to give the same influence functions as those for the straight double layer (5.16). Likewise, a single layer with curvilinear geometry is obtained from integration (5.11) using the influence functions for a single layer (5.46)

$$\Omega(\mathcal{Z}) = \sum_{n=0}^{N} \overset{\mathrm{sl}}{c}_n \sum_{q=0}^{Pn} \mathbf{d}_{nq} \frac{1}{2\pi} \int_{-1}^{1} \tilde{\mathcal{Z}}^q \ln(\mathcal{Z} - \tilde{\mathcal{Z}}) \mathrm{d}\tilde{\mathcal{Z}}, \quad \boxed{\Omega = \sum_{n=0}^{N} \overset{\mathrm{sl}}{c}_n \sum_{q=0}^{Pn} \mathbf{d}_{nq} \overset{\mathrm{sl}}{\Omega}_q(\mathcal{Z})} \tag{5.111}$$

Similarly, the vector fields for curvilinear elements take on the following forms for a double layer (5.19)

$$\boxed{v = \sum_{n=0}^{N} \overline{\overset{\mathrm{dl}}{c}_n} \sum_{q=0}^{Pn} \overline{\mathbf{d}_{nq} \overset{\mathrm{dl}}{v}_q(\mathcal{Z})}} \tag{5.112}$$

and a single layer(5.48)

$$\boxed{v = \sum_{n=0}^{N} \overline{\overset{\text{sl}}{c}_n} \sum_{q=0}^{Pn} \overline{\mathbf{d}_{nq}} \overset{\text{sl}}{v}_q(\mathcal{Z})} \tag{5.113}$$

The solution methods presented earlier in this chapter for straight singular integral equations are applicable to curvilinear elements with slight modification. First, the matrices $\mathbf{A}$ for straight line elements must be multiplied by the matrix $\mathbf{d}_{nq}$ relating the coefficients and influence functions. While the influence functions have the same form for straight and curvilinear elements, the location of the branch cut must be adjusted to fall along the curvilinear element. This may be accomplished by adding $\pm i\pi/2$ to the logarithm terms in Fig. 5.4 in the region between the curvilinear element and the straight line connecting its endpoints. Note that singularities in the conformal mappings limit the neighborhood in which Taylor series representation is valid; so computations with the near-field expansions occur inside a radius $|\mathcal{Z}| \leq R_{\text{far}}$ that excludes these singularities (Steward et al., 2008). Computations outside this neighborhood use a far-field representation of the complex potential and vector fields as previously formulated for straight elements in Section 5.4.1.

5.6.2 Curvilinear Double Layers and Single Layers

This section illustrates how to delineate curvilinear elements using standard conformal mappings. A set of elements with the geometry of a hyperbola, parabola, circle, and ellipse are shown in Fig. 5.22, where each set of elements uses the mathematics of a double layer to satisfy a boundary condition of uniform stream function. Each of these examples was placed in a string with straight elements lying along the x-axis and connected to the curvilinear elements at their endpoints. Note that all examples except the hyperbola were separated into smaller elements. This is necessary because those conformal mappings contained singularities, and each element must be made small enough so the singularity lies in the far-field, outside $|\mathcal{Z}| > R_{\text{far}}$, shown in Fig. 5.21.

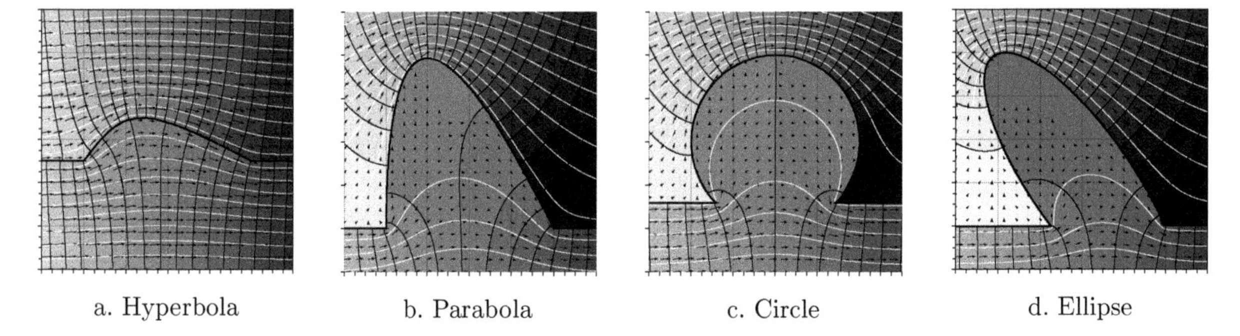

a. Hyperbola　　b. Parabola　　c. Circle　　d. Ellipse

Figure 5.22 *Curvilinear elements composed of arcs with idealized geometries with double-layer boundary conditions. (Reprinted from Steward, Le Grand, Janković, and Strack, 2008, Analytic formulation of Cauchy integrals for boundaries with curvilinear geometry,* Proceedings of The Royal Society of London, Series A, *Vol. 464, Fig. 3, used with permission of The Royal Society; permission conveyed through Copyright Clearance Center, Inc.)*

The mapping from the local curvilinear $\mathcal{Z}$-plane to the straight line in the ζ-plane is given directly for a ***hyperbola*** using

$$\zeta = a_0 \left(1 - \mathcal{Z}^2\right) + \mathcal{Z} \tag{5.114}$$

with $\zeta_{\min} = -1$ and $\zeta_{\max} = 1$ This constant a_0 controls the angles with which the curvilinear element intersects the $\mathcal{Z}$-axis at the endpoints. A similar conformal mapping between $\mathcal{Z}$- and ζ-coordinates exists for a ***parabola***

$$\mathcal{Z} = a_0 \left(1 - \zeta^2\right) + \zeta \tag{5.115}$$

Formulation of this element requires the inverse transformation $\zeta(\mathcal{Z})$, for the series (5.109). This may be obtained using Bell polynomials, with details presented later in (5.132).

Curvilinear elements with the geometry of an arc along a ***circle*** may be mapped from the physical z-plane to the real axis in $\tilde{\zeta}$-plane using the bilinear transformation

$$\tilde{\zeta} = \frac{z - z_0}{z - z_\infty} \tilde{\zeta}_a \frac{z_a - z_\infty}{z_a - z_0} \tag{5.116}$$

where z_∞, z_0, and z_a are three points on the circle that map to $\tilde{\zeta}(z_\infty) = \infty$, $\tilde{\zeta}(z_0) = 0$, and $\tilde{\zeta}(z_a) = \tilde{\zeta}_a$. The derivatives of this mapping, which are needed for the coefficients of the series (5.107), are given by

$$\frac{\mathrm{d}^p \tilde{\zeta}}{\mathrm{d}z^p} = \tilde{\zeta}_a \frac{z_a - z_\infty}{z_a - z_0} (z_0 - z_\infty)\, p! \left(\frac{-1}{z - z_\infty}\right)^{p+1} \tag{5.117}$$

A composition of conformal mappings may be used for elements lying along an ***ellipse*** centered at z_c with major and minor axes of length $r_0 + \rho$ and $r_0 - \rho$ and major axis oriented at angle ϑ. This is mapped to a circle of unit radius centered at the origin in the $\tilde{z}$-plane by

$$z = \left[\tilde{z} + \frac{\rho}{\tilde{z}}\right] \mathrm{e}^{\mathrm{i}\vartheta} + z_c \tag{5.118}$$

with inverse

$$\tilde{z} = \frac{(z - z_c)\mathrm{e}^{-\mathrm{i}\vartheta}}{2} \pm \left\{\left[\frac{(z - z_c)\mathrm{e}^{-\mathrm{i}\vartheta}}{2}\right]^2 - \rho\right\}^{1/2} \tag{5.119}$$

and the $\pm$ sign is chosen as per Steward et al. (2008). This is mapped to the real axis using the same transformation as just used for a circular arc:

$$\tilde{\zeta} = \frac{\tilde{z} - \tilde{z}_0}{\tilde{z} - \tilde{z}_\infty} \tilde{\zeta}_a \frac{\tilde{z}_a - \tilde{z}_\infty}{\tilde{z}_a - \tilde{z}_0} \tag{5.120}$$

The use of curvilinear elements as single-layer elements to match boundary conditions of specified potential is illustrated in Fig. 5.23. The first string of elements lie along a ***Bezier curve*** (Bezier, 1986), with a conformal mapping on the $\tilde{\zeta}$-axis between 0 and 1 and the element is given by the Bernstein polynomials,

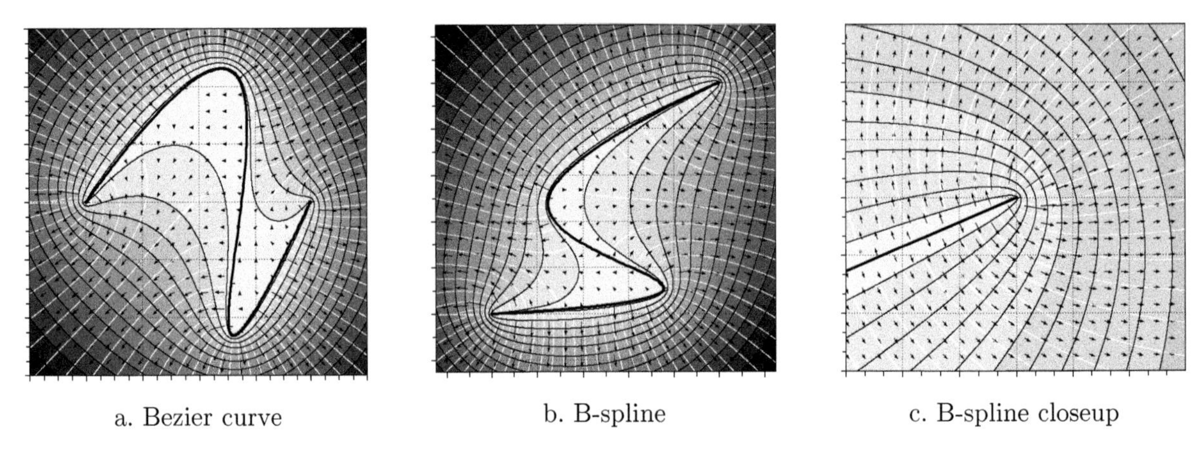

Figure 5.23 *Curvilinear elements with single-layer boundary conditions. (Reprinted from Steward, Le Grand, Janković, and Strack, 2008, Analytic formulation of Cauchy integrals for boundaries with curvilinear geometry,* Proceedings of The Royal Society of London, Series A, *Vol. 464, Figs. 4,5, used with permission of The Royal Society; permission conveyed through Copyright Clearance Center, Inc.)*

$$z = \sum_{n=0}^{K} \binom{K}{n} z_n \varsigma^n (1-\varsigma)^{K-n} \tag{5.121}$$

where z_n are the $K+1$ vertices of the characteristic polygon. A second example is illustrated for a ***B-spline***. This conformal mapping for a central Basis-spline with knots at $\varsigma - m = \{-n/2, -n/2+1, \cdots, n/2\}$ is given by (Schoenberg, 1946, 1973)

$$z = \sum_{m=-\infty}^{\infty} c_m M_n(\varsigma - m) \quad , \quad M_n(\varsigma) = \frac{1}{(n-1)!} \sum_{p=0}^{n} (-1)^p \binom{n}{p} \left(\varsigma + \frac{n}{2} - p\right)_+^{n-1} \tag{5.122}$$

where $x_+ = \max(0, x)$ is the one-sided power function. Additional details for these curvilinear elements may be found in Steward et al. (2008).

5.6.3 Bell Polynomials

The exponential partial Bell polynomials (Bell, 1934) provide a framework for organizing the series manipulations found in curvilinear elements, following Steward et al. (2008). The composition of two series, $f(\zeta(\mathcal{Z}))$, with coefficients

$$f = \sum_{n=0}^{N} f_n \frac{\zeta^n}{n!} \quad , \quad \zeta = \sum_{p=1}^{P} a_p \frac{\mathcal{Z}^p}{p!} \tag{5.123}$$

is given by the Faà di Bruno (1855) formula, which may be written in terms of Bell polynomials as (Comtet, 1974, p.137)

$$f = \sum_{q=0}^{PN} b_q \frac{\mathcal{Z}^q}{q!} \quad , \quad b_q = \sum_{n=0}^{\min(q,N)} f_n \mathbf{B}_{qn}(a_1, \cdots, a_p, \cdots, a_P,) \tag{5.124}$$

The composition used in curvilinear elements (5.109), contains the nth power of ζ, which is given by setting $f_n = n!$ and replacing a_p with $p!a_p$, in the previous expression, giving (Comtet, 1974, p.133)

$$\zeta^n = \left(\sum_{p=1}^{P} a_p \mathcal{Z}^p \right)^n = \sum_{q=n}^{Pn} \left[\frac{n!}{q!} \mathbf{B}_{qn}(p!a_p) \right] \mathcal{Z}^q \qquad (n = 0, 1, \cdots) \tag{5.125}$$

While explicit functional forms for Bell polynomials have been compiled for order $q \leq 12$ (Comtet, 1974, p. 307), values of $\mathbf{B}_{qn}$ may be computed recursively for specified values of a_p using

$$\mathbf{B}_{0n} = \begin{cases} 1 & (n = 0) \\ 0 & (n > 0) \end{cases} \tag{5.126a}$$

and, for $q \geq 1$

$$\mathbf{B}_{qn} = \begin{cases} 0 & (n = 0) \\ a_q & (n = 1) \\ \dfrac{1}{n} \displaystyle\sum_{l=n-1}^{q-1} \binom{q}{l} a_{q-l} \mathbf{B}_{l(n-1)} & (1 < n < q) \\ (a_1)^q & (n = q) \\ 0 & (n > q) \end{cases} \tag{5.126b}$$

In matrix form, this gives

$$\mathbf{B}_{qn} = \begin{bmatrix} 1 & 0 & 0 & 0 & 0 & 0 \\ 0 & a_1 & 0 & 0 & 0 & 0 \\ 0 & a_2 & (a_1)^2 & 0 & 0 & 0 \\ 0 & a_3 & 3a_1a_2 & (a_1)^3 & 0 & 0 \\ 0 & a_4 & 4a_1a_3 + 3a_2{}^2 & 6a_1{}^2a_2 & (a_1)^4 & 0 \\ \cdots & \cdots & \cdots & \cdots & \cdots & \cdots \\ 0 & a_q & \cdots & \cdots & \cdots & (a_1)^q \end{bmatrix} \tag{5.126c}$$

The series used to formulate curvilinear elements also contains the constant term a_0, whereas formulation of the composition using Bell polynomials begins at $p = 1$, (5.123). This extra term may be incorporated using the binomial theorem

$$\left[a_0 + \left(\sum_{p=1}^{P} a_p \mathcal{Z}^p \right) \right]^n = \sum_{r=0}^{n} \binom{n}{r} (a_0)^{n-r} \left(\sum_{p=1}^{P} a_p \mathcal{Z}^p \right)^r \tag{5.127}$$

The powers of this series may be evaluated recursively using (5.125) to give

$$\left(\sum_{p=0}^{P} a_p \mathcal{Z}^p \right)^n = \sum_{r=0}^{n} \binom{n}{r} (a_0)^{n-r} \sum_{q=r}^{Pr} \left[\frac{r!}{q!} \mathbf{B}_{qr}(p!a_p) \right] \mathcal{Z}^q = \sum_{q=0}^{Pn} \mathbf{d}_{nq} \mathcal{Z}^q \tag{5.128}$$

where the constant terms related to the coefficients a_p are gathered in the matrix $\mathbf{d}_{nq}$, and this representation is used in (5.109). This series expansion is a generalization of the recursive relationships presented by Henrici (1956) and Knuth (1998).

Some conformal mappings are specified as functions $\mathcal{Z}(\zeta)$ and their inverse is needed for developing the complex potential of curvilinear elements:

$$\mathcal{Z} = \sum_{p=1}^{P} a_p \frac{\zeta^p}{p!} \quad \rightarrow \quad \zeta = \sum_{p=1} a_p{}^{<-1>} \frac{\mathcal{Z}^p}{p!} \tag{5.129}$$

The coefficients for the inverse mapping may be obtained using Bell polynomials following (Comtet, 1974, p.151)

$$a_1{}^{<-1>} = 1/a_1 \quad , \quad a_p{}^{<-1>} = \sum_{n=1}^{p-1} [-p]_n (a_1)^{-p-n} \mathbf{B}_{(p-1)n}\left(\frac{a_2}{2}, \frac{a_3}{3}, \cdots\right) \tag{5.130}$$

where $[-p]_n$ is obtained by evaluating the falling factorial function

$$[-p]_n = (-p) \times [(-p) - 1] \times \cdots \times [(-p) - n + 1] \tag{5.131}$$

Inverse series with non-zero constant terms may be rearranged using

$$\mathcal{Z} = \sum_{p=0}^{P} a_p \frac{\zeta^p}{p!} \quad \rightarrow \quad (\mathcal{Z} - a_0) = \sum_{p=1} a_p \frac{\zeta^p}{p!} \quad \rightarrow \quad \zeta = \sum_{p=1} a_p{}^{<-1>} \frac{(\mathcal{Z} - a_0)^p}{p!} \tag{5.132}$$

and the terms $(\mathcal{Z} - a_0)^p$ may be expanded using (5.128).

Example 5.2 The power of the series

$$\left(1 + 2\mathcal{Z} + 3\mathcal{Z}^2\right)^n \ , \quad a_0 = 1 \ , \ a_1 = 2 \ , \ a_2 = 3$$

may be expanded using the composition equation (5.128) with the matrix of Bell polynomial coefficients from (5.126)

$$\mathbf{B}_{qn} = \begin{bmatrix} 1 & 0 & 0 & 0 & 0 & 0 & 0 \\ 0 & 2 & 0 & 0 & 0 & 0 & 0 \\ 0 & 6 & 4 & 0 & 0 & 0 & 0 \\ 0 & 0 & 36 & 8 & 0 & 0 & 0 \\ 0 & 0 & 108 & 144 & 16 & 0 & 0 \\ 0 & 0 & 0 & 1080 & 480 & 32 & 0 \\ 0 & 0 & 0 & 3240 & 6480 & 1440 & 64 \end{bmatrix}$$

For example, the first few powers of the series are given by

$$\begin{aligned}
\left(1 + 2\mathcal{Z} + 3\mathcal{Z}^2\right)^0 &= 1 \\
\left(1 + 2\mathcal{Z} + 3\mathcal{Z}^2\right)^1 &= 1 + 2\mathcal{Z} + 3\mathcal{Z}^2 \\
\left(1 + 2\mathcal{Z} + 3\mathcal{Z}^2\right)^2 &= 1 + 4\mathcal{Z} + 10\mathcal{Z}^2 + 12\mathcal{Z}^3 + 9\mathcal{Z}^4 \\
\left(1 + 2\mathcal{Z} + 3\mathcal{Z}^2\right)^3 &= 1 + 6\mathcal{Z} + 21\mathcal{Z}^2 + 44\mathcal{Z}^3 + 63\mathcal{Z}^4 + 54\mathcal{Z}^5 + 27\mathcal{Z}^6
\end{aligned}$$

Problem 5.22 Expand the powers of the following polynomials to obtain a power series in terms of $\mathcal{Z}$:

A. $(2+3\mathcal{Z}+4\mathcal{Z}^2)^2$
B. $(2+3\mathcal{Z}+4\mathcal{Z}^2)^3$
C. $(3+4\mathcal{Z}+5\mathcal{Z}^2)^2$
D. $(3+4\mathcal{Z}+5\mathcal{Z}^2)^3$
E. $(4+5\mathcal{Z}+6\mathcal{Z}^2)^2$
F. $(4+5\mathcal{Z}+6\mathcal{Z}^2)^3$
G. $(5+6\mathcal{Z}+7\mathcal{Z}^2)^2$
H. $(5+6\mathcal{Z}+7\mathcal{Z}^2)^3$
I. $(6+7\mathcal{Z}+8\mathcal{Z}^2)^2$
J. $(6+7\mathcal{Z}+8\mathcal{Z}^2)^3$
K. $(7+8\mathcal{Z}+9\mathcal{Z}^2)^2$
L. $(7+8\mathcal{Z}+9\mathcal{Z}^2)^3$

5.7 Three-Dimensional Vector Fields

A wide range of analytic elements have been developed within a two-dimensional framework, where the vector field has components within the x–y plane. Yet, these two-dimensional vector fields may be visualized within a three-dimensional domain. For example, the potential function for a two-dimensional point-sink, (3.1),

$$\Phi = \frac{Q}{2\pi}\ln\sqrt{(x-x_c)^2+(y-y_c)^2}\,, \qquad \begin{aligned} v_x &= -\frac{Q}{2\pi}\frac{x-x_c}{(x-x_c)^2+(y-y_c)^2} \\ v_y &= -\frac{Q}{2\pi}\frac{y-y_c}{(x-x_c)^2+(y-y_c)^2} \\ v_z &= 0 \end{aligned} \tag{5.133}$$

has a vector field with component v_z in the z-direction equal to zero. This vector field is visualized in Fig. 5.24 using a set of ***stream surfaces*** obtained by tracing particles that pass through points along the horizontal and vertical lines connecting points A, B, C, and D. ***Equipotential surfaces*** are visualized by connecting neighboring streamlines at points with equal value of potential, and the plan view in Fig. 5.24b is similar to that shown for the two-dimensional vector field in Fig. 3.1. The streamlines lie on horizontal planes in Fig. 5.24c since $v_z = 0$, and the three-dimensional depiction in Fig. 5.24a represents the solution for an infinitely long vertical line-sink aligned parallel to the z-axis.

This method of visualizing vector fields by tracing particles along stream surfaces is useful for studying analytic elements with three-dimensional vector fields. For example, a three-dimensional line-sink that removes a flux along a finite segment is illustrated in Fig. 5.25,

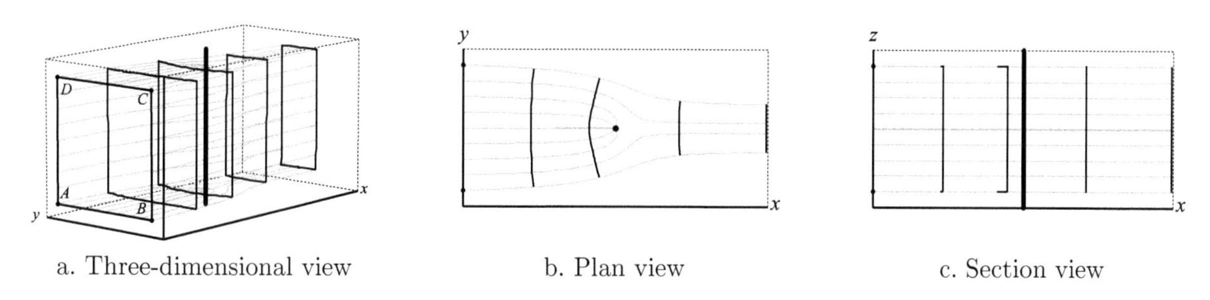

Figure 5.24 *Stream surfaces near fully penetrating sink in a uniform vector field. (Reprinted from Steward, 1998, Stream surfaces in two-dimensional and three-dimensional divergence-free flows,* Water Resources Research, *Vol. 34(5), Fig. 3, with permission from John Wiley and Sons. Copyright 1998 by the American Geophysical Union.)*

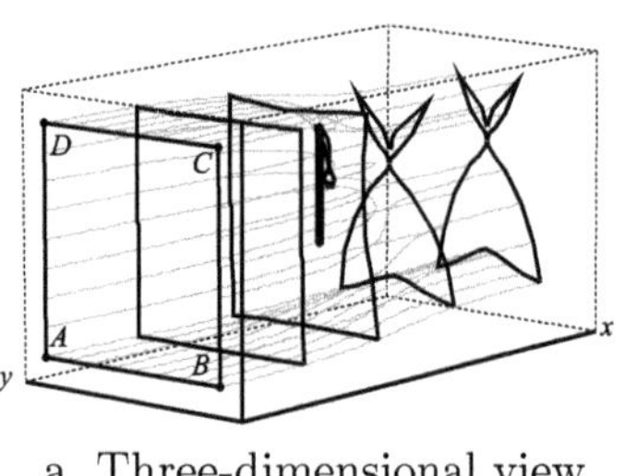

a. Three-dimensional view

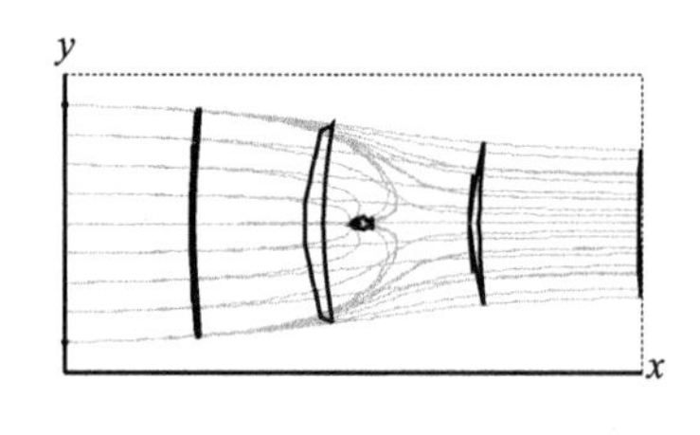

b. Plan view

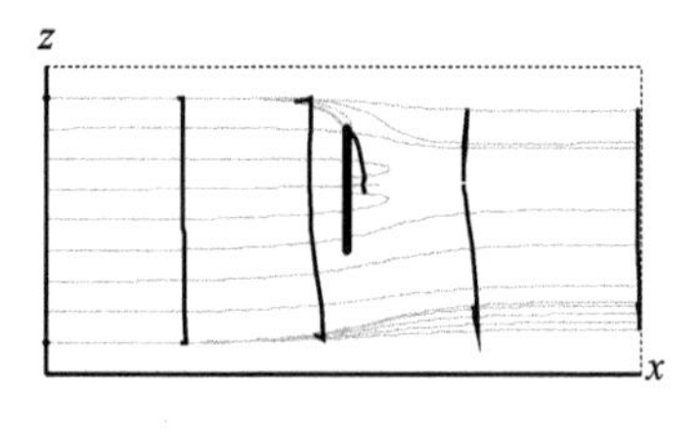

c. Section view

Figure 5.25 *Stream surfaces near partially penetrating sink. (Reprinted from Steward, 1998, Stream surfaces in two-dimensional and three-dimensional divergence-free flows,* Water Resources Research, *Vol. 34(5), Fig. 5, with permission from John Wiley and Sons. Copyright 1998 by the American Geophysical Union.)*

and the mathematical development of this element is presented shortly in (5.135). This figure illustrates how stream surfaces as a uniform vector field in the x-direction act to move particles towards the finite line-sink, where particles are tracked through the same locations between points A, B, C, and D as the two-dimensional vector field depicted in Fig. 5.24.

An important phenomenon is observed when comparing the deformation of stream surfaces in the two-dimensional and three-dimensional vector fields in Figs. 5.24 and 5.25. While a two-dimensional analytic element causes a shift in the relative position of neighboring streamlines near the element, they lie in the same relative proximity to one another downstream as they do upstream. However, a three-dimensional analytic element generates a persistent rearrangement in the relative position of neighboring streamlines as particles travel past the element. This was observed by Steward (1998), who noted that the stream surfaces with cross sections of squares take on cross sections shaped like a rosette after passing near the finite line-sink element in Fig. 5.25. Thus, ***studies of particle movement through fields of analytic elements with three-dimensional geometry require three-dimensional models, since two-dimensional models are incapable of reproducing a persistent rearrangement of the relative location of neighboring trajectories***.

This persistent rearrangement of the locations of neighboring particles as they travel along their trajectories is also evident for analytic elements that do not remove a net flux from the domain. For example, the impact of a heterogeneity with geometry of a prolate spheroid through which flux moves more readily inside than outside the element is illustrated in Fig. 5.26. The stream surfaces are visualized by tracing particles lying along the sets of interconnecting lines on plane a, and then connecting the location of neighboring trajectories as they intersect planes b and c. The solution for the prolate spheroid in Fig. 5.26a with major axis aligned parallel to the background vector field was formulated by Steward and Janković (2001) as an axisymmetric vector field. Clearly, an ***axisymmetric vector field rearranges the relative position of neighboring streamlines in the neighborhood of the element***, since flux travels more readily through the object. Yet, trajectories maintain the ***same relative position to one another between large distances upgradient and downgradient from the element***.

The three-dimensional vector field in Fig. 5.26b is generated by a spheroid placed at an oblique orientation to the background vector field following Janković et al. (2009). Clearly, ***a persistent rearrangement in the relative position of neighboring trajectories occurs in a three-dimensional non-axisymmetric vector field***, due to the preferential flux through this element. These observations led to the theory of ***advective***

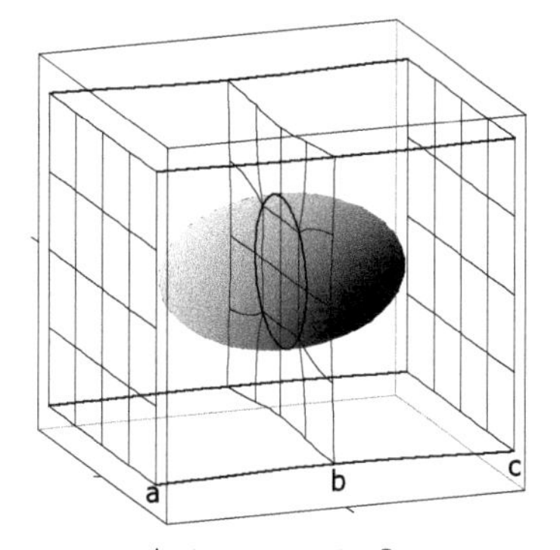

a. Axisymmetric flow (from Fig. 4.33a)

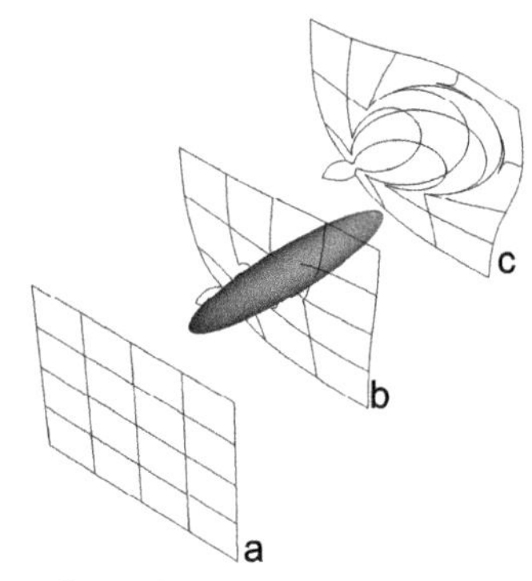

b. Uniform flow oblique to element axis (from Fig. 4.34a)

Figure 5.26 *Intersection of stream surfaces and breakthrough planes for streamlines near a highly conductive prolate spheroid.*

mixing, whereby three-dimensional heterogeneities generate a transverse displacement of the relative position of particles that causes fluids to mix as they travel through three-dimensional domains. This active area of research studies how three-dimensional analytic elements with different capacity to transmit fluxes lead to the transverse dispersion of fluids as they pass through a heterogeneous domain.

5.7.1 Point-Sinks, Line-Sinks, and the Method of Images

A set of analytic elements are presented in this section to study three-dimensional vector fields associated with objects that remove a net flux from a domain. For example, an element that that removes a flux Q at the point (x_p, y_p, z_p) in a three-dimensional domain is given by a ***point-sink*** with potential function and three-dimensional vector field

$$\Phi = -\frac{Q}{4\pi}\frac{1}{r}, \qquad r = \left[(x-x_c)^2 + (y-y_c)^2 + (z-z_c))^2\right]^{1/2} \qquad \begin{aligned} v_x &= -\frac{Q}{4\pi}\frac{x-x_c}{r^3}, \\ v_y &= -\frac{Q}{4\pi}\frac{y-y_c}{r^3}, \\ v_z &= -\frac{Q}{4\pi}\frac{z-z_c}{r^3} \end{aligned} \tag{5.134}$$

where r is the distance from the point-sink. A ***three-dimensional line-sink*** that removes a flux along a line segment in an infinite domain is obtained by integrating these functions along the segment. For example, an element that removes a constant flux along a vertical line has a potential function from (Steward, 2001, eqns 4.1–2)

$$\Phi = -\frac{\sigma_0}{4\pi}\ln\frac{r_{\min} + (z - z_c + L)}{r_{\max} + (z - z_c - L)}, \qquad \begin{aligned} r_{\min} &= \left[(x-x_c)^2 + (y-y_c)^2 + (z-z_c+L)^2\right]^{1/2} \\ r_{\max} &= \left[(x-x_c)^2 + (y-y_c)^2 + (z-z_c-L)^2\right]^{1/2} \end{aligned} \tag{5.135a}$$

where $r_{\min}$ is the distance from the endpoint at $(x_c, y_c, z_c - L)$, and $r_{\max}$ is the distance from the other endpoint at $(x_c, y_c, z_c + L)$. Its vector field has components

$$v_x = -\frac{\sigma_0}{4\pi}\frac{(x-x_c)}{(x-x_c)^2+(y-y_c)^2}\left(\frac{z-z_c+L}{r_{\min}} - \frac{z-z_c-L}{r_{\max}}\right),$$
$$v_y = -\frac{\sigma_0}{4\pi}\frac{(y-y_c)}{(x-x_c)^2+(y-y_c)^2}\left(\frac{z-z_c+L}{r_{\min}} - \frac{z-z_c-L}{r_{\max}}\right), \quad (5.135b)$$
$$v_z = \frac{\sigma_0}{4\pi}\left(\frac{1}{r_{\min}} - \frac{1}{r_{\max}}\right)$$

This element was used in the visualization of stream surfaces in Fig. 5.25.

The ***method of images*** may be used to extend these solutions to a three-dimensional domain with a bottom and top in the z-direction, and with infinite extent in the x–y plane. This is accomplished for a domain with bottom at $z = 0$ and top at $z = H$ by placing image elements outside the domain with the same x_c and y_c locations as the element but with endpoints at elevations

$$\begin{aligned} z_{\min} &= 2iH \pm z_c - L \\ z_{\max} &= 2iH \pm z_c + L \end{aligned} \qquad (i = -I, \cdots, 0, \cdots, I) \qquad (5.136)$$

For example, the element and its image elements outside the domain are shown in Fig. 5.27b for $I = 2$. The method of images results in a solution with ***impermeable planes at the bottom and top of the domain*** when each element removes a net flux Q from the domain, so $\sigma_0 = Q/(z_{\max} - z_{\min})$ for a constant strength analytic element. In theory, an infinite number I of elements is required to exactly reproduce the condition of $v_z = 0$ at the top and bottom; however, the elements at larger elevations may be approximated as two semi-infinite elements as illustrated in Fig. 5.27b with strength $\sigma_0 = Q/H$ obtained using

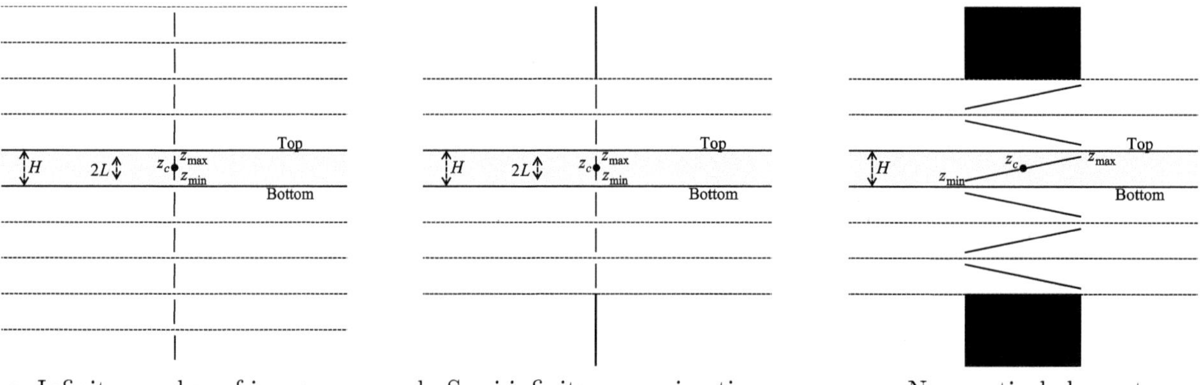

Figure 5.27 *Method of images for vertical and inclined elements with semi-infinite approximation. (Reprinted from Steward, 2001, A vector potential for a partly penetrating well and flux in an approximate method of images,* Proceedings of The Royal Society of London, Series A, *Vol. 457, Fig. 7, used with permission of The Royal Society; permission conveyed through Copyright Clearance Center, Inc.) (Reprinted from Steward and Jin, 2003, Drawdown and capture zone topology for non-vertical wells,* Water Resources Research, *Vol. 39(8), Fig. 3, with permission from John Wiley and Sons. Copyright 2003 by the American Geophysical Union.)*

a two-dimensional element minus one finite element. The method of images may also be extended to elements with non-vertical placement as illustrated in Fig. 5.27c, with details presented in Steward (1996).

The ***three-dimensional trajectories*** of particles as they travel towards elements in a domain with impermeable bottom and top are illustrated in Fig. 5.28. These examples illustrate a partially penetrating vertical line-sink, a horizontal line-sink, and a three-dimensional area-sink located at the domain. Clearly a three-dimensional vector field occurs in a local neighborhood around the element. However, the vector field approaches that of a two-dimensional element at larger horizontal distances from the element (Haitjema, 1995).

These analytic elements have been used to study the ***capture zone*** of elements that remove a flux along a line segment in a bounded domain. These capture zones are illustrated in Fig. 5.29 for a two-dimensional line-sink, and partially penetrating line-sinks with the same length but oblique or horizontal placement. These trajectories of particles were traced backwards in time from locations along the surface of the element and shown in section view at a distance upgradient from an element. Each element satisfies a condition of uniform potential along its boundary, and panels illustrate how the geometry of a capture zone changes as the net extraction increases. Details of the advanced three-dimensional analytic elements and the system of equations used to solve boundary conditions using Lagrange multipliers (2.30) are found in Steward (1996) and Steward and Jin (2003).

Problem 5.23 Calculate the potential Φ and vector (v_x, v_y, v_z) at the (x, y, z) location specified below, for a three-dimensional point-sink in an infinite aquifer with $Q = 4$ and $(x_c, y_c, z_c) = (0, 0, 6)$.

A. $(-1, -1, 0)$	C. $(-1, -1, 2)$	E. $(-1, -1, 4)$	G. $(-1, -1, 6)$	I. $(-1, -1, 8)$	K. $(-1, -1, 10)$
B. $(-1, -1, 1)$	D. $(-1, -1, 3)$	F. $(-1, -1, 5)$	H. $(-1, -1, 7)$	J. $(-1, -1, 9)$	L. $(-1, -1, 11)$

Problem 5.24 Calculate the potential Φ and vector (v_x, v_y, v_z) at the (x, y, z) location specified in Problem 5.23, for a three-dimensional vertical line-sink in an infinite aquifer with $\sigma_0 = 1$, $(x_c, y_c, z_c) = (0, 0, 6)$ and $L = 2$.

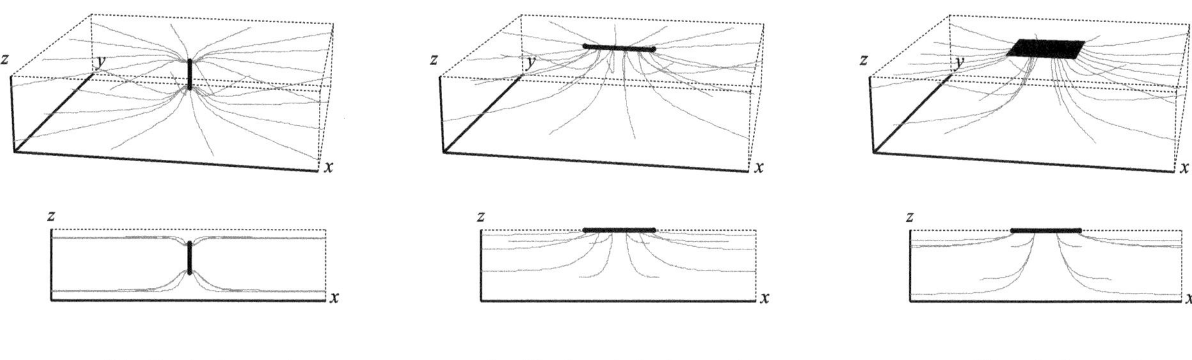

Figure 5.28 *Pathlines in three-dimensional and section views near elements. (Reprinted from Steward, 1999, Three-dimensional analysis of the capture of contaminated leachate by fully penetrating, partially penetrating, and horizontal wells,* Water Resources Research, *Vol. 35(2), Fig. 2, with permission from John Wiley and Sons. Copyright 1999 by the American Geophysical Union.)*

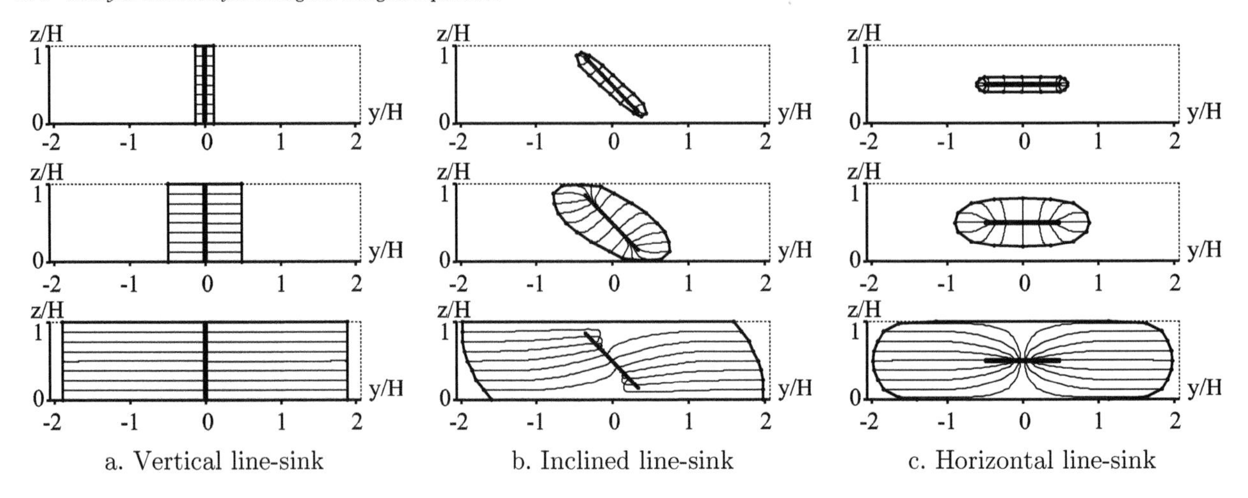

Figure 5.29 *Pathlines tracing particles in section view captured by vertical, inclined, and horizontal elements, with increasing net flux Q in top to bottom panels. (Reprinted from Steward and Jin, 2003, Drawdown and capture zone topology for non-vertical wells, Water Resources Research, Vol. 39(8), Fig. 8, with permission from John Wiley and Sons. Copyright 2003 by the American Geophysical Union.)*

5.7.2 The Vector Potential

The vector potential $\boldsymbol{\Psi}$ provides a mathematical tool useful for studying three-dimensional vector fields. It is defined such that minus its curl, (1.4), gives the vector field $\mathbf{v}$:

$$\mathbf{v} = -\nabla \times \boldsymbol{\Psi} \tag{5.137}$$

The vector potential is important in computing flux F across a surface S, since application of ***Stokes, theorem*** to the integral of the normal component of the vector field gives

$$F|_S = \iint_S \mathbf{v} \cdot \mathfrak{n} \, dS = -\oint_{\partial S} \boldsymbol{\Psi} \cdot \mathfrak{s} \, d\mathfrak{s}. \tag{5.138}$$

where the unit vector $\mathfrak{n}$ is normal to S and $\mathfrak{s}$ is tangent to its boundary ∂S. This effectively replaces a surface integral with a line integral of the tangential component of the vector potential along its boundary. And, yet, construction of a vector potential that reproduces a given vector field is difficult to achieve.

Analytic elements with two-dimensional divergence-free vector fields were previously formulated using the Lagrange (1781) stream function, (1.10):

$$v_x = -\frac{\partial \Psi}{\partial y}, \quad v_y = +\frac{\partial \Psi}{\partial x}, \quad v_z = 0 \tag{5.139}$$

The vector potential, (5.137), may also be expressed in Cartesian components (Ψ_x, Ψ_y, Ψ_z) as

$$\mathbf{v} = \left(-\frac{\partial \Psi_z}{\partial y} + \frac{\partial \Psi_y}{\partial z}\right)\hat{\mathbf{x}} + \left(-\frac{\partial \Psi_x}{\partial z} + \frac{\partial \Psi_z}{\partial x}\right)\hat{\mathbf{y}} + \left(-\frac{\partial \Psi_y}{\partial x} + \frac{\partial \Psi_x}{\partial y}\right)\hat{\mathbf{z}} \tag{5.140}$$

Steward (2002, eqn 2.9) ***developed a vector potential by multiplying the Lagrange stream function, which is constant in the z-direction, by a unit vector in the z-direction***, which gives components

$$\boldsymbol{\Psi} = \Psi \nabla z = \Psi \hat{\mathbf{z}} \quad \rightarrow \quad \begin{array}{l} \Psi_x = 0 \\ \Psi_y = 0 \\ \Psi_z = \Psi \end{array} \tag{5.141}$$

Clearly, substituting these components into the vector potential, (5.140), reproduces the vector field in (5.139). For example, the vector potential for an infinite line-sink (Steward, 2002, eqn 4.4) in a uniform vector field (Steward, 2002, eqn 4.1) is given by

$$\boldsymbol{\Psi} = \left(\frac{\sigma_0}{2\pi}\theta - v_{x0}y + v_{y0}x\right)\hat{\mathbf{z}} \tag{5.142}$$

The streamlines with constant stream function and the vector potential for an infinite line-sink are illustrated in Fig. 5.30, and minus its curl gives the vector field (5.133).

Similarly, analytic elements with axisymmetric vector fields symmetric about the z-axis in three-dimensions may be formulated using the Stokes (1842) stream function:

$$v_r = -\frac{1}{r}\frac{\partial \Psi}{\partial z}\,, \quad v_\theta = 0\,, \quad v_z = \frac{1}{r}\frac{\partial \Psi}{\partial r} \tag{5.143}$$

which was previously applied to the study of spheroids in (4.132). The vector potential, (5.137), may also be expressed as $(\Psi_r, \Psi_\theta, \Psi_z)$ components in cylindrical (r, θ, z) coordinates as

$$\mathbf{v} = \left[-\frac{1}{r}\frac{\partial \Psi_z}{\partial \theta} + \frac{1}{r}\frac{\partial (r\Psi_\theta)}{\partial z}\right]\hat{\mathbf{r}} + \left(-\frac{\partial \Psi_r}{\partial z} + \frac{\partial \Psi_z}{\partial r}\right)\hat{\boldsymbol{\theta}} + \left[-\frac{1}{r}\frac{\partial (r\Psi_\theta)}{\partial r} + \frac{1}{r}\frac{\partial \Psi_r}{\partial \theta}\right]\hat{\mathbf{z}} \tag{5.144}$$

Steward (2002, eqn 2.14) ***developed a vector potential by multiplying the Stokes stream function by the gradient of the angle*** θ, which gives vector components

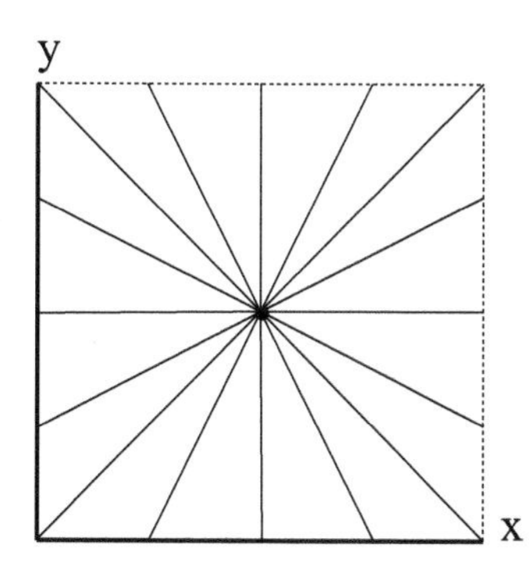

a. Streamlines with uniform Ψ

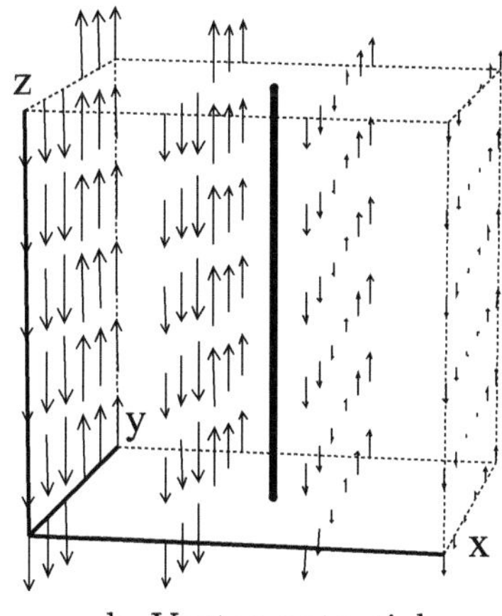

b. Vector potential

Figure 5.30 *Vector potential from the Lagrange stream function for an infinite line-sink. (Reprinted from Steward, 2002, A vector potential and exact flux through surfaces using Lagrange and Stokes stream functions,* Proceedings of The Royal Society of London, Series A, *Vol. 458, Fig. 4, used with permission of The Royal Society; permission conveyed through Copyright Clearance Center, Inc.)*

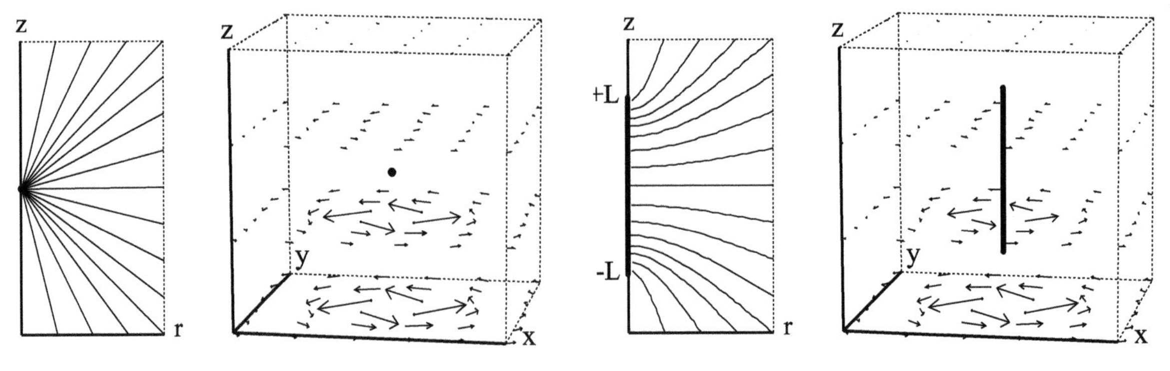

Figure 5.31 *Vector potential from the Stokes stream function for point-sink and line-sink. (Reprinted from Steward, 2002, A vector potential and exact flux through surfaces using Lagrange and Stokes stream functions,* Proceedings of The Royal Society of London, Series A, *Vol. 458, Figs. 6,7, used with permission of The Royal Society; permission conveyed through Copyright Clearance Center, Inc.)*

$$\boldsymbol{\Psi} = -\Psi\nabla\theta\,, \quad \theta = \arctan\frac{z}{y} \quad \rightarrow \quad \begin{aligned} \Psi_r &= 0 \\ \Psi_\theta &= -\frac{1}{r}\Psi \\ \Psi_z &= 0 \end{aligned} \tag{5.145}$$

Clearly, substituting these components into the vector potential, (5.144), reproduces the vector field in (5.143). The Stokes stream function is illustrated for a point-sink in Fig. 5.31a by drawing streamlines with constant Ψ. This stream function is substituted into (5.145) to give its vector potential (Steward, 2002, eqn 5.1)

$$\boldsymbol{\Psi} = \frac{Q}{4\pi}\left(1 - \frac{z}{r}\right)\nabla\left(\arctan\frac{y}{x}\right) \tag{5.146}$$

Similarly, the Stokes stream function that reproduces the vector field of a line-sink, (5.135), gives the vector potential (Steward, 2002, eqn 5.12)

$$\boldsymbol{\Psi} = -\frac{\sigma_0}{4\pi}\left(r_{\min} - r_{\max} - 2L\right)\nabla\left(\arctan\frac{y}{x}\right) \tag{5.147}$$

and these streamlines and vector potential are shown in Fig. 5.31b.

The ***vector potential for analytic elements with more complex geometry may be obtained using integrals*** developed by Goursat (1904). Here, one component of the vector potential may be chosen equal to zero, and integration of the remaining components (5.137) of the divergence-free vector field leads to (Steward, 2001, eqn 2.8)

$$\Psi_x = 0\,, \quad \Psi_y = -\int_{x_l}^{x} v_z(\tilde{x}, y, z)\,\mathrm{d}\tilde{x}\,, \quad \Psi_z = \int_{x_l}^{x} v_y(\tilde{x}, y, z)\,\mathrm{d}\tilde{x} - \int_{y_l}^{y} v_x(x_l, \tilde{y}, z)\,\mathrm{d}\tilde{y} \tag{5.148}$$

where $\boldsymbol{\Psi} = 0$ at the lower limit of integration, (x_l, y_l). The vector potential for a line-sink with uniform flux per length may be obtained by substituting its vector, (5.135), into the integrals (5.148) and integrating with the lower limit of integration at plus infinity $x_l \to \infty$, giving (Steward, 2001, eqn 4.6)

$$\Psi_x = 0\,, \quad \Psi_y = -\frac{\sigma_0}{4\pi} \ln \frac{r_{\min} + x}{r_{\max} + x}\,,$$
$$\Psi_z = -\frac{\sigma_0}{4\pi} \left[\arctan \frac{x(z+L)}{y r_{\min}} - \arctan \frac{z+L}{y} - \arctan \frac{x(z-L)}{y r_{\max}} + \arctan \frac{z-L}{y} \right] \tag{5.149}$$

where the arctan functions are each evaluated to lie within the range of $-\pi/2$ and $\pi/2$. This formulation of the vector potential for a three-dimensional line-sink is illustrated in Fig. 5.32. Note that this formulation of the vector potential clearly differs from that obtained using the Stokes stream function in (5.147), and yet, each reproduces the same vector field and trajectories of particles. Thus, ***the vector potential is not uniquely defined for a given vector field***. The method of images may also be applied to the vector potential for bounded domains with $v_z = 0$ along the top and bottom of the domain. This is illustrated in Fig. 5.33, which shows the vector potential for a line-sink in a bounded domain, and the vector field corresponds to the potential function illustrated earlier with the stream surfaces in Fig. 5.25.

The depictions of vector potentials for point-sinks and line-sinks identify the occurrences of singularities that must be accounted for in order for the vector potential to provide accurate estimates of net flux using the Stokes theorem, (5.138). A ***vortex line where $\boldsymbol{\Psi}$ is singular*** occurs in the negative z-direction from the elements in Fig. 5.31. This is analyzed for a surface that intersects this singularity in Fig. 5.34, where the small circular surface V with radius d is normal to and centered on the z-axis at z_V. The flux through V may be obtained by substituting the vector potential (5.146) into the integral (5.138),

$$F|_V = -\oint_{\partial V} \boldsymbol{\Psi} \cdot \mathfrak{s}\, ds = -\frac{Q}{4\pi} \oint_{\partial V} \left[1 - \frac{z_V}{\left(z_V{}^2 + d^2\right)^{1/2}} \right] (\nabla\theta) \cdot \mathfrak{s}\, ds \tag{5.150}$$

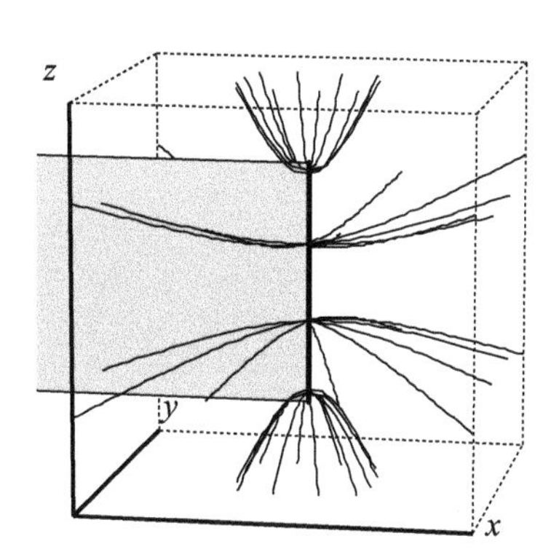

a. Streamlines and vortex sheet

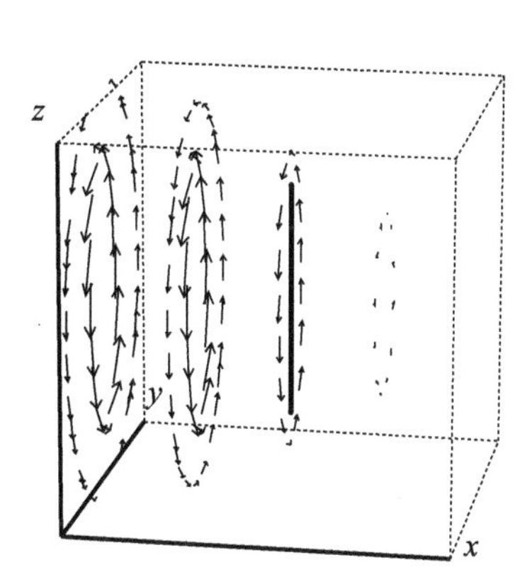

b. Vector potential

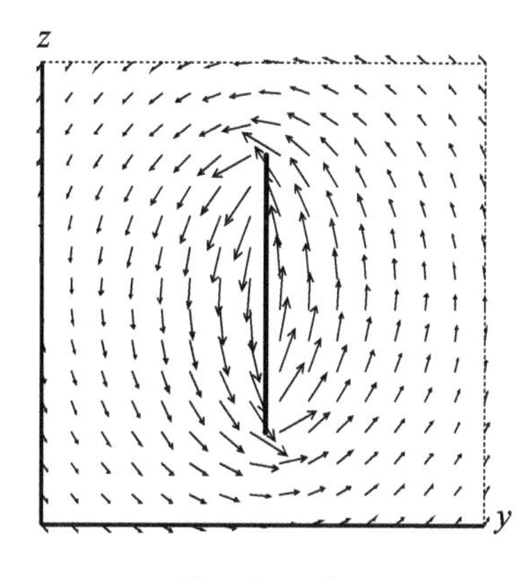

c. Section view

Figure 5.32 *Vector potential for a line-sink. (Reprinted from Steward, 2001, A vector potential for a partly penetrating well and flux in an approximate method of images,* Proceedings of The Royal Society of London, Series A, *Vol. 457, Figs. 4,5, used with permission of The Royal Society; permission conveyed through Copyright Clearance Center, Inc.)*

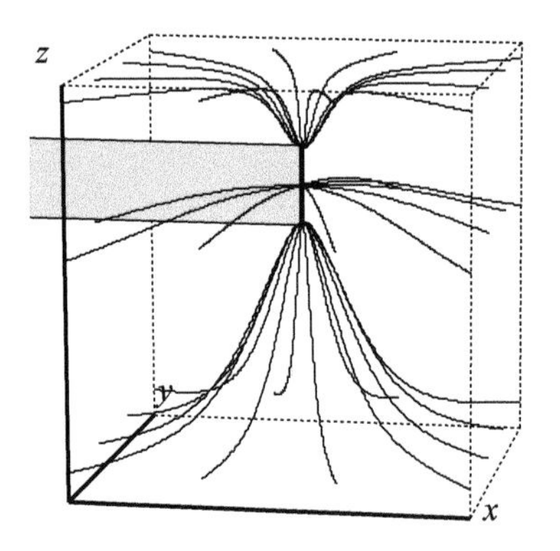

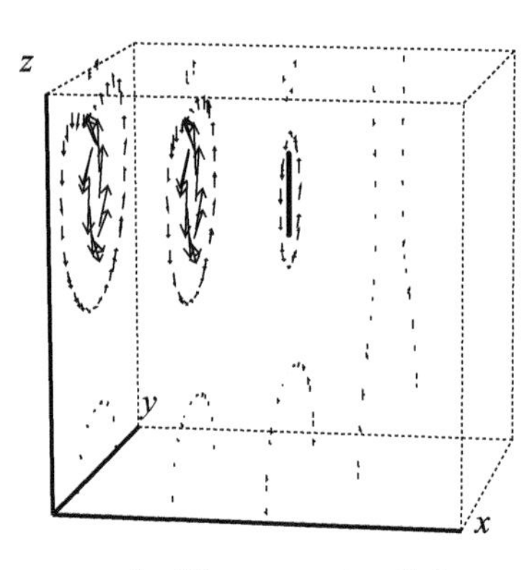

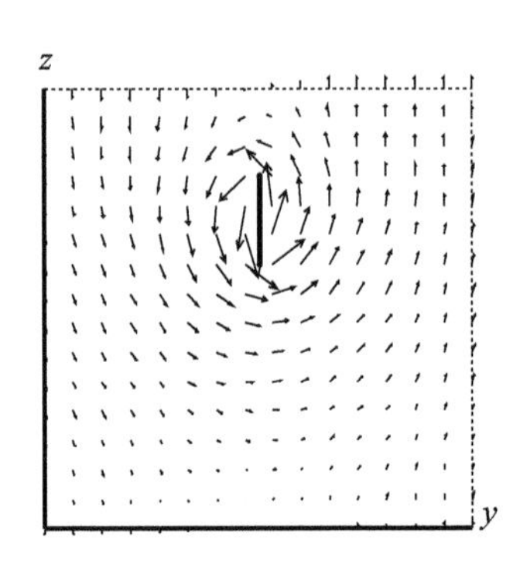

a. Streamlines and vortex sheet b. Vector potential c. Section view

Figure 5.33 *Vector potential for a line-sink in a bounded domain with impermeable base and top. (Reprinted from Steward, 2001, A vector potential for a partly penetrating well and flux in an approximate method of images,* Proceedings of The Royal Society of London, Series A, *Vol. 457, Figs. 8,9, used with permission of The Royal Society; permission conveyed through Copyright Clearance Center, Inc.)*

The variable of integration is expressed in terms of θ in the limit as the circle becomes infinitesimal

$$F|_V = -\lim_{d\to\infty} \frac{Q}{4\pi}\left[1 - \frac{z_V}{\left({z_V}^2 + d^2\right)^{1/2}}\right]\int_0^{2\pi} \mathrm{d}\theta = -\frac{Q}{2}\left(1 - \frac{z_V}{|z_V|}\right) \tag{5.151}$$

This gives a ***virtual flux*** equal to the strength of the point-sink and directed away from the element (Steward, 2001, eqn 3.12)

$$F|_V = \begin{cases} 0 & (0 < z_V) \\ -Q & (z_V < 0) \end{cases} \tag{5.152}$$

that is similar to the virtual discharge flowing through the branch cut of the stream function of a two-dimensional point-sink in Fig. 3.1. This flux is virtual in that ***it does not physically exist in the problem domain, but is an artifact of the fact that the vector potential must everywhere generate a divergence-free vector field***. Thus, the flux Q removed from the domain by the point-sink is transported outside the domain in the negative z-direction with infinite velocity through the singular vortex line. This virtual discharge must be removed from the flux obtained from the vector potential (5.138) to compute the actual flux occurring through a surface:

$$\boxed{F|_S = -\oint_{\partial S} \boldsymbol{\Psi}\cdot\mathfrak{s}\,\mathrm{d}\mathfrak{s} \pm F|_V} \tag{5.153}$$

where the $\pm$ sign is chosen by the direction in which S encircles the singular vortex line. Similarly, the singular vortex line of a line-sink generates a virtual flux along the z-axis, where the vortex sheet intersects the axis at z_V of (Steward, 2002, eqn 5.23)

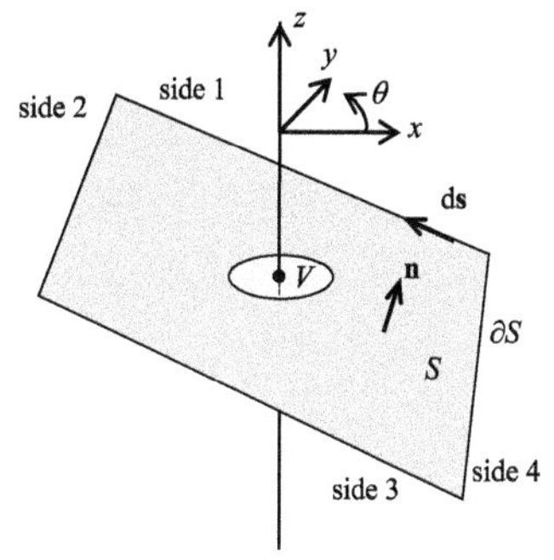

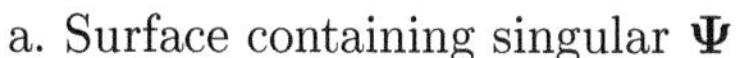
a. Surface containing singular $\boldsymbol{\Psi}$

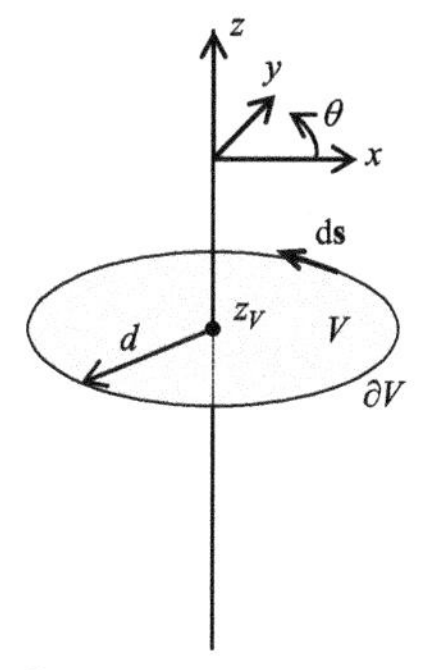

b. Singular vortex line

Figure 5.34 *Virtual flux through a singular vortex line.*

$$F|_V = \begin{cases} 0 & (L \le z_V) \\ -\sigma_0(L - z_V) & (-L < z_V < L) \\ -2\sigma_0 L & (z_V \le -L) \end{cases} \tag{5.154}$$

Discontinuities that occur along sheets must also be accounted for in computing net flux using the vector potential. A ***virtual flux through vortex sheet occurs along surfaces across which the vector potential is discontinuous,*** as illustrated for a line-sink in Fig. 5.32 and in Fig. 5.33 for a line-sink with the method of images. This flux may be computed across the surface V that intersects the discontinuous vortex sheet between elevations z_A and z_B in Fig. 5.35. A virtual flux is generated along the vortex sheet where Ψ_z is discontinuous. This discharge is obtained by the integral of the tangential component of the vector potential

$$F|_V = -\lim_{d \to 0} \left(\int_{y_A^-}^{y_A^+} \Psi_y \, \mathrm{d}y + \int_{z_A^+}^{z_B^+} \Psi_z \, \mathrm{d}z + \int_{y_B^+}^{y_B^-} \Psi_y \, \mathrm{d}y + \int_{z_B^-}^{z_A^-} \Psi_z \, \mathrm{d}z \right) \tag{5.155}$$

along the boundary of a rectangular surface V with vertices a small distance d in the y-direction from where the surface intersects the singular vortex sheet between U and D. The integrals in the y-direction are zero since Ψ_y in (5.149) is continuous, and the component Ψ_z is substituted into the two remaining integrals to give (Steward, 2001, eqn 4.9)

$$F|_V = -\int_{z_A^+}^{z_B^+} \frac{\sigma_0}{2} \, \mathrm{d}z + \int_{z_B^-}^{z_A^-} \frac{\sigma_0}{2} \, \mathrm{d}z = -\sigma_0 \left(z_B - z_A \right) \tag{5.156}$$

This virtual flux is equal to the flux removed from the domain by the line-sink between elevations z_U and z_L, and must be accounted for when computing actual discharges, as in (5.153). While closed-form expressions exist for many of the line integrals associated with the tangential component of the vector potential along a straight line (Steward, 1996), they may also be evaluated numerically, for example using the trapezoid rule:

a. Surfacecontaining discontinuous $\boldsymbol{\Psi}$

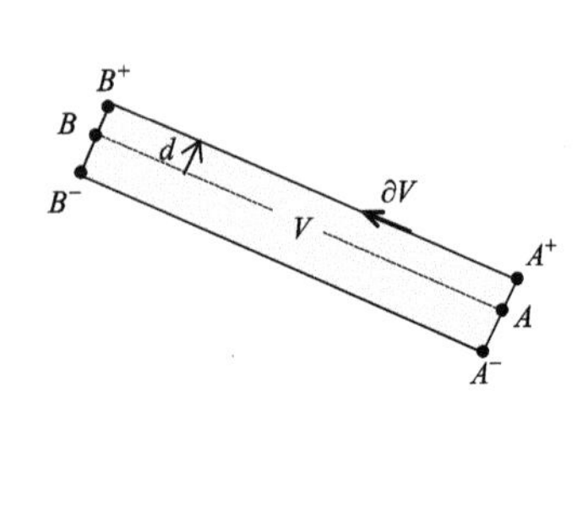

b. Discontinuous vortex sheet

Figure 5.35 *Virtual flux through discontinuous vortex sheet.*

$$\int_{\mathfrak{s}_0}^{\mathfrak{s}_I} f(\mathfrak{s})\mathrm{d}\mathfrak{s} = \left[\frac{f(\mathfrak{s}_0)}{2} + \sum_{i=1}^{I-1} f(\mathfrak{s}_i) + \frac{f(\mathfrak{s}_I)}{2}\right]\Delta\mathfrak{s} \tag{5.157}$$

or Simpson's rule:

$$\begin{aligned}\int_{\mathfrak{s}_0}^{\mathfrak{s}_I} f(\mathfrak{s})\mathrm{d}\mathfrak{s} = &\left[\frac{f(\mathfrak{s}_0)}{3} + \frac{4f(\mathfrak{s}_1)}{3} + \frac{2f(\mathfrak{s}_2)}{3} + \frac{4f(\mathfrak{s}_3)}{3} + \frac{2f(\mathfrak{s}_4)}{3} + \cdots\right.\\ &\left. + \frac{2f(\mathfrak{s}_{I-2})}{3} + \frac{4f(\mathfrak{s}_{I-1})}{3} + \frac{f(\mathfrak{s}_I)}{3}\right]\Delta\mathfrak{s}\end{aligned} \tag{5.158}$$

when the number of intervals is even.

Problem 5.25 Calculate the vector potential $\boldsymbol{\Psi}$ at four locations on a plane $(x_1, y_1, z_1) = (1, 1, z_a), (x_2, y_2, z_2) = (-1, 1, z_a), (x_3, y_3, z_3) = (-1, -1, z_a)$, and $(x_4, y_4, z_4) = (1, -1, z_a)$ with the z_a specified below, for a three-dimensional point-sink in an infinite aquifer with $Q = 4$ located at the origin.

A. $z_a = -6$ C. $z_a = -4$ E. $z_a = -2$ G. $z_a = 0$ I. $z_a = 2$ K. $z_a = 4$

B. $z_a = -5$ D. $z_a = -3$ F. $z_a = -1$ H. $z_a = 1$ J. $z_a = 3$ L. $z_a = 5$

Problem 5.26 Calculate the vector potential Ψ at the four locations specified in Problem 5.25, for a three-dimensional vertical line-sink in an infinite aquifer with $\sigma_0 = 1$ and $L = 2$ centered at the origin using the formulation with the Stokes stream function, (5.147).

Problem 5.27 Calculate the vector potential $\boldsymbol{\Psi}$ at four locations on a plane $(x_1, y_1, z_1) = (x_a, 1, -1), (x_2, y_2, z_2) = (x_a, 1, 1), (x_3, y_3, z_3) = (x_a, -1, 1)$, and $(x_4, y_4, z_4) = (x_a, -1, -1)$, with the x_a specified below, for a three-dimensional vertical line-sink in an infinite aquifer with $\sigma_0 = 1$ and $L = 2$ centered at the origin using the formulation with integration, (5.149).

A. $x_a = -6$ C. $x_a = -4$ E. $x_a = -2$ G. $x_a = 0$ I. $x_a = 2$ K. $x_a = 4$

B. $x_a = -5$ D. $x_a = -3$ F. $x_a = -1$ H. $x_a = 1$ J. $x_a = 3$ L. $x_a = 5$

Problem 5.28 Calculate the vector potential $\boldsymbol{\Psi}$ at the four locations specified in Problem 5.27, for an infinitely long vertical line-sink with $\sigma_0 = 1$ centered at the origin.

Further Reading

Section 5.1, Singular Integral Equations

- Burton and Miller (1971)
- Courant and Hilbert (1962)
- Fredholm (1903)
- Helmholtz (1858)
- Muskhelishvili (1953*a*)
- Saffman (1992)
- Steward, Le Grand, Janković, and Strack (2008)
- Strack and Haitjema (1981*a*)
- Strack and Haitjema (1981*b*)
- Strack (1989)
- Tait (1867)

Sections 5.2, 5.3, and 5.4, Line Elements and Their Applications

- Abramowitz and Stegun (1972)
- Gröbner and Hofreiter (1975)
- Helmholtz (1858)
- Janković and Barnes (1999*a*)
- Le Grand (1999)
- Muskhelishvili (1953*a*)
- Steward, Le Grand, Janković, and Strack (2008)
- Steward and Ahring (2009)
- Steward (2015)
- Strack (1989)
- Strack (2003)
- Tait (1867)

Sections 5.5 and 5.6, Polygon and Curvilinear Elements

- Bell (1934)
- Bezier (1986)
- Comtet (1974)
- Constantine and Savits (1996)
- Craig, Janković, and Barnes (2006)
- Faà di Bruno (1855)
- Fitts (2010)
- Henrici (1956)
- Janković (1997)
- Knuth (1998)
- Le Grand (1999)
- Le Grand (2003)
- Lukács (1955)
- Muskhelishvili (1953*a*)
- Schoenberg (1946)
- Schoenberg (1973)
- Steward, Le Grand, Janković, and Strack (2008)
- Steward and Ahring (2009)
- Strack and Haitjema (1981*b*)
- Strack (1989)
- Strack, Janković and Barnes (1999)
- Strack and Janković (1999)
- Wheeler (1987)

Section 5.7, Three-Dimensional Vector Fields

- Clebsch (1857)
- Goursat (1904)
- Fitts (1989)
- Haitjema (1985)
- Haitjema and Kraemer (1988)
- Haitjema (1995)
- Hess and Smith (1967)
- Janković, Steward, Barnes, and Dagan (2009)
- Lagrange (1781)
- Steward (1996)
- Steward (1998)
- Steward (2001)
- Steward and Janković (2001)
- Steward (2002)
- Steward and Jin (2003)
- Stokes (1842)

Appendix A
List of Symbols

Symbol	Variable
$\mathbf{A}$	system of equations: linear coefficients
$\mathbf{b}$	system of equations: specified conditions
$\mathbf{c}$	system of equations: unknown coefficients
E	modulus of elasticity
$E_1(u)$	exponential integral
f	function
$\mathscr{F}$	objective function
$\mathcal{F}$	Airy stress function
F_x, F_y, F_z	body force
$H_n^{(1)}(r), H_n^{(2)}(r)$	Bessel functions of third kind (Hankel function)
i, I	index and number of element being solved
$I_n(r), K_n(r)$	modified Bessel functions
$j, \mathcal{J}$	index and number of elements in additional function
$\mathcal{J}_n(r)$	Bessel function of first kind
l	iterate number for iterative solutions
L	line element length
m, M	index and number of control points
n, N	index and number of strength coefficients
$\mathfrak{n}, \hat{\mathfrak{n}}$	normal direction and unit normal vector
$P_n(\xi)$	Legendre function of the first kind
$P_{n_1}^{n_2}(\xi), Q_{n_1}^{n_2}(\xi)$	Legendre associated functions
$Q_n(\xi)$	Legendre function of the second kind
$\mathfrak{s}, \hat{\mathfrak{s}}$	tangent direction and unit tangent vector
S	tangent direction in local coordinates
u_x, u_y, u_z	displacement
$\mathbf{v}$	eigenvector

Symbol	Variable
$v = v_x + iv_y$	complex vector field
(x, y, z)	coordinate axes
$Y_n(r)$	Bessel function of second kind (Weber's function)
$z = x + iy$	complex variable
$z_{\min}, z_{\max}$	line element endpoints
$Z = X + iY$	complex variable in local coordinates
ε_m	residual error at control point
$\varepsilon_x, \varepsilon_y, \varepsilon_y$	normal strain
ϑ	line element orientation
$\gamma_{xy}, \gamma_{yz}, \gamma_{xz}$	shear strain
κ	Kolosov formula variable
λ	eigenvalue
μ	modulus of rigidity
ν	Poisson ratio
$\sigma_x, \sigma_y, \sigma_z$	normal stress (Cartesian)
σ_1, σ_2	normal stress (principal)
$\tau_{xy}, \tau_{yz}, \tau_{xz}$	shear stress (Cartesian)
τ_{max}	shear stress (max)
φ	complex wave function
ϕ, ψ	Kolosov functions
Φ	scalar potential function
Ψ	stream function
$\Omega(z) = \Phi + i\Psi$	complex function
$\overline{\Omega(z)} = \Phi - i\Psi$	complex conjugate
$\lvert\Omega(z)\rvert$	modulus of complex function
$\arg[\Omega(z)]$	argument of complex function
∇^2	Laplacian vector operator
∇^4	biharmonic vector operator
∇	gradient vector operator
$\nabla\cdot$	divergence vector operator
$\nabla\times$	curl vector operator

Appendix B
Solutions to Alternate Problem Sets

Problem 1.1

A. $v_x = -\frac{Q}{2\pi}\frac{x}{r^2},\ v_y = -\frac{Q}{2\pi}\frac{y}{r^2}$

C. $v_x = -\frac{\Gamma}{2\pi}\frac{y}{r^2},\ v_y = \frac{\Gamma}{2\pi}\frac{x}{r^2}$

E. $v_x = \frac{x}{2},\ v_y = \frac{y}{2}$

G. $$v_x = +\frac{2\pi n}{y_{\max}-y_{\min}}\frac{\cosh 2\pi n\frac{x_{\max}-x}{y_{\max}-y_{\min}}}{\sinh 2\pi n\frac{x_{\max}-x_{\min}}{y_{\max}-y_{\min}}}\cos 2\pi n\frac{y-y_{\min}}{y_{\max}-y_{\min}},$$

$$v_y = +\frac{2\pi n}{y_{\max}-y_{\min}}\frac{\sinh 2\pi n\frac{x_{\max}-x}{y_{\max}-y_{\min}}}{\sinh 2\pi n\frac{x_{\max}-x_{\min}}{y_{\max}-y_{\min}}}\sin 2\pi n\frac{y-y_{\min}}{y_{\max}-y_{\min}}$$

I. $$v_x = +\frac{2\pi n}{x_{\max}-x_{\min}}\sin 2\pi n\frac{x-x_{\min}}{x_{\max}-x_{\min}}\frac{\sinh 2\pi n\frac{y_{\max}-y}{x_{\max}-x_{\min}}}{\sinh 2\pi n\frac{y_{\max}-y_{\min}}{x_{\max}-x_{\min}}},$$

$$v_y = +\frac{2\pi n}{x_{\max}-x_{\min}}\cos 2\pi n\frac{x-x_{\min}}{x_{\max}-x_{\min}}\frac{\cosh 2\pi n\frac{y_{\max}-y}{x_{\max}-x_{\min}}}{\sinh 2\pi n\frac{y_{\max}-y_{\min}}{x_{\max}-x_{\min}}}$$

K. $$v_x = -\frac{2\pi n}{y_{\max}-y_{\min}}\frac{\cosh 2\pi n\frac{x-x_{\min}}{y_{\max}-y_{\min}}}{\sinh 2\pi n\frac{x_{\max}-x_{\min}}{y_{\max}-y_{\min}}}\cos 2\pi n\frac{y-y_{\min}}{y_{\max}-y_{\min}},$$

$$v_y = +\frac{2\pi n}{y_{\max}-y_{\min}}\frac{\sinh 2\pi n\frac{x-x_{\min}}{y_{\max}-y_{\min}}}{\sinh 2\pi n\frac{x_{\max}-x_{\min}}{y_{\max}-y_{\min}}}\sin 2\pi n\frac{y-y_{\min}}{y_{\max}-y_{\min}}$$

M. $$v_x = +\frac{2\pi n}{x_{\max}-x_{\min}}\sin 2\pi n\frac{x-x_{\min}}{x_{\max}-x_{\min}}\frac{\sinh 2\pi n\frac{y-y_{\min}}{x_{\max}-x_{\min}}}{\sinh 2\pi n\frac{y_{\max}-y_{\min}}{x_{\max}-x_{\min}}},$$

$$v_y = -\frac{2\pi n}{x_{\max}-x_{\min}}\cos 2\pi n\frac{x-x_{\min}}{x_{\max}-x_{\min}}\frac{\cosh 2\pi n\frac{y-y_{\min}}{x_{\max}-x_{\min}}}{\sinh 2\pi n\frac{y_{\max}-y_{\min}}{x_{\max}-x_{\min}}}$$

Problem 1.2

A. $h = 28$ m, $v_x = 0.12$ m/day, $v_z = -0.00132$ m/day

C. $h = 27.45$ m, $v_x = 0.1443$ m/day, $v_z = -0.001373$ m/day

E. $h = 26.79$ m, $v_x = -0.01074$ m/day, $v_z = -0.001201$ m/day

G. $h = 22.93$ m, $v_x = 0.7929$ m/day, $v_z = -0.006439$ m/day

I. $h = 24.86$ m, $v_x = 0.2317$ m/day, $v_z = 0$ m/day

K. $h = 22.93$ m, $v_x = 0.3035$ m/day, $v_z = -0.001967$ m/day

M. $h = 20$ m, $v_x = 0.408$ m/day, $v_z = -0.002587$ m/day

Problem 1.3

A. $p_A = 0.0$ m, $p_B = -1.96$ m, $p_C = -2.10$ m, $p_D = -3.21$ m
C. $p_A = 0.0$ m, $p_B = -1.90$ m, $p_C = -1.88$ m, $p_D = -2.71$ m
E. $p_A = 0.0$ m, $p_B = -1.84$ m, $p_C = -1.78$ m, $p_D = -2.46$ m
G. $p_A = 0.0$ m, $p_B = -1.79$ m, $p_C = -1.71$ m, $p_D = -2.30$ m
I. $p_A = 0.0$ m, $p_B = -1.75$ m, $p_C = -1.66$ m, $p_D = -2.18$ m
K. $p_A = 0.0$ m, $p_B = -1.68$ m, $p_C = -1.60$ m, $p_D = -2.04$ m

Problem 1.4

A. $\Delta P = -500$ N/m^2
C. $\Delta P = -1500$ N/m^2
E. $\Delta P = -2000$ N/m^2
G. $\Delta P = -2000$ N/m^2
I. $\Delta P = -6000$ N/m^2

Problem 1.5

A. $q_r(t=0) = 0, q_r(10^{-6}) = -158.80$ W/m^2, $q_r(1) = -159.15$ W/m^2
C. $q_r(t=0) = 0, q_r(10^{-6}) = -156.19$ W/m^2, $q_r(1) = -159.15$ W/m^2
E. $q_r(t=0) = 0, q_r(10^{-6}) = -103.42$ W/m^2, $q_r(1) = -159.15$ W/m^2
G. $q_r(t=0) = 0, q_r(10^{-6}) = -71.05$ W/m^2, $q_r(1) = -159.15$ W/m^2

Problem 1.6

A. $E_x(-4,-1) = 1.05, E_y(-4,-1) = 0.028, E_x(-4,0) = 1.06, E_y(-4,0) = 0, E_x(-4,1) = 1.05, E_y(-4,1) = -0.028$
C. $E_x(-2,-1) = 1.12, E_y(-2,-1) = 0.16, E_x(-2,0) = 1.25, E_y(-2,0) = 0, E_x(-2,1) = 1.12, E_y(-2,1) = -0.16$
E. $E_x(1,-1) = 1, E_y(1,-1) = -0.5, E_x(1,0) = 2, E_y(1,0) = 0, E_x(1,1) = 1, E_y(1,1) = 0.5$
G. $E_x(3,-1) = 1.08, E_y(3,-1) = -0.06, E_x(3,0) = 1.11, E_y(3,0) = 0, E_x(3,1) = 1.08, E_y(3,1) = 0.06$

Problem 1.7

A. φ =10, $-8.01 + 5.98$i, $2.84 - 9.59$i; cos[arg(φ)] =1, -0.80, 0.28
C. φ =10, $-9.98 - 0.63$i, $9.92 + 1.25$i; cos[arg(φ)] =1, -1.00, 0.99
E. φ =10, $-9.28 - 3.71$i, $7.24 + 6.90$i; cos[arg(φ)] =1, -0.93, 0.72
G. φ =10, $-9.63 - 2.70$i, $8.54 + 5.20$i; cos[arg(φ)] =1, -0.96, 0.85
I. φ =10, $-9.70 + 2.43$i, $8.82 - 4.71$i; cos[arg(φ)] =1, -0.97, 0.88

Problem 1.8

A. $\omega = 2.094/\text{sec}$, $k = 2.131/\text{m}$, $L = 2.949$ m, $C = 0.9829$ m/sec, $C_g = 0.9683$ m/sec, $\max v_x = 0.3991$ m/sec, $\max v_z = 0.08377$ m/sec

C. $\omega = 2.094/\text{sec}$, $k = 0.4473/\text{m}$, $L = 14.04$ m, $C = 4.681$ m/sec, $C_g = 2.346$ m/sec, $\max v_x = 0.08380$ m/sec, $\max v_z = 0.08377$ m/sec

E. $\omega = 1.257/\text{sec}$, $k = 0.2073/\text{m}$, $L = 30.30$ m, $C = 6.060$ m/sec, $C_g = 4.636$ m/sec, $\max v_x = 0.1618$ m/sec, $\max v_z = 0.1257$ m/sec

G. $\omega = 0.6283/\text{sec}$, $k = 0.1438/\text{m}$, $L = 43.69$ m, $C = 4.369$ m/sec, $C_g = 4.253$ m/sec, $\max v_x = 1.122$ m/sec, $\max v_z = 0.3141$ m/sec

I. $\omega = 0.6283/\text{sec}$, $k = 0.04026/\text{m}$, $L = 156.1$ m, $C = 15.61$ m/sec, $C_g = 7.804$ m/sec, $\max v_x = 0.3141$ m/sec, $\max v_z = 0.3141$ m/sec

K. $\omega = 0.3141/\text{sec}$, $k = 0.01549/\text{m}$, $L = 405.6$ m, $C = 20.28$ m/sec, $C_g = 17.13$ m/sec, $\max v_x = 0.4836$ m/sec, $\max v_z = 0.3141$ m/sec

Problem 1.9

A. $|\varphi| = 2, 0, 2$

C. $|\varphi| = 2, 1.414, 0$

E. $|\varphi| = 2, 1.247, 0.445$

G. $|\varphi| = 2, 0.868, 1.247$

Problem 1.10

A. $\sigma_1 = -7.80\text{ kN/m}^2, \sigma_2 = -18.19\text{ kN/m}^2, \tau_{\max} = 5.20\text{ kN/m}^2$

C. $\sigma_1 = -23.39\text{ kN/m}^2, \sigma_2 = -54.57\text{ kN/m}^2, \tau_{\max} = 15.59\text{ kN/m}^2$

E. $\sigma_1 = -45.85\text{ kN/m}^2, \sigma_2 = -93.90\text{ kN/m}^2, \tau_{\max} = 24.02\text{ kN/m}^2$

G. $\sigma_1 = -75.17\text{ kN/m}^2, \sigma_2 = -136.16\text{ kN/m}^2, \tau_{\max} = 30.50\text{ kN/m}^2$

Problem 1.11

A. $\sigma_1 = 4.24, \sigma_2 = -0.236, \hat{\mathbf{v}}_1 = \pm\begin{bmatrix}0.526\\0.851\end{bmatrix}, \hat{\mathbf{v}}_2 = \pm\begin{bmatrix}-0.851\\0.526\end{bmatrix}$

C. $\sigma_1 = 8.12, \sigma_2 = -0.123, \hat{\mathbf{v}}_1 = \pm\begin{bmatrix}0.615\\0.788\end{bmatrix}, \hat{\mathbf{v}}_2 = \pm\begin{bmatrix}-0.788\\0.615\end{bmatrix}$

E. $\sigma_1 = 12.08, \sigma_2 = -0.083, \hat{\mathbf{v}}_1 = \pm\begin{bmatrix}0.646\\0.763\end{bmatrix}, \hat{\mathbf{v}}_2 = \pm\begin{bmatrix}-0.763\\0.646\end{bmatrix}$

G. $\sigma_1 = 16.06, \sigma_2 = -0.062, \hat{\mathbf{v}}_1 = \pm\begin{bmatrix}0.662\\0.750\end{bmatrix}, \hat{\mathbf{v}}_2 = \pm\begin{bmatrix}-0.750\\0.662\end{bmatrix}$

I. $\sigma_1 = 20.05, \sigma_2 = -0.050, \hat{\mathbf{v}}_1 = \pm\begin{bmatrix}0.671\\0.741\end{bmatrix}, \hat{\mathbf{v}}_2 = \pm\begin{bmatrix}-0.741\\0.671\end{bmatrix}$

Problem 1.12

A. $\sigma_x = -3 \times 10^8 \text{N/m}^2$, $\sigma_y = 0$, $\tau_{xy} = 0$, $\sigma_1 = 0$, $\sigma_2 = -3 \times 10^8 \text{N/m}^2$, $\tau_{max} = 1.5 \times 10^8 \text{N/m}^2$, $u_x = 6.8750 \times 10^{-6}$m, $u_y = -1.1250 \times 10^{-5}$m

C. $\sigma_x = 0$, $\sigma_y = 0$, $\tau_{xy} = -1.5 \times 10^7 \text{N/m}^2$, $\sigma_1 = 1.5 \times 10^7 \text{N/m}^2$, $\sigma_2 = -1.5 \times 10^7 \text{N/m}^2$, $\tau_{max} = 1.5 \times 10^7 \text{N/m}^2$, $u_x = 0$, $u_y = 0$

E. $\sigma_x = 3 \times 10^8 \text{N/m}^2$, $\sigma_y = 0$, $\tau_{xy} = 0$, $\sigma_1 = 3 \times 10^8 \text{N/m}^2$, $\sigma_2 = 0$, $\tau_{max} = 1.5 \times 10^8 \text{N/m}^2$, $u_x = -6.8750 \times 10^{-6}$m, $u_y = -1.1250 \times 10^{-5}$m

G. $\sigma_x = -7.5 \times 10^7 \text{N/m}^2$, $\sigma_y = 0$, $\tau_{xy} = -1.1250 \times 10^7 \text{N/m}^2$, $\sigma_1 = 1.6511 \times 10^6 \text{N/m}^2$, $\sigma_2 = -7.6651 \times 10^7 \text{N/m}^2$, $\tau_{max} = 3.9151 \times 10^7 \text{N/m}^2$, $u_x = -1.3611 \times 10^{-4}$m, $u_y = -7.8266 \times 10^{-4}$m

I. $\sigma_x = 7.5 \times 10^7 \text{N/m}^2$, $\sigma_y = 0$, $\tau_{xy} = -1.1250 \times 10^7 \text{N/m}^2$, $\sigma_1 = 7.6651 \times 10^7 \text{N/m}^2$, $\sigma_2 = -1.6511 \times 10^6 \text{N/m}^2$, $\tau_{max} = 3.9151 \times 10^7 \text{N/m}^2$, $u_x = 1.3611 \times 10^{-4}$m, $u_y = -7.8266 \times 10^{-4}$m

K. $\sigma_x = 0$, $\sigma_y = 0$, $\tau_{xy} = 0$, $\sigma_1 = 0$, $\sigma_2 = 0$, $\tau_{max} = 0$, $u_x = -3.6813 \times 10^{-4}$m, $u_y = -2.5000 \times 10^{-3}$m

M. $\sigma_x = 0$, $\sigma_y = 0$, $\tau_{xy} = -1.5 \times 10^7 \text{N/m}^2$, $\sigma_1 = 1.5 \times 10^7 \text{N/m}^2$, $\sigma_2 = -1.5 \times 10^7 \text{N/m}^2$, $\tau_{max} = 1.5 \times 10^7 \text{N/m}^2$, $u_x = 0$, $u_y = -2.5000 \times 10^{-3}$m

O. $\sigma_x = 0$, $\sigma_y = 0$, $\tau_{xy} = 0$, $\sigma_1 = 0$, $\sigma_2 = 0$, $\tau_{max} = 0$, $u_x = 3.6813 \times 10^{-4}$m, $u_y = -2.5000 \times 10^{-3}$m

Problem 2.1

A. $c_0 = 1.6154$, $c_1 =$ -0.1133, $\mathscr{F} = 0.2332$

C. $c_0 = 1.5304$, $c_1 =$ -0.1046, $\mathscr{F} = 0.2420$

E. $c_0 = 1.6441$, $c_1 =$ -0.1157, $\mathscr{F} = 0.2282$

G. $c_0 = 1.4764$, $c_1 =$ -0.0927, $\mathscr{F} = 0.2492$

I. $c_0 = 1.6259$, $c_1 =$ -0.1068, $\mathscr{F} = 0.2584$

Problem 2.2

A. $c_0 = 1.7232$, $c_1 = 0.2369$, $c_2 =$ -0.1530, $c_3 = 0.0127$, $\mathscr{F} = 0.0316$

C. $c_0 = 1.7450$, $c_1 = 0.1376$, $c_2 =$ -0.1319, $c_3 = 0.0116$, $\mathscr{F} = 0.0277$

E. $c_0 = 1.8784$, $c_1 = 0.0787$, $c_2 =$ -0.1174, $c_3 = 0.0105$, $\mathscr{F} = 0.0347$

G. $c_0 = 1.7218$, $c_1 = 0.1293$, $c_2 =$ -0.1300, $c_3 = 0.0116$, $\mathscr{F} = 0.0208$

I. $c_0 = 1.8656$, $c_1 = 0.1255$, $c_2 =$ -0.1328, $c_3 = 0.0118$, $\mathscr{F} = 0.0268$

Problem 2.3

A. $c_0 = 0$, $c_1 = 0.1175$, $\lambda_1 = 5.0771$, $\mathscr{F} = 0.9788$, $f_1 = 0.0000$, $f_2 = 0.1175$, $f_3 = 0.2350$, $f_4 = 0.3525$, $f_5 = 0.4700$, $f_6 = 0.5875$, $f_7 = 0.7050$, $f_8 = 0.8225$, $f_9 = 0.9400$, $f_{10} = 1.0575$, $f_{11} = 1.1750$

C. $c_0 = 0$, $c_1 = 0.1140$, $\lambda_1 = 0.0000$, $\mathscr{F} = 0.9112$, $f_1 = 0.0000$, $f_2 = 0.1140$, $f_3 = 0.2280$, $f_4 = 0.3420$, $f_5 = 0.4560$, $f_6 = 0.5700$, $f_7 = 0.6840$, $f_8 = 0.7980$, $f_9 = 0.9120$, $f_{10} = 1.0260$, $f_{11} = 1.1400$

E. $c_0 = 0$, $c_1 = 0.1191$, $\lambda_1 = 5.1671$, $\mathscr{F} = 1.0005$, $f_1 = 0.0000$, $f_2 = 0.1191$, $f_3 = 0.2383$, $f_4 = 0.3574$, $f_5 = 0.4766$, $f_6 = 0.5957$, $f_7 = 0.7148$, $f_8 = 0.8340$, $f_9 = 0.9531$, $f_{10} = 1.0723$, $f_{11} = 1.1914$

G. $c_0 = 0$, $c_1 = 0.1182$, $\lambda_1 = 4.6400$, $\mathscr{F} = 0.8720$, $f_1 = 0.0000$, $f_2 = 0.1182$, $f_3 = 0.2364$, $f_4 = 0.3545$, $f_5 = 0.4727$, $f_6 = 0.5909$, $f_7 = 0.7091$, $f_8 = 0.8273$, $f_9 = 0.9454$, $f_{10} = 1.0636$, $f_{11} = 1.1818$

I. $c_0 = 0$, $c_1 = 0.1254$, $\lambda_1 = 5\ .1100$, $\mathscr{F} = 1.0137$, $f_1 = 0.0000$, $f_2 = 0.1254$, $f_3 = 0.2509$, $f_4 = 0.3763$, $f_5 = 0.5018$, $f_6 = 0.6273$, $f_7 = 0.7527$, $f_8 = 0\ .8782$, $f_9 = 1.0036$, $f_{10} = 1.1291$, $f_{11} = 1.2545$

Problem 2.4

A. $c_0 = 0.4787, c_1 = \text{-}1.9253, c_2 = 2.6213, c_3 = 2.5341, c_4 = \text{-}1.6790, c_5 = \text{-}0.7462, \mathscr{F} = 0.003314$

C. $c_0 = 0.4053, c_1 = \text{-}1.8249, c_2 = 2.6200, c_3 = 2.5670, c_4 = \text{-}1.5661, c_5 = \text{-}0.8814, \mathscr{F} = 0.002428$

E. $c_0 = 0.4730, c_1 = \text{-}1.8975, c_2 = 2.6103, c_3 = 2.8933, c_4 = \text{-}1.5862, c_5 = \text{-}1.24120, \mathscr{F} = 0.006974$

G. $c_0 = 0.4314, c_1 = \text{-}1.7159, c_2 = 2.3116, c_3 = 2.3591, c_4 = \text{-}1.2055, c_5 = \text{-}0.7161, \mathscr{F} = 0.005640$

I. $c_0 = 0.5000, c_1 = \text{-}2.0176, c_2 = 2.3974, c_3 = 3.2715, c_4 = \text{-}1.2893, c_5 = \text{-}1.4173, \mathscr{F} = 0.006229$

Problem 2.5

A. $c_0 = 1.0491, \overset{\cos}{c}_1 = 0.6802, \overset{\cos}{c}_2 = \text{-}0.0906, \overset{\cos}{c}_3 = \text{-}0.0473, \overset{\sin}{c}_1 = 0.4912, \overset{\sin}{c}_2 = 0.0445, \overset{\sin}{c}_3 = \text{-}0.0138, \mathscr{F} = 0.003220$

C. $c_0 = 1.0073, \overset{\cos}{c}_1 = 0.7134, \overset{\cos}{c}_2 = \text{-}0.0723, \overset{\cos}{c}_3 = \text{-}0.0322, \overset{\sin}{c}_1 = 0.4270, \overset{\sin}{c}_2 = 0.0193, \overset{\sin}{c}_3 = \text{-}0.0134, \mathscr{F} = 0.002442$

E. $c_0 = 1.0654, \overset{\cos}{c}_1 = 0.7072, \overset{\cos}{c}_2 = \text{-}0.1012, \overset{\cos}{c}_3 = 0.0581, \overset{\sin}{c}_1 = 0.4499, \overset{\sin}{c}_2 = 0.0482, \overset{\sin}{c}_3 = \text{-}0.0113, \mathscr{F} = 0.002806$

G. $c_0 = 1.0127, \overset{\cos}{c}_1 = 0.7146, \overset{\cos}{c}_2 = \text{-}0.0649, \overset{\cos}{c}_3 = 0.0044, \overset{\sin}{c}_1 = 0.3821, \overset{\sin}{c}_2 = \text{-}0.0138, \overset{\sin}{c}_3 = \text{-}0.0190, \mathscr{F} = 0.004484$

I. $c_0 = 1.0918, \overset{\cos}{c}_1 = 0.7342, \overset{\cos}{c}_2 = \text{-}0.0677, \overset{\cos}{c}_3 = \text{-}0.0157, \overset{\sin}{c}_1 = 0.4360, \overset{\sin}{c}_2 = \text{-}0.0030, \overset{\sin}{c}_3 = \text{-}0.0769, \mathscr{F} = 0.002521$

Problem 2.6

A. $c_0 = 2.9238, c_1 = \text{-}1.5911, \mathscr{F} = 0.006250$

C. $c_0 = 3.9370, c_1 = \text{-}2.0248, \mathscr{F} = 0.000450$

E. $c_0 = 3.0032, c_1 = \text{-}1.5851, \mathscr{F} = 0.002730$

G. $c_0 = 1.9009, c_1 = \text{-}1.1809, \mathscr{F} = 0.006028$

I. $c_0 = 2.8282, c_1 = \text{-}1.5190, \mathscr{F} = 0.007109$

Problem 2.7

A. River with specified head: Dirichlet

C. Highly conductive slit: Dirichlet or Robin

E. Edge with given inflow: Neumann

G. Impermeable wall: Neumann

I. Perfect insulator: Neumann

K. Current into domain specified: Neumann

M. Free surface boundary: Robin

O. Intensity specified: Dirichlet

Q. Displacement specified: Dirichlet

S. Given force/stress: Neumann

Problem 2.8

A. Line element with low conductivity: function, normal component

C. Area of recharge: function, normal component

E. Jump in sorptive number: function, normal component

G. Jump in fluid depth: function, tangential component

I. Line element with high conductivity: function, tangential component

K. Line with low resistance: function, normal component

M. Jump in water depth: function, normal component

O. Transmission into a softer medium: function, normal component

Q. Jump in coefficient of elasticity: function, normal and tangential components

Problem 2.9

A. $z_1 = 1.0000 + i1.0000 = 1.4142e^{i0.7854}$, $z_2 = 2.0000 + i2.0000 = 2.8284e^{i0.7854}$, $z_1 + z_2 = 3.0000 + i3.0000 = 4.2426e^{i0.7854}$, $z_1 - z_2 = -1.0000 + i - 1.0000 = 1.4142e^{i-2.3562}$, $z_1 \times z_2 = 0.0000 + i4.0000 = 4.0000e^{i1.5708}$, $z_1/z_2 = 0.5000 + i0.0000 = 0.5000e^{i0.0000}$

C. $z_1 = 5.0000 + i5.0000 = 7.0711e^{i0.7854}$, $z_2 = 6.0000 + i6.0000 = 8.4853e^{i0.7854}$, $z_1 + z_2 = 11.0000 + i11.0000 = 15.5563e^{i0.7854}$, $z_1 - z_2 = -1.0000 + i - 1.0000 = 1.4142e^{i-2.3562}$, $z_1 \times z_2 = 0.0000 + i60.0000 = 60.0000e^{i1.5708}$, $z_1/z_2 = 0.8333 + i0.0000 = 0.8333e^{i0.0000}$

E. $z_1 = 0.0000 + i1.0000 = 1.0000e^{i1.5708}$, $z_2 = 0.0000 + i2.0000 = 2.0000e^{i1.5708}$, $z_1 + z_2 = 0.0000 + i3.0000 = 3.0000e^{i1.5708}$, $z_1 - z_2 = -0.0000 + i - 1.0000 = 1.0000e^{i-1.5708}$, $z_1 \times z_2 = -2.0000 + i0.0000 = 2.0000e^{i3.1416}$, $z_1/z_2 = 0.5000 + i0.0000 = 0.5000e^{i0.0000}$

G. $z_1 = 0.0000 + i5.0000 = 5.0000e^{i1.5708}$, $z_2 = -0.0000 + i - 6.0000 = 6.0000e^{i-1.5708}$, $z_1 + z_2 = 0.0000 + i - 1.0000 = 1.0000e^{i-1.5708}$, $z_1 - z_2 = 0.0000 + i11.0000 = 11.0000e^{i1.5708}$, $z_1 \times z_2 = 30.0000 + i - 0.0000 = 30.0000e^{i-0.0000}$, $z_1/z_2 = -0.8333 + i0.0000 = 0.8333e^{i3.1416}$

Problem 2.10

A. $\Phi(1+i) = 1.0000$, $\Psi(1+i) = 1.0000$, $v_x(1+i) = -1.0000$, $v_y(1+i) = 0.0000$

C. $\Phi(1+i) = -2.0000$, $\Psi(1+i) = 2.0000$, $v_x(1+i) = -0.0000$, $v_y(1+i) = 6.0000$

E. $\Phi(1+i) = 0.5000$, $\Psi(1+i) = -0.5000$, $v_x(1+i) = 0.0000$, $v_y(1+i) = 0.5000$

G. $\Phi(1+i) = -0.2500$, $\Psi(1+i) = -0.2500$, $v_x(1+i) = -0.7500$, $v_y(1+i) = 0.0000$

I. $\Phi(1+i) = 1.0987$, $\Psi(1+i) = 0.4551$, $v_x(1+i) = -0.3884$, $v_y(1+i) = -0.1609$

K. $\Phi(1+i) = 1.0696$, $\Psi(1+i) = 0.2127$, $v_x(1+i) = -0.1603$, $v_y(1+i) = -0.1071$

M. $\Phi(1+i) = 0.7769$, $\Psi(1+i) = -0.3218$, $v_x(1+i) = 0.1138$, $v_y(1+i) = 0.2747$

O. $\Phi(1+i) = 0.8994$, $\Psi(1+i) = -0.1789$, $v_x(1+i) = 0.0901$, $v_y(1+i) = 0.1348$

Problem 3.1

A. $\Phi(1) = 0, \Psi(1) = 0, v_x(1) = 2.2, v_y(1) = 0$
$\Phi(1i) = 0, \Psi(1i) = -2.2, v_x(1i) = 0, v_y(1i) = 0$
$\Phi(-1) = 0, \Psi(-1) = 0, v_x(-1) = 2.2, v_y(-1) = 0$
$\Phi(-1i) = 0, \Psi(-1i) = 2.2, v_x(-1i) = 0, v_y(-1i) = 0$

C. $\Phi(3) = 0, \Psi(3) = 0, v_x(3) = 2.6, v_y(3) = 0$
$\Phi(3i) = 0, \Psi(3i) = -7.8, v_x(3i) = 0, v_y(3i) = 0$
$\Phi(-3) = 0, \Psi(-3) = 0, v_x(-3) = 2.6, v_y(-3) = 0$
$\Phi(-3i) = 0, \Psi(-3i) = 7.8, v_x(-3i) = 0, v_y(-3i) = 0$

E. $\Phi(5) = 0, \Psi(5) = 0, v_x(5) = 3, v_y(5) = 0$
$\Phi(5i) = 0, \Psi(5i) = -15, v_x(5i) = 0, v_y(5i) = 0$
$\Phi(-5) = 0, \Psi(-5) = 0, v_x(-5) = 3, v_y(-5) = 0$
$\Phi(-5i) = 0, \Psi(-5i) = 15, v_x(-5i) = 0, v_y(-5i) = 0$

G. $\Phi(1) = 0, \Psi(1) = 2.2, v_x(1) = 0, v_y(1) = 0$
$\Phi(1i) = 0, \Psi(1i) = 0, v_x(1i) = 0, v_y(1i) = 2.2$
$\Phi(-1) = 0, \Psi(-1) = -2.2, v_x(-1) = 0, v_y(-1) = 0$
$\Phi(-1i) = 0, \Psi(-1i) = 0, v_x(-1i) = 0, v_y(-1i) = 2.2$

I. $\Phi(3) = 0, \Psi(3) = 7.8, v_x(3) = 0, v_y(3) = 0$
$\Phi(3i) = 0, \Psi(3i) = 0, v_x(3i) = 0, v_y(3i) = 2.6$
$\Phi(-3) = 0, \Psi(-3) = -7.8, v_x(-3) = 0, v_y(-3) = 0$
$\Phi(-3i) = 0, \Psi(-3i) = 0, v_x(-3i) = 0, v_y(-3i) = 2.6$

K. $\Phi(5) = 0, \Psi(5) = 15, v_x(5) = 0, v_y(5) = 0$
$\Phi(5i) = 0, \Psi(5i) = 0, v_x(5i) = 0, v_y(5i) = 3$
$\Phi(-5) = 0, \Psi(-5) = -15, v_x(-5) = 0, v_y(-5) = 0$
$\Phi(-5i) = 0, \Psi(-5i) = 0, v_x(-5i) = 0, v_y(-5i) = 3$

Problem 3.2

A. $\Phi(2+1i) = -3.3$, $\Psi(2+1i) = -1.1$, $v_x(2+1i) = 0$, $v_y(2+1i) = 0$
$\Phi(1+2i) = -1.1$, $\Psi(1+2i) = -1.1$, $v_x(1+2i) = 2.2$, $v_y(1+2i) = 0$
$\Phi(1i) = 1.1$, $\Psi(1i) = -1.1$, $v_x(1i) = 0$, $v_y(1i) = 0$
$\Phi(1) = -1.1$, $\Psi(1) = -1.1$, $v_x(1) = 2.2$, $v_y(1) = 0$

C. $\Phi(4+1i) = -9.1$, $\Psi(4+1i) = -1.3$, $v_x(4+1i) = 0$, $v_y(4+1i) = 0$
$\Phi(1+4i) = -1.3$, $\Psi(1+4i) = -1.3$, $v_x(1+4i) = 2.6$, $v_y(1+4i) = 0$
$\Phi(-2+1i) = 6.5$, $\Psi(-2+1i) = -1.3$, $v_x(-2+1i) = 0$, $v_y(-2+1i) = 0$
$\Phi(1+-2i) = -1.3$, $\Psi(1+-2i) = -1.3$, $v_x(1+-2i) = 2.6$, $v_y(1+-2i) = 0$

E. $\Phi(6+1i) = -16.5$, $\Psi(6+1i) = -1.5$, $v_x(6+1i) = 0$, $v_y(6+1i) = 0$
$\Phi(1+6i) = -1.5$, $\Psi(1+6i) = -1.5$, $v_x(1+6i) = 3.0$, $v_y(1+6i) = 0$
$\Phi(-4+1i) = 13.5$, $\Psi(-4+1i) = -1.5$, $v_x(-4+1i) = 0$, $v_y(-4+1i) = 0$
$\Phi(1+-4i) = -1.5$, $\Psi(1+-4i) = -1.5$, $v_x(1+-4i) = 3.0$, $v_y(1+-4i) = 0$

G. $\Phi(2+1i) = -1.1$, $\Psi(2+1i) = 1.1$, $v_x(2+1i) = 0$, $v_y(2+1i) = 2.2$
$\Phi(1+2i) = -3.3$, $\Psi(1+2i) = 1.1$, $v_x(1+2i) = 0$, $v_y(1+2i) = 0$
$\Phi(1i) = -1.1$, $\Psi(1i) = 1.1$, $v_x(1i) = 0$, $v_y(1i) = 2.2$
$\Phi(1) = 1.1$, $\Psi(1) = 1.1$, $v_x(1) = 0$, $v_y(1) = 0$

I. $\Phi(4+1i) = -1.3$, $\Psi(4+1i) = 1.3$, $v_x(4+1i) = 0$, $v_y(4+1i) = 2.6$
$\Phi(1+4i) = -9.1$, $\Psi(1+4i) = 1.3$, $v_x(1+4i) = 0$, $v_y(1+4i) = 0$
$\Phi(-2+1i) = -1.3$, $\Psi(-2+1i) = 1.3$, $v_x(-2+1i) = 0$, $v_y(-2+1i) = 2.6$
$\Phi(1+-2i) = 6.5$, $\Psi(1+-2i) = 1.3$, $v_x(1+-2i) = 0$, $v_y(1+-2i) = 0$

K. $\Phi(6+1i) = -1.5$, $\Psi(6+1i) = 1.5$, $v_x(6+1i) = 0$, $v_y(6+1i) = 3.0$
$\Phi(1+6i) = -16.5$, $\Psi(1+6i) = 1.5$, $v_x(1+6i) = 0$, $v_y(1+6i) = 0$
$\Phi(-4+1i) = -1.5$, $\Psi(-4+1i) = 1.5$, $v_x(-4+1i) = 0$, $v_y(-4+1i) = 3.0$
$\Phi(1+-4i) = 13.5$, $\Psi(1+-4i) = 1.5$, $v_x(1+-4i) = 0$, $v_y(1+-4i) = 0$

Problem 3.3

A. $v_0 = 0.1111 + 0.0278i$, $\Phi_0 = 11.6667$, $\Omega(50+50i) = 4.7222 + -4.1667i$

C. $v_0 = 0.1111 + 0.0278i$, $\Phi_0 = 21.6667$, $\Omega(50+50i) = 14.7222 + -4.1667i$

E. $v_0 = 0.1250 + 0.0000i$, $\Phi_0 = 21.2500$, $\Omega(50+50i) = 15.0000 + -6.2500i$

G. $v_0 = 0.0125 + -0.0500i$, $\Phi_0 = 15.6250$, $\Omega(50+50i) = 17.5000 + -3.1250i$

Problem 3.4

A. $Q = -1$, $\overset{\text{out}}{c}_1 = 1$

C. $Q = -1$, $\overset{\text{out}}{c}_1 = 3$

E. $Q = -2$, $\overset{\text{out}}{c}_1 = 2$

G. $Q = -3$, $\overset{\text{out}}{c}_1 = 1$

I. $Q = -3$, $\overset{\text{out}}{c}_1 = 3$

K. $Q = -4$, $\overset{\text{out}}{c}_1 = 2$

Problem 3.5

A. $v_0 = 1.1429$, $\Phi_0 = 6.0345$, $Q = 4.8345$, $\overset{\text{out}}{c}_1 = 0.5714$,

C. $v_0 = 2.2857$, $\Phi_0 = 12.0777$, $Q = 10.8777$, $\overset{\text{out}}{c}_1 = 1.1429$,

E. $v_0 = -0.5714i$, $\Phi_0 = 3.0129$, $Q = 1.8129$, $\overset{\text{out}}{c}_1 = -0.2857i$,

G. $v_0 = -1.7143i$, $\Phi_0 = 9.0561$, $Q = 7.8561$, $\overset{\text{out}}{c}_1 = -0.8571i$,

I. $v_0 = -2.8571i$, $\Phi_0 = 15.0992$, $Q = 13.8992$, $\overset{\text{out}}{c}_1 = -1.4286i$,

K. $v_0 = 0.5714 - 0.5714i$, $\Phi_0 = 12.0777$, $Q = 10.8777$, $\overset{\text{out}}{c}_1 = 0.2857 - 0.2857i$,

M. $v_0 = 0.5714 - 0.5714i$, $\Phi_0 = 24.1640$, $Q = 22.9640$, $\overset{\text{out}}{c}_1 = 0.2857 - 0.2857i$,

Problem 3.6

A. $\Phi(0) = 0.2364$, $\overset{\text{in}}{c}_0 = 0.898262$, $\overset{\text{in}}{c}_4 = 0.036046$, $\overset{\text{in}}{c}_8 = -0.004082$, $\overset{\text{in}}{c}_{12} = 0.000616$, $\overset{\text{in}}{c}_{16} = -0.00010470$, $\overset{\text{in}}{c}_{20} = 0.00001896$

C. $\Phi(0) = 0.2667$, $\overset{\text{in}}{c}_0 = 0.928583$, $\overset{\text{in}}{c}_4 = 0.029793$, $\overset{\text{in}}{c}_8 = -0.002788$, $\overset{\text{in}}{c}_{12} = 0.000348$, $\overset{\text{in}}{c}_{16} = -0.00004886$, $\overset{\text{in}}{c}_{20} = 0.00000732$

E. $\Phi(0) = 0.2956$, $\overset{\text{in}}{c}_0 = 0.957525$, $\overset{\text{in}}{c}_4 = 0.024839$, $\overset{\text{in}}{c}_8 = -0.001938$, $\overset{\text{in}}{c}_{12} = 0.000201$, $\overset{\text{in}}{c}_{16} = -0.00002361$, $\overset{\text{in}}{c}_{20} = 0.00000295$

G. $\Phi(0) = 0.3233$, $\overset{\text{in}}{c}_0 = 0.985208$, $\overset{\text{in}}{c}_4 = 0.020874$, $\overset{\text{in}}{c}_8 = -0.001368$, $\overset{\text{in}}{c}_{12} = 0.000119$, $\overset{\text{in}}{c}_{16} = -0.00001177$, $\overset{\text{in}}{c}_{20} = 0.00000124$

I. $\Phi(0) = 0.3498$, $\overset{\text{in}}{c}_0 = 1.011738$, $\overset{\text{in}}{c}_4 = 0.017669$, $\overset{\text{in}}{c}_8 = -0.000980$, $\overset{\text{in}}{c}_{12} = 0.000072$, $\overset{\text{in}}{c}_{16} = -0.00000604$, $\overset{\text{in}}{c}_{20} = 0.00000054$

K. $\Phi(0) = 0.3753$, $\overset{\text{in}}{c}_0 = 1.037206$, $\overset{\text{in}}{c}_4 = 0.015056$, $\overset{\text{in}}{c}_8 = -0.000712$, $\overset{\text{in}}{c}_{12} = 0.000044$, $\overset{\text{in}}{c}_{16} = -0.00000319$, $\overset{\text{in}}{c}_{20} = 0.00000024$

Problem 3.7

A. $\Phi^+(-1) = 0.6667$, $\Phi^-(-1) = 1.3333$, $\Phi(0) = -0.0000$, $\Phi^-(1) = -1.3333$, $\Phi^+(1) = -0.6667$, $\overset{\text{cont}}{c}_1 = 0.3333$, $\overset{\text{in}}{c}_1 = -0.3333$, $\overset{\text{out}}{c}_1 = 0.3333$

C. $\Phi^+(-1) = 0.4000$, $\Phi^-(-1) = 1.6000$, $\Phi(0) = -0.0000$, $\Phi^-(1) = -1.6000$, $\Phi^+(1) = -0.4000$, $\overset{\text{cont}}{c}_1 = 0.6000$, $\overset{\text{in}}{c}_1 = -0.6000$, $\overset{\text{out}}{c}_1 = 0.6000$

E. $\Phi^+(-1) = 0.2857$, $\Phi^-(-1) = 1.7143$, $\Phi(0) = -0.0000$, $\Phi^-(1) = -1.7143$, $\Phi^+(1) = -0.2857$, $\overset{\text{cont}}{c}_1 = 0.7143$, $\overset{\text{in}}{c}_1 = -0.7143$, $\overset{\text{out}}{c}_1 = 0.7143$

G. $\Phi^+(-1) = 1.3333$, $\Phi^-(-1) = 0.6667$, $\Phi(0) = 0.0000$, $\Phi^-(1) = -0.6667$, $\Phi^+(1) = -1.3333$, $\overset{\text{cont}}{c}_1 = -0.3333$, $\overset{\text{in}}{c}_1 = 0.3333$, $\overset{\text{out}}{c}_1 = -0.3333$

I. $\Phi^+(-1) = 1.6000$, $\Phi^-(-1) = 0.4000$, $\Phi(0) = 0.0000$, $\Phi^-(1) = -0.4000$, $\Phi^+(1) = -1.6000$, $\overset{\text{cont}}{c}_1 = -0.6000$, $\overset{\text{in}}{c}_1 = 0.6000$, $\overset{\text{out}}{c}_1 = -0.6000$

K. $\Phi^+(-1) = 1.7143$, $\Phi^-(-1) = 0.2857$, $\Phi(0) = 0.0000$, $\Phi^-(1) = -0.2857$, $\Phi^+(1) = -1.7143$, $\overset{\text{cont}}{c}_1 = -0.7143$, $\overset{\text{in}}{c}_1 = 0.7143$, $\overset{\text{out}}{c}_1 = -0.7143$

Problem 3.8

A. $\Phi^+(-L_1 e^{i\vartheta}) = 0.7500$, $\Phi^-(-L_1 e^{i\vartheta}) = 1.5000$, $\Phi^-(L_1 e^{i\vartheta}) = -1.5000$, $\Phi^+(L_1 e^{i\vartheta}) = -0.7500$, $\overset{\text{cont}}{c}_1 = 0.2500 + -0.3464i$, $\overset{\text{in}}{c}_1 = -0.3750 + -0.2598i$, $\overset{\text{out}}{c}_1 = 0.2500 + -0.3464i$

C. $\Phi^+(-L_1 e^{i\vartheta}) = 0.5000$, $\Phi^-(-L_1 e^{i\vartheta}) = 2.0000$, $\Phi^-(L_1 e^{i\vartheta}) = -2.0000$, $\Phi^+(L_1 e^{i\vartheta}) = -0.5000$, $\overset{\text{cont}}{c}_1 = 0.5000 + -0.5774i$, $\overset{\text{in}}{c}_1 = -0.7500 + -0.4330i$, $\overset{\text{out}}{c}_1 = 0.5000 + -0.5774i$

E. $\Phi^+(-L_1 e^{i\vartheta}) = 0.3750$, $\Phi^-(-L_1 e^{i\vartheta}) = 2.2500$, $\Phi^-(L_1 e^{i\vartheta}) = -2.2500$, $\Phi^+(L_1 e^{i\vartheta}) = -0.3750$, $\overset{\text{cont}}{c}_1 = 0.6250 + -0.6662i$, $\overset{\text{in}}{c}_1 = -0.9375 + -0.4996i$, $\overset{\text{out}}{c}_1 = 0.6250 + -0.6662i$

G. $\Phi^+(-L_1 e^{i\vartheta}) = 1.2000$, $\Phi^-(-L_1 e^{i\vartheta}) = 0.6000$, $\Phi^-(L_1 e^{i\vartheta}) = -0.6000$, $\Phi^+(L_1 e^{i\vartheta}) = -1.2000$, $\overset{\text{cont}}{c}_1 = -0.2000 + 0.4330i$, $\overset{\text{in}}{c}_1 = 0.3000 + 0.3248i$, $\overset{\text{out}}{c}_1 = -0.2000 + 0.4330i$

I. $\Phi^+(-L_1 e^{i\vartheta}) = 1.3333$, $\Phi^-(-L_1 e^{i\vartheta}) = 0.3333$, $\Phi^-(L_1 e^{i\vartheta}) = -0.3333$, $\Phi^+(L_1 e^{i\vartheta}) = -1.3333$, $\overset{\text{cont}}{c}_1 = -0.3333 + 0.8660i$, $\overset{\text{in}}{c}_1 = 0.5000 + 0.6495i$, $\overset{\text{out}}{c}_1 = -0.3333 + 0.8660i$

K. $\Phi^+(-L_1 e^{i\vartheta}) = 1.3846$, $\Phi^-(-L_1 e^{i\vartheta}) = 0.2308$, $\Phi^-(L_1 e^{i\vartheta}) = -0.2308$, $\Phi^+(L_1 e^{i\vartheta}) = -1.3846$, $\overset{\text{cont}}{c}_1 = -0.3846 + 1.0825i$, $\overset{\text{in}}{c}_1 = 0.5769 + 0.8119i$, $\overset{\text{out}}{c}_1 = -0.3846 + 1.0825i$

Problem 3.9

A. $\Phi(\mathrm{i}) = 0.0000$, $\Psi(\mathrm{i}) = -0.4142$, $v_x(\mathrm{i}) = -0.2929$, $v_y(\mathrm{i}) = 0.0000$
$\Phi^+(\frac{1}{2}) = 0.5000$, $\Psi^+(\frac{1}{2}) = -0.8660$, $v_x{}^+(\frac{1}{2}) = -1.0000$, $v_y{}^+(\frac{1}{2}) = 0.5774$
$\Phi^-(\frac{1}{2}) = 0.5000$, $\Psi^-(\frac{1}{2}) = 0.8660$, $v_x{}^-(\frac{1}{2}) = -1.0000$, $v_y{}^-(\frac{1}{2}) = -0.5774$

C. $\Phi(\mathrm{i}) = -0.1716$, $\Psi(\mathrm{i}) = -0.0000$, $v_x(\mathrm{i}) = -0.0000$, $v_y(\mathrm{i}) = -0.2426$
$\Phi^+(\frac{1}{2}) = -0.5000$, $\Psi^+(\frac{1}{2}) = -0.8660$, $v_x{}^+(\frac{1}{2}) = -2.0000$, $v_y{}^+(\frac{1}{2}) = -1.1547$
$\Phi^-(\frac{1}{2}) = -0.5000$, $\Psi^-(\frac{1}{2}) = 0.8660$, $v_x{}^-(\frac{1}{2}) = -2.0000$, $v_y{}^-(\frac{1}{2}) = 1.1547$

E. $\Phi(\mathrm{i}) = -0.0000$, $\Psi(\mathrm{i}) = 0.0711$, $v_x(\mathrm{i}) = 0.1508$, $v_y(\mathrm{i}) = -0.0000$
$\Phi^+(\frac{1}{2}) = -1.0000$, $\Psi^+(\frac{1}{2}) = 0.0000$, $v_x{}^+(\frac{1}{2}) = -0.0000$, $v_y{}^+(\frac{1}{2}) = -3.4641$
$\Phi^-(\frac{1}{2}) = -1.0000$, $\Psi^-(\frac{1}{2}) = -0.0000$, $v_x{}^-(\frac{1}{2}) = -0.0000$, $v_y{}^-(\frac{1}{2}) = 3.4641$

G. $\Phi(\mathrm{i}) = 0.0294$, $\Psi(\mathrm{i}) = 0.0000$, $v_x(\mathrm{i}) = 0.0000$, $v_y(\mathrm{i}) = 0.0833$
$\Phi^+(\frac{1}{2}) = -0.5000$, $\Psi^+(\frac{1}{2}) = 0.8660$, $v_x{}^+(\frac{1}{2}) = 4.0000$, $v_y{}^+(\frac{1}{2}) = -2.3094$
$\Phi^-(\frac{1}{2}) = -0.5000$, $\Psi^-(\frac{1}{2}) = -0.8660$, $v_x{}^-(\frac{1}{2}) = 4.0000$, $v_y{}^-(\frac{1}{2}) = 2.3094$

I. $\Phi(\mathrm{i}) = 0.0000$, $\Psi(\mathrm{i}) = -0.0122$, $v_x(\mathrm{i}) = -0.0431$, $v_y(\mathrm{i}) = 0.0000$
$\Phi^+(\frac{1}{2}) = 0.5000$, $\Psi^+(\frac{1}{2}) = 0.8660$, $v_x{}^+(\frac{1}{2}) = 5.0000$, $v_y{}^+(\frac{1}{2}) = 2.8868$
$\Phi^-(\frac{1}{2}) = 0.5000$, $\Psi^-(\frac{1}{2}) = -0.8660$, $v_x{}^-(\frac{1}{2}) = 5.0000$, $v_y{}^-(\frac{1}{2}) = -2.8868$

K. $\Phi(\mathrm{i}) = -0.0051$, $\Psi(\mathrm{i}) = -0.0000$, $v_x(\mathrm{i}) = -0.0000$, $v_y(\mathrm{i}) = -0.0214$
$\Phi^+(\frac{1}{2}) = 1.0000$, $\Psi^+(\frac{1}{2}) = 0.0000$, $v_x{}^+(\frac{1}{2}) = 0.0000$, $v_y{}^+(\frac{1}{2}) = 6.9282$
$\Phi^-(\frac{1}{2}) = 1.0000$, $\Psi^-(\frac{1}{2}) = 0.0000$, $v_x{}^-(\frac{1}{2}) = 0.0000$, $v_y{}^-(\frac{1}{2}) = -6.9282$

Problem 3.10

A. $\Phi^+ = -0.5000$, $\Psi^+ = -1.0000$, $v_x{}^+ = 0.5000$, $v_y{}^+ = -0.5000$
$\Phi^- = -0.5000$, $\Psi^- = 0.0000$, $v_x{}^- = 0.5000$, $v_y{}^- = -0.5000$,
$c_1 = 0.5000$

C. $\Phi^+ = -1.5000$, $\Psi^+ = -3.0000$, $v_x{}^+ = 0.5000$, $v_y{}^+ = -0.5000$
$\Phi^- = -1.5000$, $\Psi^- = 0.0000$, $v_x{}^- = 0.5000$, $v_y{}^- = -0.5000$,
$c_1 = 1.5000$

E. $\Phi^+ = -2.5000$, $\Psi^+ = -5.0000$, $v_x{}^+ = 0.5000$, $v_y{}^+ = -0.5000$
$\Phi^- = -2.5000$, $\Psi^- = 0.0000$, $v_x{}^- = 0.5000$, $v_y{}^- = -0.5000$,
$c_1 = 2.5000$

G. $\Phi^+ = -3.5000$, $\Psi^+ = -7.0000$, $v_x{}^+ = 0.5000$, $v_y{}^+ = -0.5000$
$\Phi^- = -3.5000$, $\Psi^- = 0.0000$, $v_x{}^- = 0.5000$, $v_y{}^- = -0.5000$,
$c_1 = 3.5000$

I. $\Phi^+ = -4.5000$, $\Psi^+ = -9.0000$, $v_x{}^+ = 0.5000$, $v_y{}^+ = -0.5000$
$\Phi^- = -4.5000$, $\Psi^- = 0.0000$, $v_x{}^- = 0.5000$, $v_y{}^- = -0.5000$,
$c_1 = 4.5000$

K. $\Phi^+ = -5.5000$, $\Psi^+ = -11.0000$, $v_x{}^+ = 0.5000$, $v_y{}^+ = -0.5000$
$\Phi^- = -5.5000$, $\Psi^- = 0.0000$, $v_x{}^- = 0.5000$, $v_y{}^- = -0.5000$,
$c_1 = 5.5000$

Problem 3.11

A. $\Phi^+ = -0.5000$, $\Psi^+ = -0.9433$, $v_x{}^+ = 0.5623$, $v_y{}^+ = -0.4377$, $\Phi^- = -0.5000$, $\Psi^- = -0.0567$, $v_x{}^- = 0.5623$, $v_y{}^- = -0.4377$, $c_1 = 0.446488$, $c_3 = 0.003470$, $c_5 = 0.000322$, $c_7 = 0.000078$, $c_9 = 0.000062$

C. $\Phi^+ = -1.5000$, $\Psi^+ = -2.5822$, $v_x{}^+ = 0.6520$, $v_y{}^+ = -0.3480$, $\Phi^- = -1.5000$, $\Psi^- = -0.4178$, $v_x{}^- = 0.6520$, $v_y{}^- = -0.3480$, $c_1 = 1.104150$, $c_3 = 0.024342$, $c_5 = 0.002573$, $c_7 = 0.000622$, $c_9 = 0.000490$

E. $\Phi^+ = -2.5000$, $\Psi^+ = -4.0179$, $v_x{}^+ = 0.7131$, $v_y{}^+ = -0.2869$, $\Phi^- = -2.5000$, $\Psi^- = -0.9821$, $v_x{}^- = 0.7131$, $v_y{}^- = -0.2869$, $c_1 = 1.566461$, $c_3 = 0.054601$, $c_5 = 0.006429$, $c_7 = 0.001575$, $c_9 = 0.001218$

G. $\Phi^+ = -3.5000$, $\Psi^+ = -5.3322$, $v_x{}^+ = 0.7570$, $v_y{}^+ = -0.2430$, $\Phi^- = -3.5000$, $\Psi^- = -1.6678$, $v_x{}^- = 0.7570$, $v_y{}^- = -0.2430$, $c_1 = 1.910119$, $c_3 = 0.088689$, $c_5 = 0.011449$, $c_7 = 0.002854$, $c_9 = 0.002176$

I. $\Phi^+ = -4.5000$, $\Psi^+ = -6.5685$, $v_x{}^+ = 0.7900$, $v_y{}^+ = -0.2100$, $\Phi^- = -4.5000$, $\Psi^- = -2.4315$, $v_x{}^- = 0.7900$, $v_y{}^- = -0.2100$, $c_1 = 2.176198$, $c_3 = 0.123942$, $c_5 = 0.017323$, $c_7 = 0.004407$, $c_9 = 0.003321$

K. $\Phi^+ = -5.5000$, $\Psi^+ = -7.7519$, $v_x{}^+ = 0.8155$, $v_y{}^+ = -0.1845$, $\Phi^- = -5.5000$, $\Psi^- = -3.2481$, $v_x{}^- = 0.8155$, $v_y{}^- = -0.1845$, $c_1 = 2.388703$, $c_3 = 0.159016$, $c_5 = 0.023829$, $c_7 = 0.006193$, $c_9 = 0.004624$

Problem 3.12

A. $\Phi^+ = 0.0000$, $\Psi^+ = -0.5000$, $v_x{}^+ = 0.5000$, $v_y{}^+ = 0.5000$
$\Phi^- = -1.0000$, $\Psi^- = -0.5000$, $v_x{}^- = 0.5000$, $v_y{}^- = 0.5000$, $c_1 = 0.5000i$

C. $\Phi^+ = 0.0000$, $\Psi^+ = -1.5000$, $v_x{}^+ = 0.5000$, $v_y{}^+ = 0.5000$
$\Phi^- = -3.0000$, $\Psi^- = -1.5000$, $v_x{}^- = 0.5000$, $v_y{}^- = 0.5000$, $c_1 = 1.5000i$

E. $\Phi^+ = 0.0000$, $\Psi^+ = -2.5000$, $v_x{}^+ = 0.5000$, $v_y{}^+ = 0.5000$
$\Phi^- = -5.0000$, $\Psi^- = -2.5000$, $v_x{}^- = 0.5000$, $v_y{}^- = 0.5000$, $c_1 = 2.5000i$

G. $\Phi^+ = 0.0000$, $\Psi^+ = -3.5000$, $v_x{}^+ = 0.5000$, $v_y{}^+ = 0.5000$
$\Phi^- = -7.0000$, $\Psi^- = -3.5000$, $v_x{}^- = 0.5000$, $v_y{}^- = 0.5000$, $c_1 = 3.5000i$

I. $\Phi^+ = 0.0000$, $\Psi^+ = -4.5000$, $v_x{}^+ = 0.5000$, $v_y{}^+ = 0.5000$
$\Phi^- = -9.0000$, $\Psi^- = -4.5000$, $v_x{}^- = 0.5000$, $v_y{}^- = 0.5000$, $c_1 = 4.5000i$

K. $\Phi^+ = 0.0000$, $\Psi^+ = -5.5000$, $v_x{}^+ = 0.5000$, $v_y{}^+ = 0.5000$
$\Phi^- = -11.0000$, $\Psi^- = -5.5000$, $v_x{}^- = 0.5000$, $v_y{}^- = 0.5000$, $c_1 = 5.5000i$

Problem 3.13

A. $\Phi^+ = -0.0567$, $\Psi^+ = -0.5000$, $v_x{}^+ = 0.5623$, $v_y{}^+ = 0.4377$, $\Phi^- = -0.9433$, $\Psi^- = -0.5000$, $v_x{}^- = 0.5623$, $v_y{}^- = 0.4377$, $c_1 = 0.446488i$, $c_3 = 0.003470i$, $c_5 = 0.000322i$, $c_7 = 0.000078i$, $c_9 = 0.000062i$

C. $\Phi^+ = -0.4178$, $\Psi^+ = -1.5000$, $v_x{}^+ = 0.6520$, $v_y{}^+ = 0.3480$, $\Phi^- = -2.5822$, $\Psi^- = -1.5000$, $v_x{}^- = 0.6520$, $v_y{}^- = 0.3480$, $c_1 = 1.104150i$, $c_3 = 0.024342i$, $c_5 = 0.002573i$, $c_7 = 0.000622i$, $c_9 = 0.000490i$

E. $\Phi^+ = -0.9821$, $\Psi^+ = -2.5000$, $v_x{}^+ = 0.7131$, $v_y{}^+ = 0.2869$, $\Phi^- = -4.0179$, $\Psi^- = -2.5000$, $v_x{}^- = 0.7131$, $v_y{}^- = 0.2869$, $c_1 = 1.566461i$, $c_3 = 0.054601i$, $c_5 = 0.006429i$, $c_7 = 0.001575i$, $c_9 = 0.001218i$

G. $\Phi^+ = -1.6678$, $\Psi^+ = -3.5000$, $v_x{}^+ = 0.7570$, $v_y{}^+ = 0.2430$, $\Phi^- = -5.3322$, $\Psi^- = -3.5000$, $v_x{}^- = 0.7570$, $v_y{}^- = 0.2430$, $c_1 = 1.910119i$, $c_3 = 0.088689i$, $c_5 = 0.011449i$, $c_7 = 0.002854i$, $c_9 = 0.002176i$

I. $\Phi^+ = -2.4315$, $\Psi^+ = -4.5000$, $v_x{}^+ = 0.7900$, $v_y{}^+ = 0.2100$, $\Phi^- = -6.5685$, $\Psi^- = -4.5000$, $v_x{}^- = 0.7900$, $v_y{}^- = 0.2100$, $c_1 = 2.176198i$, $c_3 = 0.123942i$, $c_5 = 0.017323i$, $c_7 = 0.004407i$, $c_9 = 0.003321i$

K. $\Phi^+ = -3.2481$, $\Psi^+ = -5.5000$, $v_x{}^+ = 0.8155$, $v_y{}^+ = 0.1845$, $\Phi^- = -7.7519$, $\Psi^- = -5.5000$, $v_x{}^- = 0.8155$, $v_y{}^- = 0.1845$, $c_1 = 2.388703i$, $c_3 = 0.159016i$, $c_5 = 0.023829i$, $c_7 = 0.006193i$, $c_9 = 0.004624i$

Problem 3.14

A. $Q = -1.5000, c_1 = 0.5000$

C. $Q = -2.5000, c_1 = 1.5000$

E. $Q = -3.5000, c_1 = 2.5000$

G. $Q = -4.5000, c_1 = 3.5000$

I. $Q = -5.5000, c_1 = 4.5000$

K. $Q = -6.5000, c_1 = 5.5000$

Problem 3.15

A. $v_0 = 0.2567 + 0.0067\mathrm{i}, \Phi_0 = 3.1762, Q = 2.1762, c_1 = 0.2633$

C. $v_0 = 0.7700 + 0.0200\mathrm{i}, \Phi_0 = 7.5285, Q = 6.5285, c_1 = 0.7900$

E. $v_0 = -0.2887 + -0.2887\mathrm{i}, \Phi_0 = 1.0000, Q = 0.0000, c_1 = -0.5774$

G. $v_0 = -0.8660 + -0.8660\mathrm{i}, \Phi_0 = 1.0000, Q = 0.0000, c_1 = -1.7321$

I. $v_0 = 0.0067 + 0.2567\mathrm{i}, \Phi_0 = 3.1762, Q = 2.1762, c_1 = 0.2633$

K. $v_0 = 0.0200 + 0.7700\mathrm{i}, \Phi_0 = 7.5285, Q = 6.5285, c_1 = 0.7900$

Problem 3.16

A. $\Gamma = -3.4558, c_1 = -0.5500\mathrm{i}$

C. $\Gamma = -4.0841, c_1 = -0.6500\mathrm{i}$

E. $\Gamma = -4.7124, c_1 = -0.7500\mathrm{i}$

G. $\Gamma = -5.3407, c_1 = -0.8500\mathrm{i}$

I. $\Gamma = -5.9690, c_1 = -0.9500\mathrm{i}$

K. $\Gamma = -6.5973, c_1 = -1.0500\mathrm{i}$

Problem 3.17

A. $2L/\text{camber} = 28.25, \Gamma = 0.4449, v^+ = -1.1466, v^- = -0.8634$

C. $2L/\text{camber} = 27.56, \Gamma = 0.4559, v^+ = -1.1504, v^- = -0.8601$

E. $2L/\text{camber} = 26.91, \Gamma = 0.4669, v^+ = -1.1541, v^- = -0.8569$

G. $2L/\text{camber} = 26.29, \Gamma = 0.4779, v^+ = -1.1579, v^- = -0.8537$

I. $2L/\text{camber} = 25.70, \Gamma = 0.4890, v^+ = -1.1617, v^- = -0.8504$

K. $2L/\text{camber} = 25.13, \Gamma = 0.5000, v^+ = -1.1655, v^- = -0.8472$

Problem 3.18

A. $\Omega = x, v = -1, \nabla^2\Omega = 0, \Omega(1+\mathrm{i}) = 1.0000, v(1+\mathrm{i}) = -1.0000, \nabla^2\Omega(1+\mathrm{i}) = 0.0000$

C. $\Omega = x^3, v = -3x^2, \nabla^2\Omega = 6x, \Omega(1+\mathrm{i}) = 1.0000, v(1+\mathrm{i}) = -3.0000, \nabla^2\Omega(1+\mathrm{i}) = 6.0000$

E. $\Omega = \mathrm{i}y, v = -1, \nabla^2\Omega = 0, \Omega(1+\mathrm{i}) = 1.0000\mathrm{i}, v(1+\mathrm{i}) = -1.0000, \nabla^2\Omega(1+\mathrm{i}) = 0.0000$

G. $\Omega = \mathrm{i}y^3, v = -3y^2, \nabla^2\Omega = 6\mathrm{i}y, \Omega(1+\mathrm{i}) = 1.0000\mathrm{i}, v(1+\mathrm{i}) = -3.0000, \nabla^2\Omega(1+\mathrm{i}) = 6.0000\mathrm{i}$

I. $\Omega = r, v = -z/r, \nabla^2\Omega = 1/r, \Omega(1+\mathrm{i}) = 1.4142, v(1+\mathrm{i}) = -0.7071 - 0.7071\mathrm{i}, \nabla^2\Omega(1+\mathrm{i}) = 0.7071$

K. $\Omega = r^3, v = -3rz, \nabla^2\Omega = 9r, \Omega(1+\mathrm{i}) = 2.8284, v(1+\mathrm{i}) = -4.2426 - 4.2426\mathrm{i}, \nabla^2\Omega(1+\mathrm{i}) = 12.7279$

Problem 3.19

A. $\Omega = r^4/16, v = -r^2z/4, \nabla^2\Omega = r^2, \Omega(1+\mathrm{i}) = 0.2500, v(1+\mathrm{i}) = -0.5000 - 0.5000\mathrm{i}, \nabla^2\Omega(1+\mathrm{i}) = 2.0000$

C. $\Omega = r^8/64, v = -r^6z/8, \nabla^2\Omega = r^6, \Omega(1+\mathrm{i}) = 0.2500, v(1+\mathrm{i}) = -1.0000 - 1.0000\mathrm{i}, \nabla^2\Omega(1+\mathrm{i}) = 8.0000$

E. $\Omega = r^{12}/144, v = -r^{10}z/12, \nabla^2\Omega = r^{10}, \Omega(1+\mathrm{i}) = 0.4444, v(1+\mathrm{i}) = -2.6667 - 2.6667\mathrm{i}, \nabla^2\Omega(1+\mathrm{i}) = 32.0000$

G. $\Omega = \mathrm{i}r^4/16, v = \mathrm{i}r^2z/4, \nabla^2\Omega = \mathrm{i}r^2, \Omega(1+\mathrm{i}) = 0.2500\mathrm{i}, v(1+\mathrm{i}) = -0.5000 + 0.5000\mathrm{i}, \nabla^2\Omega(1+\mathrm{i}) = 2.0000\mathrm{i}$

I. $\Omega = \mathrm{i}r^8/64, v = \mathrm{i}r^6z/8, \nabla^2\Omega = \mathrm{i}r^6, \Omega(1+\mathrm{i}) = 0.2500\mathrm{i}, v(1+\mathrm{i}) = -1.0000 + 1.0000\mathrm{i}, \nabla^2\Omega(1+\mathrm{i}) = 8.0000\mathrm{i}$

K. $\Omega = \mathrm{i}r^{12}/144, v = \mathrm{i}r^{10}z/12, \nabla^2\Omega = \mathrm{i}r^{10}, \Omega(1+\mathrm{i}) = 0.4444\mathrm{i}, v(1+\mathrm{i}) = -2.6667 + 2.6667\mathrm{i}, \nabla^2\Omega(1+\mathrm{i}) = 32.0000\mathrm{i}$

Problem 3.20

A. $\sigma_x = -954,930, \sigma_y = -954,930, \tau_{xy} = 954,930, \sigma_1 = 391,026, \sigma_2 = -1,632,434, \tau_{\max} = 1,011,730$

C. $\sigma_x = 1,273,240, \sigma_y = 1,273,240, \tau_{xy} = -1,273,240, \sigma_1 = 2,176,579, \sigma_2 = -521,367, \tau_{\max} = 1,348,973$

E. $\sigma_x = 286,479, \sigma_y = 286,479, \tau_{xy} = -954,930, \sigma_1 = 1,632,434, \sigma_2 = -391,026, \tau_{\max} = 1,011,730$

G. $\sigma_x = -381,972, \sigma_y = -381,972, \tau_{xy} = 1,273,240, \sigma_1 = 521,367, \sigma_2 = -2,176,579, \tau_{\max} = 1,348,973$

I. $\sigma_x = -3,151,268, \sigma_y = -3,151,268, \tau_{xy} = -668,451, \sigma_1 = 782,051, \sigma_2 = -3,264,868, \tau_{\max} = 2,023,460$

K. $\sigma_x = 4,201,690, \sigma_y = 4,201,690, \tau_{xy} = 891,268, \sigma_1 = 4,353,158, \sigma_2 = -1,042,735, \tau_{\max} = 2,697,946$

Problem 3.21

A. $u_x = 0.0000032781, u_y = -0.0000004035$

C. $u_x = 0.0000043708, u_y = -0.0000005379$

E. $u_x = -0.0000004035, u_y = 0.0000032781$

G. $u_x = -0.0000005379, u_y = 0.0000043708$

I. $u_x = 0.0000042624, u_y = 0.0000034554$

K. $u_x = 0.0000056832, u_y = 0.0000046073$

Problem 3.22

A. $\sigma_x(r_0) = -0, \sigma_y(r_0) = -1,200,000, \tau_{xy}(r_0) = 0; \sigma_x\left(r_0 e^{i\pi/4}\right) = 600,000, \sigma_y\left(r_0 e^{i\pi/4}\right) = 600,000, \tau_{xy}\left(r_0 e^{i\pi/4}\right) = -600,000;$ $\sigma_x(ir_0) = 3,600,000, \sigma_y(ir_0) = 0, \tau_{xy}(ir_0) = -0$

C. $\sigma_x(r_0) = -0, \sigma_y(r_0) = -1,600,000, \tau_{xy}(r_0) = -0; \sigma_x\left(r_0 e^{i\pi/4}\right) = 800,000, \sigma_y\left(r_0 e^{i\pi/4}\right) = 800,000, \tau_{xy}\left(r_0 e^{i\pi/4}\right) = -800,000;$ $\sigma_x(ir_0) = 4,800,000, \sigma_y(ir_0) = 0, \tau_{xy}(ir_0) = 0$

E. $\sigma_x(r_0) = 0, \sigma_y(r_0) = 3,600,000, \tau_{xy}(r_0) = 0; \sigma_x\left(r_0 e^{i\pi/4}\right) = 600,000, \sigma_y\left(r_0 e^{i\pi/4}\right) = 600,000, \tau_{xy}\left(r_0 e^{i\pi/4}\right) = -600,000;$ $\sigma_x(ir_0) = -1,200,000, \sigma_y(ir_0) = -0, \tau_{xy}(ir_0) = -0$

G. $\sigma_x(r_0) = 0, \sigma_y(r_0) = 4,800,000, \tau_{xy}(r_0) = 0; \sigma_x\left(r_0 e^{i\pi/4}\right) = 800,000, \sigma_y\left(r_0 e^{i\pi/4}\right) = 800,000, \tau_{xy}\left(r_0 e^{i\pi/4}\right) = -800,000;$ $\sigma_x(ir_0) = -1,600,000, \sigma_y(ir_0) = -0, \tau_{xy}(ir_0) = -0$

I. $\sigma_x(r_0) = 0, \sigma_y(r_0) = 0, \tau_{xy}(r_0) = 0; \sigma_x\left(r_0 e^{i\pi/4}\right) = -2,400,000, \sigma_y\left(r_0 e^{i\pi/4}\right) = -2,400,000, \tau_{xy}\left(r_0 e^{i\pi/4}\right) = 2,400,000;$ $\sigma_x(ir_0) = -0, \sigma_y(ir_0) = -0, \tau_{xy}(ir_0) = 0$

K. $\sigma_x(r_0) = 0, \sigma_y(r_0) = -0, \tau_{xy}(r_0) = 0; \sigma_x\left(r_0 e^{i\pi/4}\right) = -3,200,000, \sigma_y\left(r_0 e^{i\pi/4}\right) = -3,200,000, \tau_{xy}\left(r_0 e^{i\pi/4}\right) = 3,200,000;$ $\sigma_x(ir_0) = 0, \sigma_y(ir_0) = 0, \tau_{xy}(ir_0) = 0$

Problem 3.23

A. $\sigma_x(r_0) = 1,856,000, \sigma_y(r_0) = 677,000, \tau_{xy}(r_0) = -0; \sigma_x(r_0 e^{i\pi/4}) = 1,533,000, \sigma_y(r_0 e^{i\pi/4}) = -333,000, \tau_{xy}(r_0 e^{i\pi/4}) = 323,000; \sigma_x(ir_0) = -123,000, \sigma_y(ir_0) = -10,000, \tau_{xy}(ir_0) = 0$

C. $\sigma_x(r_0) = 2,475,000, \sigma_y(r_0) = 903,000, \tau_{xy}(r_0) = -0; \sigma_x(r_0 e^{i\pi/4}) = 2,044,000, \sigma_y(r_0 e^{i\pi/4}) = -444,000, \tau_{xy}(r_0 e^{i\pi/4}) = 431,000; \sigma_x(ir_0) = -164,000, \sigma_y(ir_0) = -14,000, \tau_{xy}(ir_0) = 0$

E. $\sigma_x(r_0) = -10,000, \sigma_y(r_0) = -123,000, \tau_{xy}(r_0) = -0; \sigma_x(r_0 e^{i\pi/4}) = -333,000, \sigma_y(r_0 e^{i\pi/4}) = 1,533,000, \tau_{xy}(r_0 e^{i\pi/4}) = 323,000; \sigma_x(ir_0) = 677,000, \sigma_y(ir_0) = 1,856,000, \tau_{xy}(ir_0) = -0$

G. $\sigma_x(r_0) = -14,000, \sigma_y(r_0) = -164,000, \tau_{xy}(r_0) = 0; \sigma_x(r_0 e^{i\pi/4}) = -444,000, \sigma_y(r_0 e^{i\pi/4}) = 2,044,000, \tau_{xy}(r_0 e^{i\pi/4}) = 431,000;$ $\sigma_x(ir_0) = 903,000, \sigma_y(ir_0) = 2,475,000, \tau_{xy}(ir_0) = -0$

I. $\sigma_x(r_0) = -0, \sigma_y(r_0) = 0, \tau_{xy}(r_0) = 1,867,000; \sigma_x(r_0 e^{i\pi/4}) = 1,333,000, \sigma_y(r_0 e^{i\pi/4}) = 1,333,000, \tau_{xy}(r_0 e^{i\pi/4}) = 533,000;$ $\sigma_x(ir_0) = 0, \sigma_y(ir_0) = -0, \tau_{xy}(ir_0) = 1,867,000$

K. $\sigma_x(r_0) = -0, \sigma_y(r_0) = -0, \tau_{xy}(r_0) = 2,489,000; \sigma_x(r_0 e^{i\pi/4}) = 1,778,000, \sigma_y(r_0 e^{i\pi/4}) = 1,778,000, \tau_{xy}(r_0 e^{i\pi/4}) = 711,000;$ $\sigma_x(ir_0) = -0, \sigma_y(ir_0) = 0, \tau_{xy}(ir_0) = 2,489,000$

Problem 4.1

A. $\hat{\mathbf{r}}(0.00/\pi) = (1.0000, 0.0000, 0.0000)$,
$\hat{\mathbf{r}}(0.17/\pi) = (0.8660, 0.5000, 0.0000)$,
$\hat{\mathbf{r}}(0.25/\pi) = (0.7071, 0.7071, 0.0000)$,
$\hat{\mathbf{r}}(0.33/\pi) = (0.5000, 0.8660, 0.0000)$,
$\hat{\mathbf{r}}(0.50/\pi) = (0.0000, 1.0000, 0.0000)$

C. $\hat{\mathbf{r}}(0.00/\pi) = (1.0000, 0.0000, 0.0000)$,
$\hat{\mathbf{r}}(0.17/\pi) = (0.8660, 0.5000, 0.0000)$,
$\hat{\mathbf{r}}(0.25/\pi) = (0.7071, 0.7071, 0.0000)$,
$\hat{\mathbf{r}}(0.33/\pi) = (0.5000, 0.8660, 0.0000)$,
$\hat{\mathbf{r}}(0.50/\pi) = (0.0000, 1.0000, 0.0000)$

E. $\hat{\mathbf{r}}(0.00/\pi) = (0.0000, 0.0000, 1.0000)$,
$\hat{\mathbf{r}}(0.25/\pi) = (0.6124, 0.3536, 0.7071)$,
$\hat{\mathbf{r}}(0.50/\pi) = (0.8660, 0.5000, 0.0000)$,
$\hat{\mathbf{r}}(0.75/\pi) = (0.6124, 0.3536, -0.7071)$,
$\hat{\mathbf{r}}(1.00/\pi) = (0.0000, 0.0000, -1.0000)$

G. $\hat{\mathbf{r}}(0.00/\pi) = (0.0000, 0.0000, 1.0000)$,
$\hat{\mathbf{r}}(0.25/\pi) = (0.3536, 0.6124, 0.7071)$,
$\hat{\mathbf{r}}(0.50/\pi) = (0.5000, 0.8660, 0.0000)$,
$\hat{\mathbf{r}}(0.75/\pi) = (0.3536, 0.6124, -0.7071)$,
$\hat{\mathbf{r}}(1.00/\pi) = (0.0000, 0.0000, -1.0000)$

I. $\hat{\eta}(0.00/\pi) = (0.0000, 0.0000, 1.0000)$,
$\hat{\eta}(0.25/\pi) = (0.6890, 0.3978, 0.6059)$,
$\hat{\eta}(0.50/\pi) = (0.8660, 0.5000, 0.0000)$,
$\hat{\eta}(0.75/\pi) = (0.6890, 0.3978, -0.6059)$,
$\hat{\eta}(1.00/\pi) = (0.0000, 0.0000, -1.0000)$

K. $\hat{\eta}(0.00/\pi) = (0.0000, 0.0000, 1.0000)$,
$\hat{\eta}(0.25/\pi) = (0.3544, 0.6139, 0.7054)$,
$\hat{\eta}(0.50/\pi) = (0.5000, 0.8660, 0.0000)$,
$\hat{\eta}(0.75/\pi) = (0.3544, 0.6139, -0.7054)$,
$\hat{\eta}(1.00/\pi) = (0.0000, 0.0000, -1.0000)$

M. $\hat{\eta}(0.00/\pi) = (0.0000, 0.0000, 1.0000)$,
$\hat{\eta}(0.25/\pi) = (0.5247, 0.3029, 0.7956)$,
$\hat{\eta}(0.50/\pi) = (0.8660, 0.5000, 0.0000)$,
$\hat{\eta}(0.75/\pi) = (0.5247, 0.3029, -0.7956)$,
$\hat{\eta}(1.00/\pi) = (0.0000, 0.0000, -1.0000)$

O. $\hat{\eta}(0.00/\pi) = (0.0000, 0.0000, 1.0000)$,
$\hat{\eta}(0.25/\pi) = (0.3527, 0.6109, 0.7089)$,
$\hat{\eta}(0.50/\pi) = (0.5000, 0.8660, 0.0000)$,
$\hat{\eta}(0.75/\pi) = (0.3527, 0.6109, -0.7089)$,
$\hat{\eta}(1.00/\pi) = (0.0000, 0.0000, -1.0000)$

Problem 4.2

A. $k_1 = 0.355$, $k_1 = 0.397$, $R = 0.039$, $\tau = 1.039$, $\varphi(-10,0) = -1.036 - 0.067\text{i}$, $\varphi(0,0) = 1.039$, $\varphi(10,0) = -0.994 - 0.300\text{i}$

C. $k_1 = 0.355$, $k_1 = 0.561$, $R = 0.178$, $\tau = 1.178$, $\varphi(-10,0) = -1.176 - 0.057\text{i}$, $\varphi(0,0) = 1.178$, $\varphi(10,0) = 0.170 - 1.166\text{i}$

E. $k_1 = 0.284$, $k_1 = 0.366$, $R = 0.095$, $\tau = 1.095$, $\varphi(-10,0) = -0.848 - 0.572\text{i}$, $\varphi(0,0) = 1.095$, $\varphi(10,0) = -1.094 - 0.034\text{i}$

G. $k_1 = 0.236$, $k_1 = 0.264$, $R = 0.039$, $\tau = 1.039$, $\varphi(-10,0) = -0.477 - 0.854\text{i}$, $\varphi(0,0) = 1.039$, $\varphi(10,0) = -0.684 + 0.782\text{i}$

I. $k_1 = 0.236$, $k_1 = 0.374$, $R = 0.180$, $\tau = 1.180$, $\varphi(-10,0) = -0.542 - 0.728\text{i}$, $\varphi(0,0) = 1.180$, $\varphi(10,0) = -1.175 - 0.113\text{i}$

K. $k_1 = 0.203$, $k_1 = 0.262$, $R = 0.095$, $\tau = 1.095$, $\varphi(-10,0) = -0.201 - 0.889\text{i}$, $\varphi(0,0) = 1.095$, $\varphi(10,0) = -0.702 + 0.841\text{i}$

Problem 4.3

A. $c_1 = 0.009966, c_2 = 0.000034, p = -1.9157\text{m}$

C. $c_1 = 0.009863, c_2 = 0.000137, p = -1.7247\text{m}$

E. $c_1 = 0.199986, c_2 = 0.000014, p = -1.8161\text{m}$

G. $c_1 = 0.199945, c_2 = 0.000055, p = -1.6098\text{m}$

I. $c_1 = 1.399997, c_2 = 0.000003, p = -0.5258\text{m}$

K. $c_1 = 1.399989, c_2 = 0.000011, p = -0.4703\text{m}$

Problem 4.4

A. $(x, y) = (-10, -10)$: $\Phi = 1.0$, $v_x = 0.3141615$, $v_y = 0.0000000$; $(x, y) = (10, -10)$: $\Phi = 0.0$, $v_x = 0.0011734$, $v_y = 0.0000000$; $(x, y) = (10, 10)$: $\Phi = 0.0$, $v_x = 0.0011734$, $v_y = -0.0000000$; $(x, y) = (-10, 10)$: $\Phi = 1.0$, $v_x = 0.3141615$, $v_y = -0.0000000$

C. $(x, y) = (-10, -10)$: $\Phi = 1.0$, $v_x = 0.6283185$, $v_y = 0.0000000$; $(x, y) = (10, -10)$: $\Phi = 0.0$, $v_x = 0.0000044$, $v_y = 0.0000000$; $(x, y) = (10, 10)$: $\Phi = 0.0$, $v_x = 0.0000044$, $v_y = -0.0000000$; $(x, y) = (-10, 10)$: $\Phi = 1.0$, $v_x = 0.6283185$, $v_y = -0.0000000$

E. $(x, y) = (-10, -10)$: $\Phi = 1.0$, $v_x = 0.0000000$, $v_y = 0.3141615$; $(x, y) = (10, -10)$: $\Phi = 1.0$, $v_x = -0.0000000$, $v_y = 0.3141615$; $(x, y) = (10, 10)$: $\Phi = 0.0$, $v_x = -0.0000000$, $v_y = 0.0011734$; $(x, y) = (-10, 10)$: $\Phi = 0.0$, $v_x = 0.0000000$, $v_y = 0.0011734$

G. $(x, y) = (-10, -10)$: $\Phi = 1.0$, $v_x = 0.0000000$, $v_y = 0.6283185$; $(x, y) = (10, -10)$: $\Phi = 1.0$, $v_x = -0.0000000$, $v_y = 0.6283185$; $(x, y) = (10, 10)$: $\Phi = 0.0$, $v_x = -0.0000000$, $v_y = 0.0000044$; $(x, y) = (-10, 10)$: $\Phi = 0.0$, $v_x = 0.0000000$, $v_y = 0.0000044$

I. $(x, y) = (-10, -10)$: $\Phi = 0.0$, $v_x = -0.0011734$, $v_y = 0.0000000$; $(x, y) = (10, -10)$: $\Phi = 1.0$, $v_x = -0.3141615$, $v_y = 0.0000000$; $(x, y) = (10, 10)$: $\Phi = 1.0$, $v_x = -0.3141615$, $v_y = -0.0000000$; $(x, y) = (-10, 10)$: $\Phi = 0.0$, $v_x = -0.0011734$, $v_y = -0.0000000$

K. $(x, y) = (-10, -10)$: $\Phi = 0.0$, $v_x = -0.0000044$, $v_y = 0.0000000$; $(x, y) = (10, -10)$: $\Phi = 1.0$, $v_x = -0.6283185$, $v_y = 0.0000000$; $(x, y) = (10, 10)$: $\Phi = 1.0$, $v_x = -0.6283185$, $v_y = -0.0000000$; $(x, y) = (-10, 10)$: $\Phi = 0.0$, $v_x = -0.0000044$, $v_y = -0.0000000$

M. $(x, y) = (-10, -10)$: $\Phi = 0.0$, $v_x = 0.0000000$, $v_y = -0.0011734$; $(x, y) = (10, -10)$: $\Phi = 0.0$, $v_x = -0.0000000$, $v_y = -0.0011734$; $(x, y) = (10, 10)$: $\Phi = 1.0$, $v_x = -0.0000000$, $v_y = -0.3141615$; $(x, y) = (-10, 10)$: $\Phi = 1.0$, $v_x = 0.0000000$, $v_y = -0.3141615$

O. $(x, y) = (-10, -10)$: $\Phi = 0.0$, $v_x = 0.0000000$, $v_y = -0.0000044$; $(x, y) = (10, -10)$: $\Phi = 0.0$, $v_x = -0.0000000$, $v_y = -0.0000044$; $(x, y) = (10, 10)$: $\Phi = 1.0$, $v_x = -0.0000000$, $v_y = -0.6283185$; $(x, y) = (-10, 10)$: $\Phi = 1.0$, $v_x = 0.0000000$, $v_y = -0.6283185$

Problem 4.5

A. $\overset{0}{c} = 10.4036, \overset{x}{c} = 0.5000, \overset{y}{c} = 0.0000, \overset{xy}{c} = 0.0000, \overset{x2y2}{c} = 0.1250, \Phi(0.5, 0.5) = 10.1473$

C. $\overset{0}{c} = 11.2109, \overset{x}{c} = 1.5000, \overset{y}{c} = 0.0000, \overset{xy}{c} = 0.0000, \overset{x2y2}{c} = 0.3750, \Phi(0.5, 0.5) = 10.4420$

E. $\overset{0}{c} = 10.4036, \overset{x}{c} = 0.0000, \overset{y}{c} = 0.5000, \overset{xy}{c} = 0.0000, \overset{x2y2}{c} = -0.1250, \Phi(0.5, 0.5) = 10.1473$

G. $\overset{0}{c} = 11.2109, \overset{x}{c} = 0.0000, \overset{y}{c} = 1.5000, \overset{xy}{c} = 0.0000, \overset{x2y2}{c} = -0.3750, \Phi(0.5, 0.5) = 10.4420$

I. $\overset{0}{c} = 10.8073, \overset{x}{c} = 0.5000, \overset{y}{c} = 0.5000, \overset{xy}{c} = 0.0000, \overset{x2y2}{c} = 0.0000, \Phi(0.5, 0.5) = 10.2947$

K. $\overset{0}{c} = 12.4218, \overset{x}{c} = 1.5000, \overset{y}{c} = 1.5000, \overset{xy}{c} = 0.0000, \overset{x2y2}{c} = 0.0000, \Phi(0.5, 0.5) = 10.8841$

Problem 4.6

A. $\overset{0}{c} = 2.5000, \overset{x}{c} = -0.2604, \overset{y}{c} = -7.5000, \overset{xy}{c} = 0.0000, \overset{x2y2}{c} = 0.0000, \Phi(1.0, 0.5) = 2.5000, v_x = 0.1008, v_y = 15.0000$

C. $\overset{0}{c} = 3.5000, \overset{x}{c} = -0.2604, \overset{y}{c} = -6.5000, \overset{xy}{c} = 0.0000, \overset{x2y2}{c} = 0.0000, \Phi(1.0, 0.5) = 3.5000, v_x = 0.1008, v_y = 13.0000$

E. $\overset{0}{c} = 4.5000, \overset{x}{c} = -0.2604, \overset{y}{c} = -5.5000, \overset{xy}{c} = -0.0000, \overset{x2y2}{c} = 0.0000, \Phi(1.0, 0.5) = 4.5000, v_x = 0.1008, v_y = 11.0000$

G. $\overset{0}{c} = 5.5000, \overset{x}{c} = -0.2604, \overset{y}{c} = -4.5000, \overset{xy}{c} = -0.0000, \overset{x2y2}{c} = 0.0000, \Phi(1.0, 0.5) = 5.5000, v_x = 0.1008, v_y = 9.0000$

I. $\overset{0}{c} = 6.5000, \overset{x}{c} = -0.2604, \overset{y}{c} = -3.5000, \overset{xy}{c} = 0.0000, \overset{x2y2}{c} = 0.0000, \Phi(1.0, 0.5) = 6.5000, v_x = 0.1008, v_y = 7.0000$

K. $\overset{0}{c} = 7.5000, \overset{x}{c} = -0.2604, \overset{y}{c} = -2.5000, \overset{xy}{c} = 0.0000, \overset{x2y2}{c} = 0.0000, \Phi(1.0, 0.5) = 7.5000, v_x = 0.1008, v_y = 5.0000$

Problem 4.7

A. left: $\overset{0}{c} = 0.6572$, $\overset{x}{c} = -0.3420$, $\overset{y}{c} = -0.0502$, $\overset{xy}{c} = -0.0914$, $\overset{x2y2}{c} = 0.0014$, $\Phi(0.5, 0.5) = 0.6588$, $v_x = 0.6742$, $v_y = 0.0941$;
right: $\overset{0}{c} = 0.1651$, $\overset{x}{c} = -0.1064$, $\overset{y}{c} = -0.1165$, $\overset{xy}{c} = 0.0537$, $\overset{x2y2}{c} = 0.0855$, $\Phi(1.5, 0.5) = 0.1659$, $v_x = 0.2052$, $v_y = 0.2217$;

C. left: $\overset{0}{c} = 1.9716$, $\overset{x}{c} = -1.0261$, $\overset{y}{c} = -0.1506$, $\overset{xy}{c} = -0.2742$, $\overset{x2y2}{c} = 0.0042$, $\Phi(0.5, 0.5) = 1.9765$, $v_x = 2.0225$, $v_y = 0.2823$;
right: $\overset{0}{c} = 0.4952$, $\overset{x}{c} = -0.3191$, $\overset{y}{c} = -0.3495$, $\overset{xy}{c} = 0.1611$, $\overset{x2y2}{c} = 0.2564$, $\Phi(1.5, 0.5) = 0.4977$, $v_x = 0.6156$, $v_y = 0.6651$;

E. left: $\overset{0}{c} = 3.2861$, $\overset{x}{c} = -1.7101$, $\overset{y}{c} = -0.2511$, $\overset{xy}{c} = -0.4571$, $\overset{x2y2}{c} = 0.0070$, $\Phi(0.5, 0.5) = 3.2941$, $v_x = 3.3709$, $v_y = 0.4704$;
right: $\overset{0}{c} = 0.8253$, $\overset{x}{c} = -0.5318$, $\overset{y}{c} = -0.5826$, $\overset{xy}{c} = 0.2684$, $\overset{x2y2}{c} = 0.4274$, $\Phi(1.5, 0.5) = 0.8296$, $v_x = 1.0260$, $v_y = 1.1086$;

G. left: $\overset{0}{c} = 4.6005$, $\overset{x}{c} = -2.3942$, $\overset{y}{c} = -0.3515$, $\overset{xy}{c} = -0.6399$, $\overset{x2y2}{c} = 0.0097$, $\Phi(0.5, 0.5) = 4.6118$, $v_x = 4.7192$, $v_y = 0.6586$;
right: $\overset{0}{c} = 1.1554$, $\overset{x}{c} = -0.7445$, $\overset{y}{c} = -0.8156$, $\overset{xy}{c} = 0.3758$, $\overset{x2y2}{c} = 0.5983$, $\Phi(1.5, 0.5) = 1.1614$, $v_x = 1.4365$, $v_y = 1.5520$;

I. left: $\overset{0}{c} = 5.9149$, $\overset{x}{c} = -3.0783$, $\overset{y}{c} = -0.4519$, $\overset{xy}{c} = -0.8227$, $\overset{x2y2}{c} = 0.0125$, $\Phi(0.5, 0.5) = 5.9294$, $v_x = 6.0676$, $v_y = 0.8468$;
right: $\overset{0}{c} = 1.4855$, $\overset{x}{c} = -0.9572$, $\overset{y}{c} = -1.0486$, $\overset{xy}{c} = 0.4832$, $\overset{x2y2}{c} = 0.7692$, $\Phi(1.5, 0.5) = 1.4932$, $v_x = 1.8469$, $v_y = 1.9954$;

K. left: $\overset{0}{c} = 7.2294$, $\overset{x}{c} = -3.7623$, $\overset{y}{c} = -0.5523$, $\overset{xy}{c} = -1.0055$, $\overset{x2y2}{c} = 0.0153$, $\Phi(0.5, 0.5) = 7.2470$, $v_x = 7.4159$, $v_y = 1.0349$;
right: $\overset{0}{c} = 1.8156$, $\overset{x}{c} = -1.1699$, $\overset{y}{c} = -1.2816$, $\overset{xy}{c} = 0.5906$, $\overset{x2y2}{c} = 0.9402$, $\Phi(1.5, 0.5) = 1.8250$, $v_x = 2.2573$, $v_y = 2.4389$;

Problem 4.8

A. bottom: $\overset{0}{c} = 0.8358$, $\overset{x}{c} = -0.1166$, $\overset{y}{c} = -0.1047$, $\overset{xy}{c} = -0.0593$, $\overset{x2y2}{c} = 0.0829$, $\Phi(0.5, 0.5) = 0.8350$, $v_x = 0.2205$, $v_y = 0.2023$;
top: $\overset{0}{c} = 0.3437$, $\overset{x}{c} = -0.0503$, $\overset{y}{c} = -0.3437$, $\overset{xy}{c} = 0.0953$, $\overset{x2y2}{c} = 0.0001$, $\Phi(0.5, 1.5) = 0.3421$, $v_x = 0.0929$, $v_y = 0.6771$;

C. bottom: $\overset{0}{c} = 2.5074$, $\overset{x}{c} = -0.3497$, $\overset{y}{c} = -0.3140$, $\overset{xy}{c} = -0.1779$, $\overset{x2y2}{c} = 0.2486$, $\Phi(0.5, 0.5) = 2.5049$, $v_x = 0.6616$, $v_y = 0.6068$;
top: $\overset{0}{c} = 1.0310$, $\overset{x}{c} = -0.1509$, $\overset{y}{c} = -1.0310$, $\overset{xy}{c} = 0.2860$, $\overset{x2y2}{c} = 0.0002$, $\Phi(0.5, 1.5) = 1.0262$, $v_x = 0.2787$, $v_y = 2.0314$;

E. bottom: $\overset{0}{c} = 4.1791$, $\overset{x}{c} = -0.5828$, $\overset{y}{c} = -0.5233$, $\overset{xy}{c} = -0.2965$, $\overset{x2y2}{c} = 0.4144$, $\Phi(0.5, 0.5) = 4.1748$, $v_x = 1.1026$, $v_y = 1.0113$;
top: $\overset{0}{c} = 1.7183$, $\overset{x}{c} = -0.2515$, $\overset{y}{c} = -1.7183$, $\overset{xy}{c} = 0.4767$, $\overset{x2y2}{c} = 0.0004$, $\Phi(0.5, 1.5) = 1.7103$, $v_x = 0.4645$, $v_y = 3.3856$;

G. bottom: $\overset{0}{c} = 5.8507$, $\overset{x}{c} = -0.8159$, $\overset{y}{c} = -0.7326$, $\overset{xy}{c} = -0.4151$, $\overset{x2y2}{c} = 0.5801$, $\Phi(0.5, 0.5) = 5.8448$, $v_x = 1.5436$, $v_y = 1.4158$;
top: $\overset{0}{c} = 2.4056$, $\overset{x}{c} = -0.3521$, $\overset{y}{c} = -2.4057$, $\overset{xy}{c} = 0.6673$, $\overset{x2y2}{c} = 0.0005$, $\Phi(0.5, 1.5) = 2.3944$, $v_x = 0.6502$, $v_y = 4.7399$;

I. bottom: $\overset{0}{c} = 7.5223$, $\overset{x}{c} = -1.0490$, $\overset{y}{c} = -0.9420$, $\overset{xy}{c} = -0.5336$, $\overset{x2y2}{c} = 0.7459$, $\Phi(0.5, 0.5) = 7.5147$, $v_x = 1.9847$, $v_y = 1.8203$;
top: $\overset{0}{c} = 3.0929$, $\overset{x}{c} = -0.4528$, $\overset{y}{c} = -3.0930$, $\overset{xy}{c} = 0.8580$, $\overset{x2y2}{c} = 0.0006$, $\Phi(0.5, 1.5) = 3.0786$, $v_x = 0.8360$, $v_y = 6.0941$;

K. bottom: $\overset{0}{c} = 9.1940$, $\overset{x}{c} = -1.2822$, $\overset{y}{c} = -1.1513$, $\overset{xy}{c} = -0.6522$, $\overset{x2y2}{c} = 0.9116$, $\Phi(0.5, 0.5) = 9.1846$, $v_x = 2.4257$, $v_y = 2.2248$;
top: $\overset{0}{c} = 3.7802$, $\overset{x}{c} = -0.5534$, $\overset{y}{c} = -3.7803$, $\overset{xy}{c} = 1.0486$, $\overset{x2y2}{c} = 0.0008$, $\Phi(0.5, 1.5) = 3.7627$, $v_x = 1.0218$, $v_y = 7.4484$;

Problem 4.9

A. $|\varphi|(1) = 0.0506, \arg\varphi(1) = -0.8050,$
$|\varphi|(2) = 0.0358, \arg\varphi(2) = -0.7953,$
$|\varphi|(5) = 0.0227, \arg\varphi(5) = -0.7894,$
$|\varphi|(10) = 0.0160, \arg\varphi(10) = -0.7874$

C. $|\varphi|(1) = 0.0868, \arg\varphi(1) = 1.2540,$
$|\varphi|(2) = 0.0618, \arg\varphi(2) = -2.9089,$
$|\varphi|(5) = 0.0392, \arg\varphi(5) = -2.8917,$
$|\varphi|(10) = 0.0277, \arg\varphi(10) = 1.3030$

E. $|\varphi|(1) = 0.1104, \arg\varphi(1) = 0.3873,$
$|\varphi|(2) = 0.0794, \arg\varphi(2) = 1.6811,$
$|\varphi|(5) = 0.0506, \arg\varphi(5) = -0.8050,$
$|\varphi|(10) = 0.0358, \arg\varphi(10) = -0.7953$

G. $|\varphi|(1) = 0.1287, \arg\varphi(1) = 0.0044,$
$|\varphi|(2) = 0.0934, \arg\varphi(2) = 0.9470,$
$|\varphi|(5) = 0.0598, \arg\varphi(5) = -2.6078,$
$|\varphi|(10) = 0.0424, \arg\varphi(10) = 1.8936$

I. $|\varphi|(1) = 0.1437, \arg\varphi(1) = -0.2152,$
$|\varphi|(2) = 0.1052, \arg\varphi(2) = 0.5336,$
$|\varphi|(5) = 0.0677, \arg\varphi(5) = 2.6707,$

K. $|\varphi|(1) = 0.1564, \arg\varphi(1) = -0.3593,$
$|\varphi|(2) = 0.1154, \arg\varphi(2) = 0.2667,$
$|\varphi|(5) = 0.0746, \arg\varphi(5) = 2.0290,$
$|\varphi|(10) = 0.0530, \arg\varphi(10) = -1.3782$

Problem 4.10

A. $(x,y) = (1,0)$: $\Phi = 0.7652$, $v_x = 0.4401$, $v_y = 0.0000$, $v_r = 0.4401$, $v_\theta = 0.0000$; $(x,y) = (0,1)$: $\Phi = 0.7652$, $v_x = 0.0000$, $v_y = 0.4401$, $v_r = 0.4401$, $v_\theta = 0.0000$; $(x,y) = (1,1)$: $\Phi = 0.5591$, $v_x = 0.3850$, $v_y = 0.3850$, $v_r = 0.5445$, $v_\theta = 0.0000$

C. $(x,y) = (1,0)$: $\Phi = 0.1149$, $v_x = -0.2102$, $v_y = 0.0000$, $v_r = -0.2102$, $v_\theta = 0.0000$; $(x,y) = (0,1)$: $\Phi = -0.1149$, $v_x = 0.0000$, $v_y = 0.2102$, $v_r = 0.2102$, $v_\theta = 0.0000$; $(x,y) = (1,1)$: $\Phi = 0.0000$, $v_x = -0.2109$, $v_y = 0.2109$, $v_r = 0.0000$, $v_\theta = 0.2982$

E. $(x,y) = (1,0)$: $\Phi = 0.0000$, $v_x = 0.0000$, $v_y = -0.2298$, $v_r = 0.0000$, $v_\theta = -0.2298$; $(x,y) = (0,1)$: $\Phi = 0.0000$, $v_x = -0.2298$, $v_y = 0.0000$, $v_r = 0.0000$, $v_\theta = 0.2298$; $(x,y) = (1,1)$: $\Phi = 0.2109$, $v_x = -0.1741$, $v_y = -0.1741$, $v_r = -0.2463$, $v_\theta = 0.0000$

G. $(x,y) = (1,0)$: $\Phi = 0.0883$, $v_x = -0.7812$, $v_y = 0.0000$, $v_r = -0.7812$, $v_\theta = 0.0000$; $(x,y) = (0,1)$: $\Phi = 0.0883$, $v_x = 0.0000$, $v_y = -0.7812$, $v_r = -0.7812$, $v_\theta = 0.0000$; $(x,y) = (1,1)$: $\Phi = 0.3446$, $v_x = -0.3320$, $v_y = -0.3320$, $v_r = -0.4695$, $v_\theta = 0.0000$

I. $(x,y) = (1,0)$: $\Phi = -1.6507$, $v_x = -2.5202$, $v_y = 0.0000$, $v_r = -2.5202$, $v_\theta = 0.0000$; $(x,y) = (0,1)$: $\Phi = 1.6507$, $v_x = 0.0000$, $v_y = 2.5202$, $v_r = 2.5202$, $v_\theta = 0.0000$; $(x,y) = (1,1)$: $\Phi = 0.0000$, $v_x = 1.0086$, $v_y = -1.0086$, $v_r = 0.0000$, $v_\theta = -1.4264$

K. $(x,y) = (1,0)$: $\Phi = 0.0000$, $v_x = 0.0000$, $v_y = 3.3014$, $v_r = 0.0000$, $v_\theta = 3.3014$; $(x,y) = (0,1)$: $\Phi = 0.0000$, $v_x = 3.3014$, $v_y = 0.0000$, $v_r = 0.0000$, $v_\theta = -3.3014$; $(x,y) = (1,1)$: $\Phi = -1.0086$, $v_x = -0.6766$, $v_y = -0.6766$, $v_r = -0.9569$, $v_\theta = 0.0000$

Problem 4.11

A. ${}^{H}\overset{\cos}{c}_0 = -0.705432 - i0.455848$, ${}^{H}\overset{\cos}{c}_1 = -0.335601 - i0.057996$, ${}^{H}\overset{\cos}{c}_2 = 0.198627 - i0.598165$, ${}^{H}\overset{\cos}{c}_3 = 0.087577 + i0.003842$, ${}^{H}\overset{\cos}{c}_4 = -0.000011 + i0.004755$, ${}^{H}\overset{\cos}{c}_5 = -0.000154 - i0.000000$

C. ${}^{H}\overset{\cos}{c}_0 = -0.011201 - i0.105242$, ${}^{H}\overset{\cos}{c}_1 = -0.233112 - i0.027550$, ${}^{H}\overset{\cos}{c}_2 = 0.000010 - i0.004559$, ${}^{H}\overset{\cos}{c}_3 = 0.000029 + i0.000000$, ${}^{H}\overset{\cos}{c}_4 = -0.000000 + i0.000000$, ${}^{H}\overset{\cos}{c}_5 = -0.000000 - i0.000000$

E. ${}^{H}\overset{\cos}{c}_0 = -0.705432 - i0.455848$, ${}^{H}\overset{\cos}{c}_1 = -0.335601 - i0.057996$, ${}^{H}\overset{\cos}{c}_2 = 0.198627 - i0.598165$, ${}^{H}\overset{\cos}{c}_3 = 0.087577 + i0.003842$, ${}^{H}\overset{\cos}{c}_4 = -0.000011 + i0.004755$, ${}^{H}\overset{\cos}{c}_5 = -0.000154 - i0.000000$

G. ${}^{H}\overset{\cos}{c}_0 = -0.441088 + i0.496512$, ${}^{H}\overset{\cos}{c}_1 = 0.960674 - i1.277605$, ${}^{H}\overset{\cos}{c}_2 = 0.299243 - i0.713611$, ${}^{H}\overset{\cos}{c}_3 = 0.049955 + i1.998734$, ${}^{H}\overset{\cos}{c}_4 = -0.721544 - i0.958037$, ${}^{H}\overset{\cos}{c}_5 = -0.143129 + i0.000763$

I. ${}^{H}\overset{\cos}{c}_0 = -0.705432 - i0.455848$, ${}^{H}\overset{\cos}{c}_1 = -0.335601 - i0.057996$, ${}^{H}\overset{\cos}{c}_2 = 0.198627 - i0.598165$, ${}^{H}\overset{\cos}{c}_3 = 0.087577 + i0.003842$, ${}^{H}\overset{\cos}{c}_4 = -0.000011 + i0.004755$, ${}^{H}\overset{\cos}{c}_5 = -0.000154 - i0.000000$

K. ${}^{H}\overset{\cos}{c}_0 = -0.441088 + i0.496512$, ${}^{H}\overset{\cos}{c}_1 = 0.960674 - i1.277605$, ${}^{H}\overset{\cos}{c}_2 = 0.299243 - i0.713611$, ${}^{H}\overset{\cos}{c}_3 = 0.049955 + i1.998734$, ${}^{H}\overset{\cos}{c}_4 = -0.721544 - i0.958037$, ${}^{H}\overset{\cos}{c}_5 = -0.143129 + i0.000763$

Problem 4.12

A. $^{H\cos}c_0 = -0.625608 - i0.241998$, $^{H\cos}c_1 = -0.129671 - i0.490463$, $^{H\cos}c_2 = 0.413643 - i0.329020$, $^{H\cos}c_3 = 0.076650 + i0.037914$, $^{H\cos}c_4 = -0.001349 + i0.004550$, $^{H\cos}c_5 = -0.000150 - i0.000034$

C. $^{H\cos}c_0 = -0.149922 - i0.091385$, $^{H\cos}c_1 = -0.209498 - i0.087225$, $^{H\cos}c_2 = 0.000621 - i0.004514$, $^{H\cos}c_3 = 0.000029 + i0.000003$, $^{H\cos}c_4 = -0.000000 + i0.000000$, $^{H\cos}c_5 = -0.000000 - i0.000000$

E. $^{H\cos}c_0 = -0.625608 - i0.241998$, $^{H\cos}c_1 = -0.129671 - i0.490463$, $^{H\cos}c_2 = 0.413643 - i0.329020$, $^{H\cos}c_3 = 0.076650 + i0.037914$, $^{H\cos}c_4 = -0.001349 + i0.004550$, $^{H\cos}c_5 = -0.000150 - i0.000034$

G. $^{H\cos}c_0 = -0.465504 + i0.249494$, $^{H\cos}c_1 = 0.476755 - i1.148453$, $^{H\cos}c_2 = 0.651222 - i0.337873$, $^{H\cos}c_3 = 0.039131 + i1.458191$, $^{H\cos}c_4 = -0.909563 - i0.408058$, $^{H\cos}c_5 = -0.002375 - i0.630478$

I. $^{H\cos}c_0 = -0.625608 - i0.241998$, $^{H\cos}c_1 = -0.129671 - i0.490463$, $^{H\cos}c_2 = 0.413643 - i0.329020$, $^{H\cos}c_3 = 0.076650 + i0.037914$, $^{H\cos}c_4 = -0.001349 + i0.004550$, $^{H\cos}c_5 = -0.000150 - i0.000034$

K. $^{H\cos}c_0 = -0.465504 + i0.249494$, $^{H\cos}c_1 = 0.476755 - i1.148453$, $^{H\cos}c_2 = 0.651222 - i0.337873$, $^{H\cos}c_3 = 0.039131 + i1.458191$, $^{H\cos}c_4 = -0.909563 - i0.408058$, $^{H\cos}c_5 = -0.002375 - i0.630478$

Problem 4.13

A. $^{H\cos}c_0 = -0.572970 - i0.008506$, $^{H\cos}c_1 = 0.160900 - i0.873406$, $^{H\cos}c_2 = 0.467685 + i0.028501$, $^{H\cos}c_3 = 0.032591 + i0.073571$, $^{H\cos}c_4 = -0.003429 + i0.003218$, $^{H\cos}c_5 = -0.000123 - i0.000092$

C. $^{H\cos}c_0 = -0.328585 - i0.120069$, $^{H\cos}c_1 = -0.135044 - i0.165787$, $^{H\cos}c_2 = 0.001774 - i0.004190$, $^{H\cos}c_3 = 0.000028 + i0.000008$, $^{H\cos}c_4 = -0.000000 + i0.000000$, $^{H\cos}c_5 = -0.000000 - i0.000000$

E. $^{H\cos}c_0 = -0.572970 - i0.008506$, $^{H\cos}c_1 = 0.160900 - i0.873406$, $^{H\cos}c_2 = 0.467685 + i0.028501$, $^{H\cos}c_3 = 0.032591 + i0.073571$, $^{H\cos}c_4 = -0.003429 + i0.003218$, $^{H\cos}c_5 = -0.000123 - i0.000092$

G. $^{H\cos}c_0 = -0.480270 + i0.000755$, $^{H\cos}c_1 = -0.011001 - i1.039434$, $^{H\cos}c_2 = 0.978352 + i0.045707$, $^{H\cos}c_3 = 0.059028 + i0.947448$, $^{H\cos}c_4 = -1.118408 + i0.066563$, $^{H\cos}c_5 = 0.185941 - i1.122361$

I. $^{H\cos}c_0 = -0.572970 - i0.008506$, $^{H\cos}c_1 = 0.160900 - i0.873406$, $^{H\cos}c_2 = 0.467685 + i0.028501$, $^{H\cos}c_3 = 0.032591 + i0.073571$, $^{H\cos}c_4 = -0.003429 + i0.003218$, $^{H\cos}c_5 = -0.000123 - i0.000092$

K. $^{H\cos}c_0 = -0.480270 + i0.000755$, $^{H\cos}c_1 = -0.011001 - i1.039434$, $^{H\cos}c_2 = 0.978352 + i0.045707$, $^{H\cos}c_3 = 0.059028 + i0.947448$, $^{H\cos}c_4 = -1.118408 + i0.066563$, $^{H\cos}c_5 = 0.185941 - i1.122361$

Problem 4.14

A. $^{H\cos}c_0 = -0.106248 - i0.308155$, $^{H\cos}c_1 = 0.056772 - i0.001613$, $^{H\cos}c_2 = 0.000001 + i0.001602$, $^{H\cos}c_3 = -0.000021 + i0.000000$, $^{H\cos}c_4 = -0.000000 - i0.000000$, $^{H\cos}c_5 = 0.000000 + i0.000000$

C. $^{H\cos}c_0 = -0.833503 + i0.372526$, $^{H\cos}c_1 = -0.309442 - i0.049082$, $^{H\cos}c_2 = 0.000019 - i0.006126$, $^{H\cos}c_3 = 0.000074 + i0.000000$, $^{H\cos}c_4 = -0.000000 + i0.000001$, $^{H\cos}c_5 = -0.000000 + i0.000000$

E. $^{H\cos}c_0 = -0.513557 - i0.499816$, $^{H\cos}c_1 = 0.435857 - i0.099984$, $^{H\cos}c_2 = 0.001685 + i0.058026$, $^{H\cos}c_3 = -0.003437 + i0.000006$, $^{H\cos}c_4 = -0.000000 - i0.000114$, $^{H\cos}c_5 = 0.000002 + i0.000000$

G. $^{H\cos}c_0 = -0.489373 + i0.499887$, $^{H\cos}c_1 = 0.471348 - i1.881947$, $^{H\cos}c_2 = 0.603519 + i0.918043$, $^{H\cos}c_3 = -0.340284 + i0.059677$, $^{H\cos}c_4 = -0.001869 - i0.061110$, $^{H\cos}c_5 = 0.006349 - i0.000020$

I. $^{H\cos}c_0 = -0.084841 - i0.278645$, $^{H\cos}c_1 = -0.498030 - i0.132840$, $^{H\cos}c_2 = 0.841846 + i0.987414$, $^{H\cos}c_3 = -0.952025 + i0.693979$, $^{H\cos}c_4 = -0.913296 - i0.996235$, $^{H\cos}c_5 = -0.065655 - i0.002156$

K. $^{H\cos}c_0 = -0.090701 + i0.287175$, $^{H\cos}c_1 = 0.549143 - i0.164314$, $^{H\cos}c_2 = 0.337614 - i0.749365$, $^{H\cos}c_3 = 0.545868 + i1.837673$, $^{H\cos}c_4 = -1.633020 - i0.773164$, $^{H\cos}c_5 = 0.917793 - i0.594018$

Problem 4.15

A. ${}^{\mathcal{J}\cos}c_{0} = 0.737047 - i0.254125$, ${}^{\mathcal{J}\cos}c_{1} = 0.194613 + i6.850446$, ${}^{\mathcal{J}\cos}c_{2} = -29.673012 + i0.023770$, ${}^{\mathcal{J}\cos}c_{3} = -0.001299 - i121.893905$, ${}^{\mathcal{J}\cos}c_{4} = 493.702702 - i0.000041$, ${}^{\mathcal{J}\cos}c_{5} = 0.000002 + i1989.396063$

C. ${}^{\mathcal{J}\cos}c_{0} = 0.380288 + i0.850869$, ${}^{\mathcal{J}\cos}c_{1} = -0.284425 + i1.793197$, ${}^{\mathcal{J}\cos}c_{2} = -0.646112 - i0.001979$, ${}^{\mathcal{J}\cos}c_{3} = 0.000011 - i0.293586$, ${}^{\mathcal{J}\cos}c_{4} = 0.140698 + i0.000000$, ${}^{\mathcal{J}\cos}c_{5} = -0.000000 + i0.068651$

E. ${}^{\mathcal{J}\cos}c_{0} = 0.510230 - i0.524257$, ${}^{\mathcal{J}\cos}c_{1} = 0.579403 + i2.525769$, ${}^{\mathcal{J}\cos}c_{2} = -6.120465 + i0.177724$, ${}^{\mathcal{J}\cos}c_{3} = -0.023303 - i13.559161$, ${}^{\mathcal{J}\cos}c_{4} = 28.395265 - i0.001618$, ${}^{\mathcal{J}\cos}c_{5} = 0.000076 + i58.241001$

G. ${}^{\mathcal{J}\cos}c_{0} = -0.375163 - i0.367272$, ${}^{\mathcal{J}\cos}c_{1} = 2.228964 + i0.558261$, ${}^{\mathcal{J}\cos}c_{2} = -8.006002 + i5.263126$, ${}^{\mathcal{J}\cos}c_{3} = -8.698560 - i49.599825$, ${}^{\mathcal{J}\cos}c_{4} = 262.240385 - i8.020047$, ${}^{\mathcal{J}\cos}c_{5} = 3.951437 + i1239.010094$

I. ${}^{\mathcal{J}\cos}c_{0} = -0.848625 + i0.258387$, ${}^{\mathcal{J}\cos}c_{1} = 0.442947 - i1.660649$, ${}^{\mathcal{J}\cos}c_{2} = 1.610266 - i1.372874$, ${}^{\mathcal{J}\cos}c_{3} = 0.682256 + i0.935943$, ${}^{\mathcal{J}\cos}c_{4} = -2.093106 + i1.918849$, ${}^{\mathcal{J}\cos}c_{5} = -0.018569 + i0.565307$

K. ${}^{\mathcal{J}\cos}c_{0} = -0.874605 - i0.276225$, ${}^{\mathcal{J}\cos}c_{1} = -0.547782 - i1.831080$, ${}^{\mathcal{J}\cos}c_{2} = 1.222057 + i0.550769$, ${}^{\mathcal{J}\cos}c_{3} = -1.807004 + i0.536647$, ${}^{\mathcal{J}\cos}c_{4} = 1.624429 - i3.423955$, ${}^{\mathcal{J}\cos}c_{5} = 5.204372 + i7.902462$

Problem 4.16

A. ${}^{1D}c_{1} = 0.00134468$, ${}^{1D}c_{2} = 0.00001607$, ${}^{K\cos}c_{0} = 0.00000305$, ${}^{K\cos}c_{1} = 0$, ${}^{K\cos}c_{2} = -0.00000001$, ${}^{K\cos}c_{3} = 0$, ${}^{K\cos}c_{4} = 0$, ${}^{K\cos}c_{5} = 0$, ${}^{K\sin}c_{1} = 0.00000306$, ${}^{K\sin}c_{2} = 0$, ${}^{K\sin}c_{3} = 0$, ${}^{K\sin}c_{4} = 0$, ${}^{K\sin}c_{5} = 0$, ${}^{I\cos}c_{0} = 0.00000033$, ${}^{I\cos}c_{1} = 0$, ${}^{I\cos}c_{2} = -0.00023197$, ${}^{I\cos}c_{3} = 0$, ${}^{I\cos}c_{4} = -0.00140461$, ${}^{I\cos}c_{5} = 0$, ${}^{I\sin}c_{1} = 0.00002155$, ${}^{I\sin}c_{2} = 0$, ${}^{I\sin}c_{3} = -0.00038972$, ${}^{I\sin}c_{4} = 0$, ${}^{I\sin}c_{5} = -0.01274800$

C. ${}^{1D}c_{1} = 0.00132785$, ${}^{1D}c_{2} = 0.00004747$, ${}^{K\cos}c_{0} = 0.00000936$, ${}^{K\cos}c_{1} = 0$, ${}^{K\cos}c_{2} = -0.00000004$, ${}^{K\cos}c_{3} = 0$, ${}^{K\cos}c_{4} = 0$, ${}^{K\cos}c_{5} = 0$, ${}^{K\sin}c_{1} = 0.00000941$, ${}^{K\sin}c_{2} = 0$, ${}^{K\sin}c_{3} = 0.00000001$, ${}^{K\sin}c_{4} = 0$, ${}^{K\sin}c_{5} = 0$, ${}^{I\cos}c_{0} = 0.00000097$, ${}^{I\cos}c_{1} = 0$, ${}^{I\cos}c_{2} = -0.00067862$, ${}^{I\cos}c_{3} = 0$, ${}^{I\cos}c_{4} = -0.00446056$, ${}^{I\cos}c_{5} = 0$, ${}^{I\sin}c_{1} = 0.00006454$, ${}^{I\sin}c_{2} = 0$, ${}^{I\sin}c_{3} = -0.00107285$, ${}^{I\sin}c_{4} = 0$, ${}^{I\sin}c_{5} = -0.03574506$

E. ${}^{1D}c_{1} = 0.00131167$, ${}^{1D}c_{2} = 0.00007795$, ${}^{K\cos}c_{0} = 0.00001596$, ${}^{K\cos}c_{1} = 0$, ${}^{K\cos}c_{2} = -0.00000011$, ${}^{K\cos}c_{3} = 0$, ${}^{K\cos}c_{4} = 0$, ${}^{K\cos}c_{5} = 0$, ${}^{K\sin}c_{1} = 0.00001608$, ${}^{K\sin}c_{2} = 0$, ${}^{K\sin}c_{3} = 0.00000001$, ${}^{K\sin}c_{4} = 0$, ${}^{K\sin}c_{5} = 0$, ${}^{I\cos}c_{0} = 0.00000160$, ${}^{I\cos}c_{1} = 0$, ${}^{I\cos}c_{2} = -0.00110352$, ${}^{I\cos}c_{3} = 0$, ${}^{I\cos}c_{4} = -0.00777176$, ${}^{I\cos}c_{5} = 0$, ${}^{I\sin}c_{1} = 0.00010735$, ${}^{I\sin}c_{2} = 0$, ${}^{I\sin}c_{3} = -0.00163884$, ${}^{I\sin}c_{4} = 0$, ${}^{I\sin}c_{5} = -0.05530699$

G. ${}^{1D}c_{1} = 0$, ${}^{1D}c_{2} = 0$, ${}^{K\cos}c_{0} = 0.00000900$, ${}^{K\cos}c_{1} = 0$, ${}^{K\cos}c_{2} = -0.00000048$, ${}^{K\cos}c_{3} = 0$, ${}^{K\cos}c_{4} = 0$, ${}^{K\cos}c_{5} = 0$, ${}^{K\sin}c_{1} = 0.00000947$, ${}^{K\sin}c_{2} = 0$, ${}^{K\sin}c_{3} = -0.00000001$, ${}^{K\sin}c_{4} = 0$, ${}^{K\sin}c_{5} = 0$, ${}^{I\cos}c_{0} = 0$, ${}^{I\cos}c_{1} = 0$, ${}^{I\cos}c_{2} = 0$, ${}^{I\cos}c_{3} = 0$, ${}^{I\cos}c_{4} = 0$, ${}^{I\cos}c_{5} = 0$, ${}^{I\sin}c_{1} = 0$, ${}^{I\sin}c_{2} = 0$, ${}^{I\sin}c_{3} = 0$, ${}^{I\sin}c_{4} = 0$, ${}^{I\sin}c_{5} = 0$

I. ${}^{1D}c_{1} = 0$, ${}^{1D}c_{2} = 0$, ${}^{K\cos}c_{0} = 0.00002700$, ${}^{K\cos}c_{1} = 0$, ${}^{K\cos}c_{2} = -0.00000143$, ${}^{K\cos}c_{3} = 0$, ${}^{K\cos}c_{4} = 0$, ${}^{K\cos}c_{5} = 0$, ${}^{K\sin}c_{1} = 0.00002841$, ${}^{K\sin}c_{2} = 0$, ${}^{K\sin}c_{3} = -0.00000002$, ${}^{K\sin}c_{4} = 0$, ${}^{K\sin}c_{5} = 0$, ${}^{I\cos}c_{0} = 0$, ${}^{I\cos}c_{1} = 0$, ${}^{I\cos}c_{2} = 0$, ${}^{I\cos}c_{3} = 0$, ${}^{I\cos}c_{4} = 0$, ${}^{I\cos}c_{5} = 0$, ${}^{I\sin}c_{1} = 0$, ${}^{I\sin}c_{2} = 0$, ${}^{I\sin}c_{3} = 0$, ${}^{I\sin}c_{4} = 0$, ${}^{I\sin}c_{5} = 0$

K. ${}^{1D}c_{1} = 0$, ${}^{1D}c_{2} = 0$, ${}^{K\cos}c_{0} = 0.00004501$, ${}^{K\cos}c_{1} = 0$, ${}^{K\cos}c_{2} = -0.00000239$, ${}^{K\cos}c_{3} = 0$, ${}^{K\cos}c_{4} = 0$, ${}^{K\cos}c_{5} = 0$, ${}^{K\sin}c_{1} = 0.00004736$, ${}^{K\sin}c_{2} = 0$, ${}^{K\sin}c_{3} = -0.00000004$, ${}^{K\sin}c_{4} = 0$, ${}^{K\sin}c_{5} = 0$, ${}^{I\cos}c_{0} = 0$, ${}^{I\cos}c_{1} = 0$, ${}^{I\cos}c_{2} = 0$, ${}^{I\cos}c_{3} = 0$, ${}^{I\cos}c_{4} = 0$, ${}^{I\cos}c_{5} = 0$, ${}^{I\sin}c_{1} = 0$, ${}^{I\sin}c_{2} = 0$, ${}^{I\sin}c_{3} = 0$, ${}^{I\sin}c_{4} = 0$, ${}^{I\sin}c_{5} = 0$

Problem 4.17

A. $|\varphi|(1) = 0.0020$, $\arg\varphi(1) = -1.5708$, $|\varphi|(2) = 0.0010$, $\arg\varphi(2) = -1.5708$, $|\varphi|(5) = 0.0004$, $\arg\varphi(5) = -1.5708$, $|\varphi|(10) = 0.0002$, $\arg\varphi(10) = -1.5708$

C. $|\varphi|(1) = 0.0060$, $\arg\varphi(1) = 0.5236$, $|\varphi|(2) = 0.0030$, $\arg\varphi(2) = 2.6180$, $|\varphi|(5) = 0.0012$, $\arg\varphi(5) = 2.6180$, $|\varphi|(10) = 0.0006$, $\arg\varphi(10) = 0.5236$

E. $|\varphi|(1) = 0.0101$, $\arg\varphi(1) = -0.3142$, $|\varphi|(2) = 0.0050$, $\arg\varphi(2) = 0.9425$, $|\varphi|(5) = 0.0020$, $\arg\varphi(5) = -1.5708$, $|\varphi|(10) = 0.0010$, $\arg\varphi(10) = -1.5708$

G. $|\varphi|(1) = 0.0141$, $\arg\varphi(1) = -0.6732$, $|\varphi|(2) = 0.0071$, $\arg\varphi(2) = 0.2244$, $|\varphi|(5) = 0.0028$, $\arg\varphi(5) = 2.9172$, $|\varphi|(10) = 0.0014$, $\arg\varphi(10) = 1.1220$

I. $|\varphi|(1) = 0.0181$, $\arg\varphi(1) = -0.8727$, $|\varphi|(2) = 0.0091$, $\arg\varphi(2) = -0.1745$, $|\varphi|(5) = 0.0036$, $\arg\varphi(5) = 1.9199$, $|\varphi|(10) = 0.0018$, $\arg\varphi(10) = -0.8727$

K. $|\varphi|(1) = 0.0222$, $\arg\varphi(1) = -0.9996$, $|\varphi|(2) = 0.0111$, $\arg\varphi(2) = -0.4284$, $|\varphi|(5) = 0.0044$, $\arg\varphi(5) = 1.2852$, $|\varphi|(10) = 0.0022$, $\arg\varphi(10) = -2.1420$

Problem 4.18

A. $\overset{\text{out}}{\overset{\text{cos}}{c}}_{11} = -0.9706$, $\overset{\text{in}}{\overset{\text{cos}}{c}}_{11} = 1.9412$, $\Phi^{+}(-r_0) = 0.0294$, $\Phi^{-}(-r_0) = 2.9412$, $\Phi^{-}(r_0) = -2.9412$, $\Phi^{+}(r_0) = -0.0294$, $v_x(-r_0) = 2.9412$, $v_x(r_0) = 2.9412$

C. $\overset{\text{out}}{\overset{\text{cos}}{c}}_{11} = -0.3333$, $\overset{\text{in}}{\overset{\text{cos}}{c}}_{11} = 0.6667$, $\Phi^{+}(-r_0) = 0.6667$, $\Phi^{-}(-r_0) = 1.6667$, $\Phi^{-}(r_0) = -1.6667$, $\Phi^{+}(r_0) = -0.6667$, $v_x(-r_0) = 1.6667$, $v_x(r_0) = 1.6667$

E. $\overset{\text{out}}{\overset{\text{cos}}{c}}_{11} = -0.0769$, $\overset{\text{in}}{\overset{\text{cos}}{c}}_{11} = 0.1538$, $\Phi^{+}(-r_0) = 0.9231$, $\Phi^{-}(-r_0) = 1.1538$, $\Phi^{-}(r_0) = -1.1538$, $\Phi^{+}(r_0) = -0.9231$, $v_x(-r_0) = 1.1538$, $v_x(r_0) = 1.1538$

G. $\overset{\text{out}}{\overset{\text{cos}}{c}}_{11} = 0.0588$, $\overset{\text{in}}{\overset{\text{cos}}{c}}_{11} = -0.1176$, $\Phi^{+}(-r_0) = 1.0588$, $\Phi^{-}(-r_0) = 0.8824$, $\Phi^{-}(r_0) = -0.8824$, $\Phi^{+}(r_0) = -1.0588$, $v_x(-r_0) = 0.8824$, $v_x(r_0) = 0.8824$

I. $\overset{\text{out}}{\overset{\text{cos}}{c}}_{11} = 0.1429$, $\overset{\text{in}}{\overset{\text{cos}}{c}}_{11} = -0.2857$, $\Phi^{+}(-r_0) = 1.1429$, $\Phi^{-}(-r_0) = 0.7143$, $\Phi^{-}(r_0) = -0.7143$, $\Phi^{+}(r_0) = -1.1429$, $v_x(-r_0) = 0.7143$, $v_x(r_0) = 0.7143$

K. $\overset{\text{out}}{\overset{\text{cos}}{c}}_{11} = 0.2000$, $\overset{\text{in}}{\overset{\text{cos}}{c}}_{11} = -0.4000$, $\Phi^{+}(-r_0) = 1.2000$, $\Phi^{-}(-r_0) = 0.6000$, $\Phi^{-}(r_0) = -0.6000$, $\Phi^{+}(r_0) = -1.2000$, $v_x(-r_0) = 0.6000$, $v_x(r_0) = 0.6000$

Problem 4.19

A. $c_1 = -2893.9307$, $c_2 = 2.9527$, $\Phi^{+}(-L) = 0.0295$, $\Phi^{-}(-L) = 2.9527$, $\Phi^{-}(L) = -2.9527$, $\Phi^{+}(L) = -0.0295$, $v_z(-L) = 2.9527$, $v_z(L) = 2.9527$

C. $c_1 = -355.0217$, $c_2 = 2.9884$, $\Phi^{+}(-L) = 0.0299$, $\Phi^{-}(-L) = 2.9884$, $\Phi^{-}(L) = -2.9884$, $\Phi^{+}(L) = -0.0299$, $v_z(-L) = 2.9884$, $v_z(L) = 2.9884$

E. $c_1 = -40.9525$, $c_2 = 3.1517$, $\Phi^{+}(-L) = 0.0315$, $\Phi^{-}(-L) = 3.1517$, $\Phi^{-}(L) = -3.1517$, $\Phi^{+}(L) = -0.0315$, $v_z(-L) = 3.1517$, $v_z(L) = 3.1517$

G. $c_1 = -10.3567$, $c_2 = 3.5307$, $\Phi^{+}(-L) = 0.0353$, $\Phi^{-}(-L) = 3.5307$, $\Phi^{-}(L) = -3.5307$, $\Phi^{+}(L) = -0.0353$, $v_z(-L) = 3.5307$, $v_z(L) = 3.5307$

I. $c_1 = -3.1951$, $c_2 = 4.5901$, $\Phi^{+}(-L) = 0.0459$, $\Phi^{-}(-L) = 4.5901$, $\Phi^{-}(L) = -4.5901$, $\Phi^{+}(L) = -0.0459$, $v_z(-L) = 4.5900$, $v_z(L) = 4.5901$

K. $c_1 = -1.6367$, $c_2 = 6.3433$, $\Phi^{+}(-L) = 0.0634$, $\Phi^{-}(-L) = 6.3433$, $\Phi^{-}(L) = -6.3433$, $\Phi^{+}(L) = -0.0634$, $v_z(-L) = 6.3432$, $v_z(L) = 6.3433$

Problem 4.20

A. $c_1 = 0.1159$, $c_2 = 1.1590$, $\Phi^{+}(-L) = 0.0012$, $\Phi^{-}(-L) = 0.1159$, $\Phi^{-}(L) = -0.1159$, $\Phi^{+}(L) = -0.0012$, $v_z(-L) = 1.1590$, $v_z(L) = 1.1590$

C. $c_1 = 0.2721$, $c_2 = 1.3214$, $\Phi^{+}(-L) = 0.0026$, $\Phi^{-}(-L) = 0.2643$, $\Phi^{-}(L) = -0.2643$, $\Phi^{+}(L) = -0.0026$, $v_z(-L) = 1.3214$, $v_z(L) = 1.3214$

E. $c_1 = 0.7510$, $c_2 = 1.6350$, $\Phi^{+}(-L) = 0.0065$, $\Phi^{-}(-L) = 0.6540$, $\Phi^{-}(L) = -0.6540$, $\Phi^{+}(L) = -0.0065$, $v_z(-L) = 1.6350$, $v_z(L) = 1.6350$

G. $c_1 = 1.5416$, $c_2 = 1.9083$, $\Phi^{+}(-L) = 0.0115$, $\Phi^{-}(-L) = 1.1450$, $\Phi^{-}(L) = -1.1450$, $\Phi^{+}(L) = -0.0115$, $v_z(-L) = 1.9083$, $v_z(L) = 1.9083$

I. $c_1 = 2.7652$, $c_2 = 2.1289$, $\Phi^{+}(-L) = 0.0170$, $\Phi^{-}(-L) = 1.7031$, $\Phi^{-}(L) = -1.7031$, $\Phi^{+}(L) = -0.0170$, $v_z(-L) = 2.1289$, $v_z(L) = 2.1289$

K. $c_1 = 3.5800$, $c_2 = 2.2199$, $\Phi^{+}(-L) = 0.0200$, $\Phi^{-}(-L) = 1.9979$, $\Phi^{-}(L) = -1.9979$, $\Phi^{+}(L) = -0.0200$, $v_z(-L) = 2.2199$, $v_z(L) = 2.2199$

Problem 5.1

A. $\mathcal{Z} = 0.0000 + \mathrm{i}1.0000$, $v_{\mathfrak{s}} = 0.0000$, $v_{\mathfrak{n}} = 1.4142$

C. $\mathcal{Z} = -0.2000 + \mathrm{i}0.4000$, $v_{\mathfrak{s}} = -0.6325$, $v_{\mathfrak{n}} = 1.2649$

E. $\mathcal{Z} = -0.1538 + \mathrm{i}0.2308$, $v_{\mathfrak{s}} = -0.7845$, $v_{\mathfrak{n}} = 1.1767$

G. $\mathcal{Z} = 0.0000 - \mathrm{i}1.0000$, $v_{\mathfrak{s}} = 0.0000$, $v_{\mathfrak{n}} = -1.4142$

I. $\mathcal{Z} = 0.2000 - \mathrm{i}0.4000$, $v_{\mathfrak{s}} = 0.6325$, $v_{\mathfrak{n}} = -1.2649$

K. $\mathcal{Z} = 0.1538 - \mathrm{i}0.2308$, $v_{\mathfrak{s}} = 0.7845$, $v_{\mathfrak{n}} = -1.1767$

Problem 5.2

A. $\overset{\text{dl}}{\Omega}_0 = -0.0552 - i0.1250, \overset{\text{dl}}{v}_0 = -0.0398 - i0.0398,$
$\overset{\text{dl}}{\Omega}_1 = -0.0132 + i0.0147, \overset{\text{dl}}{v}_1 = 0.0154 - i0.0056,$
$\overset{\text{dl}}{\Omega}_2 = -0.0162 - i0.0410, \overset{\text{dl}}{v}_2 = -0.0135 - i0.0104,$
$\overset{\text{dl}}{\Omega}_3 = -0.0079 + i0.0086, \overset{\text{dl}}{v}_3 = 0.0089 - i0.0036$

C. $\overset{\text{dl}}{\Omega}_0 = 0.0552 - i0.1250, \overset{\text{dl}}{v}_0 = -0.0398 + i0.0398,$
$\overset{\text{dl}}{\Omega}_1 = -0.0132 - i0.0147, \overset{\text{dl}}{v}_1 = -0.0154 - i0.0056,$
$\overset{\text{dl}}{\Omega}_2 = 0.0162 - i0.0410, \overset{\text{dl}}{v}_2 = -0.0135 + i0.0104,$
$\overset{\text{dl}}{\Omega}_3 = -0.0079 - i0.0086, \overset{\text{dl}}{v}_3 = -0.0089 - i0.0036$

E. $\overset{\text{dl}}{\Omega}_0 = 0.0000 - i0.2500, \overset{\text{dl}}{v}_0 = -0.1592 + i0.0000,$
$\overset{\text{dl}}{\Omega}_1 = -0.0683 + i0.0000, \overset{\text{dl}}{v}_1 = 0.0000 - i0.0908,$
$\overset{\text{dl}}{\Omega}_2 = 0.0000 - i0.0683, \overset{\text{dl}}{v}_2 = -0.0225 + i0.0000,$
$\overset{\text{dl}}{\Omega}_3 = -0.0378 + i0.0000, \overset{\text{dl}}{v}_3 = 0.0000 - i0.0458$

G. $\overset{\text{dl}}{\Omega}_0 = -0.1281 + i0.1762, \overset{\text{dl}}{v}_0 = -0.0637 + i0.1273,$
$\overset{\text{dl}}{\Omega}_1 = -0.0140 - i0.0481, \overset{\text{dl}}{v}_1 = 0.0644 - i0.0148,$
$\overset{\text{dl}}{\Omega}_2 = -0.0341 + i0.0622, \overset{\text{dl}}{v}_2 = -0.0356 + i0.0311,$
$\overset{\text{dl}}{\Omega}_3 = -0.0098 - i0.0281, \overset{\text{dl}}{v}_3 = 0.0387 - i0.0045$

I. $\overset{\text{dl}}{\Omega}_0 = 0.1281 + i0.1762, \overset{\text{dl}}{v}_0 = -0.0637 - i0.1273,$
$\overset{\text{dl}}{\Omega}_1 = -0.0140 + i0.0481, \overset{\text{dl}}{v}_1 = -0.0644 - i0.0148,$
$\overset{\text{dl}}{\Omega}_2 = 0.0341 + i0.0622, \overset{\text{dl}}{v}_2 = -0.0356 - i0.0311,$
$\overset{\text{dl}}{\Omega}_3 = -0.0098 + i0.0281, \overset{\text{dl}}{v}_3 = -0.0387 - i0.0045$

K. $\overset{\text{dl}}{\Omega}_0 = 0.0000 + i0.1476, \overset{\text{dl}}{v}_0 = -0.0637 + i0.0000,$
$\overset{\text{dl}}{\Omega}_1 = -0.0231 + i0.0000, \overset{\text{dl}}{v}_1 = 0.0000 + i0.0203,$
$\overset{\text{dl}}{\Omega}_2 = 0.0000 + i0.0463, \overset{\text{dl}}{v}_2 = -0.0174 + i0.0000,$
$\overset{\text{dl}}{\Omega}_3 = -0.0135 + i0.0000, \overset{\text{dl}}{v}_3 = 0.0000 + i0.0115$

Problem 5.3

A. $z = -0.5$: $\Phi = -0.1748, \Psi^+ = -0.5000, \Psi^- = 0.5000, v_s = -0.4244, v_n{}^+ = -0.0000, v_n{}^- = 0.0000;$
$z = 0.5$: $\Phi = 0.1748, \Psi^+ = -0.5000, \Psi^- = 0.5000, v_s = -0.4244, v_n{}^+ = 0.0000, v_n{}^- = -0.0000$

C. $z = -0.5$: $\Phi = 0.1154, \Psi^+ = -0.1250, \Psi^- = 0.1250, v_s = 0.0374, v_n{}^+ = 0.5000, v_n{}^- = -0.5000;$
$z = 0.5$: $\Phi = -0.1154, \Psi^+ = -0.1250, \Psi^- = 0.1250, v_s = 0.0374, v_n{}^+ = -0.5000, v_n{}^- = 0.5000$

E. $z = -0.5$: $\Phi = 0.0819, \Psi^+ = -0.0313, \Psi^- = 0.0313, v_s = 0.2309, v_n{}^+ = 0.2500, v_n{}^- = -0.2500;$
$z = 0.5$: $\Phi = -0.0819, \Psi^+ = -0.0313, \Psi^- = 0.0313, v_s = 0.2309, v_n{}^+ = -0.2500, v_n{}^- = 0.2500$

G. $z = -0.5$: $\Phi = 0.0523, \Psi^+ = -0.0078, \Psi^- = 0.0078, v_s = 0.2033, v_n{}^+ = 0.0938, v_n{}^- = -0.0938;$
$z = 0.5$: $\Phi = -0.0523, \Psi^+ = -0.0078, \Psi^- = 0.0078, v_s = 0.2033, v_n{}^+ = -0.0938, v_n{}^- = 0.0938$

I. $z = -0.5$: $\Phi = 0.0358, \Psi^+ = -0.0020, \Psi^- = 0.0020, v_s = 0.1486, v_n{}^+ = 0.0313, v_n{}^- = -0.0313;$
$z = 0.5$: $\Phi = -0.0358, \Psi^+ = -0.0020, \Psi^- = 0.0020, v_s = 0.1486, v_n{}^+ = -0.0313, v_n{}^- = 0.0313$

K. $z = -0.5$: $\Phi = 0.0266, \Psi^+ = -0.0005, \Psi^- = 0.0005, v_s = 0.1083, v_n{}^+ = 0.0098, v_n{}^- = -0.0098;$
$z = 0.5$: $\Phi = -0.0266, \Psi^+ = -0.0005, \Psi^- = 0.0005, v_s = 0.1083, v_n{}^+ = -0.0098, v_n{}^- = 0.0098$

Problem 5.4

A. $\Phi^+ = -0.500, \Phi^- = -0.500, \Psi^+ = -1.010, \Psi^- = 0.010, \overset{\text{dl}}{c}_0 = 1.0198, \overset{\text{dl}}{c}_2 = -0.4903, \overset{\text{dl}}{c}_4 = -0.1112, \overset{\text{dl}}{c}_6 = -0.0980, \overset{\text{dl}}{c}_8 = 0.0831, \overset{\text{dl}}{c}_{10} = -0.1439$

C. $\Phi^+ = -1.500, \Phi^- = -1.500, \Psi^+ = -3.030, \Psi^- = 0.030, \overset{\text{dl}}{c}_0 = 3.0593, \overset{\text{dl}}{c}_2 = -1.4709, \overset{\text{dl}}{c}_4 = -0.3337, \overset{\text{dl}}{c}_6 = -0.2940, \overset{\text{dl}}{c}_8 = 0.2492, \overset{\text{dl}}{c}_{10} = -0.4316$

E. $\Phi^+ = -2.500, \Phi^- = -2.500, \Psi^+ = -5.049, \Psi^- = 0.049, \overset{\text{dl}}{c}_0 = 5.0988, \overset{\text{dl}}{c}_2 = -2.4515, \overset{\text{dl}}{c}_4 = -0.5561, \overset{\text{dl}}{c}_6 = -0.4900, \overset{\text{dl}}{c}_8 = 0.4153, \overset{\text{dl}}{c}_{10} = -0.7193$

G. $\Phi^+ = -3.500, \Phi^- = -3.500, \Psi^+ = -7.069, \Psi^- = 0.069, \overset{\text{dl}}{c}_0 = 7.1384, \overset{\text{dl}}{c}_2 = -3.4321, \overset{\text{dl}}{c}_4 = -0.7786, \overset{\text{dl}}{c}_6 = -0.6859, \overset{\text{dl}}{c}_8 = 0.5814, \overset{\text{dl}}{c}_{10} = -1.0070$

I. $\Phi^+ = -4.500, \Phi^- = -4.500, \Psi^+ = -9.089, \Psi^- = 0.089, \overset{\text{dl}}{c}_0 = 9.1779, \overset{\text{dl}}{c}_2 = -4.4127, \overset{\text{dl}}{c}_4 = -1.0011, \overset{\text{dl}}{c}_6 = -0.8819, \overset{\text{dl}}{c}_8 = 0.7475, \overset{\text{dl}}{c}_{10} = -1.2947$

K. $\Phi^+ = -5.500, \Phi^- = -5.500, \Psi^+ = -11.109, \Psi^- = 0.109, \overset{\text{dl}}{c}_0 = 11.2174, \overset{\text{dl}}{c}_2 = -5.3932, \overset{\text{dl}}{c}_4 = -1.2235, \overset{\text{dl}}{c}_6 = -1.0779, \overset{\text{dl}}{c}_8 = 0.9136, \overset{\text{dl}}{c}_{10} = -1.5824$

Problem 5.5

A. $\Phi^+ = -0.500$, $\Phi^- = -0.500$, $\Psi^+ = -0.949$, $\Psi^- = -0.051$, $\overset{\text{dl}}{c}_0 = 0.8987$, $\overset{\text{dl}}{c}_2 = -0.4144$, $\overset{\text{dl}}{c}_4 = -0.0932$, $\overset{\text{dl}}{c}_6 = -0.1576$, $\overset{\text{dl}}{c}_8 = 0.2040$, $\overset{\text{dl}}{c}_{10} = -0.2234$

C. $\Phi^+ = -1.500$, $\Phi^- = -1.500$, $\Psi^+ = -2.594$, $\Psi^- = -0.406$, $\overset{\text{dl}}{c}_0 = 2.1879$, $\overset{\text{dl}}{c}_2 = -0.9125$, $\overset{\text{dl}}{c}_4 = -0.2458$, $\overset{\text{dl}}{c}_6 = -0.4196$, $\overset{\text{dl}}{c}_8 = 0.5407$, $\overset{\text{dl}}{c}_{10} = -0.5933$

E. $\Phi^+ = -2.500$, $\Phi^- = -2.500$, $\Psi^+ = -4.031$, $\Psi^- = -0.969$, $\overset{\text{dl}}{c}_0 = 3.0625$, $\overset{\text{dl}}{c}_2 = -1.1617$, $\overset{\text{dl}}{c}_4 = -0.3583$, $\overset{\text{dl}}{c}_6 = -0.6342$, $\overset{\text{dl}}{c}_8 = 0.8175$, $\overset{\text{dl}}{c}_{10} = -0.8977$

G. $\Phi^+ = -3.500$, $\Phi^- = -3.500$, $\Psi^+ = -5.345$, $\Psi^- = -1.655$, $\overset{\text{dl}}{c}_0 = 3.6908$, $\overset{\text{dl}}{c}_2 = -1.2797$, $\overset{\text{dl}}{c}_4 = -0.4383$, $\overset{\text{dl}}{c}_6 = -0.8165$, $\overset{\text{dl}}{c}_8 = 1.0569$, $\overset{\text{dl}}{c}_{10} = -1.1600$

I. $\Phi^+ = -4.500$, $\Phi^- = -4.500$, $\Psi^+ = -6.581$, $\Psi^- = -2.419$, $\overset{\text{dl}}{c}_0 = 4.1615$, $\overset{\text{dl}}{c}_2 = -1.3251$, $\overset{\text{dl}}{c}_4 = -0.4925$, $\overset{\text{dl}}{c}_6 = -0.9749$, $\overset{\text{dl}}{c}_8 = 1.2711$, $\overset{\text{dl}}{c}_{10} = -1.3931$

K. $\Phi^+ = -5.500$, $\Phi^- = -5.500$, $\Psi^+ = -7.763$, $\Psi^- = -3.237$, $\overset{\text{dl}}{c}_0 = 4.5257$, $\overset{\text{dl}}{c}_2 = -1.3291$, $\overset{\text{dl}}{c}_4 = -0.5270$, $\overset{\text{dl}}{c}_6 = -1.1145$, $\overset{\text{dl}}{c}_8 = 1.4672$, $\overset{\text{dl}}{c}_{10} = -1.6042$

Problem 5.6

A. $z = -0.5$: $\Phi^+ = 0.5000$, $\Phi^- = -0.5000$, $\Psi = -0.1748$, $v_{\mathfrak{s}}{}^+ = -0.0000$, $v_{\mathfrak{s}}{}^- = 0.0000$, $v_{\mathfrak{n}} = 0.4244$;
$z = 0.5$: $\Phi^+ = 0.5000$, $\Phi^- = -0.5000$, $\Psi = 0.1748$, $v_{\mathfrak{s}}{}^+ = 0.0000$, $v_{\mathfrak{s}}{}^- = -0.0000$, $v_{\mathfrak{n}} = 0.4244$

C. $z = -0.5$: $\Phi^+ = 0.1250$, $\Phi^- = -0.1250$, $\Psi = 0.1154$, $v_{\mathfrak{s}}{}^+ = 0.5000$, $v_{\mathfrak{s}}{}^- = -0.5000$, $v_{\mathfrak{n}} = -0.0374$;
$z = 0.5$: $\Phi^+ = 0.1250$, $\Phi^- = -0.1250$, $\Psi = -0.1154$, $v_{\mathfrak{s}}{}^+ = -0.5000$, $v_{\mathfrak{s}}{}^- = 0.5000$, $v_{\mathfrak{n}} = -0.0374$

E. $z = -0.5$: $\Phi^+ = 0.0313$, $\Phi^- = -0.0313$, $\Psi = 0.0819$, $v_{\mathfrak{s}}{}^+ = 0.2500$, $v_{\mathfrak{s}}{}^- = -0.2500$, $v_{\mathfrak{n}} = -0.2309$;
$z = 0.5$: $\Phi^+ = 0.0313$, $\Phi^- = -0.0313$, $\Psi = -0.0819$, $v_{\mathfrak{s}}{}^+ = -0.2500$, $v_{\mathfrak{s}}{}^- = 0.2500$, $v_{\mathfrak{n}} = -0.2309$

G. $z = -0.5$: $\Phi^+ = 0.0078$, $\Phi^- = -0.0078$, $\Psi = 0.0523$, $v_{\mathfrak{s}}{}^+ = 0.0938$, $v_{\mathfrak{s}}{}^- = -0.0938$, $v_{\mathfrak{n}} = -0.2033$;
$z = 0.5$: $\Phi^+ = 0.0078$, $\Phi^- = -0.0078$, $\Psi = -0.0523$, $v_{\mathfrak{s}}{}^+ = -0.0938$, $v_{\mathfrak{s}}{}^- = 0.0938$, $v_{\mathfrak{n}} = -0.2033$

I. $z = -0.5$: $\Phi^+ = 0.0020$, $\Phi^- = -0.0020$, $\Psi = 0.0358$, $v_{\mathfrak{s}}{}^+ = 0.0313$, $v_{\mathfrak{s}}{}^- = -0.0313$, $v_{\mathfrak{n}} = -0.1486$;
$z = 0.5$: $\Phi^+ = 0.0020$, $\Phi^- = -0.0020$, $\Psi = -0.0358$, $v_{\mathfrak{s}}{}^+ = -0.0313$, $v_{\mathfrak{s}}{}^- = 0.0313$, $v_{\mathfrak{n}} = -0.1486$

K. $z = -0.5$: $\Phi^+ = 0.0005$, $\Phi^- = -0.0005$, $\Psi = 0.0266$, $v_{\mathfrak{s}}{}^+ = 0.0098$, $v_{\mathfrak{s}}{}^- = -0.0098$, $v_{\mathfrak{n}} = -0.1083$;
$z = 0.5$: $\Phi^+ = 0.0005$, $\Phi^- = -0.0005$, $\Psi = -0.0266$, $v_{\mathfrak{s}}{}^+ = -0.0098$, $v_{\mathfrak{s}}{}^- = 0.0098$, $v_{\mathfrak{n}} = -0.1083$

Problem 5.7

A. $\Phi^+ = 0.010$, $\Phi^- = -1.010$, $\Psi^+ = -0.500$, $\Psi^- = -0.500$, $\overset{\text{dl}}{c}_0 = 1.0198i$, $\overset{\text{dl}}{c}_2 = -0.4903i$, $\overset{\text{dl}}{c}_4 = -0.1112i$, $\overset{\text{dl}}{c}_6 = -0.0980i$, $\overset{\text{dl}}{c}_8 = 0.0831i$, $\overset{\text{dl}}{c}_{10} = -0.1439i$

C. $\Phi^+ = 0.030$, $\Phi^- = -3.030$, $\Psi^+ = -1.500$, $\Psi^- = -1.500$, $\overset{\text{dl}}{c}_0 = 3.0593i$, $\overset{\text{dl}}{c}_2 = -1.4709i$, $\overset{\text{dl}}{c}_4 = -0.3337i$, $\overset{\text{dl}}{c}_6 = -0.2940i$, $\overset{\text{dl}}{c}_8 = 0.2492i$, $\overset{\text{dl}}{c}_{10} = -0.4316i$

E. $\Phi^+ = 0.049$, $\Phi^- = -5.049$, $\Psi^+ = -2.500$, $\Psi^- = -2.500$, $\overset{\text{dl}}{c}_0 = 5.0988i$, $\overset{\text{dl}}{c}_2 = -2.4515i$, $\overset{\text{dl}}{c}_4 = -0.5561i$, $\overset{\text{dl}}{c}_6 = -0.4900i$, $\overset{\text{dl}}{c}_8 = 0.4153i$, $\overset{\text{dl}}{c}_{10} = -0.7193i$

G. $\Phi^+ = 0.069$, $\Phi^- = -7.069$, $\Psi^+ = -3.500$, $\Psi^- = -3.500$, $\overset{\text{dl}}{c}_0 = 7.1384i$, $\overset{\text{dl}}{c}_2 = -3.4321i$, $\overset{\text{dl}}{c}_4 = -0.7786i$, $\overset{\text{dl}}{c}_6 = -0.6859i$, $\overset{\text{dl}}{c}_8 = 0.5814i$, $\overset{\text{dl}}{c}_{10} = -1.0070i$

I. $\Phi^+ = 0.089$, $\Phi^- = -9.089$, $\Psi^+ = -4.500$, $\Psi^- = -4.500$, $\overset{\text{dl}}{c}_0 = 9.1779i$, $\overset{\text{dl}}{c}_2 = -4.4127i$, $\overset{\text{dl}}{c}_4 = -1.0011i$, $\overset{\text{dl}}{c}_6 = -0.8819i$, $\overset{\text{dl}}{c}_8 = 0.7475i$, $\overset{\text{dl}}{c}_{10} = -1.2947i$

K. $\Phi^+ = 0.109$, $\Phi^- = -11.109$, $\Psi^+ = -5.500$, $\Psi^- = -5.500$, $\overset{\text{dl}}{c}_0 = 11.2174i$, $\overset{\text{dl}}{c}_2 = -5.3932i$, $\overset{\text{dl}}{c}_4 = -1.2235i$, $\overset{\text{dl}}{c}_6 = -1.0779i$, $\overset{\text{dl}}{c}_8 = 0.9136i$, $\overset{\text{dl}}{c}_{10} = -1.5824i$

Problem 5.8

A. $\Phi^+ = -0.051$, $\Phi^- = -0.949$, $\Psi^+ = -0.500$, $\Psi^- = -0.500$, $\overset{\text{dl}}{c}_0 = 0.8987i$, $\overset{\text{dl}}{c}_2 = -0.4144i$, $\overset{\text{dl}}{c}_4 = -0.0932i$, $\overset{\text{dl}}{c}_6 = -0.1576i$, $\overset{\text{dl}}{c}_8 = 0.2040i$, $\overset{\text{dl}}{c}_{10} = -0.2234i$

C. $\Phi^+ = -0.406$, $\Phi^- = -2.594$, $\Psi^+ = -1.500$, $\Psi^- = -1.500$, $\overset{\text{dl}}{c}_0 = 2.1879i$, $\overset{\text{dl}}{c}_2 = -0.9125i$, $\overset{\text{dl}}{c}_4 = -0.2458i$, $\overset{\text{dl}}{c}_6 = -0.4196i$, $\overset{\text{dl}}{c}_8 = 0.5407i$, $\overset{\text{dl}}{c}_{10} = -0.5933i$

E. $\Phi^+ = -0.969$, $\Phi^- = -4.031$, $\Psi^+ = -2.500$, $\Psi^- = -2.500$, $\overset{\text{dl}}{c}_0 = 3.0625i$, $\overset{\text{dl}}{c}_2 = -1.1617i$, $\overset{\text{dl}}{c}_4 = -0.3583i$, $\overset{\text{dl}}{c}_6 = -0.6342i$, $\overset{\text{dl}}{c}_8 = 0.8175i$, $\overset{\text{dl}}{c}_{10} = -0.8977i$

G. $\Phi^+ = -1.655$, $\Phi^- = -5.345$, $\Psi^+ = -3.500$, $\Psi^- = -3.500$, $\overset{\text{dl}}{c}_0 = 3.6908i$, $\overset{\text{dl}}{c}_2 = -1.2797i$, $\overset{\text{dl}}{c}_4 = -0.4383i$, $\overset{\text{dl}}{c}_6 = -0.8165i$, $\overset{\text{dl}}{c}_8 = 1.0569i$, $\overset{\text{dl}}{c}_{10} = -1.1600i$

I. $\Phi^+ = -2.419$, $\Phi^- = -6.581$, $\Psi^+ = -4.500$, $\Psi^- = -4.500$, $\overset{\text{dl}}{c}_0 = 4.1615i$, $\overset{\text{dl}}{c}_2 = -1.3251i$, $\overset{\text{dl}}{c}_4 = -0.4925i$, $\overset{\text{dl}}{c}_6 = -0.9749i$, $\overset{\text{dl}}{c}_8 = 1.2711i$, $\overset{\text{dl}}{c}_{10} = -1.3931i$

K. $\Phi^+ = -3.237$, $\Phi^- = -7.763$, $\Psi^+ = -5.500$, $\Psi^- = -5.500$, $\overset{\text{dl}}{c}_0 = 4.5257i$, $\overset{\text{dl}}{c}_2 = -1.3291i$, $\overset{\text{dl}}{c}_4 = -0.5270i$, $\overset{\text{dl}}{c}_6 = -1.1145i$, $\overset{\text{dl}}{c}_8 = 1.4672i$, $\overset{\text{dl}}{c}_{10} = -1.6042i$

Problem 5.9

A. $\overset{\text{sl}}{\Omega}_0 = 0.2626 + i0.6397, \overset{\text{sl}}{v}_0 = 0.0552 - i0.1250,$
$\overset{\text{sl}}{\Omega}_1 = 0.0195 + i0.0420, \overset{\text{sl}}{v}_1 = 0.0132 + i0.0147,$
$\overset{\text{sl}}{\Omega}_2 = 0.0893 + i0.2112, \overset{\text{sl}}{v}_2 = 0.0162 - i0.0410,$
$\overset{\text{sl}}{\Omega}_3 = 0.0115 + i0.0251, \overset{\text{sl}}{v}_3 = 0.0079 + i0.0086$

C. $\overset{\text{sl}}{\Omega}_0 = 0.2626 + i0.3603, \overset{\text{sl}}{v}_0 = -0.0552 - i0.1250,$
$\overset{\text{sl}}{\Omega}_1 = -0.0195 + i0.0420, \overset{\text{sl}}{v}_1 = 0.0132 - i0.0147,$
$\overset{\text{sl}}{\Omega}_2 = 0.0893 + i0.1221, \overset{\text{sl}}{v}_2 = -0.0162 - i0.0410,$
$\overset{\text{sl}}{\Omega}_3 = -0.0115 + i0.0251, \overset{\text{sl}}{v}_3 = 0.0079 - i0.0086$

E. $\overset{\text{sl}}{\Omega}_0 = 0.0420 + i0.5000, \overset{\text{sl}}{v}_0 = 0.0000 - i0.2500,$
$\overset{\text{sl}}{\Omega}_1 = 0.0000 + i0.0908, \overset{\text{sl}}{v}_1 = 0.0683 + i0.0000,$
$\overset{\text{sl}}{\Omega}_2 = 0.0242 + i0.1667, \overset{\text{sl}}{v}_2 = 0.0000 - i0.0683,$
$\overset{\text{sl}}{\Omega}_3 = 0.0000 + i0.0531, \overset{\text{sl}}{v}_3 = 0.0378 + i0.0000$

G. $\overset{\text{sl}}{\Omega}_0 = 0.1140 - i0.7243, \overset{\text{sl}}{v}_0 = 0.1281 + i0.1762,$
$\overset{\text{sl}}{\Omega}_1 = 0.0470 - i0.0570, \overset{\text{sl}}{v}_1 = 0.0140 - i0.0481,$
$\overset{\text{sl}}{\Omega}_2 = 0.0394 - i0.2348, \overset{\text{sl}}{v}_2 = 0.0341 + i0.0622,$
$\overset{\text{sl}}{\Omega}_3 = 0.0275 - i0.0346, \overset{\text{sl}}{v}_3 = 0.0098 - i0.0281$

I. $\overset{\text{sl}}{\Omega}_0 = 0.1140 - i0.2757, \overset{\text{sl}}{v}_0 = -0.1281 + i0.1762,$
$\overset{\text{sl}}{\Omega}_1 = -0.0470 - i0.0570, \overset{\text{sl}}{v}_1 = 0.0140 + i0.0481,$
$\overset{\text{sl}}{\Omega}_2 = 0.0394 - i0.0986, \overset{\text{sl}}{v}_2 = -0.0341 + i0.0622,$
$\overset{\text{sl}}{\Omega}_3 = -0.0275 - i0.0346, \overset{\text{sl}}{v}_3 = 0.0098 + i0.0281$

K. $\overset{\text{sl}}{\Omega}_0 = 0.2330 - i0.5000, \overset{\text{sl}}{v}_0 = 0.0000 + i0.1476,$
$\overset{\text{sl}}{\Omega}_1 = 0.0000 - i0.0506, \overset{\text{sl}}{v}_1 = 0.0231 + i0.0000,$
$\overset{\text{sl}}{\Omega}_2 = 0.0809 - i0.1667, \overset{\text{sl}}{v}_2 = 0.0000 + i0.0463,$
$\overset{\text{sl}}{\Omega}_3 = 0.0000 - i0.0301, \overset{\text{sl}}{v}_3 = 0.0135 + i0.0000$

Problem 5.10

A. $z = -0.5$: $\Phi = -0.2767, \Psi^+ = 0.7500, \Psi^- = -0.7500, v_s = 0.1748, v_n{}^+ = -0.5000, v_n{}^- = 0.5000;$
$z = 0.5$: $\Phi = -0.2767, \Psi^+ = 0.2500, \Psi^- = -0.2500, v_s = -0.1748, v_n{}^+ = -0.5000, v_n{}^- = 0.5000$

C. $z = -0.5$: $\Phi = -0.0699, \Psi^+ = 0.1875, \Psi^- = -0.1875, v_s = -0.1154, v_n{}^+ = -0.1250, v_n{}^- = 0.1250;$
$z = 0.5$: $\Phi = -0.0699, \Psi^+ = 0.1458, \Psi^- = -0.1458, v_s = 0.1154, v_n{}^+ = -0.1250, v_n{}^- = 0.1250$

E. $z = -0.5$: $\Phi = -0.0301, \Psi^+ = 0.1031, \Psi^- = -0.1031, v_s = -0.0819, v_n{}^+ = -0.0313, v_n{}^- = 0.0313;$
$z = 0.5$: $\Phi = -0.0301, \Psi^+ = 0.0969, \Psi^- = -0.0969, v_s = 0.0819, v_n{}^+ = -0.0313, v_n{}^- = 0.0313$

G. $z = -0.5$: $\Phi = -0.0168, \Psi^+ = 0.0720, \Psi^- = -0.0720, v_s = -0.0523, v_n{}^+ = -0.0078, v_n{}^- = 0.0078;$
$z = 0.5$: $\Phi = -0.0168, \Psi^+ = 0.0709, \Psi^- = -0.0709, v_s = 0.0523, v_n{}^+ = -0.0078, v_n{}^- = 0.0078$

I. $z = -0.5$: $\Phi = -0.0110, \Psi^+ = 0.0557, \Psi^- = -0.0557, v_s = -0.0358, v_n{}^+ = -0.0020, v_n{}^- = 0.0020;$
$z = 0.5$: $\Phi = -0.0110, \Psi^+ = 0.0554, \Psi^- = -0.0554, v_s = 0.0358, v_n{}^+ = -0.0020, v_n{}^- = 0.0020$

K. $z = -0.5$: $\Phi = -0.0080, \Psi^+ = 0.0455, \Psi^- = -0.0455, v_s = -0.0266, v_n{}^+ = -0.0005, v_n{}^- = 0.0005;$
$z = 0.5$: $\Phi = -0.0080, \Psi^+ = 0.0454, \Psi^- = -0.0454, v_s = 0.0266, v_n{}^+ = -0.0005, v_n{}^- = 0.0005$

Problem 5.11

A. $\Phi^+ = 1.00, \Phi^- = 1.00, \Psi^+ = -4.32, \Psi^- = 3.32, v_s{}^+ = -0.004, v_s{}^- = -0.004, v_n{}^+ = 2.328, v_n{}^- = -3.742, \overset{\text{sl}}{c}_0 = -4.3, \overset{\text{sl}}{c}_1 = -1.1, \overset{\text{sl}}{c}_2 = -3.1, \overset{\text{sl}}{c}_3 = 1.2, \overset{\text{sl}}{c}_4 = 12.2, \overset{\text{sl}}{c}_5 = -11.7, \overset{\text{sl}}{c}_6 = -73.6, \overset{\text{sl}}{c}_7 = 26.0, \overset{\text{sl}}{c}_8 = 145.3, \overset{\text{sl}}{c}_9 = -20.4, \overset{\text{sl}}{c}_{10} = -103.3$

C. $\Phi^+ = 1.00, \Phi^- = 1.00, \Psi^+ = -8.52, \Psi^- = 5.52, v_s{}^+ = -0.004, v_s{}^- = -0.004, v_n{}^+ = 0.979, v_n{}^- = -2.393, \overset{\text{sl}}{c}_0 = -7.2, \overset{\text{sl}}{c}_1 = -3.2, \overset{\text{sl}}{c}_2 = -5.2, \overset{\text{sl}}{c}_3 = 3.5, \overset{\text{sl}}{c}_4 = 20.4, \overset{\text{sl}}{c}_5 = -35.0, \overset{\text{sl}}{c}_6 = -122.7, \overset{\text{sl}}{c}_7 = 77.9, \overset{\text{sl}}{c}_8 = 242.1, \overset{\text{sl}}{c}_9 = -61.3, \overset{\text{sl}}{c}_{10} = -172.2$

E. $\Phi^+ = 1.00, \Phi^- = 1.00, \Psi^+ = -12.72, \Psi^- = 7.72, v_s{}^+ = -0.004, v_s{}^- = -0.004, v_n{}^+ = 0.709, v_n{}^- = -2.123, \overset{\text{sl}}{c}_0 = -10.0, \overset{\text{sl}}{c}_1 = -5.4, \overset{\text{sl}}{c}_2 = -7.3, \overset{\text{sl}}{c}_3 = 5.9, \overset{\text{sl}}{c}_4 = 28.5, \overset{\text{sl}}{c}_5 = -58.3, \overset{\text{sl}}{c}_6 = -171.7, \overset{\text{sl}}{c}_7 = 129.9, \overset{\text{sl}}{c}_8 = 339.0, \overset{\text{sl}}{c}_9 = -102.2, \overset{\text{sl}}{c}_{10} = -241.1$

G. $\Phi^+ = 1.00, \Phi^- = 1.00, \Psi^+ = -16.93, \Psi^- = 9.93, v_s{}^+ = -0.004, v_s{}^- = -0.004, v_n{}^+ = 0.593, v_n{}^- = -2.008, \overset{\text{sl}}{c}_0 = -12.9, \overset{\text{sl}}{c}_1 = -7.6, \overset{\text{sl}}{c}_2 = -9.3, \overset{\text{sl}}{c}_3 = 8.3, \overset{\text{sl}}{c}_4 = 36.6, \overset{\text{sl}}{c}_5 = -81.6, \overset{\text{sl}}{c}_6 = -220.8, \overset{\text{sl}}{c}_7 = 181.9, \overset{\text{sl}}{c}_8 = 435.8, \overset{\text{sl}}{c}_9 = -143.1, \overset{\text{sl}}{c}_{10} = -310.0$

I. $\Phi^+ = 1.00, \Phi^- = 1.00, \Psi^+ = -21.13, \Psi^- = 12.13, v_s{}^+ = -0.004, v_s{}^- = -0.004, v_n{}^+ = 0.529, v_n{}^- = -1.943, \overset{\text{sl}}{c}_0 = -15.7, \overset{\text{sl}}{c}_1 = -9.7, \overset{\text{sl}}{c}_2 = -11.4, \overset{\text{sl}}{c}_3 = 10.6, \overset{\text{sl}}{c}_4 = 44.8, \overset{\text{sl}}{c}_5 = -105.0, \overset{\text{sl}}{c}_6 = -269.9, \overset{\text{sl}}{c}_7 = 233.8, \overset{\text{sl}}{c}_8 = 532.6, \overset{\text{sl}}{c}_9 = -183.9, \overset{\text{sl}}{c}_{10} = -378.9$

K. $\Phi^+ = 1.00, \Phi^- = 1.00, \Psi^+ = -25.33, \Psi^- = 14.33, v_s{}^+ = -0.004, v_s{}^- = -0.004, v_n{}^+ = 0.488, v_n{}^- = -1.903, \overset{\text{sl}}{c}_0 = -18.6, \overset{\text{sl}}{c}_1 = -11.9, \overset{\text{sl}}{c}_2 = -13.5, \overset{\text{sl}}{c}_3 = 13.0, \overset{\text{sl}}{c}_4 = 52.9, \overset{\text{sl}}{c}_5 = -128.3, \overset{\text{sl}}{c}_6 = -318.9, \overset{\text{sl}}{c}_7 = 285.8, \overset{\text{sl}}{c}_8 = 629.5, \overset{\text{sl}}{c}_9 = -224.8, \overset{\text{sl}}{c}_{10} = -447.8$

Problem 5.12

A. $\Phi^+ = -0.22$, $\Phi^- = -0.22$, $\Psi^+ = -1.02$, $\Psi^- = 0.02$, $v_{\mathfrak{s}}{}^+ = 0.574$, $v_{\mathfrak{s}}{}^- = 0.574$, $v_{\mathrm{n}}{}^+ = -0.096$, $v_{\mathrm{n}}{}^- = -1.318$, $\overset{\text{sl}}{c}_0 = -0.864$, $\overset{\text{sl}}{c}_1 = -0.287$, $\overset{\text{sl}}{c}_2 = -0.087$, $\overset{\text{sl}}{c}_3 = -0.019$, $\overset{\text{sl}}{c}_4 = -0.017$, $\overset{\text{sl}}{c}_5 = -0.009$, $\overset{\text{sl}}{c}_6 = -0.024$, $\overset{\text{sl}}{c}_7 = 0.009$, $\overset{\text{sl}}{c}_8 = 0.030$, $\overset{\text{sl}}{c}_9 = -0.011$, $\overset{\text{sl}}{c}_{10} = -0.030$

C. $\Phi^+ = -0.47$, $\Phi^- = -0.47$, $\Psi^+ = -3.75$, $\Psi^- = 0.75$, $v_{\mathfrak{s}}{}^+ = 0.409$, $v_{\mathfrak{s}}{}^- = 0.409$, $v_{\mathrm{n}}{}^+ = 0.030$, $v_{\mathrm{n}}{}^- = -1.444$, $\overset{\text{sl}}{c}_0 = -3.126$, $\overset{\text{sl}}{c}_1 = -1.841$, $\overset{\text{sl}}{c}_2 = -0.760$, $\overset{\text{sl}}{c}_3 = -0.313$, $\overset{\text{sl}}{c}_4 = -0.210$, $\overset{\text{sl}}{c}_5 = -0.211$, $\overset{\text{sl}}{c}_6 = -0.332$, $\overset{\text{sl}}{c}_7 = 0.208$, $\overset{\text{sl}}{c}_8 = 0.429$, $\overset{\text{sl}}{c}_9 = -0.256$, $\overset{\text{sl}}{c}_{10} = -0.430$

E. $\Phi^+ = -0.61$, $\Phi^- = -0.61$, $\Psi^+ = -6.96$, $\Psi^- = 1.96$, $v_{\mathfrak{s}}{}^+ = 0.312$, $v_{\mathfrak{s}}{}^- = 0.312$, $v_{\mathrm{n}}{}^+ = 0.097$, $v_{\mathrm{n}}{}^- = -1.511$, $\overset{\text{sl}}{c}_0 = -5.687$, $\overset{\text{sl}}{c}_1 = -3.909$, $\overset{\text{sl}}{c}_2 = -1.910$, $\overset{\text{sl}}{c}_3 = -0.944$, $\overset{\text{sl}}{c}_4 = -0.634$, $\overset{\text{sl}}{c}_5 = -0.835$, $\overset{\text{sl}}{c}_6 = -1.204$, $\overset{\text{sl}}{c}_7 = 0.888$, $\overset{\text{sl}}{c}_8 = 1.594$, $\overset{\text{sl}}{c}_9 = -1.076$, $\overset{\text{sl}}{c}_{10} = -1.590$

G. $\Phi^+ = -0.69$, $\Phi^- = -0.69$, $\Psi^+ = -10.44$, $\Psi^- = 3.44$, $v_{\mathfrak{s}}{}^+ = 0.250$, $v_{\mathfrak{s}}{}^- = 0.250$, $v_{\mathrm{n}}{}^+ = 0.138$, $v_{\mathrm{n}}{}^- = -1.552$, $\overset{\text{sl}}{c}_0 = -8.367$, $\overset{\text{sl}}{c}_1 = -6.138$, $\overset{\text{sl}}{c}_2 = -3.330$, $\overset{\text{sl}}{c}_3 = -1.783$, $\overset{\text{sl}}{c}_4 = -1.218$, $\overset{\text{sl}}{c}_5 = -1.985$, $\overset{\text{sl}}{c}_6 = -2.824$, $\overset{\text{sl}}{c}_7 = 2.285$, $\overset{\text{sl}}{c}_8 = 3.864$, $\overset{\text{sl}}{c}_9 = -2.715$, $\overset{\text{sl}}{c}_{10} = -3.819$

I. $\Phi^+ = -0.74$, $\Phi^- = -0.74$, $\Psi^+ = -14.08$, $\Psi^- = 5.08$, $v_{\mathfrak{s}}{}^+ = 0.207$, $v_{\mathfrak{s}}{}^- = 0.207$, $v_{\mathrm{n}}{}^+ = 0.165$, $v_{\mathrm{n}}{}^- = -1.579$, $\overset{\text{sl}}{c}_0 = -11.103$, $\overset{\text{sl}}{c}_1 = -8.410$, $\overset{\text{sl}}{c}_2 = -4.896$, $\overset{\text{sl}}{c}_3 = -2.716$, $\overset{\text{sl}}{c}_4 = -1.868$, $\overset{\text{sl}}{c}_5 = -3.705$, $\overset{\text{sl}}{c}_6 = -5.315$, $\overset{\text{sl}}{c}_7 = 4.600$, $\overset{\text{sl}}{c}_8 = 7.539$, $\overset{\text{sl}}{c}_9 = -5.348$, $\overset{\text{sl}}{c}_{10} = -7.363$

K. $\Phi^+ = -0.78$, $\Phi^- = -0.78$, $\Psi^+ = -17.83$, $\Psi^- = 6.83$, $v_{\mathfrak{s}}{}^+ = 0.176$, $v_{\mathfrak{s}}{}^- = 0.176$, $v_{\mathrm{n}}{}^+ = 0.184$, $v_{\mathrm{n}}{}^- = -1.599$, $\overset{\text{sl}}{c}_0 = -13.869$, $\overset{\text{sl}}{c}_1 = -10.682$, $\overset{\text{sl}}{c}_2 = -6.538$, $\overset{\text{sl}}{c}_3 = -3.666$, $\overset{\text{sl}}{c}_4 = -2.502$, $\overset{\text{sl}}{c}_5 = -6.017$, $\overset{\text{sl}}{c}_6 = -8.764$, $\overset{\text{sl}}{c}_7 = 7.995$, $\overset{\text{sl}}{c}_8 = 12.869$, $\overset{\text{sl}}{c}_9 = -9.094$, $\overset{\text{sl}}{c}_{10} = -12.411$

Problem 5.13

A. $v_0 = 0.2565 + 0.0065i$, $\Phi_0 = 1.2430$, $\overset{\text{sl}}{c}_0 = 0.7$, $\overset{\text{sl}}{c}_1 = -0.6$, $\overset{\text{sl}}{c}_2 = 0.5$, $\overset{\text{sl}}{c}_3 = 0.6$, $\overset{\text{sl}}{c}_4 = -2.0$, $\overset{\text{sl}}{c}_5 = -6.1$, $\overset{\text{sl}}{c}_6 = 11.9$, $\overset{\text{sl}}{c}_7 = 13.7$, $\overset{\text{sl}}{c}_8 = -23.5$, $\overset{\text{sl}}{c}_9 = -10.7$, $\overset{\text{sl}}{c}_{10} = 16.7$

C. $v_0 = 0.7695 + 0.0195i$, $\Phi_0 = 1.7291$, $\overset{\text{sl}}{c}_0 = 2.1$, $\overset{\text{sl}}{c}_1 = -1.7$, $\overset{\text{sl}}{c}_2 = 1.5$, $\overset{\text{sl}}{c}_3 = 1.9$, $\overset{\text{sl}}{c}_4 = -5.9$, $\overset{\text{sl}}{c}_5 = -18.4$, $\overset{\text{sl}}{c}_6 = 35.8$, $\overset{\text{sl}}{c}_7 = 41.0$, $\overset{\text{sl}}{c}_8 = -70.6$, $\overset{\text{sl}}{c}_9 = -32.2$, $\overset{\text{sl}}{c}_{10} = 50.2$

E. $v_0 = -0.2872 + -0.2872i$, $\Phi_0 = 1.0000$, $\overset{\text{sl}}{c}_0 = 0.0$, $\overset{\text{sl}}{c}_1 = 1.2$, $\overset{\text{sl}}{c}_2 = 0.0$, $\overset{\text{sl}}{c}_3 = -1.4$, $\overset{\text{sl}}{c}_4 = 0.0$, $\overset{\text{sl}}{c}_5 = 13.4$, $\overset{\text{sl}}{c}_6 = 0.0$, $\overset{\text{sl}}{c}_7 = -29.8$, $\overset{\text{sl}}{c}_8 = 0.0$, $\overset{\text{sl}}{c}_9 = 23.5$, $\overset{\text{sl}}{c}_{10} = 0.0$

G. $v_0 = -0.8616 + -0.8616i$, $\Phi_0 = 1.0000$, $\overset{\text{sl}}{c}_0 = 0.0$, $\overset{\text{sl}}{c}_1 = 3.7$, $\overset{\text{sl}}{c}_2 = 0.0$, $\overset{\text{sl}}{c}_3 = -4.1$, $\overset{\text{sl}}{c}_4 = 0.0$, $\overset{\text{sl}}{c}_5 = 40.2$, $\overset{\text{sl}}{c}_6 = 0.0$, $\overset{\text{sl}}{c}_7 = -89.5$, $\overset{\text{sl}}{c}_8 = 0.0$, $\overset{\text{sl}}{c}_9 = 70.4$, $\overset{\text{sl}}{c}_{10} = 0.0$

I. $v_0 = 0.0065 + 0.2565i$, $\Phi_0 = 1.2430$, $\overset{\text{sl}}{c}_0 = 0.7$, $\overset{\text{sl}}{c}_1 = -0.6$, $\overset{\text{sl}}{c}_2 = 0.5$, $\overset{\text{sl}}{c}_3 = 0.6$, $\overset{\text{sl}}{c}_4 = -2.0$, $\overset{\text{sl}}{c}_5 = -6.1$, $\overset{\text{sl}}{c}_6 = 11.9$, $\overset{\text{sl}}{c}_7 = 13.7$, $\overset{\text{sl}}{c}_8 = -23.5$, $\overset{\text{sl}}{c}_9 = -10.7$, $\overset{\text{sl}}{c}_{10} = 16.7$

K. $v_0 = 0.0195 + 0.7695i$, $\Phi_0 = 1.7291$, $\overset{\text{sl}}{c}_0 = 2.1$, $\overset{\text{sl}}{c}_1 = -1.7$, $\overset{\text{sl}}{c}_2 = 1.5$, $\overset{\text{sl}}{c}_3 = 1.9$, $\overset{\text{sl}}{c}_4 = -5.9$, $\overset{\text{sl}}{c}_5 = -18.4$, $\overset{\text{sl}}{c}_6 = 35.8$, $\overset{\text{sl}}{c}_7 = 41.0$, $\overset{\text{sl}}{c}_8 = -70.6$, $\overset{\text{sl}}{c}_9 = -32.2$, $\overset{\text{sl}}{c}_{10} = 50.2$

Problem 5.14

A. $z = -0.5$: $\Phi^+ = -0.7500$, $\Phi^- = 0.7500$, $\Psi = -0.2767$, $v_{\mathfrak{s}}{}^+ = -0.5000$, $v_{\mathfrak{s}}{}^- = 0.5000$, $v_{\mathrm{n}} = -0.1748$;
$z = 0.5$: $\Phi^+ = -0.2500$, $\Phi^- = 0.2500$, $\Psi = -0.2767$, $v_{\mathfrak{s}}{}^+ = -0.5000$, $v_{\mathfrak{s}}{}^- = 0.5000$, $v_{\mathrm{n}} = 0.1748$

C. $z = -0.5$: $\Phi^+ = -0.1875$, $\Phi^- = 0.1875$, $\Psi = -0.0699$, $v_{\mathfrak{s}}{}^+ = -0.1250$, $v_{\mathfrak{s}}{}^- = 0.1250$, $v_{\mathrm{n}} = 0.1154$;
$z = 0.5$: $\Phi^+ = -0.1458$, $\Phi^- = 0.1458$, $\Psi = -0.0699$, $v_{\mathfrak{s}}{}^+ = -0.1250$, $v_{\mathfrak{s}}{}^- = 0.1250$, $v_{\mathrm{n}} = -0.1154$

E. $z = -0.5$: $\Phi^+ = -0.1031$, $\Phi^- = 0.1031$, $\Psi = -0.0301$, $v_{\mathfrak{s}}{}^+ = -0.0313$, $v_{\mathfrak{s}}{}^- = 0.0313$, $v_{\mathrm{n}} = 0.0819$;
$z = 0.5$: $\Phi^+ = -0.0969$, $\Phi^- = 0.0969$, $\Psi = -0.0301$, $v_{\mathfrak{s}}{}^+ = -0.0313$, $v_{\mathfrak{s}}{}^- = 0.0313$, $v_{\mathrm{n}} = -0.0819$

G. $z = -0.5$: $\Phi^+ = -0.0720$, $\Phi^- = 0.0720$, $\Psi = -0.0168$, $v_{\mathfrak{s}}{}^+ = -0.0078$, $v_{\mathfrak{s}}{}^- = 0.0078$, $v_{\mathrm{n}} = 0.0523$;
$z = 0.5$: $\Phi^+ = -0.0709$, $\Phi^- = 0.0709$, $\Psi = -0.0168$, $v_{\mathfrak{s}}{}^+ = -0.0078$, $v_{\mathfrak{s}}{}^- = 0.0078$, $v_{\mathrm{n}} = -0.0523$

I. $z = -0.5$: $\Phi^+ = -0.0557$, $\Phi^- = 0.0557$, $\Psi = -0.0110$, $v_{\mathfrak{s}}{}^+ = -0.0020$, $v_{\mathfrak{s}}{}^- = 0.0020$, $v_{\mathrm{n}} = 0.0358$;
$z = 0.5$: $\Phi^+ = -0.0554$, $\Phi^- = 0.0554$, $\Psi = -0.0110$, $v_{\mathfrak{s}}{}^+ = -0.0020$, $v_{\mathfrak{s}}{}^- = 0.0020$, $v_{\mathrm{n}} = -0.0358$

K. $z = -0.5$: $\Phi^+ = -0.0455$, $\Phi^- = 0.0455$, $\Psi = -0.0080$, $v_{\mathfrak{s}}{}^+ = -0.0005$, $v_{\mathfrak{s}}{}^- = 0.0005$, $v_{\mathrm{n}} = 0.0266$;
$z = 0.5$: $\Phi^+ = -0.0454$, $\Phi^- = 0.0454$, $\Psi = -0.0080$, $v_{\mathfrak{s}}{}^+ = -0.0005$, $v_{\mathfrak{s}}{}^- = 0.0005$, $v_{\mathrm{n}} = -0.0266$

Problem 5.15

A. $\overset{\text{sl}}{c}_0 = -1.057\text{i}, \overset{\text{sl}}{c}_1 = 1.190\text{i}, \overset{\text{sl}}{c}_2 = -0.767\text{i}, \overset{\text{sl}}{c}_3 = -1.300\text{i}, \overset{\text{sl}}{c}_4 = 3.019\text{i}, \overset{\text{sl}}{c}_5 = 12.823\text{i}, \overset{\text{sl}}{c}_6 = -18.167\text{i}, \overset{\text{sl}}{c}_7 = -28.573\text{i}, \overset{\text{sl}}{c}_8 = 35.840\text{i}, \overset{\text{sl}}{c}_9 = 22.475\text{i}, \overset{\text{sl}}{c}_{10} = -25.481\text{i}$

C. $\overset{\text{sl}}{c}_0 = -1.249\text{i}, \overset{\text{sl}}{c}_1 = 1.406\text{i}, \overset{\text{sl}}{c}_2 = -0.907\text{i}, \overset{\text{sl}}{c}_3 = -1.536\text{i}, \overset{\text{sl}}{c}_4 = 3.567\text{i}, \overset{\text{sl}}{c}_5 = 15.154\text{i}, \overset{\text{sl}}{c}_6 = -21.471\text{i}, \overset{\text{sl}}{c}_7 = -33.768\text{i}, \overset{\text{sl}}{c}_8 = 42.356\text{i}, \overset{\text{sl}}{c}_9 = 26.561\text{i}, \overset{\text{sl}}{c}_{10} = -30.114\text{i}$

E. $\overset{\text{sl}}{c}_0 = -1.442\text{i}, \overset{\text{sl}}{c}_1 = 1.623\text{i}, \overset{\text{sl}}{c}_2 = -1.046\text{i}, \overset{\text{sl}}{c}_3 = -1.773\text{i}, \overset{\text{sl}}{c}_4 = 4.116\text{i}, \overset{\text{sl}}{c}_5 = 17.485\text{i}, \overset{\text{sl}}{c}_6 = -24.774\text{i}, \overset{\text{sl}}{c}_7 = -38.963\text{i}, \overset{\text{sl}}{c}_8 = 48.872\text{i}, \overset{\text{sl}}{c}_9 = 30.648\text{i}, \overset{\text{sl}}{c}_{10} = -34.747\text{i}$

G. $\overset{\text{sl}}{c}_0 = -1.634\text{i}, \overset{\text{sl}}{c}_1 = 1.839\text{i}, \overset{\text{sl}}{c}_2 = -1.186\text{i}, \overset{\text{sl}}{c}_3 = -2.009\text{i}, \overset{\text{sl}}{c}_4 = 4.665\text{i}, \overset{\text{sl}}{c}_5 = 19.817\text{i}, \overset{\text{sl}}{c}_6 = -28.077\text{i}, \overset{\text{sl}}{c}_7 = -44.158\text{i}, \overset{\text{sl}}{c}_8 = 55.389\text{i}, \overset{\text{sl}}{c}_9 = 34.734\text{i}, \overset{\text{sl}}{c}_{10} = -39.380\text{i}$

I. $\overset{\text{sl}}{c}_0 = -1.826\text{i}, \overset{\text{sl}}{c}_1 = 2.056\text{i}, \overset{\text{sl}}{c}_2 = -1.325\text{i}, \overset{\text{sl}}{c}_3 = -2.246\text{i}, \overset{\text{sl}}{c}_4 = 5.214\text{i}, \overset{\text{sl}}{c}_5 = 22.148\text{i}, \overset{\text{sl}}{c}_6 = -31.380\text{i}, \overset{\text{sl}}{c}_7 = -49.353\text{i}, \overset{\text{sl}}{c}_8 = 61.905\text{i}, \overset{\text{sl}}{c}_9 = 38.820\text{i}, \overset{\text{sl}}{c}_{10} = -44.013\text{i}$

K. $\overset{\text{sl}}{c}_0 = -2.018\text{i}, \overset{\text{sl}}{c}_1 = 2.272\text{i}, \overset{\text{sl}}{c}_2 = -1.465\text{i}, \overset{\text{sl}}{c}_3 = -2.482\text{i}, \overset{\text{sl}}{c}_4 = 5.763\text{i}, \overset{\text{sl}}{c}_5 = 24.480\text{i}, \overset{\text{sl}}{c}_6 = -34.683\text{i}, \overset{\text{sl}}{c}_7 = -54.548\text{i}, \overset{\text{sl}}{c}_8 = 68.421\text{i}, \overset{\text{sl}}{c}_9 = 42.907\text{i}, \overset{\text{sl}}{c}_{10} = -48.646\text{i}$

Problem 5.16

A. near-field: $\Omega = 0.003900 + -0.021121\text{i}, 0.007469 + -0.016392\text{i}, 0.003949 + -0.004226\text{i}$,
far-field: $\Omega = 0.003819 + -0.020465\text{i}, 0.007469 + -0.016391\text{i}, 0.003949 + -0.004226\text{i}$,
near-field: $v = -0.018484 + -0.004535\text{i}, -0.010442 + 0.005656\text{i}, -0.000117 + 0.001151\text{i}$,
far-field: $v = -0.009065 + -0.015665\text{i}, -0.010440 + 0.005651\text{i}, -0.000117 + 0.001151\text{i}$

C. near-field: $\Omega = 0.003048 + -0.017788\text{i}, 0.006189 + -0.013888\text{i}, 0.003339 + -0.003578\text{i}$,
far-field: $\Omega = 0.002972 + -0.017168\text{i}, 0.006189 + -0.013888\text{i}, 0.003339 + -0.003578\text{i}$,
near-field: $v = -0.015474 + -0.004423\text{i}, -0.008952 + 0.004566\text{i}, -0.000102 + 0.000973\text{i}$,
far-field: $v = -0.006558 + -0.014925\text{i}, -0.008950 + 0.004561\text{i}, -0.000102 + 0.000973\text{i}$

E. near-field: $\Omega = 0.002479 + -0.015357\text{i}, 0.005277 + -0.012046\text{i}, 0.002892 + -0.003103\text{i}$,
far-field: $\Omega = 0.002408 + -0.014769\text{i}, 0.005277 + -0.012046\text{i}, 0.002891 + -0.003103\text{i}$,
near-field: $v = -0.013280 + -0.004208\text{i}, -0.007830 + 0.003810\text{i}, -0.000090 + 0.000843\text{i}$,
far-field: $v = -0.004817 + -0.014149\text{i}, -0.007828 + 0.003805\text{i}, -0.000090 + 0.000843\text{i}$

G. near-field: $\Omega = 0.002076 + -0.013507\text{i}, 0.004596 + -0.010635\text{i}, 0.002549 + -0.002738\text{i}$,
far-field: $\Omega = 0.002010 + -0.012948\text{i}, 0.004596 + -0.010635\text{i}, 0.002550 + -0.002739\text{i}$,
near-field: $v = -0.011617 + -0.003965\text{i}, -0.006956 + 0.003258\text{i}, -0.000085 + 0.000748\text{i}$,
far-field: $v = -0.003562 + -0.013402\text{i}, -0.006954 + 0.003253\text{i}, -0.000080 + 0.000744\text{i}$

I. near-field: $\Omega = 0.001779 + -0.012052\text{i}, 0.004069 + -0.009519\text{i}, 0.002234 + -0.002487\text{i}$,
far-field: $\Omega = 0.001716 + -0.011519\text{i}, 0.004069 + -0.009519\text{i}, 0.002280 + -0.002452\text{i}$,
near-field: $v = -0.010315 + -0.003725\text{i}, -0.006255 + 0.002839\text{i}, 0.000155 + 0.000658\text{i}$,
far-field: $v = -0.002631 + -0.012706\text{i}, -0.006254 + 0.002835\text{i}, -0.000073 + 0.000666\text{i}$

K. near-field: $\Omega = 0.001552 + -0.010878\text{i}, 0.003649 + -0.008615\text{i}, 0.003730 + -0.003730\text{i}$,
far-field: $\Omega = 0.001492 + -0.010370\text{i}, 0.003649 + -0.008615\text{i}, 0.002063 + -0.002219\text{i}$,
near-field: $v = -0.009270 + -0.003500\text{i}, -0.005682 + 0.002512\text{i}, 0.000583 + -0.004974\text{i}$,
far-field: $v = -0.001925 + -0.012067\text{i}, -0.005681 + 0.002508\text{i}, -0.000066 + 0.000602\text{i}$

Problem 5.17

A. near-field: Ω =0.005663 + 0.031503i, 0.012707 + 0.027918i, 0.046581 + 0.023217i,
far-field: Ω =0.005695 + 0.031543i, 0.012707 + 0.027918i, 0.046581 + 0.023217i,
near-field: v =−0.003900 + −0.021121i, −0.007469 + −0.016392i, −0.003949 + −0.004226i,
far-field: v =−0.003819 + −0.020465i, −0.007469 + −0.016391i, −0.003949 + −0.004226i

C. near-field: Ω =0.004884 + 0.026804i, 0.010786 + 0.023723i, 0.039415 + 0.019655i,
far-field: Ω =0.004909 + 0.026837i, 0.010786 + 0.023723i, 0.039415 + 0.019655i,
near-field: v =−0.003048 + −0.017788i, −0.006189 + −0.013888i, −0.003339 + −0.003578i,
far-field: v =−0.002972 + −0.017168i, −0.006189 + −0.013888i, −0.003339 + −0.003578i

E. near-field: Ω =0.004294 + 0.023327i, 0.009371 + 0.020626i, 0.034160 + 0.017041i,
far-field: Ω =0.004316 + 0.023354i, 0.009371 + 0.020626i, 0.034160 + 0.017041i,
near-field: v =−0.002479 + −0.015357i, −0.005277 + −0.012046i, −0.002892 + −0.003103i,
far-field: v =−0.002408 + −0.014769i, −0.005277 + −0.012046i, −0.002891 + −0.003103i

G. near-field: Ω =0.003833 + 0.020649i, 0.008285 + 0.018245i, 0.030141 + 0.015041i,
far-field: Ω =0.003851 + 0.020671i, 0.008285 + 0.018245i, 0.030141 + 0.015041i,
near-field: v =−0.002076 + −0.013507i, −0.004596 + −0.010635i, −0.002549 + −0.002738i,
far-field: v =−0.002010 + −0.012948i, −0.004596 + −0.010635i, −0.002550 + −0.002739i

I. near-field: Ω =0.003462 + 0.018523i, 0.007425 + 0.016357i, 0.026960 + 0.013440i,
far-field: Ω =0.003477 + 0.018542i, 0.007425 + 0.016357i, 0.026969 + 0.013461i,
near-field: v =−0.001779 + −0.012052i, −0.004069 + −0.009519i, −0.002234 + −0.002487i,
far-field: v =−0.001716 + −0.011519i, −0.004069 + −0.009519i, −0.002280 + −0.002452i

K. near-field: Ω =0.003156 + 0.016794i, 0.006727 + 0.014824i, 0.024868 + 0.012197i,
far-field: Ω =0.003169 + 0.016810i, 0.006727 + 0.014824i, 0.024400 + 0.012181i,
near-field: v =−0.001552 + −0.010878i, −0.003649 + −0.008615i, −0.003730 + −0.003730i,
far-field: v =−0.001492 + −0.010370i, −0.003649 + −0.008615i, −0.002063 + −0.002219i

Problem 5.18

A. $Q = 2.6667$, $\overset{\text{dl}}{c}_0 = -2.0833$, $\overset{\text{dl}}{c}_1 = 1.0000$, $\overset{\text{dl}}{c}_2 = 0.5000$, $\overset{\text{dl}}{c}_3 = 0.3333$, $\overset{\text{dl}}{c}_4 = 0.2500$

C. $Q = 3.0667$, $\overset{\text{dl}}{c}_0 = -2.4500$, $\overset{\text{dl}}{c}_1 = 1.0000$, $\overset{\text{dl}}{c}_2 = 0.5000$, $\overset{\text{dl}}{c}_3 = 0.3333$, $\overset{\text{dl}}{c}_4 = 0.2500$, $\overset{\text{dl}}{c}_5 = 0.2000$, $\overset{\text{dl}}{c}_6 = 0.1667$

E. $Q = 3.3524$, $\overset{\text{dl}}{c}_0 = -2.7179$, $\overset{\text{dl}}{c}_1 = 1.0000$, $\overset{\text{dl}}{c}_2 = 0.5000$, $\overset{\text{dl}}{c}_3 = 0.3333$, $\overset{\text{dl}}{c}_4 = 0.2500$, $\overset{\text{dl}}{c}_5 = 0.2000$, $\overset{\text{dl}}{c}_6 = 0.1667$, $\overset{\text{dl}}{c}_7 = 0.1429$, $\overset{\text{dl}}{c}_8 = 0.1250$

G. $Q = 3.5746$, $\overset{\text{dl}}{c}_0 = -2.9290$, $\overset{\text{dl}}{c}_1 = 1.0000$, $\overset{\text{dl}}{c}_2 = 0.5000$, $\overset{\text{dl}}{c}_3 = 0.3333$, $\overset{\text{dl}}{c}_4 = 0.2500$, $\overset{\text{dl}}{c}_5 = 0.2000$, $\overset{\text{dl}}{c}_6 = 0.1667$, $\overset{\text{dl}}{c}_7 = 0.1429$, $\overset{\text{dl}}{c}_8 = 0.1250$, $\overset{\text{dl}}{c}_9 = 0.1111$, $\overset{\text{dl}}{c}_{10} = 0.1000$

I. $Q = 3.7564$, $\overset{\text{dl}}{c}_0 = -3.1032$, $\overset{\text{dl}}{c}_1 = 1.0000$, $\overset{\text{dl}}{c}_2 = 0.5000$, $\overset{\text{dl}}{c}_3 = 0.3333$, $\overset{\text{dl}}{c}_4 = 0.2500$, $\overset{\text{dl}}{c}_5 = 0.2000$, $\overset{\text{dl}}{c}_6 = 0.1667$, $\overset{\text{dl}}{c}_7 = 0.1429$, $\overset{\text{dl}}{c}_8 = 0.1250$, $\overset{\text{dl}}{c}_9 = 0.1111$, $\overset{\text{dl}}{c}_{10} = 0.1000$, $\overset{\text{dl}}{c}_{11} = 0.0909$, $\overset{\text{dl}}{c}_{12} = 0.0833$

K. $Q = 3.9103$, $\overset{\text{dl}}{c}_0 = -3.2516$, $\overset{\text{dl}}{c}_1 = 1.0000$, $\overset{\text{dl}}{c}_2 = 0.5000$, $\overset{\text{dl}}{c}_3 = 0.3333$, $\overset{\text{dl}}{c}_4 = 0.2500$, $\overset{\text{dl}}{c}_5 = 0.2000$, $\overset{\text{dl}}{c}_6 = 0.1667$, $\overset{\text{dl}}{c}_7 = 0.1429$, $\overset{\text{dl}}{c}_8 = 0.1250$, $\overset{\text{dl}}{c}_9 = 0.1111$, $\overset{\text{dl}}{c}_{10} = 0.1000$, $\overset{\text{dl}}{c}_{11} = 0.0909$, $\overset{\text{dl}}{c}_{12} = 0.0833$, $\overset{\text{dl}}{c}_{13} = 0.0769$, $\overset{\text{dl}}{c}_{14} = 0.0714$

Problem 5.19

A. $\Gamma = 2.6667$, $\overset{\text{dl}}{c}_0 = -2.0833\text{i}$, $\overset{\text{dl}}{c}_1 = 1.0000\text{i}$, $\overset{\text{dl}}{c}_2 = 0.5000\text{i}$, $\overset{\text{dl}}{c}_3 = 0.3333\text{i}$, $\overset{\text{dl}}{c}_4 = 0.2500\text{i}$

C. $\Gamma = 3.0667$, $\overset{\text{dl}}{c}_0 = -2.4500\text{i}$, $\overset{\text{dl}}{c}_1 = 1.0000\text{i}$, $\overset{\text{dl}}{c}_2 = 0.5000\text{i}$, $\overset{\text{dl}}{c}_3 = 0.3333\text{i}$, $\overset{\text{dl}}{c}_4 = 0.2500\text{i}$, $\overset{\text{dl}}{c}_5 = 0.2000\text{i}$, $\overset{\text{dl}}{c}_6 = 0.1667\text{i}$

E. $\Gamma = 3.3524$, $\overset{\text{dl}}{c}_0 = -2.7179\text{i}$, $\overset{\text{dl}}{c}_1 = 1.0000\text{i}$, $\overset{\text{dl}}{c}_2 = 0.5000\text{i}$, $\overset{\text{dl}}{c}_3 = 0.3333\text{i}$, $\overset{\text{dl}}{c}_4 = 0.2500\text{i}$, $\overset{\text{dl}}{c}_5 = 0.2000\text{i}$, $\overset{\text{dl}}{c}_6 = 0.1667\text{i}$, $\overset{\text{dl}}{c}_7 = 0.1429\text{i}$, $\overset{\text{dl}}{c}_8 = 0.1250\text{i}$

G. $\Gamma = 3.5746$, $\overset{\text{dl}}{c}_0 = -2.9290\text{i}$, $\overset{\text{dl}}{c}_1 = 1.0000\text{i}$, $\overset{\text{dl}}{c}_2 = 0.5000\text{i}$, $\overset{\text{dl}}{c}_3 = 0.3333\text{i}$, $\overset{\text{dl}}{c}_4 = 0.2500\text{i}$, $\overset{\text{dl}}{c}_5 = 0.2000\text{i}$, $\overset{\text{dl}}{c}_6 = 0.1667\text{i}$, $\overset{\text{dl}}{c}_7 = 0.1429\text{i}$, $\overset{\text{dl}}{c}_8 = 0.1250\text{i}$, $\overset{\text{dl}}{c}_9 = 0.1111\text{i}$, $\overset{\text{dl}}{c}_{10} = 0.1000\text{i}$

I. $\Gamma = 3.7564$, $\overset{\text{dl}}{c}_0 = -3.1032\text{i}$, $\overset{\text{dl}}{c}_1 = 1.0000\text{i}$, $\overset{\text{dl}}{c}_2 = 0.5000\text{i}$, $\overset{\text{dl}}{c}_3 = 0.3333\text{i}$, $\overset{\text{dl}}{c}_4 = 0.2500\text{i}$, $\overset{\text{dl}}{c}_5 = 0.2000\text{i}$, $\overset{\text{dl}}{c}_6 = 0.1667\text{i}$, $\overset{\text{dl}}{c}_7 = 0.1429\text{i}$, $\overset{\text{dl}}{c}_8 = 0.1250\text{i}$, $\overset{\text{dl}}{c}_9 = 0.1111\text{i}$, $\overset{\text{dl}}{c}_{10} = 0.1000\text{i}$, $\overset{\text{dl}}{c}_{11} = 0.0909\text{i}$, $\overset{\text{dl}}{c}_{12} = 0.0833\text{i}$

K. $\Gamma = 3.9103$, $\overset{\text{dl}}{c}_0 = -3.2516\text{i}$, $\overset{\text{dl}}{c}_1 = 1.0000\text{i}$, $\overset{\text{dl}}{c}_2 = 0.5000\text{i}$, $\overset{\text{dl}}{c}_3 = 0.3333\text{i}$, $\overset{\text{dl}}{c}_4 = 0.2500\text{i}$, $\overset{\text{dl}}{c}_5 = 0.2000\text{i}$, $\overset{\text{dl}}{c}_6 = 0.1667\text{i}$, $\overset{\text{dl}}{c}_7 = 0.1429\text{i}$, $\overset{\text{dl}}{c}_8 = 0.1250\text{i}$, $\overset{\text{dl}}{c}_9 = 0.1111\text{i}$, $\overset{\text{dl}}{c}_{10} = 0.1000\text{i}$, $\overset{\text{dl}}{c}_{11} = 0.0909\text{i}$, $\overset{\text{dl}}{c}_{12} = 0.0833\text{i}$, $\overset{\text{dl}}{c}_{13} = 0.0769\text{i}$, $\overset{\text{dl}}{c}_{14} = 0.0714\text{i}$

Problem 5.20

A. $\Phi^+(1) = -1.352$, $\Phi^-(1) = -2.704$, $\Phi^+(2+\text{i}) = -2.000$, $\Phi^-(2+\text{i}) = -4.000$, $\Phi^+(3) = -2.648$, $\Phi^-(3) = -5.296$

C. $\Phi^+(1) = -1.617$, $\Phi^-(1) = -6.468$, $\Phi^+(2+\text{i}) = -2.000$, $\Phi^-(2+\text{i}) = -8.000$, $\Phi^+(3) = -2.383$, $\Phi^-(3) = -9.532$

E. $\Phi^+(1) = -1.727$, $\Phi^-(1) = -10.365$, $\Phi^+(2+\text{i}) = -2.000$, $\Phi^-(2+\text{i}) = -12.000$, $\Phi^+(3) = -2.273$, $\Phi^-(3) = -13.635$

G. $\Phi^+(1) = -0.612$, $\Phi^-(1) = -0.306$, $\Phi^+(2+\text{i}) = -2.000$, $\Phi^-(2+\text{i}) = -1.000$, $\Phi^+(3) = -3.388$, $\Phi^-(3) = -1.694$

I. $\Phi^+(1) = -0.263$, $\Phi^-(1) = -0.066$, $\Phi^+(2+\text{i}) = -2.000$, $\Phi^-(2+\text{i}) = -0.500$, $\Phi^+(3) = -3.737$, $\Phi^-(3) = -0.934$

K. $\Phi^+(1) = -0.100$, $\Phi^-(1) = -0.017$, $\Phi^+(2+\text{i}) = -2.000$, $\Phi^-(2+\text{i}) = -0.333$, $\Phi^+(3) = -3.900$, $\Phi^-(3) = -0.650$

Problem 5.21

A. $\underset{1}{\overset{\text{dl}}{c}_0} = 1\text{i}$, $\underset{2}{\overset{\text{dl}}{c}_3} = -1\text{i}$, $\underset{3}{\overset{\text{dl}}{c}_0} = -1\text{i}$, $\underset{4}{\overset{\text{dl}}{c}_3} = 1\text{i}$, $\underset{1}{\overset{\text{sl}}{c}_0} = 3$, $\underset{3}{\overset{\text{sl}}{c}_0} = -3$

C. $\underset{1}{\overset{\text{dl}}{c}_0} = 1\text{i}$, $\underset{2}{\overset{\text{dl}}{c}_5} = -1\text{i}$, $\underset{3}{\overset{\text{dl}}{c}_0} = -1\text{i}$, $\underset{4}{\overset{\text{dl}}{c}_5} = 1\text{i}$, $\underset{1}{\overset{\text{sl}}{c}_0} = 5$, $\underset{3}{\overset{\text{sl}}{c}_0} = -5$

E. $\underset{1}{\overset{\text{dl}}{c}_2} = -1\text{i}$, $\underset{2}{\overset{\text{dl}}{c}_0} = -1\text{i}$, $\underset{3}{\overset{\text{dl}}{c}_2} = -1\text{i}$, $\underset{4}{\overset{\text{dl}}{c}_0} = -1\text{i}$, $\underset{2}{\overset{\text{sl}}{c}_0} = -2$, $\underset{4}{\overset{\text{sl}}{c}_0} = -2$

G. $\underset{1}{\overset{\text{dl}}{c}_4} = -1\text{i}$, $\underset{2}{\overset{\text{dl}}{c}_0} = -1\text{i}$, $\underset{3}{\overset{\text{dl}}{c}_4} = -1\text{i}$, $\underset{4}{\overset{\text{dl}}{c}_0} = -1\text{i}$, $\underset{2}{\overset{\text{sl}}{c}_0} = -4$, $\underset{4}{\overset{\text{sl}}{c}_0} = -4$

I. $\underset{1}{\overset{\text{dl}}{c}_0} = -1\text{i}$, $\underset{1}{\overset{\text{dl}}{c}_2} = -1\text{i}$, $\underset{2}{\overset{\text{dl}}{c}_0} = -1\text{i}$, $\underset{2}{\overset{\text{dl}}{c}_2} = -1\text{i}$, $\underset{3}{\overset{\text{dl}}{c}_0} = -1\text{i}$, $\underset{3}{\overset{\text{dl}}{c}_2} = -1\text{i}$, $\underset{4}{\overset{\text{dl}}{c}_0} = -1\text{i}$, $\underset{4}{\overset{\text{dl}}{c}_2} = -1\text{i}$, $\underset{1}{\overset{\text{sl}}{c}_0} = -2$, $\underset{2}{\overset{\text{sl}}{c}_0} = -2$, $\underset{3}{\overset{\text{sl}}{c}_0} = -2$, $\underset{4}{\overset{\text{sl}}{c}_0} = -2$

K. $\underset{1}{\overset{\text{dl}}{c}_0} = -1\text{i}$, $\underset{1}{\overset{\text{dl}}{c}_4} = -1\text{i}$, $\underset{2}{\overset{\text{dl}}{c}_0} = -1\text{i}$, $\underset{2}{\overset{\text{dl}}{c}_4} = -1\text{i}$, $\underset{3}{\overset{\text{dl}}{c}_0} = -1\text{i}$, $\underset{3}{\overset{\text{dl}}{c}_4} = -1\text{i}$, $\underset{4}{\overset{\text{dl}}{c}_0} = -1\text{i}$, $\underset{4}{\overset{\text{dl}}{c}_4} = -1\text{i}$, $\underset{1}{\overset{\text{sl}}{c}_0} = -4$, $\underset{2}{\overset{\text{sl}}{c}_0} = -4$, $\underset{3}{\overset{\text{sl}}{c}_0} = -4$, $\underset{4}{\overset{\text{sl}}{c}_0} = -4$

Problem 5.22

A. $4\mathcal{Z}^0 + 12\mathcal{Z}^1 + 25\mathcal{Z}^2 + 24\mathcal{Z}^3 + 16\mathcal{Z}^4$

C. $9\mathcal{Z}^0 + 24\mathcal{Z}^1 + 46\mathcal{Z}^2 + 40\mathcal{Z}^3 + 25\mathcal{Z}^4$

E. $16\mathcal{Z}^0 + 40\mathcal{Z}^1 + 73\mathcal{Z}^2 + 60\mathcal{Z}^3 + 36\mathcal{Z}^4$

G. $25\mathcal{Z}^0 + 60\mathcal{Z}^1 + 106\mathcal{Z}^2 + 84\mathcal{Z}^3 + 49\mathcal{Z}^4$

I. $36\mathcal{Z}^0 + 84\mathcal{Z}^1 + 145\mathcal{Z}^2 + 112\mathcal{Z}^3 + 64\mathcal{Z}^4$

K. $49\mathcal{Z}^0 + 112\mathcal{Z}^1 + 190\mathcal{Z}^2 + 144\mathcal{Z}^3 + 81\mathcal{Z}^4$

Problem 5.23

A. $\Phi = -0.0516, v_x = 0.0014, v_y = 0.0014, v_z = 0.0082$

C. $\Phi = -0.0750, v_x = 0.0042, v_y = 0.0042, v_z = 0.0167$

E. $\Phi = -0.1299, v_x = 0.0217, v_y = 0.0217, v_z = 0.0433$

G. $\Phi = -0.2251, v_x = 0.1125, v_y = 0.1125, v_z = -0.0000$

I. $\Phi = -0.1299, v_x = 0.0217, v_y = 0.0217, v_z = -0.0433$

K. $\Phi = -0.0750, v_x = 0.0042, v_y = 0.0042, v_z = -0.0167$

Problem 5.24

A. $\Phi = -0.0534, v_x = 0.0017, v_y = 0.0017, v_z = 0.0090$

C. $\Phi = -0.0800, v_x = 0.0062, v_y = 0.0062, v_z = 0.0196$

E. $\Phi = -0.1403, v_x = 0.0375, v_y = 0.0375, v_z = 0.0375$

G. $\Phi = -0.1824, v_x = 0.0650, v_y = 0.0650, v_z = 0.0000$

I. $\Phi = -0.1403, v_x = 0.0375, v_y = 0.0375, v_z = -0.0375$

K. $\Phi = -0.0800, v_x = 0.0062, v_y = 0.0062, v_z = -0.0196$

Problem 5.25

A. $\Psi_1 = (-0.3141, 0.3141, 0), \Psi_2 = (-0.3141, -0.3141, 0), \Psi_3 = (0.3141, -0.3141, 0), \Psi_4 = (0.3141, 0.3141, 0)$

C. $\Psi_1 = (-0.3092, 0.3092, 0), \Psi_2 = (-0.3092, -0.3092, 0), \Psi_3 = (0.3092, -0.3092, 0), \Psi_4 = (0.3092, 0.3092, 0)$

E. $\Psi_1 = (-0.2891, 0.2891, 0), \Psi_2 = (-0.2891, -0.2891, 0), \Psi_3 = (0.2891, -0.2891, 0), \Psi_4 = (0.2891, 0.2891, 0)$

G. $\Psi_1 = (-0.1592, 0.1592, 0), \Psi_2 = (-0.1592, -0.1592, 0), \Psi_3 = (0.1592, -0.1592, 0), \Psi_4 = (0.1592, 0.1592, 0)$

I. $\Psi_1 = (-0.0292, 0.0292, 0), \Psi_2 = (-0.0292, -0.0292, 0), \Psi_3 = (0.0292, -0.0292, 0), \Psi_4 = (0.0292, 0.0292, 0)$

K. $\Psi_1 = (-0.0091, 0.0091, 0), \Psi_2 = (-0.0091, -0.0091, 0), \Psi_3 = (0.0091, -0.0091, 0), \Psi_4 = (0.0091, 0.0091, 0)$

Problem 5.26

A. $\Psi_1 = (-0.3136, 0.3136, 0), \Psi_2 = (-0.3136, -0.3136, 0), \Psi_3 = (0.3136, -0.3136, 0), \Psi_4 = (0.3136, 0.3136, 0)$

C. $\Psi_1 = (-0.3070, 0.3070, 0), \Psi_2 = (-0.3070, -0.3070, 0), \Psi_3 = (0.3070, -0.3070, 0), \Psi_4 = (0.3070, 0.3070, 0)$

E. $\Psi_1 = (-0.2717, 0.2717, 0), \Psi_2 = (-0.2717, -0.2717, 0), \Psi_3 = (0.2717, -0.2717, 0), \Psi_4 = (0.2717, 0.2717, 0)$

G. $\Psi_1 = (-0.1592, 0.1592, 0), \Psi_2 = (-0.1592, -0.1592, 0), \Psi_3 = (0.1592, -0.1592, 0), \Psi_4 = (0.1592, 0.1592, 0)$

I. $\Psi_1 = (-0.0466, 0.0466, 0), \Psi_2 = (-0.0466, -0.0466, 0), \Psi_3 = (0.0466, -0.0466, 0), \Psi_4 = (0.0466, 0.0466, 0)$

K. $\Psi_1 = (-0.0113, 0.0113, 0), \Psi_2 = (-0.0113, -0.0113, 0), \Psi_3 = (0.0113, -0.0113, 0), \Psi_4 = (0.0113, 0.0113, 0)$

Problem 5.27

A. $\Psi_1 = (0, 0.1241, 0.3196), \Psi_2 = (0, -0.1241, 0.3196), \Psi_3 = (0, -0.1241, -0.3196), \Psi_4 = (0, 0.1241, -0.3196)$

C. $\Psi_1 = (0, 0.1202, 0.3151), \Psi_2 = (0, -0.1202, 0.3151), \Psi_3 = (0, -0.1202, -0.3151), \Psi_4 = (0, 0.1202, -0.3151)$

E. $\Psi_1 = (0, 0.1078, 0.2970), \Psi_2 = (0, -0.1078, 0.2970), \Psi_3 = (0, -0.1078, -0.2970), \Psi_4 = (0, 0.1078, -0.2970)$

G. $\Psi_1 = (0, 0.0640, 0.1619), \Psi_2 = (0, -0.0640, 0.1619), \Psi_3 = (0, -0.0640, -0.1619), \Psi_4 = (0, 0.0640, -0.1619)$

I. $\Psi_1 = (0, 0.0203, 0.0268), \Psi_2 = (0, -0.0203, 0.0268), \Psi_3 = (0, -0.0203, -0.0268), \Psi_4 = (0, 0.0203, -0.0268)$

K. $\Psi_1 = (0, 0.0079, 0.0087), \Psi_2 = (0, -0.0079, 0.0087), \Psi_3 = (0, -0.0079, -0.0087), \Psi_4 = (0, 0.0079, -0.0087)$

Problem 5.28

A. $\Psi_1 = (0, 0.0000, 0.4737), \Psi_2 = (0, 0.0000, 0.4737), \Psi_3 = (0, 0.0000, -0.4737), \Psi_4 = (0, 0.0000, -0.4737)$

C. $\Psi_1 = (0, 0.0000, 0.4610), \Psi_2 = (0, 0.0000, 0.4610), \Psi_3 = (0, 0.0000, -0.4610), \Psi_4 = (0, 0.0000, -0.4610)$

E. $\Psi_1 = (0, 0.0000, 0.4262), \Psi_2 = (0, 0.0000, 0.4262), \Psi_3 = (0, 0.0000, -0.4262), \Psi_4 = (0, 0.0000, -0.4262)$

G. $\Psi_1 = (0, 0.0000, 0.2500), \Psi_2 = (0, 0.0000, 0.2500), \Psi_3 = (0, 0.0000, -0.2500), \Psi_4 = (0, 0.0000, -0.2500)$

I. $\Psi_1 = (0, 0.0000, 0.0738), \Psi_2 = (0, 0.0000, 0.0738), \Psi_3 = (0, 0.0000, -0.0738), \Psi_4 = (0, 0.0000, -0.0738)$

K. $\Psi_1 = (0, 0.0000, 0.0390), \Psi_2 = (0, 0.0000, 0.0390), \Psi_3 = (0, 0.0000, -0.0390), \Psi_4 = (0, 0.0000, -0.0390)$

References

Abate, J. and Valkó, P. P. (2004). Multi-precision Laplace transform inversion. *International Journal for Numerical Methods in Engineering*, **60**, 979–93.

Abbott, Ira H. (1959). *Theory of wing sections Including a summary of airfoil data*. Dover Publications, New York.

Abbott, Ira H. and von Doenhoff, Albert E. (1949). *Theory of wing sections including a summary of airfoil data* (first edn). McGraw-Hill, New York.

Abramowitz, M. and Stegun, I. A. (1972). *Handbook of mathematical functions*. Dover Publications, New York.

Airy, George Biddell (1845). Tides and waves. In *Encyclopedia Metropolitana* (ed. E. Smedley, H. J. Rose, and H. J. Rose), Volume 5, pp. 241–396. B. Fellows, London.

Airy, George Biddell (1863). On the strains in the interior of beams. *Philosophical Transactions of the Royal Society of London*, **153**(1), 49–80.

Airy, George Biddell (1871). *On sound and atmospheric vibrations, with the mathematical elements of music* (second edn). Macmillan, London.

Anderson, Erik I. (2005). Modeling groundwater-surface water interactions using the Dupuit approximation. *Advances in Water Resources*, **28**, 315–27.

Anderson, Jr., John D. (1978). *Introduction to flight its engineering and history*. McGraw-Hill, New York.

Aravin, V. I. and Numerov, S. N. (1965). *Theory of fluid flow in undeformable porous media*. S. Monson, Jerusalem.

Bakker, Mark (2008). Derivation and relative performance of strings of line elements for modeling (un)confined and semi-confined flow. *Advances in Water Resources*, **31**, 906–14.

Bakker, Mark (2013). Semi-analytic modeling of transient multi-layer flow with TTim. *Hydrogeology Journal*, **21**, 935–43.

Bakker, Mark and Nieber, John L. (2004*a*). Analytic element modeling of cylindrical drains and cylindrical inhomogeneities in steady two-dimensional unsaturated flow. *Vadose Zone Journal*, **3**, 1038–49.

Bakker, Mark and Nieber, John L. (2004*b*). Two-dimensional steady unsaturated flow through embedded elliptical layers. *Water Resources Research*, **40**, W12406:1–12.

Barnes, Randal and Janković, Igor (1999). Two-dimensional flow through large number of circular inhomogeneities. *Journal of Hydrology*, **226**, 204–10.

Bascom, Willard (1980). *Waves and beaches*. Anchor Books, Garden City, New York.

Basha, H. A. (1999, January). Multidimensional linearized nonsteady infiltration with prescribed boundary conditions at the soil surface. *Water Resources Research*, **35**(1), 75–83.

Batchelor, G. K. (1967). *Introduction to fluid dynamics*. Cambridge University Press, Cambridge, UK.

Bear, Jacob (1972). *Dynamics of fluids in porous media*. Dover Publications, New York.

Bell, E. T. (1934). Exponential polynomials. *Annals of Mathematics*, **35**(2), 258–77.

Berkhoff, J. C. W. (1972, July). Computation of combined refraction-diffraction. In *Proceedings of the Thirteenth Coastal Engineering Conference*, New York, pp. 471–90. ASCE.

Berkhoff, Juri Cornelis Willem (1976). *Mathematical models for simple harmonic linear water waves, wave diffraction and refraction*. Ph.D. dissertation, Delft Hydraulics Laboratory. Publication no. 163.

Bernoulli, Danielis (1738). *Hydrodynamica, sive de viribus et motibus fluidorum commentarii*. Johannis Reinholdi Dulseckeri, Argentorati. in Latin.

Bezier, P. (1986). *The mathematical basis of the UNISURF CAD system*. Butterworth, London.

Boresi, Arthur P. (1965). *Elasticity in engineering mechanics*. Prentice-Hall, Englewood Cliffs, NJ.

Boussinesq, M. J. (1871). Theore de l'intumescence liquide, applelee onde solitaire o de translation, se propageant dans un canal rectangulaire. *Comptes rendus hebdomadaires des seances de l'Academie des sciences*, **72**, 755–9.

Braester, Carol (1973, June). Moisture variation at the soil surface and the advance of the wetting front during infiltration at constant flux. *Water Resources Research*, **9**(3), 687–94.

Brekhovskikh, L. M. and Godin, O. A. (1990). *Acoustics of layered media I plane and quasi-plane waves*. Springer-Verlag, Berlin.

Bulatewicz, Tom, Allen, Andrew, Peterson, Jeffrey M., Staggenborg, Scott, Welch, Stephen M., and Steward, David R. (2013, January). The simple script wrapper for OpenMI: Enabling interdisciplinary modeling studies. *Environmental Modelling & Software*, **39**, 283–94.

Bulatewicz, Tom, Yang, Xiaoying, Peterson, Jeffrey M., Staggenborg, Scott, Welch, Stephen M., and Steward, David R. (2010). Accessible integration of agriculture, groundwater, and economic models using the Open Modeling Interface (OpenMI): methodology and initial results. *Hydrology and Earth Systems Science*, **14**, 521–34.

Burton, A. J. and Miller, G. F. (1971). The application of integral equation methods to the numerical solution of some exterior boundary-value problems. *Proc. R. Soc. Lond. A*, **323**, 201–10.

Byerly, William Elwood (1893). *An elementary treatise on Fourier's Series and spherical, cylindrical and ellipsoidal harmonics with applications to problems in mathematical physics.* Dover Publications, New York.

Carslaw, H. S. (1921). *Introduction to the theory of Fourier's series and integrals* (second edn). MacMillan, London.

Carslaw, H. S. and Jaeger, J. C. (1959). *Conduction of heat in solids* (second edn). Oxford University Press, Oxford.

Chou, Pei Chi and Pagano, Nicholas J. (1967). *Elasticity tensor, dyadic and engineering approaches.* D. van Nostrand, Princeton, NJ.

Choudhury, M. H. (1989). *Electromagnetism.* Ellis Horwood, Chichester, West Sussex.

Churchill, R. V. and Brown, J. W. (1984). *Complex variables and applications.* McGraw Hill, New York.

Clebsch, H. A. (1857). Über eine allgemeine transformation der hydrodynamischen gleichungen. In *Journal fur die reine und angewandte mathematik* (ed. A. L. Crelle), Volume 54, pp. 293–312. W. de Gruyter, Berlin.

Comtet, L. (1974). *Advanced combinatorics; The art of finite and infinite expansions.* D. Reidel, Dordrecht, Boston.

Constantine, G. M. and Savits, T. H. (1996). A multivariate Faa di Bruno formula with applications. *Transactions of the American Mathematical Society*, **348**(2), 503–20.

Corke, Thomas C. (2003). *Design of aircraft.* Pearson Education, Upper Saddle River, New Jersey, NJ.

Courant, Richard (1950). *Dirichlet's principle, conformal mapping, and minimal surfaces.* Interscience Publishers, New York.

Courant, Richard and Hilbert, D. (1962). *Methods of mathematical physics.* Volume II. Partial differential equations. Wiley, New York.

Craig, James R., Janković, Igor, and Barnes, Randal (2006). The nested superblock approach for regional-scale Analytic Element Models. *Groundwater*, **44**(1), 76–80.

Dagan, G. (1981). Analysis of flow through heterogeneous random aquifers by the methods of embedding matrix 1. steady flow. *Water Resources Research*, **17**(1), 107–21.

Dagan, G., Fiori, A., and Janković, Igor (2003). Flow and transport in highly heterogeneous formations: 1. conceptual framework and validity of first-order approximations. *Water Resources Research*, **39**(9), SHB14:1–12.

Dagan, G., Fiori, A., and Janković, Igor (2004). Transmissivity and head covariances for flow in highly heterogeneous aquifers. *Journal of Hydrology*, **294**, 39–56.

Darcy, Henry (1856). *Les fontaines publiques de la Ville de Dijon.* Dalmont, Paris.

Davies, B. (1978). *Integral Transforms and their applications.* Springer-Verlag, New York.

de Lange, Willem J. (2006). Development of an Analytic Element ground water model of the Netherlands. *Ground Water*, **44**(1), 111–15.

Dean, Robert G. and Dalrymple, Robert A. (1984). *Water wave mechanics for engineers and scientists.* Prentice-Hall, Englewood Cliffs, NJ.

Dupuit, Jules (1863). *Études théoriques et pratiques sur le mouvement des eaux dans les canaux découverts et a travers les terrains perméables.* Dunod, Paris.

Duschek, Adalbert and Hochrainer, August (1950). *Grundzüge der tensorrechnung in analytischer darstellung II. Tensoranalysis.* Springer-Verlag, Vienna.

Ehrenmark, Ulf Torsten (1998). Oblique wave incidence on a plane beach: The classical problem revisited. *Journal of Fluid Mechanics*, **368**, 291–319.

Einstein, Albert (1995). *The collected papers of Albert Einstein*, Volume 4: *The Swiss Years; Writings, 1912–1914* (ed. M. J. Klein, A. J. Kox, J. Renn, and R. Schulmann). Princeton University Press, Princeton, NJ.

Faà di Bruno, C. F. (1855). Note sur une nouvelle formule du calcul diferentiel. *Quart. J. Math.*, **1**, 359–60.

Feynman, R. P., Leighton, R. B., and Sands, M. (1965). *The Feynman lectures on physics*, Volume II, Section 15. Addison-Wesley, Reading, MA.

Fitts, Charles R. (1989). Simple analytic functions for modeling three-dimensional flow in layered aquifers. *Water Resources Research*, **25**(5), 943–8.

Fitts, Charles R. (1990). *Modeling three-dimensional groundwater flow about ellipsoids of revolution using analytic functions.* Ph.D. thesis, University of Minnesota, Minneapolis, Minnesota.

Fitts, Charles R. (1991, May). Modeling three-dimensional flow about ellipsoidal inhomogeneities with application to flow to a gravel-packed well and flow through lens-shaped inhomogeneities. *Water Resources Research*, **27**(5), 815–24.

Fitts, Charles R. (2010). Modeling aquifer systems with analytic elements and subdomains. *Water Resources Research*, **46**, W07521.

Fourier, Jean Baptiste Joseph (1808, March). Mémoire sur la propagation de la chaleur dans les corps solides. *Nouveau Bulletin des Sciences par la Société Philomathique de Paris*, **I**, 112–116. Reprinted in Oeuvres complètes, tome 2. pp. 215–21.

Fourier, Joseph (1878). *The analytical theory of heat* (Translated with Notes by Alexander Freeman). Cambridge University Press, London.

Fredholm, E. I. (1903). Sur une classe d'equations fonctionnelles. *Acta Mathematica*, **27**, 365–90.

Freeze, R. A. and Cherry, J. A. (1979). *Groundwater*. Prentice Hall, Englewood Cliffs, NJ.

Furman, Alex and Neuman, Shlomo P. (2003). Laplace-transform analytic element solution of transient flow in porous media. *Advances in Water Resources*, **26**, 1229–37.

Gardner, W. R. (1958). Some steady-state solutions of the unsaturated moisture flow equation with application to evaporation from a water table. *Soil Science*, **85**, 228–32.

Gauss, Carl Friedrich (1809). *Theoria Motus Corporum Coelestium in sectionibus conicis solem ambientium* (translated in 1857 by Carles Henry Davis). Little, Brown, Boston.

Goodier, J. N. and Hodge, Jr., P. G. (1958). *Elasticity and plasticity*. Wiley, London.

Goodman, Joseph W. (1996). *Introduction to Fourier optics* (second edn). McGraw Hill, New York.

Goursat, E. (1898). Sur l'équation $\Delta\Delta u = 0$. *Bulletin de la Société Mathématique de France*, **26**, 236–7. `archive.org/stream/bulletinsocit26soci#page/236`.

Goursat, E. (1904). *A course in mathematical analysis* (Volume I, translated by E. R. Hedrick). Ginn and Company, Boston.

Gradshteyn, I. S. and Ryzhik, I. M. (1980). *Table of integrals, series, and products*. Academic Press, Orlando, FL.

Grant, I. S. and Phillips, W. R. (1990). *Electromagnetism* (second edn). Wiley, Chichester, West Sussex.

Green, A. E. and Zerna, W. (1968). *Theoretical elasticity* (second edn). Oxford University Press, London.

Gröbner, W. and Hofreiter, N. (1975). *Integraltafel*. Springer-Verlag, Vienna.

Haitjema, H. M. (1985). Modeling three-dimensional flow in confined aquifers by superposition of both two- and three-dimensional analytic functions. *Water Resources Research*, **21**(10), 1557–66.

Haitjema, Hendrik M. (1995). *Analytic element modeling of groundwater Flow*. Academic Press, San Diego.

Haitjema, H. M. and Kraemer, S. P. (1988). A new analytic function for modeling partially penetrating wells. *Water Resources Research*, **24**(5), 683–90.

Hammond, P. (1986). *Electromagnetism for engineers* (third edn). Pergamon Press, Oxford.

Helmholtz, H. (1858). Über integrale der hydrodynamischen gleichungen, welche den wirbelbewegungen entsprechen. *Crelles (Journal für die reine und angewandte Mathematik*, **55**, 25–55.

Helmholtz, H. (1881). *Popular lectures on scientific subjects* (second edn). Longman's, Green, and Co., London.

Henrici, P. (1956). Automatic computations with power series. *JACM*, **3**, 10–15.

Hess, J. L. (1990). Panel method in computational fluid dynamics. *Annual Review of Fluid Mechanics*, **22**, 255–74.

Hess, J. L. and Smith, A. M. O. (1967). Calculation of potential flow about arbitrary bodies. In *Progress in Aeronautical Sciences* (ed. D. Küchemann, P. Carrière, B. Etkin, W. Fiszdon, N. Rott, J. Smolderen, I. Tani, and W. Wuest), Volume 8, pp. 1–138. Pergamon Press, Oxford.

Hille, Einar (1982). *Analytic function theory: Volume 1*. Chelsea Publishing, New York.

Homma, S. (1950). On the behaviour of seismic sea waves around circular island. *Geophysical Magazine*, **21**(3), 199–208.

Hooke, Robert (1678). *Lectures de potentia restitutiva, Or of spring explaining the power of springing bodies*. John Martyn, London.

Jaeger, John Conrad (1969). *Elasticity, fracture and flow with engineering and geological applications* (third edn). Methuen, London.

Jaeger, John Conrad and Cook, N. G. W. (1976). *Fundamentals of rock mechanics* (second edn). Chapman and Hall, London.

Janković, Igor (1997). *High-order Analytic Elements in modeling groundwater flow*. Ph.D. dissertation, University of Minnesota, Minneapolis. Department of Civil Engineering.

Janković, Igor and Barnes, Randal J. (1999*a*). High-order line elements in modeling two-dimensional groundwater flow. *Journal of Hydrology*, **226**(3–4), 211–23.

Janković, Igor and Barnes, Randal J. (1999*b*). Three-dimensional flow through large numbers of spheroidal inhomogeneities. *Journal of Hydrology*, **226**(3–4), 224–33.

Janković, Igor, Fiori, A., and Dagan, G. (2003). Flow and transport in highly heterogeneous formations: 3. numerical simulations and comparison with theoretical results. *Water Resources Research*, **39**(9), SBH16:1–13.

Janković, Igor, Fiori, Aldo, and Dagan, Gedeon (2006). Modeling flow and transport in highly heterogeneous three-dimensional aquifer: Ergodicity, Gaussianity, and anomalous behavior - 1. conceptual issues and numerical simulations. *Water Resources Research*, **42**, W06D12:1–9.

Janković, Igor, Steward, David R., Barnes, Randal J., and Dagan, Gedeon (2009). Is transverse macrodispersivity in three-dimensional transport through isotropic heterogeneous formation equal to zero? A counterexample. *Water Resources Research*, **45**(8), W08415:1–10.

Jeffreys, Harold and Jeffreys, Bertha Swirles (1956). *Methods of mathematical physics* (third edn). Cambridge University Press, Cambridge, UK.

Jonsson, Ivar G., Skovgaard, Ove, and Brink-Kjaer, Ole (1976). Diffraction and refraction calculations for waves incident on an island. *Journal of Marine Research*, **34**, 469–496.

Joukowsky, Nikolai (1910, November). Über die konturen der tragflächen der drachenflieger. *Zeitschrift für Flugtechnik und Motorluftschiffahrt*, **1**(22), 281–285.

Joukowsky, Nikolai (1912, March). Über die konturen der tragflächen der drachenflieger. *Zeitschrift für Flugtechnik und Motorluftschiffahrt*, **3**(6), 81–6.

Kellogg, Oliver Dimon (1929). *Foundations of potential theory*. Verlag von Julius Springer, Berlin.

Kirchhoff, G. (1894). *Vorlesungen über der theorie der wärme*. Leipzig, Barth.

Knight, John H., Philip, John R., and Waechter, R. T. (1989, January). The seepage exclusion problem for spherical cavities. *Water Resources Research*, **25**(1), 29–37.

Knuth, D. E. (1998). *Art of computer programming*, volume 2: *Seminumerical algorithms* (third edn). Addison Wesley, Boston.

Kraemer, Stephen R. (2007, July-August). Analytic Element ground water modeling as a research program (1980 to 2006). *Ground Water*, **45**(4), 402–8.

Kraus, Werner (1978). *Panel method in aerodynamics*, Chapter 4, pp. 237–97. Numerical methods in fluid dynamics. Hemisphere, Washington, DC.

Kuhlman, Kristopher L. and Neuman, Shlomo P. (2009, June). Laplace-transform analytic-element method for transient porous-media flow. *Journal of Engineering Mathematics*, **64**(2), 113–30.

Kuttruff, Heinrich (2007). *Acoustics an introduction*. Taylor & Francis, London.

Lagrange, Joseph Louis (1768). Nouvelle méthode pour résoudre les équations littérales par le moyen des séries. *Histoire de l'Académie Royale des Sciences et des Belles Lettres de Berlin*, **24**, 251–326.

Lagrange, Joseph Louis (1781). Mémoire sur la théorie du mouvement des fluides. In *Nouveaux Mémoires de l'Académie Royale des Sciences et Belles-Lettres*, pp. 150–98. George Jacques Decker, Berlin.

Lamb, Horace (1879). *A treatise on the mathematical theory of the motion of fluids*. Cambridge University Press, London.

Lamb, H. (1916). *Hydrodynamics* (fourth edn). Cambridge University Press, London.

Legendre, Adrien-Marie (1806). *Nouvelles méthodes pour la détermination des orbites des comètes*. Courcier, Paris.

Le Grand, Philippe (1999). Analytic elements of high degree along bezier spline curves for the modeling of two-dimensional groundwater flow. M.S. thesis, University of Minnesota, Minneapolis. Department of Civil Engineering.

Le Grand, Philippe (2003). *Formes curvilinéaires avancées pour la modélisation centrée objet des écoulements souterrains par la méthode des éléments analytiques*. Ph.D. dissertation, Ecole Nationale Supérieure des Mines, Saint-Etienne, France. SITE Division.

Levenberg, Kenneth (1944). A method for the solution of certain non-linear problems in least squares. *Quart. Appl. Math*, **2**(2), 164–8.

Liu, Huan-Wen, and Li, Yan-Bao (2007). An analytical solution for long-wave scattering by a submerged circular truncated shoal. *Journal of Engineering Mathematics*, **57**, 133–44.

Liu, Huan-Wen, Lin, Pengzhi, and Shankar, N. Jothi (2004). An analytical solution of the mild-slope equation for waves around a circular island on a paraboloidal shoal. *Coastal Engineering*, **51**(5–6), 421–37.

Longuet-Higgins, M. S. (1967). On the trapping of wave energy round islands. *Journal of Fluid Mechanics*, **29**(4), 781–821.

Love, A. E. H. (1927). *A treatise on the mathematical theory of elasticity* (fourth edn). Cambridge University Press, London.

Lukács, E. (1955). Applications of Faà di Bruno's formula in mathematical statistics. *Amer. Math. Monthly*, **62**, 340–8.

MacCamy, R. C. and Fuchs, R. A. (1954, December). Wave forces on piles: A diffraction theory. Technical Memorandum 69, Beach Erosion Board Corps of Engineers.

McCormick, Barnes W. (1995). *Aeronautics, and flight mechanics* (second edn). Wiley, New York.

MacRobert, T. M. (1967). *Spherical harmonics: An elementary treatise on harmonic functions with applications* (third edn). Pergamon Press, Oxford.

Marquardt, Donald W. (1963, June). An algorithm for least-squares estimation of nonlinear parameters. *SIAM Journal on Applied Mathematics*, **11**(2), 431–41.
Maxwell, J. C. (1865). A dynamical theory of the electromagnetic field. *Philos. Trans.*, **155**, 459–512.
Maxwell, J. C. (1870). On the displacement in a case of fluid motion. *Proc. Lond. Math. Soc.*, **3**, 82–97.
Maxwell, J. C. (1881). *A treatise on electricity and magnetism* (second edn), Volume 2. Oxford University Press, London.
Maxwell, James Clerk (1890*a*). *The scientific papers of James Clerk Maxwell*, Volume 1. Dover Publications, New York.
Maxwell, James Clerk (1890*b*). *The scientific papers of James Clerk Maxwell*, Volume 2. Dover Publications, New York.
Maxwell, James Clerk (1904*a*). *Theory of heat*. Longmans, Green, New York.
Maxwell, James Clerk (1904*b*). *A treatise on electricity and magnetism* (third edn). Oxford University Press, London.
Mei, Chiang C. (1989). *The applied dynamics of ocean surface waves*. World Scientific, London.
Mogilevskaya, S. G. and Crouch, S. L. (2001). A Galerkin boundary integral method for multiple circular elastic inclusions. *International Journal for Numerical Methods in Engineering*, **52**, 1069–106.
Moon, Parry and Spencer, Domina Eberle (1960). *Foundations of electrodynamics*. Van Nostrand, Princeton, NJ.
Moon, Parry and Spencer, Domina Eberle (1961*a*). *Field theory for engineers*. Van Nostrand, Princeton, NJ.
Moon, Parry and Spencer, Domina Eberle (1961*b*). *Field theory handbook including coordinate systems differential equations and their solutions*. Springer-Verlag, Berlin.
Morse, Philip M. (1936). *Vibration and sound* (first edn). McGraw Hill, New York.
Morse, Philip M. and Feshbach, Herman (1953). *Methods of theoretical physics*. McGraw-Hill, New York.
Morse, Philip M. and Ingard, K. Uno (1968). *Theoretical acoustics*. McGraw Hill, New York.
Mualem, Yechezkel (1976, July). A catalogue of hydraulic properties of unsaturated soils. Technion Israel Institute of Technology, Haifa, Israel.
Muskat, M. (1932). Potential distributions in large cylindrical disks with partially penetrating electrodes. *Physics*, **2**, 329–64.
Muskat, M. (1937). *The flow of homogeneous fluids through porous media*. McGraw Hill, New York.
Muskhelishvili, N. I. (1953*a*). *Singular integral equations*. P. Noordhoof N. V., Groningen, The Netherlands.
Muskhelishvili, N. I. (1953*b*). *Some basic problems of the mathematical theory of elasticity* (third edn). P. Noordhoof N. V., Groningen, The Netherlands.
Oberhettinger, F. (1973). *Tables of Laplace transforms*. Springer-Verlag, New York.
Ohm, Georg Simon (1827). *Die galvanische kette, mathematische bearbeitet*. T. H. Riemann, Berlin.
Philip, John R. (1968, October). Steady infiltration from buried point sources and spherical cavities. *Water Resources Research*, **4**(3), 1039–47.
Philip, John R., Knight, John H., and Waechter, R. T. (1989, January). Unsaturated seepage and subterranean holes: Conspectus, and exclusion problem for circular cylindrical cavities. *Water Resources Research*, **25**(1), 16–28.
Planck, Max (1903). *Treatise on thermodynamics*. Longmans, Green, New York.
Polubarinova-Kochina, P. Y. (1962). *Theory of ground water movement*. Princeton University Press, Princetonm, NJ.
Press, W. H., Teukolsky, S. A., Vetterling, W. T., and Flannery, B. P. (1992). *Numerical recipes in C: The art of scientific computing* (second edn). Cambridge University Press, Cambridge, UK.
Pullan, A. J. (1990, June). The quasilinear approximation for unsaturated porous media flow. *Water Resources Research*, **26**(6), 1219–34.
Raats, Peter A. C. (1970, September). Steady infiltration from line sources and furrows. *Soil Science Society of America Journal*, **34**(5), 709–14.
Raats, Peter A. C. (1971, September). Steady infiltration from point sources, cavities, and basins. *Soil Science Society of America Journal*, **35**(5), 689–94.
Raats, Peter A. C. and Gardner, W. R. (1974). *Movement of water in the unsaturated zone near a water table*, Chapter 13, pp. 311–57. Agronomy Monogram 17. American Society of Agronomy, Madison, WI.
Remmert, Reinhold (1991). *Theory of complex functions*. Translated d by Robert B. Burckel. Springer-Verlag, New York.
Richards, Lorenzo A. (1931, November). Capillary conduction of liquids through porous mediums. *Physics*, **1**, 318–33.
Saffman, P. G. (1992). *Vortex dynamics*. Cambridge University Press, Cambridge, UK.
Sampson, R. A. (1891). On stoke's current function. *Philosophical Transactions of the Royal Society of London, Series A*, **182**, 449–518.
Sanford, R. J. (2003). *Principles of fracture mechanics*. Pearson Education, Upper Saddle River, NJ.
Sarpkaya, Turgut and Isaacson, Michael (1981). *Mechanics of wave forces on offshore structures*. Van Nostrand Reinhold, New York.

Schoenberg, I. J. (1946). Contributions to the problem of approximation of equidistant data by analytic functions. Part A.–on the problem of smoothing or graduation. a first class of analytic approximation formulae. *Quarterly of Applied Mathematics*, **4**, 45–99.

Schoenberg, I. J. (1973). *Cardinal spline interpolation*. SIAM, Philadelphia.

Selby, S. M. (1975). *Standard mathematical tables*. CRC Press, Cleveland.

Selvadurai, A. P. S. (2000). *Partial differential equations in mechanics 2: The biharmonic equation, Poisson equation*. Springer-Verlag, Berlin.

Slater, John C. and Frank, Nathaniel H. (1947). *Electromagnetism* (First edn). McGraw-Hill, New York.

Smith, Ronald and Sprinks, T. (1975, November). Scattering of surface waves by a conical island. *Journal of Fluid Mechanics*, **72**(2), 373–84.

Sokolnikoff, I. S. and Sokolnikoff, E. S. (1941). *Higher mathematics for engineers and physicists*. McGraw Hill, New York.

Sommerfeld, Arnold (1949). *Partial differential equations in physics*. Academic Press, New York.

Sommerfeld, Arnold (1952). Electrodynamics. In *Lectures on theoretical physics*, Volume 3. Academic Press, New York.

Sommerfeld, Arnold (1972). Partial differential equations in physics. In *Lectures on theoretical physics* (fifth edn), Volume 6. Academic Press, New York.

Stehfest, Harald (1970, January). Algorithm 368 numerical inversion of Laplace transforms. *Communications of the ACM*, **13**(1), 47–54.

Stephens, R. W. B. and Bate, A. E. (1966). *Acoustics and vibrational physics* (Second edn). Edward Arnold, London.

Stevenson, A. F. (1954). Note on the existence and determination of a vector potential. *Quart. Appl. Math*, **12**(2), 194–8.

Steward, David R. (1991). Analytic boundary integral equation method for steady-state and transient groundwater flow. M.S. Thesis, University of Minnesota, Department of Civil Engineering, Minneapolis.

Steward, David R. (1996). *Vector potential functions and stream surfaces in three-dimensional groundwater flow*. Ph.D. Dissertation, University of Minnesota, Department of Civil Engineering, Minneapolis.

Steward, David R. (1998). Stream surfaces in two-dimensional and three-dimensional divergence-free flows. *Water Resources Research*, **34**(5), 1345–50.

Steward, David R. (1999). Three-dimensional analysis of the capture of contaminated leachate by fully penetrating, partially penetrating, and horizontal wells. *Water Resources Research*, **35**(2), 461–8.

Steward, David R. (2001). A vector potential for a partly penetrating well and flux in an approximate method of images. *Proceedings of the Royal Society of London, Series A, Mathematical, Physical and Engineering Sciences*, **457**, 2093–111.

Steward, David R. (2002). A vector potential and exact flux through surfaces using Lagrange and Stokes stream functions. *Proceedings of the Royal Society of London, Series A, Mathematical, Physical and Engineering Sciences*, **458**(2018), 489–509.

Steward, David R. (2007). Groundwater response to changing water-use practices in sloping aquifers. *Water Resources Research*, **43**(5), W05408.

Steward, David R. (2015, November). Analysis of discontinuities across thin inhomogeneities, groundwater/surface water interactions in river networks, and circulation about slender bodies using slit elements in the Analytic Element Method. *Water Resources Research*, **51**(11), 8684–703.

Steward, David R. (2016, November). Analysis of vadose zone inhomogeneity toward distinguishing recharge rates: Solving the nonlinear interface problem with Newton method. *Water Resources Research*, **52**(11), 8756–74.

Steward, David R. (2018, July). Wave resonance and dissipation in collections of partially reflecting vertical cylinders. *Journal of Waterways, Ports, Coastal, and Ocean Engineering (ASCE)*, **144**(4), 04018004–1–16.

Steward, David R. (2020). Waves in collections of circular shoals and bathymetric depressions. *Journal of Waterways, Ports, Coastal, and Ocean Engineering (ASCE)*, 146(4), 04020018.

Steward, David R. and Ahring, Trevor (2009, June). An analytic solution for groundwater uptake by phreatophytes spanning spatial scales from plant to field to regional. *Journal of Engineering Mathematics*, **64**(2), 85–103.

Steward, David R. and Allen, Andrew J. (2013, October). The Analytic Element Method for rectangular gridded domains, benchmark comparisons and application to the High Plains Aquifer. *Advances in Water Resources*, **60**, 89–99.

Steward, David R. and Allen, Andrew J. (2016, May). Peak groundwater depletion in the High Plains Aquifer, projections from 1930 to 2110. *Agricultural Water Management*, **170**, 36–48.

Steward, David R. and Bernard, Eric A. (2006). The synergistic powers of AEM and GIS geodatabase models in water resources studies. *Ground Water*, **44**(1), 56–61.

Steward, David R., Bruss, Paul J., Yang, Xiaoying, Staggenborg, Scott A., Welch, Stephen M., and Apley, Michael D. (2013, September 10). Tapping unsustainable groundwater stores for agricultural production in the High Plains Aquifer of Kansas, projections to 2110. *Proceedings of the National Academy of Sciences of the United States of America*, **110**(37), E3477–86.

Steward, David R. and Janković, Igor (2001). Deformation of stream surfaces in steady axisymmetric flow. *Water Resources Research*, **37**(2), 307–15.

Steward, David R. and Jin, Wei (2001). Gaining and losing sections of horizontal wells. *Water Resources Research*, **37**(11), 2677–85.

Steward, David R. and Jin, Wei (2003). Drawdown and capture zone topology for nonvertical wells. *Water Resources Research*, **39**(8), 9:1–11.

Steward, David R., Le Grand, Philippe, Janković, Igor, and Strack, Otto D. L. (2008). Analytic formulation of Cauchy integrals for boundaries with curvilinear geometry. *Proceedings of the Royal Society of London, Series A, Mathematical, Physical and Engineering Sciences*, **464**, 223–48.

Steward, David R. and Panchang, Vijay G. (2000). Improved coastal boundary condition for surface water waves. *Ocean Engineering*, **28**(1), 139–57.

Steward, David R., Peterson, Jeffrey M., Yang, Xiaoying, Bulatewicz, Tom, Herrera-Rodriguez, Mauricio, Mao, Dazhi, and Hendricks, Nathan (2009, May). Groundwater economics: An object oriented foundation for integrated studies of irrigated agricultural systems. *Water Resources Research*, **45**, W05430.

Stewart, George Walter and Lindsay, Robert Bruce (1930). *Acoustics a text on theory and applications*. Van Nostrand, New York.

Stokes, George Gabriel (1842). On the steady motion of incompressible fluids. *Transactions of the Cambridge Philosophical Society*, **7**(3), 439–53. (Also in *Mathematical and Physical Papers 1*, 1–16, 1880).

Stokes, George Gabriel (1847). On the theory of oscillatory waves. *Transactions of the Cambridge Philosophical Society*, **8**, 441–55. (Also in *Mathematical and Physical Papers 1*, 197–229, 1880).

Stokes, George Gabriel (1880). *Mathematical and physical papers*, Volume 1. Cambridge University Press, London.

Strack, Otto D. L. (1981). Flow in aquifers with clay laminae 1. the comprehensive potential. *Water Resources Research*, **17**(4), 985–92.

Strack, Otto D. L. (1984). Three-dimensional streamlines in Dupuit-Forchheimer models. *Water Resources Research*, **20**(7), 812–22.

Strack, Otto D. L. (1989). *Groundwater Mechanics*. Prentice Hall, Englewood Cliffs, NJ.

Strack, Otto D. L. (1999). Principles of the analytic element method. *Journal of Hydrology*, **226**, 128–38.

Strack, Otto D. L. (2003). Theory and applications of the Analytic Element Method. *Rev. Geophys.*, **41**(2), 1(1–19).

Strack, Otto D. L. (2005). Comment on 'steady two-dimensional groundwater flow through many elliptical inhomogeneities' by Raghavendra Suribhatla, Mark Bakker, Karl Bandilla, and Igor Janković. *Water Resources Research*, **41**, W11601:1–2.

Strack, Otto D. L. (2009*a*, June). The generating analytic element approach with application to the modified Helmholtz equation. *Journal of Engineering Mathematics*, **64**(2), 163–91.

Strack, Otto D. L. (2009*b*). Using Wirtinger calculus and holomorphic matching to obtain the discharge potential for an elliptical pond. *Water Resources Research*, **45**(1), W01409:1–9.

Strack, Otto D. L. and Haitjema, H. M. (1981*a*). Modeling double aquifer flow using a comprehensive potential and distributed singularities 1. solution for homogeneous permeability. *Water Resources Research*, **17**(5), 1535–49.

Strack, O. D. L. and Haitjema, H. M. (1981*b*). Modeling double aquifer flow using a comprehensive potential and distributed singularities 2. solution for inhomogeneous permeabilities. *Water Resources Research*, **17**(5), 1551–60.

Strack, Otto D. L. and Janković, Igor (1999). A multi-quadric area-sink for Analytic Element Modeling of groundwater flow. *Journal of Hydrology*, **226**(3–4), 188–96.

Strack, Otto D. L., Janković, Igor, and Barnes, Randal J. (1999). The superblock approach for the Analytic Element Method. *Journal of Hydrology*, **226**(3–4), 179–87.

Strack, Otto D. L. and Namazi, Taha (2014). A new formulation for steady multiaquifer flow: An analytic element for piecewise constant infiltration. *Water Resources Research*, **50**, 7839–956.

Strang, Gilbert (1980). *Linear algebra and its applications* (Second edn). Academic Press, Orlando, FL.

Strang, Gilbert (1986). *Introduction to applied mathematics*. Wellesley-Cambridge Press, Wellesley, MA.

Strang, Gilbert (2007). *Computational science and engineering*. Wellesley-Cambridge Press, Wellesley, MA.

Sunada, Shigeru S. (1997, March). Airfoil section characteristics at a low Reynolds number. *Journal of Fluids Engineering*, **119**(1), 129–35.

Suribhatla, Raghavendra, Bakker, Mark, Bandilla, Karl, and Janković, Igor (2004). Steady two-dimensional groundwater flow through many elliptical inhomogeneities. *Water Resources Research*, **40**, W042002:1–10.

Tait, P. G. (1867). Translation of 'On integrals of the hydrodynamical equations which express vortex-motion by H. Helmholtz'. *Philosophical Magazine*, **4**(33), 485–512.

Theis, Charles V. (1935). The relation between the lowering of the piezometric surface and the rate and duration of discharge of a well using ground-water storage. In *Transactions, American Geophysical Union*, 16th Annual Meeting, pp. 519–24.

Thiem, G. (1906). *Hydrologische methoden*. J. M. Gebhardt, Leipzig.

Thomson, J. J. (1904). *Electricity and matter*. Yale University Press, New Haven, CT.

van Genuchten, M. Th. (1980). A closed-form equation for predicting the hydraulic conductivity of unsaturated soils. *Soil Science Society of America Journal*, **44**, 892–8.

Verhulst, Pierre-François (1838). Notice sur la loi que la population suit dans son accroissement. *Correspondance Mathématique et Physique*, **10**, 113–21.

Verruijt, Arnold (1982). *Theory of groundwater flow* (second edn). MacMillan Press, London.

von Mises, Richard (1945). *Theory of flight* (first edn). McGraw-Hill, New York.

Wait, James R. (1970). *Electromagnetic waves in stratified media* (second edn). Pergamon Press, Oxford.

Warrick, Arthur W. (1974). Time-dependent linearized infiltration. I. point sources. *Soil Science Society of America Journal*, **38**(3), 383–6.

Warrick, Arthur W. and Knight, John H. (2002, July). Two-dimensional unsaturated flow through a circular inclusion. *Water Resources Research*, **38**(7), 18:1–6.

Warrick, Arthur W. and Knight, John H. (2004, May). Unsaturated flow through a spherical inclusion. *Water Resources Research*, **40**(5), W05101:1–6.

Watson, G. N. (1914). *Complex integration and Cauchy's theorem*. Cambridge University Press, Cambridge, UK.

Weatherburn, C. E. (1960). *Advanced vector analysis*. G. Bell and Sons, London.

Westergaard, H. M. (1952). *Theory of elasticity and plasticity*. Harvard University Press, Cambridge, MA.

Wheeler, F. S. (1987). Bell polynomials. *ACM SIGSAM Bulletin*, **21**(3), 44–53.

Wirtinger, W. (1927). Zur formalen theorie der funktionen von mehrenen komplexen veranderlichen. *Mathematischen Annalen*, **97**, 357–75.

Wooding, R. A. (1968, November). Steady infiltration from a shallow circular pond. *Water Resources Research*, **4**(6), 1259–73.

Xu, Bingyi, Panchang, Vijay, and Demirbilek, Zeki (1996, May–June). Exterior reflections in elliptic harbor wave models. *Journal of Waterway, Port, Coastal and Ocean Engineering*, **122**(3), 118–26.

Zaadnoordijk, Willem J. (1998, December). Transition from transient Theis wells to steady Thiem wells. *Hydrological Sciences Journal*, **43**(6), 859–73.

Zaadnoordijk, W. J. and Strack, Otto D. L. (1993). Area sinks in the analytic element method for transient groundwater flow. *Water Resources Research*, **29**(12), 4121–30.

Index